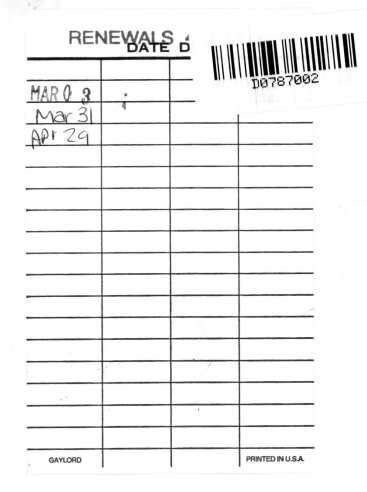

RENEWALS
DATE D

| MAR 0 3 | | | |
| Mar 31 | | | |
| Apr 29 | | | |
| | | | |
| | | | |
| | | | |
| | | | |
| | | | |
| | | | |
| | | | |
| | | | |
| | | | |
| | | | |
| | | | |
| | | | |
| | | | |
| | | | |
| GAYLORD | | | PRINTED IN U.S.A. |

X-RAY DIFFRACTION

# INTERNATIONAL SERIES IN PURE AND APPLIED PHYSICS

ADLER, BAZIN, AND SCHIFFER: Introduction to General Relativity
ALLIS AND HERLIN: Thermodynamics and Statistical Mechanics
AZÁROFF (Editor): X-Ray Spectroscopy
AZÁROFF, KAPLOW, KATO, WEISS, WILSON, AND YOUNG: X-Ray Diffraction
BECKER: Introduction to Theoretical Mechanics
BJORKEN AND DRELL: Relativistic Quantum Fields
BJORKEN AND DRELL: Relativistic Quantum Mechanics
CHODOROW AND SUSSKIND: Fundamentals of Microwave Electronics
CLARK: Applied X-Rays
COLLIN: Field Theory of Guided Waves
EVANS: The Atomic Nucleus
FETTER AND WALECKA: Quantum Theory of Many-Particle Systems
FEYNMAN AND HIBBS: Quantum Mechanics and Path Integrals
HALL: Introduction to Electron Microscopy
HARDY AND PERRIN: The Principles of Optics
HARNWELL: Electricity and Electromagnetism
HARNWELL AND LIVINGOOD: Experimental Atomic Physics
HARRISON: Solid State Theory
JAMMER: The Conceptual Development of Quantum Mechanics
KRALL AND TRIVELPIECE: Principles of Plasma Physics
LEIGHTON: Principles of Modern Physics
LINDSAY: Mechanical Radiation
LIVINGSTON AND BLEWETT: Particle Accelerators
MORSE: Vibration and Sound
MORSE AND FESHBACH: Methods of Theoretical Physics
MORSE AND INGARD: Theoretical Acoustics
NEWTON: Scattering Theory of Waves and Particles
PARK: Introduction to the Quantum Theory
PRESENT: Kinetic Theory of Gases
READ: Dislocations in Crystals
RICHTMYER, KENNARD, AND COOPER: Introduction to Modern Physics
ROSSI AND OLBERT: Introduction to the Physics of Space
SCHIFF: Quantum Mechanics
SCHWARTZ: Introduction to Special Relativity
SCHWARTZ: Principles of Electrodynamics
SLATER: Quantum Theory of Atomic Structure, Vol. I
SLATER: Quantum Theory of Matter
SLATER: Electronic Structure of Molecules: Quantum Theory of Molecules and Solids, Vol. 1
SLATER: Insulators, Semiconductors, and Metals: Quantum Theory of Molecules and Solids, Vol. 3
SLATER: The Self-Consistent Field for Molecules and Solids: Quantum Theory of Molecules and Solids, Vol. 4
SLATER AND FRANK: Introduction to Theoretical Physics
SMYTHE: Static and Dynamic Electricity
STRATTON: Electromagnetic Theory
TINKHAM: Group Theory and Quantum Mechanics
TOWNES AND SCHAWLOW: Microwave Spectroscopy
WANG: Solid-State Electronics
WHITE: Introduction to Atomic Spectra

McGRAW-HILL
BOOK COMPANY
New York
St. Louis
San Francisco
Düsseldorf
Johannesburg
Kuala Lumpur
London
Mexico
Montreal
New Delhi
Panama
Paris
São Paulo
Singapore
Sydney
Tokyo
Toronto

## LEONID V. AZÁROFF

*University of Connecticut*
*Storrs, Connecticut*

## ROY KAPLOW

*Massachusetts Institute of Technology*
*Cambridge, Massachusetts*

## N. KATO

*Nagoya University*
*Chikusa-ku, Nagoya, Japan*

## RICHARD J. WEISS

*Army Materials and Mechanics Research Center*
*Watertown, Massachusetts*

## A. J. C. WILSON

*The University of Birmingham*
*Birmingham, Great Britain*

## R. A. YOUNG

*Georgia Institute of Technology*
*Atlanta, Georgia*

# X-Ray Diffraction

This book was set in Times New Roman.
The editors were Jack L. Farnsworth and James W. Bradley;
the production supervisor was Sam Ratkewitch.
The drawings were done by John Cordes, J & R Technical Services, Inc.
The printer and binder was The Maple Press Company.

**Library of Congress Cataloging in Publication Data**
Main entry under title:

X-ray diffraction.

(International series in pure and applied physics)
Bibliography:    p.
1.   X-rays—Diffraction.    2    X-ray crystallography.
I. Azároff, Leonid V.
QC482.D5X7    1974                544'.66                73–21547
ISBN 0–07–002672–6

## X-RAY DIFFRACTION

1 2 3 4 5 6 7 8 9 0 M A M M 7 9 8 7 6 5 4

# CONTENTS

**Preface**                                                                    xv

**1   Scattering by atoms**                                                     1

I.   Elastic scattering theory                                                  1

   A.   Introduction                                             1

   B.   Theory                                                    2
      1. Hydrogenic solutions; 2. Expectation values; 3. Many-electron
      atoms; 4. Molecules; 5. Accuracy of wavefunctions.

   C.   Evaluation of the scattering factor $f_0$                 12
      1. Direct evaluation; 2. Total scattering from free atoms.

   D.   Anomalous dispersion                                      16

II.  Inelastic scattering theory                                                18

   A.   Compton effect                                            18
      1. Momentum space.

   B.   Compton profile                                           21
      1. Hydrogenic solution; 2. Many-electron atom; 3. Analysis of data

in the impulse approximation; 4. Relativistic corrections in the impulse approximation; 5. Nonspherical distributions.

C.   Approximate relationship between the scattering factor and the Compton profile ............................................. 33

D.   Total Compton cross sections .................................. 35

III.   Experiments ..................................................... 36

A.   Introduction .................................................. 36

B.   Experimental techniques ...................................... 37
1. Scattering factors; 2. Compton profile.

C.   Experimental results ......................................... 43
1. Inert gases He, Ne, Ar, Kr, and Xe; 2. $H_2$ molecule; 3. Lithium and sodium; 4. Magnesium and LiMg alloy; 5. Carbon (diamond, graphite, carbon black); 6. LiH and LiF; 7. B, BN, $B_4C$, and BeO; 8. Benzene, cyclohexane, and polyethylene; 9. Anisotropy in vanadium and iron; 10. Aluminum metal; 11. Titanium (polycrystalline).

D.   Conclusion ................................................... 66

**2   Kinematical theory** ............................................ 68

I.   Fundamental theory ............................................. 69

A.   Introduction .................................................. 69

B.   Restrictions in the theory .................................... 70
1. Small scattering amplitude; 2. Plane-wave approximation; 3. Coherent scattering.

C.   Directional variation of the amplitude and intensity of the scattered wave ........................................................ 72
1. The phase relationships among scattered rays; 2. The diffraction vector **K**; 3. The Fourier relationship between the set of inter-scatterer vectors and the intensity of the scattered waves.

D.   The three-dimensional pair distribution; retrieving the structure from measured intensities of scattering ..................... 75

II.   Spherically averaged samples: gases, liquids, glasses, and other amorphous solids, and polycrystalline aggregates ................. 79

A.   Introduction .................................................. 79

B.   The spherical average; reduction to one-dimensional functions .. 80

C.   Finite samples and particle size effects ...................... 85

D.   Forward scattering; the contribution of the mean density            87

E.   The finiteness of data; practical analysis for monatomic samples     95

F.   Atomic vibrations and other displacements                          i01
     1. The **r**-space broadening functions; 2. The general effects in $K$
     space; 3. Thermal diffuse scattering in polycrystalline aggregates;
     4. Anharmonic vibrations and asymmetric motions.

G.   Atomic charge distributions; the scattering factor                  116

H.   Scattering by two or more elements                                  118
     1. Partial distribution functions; 2. General analysis techniques;
     3. Short-range order, particularly in polycrystalline samples

I.   The use of pair distributions in the determination of crystal
     structures and the study of structural defects                      139

III.  Single crystals                                                    140

A.   The geometry of crystals                                            141
     1. Unit cells; 2. Transformations.

B.   The reciprocal lattice; the diffraction pattern for a single crystal  144

C.   The structure factor                                                146

D.   Geometrical constructs in reciprocal space                          148
     1. Brillouin zones and the Bragg law; 2. Diffraction of a continuum
     of wavelengths; the Laue pattern; 3. Simultaneous multiple diffrac-
     tion; 4. Sampling region.

E.   Particle size effects                                               154

F.   Lattice parameter variations; the simplest model for **K**-dependent
     perturbations                                                       156

G.   Numerical calculation of the diffraction by localized defects       159
     1. Theory; 2. Sample calculations and particular effects.

**3   Diffraction by crystals**                                         176

I.   Fundamentals of crystal diffraction                                 176

A.   Maxwell equations                                                   176

B.   Energy flux and energy density associated with the electromagnetic
     field                                                               179

C.   The x-ray wave fields in matter                                     180
     1. Dielectric constant for x rays; 2. Differential form of the field
     equation; 3. Integral form of the field equation.

D.  The propagation of waves                                         184
    1. Plane waves of complex variables; 2. Plane-wave solution of real
    variables; 3. Energy flow of the plane waves; 4. Absorption.

E.  Reflection and refraction                                        189
    1. Tangential continuity of the wave vectors; 2. The continuity of
    the amplitudes; 3. Total reflection.

II. Kinematical theory of crystal diffraction                        195

A.  Fraunhofer equation for three-dimensional diffraction            195
    1. Energy flow in the spherical wave; 2. Thomson scattering;
    3. Coherence of scattered waves; 4. Polarization.

B.  The form factor for Laue-Bragg reflection                        199
    1. Atomic form factor $f$; 2. Crystal-structure factor $F$; 3. Crystal
    shape function $S$; 4. Laue-Bragg reflection and the Ewald con-
    struction; 5. A remark on the crystal-structure factor.

C.  Imperfect crystals                                               204
    1. Finiteness of the crystal; 2. The direct image and Fraunhofer
    diffraction; 3. Planar defects; 4. Continuously bent lattice plane.

    Appendix 3A   Stationary-phase method                            217

I.   One-dimensional cases                                           217
II.  Two-dimensional cases                                           220
III. Three-dimensional cases                                         220
IV.  Fraunhofer diffraction                                          221
V.   Remarks on application to relations                             221

4   Dynamical theory for perfect crystals                            222

I.  Plane-wave theory                                                222

A.  Historical introduction                                          222

B.  Propagation of waves in crystalline media                        223
    1. Fundamental equations; 2. Single-beam cases; 3. Two-beam
    cases; 4. Dispersion surface; 5. Approximate forms of the dispersion
    surface and amplitude ratio; 6. Poynting vector and the wave
    packet; 7. Attenuation of the Bloch wave.

C.  Wave fields in two-beam cases                                    234
    1. Laue and Bragg cases; 2. Wave vectors in crystals; 3. Wave fields
    in vacuum and in crystals.

D.  Experimental aspects                                                                 262
    1. Double refraction and amplitude region; 2. *Pendellösung* fringes;
    3. Double refraction of the rays; 4. Intensity profiles of *O* and *G*
    beams; 5. Borrmann absorption and *Netzebene* fringes; 6. Rocking
    curves; 7. Integrated power.

II.  Spherical-wave theory                                                               295

    A.  Necessity for a spherical-wave theory                                            296
        1. Divergence of the wave vectors; 2. Divergence of the rays in the
        crystal; 3. Spherical form of the coherent wave front.

    B.  The spherical-wave theory for Laue cases                                         298
        1. The incident wave; 2. Crystal wave fields; 3. Approximate
        behavior of the wave fields; 4. Transmitted wave fields (Laue-Laue
        case).

    C.  Ray considerations for the crystal wave fields                                   306
        1. Rays associated with the wave fields given by (4-202); 2. Effect
        of the spherical form of the wave front of the incident wave.

    D.  Laue-Bragg case                                                                  311
        1. Type I; 2. Type II.

    E.  Bragg case                                                                       317

    F.  Experimental verification of the spherical-wave theory                           320
        1. Topographic observations; 2. Integrated power in the Laue-Laue
        case; 3. *Pendellösung* fringes in Laue (or Laue-Laue) cases; 4. Fringe
        phenomena in other cases; 5. Accurate determination of the
        structure factor.

**Appendix 4A   Mathematical structure of $Z = z \pm (z^2 + U^2)^{1/2}$**                **339**

  I.  Properties of the Fourier coefficients $\chi_g$ and $\chi_{-g}$                     339
 II.  Rigorous expressions of $Z$ for $z = u$                                             340
III.  Proof that $v \geq |V|$ in the Bragg case                                           341
 IV.  Proof that $|Z(+)| > |U|$ and $|Z(-)| < |U|$ in the Bragg case                      341
  V.  Extremal point of $|Z|$ on the line $u = w + iv$ in the Bragg case                  342
 VI.  Proof that $\alpha = |\chi_g/\chi_{-g}||u - (u^2 + U^2)^{1/2}|/|U|^2 < 1$ in the Bragg case   343
VII.  Imaginary part of *Anpassung* $\delta_e$ (4-56c)                                    343

**Appendix 4B   Interference**                                                           **344**

**Appendix 4C   Fourier integrals required for the spherical-wave theory**   **344**

I.   Laue-(Bragg)$^n$ cases; Type I   344

II.   Laue-(Bragg)$^n$ cases; Type II   347

III.   (Bragg)$^n$ cases   348

**5   Dynamical theory for imperfect crystals**   **350**

I.   Finiteness of the crystal and planar defects   351

   A.   Finite polyhedral crystals   351

   B.   Crystals containing a fault plane   354
       1. Plane-wave theory; 2. Spherical-wave theory.

   C.   Stacking faults   358
       1. Plane-wave solution; 2. Spherical-wave theory.

   D.   Twinned crystals   371
       1. Plane-wave theory; 2. Spherical-wave theory.

   E.   Misfit boundaries   380
       1. Plane-wave theory; 2. Spherical-wave theory.

II.   Continuously deformed crystals   389

   A.   Wave-optical basis of ordinary geometrical optics   389
       1. Eikonal theory; 2. Fermat's principle; 3. Variational principle and the ray equation; 4. Intensity and amplitude; 5. Absorption; 6. Reciprocity.

   B.   Eikonal theory in x-ray diffraction   400
       1. Polarizability in distorted crystals; 2. Solution of Maxwell's equation; 3. Fermat's principle and the ray equation; 4. Phases and *Pendellösung* fringes; 5. Intensity and amplitude.

   C.   Case of constant strain gradient   413
       1. Ray trajectory; 2. The Eikonal; 3. Intensity; 4. Absorption; 5. Integrated intensity; 6. Departure from Friedel's law; 7. Validity of present theory.

   D.   Lattice distortion caused by an oxide film   427
       1. Analysis of the fringe positions; 2. Symmetrical properties of the section topographs; 3. Other studies.

   E.   Related analyses   433

III.   Multicrystal systems   434

   A.   X-ray interferometer (three-crystal system)   434

   B.   Moiré fringes (two-crystal system)   436

**6   Powder diffractometry**                                                              **439**

I.   Introduction                                                                          439

II.  Geometry of powder diffractometers                                                    440

    A.   Bragg-Brentano focusing                                       440

    B.   Practical considerations                                       441

    C.   The Seemann-Bohlin arrangement                                 442

    D.   The symmetrical Bragg-Brentano arrangement                     443

    E.   Other arrangements without monochromator                       444

    F.   Arrangements with monochromator                                444
        1. Plane-crystal arrangements; 2. Curved-crystal arrangements.

III. Geometrical aberrations in the general Bragg-Brentano arrangement                      449

    A.   General calculation                                            451

    B.   Specific aberrations                                           454
        1. Finite width of source; 2. Finite width of receiving slit; 3. Specimen displacement and transparency; 4. Specimen tilt; 5. Equatorial divergence; 6. Equatorial cross terms; 7. Axial divergence.

IV.  Geometrical aberrations in monochromator arrangements                                  463

    A.   Plane-crystal monochromators                                   463

    B.   Curved-crystal monochromators in reflection                    463

    C.   Curved-crystal monochromators in transmission                  464

    D.   Effect on wavelength distribution                              464

V.   Physical aberrations in powder diffractometry                                         464

    A.   The effective spectrum                                         464

    B.   Refraction                                                     465

    C.   Dispersion                                                     466

    D.   Response variations                                            467
        1. Polarization factor; 2. Lorentz and other trigonometrical factors; 3. Absorption and extinction; 4. Quantum-counting efficiency; 5. Pulse-size discrimination.

VI.  Experimental determination of intrinsic line profiles                                 471

    A.   Origin of the observed profile                                 471

    B.   Properties of convoluted functions                             472

    C.   Synthesis of line profiles                                     474

D.   Analysis of line profiles     474
1. The available methods; 2. Fourier methods; 3. Variance methods; 4. Other methods.

VII.   Intensity analysis     480

A.   Statistical matters     481
1. Counting statistics; 2. Random setting errors; 3. Statistical effects in particle orientation; 4. Statistical absorption effects.

B.   Measurement of Bragg intensity     486
1. Correction for background; 2. Correction for experimental effects; 3. Propagation of errors.

C.   Optimization of effort     493
1. Minimization of the variance of a single parameter; 2. Minimization of an agreement index; 3. Optimization of many parameters.

**7   Single-crystal intensities**     **500**

I.   Introduction     500

II.   Single-crystal diffractometry     501

A.   Reciprocal-space sampling region     501

B.   Task of the diffractometer     508

C.   Diffractometer motions     509
1. The two general types of instruments; 2. Mechanical analog instruments; 3. Independent-axes diffractometers; 4. Axis-positioning and alignment considerations.

III.   Wavelength discrimination     529

A.   Need for discrimination     529

B.   Discrimination methods     530
1. Pulse-height analysis; 2. Balanced filters; 3. Fluorescent discriminators; 4. Total reflection; 5. Crystal monochromators.

IV.   Background control     548

A.   Degree of control needed     548

B.   Unnecessary background intensity     550

C.   Systematic errors from extraneous-wavelength radiation     554

D.   Background structure     559

E.   Thermal diffuse scattering     562

V.   Other effects                                              569

   A.   Simultaneous diffraction                               569

   B.   Extinction corrections                                 573

   C.   Effective integration of intensities                   578

**Appendixes**

   **A.**   Compton Total Cross Sections                        581

   **B.**   Compton Profiles for Hartree-Fock Wavefunctions     594

   **C.**   Scattering Factors for Hartree-Fock Wavefunctions   609

**References**                                                  619

**Indexes**                                                     651

   **Name Index**
   **Subject Index**

The field of x-ray diffraction owes its inception to a momentous discovery by Max Laue in 1913. As is often the case, the discovery resulted from a confluence in one place of several ideas, and, more importantly, from the proximity of several ingenious individuals, including the discoverer of x radiation, Roentgen, the outstanding theoretical physicist, Sommerfeld, and a leading crystallographer, Groth. [A most interesting recounting of the exciting days that led to this discovery has been presented by one of the protagonists, Paul Ewald, in *Fifty years of x-ray diffraction* (N. V. A. Oosthoek's Uitgeversmaatschappij, Utrecht, The Netherlands, 1962).] Within two years of this discovery, a most comprehensive theory explaining the basis of x-ray diffraction was developed by C. G. Darwin, who spent the next several years expanding this theory, before turning his attention to other problems in science.

The importance of x-ray diffraction to the development of modern science can be appreciated best by considering its widespread influence on chemistry, metallurgy, mineralogy, physics, and related sciences. First of all, x-ray diffraction confirmed the speculations of leading crystallographers and chemists that crystals were composed of atoms arranged in an orderly and periodic array. This enabled the development of the band theory of solids on which all the developments of solid-state physics are based, for example, the theoretical basis for classifying solids into conductors (metals), semiconductors, and nonconductors (insulators), and the subsequent discovery of transistors, lasers, and other solid-state devices so essential to the modern technology of computers, communication devices, space electronics, etc. Similarly, the modern understanding of metals and alloys, their structures, defects, and various properties, would not be possible if their crystal structures had not been previously revealed by x-ray diffraction studies.

The importance of the knowledge of the atomic arrangement is not limited to inorganic compounds, however, even though most of the concepts evolved in modern chemistry are based on the new knowledge thus gained regarding the nature of bonding in chemical compounds and minerals. It is not limited to the vistas it opened in the understanding of the mechanical properties of solids such as twinning, gliding, cleavage, and other consequences of deforming such solids, or the diffusion of atoms in crystals, or the growth of crystals, etc. It has also served to provide many important clues of more direct importance to man, such as the structure of DNA, RNA, and other life-controlling molecules. The determination of the crystal structure of penicillin during World War II enabled the purification of this important antibiotic, while the determination of the crystallite size and $SiO_2$ polymorphs in silica dust has saved the lives of countless workers exposed to such hazards. In fact, it is not an exaggeration to claim that x-ray diffraction has affected, directly or indirectly, virtually every terrestrial science, while making possible the technology that has sent men flying successfully in outer space.

An important ingredient in the advancement of any science is the availability of authoritative texts that can be used by all students coming to the field for the first time. A tremendous service was rendered to the scientific community by A. H. Compton and S. K. Allison when they published their volume *X-rays in theory and experiment* (D. Van Nostrand Company, Inc., Princeton, N.J., 1926) in which the early developments of x-ray diffraction theory and all the pioneering work of x-ray physicists that enabled these developments were described. Similarly, W. H. Zachariasen, in *Theory of x-ray diffraction in crystals* (John Wiley & Sons, Inc., New York, 1945), R. W. James, in *The optical principles of the diffraction of x-rays* (G. Bell & Sons, Ltd., London, 1950) and M. von Laue, in *Röntgenstrahl-Interferenzen* (Akademische Verlagsgesellschaft, Frankfurt, 1960), have expounded the basis of modern x-ray diffraction theory and its application to the characterization of crystals. Of equal importance have been the books devoted to the development of various methods of crystal-structure analysis, the most comprehensive of which is M. J. Buerger's *Crystal-structure analysis* (John Wiley & Sons, Inc., New York, 1960). These books are well known to all serious students of x-ray diffraction.

In preparing the present book, therefore, the authors have assumed a prior familiarity on the reader's part with the contents of the above-cited classic books. The reason for this undertaking was that a similarly authoritative treatment of more recent developments in x-ray diffraction was not available in a single volume. Because of the high degree of specialization that has resulted in the wake of the rapid proliferation of x-ray diffraction, it is no longer possible to find a single person who has expertise in all its ramifications, and a collaboration of several authors was necessary. Thus, R. J. Weiss, who has spent years studying the interaction of x-rays with atoms, prepared the first chapter in this book. Roy Kaplow, well known for his studies of x-ray scattering from liquids and polycrystalline aggregates, wrote the second chapter. The really important advances in the theory of x-ray diffraction by single crystals have come in the dynamical theory, and so N. Kato, who is actively following in the pioneering footsteps of Darwin, Ewald, and von Laue, wrote the next three chapters. In order to present the experimental aspects of x-ray diffraction, especially the analytical procedures necessary for interpreting actual observations in terms of the available theory, two well-known x-ray crystallographers, A. J. C. Wilson and R. A. Young, collaborated in preparing the last two chapters in this book. The undersigned has been concerned, primarily, with coordinating their efforts, lending support to them to complete their respective tasks, styling the manuscript in order to make it more uniformly comprehensible to the reader, and writing this preface to protect these authors from the reviewer's criticisms.

This book contains seven chapters that attempt to present up-to-date expositions of the developments in x-ray diffraction during the years that followed the publication of Compton and Allison's second edition of their monumental work in 1935. The elastic and inelastic scattering of x rays by atoms and the information derived therefrom regarding momentum densities and atomic wavefunctions is described in the first chapter. A novel approach to the kinematical theory of x-ray diffraction through the use of pair-distribution functions and the application of high-speed computers enabling more exact calculations is presented in Chap. 2. This chapter is primarily concerned with diffraction by noncrystalline and polycrystalline aggregates and with imperfections in single crystals. Still a different presentation of the kinematical theory is given in Chap. 3. The theory is first derived from Maxwell's equations, and the groundwork is laid for the dynamical theory which is fully expounded in Chap. 4. Both the plane-wave and the spherical-wave theories are considered in Chap. 4 so that it constitutes the most comprehensive coverage yet presented in a book, particularly when combined with the dynamical theory for imperfect crystals developed in Chap. 5. Modern advances in experimentation are presented in the last two chapters. Here, emphasis is placed on the interpretation of experimental results rather than on instrumentation. Chapter 6 considers the geometries and analyses required in powder diffractometry, while Chap. 7 does the same for single-crystal diffractometry, emphasizing the case of small single crystals bathed entirely in the incident x-ray beam. For the most part, the chapters in this book do not strive for complete coverage of all possible topics but, rather, select the important new developments for emphasis.

This has been done to limit the size of this book as well as to assure its maximum utility to the reader. A companion volume devoted to x-ray spectroscopy presents a discussion of two-crystal spectrometers as well as the recent developments in this field of x-ray physics and their applications to various problems in science.

Several important topics are omitted entirely from this book, including the subjects of crystal-structure analysis, x-ray diffraction instrumentation, x-ray topography, and computational methods in x-ray diffraction. This has been done deliberately in order to keep the size of the book within reasonable bounds. Excellent books, monographs, and review articles on these topics are already available so that it was felt that their exclusion here would not inconvenience the interested reader. References to these works can be found in the literature lists collected at the end of this book, and specific citations are included in appropriate places in the text.

A note is in order about the preparation of a manuscript such as this. Advances in science, and books describing them, are made by individuals of disparate tempers and interests. In order to exhort them to the necessary exertions in view of the limited emoluments that accrue to writers on highly technical subjects, the authors were enticed to confabulations that took place on the shores of the Atlantic Ocean in Cape Cod, the Alpine Village of Aussois in France, and on the rolling hills of Storrs, Connecticut. Discussions were lubricated by ample quantities of the wine of the land, and finally, appeals were made to their responsibility to the scientific community. A not insignificant factor in all this was an award of moneys made available to the University of Connecticut[1] by the U.S. Air Force Materials Laboratory through the personal efforts of Mr. William Baun. I trust that the readers of this book will share with me a deep appreciation of all their efforts as well as those of the typists and others who made the final book a reality.

LEONID V. AZÁROFF

[1] The preparation of the manuscript for this book was supported, in part, by Contract F33615-68-C-1602 issued by U.S. Air Force Systems Command, Aeronautical Systems Division, Wright-Patterson AFB, Ohio.

**X-RAY DIFFRACTION**

# SCATTERING BY ATOMS

## I. ELASTIC SCATTERING THEORY

### A. Introduction

It is perhaps surprising that the theory of x-ray scattering by atoms is still undergoing significant revision. Unfortunately experiments have not explored the fundamental phenomena with sufficient accuracy to assess the uncertainty in the theory. Thus there has been little interplay between theory and experiment. In scattering theory it is convenient to treat elastic and inelastic scattering separately. The fundamental problem in elastic scattering theory arises from the binding of the electrons to the atom. One difficulty is the correlation problem, i.e., the Coulomb repulsion between the electrons. Only approximate solutions to the Schrödinger equation are possible for the many-electron atom, although it is believed that the Hartree-Fock approximation can describe the x-ray elastic scattering from atoms to an error of only 1%. The second difficulty associated with the binding of the electrons is the so-called anomalous dispersion effect. As the photon energy approaches the electron excitation energies, the electrons couple very strongly to the x rays, and large variations in the elastic scattering cross section may occur. A third difficulty arises from relativistic effects, a problem only recently considered for bound electrons. Corrections of several percent may be involved, but only for the heavier atoms.

Fortunately it is sometimes possible to select experimental conditions under which both the anomalous dispersion and relativistic corrections are small, so that comparison between theory and experiment is focused on the accuracy of the solution to the Schrödinger equation. Significance can be placed on differences of about 1%.

The theory of x-ray inelastic scattering by atoms has not been explored as extensively as elastic scattering since the experimental efforts in this field have been limited. For free electrons initially at rest there is the 40-year-old Klein-Nishina (1929) formula which gives the differential cross section (including relativistic effects) both for the angular distribution and for the energy loss of the photon. Only recently has the theory been extended to consider free electrons not initially at rest (Jauch and Rohrlich, 1955), but the most difficult problem of inelastic scattering by bound electrons has only been briefly considered (Platzman and Tzoar, 1965; Eisenberger and Platzman, 1970; Currat, DeCicco, and Weiss, 1971).

Inelastic scattering by bound electrons is undergoing renewed interest due to the revival of experiments measuring the differential cross section for photon energy loss (Compton profiles). These experiments provide information about the initial (ground-state) momentum of the electron before the inelastic collision. The integrated Compton profiles, i.e., the integral over all photon energy losses, was considered in a well-known theoretical paper by Waller and Hartree (1929), but more recently Bonham (1965) has calculated the corrections to the Waller-Hartree theory and has found them to be large $\sim 25\%$. Currat et al. (1971) have suggested another method for calculating the integrated Compton profiles (Section II D) and find differences of as much as a factor of 3 compared to the Waller-Hartree theory. Experimental results on integrated Compton profiles (total Compton scattering) are very sparse.

It appears at present that elastic scattering is well suited for the determination of spatial distributions of electrons on free atoms and simple molecules and inelastic scattering to their linear momentum distributions. When atoms condense to form solids it becomes difficult to determine the spatial distributions (particularly of the valence electrons) from elastic scattering because of the coherent effects between atoms. However, the positions of the atoms as a whole are readily obtainable even in very complicated crystals. Fortunately the momentum distributions are still readily obtained in solids from the inelastic scattering since the Compton process is incoherent.

At the present writing, theory and experiment for both elastic and inelastic x-ray scattering only agree occasionally.

## B.  Theory

The theory of elastic scattering of x rays by atoms requires solutions of the Schrödinger equation in the vector potential of the photoelectric field. If the energy of the photon is not close to an absorption edge or to an excited state of the atom, its interaction can be treated as a small perturbation. The Schrödinger equation is then

$$\sum_i H_0 \Psi + H_1 \Psi = i\hbar \frac{\partial \Psi}{\partial t} \qquad (1\text{-}1)$$

where $H_0$ is the Hamiltonian for the electrons in the Coulomb field of the nucleus and the other electrons and $H_1$ is the Hamiltonian for the electrons in the vector potential of the electromagnetic field. $\Psi$ is the wavefunction, and the summation is over the $i$ electrons.

$$H_0 = \sum_i \left[ -\frac{\hbar^2}{2m} \nabla_i^2 + \sum_{j \neq i} \left( \frac{e^2}{2r_{ij}} \right) - \frac{Ze^2}{r_i} \right] \qquad (1\text{-}2)$$

The first term on the right in (1-2) is the kinetic energy operator; the second, the inter-electron Coulomb repulsion; and the third, the attractive nuclear potential.

$$H_1 = -\frac{i\hbar e}{mc} (\mathbf{A}_i \cdot \nabla_i) + \frac{e^2}{2mc^2} A_i^2 \qquad (1\text{-}3)$$

where $\mathbf{A}$ is the vector potential of the electromagnetic field. The effect of the perturbation is to give rise to elastically and inelastically (Compton) scattered waves. The amplitude of the elastic wave can be written as

$$\frac{e^2}{mc^2} K(f_0 + \Delta f' + i\,\Delta f'') \equiv \frac{e^2}{mc^2} Kf \qquad (1\text{-}4)$$

where $e^2/mc^2 = 0.282 \times 10^{-12}$ cm is the fundamental photon-electron scattering length, and $K$, the polarization, equals either unity or $\cos 2\theta$ ($2\theta$ is the scattering angle). $f_0$ is the scattering factor

$$f_0 = \int \Psi_0^* \sum_i e^{i\mathbf{S} \cdot \mathbf{r}} \Psi_0 \, d\mathbf{r} \qquad (1\text{-}5)$$

where $\Psi_0$ is the ground-state wavefunction, i.e., the solution to the unperturbed Hamiltonian $H_0$ in $(1-1)$, and $\mathbf{S}$ is the scattering vector.

$$\mathbf{S} = \mathbf{k} - \mathbf{k}_0 \qquad (1\text{-}6)$$

$\mathbf{k}_0$ and $\mathbf{k}$ are, respectively, the incoming and scattered photon wave vectors, $|k| = 2\pi/\lambda$. The sum $i$ in equation (1-5) is over the $i$ electrons. $\Delta f'$ and $\Delta f''$ are the so-called anomalous dispersion corrections arising from the interaction between the perturbing electromagnetic field and the excited states (both real and virtual and including the continuum).

The theoretical problem in x-ray elastic scattering is essentially threefold:

*1*  Solving (1-1) for the unperturbed Hamiltonian and determining the ground-state wavefunction $\Psi_0$

*2*  Employing $\Psi_0$ in (1-5) to determine the scattering factor $f_0$

*3*  Calculating $\Delta f'$ and $\Delta f''$

Step 2 is trivial once $\Psi_0$ is determined from step 1 but, unfortunately, $\Psi_0$ can rarely be calculated exactly. Step 3 is the most difficult since it involves all the excited-state wavefunctions. However, when the photon energy is far from any excitation energies, $\Delta f'$ and $\Delta f''$ can be very small compared to $f_0$.

The solution of the unperturbed Schrödinger equation has attracted considerable interest. Since the Hamiltonian $H_0$ does not contain the time, the solutions are of the form

$$\Psi = \Psi_0 e^{-iEt/\hbar} \tag{1-7}$$

which, when substituted in (1-1) ($H_1 = 0$),

$$\sum_i H_0 \Psi = i\hbar \frac{\partial \Psi}{\partial t} \tag{1-8}$$

yields the time-independent Schrödinger equation

$$\sum_i H_0 \Psi_0 = E\Psi$$

$$\sum_i \left[ -\frac{h^2}{2m} \nabla_i^2 + \sum_{j \neq i} \left( \frac{e^2}{2r_{ij}} \right) - \frac{Ze^2}{V_i} \right] \Psi_0 = E\Psi_0 \tag{1-9}$$

where $E$ is the total energy and $Z$ the nuclear charge. The presence of the Coulomb repulsive term between electrons $\sum_{j \neq i} \frac{1}{2} e^2 / r_{ij}$ prevents a simple analytic solution.

**1.** *Hydrogenic solutions* Only in the case of the hydrogenic atoms (single electron) can (1-9) be solved in a simple analytic form, for in these cases the $e^2/r_{ij}$ terms are absent. Since these hydrogenic solutions are frequently employed as basis sets for the approximate solutions of the many-electron atom, we give them below. Since the nuclear potential $(-Ze^2/r)$ is spherically symmetric, the solution in spherical coordinates $r$, $\theta$, $\varphi$ can be written as a product of wavefunctions dependent respectively on $r$, $\theta$, and $\varphi$ only.

$$\Psi_0 = R(r)\Theta(\theta)\Phi(\varphi) \tag{1-10a}$$

$$R(r) = \left\{ \frac{4[(n - l - 1)!]Z^3}{[(n + l)!]^3 n^4} \right\}^{1/2} \left( \frac{2Zr}{n} \right) \exp\left( -\frac{Zr}{n} \right) \left[ L_{n+1}^{2l+1} \left( \frac{2Zr}{n} \right) \right] \tag{1-10b}$$

$$\Theta(\theta) = \left\{ \frac{(2l + 1)[(l - |m_l|)!]}{2[(l + |m_l|)!]} \right\}^{1/2} \sin^{|m_l|}\theta \left[ P_l^{|m_l|}(\cos\theta) \right] \tag{1-10c}$$

$$\Phi(\varphi) = \left( \frac{1}{2\pi} \right)^{1/2} \exp(im_l\varphi) \tag{1-10d}$$

where $n$, $l$, and $m_l$ are the quantum numbers. Atomic units of length are used throughout (1 au $= h^2/me^2 = 0.529 \times 10^{-8}$ cm). Values of the Laguerre polynomials $L_{n+1}^{2l+1}(2Zr/n)$ and Legendre functions $P_l^{|m_l|}(\cos\theta)$ are given in Table 1-1.

In atomic units, the energy $E$ is given by ($1$ au $= me^4/h^2 = 27.196$ eV)

$$E = -\frac{Z^2}{2n^2} \quad (1\text{-}11)$$

The wavefunction in (1-10) is normalized so that

$$\int \Psi_0 \Psi_0^* \, d\mathbf{r} = 1$$

$$\int_0^\infty RR^* r^2 \, dr = 1$$

$$\int_0^\pi \Theta\Theta^* \sin\theta \, d\theta = 1 \quad (1\text{-}12)$$

$$\int_0^{2\pi} \Phi\Phi^* \, d\varphi = 1$$

It is also a property of these solutions that they are orthogonal

$$\int \Psi_0 \Psi_0' \, d\mathbf{r} = \begin{cases} 1 & \text{for } n = n', l = l', \text{ and } m_l = m_l' \\ 0 & \text{for any primed and unprimed not equal} \end{cases} \quad (1\text{-}13)$$

**Table 1-1  VALUES OF $L_{n+1}^{2l+1}(2Zr/n)$ AND $P_l^{|m_l|}(\cos\theta)$**

| $n$ | $l$ | Desig-nation | $L_{n+1}^{2l+1}(2Zr/n)$ | $l$ | $m_l$ | $P_l^{|m_l|}(\cos\theta)$ |
|---|---|---|---|---|---|---|
| 1 | 0 | $1s$ | $-1$ | 0 | 0 | 1 |
| 2 | 0 | $2s$ | $2Zr - 4$ | 1 | $1, -1$ | 1 |
| 3 | 0 | $3s$ | $-\dfrac{4Z^2r^2}{3} + 12Zr - 18$ | 1 | 0 | $\cos\theta$ |
| 4 | 0 | $4s$ | $\dfrac{Z^3r^3}{2} - 12Z^2r^2 + 72Zr - 96$ | 2 | $2, -2$ | 3 |
| 2 | 1 | $2p$ | $-6$ | 2 | $1, -1$ | $3\cos\theta$ |
| 3 | 1 | $3p$ | $16Zr - 96$ | 2 | 0 | $\dfrac{3\cos^2\theta - 1}{2}$ |
| 4 | 1 | $4p$ | $-15Z^2r^2 + 300Zr - 1200$ | 3 | $3, -3$ | 15 |
| 3 | 2 | $3d$ | $-120$ | 3 | $2, -2$ | $15\cos\theta$ |
| 4 | 2 | $4d$ | $360Zr - 5760$ | 3 | $1, -1$ | $\dfrac{3(5\cos^2\theta - 1)}{2}$ |
| 5 | 2 | $5d$ | $\dfrac{2016}{5} Z^2r^2 + 35,280Zr - 105,840$ | 3 | 0 | $\dfrac{5\cos^3\theta - 3\cos\theta}{2}$ |
| 4 | 3 | $4f$ | $-5040$ | | | |
| 5 | 3 | $5f$ | $16,128Zr - 322,560$ | | | |

**2.   *Expectation values***   There are many measurable ground-state properties of the atom, and these can be calculated from the wavefunction and the appropriate operator $Q$. These are called expectation values and are denoted by $\langle Q \rangle$

$$\langle Q \rangle = \int \Psi_0^* \, Q\Psi_0 \, d\mathbf{r} \qquad (1\text{-}14)$$

Various measurable properties and the appropriate operators to employ in (1-14) are listed in Table 1-2.

**3.   *Many-electron atoms***   When the atom or molecule has more than one electron, the presence of the Coulomb repulsion between electrons presents formidable mathematical difficulties. An entirely new approach employing trial wavefunctions is taken and one makes use of the very powerful variational principle. Since the ground-state energy of an atom or molecule is the lowest possible energy, any wavefunction that is not the exact solution of the Schrödinger equation will give an expectation value of energy $\langle E \rangle$ higher than the ground state. A trial wavefunction (generally expanded in some convenient mathematical series like $\sum_n a_n r^m e^{-b}$) is employed to evaluate $\langle E \rangle$, minimizing the energy by varying the parameters employed in the trial wavefunction.

**Table 1-2   VARIOUS GROUND-STATE PROPERTIES AND THE APPROPRIATE OPERA-TORS IN EVALUATING THEM [EQUATION (1-14)]**

| Property | Operator $Q$ |
|---|---|
| Kinetic energy $\langle KE \rangle$ | $\sum_i \left\{ -\dfrac{\hbar^2}{2m} \left[ \dfrac{1}{r_i^2} \dfrac{\partial}{\partial r_i} \left( r_i^2 \dfrac{\partial}{\partial r_i} \right) + \dfrac{1}{r_i^2 \sin^2 \theta_i} \dfrac{\partial^2}{\partial \varphi_i^2} \right.\right.$ $\left.\left. + \dfrac{1}{r_i^2 \sin \theta_i} \dfrac{\partial}{\partial \theta_i} \left( \sin \theta_i \dfrac{\partial}{\partial \theta_i} \right) + \dfrac{1}{r_i^2 \sin^2 \theta_i} \dfrac{\partial^2}{\partial \varphi_i^2} \right] \right\}$ |
| Potential energy $\langle V \rangle$ | $\sum_i \left[ \sum_{j \neq i} \left( \dfrac{e^2}{2r_{ij}} \right) - \dfrac{Ze^2}{r_i} \right]$ |
| Total energy $\langle H_0 \rangle$ | Sum of $\langle KE \rangle$ and $\langle V \rangle$. For the hydrogenic solutions (10) the virial theorem is satisfied; i.e., $\langle V \rangle = -2\langle KE \rangle$ |
| X-ray scattering factor $f_0(S)$ | $\sum_i \exp(i\mathbf{S} \cdot \mathbf{r}_i)$ |
| Charge density $\rho(\mathbf{r})$ | $\sum_i \delta(\mathbf{r} - \mathbf{r}_i)$ |
| $\langle r \rangle$ $\langle r^n \rangle$ | $r$ $r^n$ |
| Components of linear momentum | |
| $\langle p_x \rangle$; $\langle p_y \rangle$; $\langle p_z \rangle$ | $-i\hbar \dfrac{\partial}{\partial x}$ ; $-i\hbar \dfrac{\partial}{\partial y}$ ; $-i\hbar \dfrac{\partial}{\partial z}$ |

There are two overall approaches in selecting a trial wavefunction. In one, the trial wavefunction contains terms that specifically include the interelectronic separation, thus allowing for correlation between electrons; i.e., it tends to keep the electrons apart due to their Coulomb repulsion. Wavefunctions which include correlation, however, are extremely cumbersome to evaluate and, even though they give the best results, their difficulty has limited their use to systems containing very few electrons ($<10$). A much more widely used but less accurate trial wavefunction neglects terms containing the interelectronic coordinates and assumes that the wavefunction of the atom can be written as a product of one-electron wavefunctions; i.e., each electron has its own wavefunction

$$\Psi_0 = \prod_i \varphi_i(\mathbf{r}_i) \qquad (1\text{-}15)$$

In addition, the Pauli exclusion principle requires that a spin function $\sigma_i$ be included for each electron. The most successful such one-electron scheme is the Hartree-Fock method which modifies (1-15) by writing the product of one-electron wavefunctions (including spin functions) as a determinant.

$$\Psi_0 = \begin{vmatrix} \varphi_1(r_1)\sigma_1(r_1) & \cdots & \varphi_1(r_n)\sigma_1(r_n) \\ \varphi_2(r_1)\sigma_2(r_1) & \cdots & \varphi_2(r_n)\sigma_2(r_n) \\ \cdots\cdots\cdots\cdots\cdots\cdots\cdots\cdots \\ \varphi_n(r_1)\sigma_n(r_1) & \cdots & \varphi_n(r_n)\sigma_n(r_n) \end{vmatrix} \qquad (1\text{-}16)$$

Such a determinantal form assures that the wavefunction is antisymmetric; i.e., swapping the coordinates of any pair of electrons reverses the sign of the wavefunction, since interchanging any pair of columns in the determinant has this property. A subsidiary condition of orthogonality [equation (1-13)] for all one-electron wavefunctions is also imposed on (1-15). It is also generally assumed in the Hartree-Fock method that the radial and angular functions are separable [equation (1-10a)]. The individual one-electron radial wavefunctions in (1-16) are generally expanded in a series of exponentials so that we have

$$\varphi_i(\mathbf{r}_i) = R_i(r_i)\Theta_i(\theta_i)\Phi_i(\varphi_i)$$
$$= \sum_n a_n r_i{}^m \exp(-b_n r_i)\Theta_i(\theta_i)\Phi_i(\varphi_i) \qquad (1\text{-}17)$$

where the angular functions are just the hydrogenic functions given by (1-10c) and (1-10d). The functions in (1-17) are called the basis sets, and it is at the discretion of the theorist to decide the form and extent of the basis set. The Hartree-Fock method yields minimum total energies $\langle E \rangle$ approximately $\frac{3}{4}\%$ too high (compared to experiment). A correlated wavefunction, though, can reduce the difference appreciably. In helium this difference has been reduced to one part in $10^{10}$. (The difference between the Hartree-Fock energy and the experimental energy has come to be called the *correlation energy*.)

In order to improve the Hartree-Fock energy without introducing correlation terms in the trial wavefunction, a procedure called *configuration interaction* is employed in which a sum of determinants like (1-16), each with a different basis set [i.e., a different selection of the angular functions in (1-17)], is used. This, of course, increases the complexity of the calculation but, in some cases, has reduced the total energy $\langle E \rangle$ to within the correct value to one part in $\sim 10^4$.

**4.   *Molecules*** For molecules, the mathematical problem is complicated by the presence of the additional nuclear centers. A variety of procedures are employed to construct trial wavefunctions such as forming a determinant of wavefunctions, each centered on the various nuclei (Heitler-London scheme), or constructing new one-electron wavefunctions from sums or differences of atomic functions (Hund-Mullikan scheme). Both exponential and Gaussian basis sets have been tried, but the types of functions and the methods of approach seem inexhaustible, limited only by the in-genuity of the theorist and the capability of the computer.

**5.   *Accuracy of wavefunctions*** Theorists have expended most of their effort in devising wavefunctions which give good energies. In most cases this is the only criterion employed in the choice of a wavefunction. It is, in fact, the only criterion which is bounded; i.e., one cannot reduce the energy below the observed energy. This in itself is of minor interest to the x-ray scattering calculations, however, since it is the accuracy of the wavefunction that is important in evaluating $f_0$ in equation (1-5). There is a general rule that the error in the wavefunction $\Delta\Psi_0$ is proportional to the square root of the error in the energy

$$\frac{\Delta\Psi_0}{\Psi_0} \sim \left(\frac{\Delta E}{E}\right)^{1/2} \qquad (1\text{-}18)$$

Thus a 1% error in the energy will yield a 10% error in the wavefunction. This is not the case for a Hartree-Fock wavefunction (single determinant), however, for which the average error in the wavefunction is of the same order as the error in the energy (Weiss, 1966, p. 168).

At the present time, there exists very limited theoretical (or experimental) evidence for assessing the true errors in the Hartree-Fock wavefunctions (or in the charge density). For the helium atom one can compare the Hartree-Fock wave-function with a very accurate wavefunction including correlation (Weiss, 1966, p. 187), and we find that the scattering factors differ at most by $\sim 1\frac{1}{2}\%$ (and $\langle r^2 \rangle$ by 1%), whereas the error in the Hartree-Fock energy is $\sim\frac{3}{4}\%$. This is probably the most ideal case. For beryllium one can compare the Hartree-Fock wavefunction with a relatively accurate 55 configuration interaction wavefunction. At most, the scattering factors differ by less than a percent, although $\langle r^2 \rangle$ differs by $\sim 5\%$. Here again the Hartree-Fock energy is about $\frac{3}{4}\%$ too high. For the best molecular wavefunctions (not including correlation) the error in the x-ray scattering factor is probably at least several percent.

It is instructive to consider two illustrative cases to see how errors in the wavefunction are reflected in both the x-ray scattering factor $f_0$ and the Compton profiles $J(z)$. A further discussion of the latter is given in Section II.

*a. Hydrogen atom* Consider a one-electron atom with nuclear charge $Z$. The exact ground-state solution ($1s$) is given by equations (1-10) ($n = 1$, $l = 0$, $m_l = 0$)

$$\Psi_0 = Z^{3/2} \frac{\exp(-Zr)}{\sqrt{\pi}}$$

$$E = \frac{-Z^2}{2}$$

(1-19)

Consider now a trial wavefunction which is the sum of Gaussian functions

$$\Psi_0 = A \sum_n \exp(-b_n r^2)$$

(1-20)

and observe how the errors in the energy and scattering factor $f_0$ vary as more terms are added to (1-20) to improve the fit to the correct wavefunction. Table 1-3 gives selected values of $f_0$, $J(z)$, and $\Psi_0$ for a one-, two-, and three-term expansion of (1-20) for both $Z = 1$ and $Z = 4$.

There are several points worth noting in Table 1-3. In general, the errors in the wavefunction itself are given approximately by (1-18), but the magnitude of the errors for $f_0$ are considerably less over the range of interest, say, from $\sin\theta/\lambda = 0$ to the value at which $f_0$ falls to one-tenth its maximum. This is due to the fact that all the $f_0$ values agree at $\sin\theta/\lambda = 0$ since $f_0 = Z$. Hence the errors are very small at small values of $\sin\theta/\lambda$. Actually the values of $Z - f_0$ more nearly evidence errors of a magnitude given by (1-18). Thus we see that, over the range of values of general interest, even a poor wavefunction yields a scattering factor that is reasonably accurate. More specifically,

$$\frac{\Delta f_0}{f_0} \sim \frac{\Delta E}{E} \qquad \text{from } Z > f_0 > 0.2Z$$

(1-21)

On the other hand, the Compton profile over the range of interest, i.e., from $J(0)$ to $J(0)/5$, shows significantly larger departure than (1-20), namely,

$$\frac{\Delta J}{J} \sim \left(\frac{\Delta E}{E}\right)^{1/2} \qquad \text{from } J(0) > J(z) > \frac{J(0)}{5}$$

(1-22)

*b.* *Helium atom* The ground state of the helium atom (two electrons) is the simplest case for which electron-electron correlation enters. As an approximate wavefunction that specifically includes interelectronic correlation, we take

$$\Psi_0 = A \exp(-ar_1) \exp(-ar_2)(1 + br_{12}^2) \qquad (1\text{-}23)$$

**Table 1-3** SELECTED VALUES OF THE WAVEFUNCTION $\Psi_0$, SCATTERING FACTOR $f_0$, AND COMPTON PROFILE $J(z)$ FOR A ONE-, TWO-, AND THREE-TERM GAUSSIAN EXPANSION (1-20) OF THE HYDROGENIC ATOM (SINGLE ELECTRON) FOR $Z = 1$ AND $Z = 4$. ALSO INCLUDED ARE THE MINIMUM ENERGY VALUES OF THE PARAMETERS AS WELL AS THE EXACT VALUES OF THE OBSERVABLES

| | $Z = 1$ | | | | $Z = 4$ | | | |
|---|---|---|---|---|---|---|---|---|
| | Exact | 1-term | 2-term | 3-term | Exact | 1-term | 2-term | 3-term |
| $A^2$ | | 0.07643 | 0.04284 | 0.02460 | | 4.8927 | 2.7548 | 1.5809 |
| $b_1$ | | 0.2829 | 0.231 | 0.192 | | 4.527 | 3.701 | 3.073 |
| $b_2$ | | | 1.6 | 0.798 | | | 25.31 | 12.707 |
| $b_3$ | | | | 3.67 | | | | 58.50 |
| $E$ | $-0.500$ | $-0.425$ | $-0.484$ | $-0.494$ | $-8.000$ | $-6.791$ | $-7.742$ | $-7.905$ |

| $r$ | $\Psi_0$ (exact) | $\Psi_0$ (1-term) | $\Psi_0$ (2-term) | $\Psi_0$ (3-term) | $\Psi_0$ (exact) | $\Psi_0$ (1-term) | $\Psi_0$ (2-term) | $\Psi_0$ (3-term) |
|---|---|---|---|---|---|---|---|---|
| 0.0 | 0.564 | 0.276 | 0.414 | 0.471 | 4.52 | 2.22 | 3.32 | 3.78 |
| 0.5 | 0.342 | 0.258 | 0.335 | 0.341 | 0.611 | 0.713 | 0.661 | 0.636 |
| 1.0 | 0.208 | 0.208 | 0.207 | 0.205 | 0.083 | 0.024 | 0.041 | 0.058 |
| 1.5 | 0.126 | 0.146 | 0.129 | 0.129 | 0.011 | 0.0000 | 0.0004 | 0.001 |
| 2.0 | 0.076 | 0.089 | 0.082 | 0.080 | 0.0015 | 0.0000 | 0.0000 | 0.0000 |
| 3.5 | 0.017 | 0.009 | 0.012 | 0.015 | | | | |
| 5.1 | 0.0034 | 0.0002 | 0.0005 | 0.0011 | | | | |

| $\dfrac{\sin\theta}{\lambda}$ | $f_0$ (exact) | $f_0$ (1-term) | $f_0$ (2-term) | $f_0$ (3-term) | $f_0$ (exact) | $f_0$ (1-term) | $f_0$ (2-term) | $f_0$ (3-term) |
|---|---|---|---|---|---|---|---|---|
| 0.0 | 1.000 | 1.000 | 1.000 | 1.000 | 1.000 | 1.000 | 1.000 | 1.000 |
| 0.2 | 0.481 | 0.458 | 0.484 | 0.480 | 0.947 | 0.952 | 0.953 | 0.951 |
| 0.4 | 0.131 | 0.044 | 0.115 | 0.123 | 0.811 | 0.811 | 0.823 | 0.819 |
| 0.6 | 0.040 | 0.001 | 0.034 | 0.039 | 0.641 | 0.645 | 0.656 | 0.648 |
| 1.0 | 0.0069 | 0.0000 | 0.0017 | 0.0061 | 0.350 | 0.295 | 0.340 | 0.342 |
| 1.5 | 0.0015 | 0.0000 | 0.0000 | 0.0004 | 0.153 | 0.064 | 0.136 | 0.145 |

| $z$ | $J$ (exact) | $J$ (1-term) | $J$ (2-term) | $J$ (3-term) | $J$ (exact) | $J$ (1-term) | $J$ (2-term) | $J$ (3-term) |
|---|---|---|---|---|---|---|---|---|
| 0.1 | 0.834 | 0.737 | 0.753 | 0.777 | 0.212 | 0.187 | 0.192 | 0.198 |
| 0.5 | 0.435 | 0.482 | 0.471 | 0.463 | 0.203 | 0.182 | 0.186 | 0.192 |
| 1.0 | 0.106 | 0.128 | 0.118 | 0.110 | 0.177 | 0.168 | 0.170 | 0.173 |
| 4.0 | 0.0002 | 0.0000 | 0.0000 | 0.0001 | 0.027 | 0.032 | 0.030 | 0.027 |

where $r_{12}$ is the distance between the two electrons and $r_1$ and $r_2$ are the distances, respectively, from electrons one and two to the nucleus. Such a wavefunction is typical of the sort of solution one expects. A cusp develops around the point $r_{12} = 0$; i.e., when the two electrons occupy the same spot, the wavefunction is a minimum. It is known that the actual form of the cusp, when $r_{12}$ is small, is exponential in helium $[\exp(\frac{1}{2}r_{12}{}^2)]$ rather than of the form $1 + br_{12}{}^2$, but the latter form is used for mathematical convenience to demonstrate how this cusp is related to the Compton profiles (Section II A 1). (The exponential form prevents an analytic evaluation of the Compton profile.) The error in the energy for the wavefunction (1-23) is $\sim 0.9\%$, which has a similar magnitude to the error in the Hartree-Fock energy, $\sim 1.3\%$. In Table 1-4a, the scattering factors $f_0$ obtained from the wavefunction of (1-23), from the Hartree-Fock wavefunction, and from a highly accurate correlated wavefunction are compared. It is interesting that the Hartree-Fock wavefunction yields more accurate scattering factors than those calculated from (1-23) even though the error in the energy is about 50% smaller for the function in (1-23). But in both cases, the error in $f_0$ over the range of interest, i.e., from $Z > f > 0.2Z$, is not more than 1%.

The Compton profile is given in Table 1-4b and, here again, the Hartree-Fock wavefunction gives more accurate values than those calculated from the wavefunction of (1-23). Over the range of principal interest the errors do not exceed $\sim 1\%$, but this may be due to the particularly simple wavefunction in helium. This is not the case for the beryllium ground state where the Hartree-Fock and configuration interaction wavefunctions yield values of $J$ that differ by $\sim 8\%$ although the scattering factors agree to 1%.

Table 1-4a    THE SCATTERING FACTOR $f_0$ FOR THE HELIUM ATOM GROUND STATE BASED ON A HARTREE-FOCK WAVEFUNCTION, ON AN APPROXIMATE WAVEFUNCTION (1-23), AND FROM AN ACCURATE CORRELATED WAVEFUNCTION ASSUMED FOR PRESENT PURPOSES TO BE "EXACT"

| $\dfrac{\sin\theta}{\lambda}$ (A$^{-1}$) | $f_0$ (H-F) | $f_0$ (1-23) | $f_0$ ("exact") |
|---|---|---|---|
| 0 | 2.000 | 2.000 | 2.000 |
| 0.1 | 1.837 | 1.844 | 1.836 |
| 0.3 | 1.060 | 1.075 | 1.058 |
| 0.5 | 0.509 | 0.512 | 0.509 |
| 1.0 | 0.095 | 0.091 | 0.096 |
| 1.5 | 0.026 | 0.024 | 0.026 |

Benesch and Smith (1970) have evaluated the momentum density for the lithium-free atom employing many different wavefunctions, including an "exact" Hylleras-type function ($r_{ij}$ terms specifically included).   Compared to the Hartree-Fock free-atom wavefunctions of Clementi, the maximum difference in momentum density was less than 2% and in $J(z)$ less than 1%.

## C.   Evaluation of the scattering factor $f_0$

1.   *Direct evaluation*   Equation (1-5) defines the x-ray scattering factor in terms of the wavefunctions.   To date, the vast bulk of wavefunctions calculated for atoms, molecules, and solids have relied on the separability of the wavefunction into radial and angular functions in (1-10a).   The angular solutions are then taken to be the hydrogenic functions in (1-10c) and (1-10d).   The scattering factor then becomes

$$f_0(\mathbf{S}) = \sum_i \int_0^\infty \int_0^\pi \int_0^{2\pi} R_i^*(r_i)\Theta_i^*(\theta_i)\Phi_i^*(\varphi_i) \exp (i\mathbf{S} \cdot \mathbf{r}_i)R_i(r_i)\Theta_i(\theta_i)\Phi_i(\varphi_i)r_i^2$$
$$\times \sin \theta_i \, dr_i \, d\theta_i \, d\varphi_i \tag{1-24}$$

If the continuous set of indices $h$, $k$, $l$ is used to define the direction of the scattering vector $\mathbf{S}$ (i.e., $h$ is the magnitude of the component along the $z$ axis, $k$ along the $x$ axis, and $l$ along the $y$ axis), it is possible to write

$$\mathbf{S} \cdot \mathbf{r} = Sr \frac{h \cos \theta + k \sin \theta \cos \varphi + l \sin \theta \sin \varphi}{(h^2 + k^2 + l^2)^{1/2}} \tag{1-25}$$

Making the substitution

$$\varphi = \varphi' + \tan^{-1} \frac{l}{k} \tag{1-26}$$

Table 1-4b   THE COMPTON PROFILE $J(z)$ FOR THE HELIUM ATOM BASED ON THE HARTREE-FOCK WAVEFUNCTION, AN APPROXIMATE WAVEFUNCTION (1-23), AND FROM AN ACCURATE CONFIGURATION INTERACTION WAVEFUNCTION ASSUMED FOR PRESENT PURPOSES TO BE "EXACT"

| $z$ | $J$ (H-F) | $J$ (1-23) | $J$ ("exact") |
|---|---|---|---|
| 0.0 | 1.071 | 1.080 | 1.068 |
| 0.2 | 1.017 | 1.028 | 1.015 |
| 0.5 | 0.791 | 0.796 | 0.789 |
| 1.0 | 0.382 | 0.400 | 0.381 |
| 2.0 | 0.068 | 0.072 | 0.069 |
| 3.0 | 0.015 | 0.019 | 0.015 |

so that the scattering factor $f_0$ becomes

$$f_0(\mathbf{S}) = \sum_i \int_0^\infty |R_i|^2 r_i^2 \, dr_i \, \frac{(2l+1)[(l-|m_l|)!]}{2[(l+|m_2|)!]} \frac{1}{2\pi}$$

$$\times \int_0^\pi \int_0^{2\pi} \exp\left[\frac{iSr_i h \cos\theta_i}{(h^2+k^2+l^2)^{1/2}}\right] \sin^{2|m_l|}\theta_i [P_l^{m_l}(\cos\theta_i)]^2$$

$$\times \exp\left(\frac{iSr_i \sin\theta_i \cos\varphi_i(l^2+k^2)}{(h^2+k^2+l^2)^{1/2}}\right) \sin\theta_i \, d\theta_i \, d\varphi_i \qquad (1\text{-}27)$$

Substituting

$$c = \frac{Sr(l^2+k^2)^{1/2}\sin\theta}{(h^2+k^2+l^2)^{1/2}} \qquad (1\text{-}28)$$

and employing the integral

$$\int_0^{2\pi} \exp(\pm ic \cos\varphi) \sin^{2n}\varphi \, d\varphi = \frac{2J_n(c)\Gamma(n+\tfrac{1}{2})\Gamma(\tfrac{1}{2})}{(c/2)^n} \qquad (1\text{-}29)$$

where $J_n(c)$ is the Bessel function and $\Gamma$ is the gamma function. Integrating (1-27) over $\varphi$ gives

$$f_0(\mathbf{S}) = \sum_i \int_0^\infty |R_i|^2 r_i^2 \, dr_i \, \frac{(2l+1)[(l-|m_l|)!]}{2[(l+|m_l|)!]} \frac{1}{2\pi}$$

$$\times \int_0^\pi 2\pi J_0(c) \exp\left[\frac{iSrh \cos\theta_i}{(h^2+k^2+l^2)^{1/2}}\right] \sin^{2|m_l|}\theta_i [P_l^{m_l}(\cos\theta_i)]^2$$

$$\times \sin\theta_i \, d\theta_i \qquad (1\text{-}30)$$

The integral over $\theta$ is accomplished by expanding

$$\sin^{2|m_l|}\theta_i [P_l^{m_l}(\cos\theta_i)]^2 = \sum_{n=0} a_n P_{2n}^0(\cos\theta_i) \qquad (1\text{-}31)$$

Making the further substitutions

$$\frac{h}{(h^2+k^2+l^2)^{1/2}} = \cos\xi$$

$$\frac{(k^2+l^2)^{1/2}}{(h^2+k^2+l^2)^{1/2}} = \sin\xi \qquad (1\text{-}32)$$

$$z = Sr$$

and using the well-known relationship

$$\int_0^\pi \exp(iz \cos\xi \cos\theta) J_m(z \sin\xi \sin\theta) P_{2n}^m(\cos\theta) \sin\theta \, d\theta$$

$$= 2i^{(2n-m)} P_{2n}^m(\cos\xi) j_{2n}(z) \qquad (1\text{-}33)$$

where $j_n(z)$ is the spherical Bessel function.  Finally,

$$f_0(\mathbf{S}) = \sum_i \int_0^\infty |R_i|^2 r_i^2 \frac{(2l + 1)[(l - |m_l|)!]}{2[(l + |m_l|)!]} \sum_n a_n i^{2n} P_{2n}{}^0(\cos \xi) j_{2n}(z) \, dr_i \qquad (1\text{-}34)$$

As a specific example of (1-34) consider a hydrogen atom in the $2p$ state $l = 1$, $m_l = 0$.  Then the coefficients $a_n$ in (1-31) are $a_0 = \frac{1}{3}$, $a_1 = \frac{2}{3}$, and the radial wave-function is $R = r \exp (-r/2)/2\sqrt{6}$ from (1-10$b$).  Integration over $r$ gives

$$f_0 = \frac{1 - S^2}{(1 + S^2)^4} - (3 \cos^2 \xi - 1) \frac{3S^2}{(1 + S^2)^4} \qquad (1\text{-}35)$$

where $\xi$ according to (1-32) is the angle between the x-ray scattering vector $\mathbf{S}$ and the $z$ axis of the charge density.  For a gas of hydrogen atoms, the scattering cross section $\sigma$ is proportional to $f_0{}^2$ and, since all orientations of the atoms are possible, one must average over $\xi$.

$$\sigma \sim f_0{}^2 = (1 + 2S^2)^2 - 18S^2(1 + 2S^2) \cos^2 \xi + \frac{81S^4 \cos^4 \xi}{(1 + S^2)^8} \qquad (1\text{-}36)$$

Denoting by $\overline{f_0{}^2}$ the average value

$$\overline{f_0{}^2} = \int_0^\pi f_0{}^2 \frac{\sin \xi}{2} \, d\xi$$

$$= \frac{1 - 2S^2 + \frac{41}{5}S^4}{(1 + S^2)^8} \qquad (1\text{-}37)$$

It is interesting to compare $\overline{f_0{}^2}$ with $(\overline{f_0})^2$ since tabulated scattering factors are for $\overline{f_0}$, not $f_0$; i.e., the angular factors in the final wavefunctions are neglected.  For atoms in $S$ states (having no total angular momentum), $f_0$ and $\overline{f_0}$ are identical, since the charge density is spherically symmetric.  For the hydrogen $2p$ state,

$$\overline{f_0} = \int_0^\pi f_0 \frac{\sin \xi}{2} \, d\xi = \frac{1 - S^2}{(1 + S^2)^4}$$

$$(\overline{f_0})^2 = \frac{1 - 2S^2 + S^4}{(1 + S^2)^8} \qquad (1\text{-}38)$$

$$S = \frac{4\pi \sin \theta}{\lambda} \qquad \text{(atomic units)}$$

so that the ratio

$$\frac{\overline{f_0{}^2}}{(\overline{f_0})^2} = \frac{1 - 2S^2 + \frac{41}{5}S^4}{1 - 2S^2 + S^4} \qquad (1\text{-}39)$$

is significantly different from unity for the larger values of $S$.

In solids it is possible to "freeze in" specific directions of the charge density relative to the scattering vector.  The departures of the charge density from spherical symmetry are still treated on a "quasi-atomic" basis in that one chooses linear combinations of the hydrogenic solutions of (1-10$c$) and (1-10$d$) that have the local point symmetry. For example, if the local symmetry is cubic, it is possible to combine the five $d$ functions ($l = 2$) to form two cubic charge densities called $e_g$ and $t_{2g}$, respectively. The two $e_g$ wavefunctions are

$$R_1(r_1)\Theta_1(\theta_1)\Phi_1(\varphi_1)$$

$$= R_1(r_1)\left(\frac{5}{16\pi}\right)^{1/2}(3\cos^2\theta_1 - 1) \qquad m_l = 0$$

$$R_2(r_2)\Theta_2(\theta_2)\Phi_2(\varphi_2)$$

$$= R_2(r_2)\left(\frac{15}{16\pi}\right)^{1/2}(\sin^2\theta_2\cos 2\varphi_2) \qquad m_l = +2 \text{ plus } m_l = -2$$

(1-40)

Assuming $R_1 = R_2$, the total charge density $\rho$ for the two electrons is (Table 1-2)

$$\rho(e_g) = R^2\frac{5}{16\pi}(3\sin^4\theta\cos^2 2\varphi + 9\cos^4\theta - 6\cos^2\theta + 1)$$

$$= R^2\frac{5}{4\pi}\left[\frac{x^4 + y^4 + z^4 - x^2y^2 - x^2z^2 - y^2z^2}{(x^2 + y^2 + z^2)^2}\right]$$

(1-41)

The $e_g$ charge density has maxima along the cube axes $x$, $y$, $z$. In a similar way one can construct the three $t_{2g}$ wavefunctions.

$$R_3(r_3)\Theta_3(\theta_3)\Phi_3(\varphi_3)$$

$$= R_3(r_3)e^{i\varphi_3}\left(\frac{15}{8\pi}\right)^{1/2}\sin\theta_3\cos\theta_3 \qquad m_l = +1$$

$$R_4(r_4)\Theta_4(\theta_4)\Phi_4(\varphi_4)$$

$$= R_4(r_4)e^{i\varphi_4}\left(\frac{15}{8\pi}\right)^{1/2}\sin\theta_4\cos\theta_4 \qquad m_l = -1$$

(1-42)

$$R_5(r_5)\Theta_5(\theta_5)\Phi_5(\varphi_5)$$

$$= R_5(r_5)i\left(\frac{15}{8\pi}\right)^{1/2}\sin 2\varphi_5\sin^2\theta_5 \qquad m_l = +2 \text{ minus } m_l = -2$$

The total $t_{2g}$ charge density, assuming $R_3 = R_4 = R_5$, is

$$\rho(t_{2g}) = \frac{15}{16\pi}R^2\sin^2\theta(4\cos^2\theta + \sin^2\theta\sin^2 2\varphi)$$

$$= \frac{15}{4\pi}R^2\left[\frac{(xy)^2 + (xy)^2 + (yz)^2}{(x^2 + y^2 + z^2)^2}\right]$$

(1-43)

The $t_{2g}$ charge density has maxima along the body diagonals of the cube. If all five of the wavefunctions are employed, the total charge density is spherically symmetric. Thus for $R_1 = R_2 = R_3 = R_4 = R_5$,

$$\rho(e_g) + \rho(t_{2y}) = \frac{5}{4\pi}R^2$$

(1-44)

Using (1-34) to find the scattering factors for the $e_g$ and $t_{2g}$ charge density

$$f_0(e_g) = \langle j_0(Sr) \rangle + \tfrac{1}{2}A \langle j_4(Sr) \rangle \qquad \text{per electron}$$
$$f_0(t_{2g}) = \langle j_0(Sr) \rangle - \tfrac{1}{3}A \langle j_4(Sr) \rangle \qquad \text{per electron}$$

(1-45)

where

$$\langle j_n(Sr) \rangle = \int_0^\infty R^2 r^2 j_n(Sr)\, dr \qquad (1-46)$$

and

$$A = \frac{3(h^4 + k^4 + l^4) - 9(h^2 K^2 + h^2 l^2 + k^2 l^2)}{(h^2 + k^2 + l^2)^2} \qquad (1-47)$$

**2.   Total scattering from free atoms**   The x-ray scattered intensity $I$, from a gas of free atoms in the low-density limit, is given by

$$I = I_0 N \left( \frac{e^2}{mc^2} \right)^2 t \exp(-\mu t) \Omega K^2 f^2 \qquad (1-48)$$

where $I_0$ is the incident intensity, $N$ the number of atoms per cubic centimeter, $e^2/mc^2$ the fundamental scattering length, $t$ the x-ray path length, $\mu$ the linear absorption coefficient, $\Omega$ the solid angle of the detector, $K$ the polarization, and $f$ the atomic scattering factor from (1-4).

In the forward direction the scattering factor per atom of a gas approaches $Z$, the total number of electrons per atom. Interference effects between the atoms, however, reduce the scattering by a factor that depends on the ratio of the average distance between atoms to the "size" of the atoms (Weiss, 1966, p. 12).

## D.   Anomalous dispersion

The fundamental theoretical calculation of the anomalous dispersion corrections $\Delta f'$ and $\Delta f''$ is quite difficult, and the problem has been approached in many approximate ways. We do not know the accuracy of these calculations since measurements of $\Delta f'$ are sparse and accurate measurements of $\Delta f''$ are limited to a small part of the energy spectrum.

It is convenient to introduce the so-called oscillator density $dg/d\omega$, which is a measure of the strength of the interaction between a photon of frequency $\omega$ and the excitation of the atom from the ground state to an excited state of energy $\hbar\omega$. It involves a knowledge of the ground-state and excited-state wavefunctions. The oscillator density is thus directly related to the linear absorption coefficient $\mu$.

$$\frac{dg}{d\omega} = \frac{mc\mu}{2\pi^2 e^2 N} \qquad (1-49)$$

where $N$ is the number of atoms per cubic centimeter. The oscillator density is further subdivided into transitions involving excitation of a $1s$ electron ($K$ oscillator density) or 2s, 2p electron (L oscillator density), etc. The integral of the oscillator density over all frequencies is called the oscillator strength $g$. Thus for the 1s excitation, the K oscillator strength is

$$g_K = \int_0^\infty \left(\frac{dg}{d\omega}\right)_K d\omega \qquad (1\text{-}50)$$

where the subscript denotes the particular shell. Due to conservation of energy, however, $dg/d\omega$ is zero when the photon energy is less than the K absorption energy $\hbar\omega$. Thus the integral (1-50) can be written

$$g_K = \int_{\omega_K}^\infty \left(\frac{dg}{d\omega}\right)_K d\omega \qquad (1\text{-}51)$$

For each electron subshell, $\Delta f'$ and $\Delta f''$ are given as a function of frequency $\omega_i$

$$\Delta f_K'(\omega_i) = \int_{\omega_K}^\infty \frac{\omega^2 (dg/d\omega)_K}{\omega_i{}^2 - \omega^2} d\omega \qquad (1\text{-}52a)$$

$$\Delta f_K''(\omega_i) = \frac{\pi}{2} \omega_i \left(\frac{dg}{d\omega}\right)_K = \frac{\pi}{2} \frac{\omega_i mc\mu}{2\pi^2 e^2 N} \qquad (1\text{-}52b)$$

The observed values of $\Delta f'$, $\Delta f''$, and $\mu$ are the sum of the contributions from the K, L, M, etc., shells. It is a common assumption to describe the frequency dependence of $\mu$ for each subshell as

$$\mu_K(\omega) = \left(\frac{\omega_K}{\omega}\right)^n \mu_K(\omega_R) \qquad \text{for } \omega \geq \omega_K$$

$$\mu_K(\omega) = 0 \qquad \text{for } \omega < \omega_K \qquad (1\text{-}53)$$

a relationship only approximately correct. The exponent $n$ has a value of $\sim 2.7$ and may differ for each subshell. Substitution of (1-53) into equations (1-51) and (1-52), respectively, gives

$$g_K = \frac{mc\mu_K(\omega_K)\omega_K}{(n-1)2\pi^2 e^2 N}$$

$$\Delta f'(\omega_i) = g_K(n-1)\left(\frac{\omega_K}{\omega_i}\right)^{n-1} \int_0^{\omega_i/\omega_K} \frac{x^{n-1}}{x^2 - 1} dx \qquad (1\text{-}54)$$

$$\Delta f''(\omega_i) = \frac{\pi}{2} \frac{n-1}{(\omega_i/\omega_K)} g_K$$

There are several possible routes to follow:

1   A completely fundamental calculation of $dg/d\omega$, such as by Hönl (1933) or Wagenfeld (1966). Some simplification of the wavefunctions is necessary since Hartree-Fock functions for the excited states are not available.

2   A semiempirical approach combining a fundamental calculation of $g_K$, $g_L$, $g_M$, etc., from the wavefunctions (Cromer, 1965) and an experimental value

of $n$ (1-53). The calculation of the oscillator strengths is not as difficult as $dg/d\omega$ since certain sum rules can be used that do not require the excited-state wavefunctions.

*3* Use experimental values of $\mu$ over a large enough frequency range to integrate (1-52a) directly.

*4* Measure $\Delta f'$ directly (very little such data are available).

Some selected values of $\Delta f'$ and $\Delta f''$ from the four methods are given in Table 1-5.


## II.   INELASTIC SCATTERING THEORY

### A.   Compton effect

Detailed studies of the inelastically scattered x rays associated with the ejection of a single electron from the ground state of an atom (Compton effect) have undergone renewed interest. The important quantity that can be deduced from these measurements is the electron momentum distribution as compared to the electron spatial distribution deduced from the scattering factor (elastic scattering).

**1.  *Momentum space***   Position and momentum are complementary quantities in quantum systems. It is possible to solve the Schrödinger equation directly in momentum space, but it becomes an integral equation and is unwieldy. For a single electron atom it is

$$\frac{p^2}{2}\,\chi(\mathbf{p}) + \int f(\mathbf{p},\,\mathbf{p}')\chi(\mathbf{p}')\,d\mathbf{p}' = E\chi(\mathbf{p})$$

$$f(\mathbf{p},\,\mathbf{p}') = \frac{1}{(2\pi)^{3/2}} \int V(\mathbf{r}) \exp\left[-i(\mathbf{p} - \mathbf{p}')\cdot\mathbf{r}\right]\,d\mathbf{r}$$

(1-55)

**Table 1-5**  SELECTED VALUES OF $\Delta f'$ AND $\Delta f''$ OBTAINED BY THE FOUR METHODS DESCRIBED ABOVE

Values are for Mo $K\alpha$ ($\lambda = 0.71$ A). The variation in the three methods is of the order $\sim 10$–$50\%$ in $\Delta f'$ and $\sim 5\%$ in $\Delta f''$

| Element | Method 1 | | Method 2 | | Method 3 | | Method 4 | |
|---------|----------|----------|----------|----------|----------|----------|----------|----------|
| | $\Delta f'$ | $\Delta f''$ | $\Delta f'$ | $\Delta f''$ | $\Delta f'$ | $\Delta f''$ | $\Delta f'$ | $\Delta f''$ |
| Neon | $\sim 0.03$ | 0.02 | $+0.03$ | 0.02 | $+0.04$ | 0.02 | $\sim 0$ | 0.02 |
| Argon | $+0.22$ | 0.26 | $+0.18$ | 0.24 | $+0.12$ | 0.20 | $+0.10$ | 0.20 |
| Krypton | $-0.37$ | | $-0.47$ | 2.96 | $-0.88$ | 2.8 | $-0.81$ | 2.8 |
| Xenon | $+0.14$ | | $-0.59$ | 2.30 | | 2.2 | $-0.40$ | 2.2 |

where $\chi(\mathbf{p})$ is the wavefunction in momentum space, $\mathbf{p}$ the momentum, and $V(\mathbf{r})$ the potential energy in position space. The first term in (1-55) is the kinetic energy, and the second term the potential energy. It is the potential energy term that presents mathematical difficulties. If the wavefunction $\Psi_0(\mathbf{r})$ is calculated in position space, however, one can transform the wavefunction to momentum space $\chi(\mathbf{p})$ (and vice versa) by the well-known Dirac transformations.

$$\chi(\mathbf{p}) = \frac{1}{(2\pi)^{3/2}} \int \Psi(\mathbf{r}_i) \prod_i \exp\left(-i\mathbf{p}_i \cdot \mathbf{r}_i\right) d\mathbf{r}_i$$

$$\Psi(\mathbf{r}) = \frac{1}{(2\pi)^{3/2}} \int \chi(\mathbf{p}_i) \prod_i \exp\left(i\mathbf{p}_i \cdot \mathbf{r}_i\right) d\mathbf{p}_i$$

(1-56)

For one-electron wavefunctions, the transformation is straightforward since the product of exponentials becomes a sum in the exponent, but for correlated wavefunctions the transformation becomes cumbersome due to the $r_{ij}$ terms. It can be shown that, for one-electron wavefunctions separable into radial and angular functions (1-10a), the one-electron momentum density is given by

$$|\chi_l|^2 = \frac{2}{\pi} |\Theta(\theta')|^2 |\Phi(\varphi')|^2 \left| \int R(r) j_l(pr) r^2 \, dr \right|^2 \qquad (1\text{-}57)$$

where $\theta'$ and $\varphi'$ are the corresponding angular variables in momentum space, i.e., the $x$, $y$, $z$ axes in position space and the $x'$, $y'$, $z'$ axes in momentum space superimpose. $l$ is the one-electron orbital quantum number $l = 0, 1, 2, 3$, etc., for $s$, $p$, $d$, $f$, etc., electrons. Some examples of space wavefunctions and their corresponding momentum wavefunctions are given in Table 1-6 for $s$ and $p$ electrons. Exponential

Table 1-6    POSITION WAVEFUNCTIONS AND THEIR CORRESPONDING MOMENTUM WAVEFUNCTIONS FOR A HYDROGENIC $1s$, $2p(m_l = \pm 1)$ AND $2p(m_l = 0)$ ELECTRON AND FOR A GAUSSIAN FUNCTION

| Electron | Position wavefunction | Momentum wavefunction |
|---|---|---|
| $1s$ | $\Psi = \dfrac{2b^{3/2} \exp\left(-br\right)}{\sqrt{4\pi}}$ | $\chi = \dfrac{4b^{5/2}}{\sqrt{2}\pi(b^2 + p^2)^2}$ |
| $2p(m_l = \pm 1)$ | $\Psi = \dfrac{a^{5/2}r \exp\left(-ar\right)\sin\theta\exp\left(\pm i\varphi\right)}{\sqrt{2\pi}}$ | $\chi = \dfrac{ia^{5/2}(\sin\theta\cos\pm i\sin\theta\sin\varphi)8ap}{\pi(a^2 + p^2)^3}$ |
| $2p(m_l = 0)$ | $\Psi = \dfrac{a^{5/2}r \cos\theta \exp\left(-ar\right)}{\sqrt{\pi}}$ | $\chi = \dfrac{ia^{5/2}(\cos\theta)8ap}{\sqrt{2}\,\pi(a^2 + p^2)^3}$ |
| $1s$ (Gaussian) | $\Psi = \dfrac{2a^{3/4} \exp\left(-ar^2\right)}{\pi^{1/4}\sqrt{4\pi}}$ | $\chi = \dfrac{2a^{3/4} \exp\left(-p^2/4a\right)}{\pi^{1/4}\sqrt{4\pi}\,(2a)^{3/2}}$ |

functions in position space transform as Lorentzians in momentum space, but Gaussians in position space transform as Gaussians.

It is a further property of the transformations (1-56) that for a determinantal wavefunction, orthogonality of the wavefunctions is preserved. Thus, including spin $\sigma$, the transformation is

$$\chi = \frac{1}{\sqrt{n!}} \begin{vmatrix} \chi_1(\mathbf{p}_1, \sigma_1) \cdots \chi_1(\mathbf{p}_n, \sigma_n) \\ \chi_2(\mathbf{p}_1, \sigma_1) \cdots \chi_2(\mathbf{p}_n, \sigma_n) \\ \cdots\cdots\cdots\cdots\cdots \\ \chi_n(\mathbf{p}_1, \sigma_1) \cdots \chi_n(\mathbf{p}_n, \sigma_n) \end{vmatrix}$$

$$= \frac{(2\pi)^{-3n/2}}{\sqrt{n!}} \begin{vmatrix} \int \exp\left[-i\mathbf{p}_1 \cdot \mathbf{r}_1\right]\psi_1(\mathbf{r}_1, \sigma_1)\, d\mathbf{r}_1 \cdots \int \exp\left[-i\mathbf{p}_n \cdot \mathbf{r}_n\right]\psi_1(\mathbf{r}_n, \sigma_n)\, d\mathbf{r}_n \\ \cdots\cdots\cdots\cdots\cdots\cdots\cdots\cdots\cdots \\ \cdots\cdots\cdots\cdots\cdots\cdots\cdots\cdots\cdots \\ \cdots\cdots\cdots\cdots\cdots\cdots\cdots\cdots\cdots \\ \int \exp\left[-i\mathbf{p}_1 \cdot \mathbf{r}_1\right]\psi_n(\mathbf{r}_1, \sigma_1)\, d\mathbf{r}_1 \cdots \int \exp\left[-i\mathbf{p}_n \cdot \mathbf{r}_n\right]\psi_n(\mathbf{r}_n, \sigma_n)\, d\mathbf{r}_n \end{vmatrix}$$

$$(1\text{-}58)$$

Referring back to the approximate correlated wavefunction for helium [equation (1-23)], one can see how such correlation is manifested in momentum space. The distance between the two electrons $r_{12}$ can be written

$$r_{12} = (r_1^2 + r_2^2 - 2r_1 r_2 \cos \alpha)^{1/2} \qquad (1\text{-}59)$$

where $\alpha$ is the angle between the lines from the electrons to the nuclei at $\mathbf{r}_1$ and $\mathbf{r}_2$. From (1-56) it follows that

$$\chi(p_1, p_2) = \frac{8Aa^2}{\pi(a^2 + p_1^2)^2(a^2 + p_2^2)^2}$$

$$\times \left\{ 1 + b\left[ \frac{12(a^2 - p_1^2)}{(a^2 + p_1^2)^2} + \frac{12(a^2 - p_2^2)}{(a^2 + p_2^2)^2} \right.\right.$$

$$\left.\left. - 2\cos\alpha \frac{(3a^2 - p_1^2)(3a^2 - p_2^2)}{a^2(a^2 + p_1^2)(a^2 + p_2^2)} \right] \right\} \qquad (1\text{-}60)$$

When the angle $\alpha$ between $\mathbf{r}_1$ and $\mathbf{r}_2$ is small, both the position and momentum wavefunctions tend toward a minimum, so that the effect of correlation is manifested in momentum space. When the two electrons are at identical positions in position space, however, $\mathbf{r}_1 = \mathbf{r}_2$ and $\cos \alpha = 1$, so that the term in $b$ is zero, but when the two electrons have identical momenta $\mathbf{p}_1$ and $\mathbf{p}_2$, $\cos \alpha = 1$, the term in $b$ is still finite.

The presence of the cusp in position space near $r_{12} = 0$ manifests itself by introducing higher momentum components in momentum space. A conceptual picture of correlation can be obtained by noting that, as the electrons approach each other, they speed up, thus reducing the time they spend near each other.

Recently, Benesch (1972) has shown how to obtain the momentum density of correlated wavefunctions by using density matrices. The radial momentum density $I_0(p)$ is given by

$$I_0(p) = p^2 \int_0^{2\pi} \int_0^{\pi} \hat{\gamma}(\mathbf{p} \mid \mathbf{p}) \sin \alpha \, d\alpha \, d\beta$$

The term $\hat{\gamma}(\mathbf{p} \mid \mathbf{p})$ is the diagonal momentum space first-order density matrix

$$\hat{\gamma}(\mathbf{p} \mid \mathbf{p}) = (2\pi)^{-3} \int e^{-i\mathbf{p}\cdot(\mathbf{r}-\mathbf{r}')}\gamma(\mathbf{r} \mid \mathbf{r}') \, d\mathbf{r} \, d\mathbf{r}'$$

where $\mathbf{p} = (p, \alpha, \beta)$ and $\mathbf{r} = (r, \theta, \psi)$. $\gamma(\mathbf{r} \mid \mathbf{r}')$ is the nondiagonal (spin-free) first-order density matrix in position space

$$\hat{\gamma}(\mathbf{r} \mid \mathbf{r}') = N \int \psi(x_1, x_2, \ldots, x_n)\psi^*(x_1', x_2', \ldots, x_n') \, ds_1 \, dx_i$$

where $N$ is the number of electrons and $dx_i$ indicates integration over the space and spin coordinates of electrons 2 to $N$, while $ds_1$ is the integration over the spin coordinates of electron 1. Benesch has shown explicitly how to evaluate $\gamma(\mathbf{r} \mid \mathbf{r}')$ in closed form for the two-electron case when the wavefunction $\psi$ specifically contains $r_{12}$ terms and has indicated how to extend the calculation to the many-electron case. It is interesting that, even though the wavefunction in position space cannot, itself, be transformed in closed form, the density matrix can.

## B.  Compton profile

In the Compton effect the x ray transfers energy and momentum to a single electron. For an unbound electron initially at rest, the energy transferred is

$$\Delta E = \frac{[2(h\nu_0)^2/mc^2] \sin^2 \theta_c}{[1 + (2h\nu_0 \sin^2 \theta_c/mc^2)]} \qquad (1\text{-}61)$$

where $h\nu_0$ is the initial x-ray energy and $\theta_c$ is one-half the scattering angle. In a typical experiment employing Mo $K\alpha$ ($\lambda = 0.71$ A) scattered at $160°$ ($2\theta_c$), the energy gain for a valence electron is about 1000 eV, so that the final electron state lies well into the continuum. As will be seen, the final-state wavefunction cannot be treated simply as a plane wave, without making some compensating approximations.

**1.** **_Hydrogenic solution_**   _a._   _Exact solution_   Only in the case of the hydrogenic atom is an "exact" solution to the Compton profile possible. The differential cross section for the Compton profile neglecting the $(\mathbf{A}_i \cdot \mathbf{V}_i)$ term of equation (2-3) is

$$\frac{d\sigma}{d\Omega \, d\omega} = \left(\frac{e^2}{mc^2}\right)^2 K^2 \frac{v}{v_0} \sum_f \sum_i |\langle \psi_f | e^{i(\mathbf{k} - \mathbf{k}_0 \cdot \mathbf{r})} | \psi_0 \rangle|^2 \, \delta(E_f - E_i - \omega) \tag{1-62}$$

where $K$ is the polarization factor, $\Omega$ the solid angle, $(e^2/mc^2)^2$ the Thomson cross section, $v$ and $v_0$ the final and initial x-ray frequencies, $\psi_f$ and $\psi_0$ the final and initial wavefunctions, $E_f$ and $E_i$ the final and initial electron energies, and $\omega$ the x-ray energy loss. The delta function is used to indicate energy conservation in the process. Neglect of the $(\mathbf{A}_i \cdot \mathbf{V}_i)$ term assumes that the x-ray energy is far from any absorption edge. The final-state wavefunction for the hydrogenic continuum state is

$$\psi_f = \left(\frac{2\pi}{pa}\right)^{1/2} (1 - e^{-2\pi/pa})^{-1/2} e^{i\mathbf{p} \cdot \mathbf{r}} F\left[\frac{i}{pa}, 1, i(pr - \mathbf{p} \cdot \mathbf{r})\right] \tag{1-63}$$

where the energy is $E_f = p^2/2m$ and $a = \hbar^2/Zme^2$. $F(a, |c|, z)$ is the confluent hypergeometric series

$$F(a, |c|, z) = 1 + \frac{az}{c} + \frac{a(a + 1)z^2}{2! \, c(c + 1)} + \frac{a(a + 1)(a + 2)z^3}{3! \, c(c + 1)(c + 2)} + \cdots \tag{1-64}$$

which for $z$ large

$$F(a, |c|, z) \xrightarrow[z \to \infty]{} \frac{\Gamma(c)}{\Gamma(A)} z^{a-c} e^z \tag{1-65}$$

so that for $r$ and $p$ large

$$\psi_f \to (2\pi)^{1/2} (pa)^{-1/2} e^{ipr} \tag{1-66}$$

The initial-state wavefunction is the $1s$ hydrogenic ground state with energy $E_0 = Z^2 me^2/2\hbar^2$. Equation (1-62) has been evaluated by Gummel and Lax (1957) for the ground-state hydrogenic $1s$ wavefunction. When averaged over all final directions for $\mathbf{p}$, the differential cross section is

$$\frac{d\sigma}{d\Omega \, d\omega} = \left(\frac{e^2}{mc^2}\right)^2 K^2 \frac{v}{v_0} 256a^2 m(1 - e^{-2\pi/pa})^{-1}$$

$$\times \exp\left(-\frac{2}{pa} \tan^{-1} \frac{2pa}{1 + S^2 a^2 - p^2 a^2}\right)$$

$$\times \frac{[S^4 a^4 + S^2 a^2(1 + p^2 a^2)/3]}{[(S^2 a^2 + 1 - p^2 a^2)^2 + 4p^2 a^2]^3} \tag{1-67}$$

where

$$S = \mathbf{k} - \mathbf{k}_0$$

Since the energy of the continuum states is $E_f = p^2/2m$, it suggests that $p$ is essentially the electron momentum of the final state. While this is approximately correct and appears similar to the relationship between energy and momentum for a free electron (plane wave), the use of a plane wave for the final state leads to erroneous results for the matrix elements in (1-67). The reason is the critical overlap region between the initial- and final-state wavefunctions. In the overlap region (which is close to the nucleus) the continuum wavefunction (1-63) departs significantly from a plane wave.

b. *Plane-wave final state*  Approximating the final-state wavefunction with a plane wave $(\psi_f = e^{i\mathbf{p}\cdot\mathbf{r}})$ where $\mathbf{p}$ is the momentum of the electron and $p^2/2m$ is the energy, the matrix elements in (1-62) can be readily evaluated

$$\frac{d\sigma}{d\Omega\,d\omega} = \left(\frac{e^2}{mc^2}\right)^2 K^2 \frac{v}{v_0} \frac{2\pi}{|\mathbf{k} - \mathbf{k}_0|} \int_{|y|}^{|y|+p} |\chi(p_0)|^2 p_0 \, dp \qquad (1\text{-}68)$$

where $y$ is the initial momentum of the electron before the collision

$$y = |\mathbf{k} - \mathbf{k}_0| - p = |\mathbf{k} - \mathbf{k}_0| - \sqrt{2(\omega - B)} \qquad (1\text{-}69)$$

$B$ is the binding energy and $\chi(p_0)$ is the ground-state-momentum wavefunction. For the hydrogenic $1s$ ground state $(Z = $ atomic number), equation (1-68) becomes

$$\frac{d\sigma}{d\Omega\,d\omega} = \left(\frac{e^2}{mc^2}\right)^2 K^2 \frac{v}{v_0} \frac{8}{3\pi|\mathbf{k} - \mathbf{k}_0|}$$

$$\times \left\{ \left[ \left(\frac{|y| + 2p}{Z}\right)^2 + 1 \right]^{-3} - \left(\frac{y^2}{Z^2} + 1\right)^{-3} \right\}$$

$$= \left(\frac{e^2}{mc^2}\right)^2 K^2 \frac{v}{v_0} \frac{J(y)}{|\mathbf{k} - \mathbf{k}_0|} \qquad (1\text{-}70)$$

In Fig. 1-1 are plotted $(d\sigma/d\Omega\,d\omega)/(e^2 mc^2)^2 K^2(v/v_0)$ for both the exact (1-67) and plane-wave approximations (1-70) for $Z = 2$, 5, and 10. It is seen that there are significant differences between these two curves because of the poor approximation of the plane-wave final-state wavefunction to the continuum function in the $1s$ overlap region.

c. *Impulse approximation*  In the impulse approximation the final-state wave-function is still assumed to be a plane wave, but in addition the *potential* energy difference between the initial and final states is neglected (Platzman and Eisenberger, 1971). It is just the region where the potential energy is large (near the nucleus) that the plane-wave and continuum-state wavefunctions differ. The reason that the potential energy is neglected is that the interaction between the x ray and electron

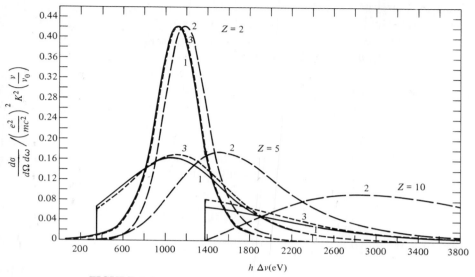

**FIGURE 1-1**
Compton profiles for the hydrogenic atoms $Z = 2$, 5, and 10. The three curves for each $Z$ are the exact calculation (1-67) (curves 1), the plane-wave approximation (1-70) (curves 2), and the impulse approximation (1-76) (curves 3). The scattering angle was 160°.

occurs so quickly that energy and momentum are exchanged in a "constant" potential field so that only momentum and kinetic energy need be considered—in effect a classical billiard ball collision.[1]

Taking the electron energy difference between the initial and final states as $p^2/2m - p_0^2/2m$ where $p$ and $p_0$ are final and initial momenta and approximating the

[1] The basis for the impulse approximation can be shown as follows: Starting from equation (1-62), assume a spherically symmetric potential and consider a thin spherical volume element in which the potential can be taken as constant. In such a constant potential the solutions to the Schrödinger equation for both $\psi_f$ and $\psi_0$ are of the form $\exp(i\mathbf{p}\cdot\mathbf{r}) = \exp(ipr\cos\phi)$, where $|\mathbf{p}| = [2m(E - V)]^{1/2}$. Integration over the angular coordinates in this spherical volume element gives $4\pi j_0(|\mathbf{p} - (\mathbf{k} - \mathbf{k}_0) + \mathbf{p}_0|r)$, which is peaked at $|\mathbf{p} - (\mathbf{k} - \mathbf{k}_0) + \mathbf{p}_0| = 0$ and expresses the conservation of momentum, since $p^2 = 2m[E_f - V(r)]$, $p_0^2 = 2m[E_i - V(r)]$ and $E_f - E_i = h\,\Delta\nu$. If $V(r)$ is taken to be the same for both the initial and final states of the electron, this yields (1-72) and (1-73). While the initial $(E_i)$ and the final $(E_f)$ total energies of the electron are different, as long as the potential $V(r)$ is the same, integration over the angular variables yields the momentum transform of the ground-state wavefunctions (1-71). Of course, if $V(r)$ differs for the initial and final states of the electron, then the equations for conservation of energy and momentum (1-72) will specifically contain the potential energy and the impulse approximation breaks down.

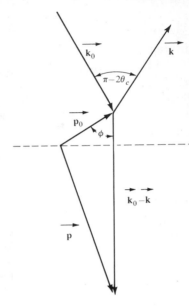

**FIGURE 1-2**
Diagram of the Compton process show-
ing momentum conservation. $p_0$ and $\mathbf{p}$
are initial and final momenta of the electron, $\mathbf{k}_0$ and $\mathbf{k}$ of the x ray.

final-state wavefunction as a plane wave, (1-62) becomes

$$\frac{d\sigma}{d\Omega\,d\omega} = \left(\frac{e^2}{mc^2}\right)^2 \frac{K^2}{|\mathbf{k} - \mathbf{k}_0|} \frac{v}{v_0} \sum_f |\chi(p_0)|^2 \, \delta\left(\frac{p^2}{2m} - h\,\Delta v\right)\left(\frac{p^2}{2m} - \frac{p_0^2}{2m} - h\,\Delta v\right)$$

(1-71)

for $h\,\Delta v \geq E_B$. The equations for conservation of energy and momentum (Fig. 1-2)
are

$$p^2 = p_0^2 + |\mathbf{k} - \mathbf{k}_0|^2 - 2p\cos\phi\,|\mathbf{k} - \mathbf{k}_0| \qquad \text{conservation of momentum}$$
$$p^2 = p_0^2 + 2mh\,\Delta v \qquad\qquad\qquad\qquad \text{conservation of energy}$$

(1-72)

where $\cos\phi$ is the angle between the electron initial momentum and the x-ray scattering vector. Eliminating $p^2$, (1-72) becomes

$$p_0\cos\phi = \frac{|\mathbf{k} - \mathbf{k}_0|}{2} - \frac{mh\,\Delta v}{|\mathbf{k} - \mathbf{k}_0|} \equiv -z$$

(1-73)

Thus under experimental conditions in which $\mathbf{k}$ and $\mathbf{k}_0$ are fixed, $p_0\cos\phi$ is uniquely
determined; i.e., for any electron whose initial component of momentum along the
scattering vector is $p_0\cos\phi$. This defines a plane in momentum space shown by the
dotted line in Fig. 1-2. Replacing the sum in (1-71) by an integral and integrating
over all final states $p$ terminating on the plane,

$$\frac{d\sigma}{d\Omega\,d\omega} = \frac{1}{|\mathbf{k} - \mathbf{k}_0|}\left(\frac{e^2}{mc^2}\right)^2 K^2 \frac{v}{v_0} \int_{|z|}^{\infty} |\chi(p_0)|^2 \, dp_x\,dp_y$$

(1-74)

where $z = -p \cos \phi$ and $dp_x$ and $dp_y$ are the elements of integration normal to the $z$ direction. For the hydrogenic $1s$ ground state (1-74) yields the curves shown in Fig. 1-1, evidencing very good agreement with the "exact" solution. For valence electrons which have a binding energy of the order of 10 eV, the error in the impulse approximation is only about one part in a thousand for the peak position.

It is convenient to express the differential cross section in terms of the initial electron momentum. From (1-73),

$$\omega = z|\mathbf{k} - \mathbf{k}_0| + \frac{|\mathbf{k} - \mathbf{k}_0|^2}{2}$$
$$d\omega = |\mathbf{k} - \mathbf{k}_0| \, dz \tag{1-75}$$

which gives for (1-74)

$$\frac{d\sigma}{d\Omega \, dz} = \frac{d\sigma}{d\Omega \, d\omega} |\mathbf{k} - \mathbf{k}_0| = \left(\frac{e^2}{mc^2}\right)^2 K^2 \frac{v}{v_0} \int_{|z|}^{\infty} |\chi(p_0)|^z \, dp_x \, dp_y$$

$$\equiv \left(\frac{e^2}{mc^2}\right)^2 K^2 \frac{v}{v_0} J(z) \tag{1-76}$$

Rewriting the coefficient of $J(z)$ in (1-76),

$$\frac{v}{v_0} = \left(1 - \frac{\Delta v}{v_0}\right) \cong \left(1 - \frac{\Delta\lambda}{\lambda_0}\right) \tag{1-77}$$

For large scattering angles ($\sin \theta_c \to 1$),

$$\frac{\Delta\lambda}{\lambda_0} \cong \frac{0.0485}{\lambda_0} \tag{1-78}$$

Thus, in a typical experiment (Mo $K\alpha$ scattered 160°), the x-ray energy dependent correction is less than 20% over the entire range of the Compton profile.

**2.   Many-electron atom**   For the many-electron atom, the differential cross section in equation (1-62) is now given by

$$\frac{d^2\gamma}{d\Omega \, d\omega} = \left(\frac{e^2}{mc^2}\right)^2 K^2 \frac{v}{v_0} \int \left| \left\langle \psi_f \left| \sum_{i=1}^{N} e^{i(\mathbf{k} - \mathbf{k}_0 \cdot \mathbf{r}_i)} \right| \psi_0 \right\rangle \right|^2 \delta(E_f - E_i - \omega) \, d\mathbf{f} \tag{1-79}$$

where the summation is over the $i$ electrons and the integration of $d\mathbf{f}$ is over all final-state wavefunctions. $\psi_f$ and $\psi_0$ are total final- and ground-state (initial) wavefunctions, respectively. The main problem lies with the total final wavefunction when one of the electrons has been given sufficient energy to eject it from the atom. The whole problem of excited-state wavefunctions presents great difficulties for the theoretician, particularly when a hole exists in an inner shell. One approximate approach is to assume that all one-electron wavefunctions are the same in the ground and excited states except for the excited electron. Such a simplification reduces

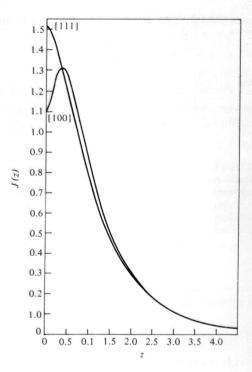

**FIGURE 1-3**
The calculated Compton-line shape for vanadium metal based on a $t_{2g}$ orbital population of 81 % (remainder $e_g$) for the approximately four $3d$ electrons. The directions [100] and [111] are the directions of the x-ray scattering vector relative to the crystal axes.

equation (1-79) to a one-electron problem involving matrix elements between the initial (ground) state and final (continuum) state of the excited electron. Currat, DeCicco, and Weiss (1971) employed numerical solutions to the final-state wavefunction in the potential of the remaining electrons. This is called the excited-state approximation. Figure 1-3 gives both the impulse and excited-state approximations for the Compton profile of the carbon $1s$ electron (binding energy 280 eV) and for the core profile $2s^2\, 2p^6$ for aluminum. In the latter case, both the $2s^2$ and $2p^6$ contributions are evaluated separately. The very good agreement with the impulse approximation provides further support for the use of the impulse approximation.

**3. *Analysis of data in the impulse approximation*** For spherically symmetric momentum distributions such as those from closed shells, (1-76) is further simplified. $J(z)$ can be written

$$J(z) = 2\pi \int_{|z|}^{\infty} |\chi(p_0)|^2 p_0 \, dp_0 \qquad (1\text{-}80)$$

Using hydrogenic solutions for the angular functions, one obtains from (1-57), for spherically symmetric distributions

$$|\chi(p_0)|^2 = \frac{2}{\pi}\left[\int R(r)j_l(pr)r^2\,dr\right]^2 \qquad (1\text{-}81)$$

where $l$ is the orbital quantum 0, 1, 2, 3, etc., for $s, p, d, f$, respectively. From (1-73) one can relate the x-ray wavelength increase to the component of initial momentum. For $p_0\cos\phi = 0$, (1-73) yields the well-known Compton shift

$$\Delta\lambda = \frac{2h}{mc}\sin^2\theta_c$$

$$= 0.04852\sin^2\theta_c\ \text{A} \qquad (1\text{-}82)$$

where $\Delta\lambda$ is the wavelength increase and $\theta_c$ is one-half the scattering angle. For $p_0\cos\phi$ finite, (1-73) can be rewritten

$$z = p_0\cos\phi$$

$$= mc\,\frac{h\nu_0\sin\theta_c}{mc^2}\left\{\frac{mc^2\,\Delta\nu/\nu_0}{2h\nu_0\sin^2\theta_c[1 - \Delta\nu/\nu_0 + (\Delta\nu/\nu_0)^2(1/4\sin^2\theta_c)]^{1/2}}\right.$$

$$\left. -\left[1 - \frac{\Delta\nu}{\nu_0} + \left(\frac{\Delta\nu}{\nu_0}\right)^2\frac{1}{4\sin^2\theta_c}\right]^{1/2}\right\} \qquad (1\text{-}83)$$

where $\nu_0$ is the incident photon frequency. Equation (1-83) enables one to convert the x-ray wavelength scale to an electron momentum scale. It is convenient to introduce a variable $q$ which is proportional to the wavelength shift $l$ from the center of the Compton line [which is at $\lambda_0 + \Delta\lambda$ according to (1-82)].

$$q = \frac{mcl}{2(\lambda\lambda_0)^{1/2}\sin\theta_c} \cong z$$

$$l = \lambda - \left(\lambda_0 + \frac{2h}{mc}\sin^2\theta_c\right) \qquad (1\text{-}84)$$

where $\lambda_0$ and $\lambda$ are the x-ray wavelengths before and after scattering. In atomic units $mc = 137$ (reciprocal of fine-structure constant). The advantage of (1-84) is that it provides a simpler relationship than (1-83) between the electron momentum and the x-ray wavelength and is good to a fraction of a percent.

**4.   Relativistic corrections in the impulse approximation**   The relativistic solution for the Compton effect has only been derived for unbound electrons (Jauch and Rohrlich, 1955) but is probably quite good for bound electrons in the impulse approximation, where binding effects can be ignored. The relativistic corrections are not large. They affect both the differential cross section and the relationship between the x-ray wavelength scale and the electron momentum scale.

The planar integration (1-74) is now slightly distorted but only at the largest values of momentum. The wavelength shift now is given by

$$\Delta\lambda = \frac{2h \sin^2 \theta_c}{\gamma mc} - \frac{2\lambda_0 p_0 \cos \phi \sin \theta_c}{\gamma mc}\left[1 + \frac{\Delta\lambda}{\lambda_0} + \left(\frac{\Delta\lambda}{\lambda_0}\right)^2 \frac{1}{4 \sin^2 \theta_c}\right]^{1/2}$$

$$\gamma = \left[1 - \frac{p_0^2}{(mc)^2}\right]^{-1/2}$$

(1-85)

Since measurements are rarely made out to values of $p_0 \sim 15$, $\gamma$ can be treated as unity. To within a fraction of a percent then, one can write

$$z = -p_0 \cos \phi = \frac{mcl}{2 \sin \theta_c \lambda_0 [1 + \Delta\lambda/\lambda_0 + (\Delta\lambda/\lambda_0)^2 (1/4 \sin^2 \theta_c)]^{1/2}} \cong q$$

(1-86)

The differences between (1-83) (nonrelativistic), (1-84), and (1-86) can generally be neglected for x rays. It is in the cross sections that the correction is more significant. For relatively high scattering angles, it can be shown (Jauch and Rohrlich, 1955) that the nonrelativistic cross section must be corrected by a term $\beta$ to account for the more rapid decrease in intensity as the energy loss increases.

$$\beta \sim \left(1 - \frac{2z}{mc}\right)^{-1}$$

(1-87)

**5.** *Nonspherical distributions*  Consider a hydrogenic $2p$ electron, $l = 1$, $m_l = +1$, whose space wavefunction is

$$\Psi_0 = R(r)\Theta(\theta)\Phi(\varphi)$$

$$R(r) = \frac{2}{\sqrt{3}} a^{5/2} r \exp(-ar)$$

$$\Theta(\theta) = \frac{\sqrt{3}}{2} \sin \theta$$

(1-88)

$$\Phi(\varphi) = \frac{1}{(2\pi)^{1/2}} \exp(\pm i\varphi)$$

On transforming to momentum space (1-56) and following the notation of (1-24),

$$\chi(\mathbf{p}) = \frac{a^{5/2}}{2\pi} \int_0^\infty \int_0^\pi \int_0^{2\pi} \exp\left[\frac{-ipr(h \cos \theta + k \sin \theta \cos \varphi + l \sin \theta \sin \varphi)}{(h^2 + k^2 + l^2)^{1/2}}\right]$$

$$\times r^3 \exp(-ar) \sin^2 \theta \exp(\pm i\varphi) \, d\theta \, dr \, d\varphi$$

$$= i \frac{a^{5/2}}{\pi} \frac{k \pm il}{(h^2 + k^2 + l^2)^{1/2}} \frac{8ap}{(a^2 + p^2)^3}$$

(1-89)

The momentum density is

$$|\chi(\mathbf{p})|^2 = \frac{a^5}{\pi^2} \sin^2 \theta \, \frac{64a^2 p^2}{(a^2 + p^2)^6} \tag{1-90}$$

where $\theta$ is the angle with the $z$ axis; i.e., $\sin^2 \theta = (k^2 + l^2)/(h^2 + k^2 + l^2)$. The Compton profile in the impulse approximation requires integration of $|\chi|^2$ over a plane normal to the scattering vector. For any angle $\gamma$ between the x-ray scattering vector and the $z$ axis of the momentum density, one can write the momentum density by rotating about the $x$ or $y$ axis (the momentum density has cylindrical symmetry about the $z$ axis). Introducing a new set of indices

$$\frac{k}{(h^2 + k^2 + l^2)^{1/2}} \rightarrow \frac{k' \cos \gamma}{(h'^2 + k'^2 + l'^2)^{1/2}} - \frac{h' \sin \gamma}{(h'^2 + k'^2 + l'^2)^{1/2}}$$

$$\frac{l}{(h^2 + k^2 + l^2)^{1/2}} \rightarrow \frac{l'}{(h'^2 + k'^2 + l'^2)^{1/2}} \tag{1-91}$$

the transformed momentum density becomes

$$|\chi(p)|^2 = \frac{a^5}{\pi^2} (\cos^2 \gamma \sin^2 \alpha \cos^2 \alpha + \sin^2 \gamma \cos^2 \alpha$$

$$- 2 \sin \gamma \cos \gamma \sin \alpha \cos \varphi + \sin^2 \alpha \sin^2 \varphi) \frac{64a^2 p^2}{(a^2 + p^2)^6} \tag{1-92}$$

where $\alpha$ is the angle with the new $z$ axis and $\varphi$ is the new azimuthal angle. Integration over a plane normal to the scattering vector gives

$$J(z) = \frac{1}{2} \int_0^{2\pi} \int_0^{\pi/2} \int_{|z|}^{\infty} |\chi(p)|^2 p^2 \, \frac{\sin \alpha}{\cos \alpha} \, d\varphi \, d\alpha \, dp \tag{1-93}$$

Integrating over $\varphi$ and noting that $\cos \alpha = z/p$, one finally obtains the Compton profile as a function of the angle $\alpha$ between the scattering vector and the polar axis of the momentum density.

$$J(z) = \frac{64a^7}{2\pi} \int_{|z|}^{\infty} \left[ (1 + \cos^2 \gamma) + (1 - 3 \cos^2 \gamma) \frac{z^2}{p^2} \right] \frac{p^3}{(a^2 + p^2)^6} \, dp$$

$$= \frac{32a^7}{\pi} \left[ \frac{(1 - \cos^2 \gamma)^2 z^2}{5(a^2 + z^2)^5} + \frac{1 + \cos^2 \gamma}{40(a^2 + z^2)^4} \right] \tag{1-94}$$

If the measurement is made on a free atom, all orientations of the atom are possible and one must average over $\gamma$. This gives for the average of $J(z)$,

$$\overline{J(z)} = \frac{32a^7(a^2 + 5z^2)}{15\pi(a^2 + z^2)^5} \tag{1-95}$$

which is identical to that obtained by first averaging over the polar axis for $|\chi|^2$.

This is unlike the square of the scattering factor for which the two averages differ in equations (1-37) and (1-38).

For free-atom $d$ electrons, Fukamachi and Hosoya (1970) have obtained the following profiles:

$$J(z)_{m_l=0} = \frac{5}{8\pi}(A_0 I_0 + B_0 I_2 + C_0 I_4)$$

$$J(z)_{m_l=\pm 1} = \frac{15}{4\pi}(A_{\pm 1} I_0 + B_{\pm 1} I_2 + C_{\pm 1} I_4) \tag{1-96}$$

$$J(z)_{m_l=\pm 2} = \frac{15}{16\pi}(A_{\pm 3} I_0 + B_{\pm 2} I_2 + C_{\pm 2} I_4)$$

where

$$I_v = \int_{|z|}^{\infty} \langle j_l(z)\rangle^2 \left(\frac{z}{p_0}\right)^v p_0 \, dp_0$$

and the coefficients are

$$A_0 = \tfrac{11}{4} - \tfrac{15}{2}\cos^2\gamma + \tfrac{27}{4}\cos^4\gamma$$
$$B_0 = -\tfrac{15}{2} + 63\cos^2\gamma - \tfrac{135}{2}\cos^4\gamma$$
$$C_0 = 9(\tfrac{3}{4} - \tfrac{15}{2}\cos^2\gamma + \tfrac{35}{4}\cos^4\gamma)$$
$$A_{\pm 1} = \tfrac{1}{4} + \tfrac{1}{2}\cos^2\gamma - \tfrac{3}{4}\cos^4\gamma$$
$$B_{\pm 1} = \tfrac{1}{2} - 6\cos^2\gamma + \tfrac{15}{2}\cos^4\gamma$$
$$C_{\pm 1} = -\tfrac{3}{4} + \tfrac{15}{2}\cos^2\gamma - \tfrac{35}{4}\cos^4\gamma$$
$$A_{\pm 2} = \tfrac{3}{4} + \tfrac{1}{2}\cos^2\gamma + \tfrac{3}{4}\cos^4\gamma$$
$$B_{\pm 2} = \tfrac{1}{2} + 3\cos^2\gamma - \tfrac{15}{2}\cos^4\gamma$$
$$C_{\pm 2} = \tfrac{3}{4} - \tfrac{15}{2}\cos^2\gamma + \tfrac{35}{4}\cos^4\gamma$$

In crystals it becomes possible to fix the direction of the nonspherical charge or momentum distribution relative to the x-ray scattering vector. For the doubly and triply degenerate $d$ orbitals in cubic crystals, treated earlier for the scattering factor, we find upon transformation of equations (1-40) and (1-42) to momentum space,

$$|\chi_1|^2 + |\chi_2|^2 = \frac{5}{2\pi^2}[\widehat{j_2(pr)}]^2 \frac{x^4 + y^4 + z^4 - x^2 y^2 - x^2 z^2 - y^2 z^2}{(x^2 + y^2 + z^2)^2}$$

$e_g$ momentum density $\qquad$ (1-97)

$$|\chi_3|^2 + |\chi_4|^2 + |\chi_5|^2 = \frac{5}{2\pi^2}[\widehat{j_2(pr)}]^2 \frac{3(x^2 y^2 + x^2 z^2 + y^2 z^2)}{(x^2 + y^2 + z^2)^2}$$

$t_{2g}$ momentum density $\qquad$ (1-98)

$$\widehat{j_n(pr)} = \int_0^{\infty} R(r) j_n(pr) r^2 \, dr$$

Thus the charge and momentum density show the same anisotropy; cf equations (1-41), (1-43), (1-97), (1-98). In a similar way to the procedure followed above for calculating the Compton profile from nonspherical distributions, one obtains, per electron

$$J(z) = \frac{5}{2\pi} \int_{|z|}^{\infty} [j_2(pr)]^2 \left( A + \frac{Bz^2}{p^2} + \frac{Cz^4}{p^4} \right) p \, dp \qquad (1\text{-}99)$$

where the coefficients $A$, $B$, $C$, for any direction $u$, $v$, $0$ or $u$, $u$, $w$ of the scattering vector relative to the cube axis, are given in Table 1-7 for the $e_g$ and $t_{2g}$ orbitals. To evaluate (1-99), the number of electrons per atom in the $e_g$ and $t_{2g}$ orbitals is required

$$\begin{pmatrix} A \\ B \\ C \end{pmatrix} = n(t_{2g}) \begin{pmatrix} A(t_{2g}) \\ B(t_{2g}) \\ C(t_{2g}) \end{pmatrix} + n(e_g) \begin{pmatrix} A(e_g) \\ B(e_g) \\ C(e_g) \end{pmatrix} \qquad (1\text{-}100)$$

The integrals in (1-99),

$$\int_{|z|}^{\infty} [j_2(pr)]^2 \left( \frac{z}{p} \right)^n p \, dp \qquad n = 0, 2, 4 \qquad (1\text{-}101)$$

have been evaluated using Hartree-Fock free-atom $3d$ radial wavefunctions (Weiss, 1966, p. 185). While applicability is limited to crystals where the functions can be fixed spatially, and one should then employ crystal radial functions, the use of free-atom radial functions is probably a good approximation. At the present time, crystal wavefunctions are not generally available.

As an example of the use of (1-99), Fig. 1-3 shows the calculated line shape for the $3d$ electrons in vanadium which appears to have a $t_{2g}$ population of $\sim 3.2$ electrons and an $e_g$ population of 0.8 electron (DeMarco and Weiss, 1965). A very sizable anisotropy, $\sim 40\%$ at $z = 0$, is predicted for such a distribution of orbitals. Recent experiments of Phillips and Weiss (1972) show negligible anisotropy in the Compton profile suggesting the inadequacy of the $e_g$, $t_{2g}$ description.

Table 1-7   THE COEFFICIENTS PER ELECTRON $A(t_{2g})$, $B(t_{2g})$, $C(t_{2g})$, $A(e_g)$, $B(e_g)$, AND $C(e_g)$ AS A FUNCTION OF THE DIRECTION $u$, $v$, $0$ OR $u$, $u$, $w$ OF THE X-RAY SCATTERING VECTOR RELATIVE TO THE CUBIC AXES OF CRYSTAL

|  | $A(t_{2g})$ | $B(t_{2g})$ | $C(t_{2g})$ |
|---|---|---|---|
| $u, v, 0$ | $\dfrac{1}{4} + \dfrac{3u^2v^2}{4(u^2 + v^2)^2}$ | $\dfrac{3}{2} - \dfrac{15u^2v^2}{2(u^2 + v^2)^2}$ | $\dfrac{35u^2v^2}{4(u^2 + v^2)^2} - \dfrac{7}{4}$ |
| $u, u, w$ | $\dfrac{3}{16} + \dfrac{16u^4 + w^4 + 28u^2w^2}{16(2u^2 + w^2)^2}$ | $\dfrac{15w^4 - 60u^2w^2}{8(2u^2 + w^2)^2} - \dfrac{3}{8}$ | $\dfrac{3}{16} + \dfrac{16u^4 - 31w^4 + 156u^2w^2}{16(2u^2 + w^2)^2}$ |
|  | $A(e_g) = 1 - \tfrac{3}{2}A(t_{2g})$ | $B(e_g) = -\tfrac{3}{2}B(t_{2g})$ | $C(e_g) = -\tfrac{3}{2}C(t_{2g})$ |

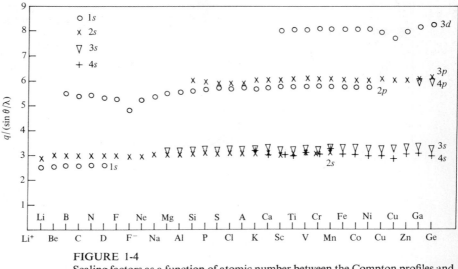

FIGURE 1-4

Scaling factors as a function of atomic number between the Compton profiles and the x-ray scattering factors. The ordinates give the ratio of $q$ (in atomic units) to $\sin\theta/\lambda$ ($A^{-1}$) at which both curves fall to half-height.

## C. Approximate relationship between the scattering factor and the Compton profile

The more the charge density is extended in space the more rapidly the scattering factor decreases with increasing $\sin\theta/\lambda$. This also implies a decrease in the average momenta of the electrons since they are less tightly bound as they move away from the nucleus. Thus the Compton profile will also decrease more rapidly with increasing $z$. It is interesting that there is considerable similarity in shape for both $f(S)$ per electron and $J(z)/J(0)$ [the latter is normalized to unity at $z = 0$ by dividing by $J(0)$]. Consider the hydrogenic functions in relations (1-10a) to (1-10d) where $Z$ is replaced by an effective charge $Z^*$. Also assume that the deviations from spherical symmetry are small. In Table 1-8 the scattering factor and reduced Compton profile are calculated for the $1s$, $2p$, and $3d$ hydrogenic functions. If the scattering factors and corresponding reduced Compton profiles are scaled by matching them at half-height, the appropriate scaling factors $z/k$ are independent of the effective charge since $Z^*$ cancels out in the ratio $\beta/\alpha$. The scaling factors are given in Table 1-8. In addition, the scaling factors were determined from $f(s)$ and $J(z)/J(0)$ calculated from Hartree-Fock wavefunctions. These are also given in Table 1-8 and evidence a marked similarity to the hydrogenic functions. The scaling factors are approximately $2.6l$ where $l$ is the orbital quantum number. Figure 1-4 plots the variation of the Hartree-

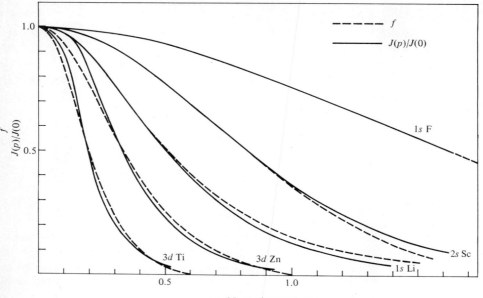

*sin θ/λ or q/scaling factor*

**FIGURE 1-5**
Employing the scaling factors in Fig. 1-4, a few selected Compton-profile and
x-ray scattering factor curves are plotted. The scattering factor curves are per
electron as a function of $\sin \theta/\lambda$ $(\text{A}^{-1})$. The Compton profiles are the normalized
curves (i.e., adjusted to unity at $q = 0$) vs. $q$ (electron momentum) divided by
the scaling factor.

Fock scaling factors with atomic number showing the relative independence with
atomic number. Figure 1-5 is a plot of some representative scattering factors and
reduced Compton profiles adjusting the momentum scale in the latter by the scaling
factor. The curves show an unexpected similarity.

**Table 1-8** **X-RAY SCATTERING FACTORS AND REDUCED COMPTON PROFILES FOR
HYDROGENIC 1s, 2p, AND 3d ELECTRONS**

$\alpha = nk/2Z^*$; $\beta = nz/Z^*$, where $Z^*$ is the effective charge and $n$ the principal quantum
number. The scaling factors for the hydrogenic and Hartree-Fock functions give the ratio of $z$
(in atomic units) to $\sin \Theta/\lambda$ (in $\text{A}^{-1}$) at which the curves fall to half-height

| Hydrogenic function | $f(s)$ | $J(z)/J(0)$ | Scaling factors Hydrogenic | H-F |
|---|---|---|---|---|
| 1s | $(1 + \alpha^2)^{-2}$ | $(1 + \beta^2)^{-3}$ | 2.64 | ~2.6 |
| 2p | $(1 - \alpha^2)(1 + \alpha^2)^{-4}$ | $(1 + 5\beta^2)(1 + \beta^2)^{-5}$ | 5.98 | ~5.6 |
| 3d | $(1 - \tfrac{10}{3}\alpha^2 + \alpha^4)(1 + \alpha^2)^{-6}$ | $(1 + 7\beta^2 + 21\beta^4)(1 + \beta^2)^{-7}$ | 9.35 | ~8.0 |

The independence of scaling factor with atomic number and the similarity in shape permits some comparison to be made between $f(S)$ and $J(z)$ measurements on the same substance, particularly when the departures of the measurements from calculated Hartree-Fock functions are small; i.e., both the magnitude and direction of the departures should be similar. This comparison, though, is only valid for inner or core electrons in solids since the valence electrons are frequently not well described with free-atom-like Hartree-Fock functions.

## D.   Total Compton cross sections

Total Compton cross sections based on the Waller-Hartree expression (1929) have been evaluated by Freeman and are listed in the International Tables (1962). These total cross sections are calculated directly from the position wavefunctions using a sum rule to account for the final electron states in the continuum. Neither binding effects nor time dependence (i.e., impulse approximation) were considered in the Waller-Hartree derivation. The relativistic effects are assumed to be given by the relativistic correction in the Klein-Nishina formula (1929) (electrons at rest). The Waller-Hartree expression for the total cross section is given as

$$\frac{d\sigma}{d\Omega} = \left(\frac{e^2}{mc^2}\right)^2 K^2 \left(Z - \sum_i \sum_j f_{ij}^2\right)\left(\frac{v}{v_0}\right)^2 \qquad (1\text{-}102)$$

where

$$f_{ij} = \int_0^\infty \Psi_i \exp\left[i(\mathbf{k} - \mathbf{k}_0)\cdot\mathbf{r}\right]\Psi_j \, d\mathbf{r} \qquad (1\text{-}103)$$

and $\Psi_i$ and $\Psi_j$ are the one-electron functions employed in a determinantal wavefunction. Since the Waller-Hartree formulation does not specifically consider the line profile, the value of the final frequency $v$ is generally taken to be the value at the center of the line, i.e., from an electron initially at rest (1-61).

Table 1-9   TOTAL COMPTON CROSS SECTIONS FOR THE HYDROGENIC ATOM $Z = 2, 5, 10$, BASED ON THE WALLER-HARTREE EXPRESSION, ON THE EXACT SOLUTION, AND ON THE IMPULSE APPROXIMATION

| $Z$ | Waller-Hartree | Exact | Impulse |
|---|---|---|---|
| 2 | 0.879 | 0.881 | 0.879 |
| 5 | 0.796 | 0.765 | 0.779 |
| 10 | 0.455 | 0.280 | 0.318 |

We can compare the Waller-Hartree total cross sections for the hydrogenic case with the exact cross section and the impulse approximation cross section by integrating equations (1-67) and (1-74) over x-ray energy. For Mo $K\alpha$ ($\lambda = 0.71$ A) scattered at 160°, the total cross sections $d\sigma/d\Omega$ in units of $(e^2/mc^2)^2K^2$ are given in Table 1-9.

As seen in Table 1-9, the Waller-Hartree expression is in serious error, $>40\%$ for $Z = 10$. Even for $Z = 2$ the Waller-Hartree expression is in error by $4\%$. In the absence of accurate calculations for many-electron atoms, it appears that the impulse approximation [for which $J(z)$ has been tabulated in the Hartree-Fock approximation] currently provides the most accurate method for evaluating total Compton cross sections. By combining (1-76) with (1-75), the total cross section becomes

$$\frac{d\sigma}{d\Omega} = \left(\frac{e^2}{mc^2}\right)^2 K^2 \sum_i \int_{z_i^*}^{\infty} \frac{v}{v_0} J_i(z)\ dz_i \qquad (1\text{-}104)$$

where $z_i^*$ is the value of $z$ corresponding to the binding energy of the $i$th electron and the sum is over the $i$ electrons. Tables of the total cross sections are given in Appendix A and are compared to the Waller-Hartree values. In some cases differences of a factor of 3 occur for small $\sin\theta/\lambda$, indicating the need for more experimental measurements.

## III.   EXPERIMENTS

### A.   Introduction

For many years, measurements of electron distributions in position space have been reported but, as the experimental problems inherent in such measurements became more fully appreciated, the great difficulty in obtaining accurate distributions became apparent, particularly for valence electrons. A review of the subject has been given by Weiss (1966), and an I.U.Cr. conference in 1968 has underscored many of the difficulties. Since the measurement of accurate scattering factors is difficult, one must rely heavily on an interchange between theory and experiment. Unfortunately there are very few precise measurements on systems for which the theory is reasonably accurate, i.e., on relatively low-atomic-number free atoms and diatomic molecules. Most of the measurements are on solids for which accurate theoretical scattering factors are not currently available. Furthermore, the experimental problems in measuring scattering factors are more prodigious in solids than in gases.

Relatively accurate measurements in momentum space (Compton profiles) have been available in recent years, all measurements being on condensed matter. In only very few calculations of molecules or solids have position wavefunctions been transformed to momentum space, so that comparisons between theory and experiment are limited. The Compton profiles are particularly sensitive to valence electrons (in contrast to scattering factors), so that one might anticipate a considerable increase in the number of such measurements and calculations.

Since the position and momentum wavefunctions are transforms of each other (1-56), it is possible to combine the discussion of experimental results on charge and momentum density. The two measurements complement each other and, when scattering factors and Compton profiles are available for the same material, the theorist is put to a severe test.

## B.  Experimental techniques

**1.  *Scattering factors***  The experimental techniques in scattering factor measurements are widely discussed in the literature (Weiss, 1966). There is little more that can be added here except to underscore some of the sources of error. A list of these is given in Table 1-10 together with an estimate of the uncertainty for both theoretical and experimental work. Since most scattering factor measurements are made on solids with virtually no theoretical calculations for comparison, it has been the custom to compare the experimental results on solids with theoretical free-atom Hartree-Fock scattering factors. Since these are believed accurate to $\sim 1\%$, any departures significantly greater than $1\%$ in the experimental result could be interpreted as a solid-state effect, i.e., due to the change in charge density when the free atoms form the solid. Actually the differences are seldom more than a few percent, and it is rare that the experimental sources of uncertainty listed in Table 1-10 have been reduced to $1\%$.

**2.  *Compton profile***  The experimental technique employed in Compton-profile measurements is shown schematically in Fig. 1-6. The entire spectrum from a Mo target x-ray tube is collimated with a set of Soller slits $\pm 1.1°$ and impinges on the sample. The radiation scattered through some fixed angle (generally between 120 and 160°) passes through a second set of very fine Soller slits $\pm 0.07°$, and the spectra are analyzed using a LiF crystal [reflections from (400) or (600) planes] and a third set of Soller slits $\pm 0.07°$.[1] If the scattered spectrum is analyzed in the vicinity of the

---

[1] Nondispersive detection (Li-drifted Ge or Si detectors with multichannel analyzers) provides a second method for Compton-profile measurements, but it requires a relatively high-energy photon source (such as 59-keV gamma rays from a radioactive americium source) to give acceptable resolution. While resolution may be poorer than for x-ray methods, it has the advantage of eliminating $\alpha_1 - \alpha_2$ separation and the Bremsstrahlung background. Furthermore, for heavier elements ($Z > 20$), there is a smaller ratio of photoelectric absorption to Compton scattering. This method appears to be superseding the dispersive x-ray method except for light elements or whenever high resolution is desired.

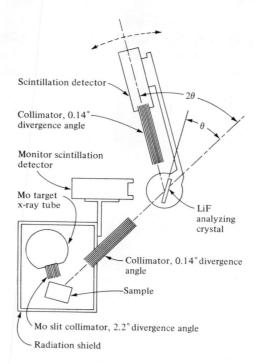

FIGURE 1-6
Experimental arrangement for Compton-profile measurements.

Mo $K\alpha$ line, an elastic component (Mo $K\alpha_1$, Mo $K\alpha_2$) is found due to thermal diffuse and possibly Bragg scattering from the specimen. The superposed Compton lines from the Mo $K\alpha_1$ and Mo $K\alpha_2$ lines in the incident beam are shifted to longer wavelengths with a maximum intensity at a wavelength increase given by (1-82). The Bremsstrahlung continuum also gives rise to an elastic and a Compton component but this is a continuous and slowly varying background that can be separated from the Compton component. While it would simplify matters somewhat to employ a monchromatic x-ray beam, the intensity lost in achieving these experimental conditions would be generally prohibitive. Occasionally extraneous peaks appear in the analyzed spectrum due to fluorescence or Bragg scattering from the sample, but these sources of scattering can often be eliminated.

There are several steps involved in analyzing Compton-profile data:

*1*   Subtraction of background, impurity peaks, and Bragg peaks
*2*   Separation of $\alpha_1$ and $\alpha_2$ components
*3*   Wavelength-dependent corrections:
   *a*   Relativistic
   *b*   Absorption
   *c*   Crystal and detector efficiency

4 Corrections for geometric resolution

5 Conversion from an x-ray wavelength to an electron momentum scale and normalization of the resulting $J(z)$ curve

After completion of these five steps, one has the total $J(z)$ curve per atom (1-76). For samples with negligible anisotropy, equation (1-80) can be inverted to obtain the total momentum density $|\chi|^2$. The core- and valence-electron contributions can generally be separated by employing core profiles calculated from Hartree-Fock wavefunctions in the impulse approximation.

In Fig. 1-7 is shown the experimental curve from lithium metal analyzed with a LiF (600) crystal. Mo $K\alpha$ radiation and a scattering angle of 120° were employed.

Table 1-10   SOME SOURCES OF UNCERTAINTY IN STRUCTURE FACTOR MEASURE-MENTS

|  | | Uncertainty | Reference and comments |
|---|---|---|---|
| 1. | Hartree-Fock free-atom scattering factor (theory) | ~1% | Very few measured. (Hall, 1966; Goodisman and Klemperer, 1963) |
| 2. | Molecular scattering factors (theory) | ~2% | Very few calculated |
| 3. | Crystal structure factors (theory) | ~3% | Very few calculated |
| 4. | Debye-Waller correction $(1 - e^{-M})$ | >10% | Especially difficult at high temperatures or when anharmonicity is large. (Chipman, 1960; Batterman and Chipman, 1962) |
| 5. | Thermal diffuse scattering correction | >10% | Very few measurements. (Nilsson, 1957; Schwartz, 1964; Chipman and Paskin, 1959; Jennings, 1968) |
| 6. | Dispersion corrections | >10% | Very few direct measurements. Reliance placed on Kramers-Kronig relations |
| 7. | Extinction corrections | >20% | Needs considerable experimental study |
| 8. | Absorption coefficients | 1–10% | International Tables unreliable |
| 9. | *Umweganregung* | ? | Needs experimental study |
| 10. | Dead-time error corrections | ~20% | |
| 11. | Porosity correction | ~20% | Needs more experimental work. (Cooper, 1965) |
| 12. | Polarization correction $(1 - K^2)$ | ~20% | Depends on the extinction in the monochromator and the sample. (Jennings, 1968) |
| 13. | Preferred orientation in powders | ? | Extreme care required to reduce to less than 1% |
| 14. | Size-effect correction in disordered alloys | ? | Very little experimental work reported. (Borie, 1957) |
| 15. | Geometrical factors | ? | Beam homogeneity, sample shape, absorption factors, etc. (Svortti and Jennings, 1971) |

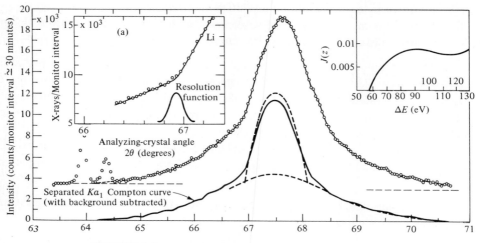

**FIGURE 1-7**

Measured Compton profile of lithium employing a LiF (600) analyzer (open circles). The solid lower curve is the separated Mo $K\alpha_1$ component after subtracting the straight-line background shown by the dashed line. Superposed on the separated $\alpha_1$ component is the calculated curve deduced from free-electron theory and the $1s^2$ Hartree-Fock free-atom-like core (dashed lines). Upper-left insert shows the region of the excitation energy per $1s$ electron (57 eV) after separating the $\alpha_1$ component and deconvoluting the resolution function.

Both the elastic Mo $K\alpha_1$, Mo $K\alpha_2$ peaks and the superposed Compton line are clearly seen (open circles). The dashed lines indicate the straight-line background subtracted in the subsequent analysis. Since the wavelength separation and the relative intensities of the $\alpha_1 - \alpha_2$ components are known, it is possible to employ a computer program to separate the $\alpha_1$ component from the total. The separated $\alpha_1$ component, after corrections are made for relativistic effects [using (1-87)], sample absorption (i.e., the wavelength dependence of the absorption coefficient), and crystal efficiency, is given as the solid curve.

The wavelength dependence of the analyzing crystal efficiency can be estimated from its rocking curve and its structure factor. If the rocking curve is considerably broader than expected for a perfect crystal, secondary extinction may be important. For a thick crystal in Bragg reflection the reflecting power for a Gaussian rocking curve is

$$\frac{P_H}{P_0} = 1 + x\left[1 - \left(1 + \frac{2}{x}\right)^{1/2}\right]$$

$$x = \frac{\sin 2\theta_B \mu V^2}{(e^2/mc^2)^2 \lambda^3 K^2 |F|^2 \exp(-2M)\sqrt{2g}\exp(-2\pi \Delta^2 g^2)}$$

(1-105)

where $0.664/g$ is the full width at half-maximum of the rocking curve, and $\Delta$ is the deviation of the incident beam from the peak of the rocking curve. If the rocking curve is wider than the Soller slits, it is a good approximation to use the peak reflecting power ($\Delta = 0$). The absorption coefficient $\mu$ can generally be written as $\mu = A\lambda^{2.7}$ so that the wavelength dependence of the argument $x$ in (1-105) is

$$x \propto \frac{1 + [1 - 2(\lambda/2d)^2]^2}{\lambda^{0.7}[1 - (\lambda/2d)^2]^{1/2}} \qquad (1\text{-}106)$$

where $2d$ is the interplanar spacing of the crystal reflecting planes. In a typical case, such as the 400 reflection of LiF with Mo $K\alpha$, the crystal efficiency (peak reflecting power) is reduced by $\sim 8\%$ over the range of interest from 0.71 to 0.81 A.

The geometrical resolution function of the system $Y(2\alpha)$ (i.e., Soller slits and analyzing crystal) can be determined from the shape of the $K\alpha_1$ line since the true line is considerably narrow. For Mo $K\alpha$, the true width of the $\alpha_1$ lines is $\sim \frac{1}{20}$ of the $\alpha_1 - \alpha_2$ separation. The measured resolution function $Y(2\alpha)$ broadens the true Compton profile $J(2\theta)$ to give the measured curve $G(2\theta)$.

$$G(2\theta) = \int_{-\infty}^{+\infty} J(2\theta)Y(2\theta - 2\alpha)\, d2\alpha \qquad (1\text{-}107)$$

A trial-and-error method can be employed to solve for $J(2\theta)$. For example, the measured curve can be used as a trial function which, when compounded with the resolution function, gives a broader curve than measured. The differences between the calculated curve and the measured curve are used to correct the first trial function. This gives a second trial function, and the process is repeated until a trial function $J(2\theta)$ is found which reproduces the measured curve $G(2\theta)$.

Once $J(2\theta)$ is found, equation (1-86) can be used to convert from a wavelength scale to an electron momentum scale $J(z)$. Normalization of $J(z)$ requires that

$$\int_{-\infty}^{+\infty} J(z)\, dz = Z \qquad \text{(total number of electrons per atom)} \qquad (1\text{-}108)$$

provided, of course, that the impulse approximation is valid over the range $-\infty$ to $+\infty$. The experimental data, however, only can be analyzed up to some finite upper limit on the long-wavelength side of the Compton profile ($z$ positive), since the contribution of the profile may become very small compared to the background. Furthermore, on the short-wavelength side of the Compton profile ($z$ negative), integration can be carried on only out to values of $z$ corresponding to the binding energy. The practical solution to this problem is to select some convenient range from $-z$ to $+z$ that easily encompasses all the valence-electron contributions, say, from $z = -4$ to $z = +4$. The integral in equation (1-108) then can be separated into various electron groups with the total integral equal to the number of valence

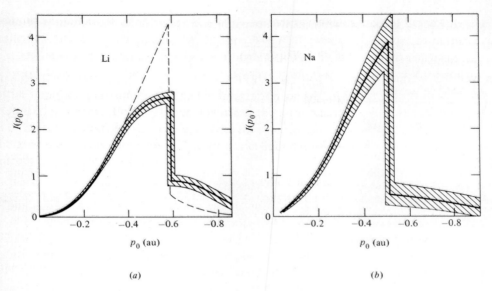

(a)                                   (b)

FIGURE 1-8

Momentum density $I(p_0)$ for the valence electron. (a) Lithium. Note the discontinuity at the Fermi momentum at $z = 0.59$ (shaded area). The dashed curve was calculated for an interacting electron gas. (b) Sodium.

electrons plus the core contribution from $z = -4$ to $z = +4$. The core contributions then can be summed by integrating the calculated free-atom Compton profiles (Appendix B) over this range. Integration of the calculated free-atom Compton profiles is, of course, subject to the additional condition that the integral is zero for values of $z$ below the binding energy. Such a procedure provides an absolute standardization that is generally accurate to within 2%.

Since single-crystal measurements on lithium failed to reveal any anisotropy (less than 2%), relation (1-80) can be inverted to yield $|\chi|^2$. Differentiation gives

$$|\chi(p_0)|^2 = \left| \frac{1}{2\pi z} \frac{dJ}{dz} \right| \qquad (1\text{-}109)$$

or for a spherical volume element

$$4\pi |\chi(p_0)|^2 p_0{}^2 \equiv I(p_0) = \left| 2z \frac{dJ}{dz} \right| \qquad (1\text{-}110)$$

For the lithium data in Fig. 1-7, application of equation (1-10b) (after using the Hartree-Fock functions in the impulse approximation to subtract the $1s^2$ core) gives the $I(p_0)$ shown in Fig. 1-8. The dashed curve is from a theoretical approximation discussed below. The discontinuity at the Fermi momentum $z = 0.59$ is clearly seen.

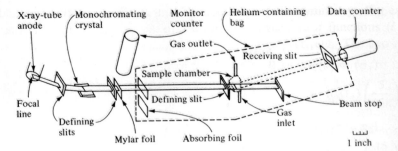

FIGURE 1-9

A schematic diagram of the apparatus used for an absolute measurement of the inert gas scattering factors. The gas was maintained at 1-atm pressure by a continuous flow, and the air (comprising most of the x-ray path length) was replaced with helium to reduce scattering. A LiF (200) monochromating crystal was employed to select the Mo $K\alpha$ line.

This discontinuity arises from the discontinuous change in slope in $J(z)$ in the vicinity of 66.96° ($2\theta$) seen in the inset of Fig. 1-7. The shaded area in Fig. 1-8 gives the probable error.

## C. Experimental results

The results discussed below were chosen to illustrate the kind of experiments that have yielded information about electron distributions in position and momentum space.

**1. Inert gases He, Ne, Ar, Kr, and Xe**[1]  Mo $K\alpha$ x-rays were selected with a singly bent LiF (200) monochromator and scattered from the gas sample, which was kept at 1-atm pressure in a container with parallel aluminum windows. The remainder of the x-ray path length was filled with helium to reduce air scattering. A schematic arrangement of the experiment is shown in Fig. 1-9. The 0.0005-in. aluminum windows showed considerable preferred orientation so that in transmission the 200 and 220 reflections were virtually eliminated at the expense of an enhanced 111 reflection. Measurements were made over the range $\sin \theta/\lambda = 0.037$ to $0.39$ A$^{-1}$, with the range in the vicinity of 0.21 A$^{-1}$ eliminated due to the interference of the 111 reflection from the aluminum windows. By extrapolating to $\sin \theta/\lambda = 0$ their absolute scattering factors yielded $f = Z + \Delta f'$ so that $\Delta f'$ could be determined. The measured values of $\Delta f'$ are given in Table 1-5 (method 4) and reveal significant

[1] Chipman and Jennings (1963).

differences with theory (methods 1 and 2). In neon, the dispersion correction at $\sin \theta/\lambda = 0$ is extremely small for Mo $K\alpha$ ($\Delta f'/Z \simeq 10^{-4}$) so that the experiment corroborated the fundamental photon-electron interaction as accurately given by $(e^2/mc^2)$ to within their experimental error ($\sim \frac{1}{2}\%$). The relativistic correction (Weiss, 1966, p. 11; Goldberger and Low, 1968) to $e^2/mc^2$ is less than $10^{-4}$.

Up to $\sin \theta/\lambda \sim 0.4$ $A^{-1}$, the measured values of $f_0$ agreed with Hartree-Fock scattering factors to 1%. It was necessary, however, to make a total Compton cross section (Waller-Hartree theory) which amounted to $\sim 20\%$ in neon and 12% in argon at the highest values of $\sin \theta/\lambda$. The total Compton cross sections (Appendix A) appear to have significant uncertainty in this range, so that agreement between theory and experiment did not fall within 1% over the range of their measurements. If these experiments were repeated with Ag $K\alpha_1$ radiation it would be possible to separate experimentally the elastic and inelastic (Compton) components with a ruthenium filter, since an energy loss of $\sim 50$ eV is sufficient to cross the ruthenium absorption edge.

Compton-profile measurements for He, Ne, and Ar have recently been reported by Eisenberger (1970) and agree with Hartree-Fock theory to within 1%. For helium and neon, more accurate wavefunctions reveal only minor departures ($<1\%$) of the Compton profile from the profile calculated from Hartree-Fock wavefunctions.

**2. $H_2$ molecule** One of the simplest systems to demonstrate the effect of bonding is the hydrogen molecule, the binding energy 4.72 eV being about 15% of the total energy (31.9 eV). The Compton-profile measurements of Eisenberger (1970) were made on $H_2$ liquid but, since the additional binding energy of the liquid (heat of vaporization) is only of the order of $10^{-2}$ eV due to the weak Van der Waals forces, the measurements probably give an accurate representation of the free molecule. The liquid was used in order to gain more scattering intensity. The results of the measurements are given in Table 1-11, and they clearly show the effect of bonding. Compared to the superposition of two hydrogen atoms, $J(0)$ is reduced by $\sim 15\%$.

Table 1-11   COMPTON-PROFILE MEASUREMENTS FOR $H_2$

| $z$ | Experiment | 2H atoms | $H_2$ (theory) |
|-----|-----------|----------|----------------|
| 0 | 1.51 | 1.70 | 1.53 |
| 0.2 | 1.38 | 1.51 | 1.39 |
| 0.5 | 0.89 | 0.87 | 0.89 |
| 1.0 | 0.25 | 0.21 | 0.25 |
| 2.0 | 0.015 | 0.013 | 0.017 |

Compared to the molecular wavefunctions of Liu (Brown and Smith, 1971), the agreement is quite good ($<2\%$). It has been suggested by Ulsh, Bonham, and Bartell (1972) that the zero-point motion of $H_2$ requires $\sim 1\%$ theoretical correction to the Compton profile at $J(0)$. Measurements of the scattering factor are not available for comparison.

**3. *Lithium and sodium*** Lithium and sodium are two of the simplest metals. Each has a free-atom-like core ($1s^2$ in lithium and $1s^2 2s^2 2p^6$ in sodium) with a single valence electron per atom responsible for its metallic properties. The wavefunctions for these valence electrons have been calculated in several approximations.

*a. Free-electon theory.* The valence electrons are assumed to be in a constant potential yielding a constant valence-electron spatial density; i.e., 1 electron per atomic volume. Correlation between electrons is neglected so that the momentum density in momentum space is constant from $p_0 = 0$ to $p_F$, the Fermi momentum. In atomic units

$$p_F = \pi \left(\frac{3\rho}{\pi}\right)^{1/3} \qquad (1\text{-}111)$$

where $\rho$ is the number of valence electrons per unit volume. The momentum density is also constant up to the Fermi momentum,

$$|\chi(p_0)|^2 = \frac{3}{4\pi p_F{}^3} \qquad 0 \le p_0 \le p_F$$
$$= 0 \qquad p_0 > p_F \qquad (1\text{-}112)$$

*b. Orthogonalized plane waves (OPW)*[1] In the orthogonalized-plane-wave method the valence-electron wavefunctions are made orthogonal to the atomlike $1s^2$ core electrons. Transformation of these wavefunctions to momentum space reveals only minor differences from the free-electron theory.

*c. Interacting electron gas*[2] In this model, the electrons are considered free from any nuclear or core-electron interactions, but the Coulomb repulsion between the valence electrons is specifically included in the calculation. When the density of the electron gas is made appropriate to the metal under consideration, the momentum

[1] Melngailis and DeBenedetti (1966).
[2] Daniel and Voski (1960).

density still evidences a sharp discontinuity at the Fermi momentum (1-110), but some momentum states below the Fermi momentum are shifted to values above $p_F$, giving rise to a "high-momentum tail." This transfer of momentum states is due to the increased kinetic energy required to keep the electrons apart.

In Fig. 1-8 there is plotted the theoretical $I(p_0)$ as calculated from the inter-acting electron gas (dashed line). The free-electron theory and the OPW calculations are not shown in Fig. 1-8 but they both would be essentially quadratic; i.e.,

$$I(p_0) = 4\pi|\chi(p_0)|^2 = \frac{3p_0^2}{p_F^3} \qquad 0 \le p_0 \le p_F$$

$$I(p_0) = \frac{3}{p_F} \qquad\qquad\qquad p_0 = p_F \qquad (1\text{-}113)$$

$$I(p_0) = 0 \qquad\qquad\qquad p_0 > p_F$$

thus rising to a maximum value of 5.1 in Li for which $p_F = 0.59$ (Fig. 1-8). The measured curve (solid line) gives clear evidence of the high-momentum tail and provides the clearest experimental evidence of the effect of correlation in a metal. The measured momentum density for sodium is shown in Fig. 1-8, and it is quite similar in detail to that for lithium. Again, free-electron theory (1-113) would give a quadratic distribution in $I(p_0)$ with a maximum value of 5.9 at $p_F = 0.51$.

At the present time, x-ray scattering factor measurements are unavailable for lithium or sodium, but it is unlikely that such measurements would yield any signifi-cant information about the valence-electron spatial distribution since their contribu-tion to the Bragg peaks ($<1\%$) is too small to measure with presently attainable accuracies.

The upper-right inset in Fig. 1-8 shows the region in the vicinity of the $1s$ excitation energy for Li (57 eV). This is typical of the Compton profiles near absorp-tion edges. There is a sudden increase in intensity at the excitation energy, and it peaks at $\sim 30$ eV above the excitation energy. The curve then dips to a minimum and, at about 100 eV above the excitation energy, appears to take on the values given by the impulse approximation. The origin of this 30-eV peak is not known, but energies of the order of 30 eV are typical of plasma energies.

**4.   *Magnesium and LiMg alloy***   Both the Compton profile (Phillips and Weiss, 1972) and absolute structure factors (Weiss, 1967) have been measured for magnesium. Only the core-electron contribution ($1s^22s^22p^6$) is significant to the structure factor measurements, and agreement with Hartree-Fock theory was obtained. The Compton profile for the valence electrons in Mg (theoretical Hartree-Fock core contribution subtracted) is plotted in Fig. 1-10. Compared to lithium, the $J(z)$ curve shows some structure which is similar to the structure expected for some $p$-like contribution to the

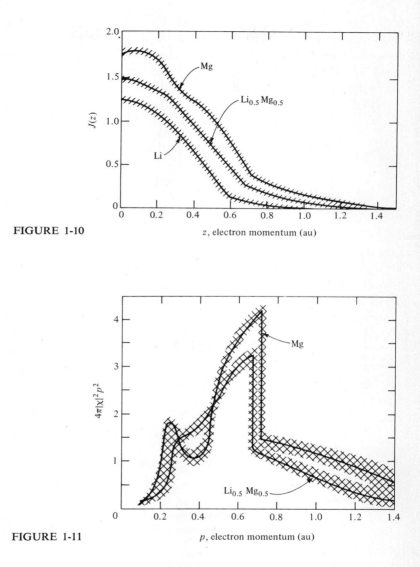

FIGURE 1-10

$z$, electron momentum (au)

FIGURE 1-11

$p$, electron momentum (au)

two valence-electron wavefunctions. This is also seen in Fig. 1-11, in which $I(p)$ is plotted showing a peak in the vicinity of 0.25 au of momentum. The discontinuity at the Fermi momentum, $p_F = 0.72$, is clearly seen to be in good agreement with the value derived from free-electron theory (1-110). This is interesting in that the significant $p$-like contribution to the wavefunction has not significantly affected the value of the Fermi momentum.

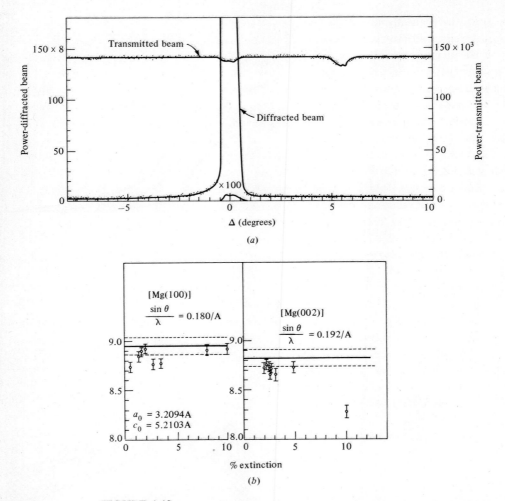

FIGURE 1-12
Experimental intensity measurements. (*a*) The point-to-point (every 3 minutes of arc) power in the transmitted (top curve) and diffracted (bottom curve) beams in a typical run. The dip in the transmitted beam in the vicinity of the diffracted peak is due to the power loss to the peak being studied, while the second dip in the transmitted beam is due to the power loss to some other peak not detected by the counter. (*b*) The experimental values of $F$ (after correction for secondary extinction) of the 100 and 002 reflections as a function of the secondary extinction correction measured at different spots on the crystal. The two points at 8 and 10% extinction (100 reflection) exhibited large extinction because of the super-position of two reflections in the transmitted beam. The solid line is the theoretical Hartree-Fock "neon" core value while the dashed lines are the probable limits of uncertainty in the theoretical value.

The scattering factor measurements (Weiss, 1967) were made on a thin single crystal, 0.012 in. thick, with the $a$ axis normal to the surface. Measurements were made, in transmission, for the 100 and 002 reflections and two counters simultaneously recorded the transmitted and diffracted beams as the crystal was rocked through the Bragg peak. A typical set of data is shown in Fig. 1-12$a$ where the transmitted beam displays the power lost due to diffraction. The observance of other dips in the transmitted beam power is evidence of *Aufhellung*; i.e., other Bragg peaks coming into the reflecting position but not detected by the counter in its position to measure the specific reflection being studied. The monochromatic Mo $K\alpha$ x-ray beam was finely collimated so that its size $\sim 0.010$ in. and divergence of about 1.8 minutes of arc were both small compared to the spatial and angular variations of the "mosaic blocks." Each point on the diffraction curve has been corrected by the corresponding departure of the transmitted beam from the smooth line due solely to true absorption. This is the correction for secondary extinction (Weiss, 1967). When secondary extinction is small, the integrated reflecting power is given by

$$\int_{-\infty}^{\infty} \frac{P_H/P_0}{P_T/P_0} \, T \, d\Delta = \frac{K^2 (e^2/mc^2)^2 |F|^2 \lambda^3 T \exp(-2M)}{V^2 \sin 2\theta_B \gamma_0} \qquad (1\text{-}114)$$

where $P_H$, $P_T$, and $P_0$ are the powers in the diffracted, transmitted, and incident beams, respectively. Measurements were made at different spots on the crystal, each requiring a different secondary extinction correction, and the results after correcting for secondary extinction and anomalous dispersion are given in Fig. 1-12$b$. The variation in the values of $f_0$ are probably due to primary extinction, for which no correction was feasible at a single x-ray wavelength. The maximum values, however, agree with the theoretical calculations for the Hartree-Fock core values.

The LiMg alloy (50–50 atomic percent) had an essentially disordered body-centered cubic structure, and comparison with the results shown in Figs. 1-10 and 1-11 suggests a valence-electron momentum distribution somewhat midway between Li and Mg. In addition, the results for the Fermi momentum ($p_F = 0.67$) were consistent with a value given by the *average* valence-electron density per atom; i.e., one and one-half electrons per atom in (1-110). This is good evidence for the collective behavior of the valence electrons.

**5.  *Carbon (diamond, graphite, carbon black)*** Elemental carbon occurs in three forms: The diamond structure, the graphite structure, and a carbon black structure consisting of very small particles comprised of parallel graphitic layers having very little $a$-axis correlation. Scattering factor measurements on diamond (Weiss, 1966, p. 181; Göttlicher and Wölfel, 1959) give evidence for a considerable departure of the valence electrons from spherical symmetry. There have been a considerable number of band calculations for diamond (Goroff and Kleinman, 1968) which have had some success in calculating the anisotropy of the charge density, but none of these calcula-tions have been extended into momentum space for comparison with Compton-profile measurements. Recently, a band calculation of diamond by Euwema, Wilhite, and Surratt (1973) has been transformed to momentum space by Wepfer, Euwema, Surratt, and Wilhite (1973) and gives good agreement with both the scattering factors and the Compton profile.

Each atom in diamond is surrounded by four nearest neighbors in tetrahedral coordination. The next-nearest neighbors are considerably farther apart. One generally assumes the $1s^2$ core electrons to be essentially free-atom-like, and Table 1-12 gives the measured total scattering factors separated into core- and valence-electron contributions. The (222) structure factor is zero in the diamond crystal for a spherically symmetric electron distribution. Thus the measurement of a finite (222) structure factor is direct evidence for anisotropy in the electron spatial distribution. Following the procedure in Section I C, one can expand the valence-electron wave-function in terms of functions having the appropriate local tetrahedral symmetry. The hydrogenic one-electron functions, (1-10c) and (1-10d), cannot be combined to produce local tetrahedral symmetry in the charge density. The leading term in such an expansion is

$$\Psi_0 = A\left[R_1(r) + \alpha\,\frac{xyz}{r^3}\,R_2(r)\right] \qquad (1\text{-}115)$$

where $R_1$ and $R_2$ are radial functions, $A$ is the normalization constant, and $\alpha$ an adjustable parameter. Equation (1-115) yields a scattering factor

$$f_0 = \langle j_0(Sr)\rangle - \frac{2i\alpha hkl}{(h^2 + k^2 + l^2)^{3/2}}\,\langle j_3(Sr)\rangle + \text{terms in } \alpha^2$$

$$\langle j_0(Sr)\rangle = \int R_1{}^2(r)j_0(Sr)r^2\,dr \qquad (1\text{-}116)$$

$$\langle j_3(Sr)\rangle = \int R_1(r)R_2(r)j_3(Sr)r^2\,dr$$

**Table 1-12   THE OBSERVED SCATTERING FACTORS FOR DI-AMOND, THE CALCULATED HARTREE-FOCK CORE CONTRIBUTIONS $1s^2$, AND THE SEPARATED VALENCE-ELECTRON CONTRIBUTION**

|  | $\sin\Theta/\lambda$ | $f$ (observed) | $f$ (core)$1s^2$ | $f$ (valence) |
|---|---|---|---|---|
| 111 | 0.2428 | 3.32 $\pm$ 0.06 | 1.92 | 1.40 $\pm$ 0.06 |
| 220 | 0.3964 | 1.98 $\pm$ 0.06 | 1.79 | 0.19 $\pm$ 0.06 |
| 311 | 0.4649 | 1.66 $\pm$ 0.05 | 1.72 | $-0.06 \pm 0.05$ |
| 222 | 0.4856 | 0.144 $\pm$ 0.015 | 0 | 0.14 $\pm$ 0.015 |
| 400 | 0.5607 | 1.48 $\pm$ 0.05 | 1.62 | $-0.14 \pm 0.05$ |
| 331 | 0.6110 | 1.58 $\pm$ 0.05 | 1.56 | $+0.02 \pm 0.05$ |

If we assume $R_1$ is given by the radial part of the carbon $2s2p^3$ Hartree-Fock wavefunction, we can evaluate $\langle j_0(Sr) \rangle$ and $\langle j_3(Sr) \rangle$. Since the $\langle j_0(Sr) \rangle$ term does not contribute to the (222) structure factor, one can determine $\alpha$ from (1-116) and the measured (222) scattering factor in Table 1-12. Neglecting terms in $\alpha^2$, we obtain $\alpha = 0.085$ and, employing this value of $\alpha$, the (111) valence-electron contribution is $\sim 1.10$, considerably less than the measured value of 1.40. As can be seen below, however, the Compton-profile measurements do not fall as rapidly as those calculated from the Hartree-Fock $2s2p^3$ wavefunctions but more closely agree with a distribution halfway between the carbon and nitrogen free-atom distributions. Because of the independence of the scaling factors from atomic number, one can determine a more appropriate $\langle j_0(Sr) \rangle$ by averaging the values for carbon and nitrogen given in Table 1-13. Using the average value for carbon and nitrogen, one obtains $\langle j_0(Sr) \rangle = 1.40$ for (111), in good agreement with the value in Table 1-12. Similarly, for (220) one obtains $\langle j_0(Sr) \rangle = 0.32$, slightly higher than the measured value.

If the position wavefunction (1-115) is transformed to momentum space, the momentum wavefunction becomes

$$\chi(p_0) = \frac{4\pi A}{(2\pi)^{3/2}} \left[ \widehat{j_0(p_0 r)} - \frac{i\alpha hkl}{(h^2 + k^2 + l^2)^{3/2}} \widehat{j_3(p_0 r)} \right]$$

$$\widehat{j_0(p_0 r)} = \int R_1(r) j_0(p_0 r) r^2 \, dr \qquad (1\text{-}117)$$

$$\widehat{j_3(p_0 r)} = \int R_2(r) j_3(p_0 r) r^2 \, dr$$

While the charge density $|\Psi_0|^2$ (1-115) has maxima in the directions of the four nearest neighbors, the momentum density $|\chi(p_0)|^2$ is centrosymmetric and has eight maxima

Table 1-13  HARTREE-FOCK VALUES OF $\langle j_0(Sr) \rangle$ PER ELECTRON FOR THE CARBON AND NITROGEN $2s$ AND $2p$ ELECTRONS

| sin $\Theta/\lambda$ | C | | N | |
|---|---|---|---|---|
| | $2s$ | $2p$ | $2s$ | $2p$ |
| 0 | 1.00 | 1.00 | 1.00 | 1.00 |
| 0.05 | 0.946 | 0.934 | 0.961 | 0.955 |
| 0.10 | 0.802 | 0.769 | 0.855 | 0.835 |
| 0.15 | 0.614 | 0.570 | 0.707 | 0.676 |
| 0.20 | 0.428 | 0.392 | 0.546 | 0.516 |
| 0.25 | 0.271 | 0.254 | 0.395 | 0.377 |
| 0.30 | 0.155 | 0.158 | 0.268 | 0.265 |
| 0.35 | 0.076 | 0.094 | 0.169 | 0.182 |
| 0.40 | 0.027 | 0.053 | 0.097 | 0.121 |

along the body diagonals. (It seems obvious that momentum density in a crystal must be centrosymmetric if the crystal is stationary.) The presence of $i$ in (1-117) gives rise only to terms in $\alpha^2$ in the momentum density, and this appreciably reduces the magnitude of the asphericity in momentum space. The Compton profile calculated from (1-117) has negligible anisotropy ($<1\%$).

The experimental results for $J(z)$ given in Table 1-14 were measured in five directions for diamond with anisotropies in $J(z)$ of $\sim 5\%$. Since the five directions [001], [110], [111], [112], [221] are all normal to the [1$\bar{1}$0] direction and since the anisotropy was small, it is a reasonable approximation to average the results over these directions to obtain the Compton profile equivalent to continuously rotating the crystal about [1$\bar{1}$0] during the measurement. The differences $\Delta J(z)$ between the average and the observed $J(z)$ values in Table 1-14 are shown in Fig. 1-13. Essentially, this is a difference-momentum-density projection on the (1$\bar{1}$0) plane. In principle, it is possible to determine the three-dimensional momentum density map from the data although, in practice, this is not a straightforward procedure. The limitation arises partly from the finite sampling of the various directions. Mijnarends (1967) has suggested a technique based on expanding the momentum density in functions having the appropriate crystal symmetry. The functions employed by Mijnarends assumed separability of the momentum density into a product of radial and angular functions, but this may not be a good assumption.

By trial and error, Phillips and Weiss (1968) deduced a three-dimensional-difference density that reproduced the observed difference profiles. Some momentum density in the region 1.3–2.3 au along the [211] direction is transferred to similar regions along the [200] direction. This is shown in the inset in Fig. 1-13, and it is clear that the anisotropy cannot be separated into a product of radial and angular functions. The measured Compton profile for diamond powder is given in Table 1-14, and since the powder provides an average over the angular variables, equations (1-109) and (1-110) can be used to convert to a momentum density. This is shown in Fig. 1-14 together with the momentum density calculated for a superposition of Hartree-Fock free-atom momentum densities. The measured momentum densities are shifted to higher values and, in addition, a very high momentum component $p_0 > 4$ au is seen which may be due to electron-electron correlation. By employing his calculational method Mijnarends (1967) concludes that a deficiency of momentum density occurs in the [110] direction with maxima along [100] and [111]. In summary, the structure factors and Compton profiles for diamond cannot be explained by employing a simple wavefunction such as (1-115). Only a band calculation can give accurate values of both.

The Compton profiles for graphite and carbon black are given in Table 1-14, and the two profiles are quite similar. This is consistent with the results of Biscoe and Warren (1942), who have shown that carbon black is composed of graphite

# Table 1-14  EXPERIMENTAL $J(z)$ VALUES ($z$ IS THE ELECTRON MOMENTUM IN ATOMIC UNITS) FOR CARBON BLACK; PYROLYTIC GRAPHITE AVERAGED OVER THE THREE MEASURED DIRECTIONS; DIAMOND POWDER; DIAMOND AVERAGED AROUND THE [110] AXIS; DIAMOND [110], [111], [100], [211], AND [221]; PYROLYTIC GRAPHITE $c$ AXIS AND 90° AND 45° TO THE $c$ AXIS. THE FINAL COLUMN GIVES THE HARTREE-FOCK $J(z)$ VALUES FOR THE FREE-ATOM $1s^2 2s 2p^3$. THE EXPERIMENTAL VALUES OF $J$ BEYOND $z = 1.6$ au ARE ESSENTIALLY IDENTICAL FOR ALL CASES. THE ABSOLUTE UNCERTAINTY IN THE EXPERIMENTAL VALUE OF $J$ IS $\pm 0.02$ (au)$^{-1}$ FOR ALL VALUES OF $z$. THE RELATIVE UNCERTAINTY IS $\pm 0.01$ (au)$^{-1}$ (1 au = $mc/137$)

| $z$ | Carbon black | Graphite ave. | Diamond powder | Diamond ave. | Diamond [110] | [111] | [100] | [211] | [221] | Graphite $c$ axis | 90° | 45° | HF $1s^2 2s 2p^3$ |
|---|---|---|---|---|---|---|---|---|---|---|---|---|---|
| 0.00 | 2 23 | 2.16 | 2.08 | 2.05 | 2.11 | 2.05 | 2.09 | 2.04 | 2.02 | 2.14 | 2.15 | 2.18 | 2.55 |
| 0.04 | 2.20 | 2.17 | 2.09 | 2.05 | 2.10 | 2.03 | 2.11 | 2.05 | 2.02 | 2.15 | 2.18 | 2.18 | |
| 0.08 | 2.19 | 2.17 | 2.09 | 2.05 | 2.08 | 2.00 | 2.12 | 2.06 | 2.03 | 2.16 | 2.18 | 2.17 | |
| 0.12 | 2.18 | 2.17 | 2.06 | 2.05 | 2.08 | 1.99 | 2.10 | 2.06 | 2.05 | 2.16 | 2.18 | 2.16 | |
| 0.16 | 2.18 | 2.15 | 2.05 | 2.04 | 2.08 | 2.00 | 2.07 | 2.05 | 2.05 | 2.14 | 2.16 | 2.14 | |
| 0.20 | 2.15 | 2.12 | 2.04 | 2.03 | 2.05 | 2.00 | 2.07 | 2.04 | 2.01 | 2.12 | 2.12 | 2.12 | 2.46 |
| 0.24 | 2.11 | 2.10 | 2.03 | 2.01 | 2.01 | 1.98 | 2.08 | 2.02 | 1.97 | 2.11 | 2.09 | 2.09 | |
| 0.28 | 2.08 | 2.07 | 2.01 | 1.99 | 1.99 | 1.99 | 2.08 | 2.00 | 1.96 | 2.08 | 2.07 | 2.07 | |
| 0.32 | 2.05 | 2.05 | 1.99 | 1.97 | 1.97 | 1.96 | 2.01 | 1.98 | 1.94 | 2.04 | 2.04 | 2.06 | |
| 0.36 | 2.03 | 2.02 | 1.96 | 1.94 | 1.97 | 1.92 | 1.94 | 1.94 | 1.93 | 2.02 | 2.01 | 2.05 | |
| 0.40 | 1.99 | 1.99 | 1.93 | 1.91 | 1.93 | 1.89 | 1.91 | 1.91 | 1.92 | 1.99 | 1.98 | 2.00 | 2.18 |
| 0.44 | 1.94 | 1.94 | 1.89 | 1.89 | 1.87 | 1.89 | 1.90 | 1.89 | 1.89 | 1.96 | 1.95 | 1.93 | |
| 0.48 | 1.89 | 1.89 | 1.85 | 1.86 | 1.83 | 1.86 | 1.89 | 1.86 | 1.85 | 1.92 | 1.90 | 1.89 | |
| 0.52 | 1.84 | 1.86 | 1.81 | 1.82 | 1.79 | 1.84 | 1.84 | 1.82 | 1.83 | 1.87 | 1.86 | 1.85 | |
| 0.56 | 1.80 | 1.81 | 1.78 | 1.79 | 1.74 | 1.80 | 1.78 | 1.79 | 1.79 | 1.82 | 1.81 | 1.81 | |
| 0.60 | 1.73 | 1.76 | 1.74 | 1.74 | 1.69 | 1.77 | 1.73 | 1.75 | 1.74 | 1.77 | 1.75 | 1.75 | 1.77 |
| 0.64 | 1.67 | 1.70 | 1.68 | 1.70 | 1.64 | 1.72 | 1.68 | 1.70 | 1.70 | 1.72 | 1.68 | 1.69 | |
| 0.68 | 1.61 | 1.64 | 1.63 | 1.65 | 1.60 | 1.68 | 1.63 | 1.65 | 1.65 | 1.65 | 1.63 | 1.63 | |
| 0.72 | 1.56 | 1.57 | 1.58 | 1.60 | 1.54 | 1.62 | 1.57 | 1.60 | 1.61 | 1.58 | 1.56 | 1.58 | |
| 0.76 | 1.50 | 1.51 | 1.52 | 1.54 | 1.50 | 1.57 | 1.51 | 1.55 | 1.55 | 1.53 | 1.49 | 1.51 | |
| 0.80 | 1.44 | 1.44 | 1.47 | 1.48 | 1.44 | 1.50 | 1.46 | 1.48 | 1.49 | 1.45 | 1.43 | 1.44 | 1.34 |
| 0.84 | 1.36 | 1.36 | 1.42 | 1.42 | 1.39 | 1.45 | 1.39 | 1.42 | 1.43 | 1.37 | 1.36 | 1.36 | |
| 0.88 | 1.30 | 1.29 | 1.36 | 1.37 | 1.35 | 1.39 | 1.33 | 1.37 | 1.38 | 1.29 | 1.28 | 1.31 | |
| 0.92 | 1.24 | 1.23 | 1.30 | 1.31 | 1.30 | 1.32 | 1.27 | 1.30 | 1.33 | 1.23 | 1.23 | 1.24 | |
| 0.96 | 1.18 | 1.17 | 1.25 | 1.25 | 1.25 | 1.25 | 1.22 | 1.25 | 1.28 | 1.15 | 1.17 | 1.17 | |
| 1.00 | 1.10 | 1.10 | 1.18 | 1.19 | 1.20 | 1.18 | 1.16 | 1.19 | 1.22 | 1.09 | 1.10 | 1.10 | 0.99 |
| 1.10 | 0.96 | 0.95 | 1.04 | 1.04 | 1.06 | 1.04 | 1.00 | 1.03 | 1.06 | 0.93 | 0.96 | 0.95 | |
| 1.20 | 0.82 | 0.82 | 0.88 | 0.88 | 0.91 | 0.88 | 0.86 | 0.87 | 0.89 | 0.80 | 0.83 | 0.82 | 0.74 |
| 1.30 | 0.70 | 0.71 | 0.76 | 0.76 | 0.79 | 0.76 | 0.74 | 0.75 | 0.77 | 0.70 | 0.72 | 0.71 | |
| 1.40 | 0.59 | 0.61 | 0.64 | 0.65 | 0.66 | 0.65 | 0.64 | 0.64 | 0.64 | 0.61 | 0.61 | 0.61 | 0.56 |
| 1.50 | 0.52 | 0.54 | 0.55 | 0.55 | 0.56 | 0.55 | 0.54 | 0.55 | 0.55 | 0.54 | 0.53 | 0.54 | |
| 1.60 | 0.46 | 0.47 | 0.47 | 0.47 | 0.47 | 0.47 | 0.47 | 0.47 | 0.47 | 0.47 | 0.47 | 0.47 | 0.45 |
| 1.8 | | 0.37 | | | | | | | | | | | 0.36 |
| 2.0 | | 0.31 | | | | | | | | | | | 0.31 |
| 2.2 | | 0.27 | | | | | | | | | | | 0.27 |
| 2.4 | | 0.24 | | | | | | | | | | | 0.24 |
| 2.6 | | 0.23 | | | | | | | | | | | |
| 2.8 | | 0.22 | | | | | | | | | | | |
| 3.0 | | 0.20 | | | | | | | | | | | 0.16 |
| 3.5 | | 0.17 | | | | | | | | | | | 0.13 |
| 4.0 | | 0.13 | | | | | | | | | | | 0.09 |
| 4.5 | | 0.10 | | | | | | | | | | | |
| 5.0 | | 0.08 | | | | | | | | | | | 0.06 |
| 5.5 | | 0.05 | | | | | | | | | | | |
| 6.0 | | 0.04 | | | | | | | | | | | 0.03 |
| 6.5 | | 0.02 | | | | | | | | | | | |
| 7.0 | | 0.01 | | | | | | | | | | | 0.02 |

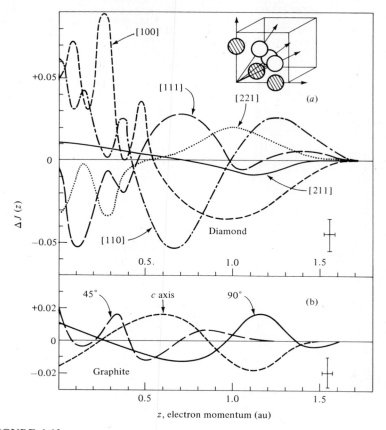

**FIGURE 1-13**

Experimental difference curves $\Delta J(z)$. (a) Between the $J(z)$ for five directions in diamond [110], [111], [100], [211], and [211], and [221] and the $J(z)$ curve averaged around the [110] axis; (b) between the $J(z)$ for the three directions in pyrolytic graphite (c axis, 45 and 90° to the c axis) and the $J(z)$ curve averaged over these directions. The inset shows an approximate distribution of negative (unshaded) and positive (shaded) momentum density which yields the gross features of the $\Delta J(z)$ curves for diamond. The error bars are in the lower-right-hand corners.

layers with little a-axis correlation. Thus the nearest-neighbor configurations are essentially identical, and this would indicate similar electron wavefunctions. The measured anisotropies in graphite were small, and the $\Delta J(z)$, averaged around an axis normal to the c axis, are shown in Fig. 1-13. No attempt was made to deduce a three-dimensional momentum density map for graphite. Accurate structure factor measurements are unavailable for graphite and carbon black.

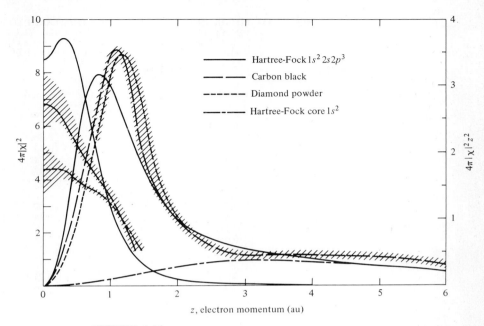

FIGURE 1-14
Spherically averaged momentum densities per atom $4\pi|\chi|^2$ for diamond powder, carbon black, and a Hartree-Fock $1s^22s2p^3$ (left-hand curves). The curves to the right are the radial momentum density $I(z) = 4\pi|\chi|^2z^2$, for which the calculated Hartree-Fock core $1s^2$ is also plotted. Because of the large uncertainty in determining $|\chi|^2$ near the origin, the data have been extrapolated to $z = 0$ (thin lines). The shading indicates the estimated experimental uncertainty.

**6. LiH and LiF**  Both Compton-profile and structure factor measurements have been reported for LiH and LiF. While chemists generally treat these as ionic crystals, i.e., a superposition of $Li^+$ ions and $F^-$ or $H^-$ ions, respectively, the Compton-profile measurements indicate that this treatment is qualitatively correct for LiF but not LiH. The measured Compton profile (Weiss, 1970) and structure factors (Merisalo and Inkinen, 1966) for LiF are given in Table 1-15 together with the theoretical Hartree-Fock values for a superposition of $Li^+$ and $F^-$ and neutral Li and F. It is clear from the large differences in Table 1-15 that the ionic models are much closer to actual observations. Berggren (1971) has improved the Compton-profile calculation by considering overlap of the ion wavefunctions in LiF and has brought theory and experiment in agreement to $\sim 1\%$.

For LiH, however, the difference in $J(z)$ between theory and experiment is quite large for both the ionic $Li^+ + H^-$ and neutral model $Li^0 + H^0$. In Table 1-16 are listed the measured Compton-profile values of LiH (Phillips and Weiss, 1969) after subtracting the $1s^2$ core contribution for Li.

Table 1-15  MEASURED AND CALCULATED COMPTON
PROFILES $J(z)$ PER MOLECULE AND STRUC-
TURE FACTORS FOR LiF. THE CALCULATED
VALUES ARE FOR BOTH $Li^+F^-$ AND $Li^0F^0$

$J(z)$

| $z$ | $Li^+F^-$ | $Li^0F^0$ | Experiment |
|---|---|---|---|
| 0 | 4.10 | 5.34 | 3.78 |
| 0.2 | 4.02 | 4 49 | 3.69 |
| 0.4 | 3.76 | 3.45 | 3.46 |
| 0.6 | 3.31 | 2.93 | 3.13 |
| 0.8 | 2.78 | 2.54 | 2.71 |
| 1.0 | 2.27 | 2.15 | 2.22 |
| 1.4 | 1.47 | 1.45 | 1.37 |
| 1.8 | 1.18 | 1.17 | 0.95 |
| 2.0 | 0.77 | 0.78 | 0.80 |
| 3.0 | 0 33 | 0.33 | 0.38 |
| 5.0 | 0.11 | 0.11 | 0.19 |
| 7.0 | 0.05 | 0.05 | 0.08 |

| $hkl$ | | $\mid F \mid$ | |
|---|---|---|---|
| 111 | 4.76 | 4.62 | 4.71 |
| 200 | 7.26 | 7.15 | 7.35 |
| 311 | 2.18 | 2.19 | 2.17 |

Table 1-16  MEASURED AND CALCULATED VALENCE-ELECTRON COMPTON PRO-
FILES $J(z)$ PER MOLECULE

The six calculated values of $J(z)$ are a superposition of ions $Li^+ + H^-$, a superposition of
neutral atoms $Li^0 + H^0$, a point-ion crystal-field calculation assuming spherical symmetry
for $H^-$ (Hurst, 1959), a point-ion crystal-field calculation with allowance for asphericity in
$H^-$ (Phillips and Weiss, 1969), and two band calculations (Brandt, 1970; Berggren and
Martino, 1971). The structure factors per molecule are also given for three theoretical models

$J(z)$

| | Hartree-Fock | | Crystal field | | | | |
|---|---|---|---|---|---|---|---|
| $z$ | $Li^+H^-$ | $Li^0H^0$ | Spherical[1] $Li^+H^-$ | Aspherical[2] $Li^+H^-$ | Band calculations[3,4] | | Experiment[2] |
| 0 | 3.75 | 2.81 | 2.32 | 1.46 | 2.01 | 1.5 | 1.47 |
| 0.2 | 1.75 | 1.96 | 1.83 | 1.34 | 1.88 | 1.4 | 1.39 |
| 0.4 | 0.68 | 0.84 | 1.02 | 1.04 | 1.42 | 1.2 | 1.11 |
| 0.6 | 0.35 | 0.38 | 0.50 | 0.72 | 0.67 | 0.8 | 0.77 |
| 0.8 | 0.18 | 0.20 | 0.24 | 0.46 | 0.07 | 0.4 | 0.45 |
| 1.0 | 0.11 | 0.12 | 0.12 | 0.28 | 0.03 | 0.2 | 0.24 |
| 1.4 | 0.03 | 0.04 | 0.03 | 0.10 | 0.02 | 0.05 | 0.07 |
| 1.8 | 0.01 | 0.01 | 0.01 | 0.04 | 0.01 | 0.02 | 0.03 |

| $hkl$ | | | | $\mid F \mid$ | | | |
|---|---|---|---|---|---|---|---|
| 111 | 0.48 | | 0.62 | 0.61 | | | 0.62 |
| 200 | 0.38 | | 0.47 | 0.57 | | | 0.52 |
| 220 | 0.18 | | 0.22 | 0.19 | | | 0.27 |
| 331 | 0.14 | | 0.14 | 0.16 | | | 0.14 |
| 222 | 0.11 | | 0.13 | 0.07 | | | 0.12 |

[1] Hurst (1959); [2] Phillips and Weiss (1969); [3] Brandt (1970) on left and [4] Berggren and Martino
(1971) on right.

A superposition of Hartree-Fock functions in either the neutral or the ionic case gives a poor fit to the experiment. The first attempt to improve the wavefunction for the solid was made by Hurst (1959). He chose an ionic model, $Li^+$ and $H^-$, and assumed the $1s^2$ core electrons on $Li^+$ to be sufficiently close to the nucleus so that the $Li^+$ ion could be replaced by a single positive point change. The $H^-$ electron wavefunctions were assumed spherically symmetric and a simple singlet combination of $1s$-like exponentials. The total energy of the two $H^-$ electrons was evaluated in the Coulomb potential of all the $Li^+$ and $H^-$ ions in the crystal, and this total energy was minimized by varying the parameters in the exponentials. The minimum-energy wavefunction was used to calculate the $J(z)$ and $|F|$ values in Table 1-16 and, while agreeing quite well with the measured structure factors (Calder et al., 1962), they were at variance with the experimental $J(z)$ values by $\sim 80\%$. An attempt was made by Phillips and Weiss (1969) to improve the $H^-$ wavefunction within the framework of the point-ion model. The wavefunction for the two electrons was chosen to be a singlet determinant with an admixture of $s$ and $d$ functions.

$$\Psi_1(r_1) = \frac{1}{[\pi(1 + \alpha^2)]^{1/2}} \left[ a^{3/2} \exp(-ar_1) + \frac{i\alpha b^{7/2}r_1{}^2}{\sqrt{6}} \right.$$

$$\left. \times \exp(-br_1) \sin^2 \theta_1 \cos 2\varphi_1 \right]$$

$$(1\text{-}118)$$

$$\Psi_2(r_2) = \frac{1}{[\pi(1 + \alpha^2)]^{1/2}} \left[ a^{3/2} \exp(-ar_2) + \frac{i\alpha b^{7/2}r_2{}^2}{3\sqrt{2}} \right.$$

$$\left. \times \exp(-br_2)(3 \cos^2 \theta_2 - 1) \right]$$

The charge density has maxima in the direction of the six $Li^+$ nearest neighbors, and the parameter $\alpha$ gives the anisotropic admixture. The parameters $a$, $b$, and $\alpha$ were chosen to give the best compromise fit to both $J(z)$ and $|F|$ ($a = 0.94$, $b = 1.5$, $\alpha = 0.6$). While the wavefunction (1-117) has improved the agreement with the measured Compton profile and structure factor, the total energy is not as favorable as that obtained from the Hurst wavefunction. More recently Brandt (1970) and Berggren and Martino (1971) have done band calculations for LiH and, as can be seen in Table 1-16, the improvement over the ionic model is small for the former but considerable for the latter.

**7. B, BN, B₄C, and BeO**  The materials B, BN, $B_4C$, and BeO are highly refractory, very hard, and have high melting points. This suggests a very strong coupling between atoms. Table 1-17 gives the measured values of $J(z)$ after the $1s^2$ core contribution is

# Table 1-17  VALENCE ELECTRON $J(z)$ PER ATOM FOR B, B$_4$C, BeO, AND BN

The calculated values are for a superposition of Hartree-Fock free-atom Compton profiles

| z | B(rhombohedral) Exp. | $2s^2\,2p$ | $sp^3$ | B$_4$C Exp. | B$_4$C Free-atom theory | BeO Exp. | BeO Free-atom theory | BN Exp. | BN Free-atom theory |
|---|---|---|---|---|---|---|---|---|---|
| 0 | 1.538 | 2.617 | 1.843 | 1.572 | 2.608 | 1.553 | 2.617 | 1.968 | 2.379 |
| 0.04 | 1.549 | | | 1.478 | | 1.553 | | 1.961 | |
| 0.08 | 1.556 | | | 1.443 | | 1.549 | | 1.949 | |
| 0.12 | 1.553 | | | 1.458 | | 1.543 | | 1.932 | |
| 0.16 | 1.533 | | | 1.481 | | 1.532 | | 1.918 | |
| 0.20 | 1.500 | 2.321 | 1.818 | 1.467 | 2.339 | 1.520 | 2.267 | 1.895 | 2.183 |
| 0.24 | 1.483 | | | 1.392 | | 1.506 | | 1.881 | |
| 0.28 | 1.453 | | | 1.346 | | 1.487 | | 1.847 | |
| 0.32 | 1.407 | | | 1.345 | | 1.467 | | 1.814 | |
| 0.36 | 1.365 | | | 1.348 | | 1.447 | | 1.769 | |
| 0.40 | 1.349 | 1.637 | 1.598 | 1.318 | 1.706 | 1.428 | 1.631 | 1.710 | 1.708 |
| 0.44 | 1.324 | | | 1.275 | | 1.402 | | 1.656 | |
| 0.48 | 1.288 | | | 1.238 | | 1.372 | | 1.612 | |
| 0.52 | 1.244 | | | 1.203 | | 1.342 | | 1.564 | |
| 0.56 | 1.202 | | | 1.162 | | 1.312 | | 1.523 | |
| 0.60 | 1.157 | 0.960 | 1.179 | 1.120 | 1.056 | 1.279 | 1.158 | 1.477 | 1.188 |
| 0.64 | 1.110 | | | 1.070 | | 1.243 | | 1.431 | |
| 0.68 | 1.058 | | | 1.023 | | 1.204 | | 1.385 | |
| 0.72 | 1.010 | | | 0.975 | | 1.162 | | 1.332 | |
| 0.76 | 0.957 | | | 0.923 | | 1.117 | | 1.284 | |
| 0.80 | 0.900 | 0.504 | 0.768 | 0.860 | 0.569 | 1.068 | 0.859 | 1.228 | 0.777 |
| 0.84 | 0.833 | | | 0.808 | | 1.019 | | 1.179 | |
| 0.88 | 0.775 | | | 0.753 | | 0.975 | | 1.128 | |
| 0.92 | 0.720 | | | 0.694 | | 0.936 | | 1.063 | |
| 0.96 | 0.657 | | | 0.642 | | 0.888 | | 1.004 | |
| 1.00 | 0.593 | 0.257 | 0.471 | 0.586 | 0.328 | 0.838 | 0.650 | 0.948 | 0.498 |
| 1.1 | 0.46 | | | 0.466 | | 0.732 | | 0.811 | |
| 1.2 | 0.352 | 0.136 | 0.283 | 0.359 | 0.185 | 0.634 | 0.490 | 0.681 | 0.321 |
| 1.3 | 0.275 | | | 0.288 | | 0.546 | | 0.549 | |
| 1.4 | 0.207 | 0.080 | 0.177 | 0.226 | 0.112 | 0.469 | 0.367 | 0.456 | 0.213 |
| 1.5 | 0.160 | | | 0.185 | | 0.395 | | 0.382 | |
| 1.6 | 0.128 | 0.053 | 0.105 | 0.155 | 0.074 | 0.340 | 0.282 | 0.311 | 0.145 |
| 1.7 | 0.109 | | | 0.133 | | 0.285 | | 0.254 | |
| 1.8 | 0.095 | 0.040 | 0.064 | 0.120 | 0.053 | 0.248 | 0.205 | 0.209 | 0.103 |
| 1.9 | 0.085 | | | 0.110 | | 0.217 | | 0.175 | |
| 2.0 | 0.074 | 0.031 | 0.040 | 0.106 | 0.041 | 0.195 | 0.155 | 0.145 | 0.076 |
| 2.1 | 0.065 | | | 0.100 | | 0.177 | | 0.128 | |
| 2.2 | 0.058 | | | 0.097 | | 0.163 | | 0.112 | |
| 2.3 | 0.052 | | | 0.093 | | 0.152 | | 0.096 | |
| 2.4 | 0.048 | | | 0.087 | | 0.140 | | 0.083 | |
| 2.5 | 0.042 | 0.0201 | 0.015 | 0.085 | 0.024 | 0.130 | 0.081 | 0.071 | 0.040 |
| 2.6 | 0.038 | | | 0.082 | | 0.122 | | 0.063 | |
| 2.7 | 0.031 | | | 0.080 | | 0.110 | | 0.054 | |
| 2.8 | 0.027 | | | 0.079 | | 0.102 | | 0.046 | |
| 2.9 | 0.023 | | | 0.078 | | 0.095 | | 0.041 | |
| 3.0 | 0.020 | 0.0132 | 0.005 | 0.077 | 0.016 | 0.087 | 0.045 | 0.037 | 0.025 |
| 3.5 | 0.016 | 0.0088 | 0.002 | 0.067 | 0.010 | 0.060 | 0.028 | 0.014 | 0.016 |
| 4.0 | 0.012 | 0.0057 | 0.001 | 0.056 | 0.007 | 0.052 | 0.018 | 0.000 | 0.011 |
| 4.5 | 0.007 | | | 0.035 | | 0.037 | | | |
| 5.0 | 0.002 | 0.0025 | ... | 0.015 | 0.003 | 0.028 | 0.008 | | 0.005 |
| 5.5 | 0.000 | | | 0.001 | | 0.020 | | | |
| 6.0 | | 0.0012 | ... | 0.000 | 0.002 | 0.017 | 0.004 | | 0.003 |

subtracted. In each case, the experimental values at $z = 0$ are significantly lower than for a superposition of free-atom ground-state Hartree-Fock Compton profiles. Better agreement is obtained by altering the free-atom electron configurations from $s$-like to $p$-like. For boron, the free-atom ground state is $2s^2 2p$ for the valence electrons, but the Compton profile for the $2p^3$ configuration given in Table 1-17 is in much better agreement. This, of course, is only a first approximation, and the results must await solid-state wavefunctions. Nonetheless it does yield a conceptual idea about the electron configuration in the solid state. Structure factor measurements are not available for comparison.

**8.** *Benzene, cyclohexane, and polyethylene* These organic substances are typical of many hydrocarbons for which wavefunctions are available. The atomic arrangements are well known, but structure factor data sufficiently accurate to determine electron distributions is not available. It would, in fact, be an extremely difficult problem to determine the scattering factors in the condensed state, since the interference effects between molecules in the liquids would be difficult to separate from the molecular structure factors. In addition, thermal effects are difficult to estimate in the liquid. Structure factor measurements could be made on gases, but they have not been reported. The Compton profile, however, provides a method for determining the accuracy of the wavefunctions, since there are not comparable interference effects (the process is incoherent). In addition, the weak Van der Waals forces only produce minor perturbations on the molecular wavefunctions. In a series of theoretical papers about 30 years ago (Coulson, 1941a, b; Coulson and Duncanson, 1941; Duncanson and Coulson, 1941), Coulson and Duncanson actually evaluated Compton profiles for a series of hydrocarbons, following a relatively simple model for constructing the molecular wavefunction that is still popular among chemists.

The benzene molecule consists of six carbon atoms in a flat hexagon with one hydrogen atom attached to each carbon atom along a line passing through the center of the hexagon and each carbon atom. The net effect is a "mariner's wheel" with the hydrogen atoms the spokes. The $1s^2$ core electrons in carbon are considered to be Hartree-Fock free-atom-like. The four valence electrons on each carbon atom are then arranged into independent bonds, one with each of the two carbon nearest neighbors, one with the hydrogen, and one considered to be relatively free. Each of the first three (called Heitler-London bonds) contains a pair of electrons with opposite spins so that, when coupled with the hydrogen electrons, each molecule has 12 electrons in Heitler-London carbon-carbon bonds, 12 electrons in Heitler-London carbon-hydrogen bonds, and 6 "free electrons" called $\pi$ bonds. Each Heitler-London bond (two electrons) is constructed from atomic-like functions $\psi_a$ and $\psi_b$:

$$\Psi(\mathbf{r}_1, \mathbf{r}_2) = \frac{\psi_a(\mathbf{r}_1)\psi_b(\mathbf{r}_2) + \psi_a(\mathbf{r}_2)\psi_b(\mathbf{r}_1)}{[2(1 + S_{ab}{}^2)]^{1/2}} \qquad (1\text{-}119)$$

where $S_{ab} = \int \psi_a(\mathbf{r})\psi_b(\mathbf{r}) \, d\mathbf{r}$ is the overlap integral. Because the carbon atom has $2s$- and $2p$-like states in the free atom, the carbon atom wavefunction in (1-118) is taken as a mixture of $s$ and $p$

$$\psi = \frac{\psi_s + \sigma\psi_p}{(1 + \sigma^2)^{1/2}} \qquad (1\text{-}120)$$

where $\sigma$ is the coefficient of $p$ mixing, generally taken to be $\sqrt{3}$. For hydrogen, a simple $1s$-like function is chosen to be used in (1-119). For the $\pi$ bonds, the wave-function for each electron is taken as a sum of six $p$-like functions centered on each carbon atom, thus giving equal probability of finding the electron on all six carbon atoms. These $\pi$ electrons can be envisaged as free or mobile.

Cyclohexane also consists of a ring of six carbon atoms (not planar) but with two hydrogen atoms attached to each carbon. This gives 24 electrons in C-H bonds, 12 electrons in C-C bonds, and no $\pi$ bonds. In terms of the Coulson-Duncanson approach, polyethylene would be similar to cyclohexane with the carbon atoms now forming a long sawtooth chain, each with two hydrogen atoms. Neglecting the end of the chain, this gives the same ratio of C-C to C-H bonds.

Figure 1-15 shows the momentum densities $|\chi|^2$ and $I(p_0)$ determined from the measured Compton profiles together with the calculated $I(p_0)$ from the Coulson-Duncanson theory. It is evident that their simple treatment of independent bonds is not adequate. The experimental valence-electron profiles are given in Table 1-18, and the difference between theory and experiment at $z = 0$ is $\sim 40\%$. More recently, Epstein (1971) has employed a linear combination of self-consistent field Slater-type atomic orbitals to construct his molecular wavefunctions. These wavefunctions are now shared by all atoms in the molecule and represent an improvement over the simple pair-type functions of Coulson and Duncanson. The results of these calcula-tions are in better agreement with experiment (Table 1-18).

**9.  *Anisotropy in vanadium and iron*** Vanadium and iron are both body-centered cubic metals, and it is possible to observe significant departures from spherical sym-metry in the charge density by measuring the "paired" reflections from (330)–(411) and (600)–(442) planes. These pairs occur at the same $\sin \theta/\lambda$ value so that the spherical core contribution to the scattering factor would be identical for each reflection of a pair. Furthermore, the Debye-Waller factors and the thermal diffuse scattering are expected to be essentially identical, so that the only source of asphericity is the valence-electron charge density. With a thin [110] crystal, both reflections in each pair can be obtained in transmission by a rotation about this axis. Furthermore, if the crystal is $\sim 1/\mu$ thick ($\mu$ is the linear absorption coefficient), variations in thick-ness are canceled to first order, since the diffracted intensity is proportional to the thickness and the absorption proportional to $\exp(-\mu t)$.

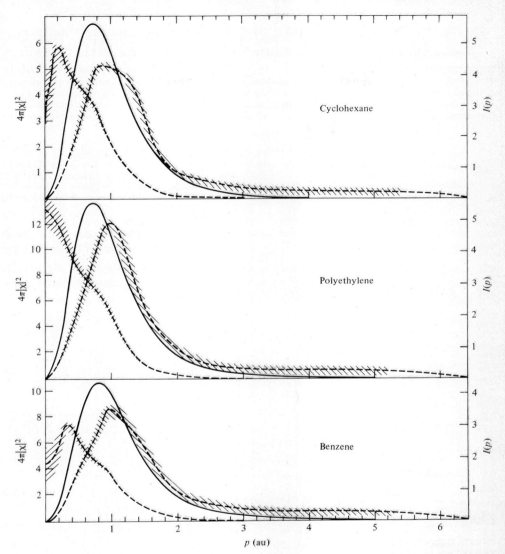

**FIGURE 1-15**

The valence-electron momentum densities $4\pi|\chi|^2$ and $4\pi|\chi|^2 p^2 = I(p)$ ($p =$ electron momentum in au) deduced from the experimental $J(z)$ curves for cyclohexane, polyethylene, and benzene. The area under the $I(p)$ curves is equal to the number of valence electrons for the smallest equivalent molecular unit, $CH_2$ (6 electrons) for cyclohexane and polyethylene, and $CH$ (5 electrons) for benzene. The theoretical $I(p)$ curves (solid lines) are based on the Coulson-Duncanson theory of independent, single-electron bonds. The higher momentum tails in $I(p)$ in the vicinity of $p \sim 3$ to 6 au may be due to valence electron-electron correlation. The $I(p)$ scale is on the right. (The shaded regions indicate the approximate uncertainty.)

**Table 1-18   THE EXPERIMENTAL AND TWO THEORETICAL VALENCE-ELECTRON COMPTON PROFILES $J(z)$ ($z$ IS ELECTRON MOMENTUM IN ATOMIC UNITS FOR CYCLOHEXANE, POLYETHYLENE, AND BENZENE**

The curves are normalized to six electrons for cyclohexane and polyethylene and five electrons for benzene. The data for polyethylene are for the x-ray scattering vector normal to the carbon chains, denoted $J(z)_\perp$. A separate column lists the small differences observed between the perpendicular and parallel orientations. The absolute experimental uncertainty in $J(z)$ is $\pm0.05$ au for all values of $z$: the relative uncertainty is $\pm0.02$ au

| | Cyclohexane | | | Polyethylene | | | Benzene | | |
|---|---|---|---|---|---|---|---|---|---|
| $z$ | Exp. | Coulson theory | Epstein theory | Exp. | Coulson theory | $J(z)_\perp$ $-J(z)_\parallel$ | Exp. | Coulson theory | Epstein theory |
| 0.00 | 2.92 | 4.08 | 3.30 | 3.00 | 4.08 | 0.04 | 2.31 | 3.07 | 2.64 |
| 0.04 | 2.92 | 4.07 | | 2.99 | 4.07 | 0.03 | 2.29 | 3.06 | |
| 0.08 | 2.91 | 4.03 | | 2.97 | 4.03 | 0.03 | 2.28 | 3.04 | |
| 0.12 | 2.89 | 3.97 | | 2.95 | 3.97 | 0.03 | 2.26 | 3.00 | |
| 0.16 | 2.86 | 3.89 | | 2.91 | 3.89 | 0.03 | 2.26 | 2.96 | |
| 0.20 | 2.80 | 3.80 | 3.18 | 2.86 | 3.80 | 0.01 | 2.23 | 2.91 | 2.56 |
| 0.24 | 2.75 | 3.69 | | 2.81 | 3.69 | 0.00 | 2.21 | 2.84 | |
| 0.28 | 2.69 | 3.57 | | 2.74 | 3.57 | 0.00 | 2.17 | 2.77 | |
| 0.32 | 2.63 | 3.42 | | 2.67 | 3.42 | 0.01 | 2.14 | 2.68 | |
| 0.36 | 2.58 | 3.26 | | 2.62 | 3.26 | 0.03 | 2.09 | 2.58 | |
| 0.40 | 2.51 | 3.10 | 2.82 | 2.54 | 3.10 | 0.02 | 2.02 | 2.48 | 2.29 |
| 0.44 | 2.44 | 2.92 | | 2.47 | 2.92 | 0.02 | 1.97 | 2.37 | |
| 0.48 | 2.34 | 2.74 | | 2.38 | 2.74 | 0.03 | 1.91 | 2.25 | |
| 0.52 | 2.26 | 2.56 | | 2.29 | 2.56 | 0.02 | 1.84 | 2.13 | |
| 0.56 | 2.16 | 2.38 | | 2.22 | 2.38 | 0.04 | 1.80 | 2.00 | |
| 0.60 | 2.07 | 2.21 | 2.31 | 2.15 | 2.21 | 0.06 | 1.71 | 1.88 | 1.92 |
| 0.64 | 1.98 | 2.04 | | 2.04 | 2.04 | 0.03 | 1.65 | 1.76 | |
| 0.68 | 1.89 | 1.87 | | 1.96 | 1.87 | 0.03 | 1.58 | 1.64 | |
| 0.72 | 1.80 | 1.72 | | 1.85 | 1.72 | 0.00 | 1.51 | 1.52 | |
| 0.76 | 1.69 | 1.57 | | 1.75 | 1.57 | 0.00 | 1.45 | 1.41 | |
| 0.80 | 1.58 | 1.43 | 1.75 | 1.65 | 1.43 | −0.01 | 1.38 | 1.30 | 1.48 |
| 0.84 | 1.49 | 1.31 | | 1.54 | 1.31 | −0.05 | 1.31 | 1.20 | |
| 0.88 | 1.39 | 1.19 | | 1.44 | 1.19 | −0.06 | 1.23 | 1.10 | |
| 0.92 | 1.30 | 1.08 | | 1.36 | 1.08 | −0.04 | 1.14 | 1.01 | |
| 0.96 | 1.21 | 0.98 | | 1.24 | 0.98 | −0.07 | 1.06 | 0.93 | |
| 1.00 | 1.13 | 0.89 | 1.21 | 1.14 | 0.89 | −0.08 | 0.99 | 0.85 | 1.05 |
| 1.04 | 1.04 | 0.80 | | 1.06 | 0.80 | −0.08 | 0.93 | 0.77 | |
| 1.08 | 0.97 | 0.73 | | 0.98 | 0.73 | −0.06 | 0.86 | 0.70 | |
| 1.12 | 0.90 | 0.66 | | 0.88 | 0.66 | −0.06 | 0.81 | 0.64 | |
| 1.16 | 0.83 | 0.59 | | 0.81 | 0.59 | −0.05 | 0.75 | 0.58 | |
| 1.20 | 0.77 | 0.54 | 0.78 | 0.74 | 0.54 | −0.06 | 0.71 | 0.53 | 0.69 |
| 1.24 | 0.70 | 0.48 | | 0.68 | 0.48 | −0.05 | 0.65 | 0.48 | |
| 1.28 | 0.65 | 0.44 | | 0.61 | 0.44 | −0.05 | 0.60 | 0.44 | |
| 1.32 | 0.60 | 0.40 | | 0.56 | 0.40 | −0.05 | 0.56 | 0.40 | |
| 1.36 | 0.54 | 0.36 | | 0.51 | 0.36 | −0.05 | 0.52 | 0.36 | |
| 1.40 | 0.50 | 0.32 | 0.46 | 0.47 | 0.32 | −0.04 | 0.48 | 0.33 | 0.42 |
| 1.44 | 0.46 | 0.29 | | 0.44 | 0.29 | −0.03 | 0.45 | 0.30 | |
| 1.48 | 0.42 | 0.27 | | 0.40 | 0.27 | −0.03 | 0.42 | 0.27 | |
| 1.52 | 0.39 | 0.24 | | 0.38 | 0.24 | −0.01 | 0.40 | 0.25 | |
| 1.56 | 0.36 | 0.22 | | 0.36 | 0.22 | −0.01 | 0.36 | 0.23 | |

TABLE 1-18 (*Continued*)

| | Cyclohexane | | | Polyethylene | | | Benzene | | |
|---|---|---|---|---|---|---|---|---|---|
| $z$ | Exp. | Coulson theory | Epstein theory | Exp. | Coulson theory | $J(z)_\perp - J(z)_\parallel$ | Exp. | Coulson theory | Epstein theory |
| 1.60 | 0.34 | 0.20 | 0.27 | 0.32 | 0.20 | $-0.01$ | 0.34 | 0.21 | 0.24 |
| 1.64 | 0.32 | 0.18 | | 0.30 | 0.18 | $-0.01$ | 0.31 | 0.19 | |
| 1.68 | 0.29 | 0.16 | | 0.28 | 0.16 | 0.00 | 0.29 | 0.17 | |
| 1.72 | 0.27 | 0.15 | | 0.26 | 0.15 | 0.00 | 0.28 | 0.16 | |
| 1.76 | 0.25 | 0.13 | | 0.24 | 0.13 | 0.00 | 0.26 | 0.14 | |
| 1.80 | 0.24 | 0.12 | 0.16 | 0.22 | 0.12 | | 0.24 | 0.13 | 0.14 |
| 1.84 | 0.23 | 0.11 | | 0.21 | 0.11 | | 0.23 | 0.12 | |
| 1.88 | 0.22 | 0.10 | | 0.20 | 0.10 | | 0.22 | 0.11 | |
| 1.92 | 0.21 | 0.09 | | 0.18 | 0.09 | | 0.21 | 0.10 | |
| 1.96 | 0.20 | 0.09 | | 0.17 | 0.09 | | 0.20 | 0.09 | |
| 2.00 | 0.20 | 0.08 | 0.11 | 0.17 | 0.08 | | 0.19 | 0.08 | 0.10 |
| 2.1 | 0.18 | 0.06 | | 0.15 | 0.06 | | 0.17 | 0.07 | |
| 2.2 | 0.16 | 0.05 | | 0.13 | 0.05 | | 0.16 | 0.05 | |
| 2.3 | 0.15 | 0.04 | | 0.12 | 0.04 | | 0.14 | 0.04 | |
| 2.4 | 0.13 | 0.03 | | 0.11 | 0.03 | | 0.13 | 0.04 | |
| 2.5 | 0.13 | 0.03 | | 0.10 | 0.03 | | 0.12 | 0.03 | |
| 2.6 | 0.12 | 0.02 | | 0.09 | 0.02 | | 0.11 | 0.02 | |
| 2.7 | 0.12 | 0.02 | | 0.08 | 0.02 | | 0.10 | 0.02 | |
| 2.8 | 0.11 | 0.01 | | 0.07 | 0.01 | | 0.09 | 0.01 | |
| 2.9 | 0.11 | 0.01 | | 0.07 | 0.01 | | 0.09 | 0.01 | |
| 3.0 | 0.10 | 0.01 | 0.02 | 0.06 | 0.01 | | 0.08 | 0.01 | 0.02 |
| 3.5 | 0.08 | 0.00 | | 0.05 | 0.00 | | 0.05 | 0.00 | |
| 4.0 | 0.07 | | | 0.05 | | | 0.05 | | |
| 4.5 | 0.06 | | | 0.04 | | | 0.04 | | |
| 5.0 | 0.04 | | | 0.03 | | | 0.03 | | |
| 5.5 | 0.03 | | | 0.02 | | | 0.02 | | |
| 6.0 | 0.01 | | | 0.01 | | | 0.01 | | |

Measurements (DeMarco and Weiss, 1965; Weiss and DeMarco, 1965) were made on thin crystals in transmission following the method described in Section I C 4 to correct for secondary extinction. Since the scattering factors were small, primary extinction was considered negligible. Equations (1-45) to (1-47) were used to analyze the differences, and it was determined that of the four $3d$-like electrons in vanadium, about $81 \pm 6\%$ were $t_{2g}$ (remainder $e_g$) compared to 60% $t_{2g}$ 40% $e_g$ for spherical symmetry. In iron, the results were 70% $t_{2g}$ 30% $e_g$. In both cases, the charge density has maxima in the direction of the nearest neighbors along [[111]]. X-ray Compton-profile calculations, as seen in Fig. 1-3, would predict a large anisotropy for vanadium if the $t_{2g} - e_g$ description is accurate in momentum space. Measurements by Phillips and Weiss (1972) show, however, that the Compton profiles of vanadium and iron have little anisotropy ($<2\%$), indicating the inadequacy of the $t_{2g} - e_g$ description.

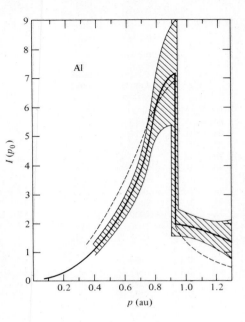

**FIGURE 1-16**
Momentum density for the three valence
electrons in aluminum deduced from the
Compton profile (solid curve). The theoretical momentum density for the interacting electron gas is shown in the
dashed line. The shading indicates the
experimental uncertainty.

**10.** *Aluminum metal* A good example of the relative sensitivities and difficulties of
scattering factor and Compton-profile measurements is provided by the work on
aluminum. There is essentially no contribution from the valence electrons to the
scattering factor so that such measurements in the metal can only determine the core
charge density. There have been five different measurements of recent vintage on both
powders and single crystals (Batterman et al., 1961; Bensch et al., 1955; DeMarco,
1967; Raccah and Henrich, 1969; Inkinen et al., 1969) and variations of $1\frac{1}{2}\%$ exist
in the reported values of $f_0$ for the 111 reflection. All results, however, for the 111
and 200 reflections are lower by more than $1\%$ than the calculated free-atom core
scattering factors. A theoretical attempt was made by Arlinghaus (1967) to seek
changes in the aluminum core due to binding effects in the solid, but no such change
was found.

The Compton-profile results for the three valence electrons in aluminum are
shown in Fig. 1-16. The Fermi momentum is clearly seen, as well as the high-momentum tail attributed to correlation effects. The dashed line is the theoretical momentum
density calculated for the interacting electron gas (Daniel and Voski, 1960). The free-
electron behavior of the valence electrons confirms the assumption in scattering factor
measurements that the valence electrons do not contribute appreciably to the scattering factor at the Bragg peaks.

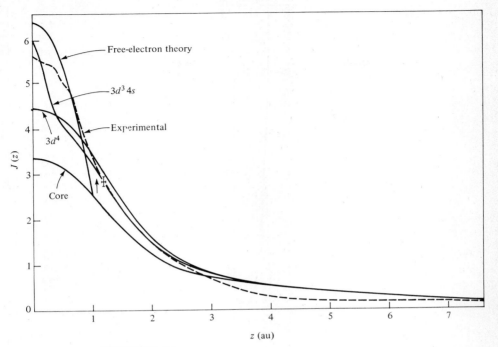

FIGURE 1-17
The measured Compton profile of polycrystalline titanium. The curves marked $3d^4$ and $3d^34s$ are calculated from Hartree-Fock wavefunctions including the $2s^22p^63s^23p^6$ core. The curve marked free-electron theory is calculated for four electrons per atom plus a Hartree-Fock core $2s^22p^63s^23p^6$. The lowest curve is calculated for the core alone. All calculated curves are evaluated in the impulse approximation.

**11.** *Titanium (polycrystalline)* The Compton-profile results for polycrystalline titanium (four valence electrons per atom) are shown in Fig. 1-17 together with three theoretical curves constructed by using:

> *1* Free-electron theory for the four valence electrons plus the Hartree-Fock core $2s^22p^63s^23p^6$ ($1s^2$ not excited)
> *2* Free-atom valence configuration $3d^34s$ plus free-atom core $2s^22p^63s^23p^6$
> *3* Free-atom valence configuration $3d^4$ plus free-atom core $2s^22p^63s^23p^6$

None of these crude theories agree very well with the data. At present, no band theory in momentum space is available for comparison. Inversion of the Compton profiles yields the momentum density $4\pi|\chi|^2$ shown in Fig. 1-18. The structure in the experimental curves is presumably a solid-state effect. The experimental discontinuity

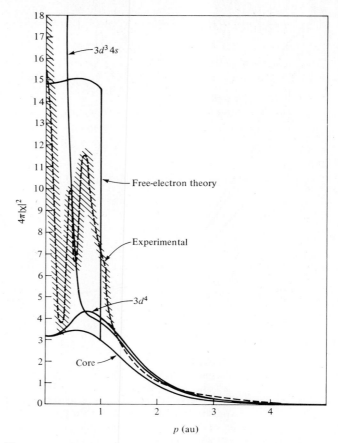

**FIGURE 1-18**
The normalized momentum density $4\pi|\chi|^2$ per unit volume in momentum $p$ is in atomic units. The curves calculated from free-atom Hartree-Fock wavefunctions are shown for the core alone $2s^2 2p^6 3s^2 p^6$ and for the core plus the valence-electron configurations $3d^3 4s$ and $3d^4$ and for the (valence-electron) free-electron theory. The value of $4\pi|\chi|^2$ at $p = 0$ for the $3d^3 4s$ configuration is 90 au.

in momentum density at 1.08 au (Fig. 1-18) may be associated with the Fermi momentum. X-ray scattering factor data are unavailable for comparison.

## D.   Conclusion

From the limited measurements of Compton profiles and scattering factors, it appears that the Compton profile is a much more powerful tool for testing the accuracy of valence-electron wavefunctions in solids and liquids. For gases, there are insufficient measurements to compare the sensitivity of the two techniques. The Compton measurements have the added advantage that problems such as termination errors, thermal effects, extinction corrections, and absolute scale are not encountered. The only major problem encountered in x-ray Compton-profile measurements is that of inadequate intensity but, with present-day equipment, this limitation is no longer severe for elements up to atomic number $\sim 20$. With the development of nondispersive gamma-ray methods, this limitation has been removed for heavier elements as well. It is expected that theoreticians will see to it that their position wavefunctions are transformed to momentum space so that both theory and experiment can be compared for both scattering factors and Compton profiles.

# 2

# KINEMATICAL THEORY

While there have been no real alterations in the fundamentals of kinematical x-ray scattering theory in recent years, this chapter is included in the present text for more than the sake of completeness. Indeed, completeness has not been an issue, and many specific applications of the theory have not been included. The viewpoint which has been taken in the present text is different from that ordinarily used, and consideration has been given to a number of subtleties which have received only slight treatment previously. Two aspects of the application of kinematical theory have been emphasized. First, pair-distribution functions are obtained by Fourier transformation of the continuous distribution of scattered intensity. Second, the study of localized defects in single crystals is developed through analysis of the intensity distributions at or near the reciprocal lattice points by direct numerical calculation of the scattering amplitudes. Throughout the chapter, the approach is based on the recognition that computers and new analytic procedures allow a more straightforward and less approximate application of the theory to specific problems than has been possible previously.

There are, of course, a number of textbooks in which closely related subject

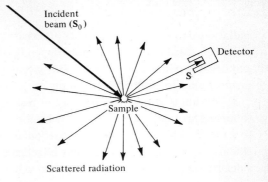

FIGURE 2-1
The scattering of a collimated beam.

material has been presented.  To name only a few, in addition to the aforementioned book by James, the texts by Guinier and Fournet, Krivoglaz, and Warren listed in the bibliography may provide coverage of particular problems not treated here, or a usefully different view of an included topic.

# I.  FUNDAMENTAL THEORY

## A.  Introduction

In this chapter, simple geometrical aspects of the scattering of x rays are considered, which can be usefully applied to the measurement of relative atomic positions in matter.  In most respects these considerations are independent of the details of the actual scattering mechanisms and will often apply to electron and neutron radiation as well as x rays.

The basic diffraction experiment is indicated schematically in Fig. 2-1.  A collimated beam of radiation (wave vector $S_0$) is incident on a specimen which scatters the beam in various directions, generally not isotropically.  A portion of the beam will be scattered without energy loss or gain (usually a small portion), and the distribution of that scattered intensity is related to the distribution of scattering material in the specimen.  Qualitatively, one may say that the more orderly the arrangement of scatterers, the more the intensity fluctuations will be regular, well defined, and intense. It is through a quantitative understanding of that relationship that radiation-scattering experiments can elucidate the internal structure of materials.

The theory which is developed here is often called the *kinematical theory*.  In formulating it, a physical model is used in which: a photon is represented as a wave that traverses the entire sample; scattering occurs at points within the sample and the radiation travels in straight lines otherwise; the amplitude phase difference between radiation scattered at two points depends on the difference in the length of the two geometrical paths; and the probability that a photon is detected at a (distant) point is proportional to the complex conjugate product of the wave amplitude.

Within that model, a number of restrictions are invoked, which are listed in the following section. Certain of the restrictions may be violated in practical experiments to a degree which would modify particular aspects of the predictions to a measurable extent. Therefore, restrictions may be relaxed for certain circumstances in order to consider particular effects, which are then usually treated as corrections or perturbations.

## B.   Restrictions in the theory

**1.   *Small scattering amplitude***   The total radiation diffracted into any one direction is assumed to be a very small portion of the incident beam. The interaction between the scattered and incident radiation is ignored. It is also assumed that beams are diminished slowly in passing through the sample by an isotropic absorption-plus-scattering factor, which causes a uniform exponential decay, $\exp(-\mu l)$. *Comment:* This is a central approximation, which is violated, however, whenever the sample is reasonably perfectly crystalline. In particular, if the diffracted beam should be comparable in intensity to the incident radiation, interactions between these waves dominate the process and the dynamical theory of Chap. 3 becomes more appropriate. Within the framework of the present chapter, violation of the above restriction is of quantitative importance to the *integrated intensity* of the diffraction maxima that occur with crystalline specimens. Of particular concern in that regard is the comparison of the relative intensities of different maxima. In the kinematical theory, the effect is usually called the *extinction effect* and is considered to be a perturbation which sometimes needs to be accounted for in the analysis of experimental data. A discussion of extinction, from that point of view, is given in Chap. 7.

The assumed weakness of the scattered radiation carries the further implication that *multiple* scattering is negligible in general. To first order, the model does make that presumption, but corrections are considered. The latter may be important in particular circumstances:

> *1*   In samples which yield intensity maxima in a discrete set of scattering directions, two successive diffractions can yield "anomalous" intensity in particular sum-or-difference directions. The effect is particularly noticeable with single crystals. With polycrystalline samples the effect may be seen most strongly at small angles, due to the summation of all those instances in which the second diffraction is the negative of the first.

*2*  When the primary scattering is not exceedingly localized to specific directions, the multiply scattered radiation is relatively structureless and of low intensity, but not necessarily negligible if absorption in the specimen is small.

**2.  *Plane-wave approximation***  The source of radiation and the detector (or film) are far from the sample, compared to the dimensions of the regions of structural regularity within the irradiated portion of the sample.  This allows the assumption that the incident beam has a plane (rather than spherical) wave front and that rays scattered to a point on the detector travel parallel paths.

*Comments:* Again, the restriction may be violated by crystal specimens in which precise spatial regularity (and hence the phase relationship) is maintained over large regions of the sample.  If one nonetheless uses the kinematical theory, the exact path differences should be used for the phase relationships in considering the precise *shapes* of the diffraction maxima, which will occur.

Regardless of the arrangement of scattering matter in the sample, *all* points scatter in phase in the forward direction in the plane-wave approximation.  Accordingly, that simplifying assumption should be questioned whenever small-angle scattering is considered.  A result of the non-plane-wave attributes of the radiation, for example, is the persistence of the tail of the "forward" scattering to larger angles than the simple model predicts.  This particular correction has been derived by Warren (1969, chap. 10).

**3.  *Coherent scattering***  Changes in the wavelength of the radiation on scattering are not explicitly considered.  *Comment:* As shown in Chap. 1, some of the scattered radiation is of altered energy (and, hence, wavelength) because some of the incident photons excite electrons in the sample to higher energy states.  Such "Compton"-type scattering is entirely ignored in this section, it being assumed that it will have been removed either by experimental means (e.g., postscattering monochromatization) or subtracted according to theoretical estimates, before further analysis.

The very small energy losses (or gains) that occur when a scattered photon exchanges momentum and energy with nuclei are also ignored.  Since the energy involved in such exchanges is small for x radiation ($\Delta E/E \approx 10^{-5}$), even in comparison to the natural energy spread of an x-ray "characteristic line" source, it is not feasible to separate such scatterings.  By the same token, the smallness of the energy transfer ($\sim kT$) compared to the binding energy of an atom precludes the possibility that the energy is transferred to one nucleus only.  Rather, the excitation must in principle involve all the atoms in a sample such that the coherence of the scattering (which is related to the *positions* of the atoms) is maintained for each photon, even though successive photons may differ in energy by slight amounts.

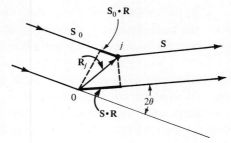

**FIGURE 2-2**
Illustration of the path-length difference
in the plane-wave approximation.

## C.   Directional variation of the amplitude and intensity of the scattered wave

**1.   *The phase relationships among scattered rays***   Consider a well-defined beam of
radiation of wavelength $\lambda$, frequency $v$, and traveling in a direction indicated by unit
vector $\mathbf{S_0}$. In standard notation, one may describe the variation of amplitude of such
radiation in space and time as a wavefunction:

$$\psi(\mathbf{R}, t) = A \exp\left[-i2\pi\left(\frac{\mathbf{S_0} \cdot \mathbf{R}}{\lambda} - vt + \alpha\right)\right] \qquad (2\text{-}1)$$

where $i = \sqrt{-1}$, $\alpha$ is an arbitrary phase, and $\mathbf{R}$ is the vector from an (arbitrary)
origin to a space point. Consider further, a single point scatterer, placed at the
arbitrary origin in the beam, of which it scatters a portion into all directions $\mathbf{S}$ (where
$\mathbf{S}$ is also a unit vector), with wavefunction given by

$$\psi_s(\mathbf{R}, t) = a'(\mathbf{S} - \mathbf{S_0})A \exp\left[-i2\pi\left(\frac{\mathbf{S} \cdot \mathbf{R}}{\lambda} - vt + \alpha + \beta\right)\right] \qquad (2\text{-}2)$$

The physics of the scattering process has been bypassed (but see Chap. 1) and summarized in the two parameters: $a'(\mathbf{S} - \mathbf{S_0})$, the scattering probability amplitude,
possibly (as indicated) a function of the change in direction, and $\beta$, a possible change
in phase on scattering. Indeed, the two parameters can be combined into a single
complex function

$$a(\mathbf{S} - \mathbf{S_0}) = a'(\mathbf{S} - \mathbf{S_0})e^{i2\pi\beta} \qquad (2\text{-}3)$$

and

$$\psi_s(\mathbf{R}, t) = Aa(\mathbf{S} - \mathbf{S_0}) \exp\left[2\pi i(vt + \alpha)\right] \exp\left(-2\pi i\frac{\mathbf{S} \cdot \mathbf{R}}{\lambda}\right) \qquad (2\text{-}4)$$

In Fig. 2-2, part of a plane wave ($\mathbf{S_0}$) is depicted as being scattered into direction $\mathbf{S}$, by
each of *two* point scatterers, one at the arbitrary origin and the other (point $j$) at $\mathbf{R}_j$.

   Initially, consider identical scatterers, such that the (complex) scattering amplitudes are the same. The phase difference between the two scattered rays is then due

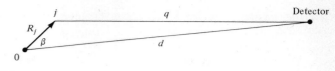

**FIGURE 2-3**
The geometry of the path lengths to a point on the detector.

only to the relative positions of the two scatterers. The path difference (from source to detector) is $\mathbf{S} \cdot \mathbf{R}_j - \mathbf{S}_0 \cdot \mathbf{R}_j$, in the limit that the distances to both source and detector are very large compared to $R_j$. This may be seen with reference to Fig. 2-3, which shows the beam paths from the two scatterers to the detector. These distances are labeled $d$ and $q$, respectively, and the path difference $(d - q)$ may be derived as follows, for the limit $d/R_j \ggg 1$:

$$q^2 = d^2 + R_j^2 - 2dR_j \cos \beta \qquad (2\text{-}5a)$$

$$q = d\left[1 + \left(\frac{R_j^2}{d^2} - \frac{2R_j \cos \beta}{d}\right)\right]^{1/2} \qquad (2\text{-}5b)$$

Expanding the square root yields the series

$$q = d\left[1 + \frac{1}{2}\left(\frac{R_j^2}{d^2} - \frac{2R_j \cos \beta}{d}\right) - \frac{1}{8}\left(\frac{R_j^2}{d^2} - \frac{2R_j \cos \beta}{d}\right)^2 + \cdots\right] \qquad (2\text{-}6)$$

Therefore, $q - d \approx -R_j \cos \beta$, and, since $\mathbf{S}$ is a unit vector, $d - q \approx \mathbf{S} \cdot \mathbf{R}_j$. The same geometry holds on the source side, with only a difference in sign, so that the total path difference would be $(\mathbf{S} - \mathbf{S}_0) \cdot \mathbf{R}_j$.

Reasonable values for $d$ and $\lambda$ are $10^9$ A (10 cm) and 1 A, respectively, and it may be seen that the approximation is inaccurate if $R_j$ is much greater than $10^3$ A. Therefore, whenever the situation is such that coherent contributions to the amplitude appear to arise from more widely separated portions of the sample, the more complete form should be considered.

Thus the total amplitude, at the detector, of rays scattered into direction $\mathbf{S}$ by $N$ identical scatterers at $N$ different positions will be

$$\psi^{\text{total}} = Aa(\mathbf{S} - \mathbf{S}_0)e^{2\pi i(vt + \alpha)} \sum_{j=1}^{N} e^{2\pi i \frac{(\mathbf{S} - \mathbf{S}_0) \cdot \mathbf{R}_j}{\lambda}} \qquad (2\text{-}7)$$

where $\alpha$ is now a different but still arbitrary phase because of the additional path from the origin to the detector.

**2.  *The diffraction vector* K**   For convenience, a *diffraction vector* is defined $\mathbf{K} \equiv 2\pi(\mathbf{S} - \mathbf{S}_0)/\lambda$.

$$\psi^{\text{total}} = Aa(\mathbf{K})e^{2\pi i(vt+\alpha)} \sum_{j=1}^{N} e^{i\mathbf{K}\cdot\mathbf{R}_j} \qquad (2\text{-}8)$$

**3.  *The Fourier relationship between the set of interscatterer vectors and the intensity of the scattered waves***   In an actual measurement, the detector does not respond to wave amplitude directly; rather it must extract energy from the scattered beam, which it can do only in quantized photons, of energy $hv$. The probable occurrence of a detectable photon is not proportional to $\psi$ but to the complex conjugate product $\psi\psi^*$.[1]

Defining a relative intensity $Q \equiv \psi\psi^*/AA^*$,

$$Q = aa^* \left[ \sum_{j=1}^{N} e^{i\mathbf{K}\cdot\mathbf{R}_j} \right]\left[ \sum_{j=1}^{N} e^{-i\mathbf{K}\cdot\mathbf{R}_j} \right] \qquad (2\text{-}9)$$

Expanding the squared summation,

$$Q = aa^* \sum_{n,m=1}^{N} e^{i\mathbf{K}\cdot(\mathbf{R}_m - \mathbf{R}_n)} \qquad (2\text{-}10)$$

It should be noted that in the given definition of $Q$ as a relative intensity, a number of factors have been ignored which are independent of the arrangement of scatterers, but which do affect the absolute probability of detecting a photon, and even the $\mathbf{K}$ dependence. These relate in part to experimental details such as slit widths, detector efficiency, and incident beam size, and to more intrinsic matters such as the electron charge, the photoelectric absorption probability, and the polarization characteristics of the incident beam. Generally speaking, these factors can be expressed as the product of three terms, $C$, $A(\mathbf{K})$, and $PL(\mathbf{K})$. $C$ is a normalization constant, most easily determined by fitting theoretical intensities to experimental counting rates. $C$ can also be computed, in principle, on the basis of direct measurements of the incident beam power, of all sizes and distances relevant to the experiment, of the sample's absorption coefficient, and of the detector efficiency. $A(\mathbf{K})$ is an absorption-dependent function, which varies with the shape of the sample and the geometry of the diffraction paths. $PL(\mathbf{K})$ accounts for polarization effects.

Continuing from equation (2-10) and extracting the ($m = n$) terms, one obtains

$$Q = aa^* \left( N + \sum_{n,m} e^{i\mathbf{K}\cdot(\mathbf{R}_m - \mathbf{R}_n)} \right) \qquad (2\text{-}11)$$

Notice that, if the double summation were equal to zero, the experimental

---

[1] It will be recognized that this discussion applies to particle "radiation" as well, although the initial wavefunctions have the same form only if the particles are traveling free.

intensity would be just the sum of individual intensities due to $N$ totally independent scatterers; this suggests the possible utility of the reduced function:

$$P(\mathbf{K}) \equiv \left(\frac{Q}{Naa^*} - 1\right) \qquad (2\text{-}12)$$

$$P(\mathbf{K}) = \frac{1}{N} \sum_m \sum_n{}' e^{i\mathbf{K} \cdot \mathbf{R}_{mn}} \qquad (2\text{-}13)$$

where $\mathbf{R}_{mn} = \mathbf{R}_m - \mathbf{R}_n$. For each $\mathbf{R}_{mn}$ in the double sum, there is a corresponding $\mathbf{R}_{nm} = -\mathbf{R}_{mn}$. The summations may therefore be replaced by a single sum over all pairs (each counted once).

$$P(\mathbf{K}) = \frac{1}{N} \sum_{\mathbf{R}_{mn}} (e^{i\mathbf{K} \cdot \mathbf{R}_{mn}} + e^{-i\mathbf{K} \cdot \mathbf{R}_{mn}}) \qquad (2\text{-}14a)$$

$$P(\mathbf{K}) = \frac{2}{N} \sum_{\mathbf{R}_{mn}} \cos (\mathbf{K} \cdot \mathbf{R}_{mn}) \qquad (2\text{-}14b)$$

The last sum contains $(N^2 - N/2)$ terms. $P(\mathbf{K})$ and the measured intensity (photons per sec) are both real and even functions of $\mathbf{K}$, regardless of the arrangements of scatterers.

## D. The three-dimensional pair distribution; retrieving the structure from measured intensities of scattering

The interscatterer vector set, $\mathbf{R}_{nm}$, can be retrieved, in principle, through a three-dimensional Fourier inversion. Multiplying both sides by $\cos (\mathbf{K} \cdot \mathbf{r})$ and integrating over $\mathbf{K}$, one finds:

$$\int P(\mathbf{K}) \cos (\mathbf{K} \cdot \mathbf{r}) \, d\mathbf{K} = \frac{2}{N} \sum_{\text{pairs}} \int \cos \mathbf{K} \cdot \mathbf{R}_{nm} \cos \mathbf{K} \cdot \mathbf{r} \, d\mathbf{K} \qquad (2\text{-}15)$$

The equation now represents a function in $\mathbf{R}$(real) space, rather than $\mathbf{K}$(reciprocal) space. Because of the orthogonality of the cosine functions the latter integrals reduce to $\delta$ functions at $\mathbf{r} = \mathbf{R}_{nm}$, if the integrals are extended to infinity:

$$\int_{-\infty}^{+\infty} P(\mathbf{K}) \cos (\mathbf{K} \cdot \mathbf{r}) \, d\mathbf{K} = \frac{2}{N} \sum_{\text{pairs}} \delta(\mathbf{r} - \mathbf{R}_{nm}) \qquad (2\text{-}16)$$

The prescription specified by equation (2-16) for a single-scatterer-type specimen may be summarized as follows:

*1* Measure the intensity scattered by the sample as a function of $\mathbf{K} = (2\pi/\lambda) \times (\mathbf{S} - \mathbf{S}_0)$ for all $\mathbf{K}$.[1]

---

[1] Since it is never possible to obtain data for all $\mathbf{K}$, equation (2-16) is an idealization in that respect also. See Sections II E and II F 2.

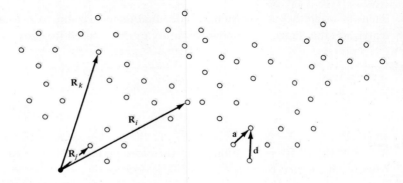

**FIGURE 2-4**
A two-dimensional random arrangement of oriented square molecules.

*2* Reduce the data to absolute units, i.e., to a multiple of the intensity which would be expected theoretically from an identical number of independent scatterers, and subtract 1.0 to obtain $P(\mathbf{K})$.

*3* Compute and normalize the three-dimensional Fourier transform of $P(\mathbf{K})$. The result,

$$\rho(\mathbf{r}) \equiv \frac{2}{N} \sum_{pairs} \delta(\mathbf{r} - \mathbf{R}_{mn}) \qquad (2\text{-}17)$$

will be a three-dimensional function of $\mathbf{r}$ which, ideally, will be equal to zero everywhere except at specific values $\mathbf{R}_{mn}$ which correspond to interscatterer vectors in the specimen. At those points the magnitude[1] of $\rho(\mathbf{r})$ will be equal to the number of such pair distances, per scatterer, in the sample.

The relationship between $\rho(\mathbf{r})$ and the structure may not be self-evident. For purposes of illustration, consider a hypothetical molecule-like substance, in which the molecules retain a specific shape and orientation, but are otherwise quite randomly distributed. Figure 2-4 shows a portion of such a (two-dimensional) structure, in which the molecules happen to be four-point "squares." The positions of the points can be specified with respect to an arbitrary origin; three are shown in the figure as $\mathbf{R}_i$, $\mathbf{R}_j$, and $\mathbf{R}_k$. In general, it might be convenient to speak of the density distribution of scatterers; for fixed-point scatterers:

$$p(\mathbf{R}) = \sum_{k=1}^{N} \delta(\mathbf{R} - \mathbf{R}_k) \qquad (2\text{-}18)$$

where $N$ is the total number of scatterers.

[1] That is, the "area" of the $\delta$ function.

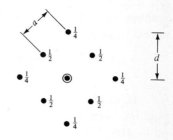

**FIGURE 2-5**
The two-dimensional pair-distribution
function corresponding to the random
square molecules in Fig. 2-4.

To construct the corresponding pair function $\rho(\mathbf{r})$, one simply makes a list of all the interscatterer vectors which occur, and of how many times each occurs. Designating the different vectors in the list as $\mathbf{r}_j$ and the occurrence frequency as $n_j$:

$$\rho(\mathbf{r}) = \frac{1}{N} \sum_j n_j \, \delta(\mathbf{r} - \mathbf{r}_j) \qquad (2\text{-}19)$$

Figure 2-5 shows the $\rho(\mathbf{r})$ corresponding to the structure in Fig. 2-4. In this instance the six repetitive "bond" vectors (an average of 1.5 per scatterer) are represented by eight $\mathbf{r}_j$'s. All the other interscatterer vectors are assumed to occur only once (or a few times) such that other $n_j/N \ll \frac{1}{8}$, and are therefore too weak to appear in Fig. 2-5. It may be noticed (as a generality) that $\rho(\mathbf{r})$ portrays the average configuration around a scatterer. Thus, except in the special circumstances that one knows that every scatterer sees an identical environment,[1] $\rho(\mathbf{r})$ itself may provide a picture which is somewhat obscure, and even ambiguous. Although $\rho(\mathbf{r})$ and other directly derived constructions can be very informative concerning the configurations of scatterers, it may be useful to stress at the outset that the structure [e.g., $p(\mathbf{R})$] is rarely derived directly from the diffraction data; rather, in the end one attempts to deduce (and systematically improve) a model which is then shown to be consistent with the experiment. In Fig. 2-5, all $\mathbf{r}_j$ were ignored whose occurrence frequency was not of the order $N$. In actual fact, that would not be a very sensible approach to a real material in which the probability that any vector would be *exactly* repeated would be small (if only because of the uncertainty principle); the list of $\mathbf{r}$'s, if precise enough, would contain a separate entry for nearly every one of the $(N^2 - N)/2$ interscatterer vectors. Obviously, it is more reasonable to use a histogram approach, in which one lumps together all the vectors which fall into each of the uniformly spaced $\Delta\mathbf{r}$ volume-element "boxes." For practical specimens, $N^2$ will be so large and $\rho(\mathbf{r})$ sufficiently irregular that the boxes can be very narrow—so narrow, in fact, can they be that $\rho(\mathbf{r})$ can be considered to be a continuous function, of which the $\delta$-function form is a

---

[1] This condition is satisfied by the points on a Bravais lattice, a topic which is discussed in Section III A.

limiting case useful for idealized models. The continuous function requires normalization:

$$\int \rho(\mathbf{r}) \, d\mathbf{r} = N - 1 \qquad (2\text{-}20)$$

If, in the example of Fig. 2-4, the relative positions of the molecules were fully random, then the value of $\rho(\mathbf{r})$ away from the prevalent vectors would be $\rho_c = (N - 4)/V$.[1] The numbers given in Fig. 2-5 at the "peaks" would correspond to $\int_{\mathbf{r}_j \pm \delta} \rho(\mathbf{r}) \, d\mathbf{r}$, with a negligibly small volume element $\delta$.

While considering a continuous $\rho(\mathbf{r})$, it is also useful to examine the ramifications of a continuous $p(\mathbf{R})$ as regards the amplitude of scattering as well as the intensity. In place of the summation in equation (2-8), one obtains

$$\psi_s^{\text{total}} = Ae^{2\pi i(vt+\alpha)}a(\mathbf{K}) \int p(\mathbf{R})e^{i\mathbf{K} \cdot \mathbf{R}} \, d\mathbf{R} \qquad (2\text{-}21)$$

Equation (2-21), perhaps more clearly than (2-7), shows that the scattered amplitude is the Fourier transform of the scatterer density, which is designated $\tau[p(\mathbf{R})]$. Using that notation

$$\psi_s = Ae^{2\pi i(vt+\alpha)}a(\mathbf{K})\tau[p(\mathbf{R})] \qquad (2\text{-}22a)$$

$$\psi_s^* = A^*e^{-2\pi i(vt+\alpha)}a^*(\mathbf{K})\tau[p^*(-\mathbf{R})] \qquad (2\text{-}22b)$$

$$Q(\mathbf{K}) = aa^*\tau[p(\mathbf{R})]\tau[p^*(-\mathbf{R})]\dagger \qquad (2\text{-}23)$$

The well-known convolution theorem states that the product of the transforms of two functions is the transform of the convolution of the two functions. Its application here yields

$$\tau^{-1}\left[\frac{Q(\mathbf{K})}{aa^*}\right] = \int_{-\infty}^{+\infty} p(\mathbf{u}) p^*(\mathbf{u} + \mathbf{r}) \, d\mathbf{u} \qquad (2\text{-}24)$$

The inverse transform has been defined earlier as

$$\tau^{-1}\left[\frac{Q(\mathbf{K})}{Naa^*} - 1\right] \equiv \rho(\mathbf{r}) \qquad (2\text{-}25)$$

---

[1] In general, for that quantity

$$\rho_c = \rho_0 \frac{1 - n/N}{1 - nV_e/V}$$

where $\rho_0 = N/V$, $n$ is the number of scatterers whose positions are specified as being nonrandom (per scatterer), and $nV_e$ is the volume excluded by those $n$ scatterers, as a unit. If $nV_e$ is zero, the substance is an *ideal* gas. There is rarely any reason to note the difference between $\rho_0$ and $\rho_c$, except when considering limits.

† Since only one type of scatterer is still being considered, $\mathbf{p}(\mathbf{R})$ must be real. Otherwise, the $a(\mathbf{S} - \mathbf{S}_0)$ might be a function of $\mathbf{R}$ and would have to be included in the integral (or sums). The conjugate symbol is left here as a reminder of that fact.

Thus, since $\tau^{-1}[1] = \delta(\mathbf{r} - 0)$,

$$\rho(\mathbf{r}) \equiv \frac{1}{N} \int_{-\infty}^{'+\infty} p(\mathbf{u}) p^*(\mathbf{u} + \mathbf{r}) \, d\mathbf{u} \qquad (2\text{-}26)$$

where the prime on the integral means that the $\mathbf{r} = 0$ term should be excluded [see equation (2-11)]. The integral (without the prime) is known as an *autocorrelation function*; this form may give a clearer visualization of the relationship between the interscatterer distribution $\rho(\mathbf{r})$ and the spatial configuration $p(\mathbf{R})$ than was afforded by the earlier discussion based on equations (2-16) and (2-19).

It should be pointed out that the equations (2-21) to (2-26) become the analogous summation equations if $\rho(\mathbf{r})$ is the sum of $\delta$ functions defined in (2-19). Indeed, the two sets of equations are formally identical if $n_j$ [in (2-19)] is allowed to be non-integral.

While the formalism of the preceding sections may possess a certain elegance, there is an obvious impracticality inherent in its direct application, totally apart from ambiguities which have been mentioned. Although the analysis (i.e., Fourier transformation) of three-dimensional, high-resolution data is a reasonably practical matter,[1] few experimentalists will be enthusiastic enough to gather the data, even with automatic experimental machinery.[2] However, while it may require millions of data points to fully detail $Q(\mathbf{K})$ (even though only to a finite $\mathbf{K}_{max}$), one might suspect that such coverage is not always necessary. Working within the framework of a particular model for the structure of the sample material, it is generally possible to work with a less extensive array of data.

## II.  SPHERICALLY AVERAGED SAMPLES: GASES, LIQUIDS, GLASSES AND OTHER AMORPHOUS SOLIDS, AND POLYCRYSTALLINE AGGREGATES

### A.  Introduction

A particular specialization that can be introduced in order to reduce the amount of data needed is spherical symmetry. Assume a model in which all interscatterer vectors of a given magnitude $|\mathbf{r}_j|$ occur in all orientations with uniform probability, averaged over the sample. Regarding the four-point molecules of Fig. 2-4, for example, such a model would apply under a variety of circumstances: (1) the molecules

---

[1] There are readily available fast, Fourier summation algorithms, implemented for high-speed, large-storage digital computers, as well as specialized "analog" data processors.

[2] If one decided to use $\Delta K = |\mathbf{K}_{max}|/100$ as a mesh interval in all three dimensions, then the resulting $10^6$ or more data values would require a year or more of measurement, at the optimistic rate of 30 sec per point.

are oriented (statically) in a completely helter-skelter manner, or (2) sections of the sample, each highly or partially oriented internally, are randomly oriented with respect to each other, or (3) random motion of the molecules causes a spherical averaging. It should be distinctly noted, however, that the assumed condition does *not* imply that each scatterer sees a spherically symmetrical environment; indeed, nothing whatever is implied about local symmetry. The model applies to a surprisingly wide range of substances, including gases, liquids, and amorphous solids. It can also be simulated with artificially randomized samples of more oriented materials (e.g., powder briquets).

In many instances, the spherically averaged sample can provide all the desired structural information about the material. Other times, the results can provide a useful basis for more detailed investigation of oriented arrangements, such as obtains in polymeric materials.

In an actual measurement the experimental intensities represent an average over the time necessary to obtain a data point, as well as an average in space over the irradiated regions of the sample. In applying equations (2-14) or (2-24) to any model, therefore, it is necessary to simulate the averaging appropriate to the model and the particular experiment.

## B. The spherical average; reduction to one-dimensional functions

Regarding the use of equations for the intensity [(2-14) or (2-24)], to obtain the form specific to spherical symmetry, it is only necessary to consider that $\rho(\mathbf{r})$ depends only on $|\mathbf{r}|$ and to average $\cos(\mathbf{K} \cdot \mathbf{r})$ or, more generally, $\exp(\pm i\mathbf{K} \cdot \mathbf{r})$ over all possible values of the $\mathbf{K}, \mathbf{r}$ included angle, $\alpha$.

$$\langle \exp(\pm i\mathbf{K} \cdot \mathbf{r}) \rangle_\alpha = \frac{\int \exp(\pm iKr \cos \alpha)\, dV}{\int dV} \tag{2-27}$$

Referring to Fig. 2-6, a convenient volume element for vector magnitudes between $r$ and $r + dr$ is

$$dV = 2\pi r \sin \alpha\, dr\, r\, d\alpha \tag{2-28}$$

$$\langle \exp(\pm i\mathbf{K} \cdot \mathbf{r}) \rangle_\alpha = \frac{\int_0^\pi 2\pi r^2\, dr \exp(\pm iKr \cos \alpha) \sin \alpha\, d\alpha}{\int_0^\pi 2\pi r^2\, dr \sin \alpha\, d\alpha} \tag{2-29}$$

The value of the normalization integral in the denominator is $4\pi r^2\, dr$. Letting $y = \cos \alpha$,

$$\langle \exp(\pm i\mathbf{K} \cdot \mathbf{r}) \rangle_\alpha = \pm \frac{1}{2} \int_{-1}^{+1} \exp(\pm iKry)\, dy \tag{2-30}$$

$$\langle \exp(\pm i\mathbf{K} \cdot \mathbf{r}) \rangle_\alpha = \pm \left[ \frac{e^{\pm iKr} - e^{\mp iKr}}{2iKr} \right] \tag{2-31}$$

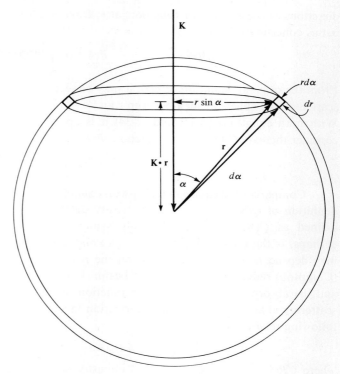

FIGURE 2-6
The integration volume for the spherical average.

Therefore,

$$\langle \exp(i\mathbf{K} \cdot \mathbf{r}) \rangle_\alpha = \frac{\sin Kr}{Kr} \qquad (2\text{-}32a)$$

$$\langle \exp(-i\mathbf{K} \cdot \mathbf{r}) \rangle_\alpha = \frac{\sin Kr}{Kr} \qquad (2\text{-}32b)$$

$$\langle \cos(\mathbf{K} \cdot \mathbf{r}) \rangle_\alpha = \frac{\sin Kr}{Kr} \qquad (2\text{-}32c)$$

Rewriting equations (2-14) in the particularized form for a spherically symmetrical $\rho(\mathbf{r})$

$$P(K) = \sum_{j=1}^{\infty} \frac{c_j \sin(Kr_j)}{Kr_j} \qquad (2\text{-}33)$$

where $r_j$ is the magnitude of an interscatterer vector $\mathbf{r}_j$, and $c_j$ is the sum over all directions of the number of neighbors at $r_j$ from an average scatterer. For the analogous continuous form (2-24)

$$P(K) = \int_0^\infty 4\pi r^2 \rho(r) \frac{\sin Kr}{Kr} dr \qquad (2\text{-}34)$$

From the original definition of $\rho(\mathbf{r})$, it may be seen that $4\pi r^2 \rho(r) \, dr$ is the number of scatterers in a spherical shell between $r$ and $r + dr$ centered on an average scatterer. With both equations (2-33) and (2-34) it is appropriate to multiply by $K$, and to consider thereafter the reduced function, $F(K) = KP(K)$.

$$F(K) \equiv \left( \frac{I(K)}{aa^*} - 1 \right) K \qquad (2\text{-}35)$$

Comparison of equation (2-35) with equations (2-14) and (2-9) and the previous definition of $Q(\mathbf{K})$ as the relative intensity $\psi\psi^*/AA^*$ will show that $I(K)$ has been defined as $Q(\mathbf{K})/N$ for a spherically symmetrical experiment. More important, perhaps, is the relationship between $I(K)$ and the measured intensity. In fact, this may depend to a significant extent on the relative amount of modified, incoherent (Compton) radiation in the detected beam. Fortunately, the normalization constant and the absorption and polarization functions apply to the Compton and coherent scattering alike. Therefore, those correction factors may be expressed such that the following relationship holds:

$$I(K) = [CR(K) - CR_B(K)]C \, A(K)PL(K) + q_c(K)I_c(K) \qquad (2\text{-}36)$$

where $CR(K)$ is the total measured intensity (or counting rate) in arbitrary units, $CR_B(K)$ is the background or noise contribution to the total, $C$ is the normalization constant, $A(K)$ is the theoretical, absorption-dependent, geometrical function, $PL(K)$ is the theoretical polarization correction, $I_c(K)$ is the total theoretical Compton-scattering contribution, and $q_c(K)$ is the fraction of the Compton intensity which is actually detected. The last function depends on the wavelength sensitivity of the detection system and may not be equal to 1.0 even if no purposeful attempt is being made to eliminate the $I_c(K)$ contribution.

For a diffraction geometry in which the incident and detected beams are symmetrical to the normal of a flat sample and cross the same face, $A(K)$ is a constant.[1]

---

[1] A volume element in a specimen contributes to the measured intensity if it is both irradiated by the incident beam and seen by the detector within its divergence collimators. The contribution of each element in the active volume is weighted by its appropriate $\exp(-\mu x)$, where $\mu$ is the linear absorption coefficient and $x$ the path length in the sample to and from the volume element. $A(\mathbf{K})$ is simply the measure of the variation of the weighted, active volume as $\mathbf{K}$ is varied. The nominal invariance of $A(\mathbf{K})$ in flat-specimen reflection geometry is not valid if the detector collimators exclude a varying portion of the effectively irradiated volume, as may occur if $1/\mu$ is not small compared to the beam widths. When the detecting system sees exactly the irradiated area at the sample surface, the active volume function, expressed in terms of the scattering angle, is

$PL(K)$ assumes one of the forms below, depending on the number of crystal monochromators in the beam:

$$PL_0(K) = 1 + \cos^2 2\theta \qquad (2\text{-}37a)$$

$$PL_1(K) = 1 + \cos^2 2\theta_m \cos^2 2\theta \qquad (2\text{-}37b)$$

$$PL_2(K) = 1 + \cos^4 2\theta_m \cos^2 2\theta \quad \text{(identical monochromators)} \qquad (2\text{-}37c)$$

where $2\theta$ and $2\theta_m$ are the scattering angles at the specimen and monochromators, respectively. Note that $K = 4\pi \sin \theta/\lambda$.

From equation (2-35), it may be seen that the units of $I(K)$ [and also of $I_C(K)$ as used in this reduction] must be the same as the units of the product $aa^*$. Thus far, however, the particular units of scattering matter, whose scattering amplitude is $a(K)$ and whose distribution is being sought, have not been specified. With x rays, one normally thinks in terms of electrons, or atoms, or even molecules, for which $a(K)$ is usually expressed in *electron units*, such that the magnitude of $a(K)$ at $K = 0$ is equal to the number of electrons in the scattering unit (apart from dispersion effects). Ordinarily, it is the atomic distribution that one deals with, and $a(K)$ is called the atomic form factor $f(K)$ and is derived by Fourier transformation of the electron distribution associated with each atom (see Section II G). Thus, it may be seen from equation (2-36) that if the scattering from the sample approaches that expected for a collection of independent scatterers, the normalization constant $C$ may be determined from a fit of the measured (partially reduced) intensity to the theoretical values, $f(K)f^*(K) + q_C(K)I_C(K)$. For many sorts of specimens the necessary condition is approached at large $K$ values, but well within the range of measurement.

In terms of $F(K)$, as defined in equation (2-35), equations (2-33) and (2-34) can be rewritten in somewhat simpler form:

$$F(K) = \sum_{i=1}^{M} \frac{c_i}{r_i} \sin Kr_i \qquad (2\text{-}39)$$

---

$$A(\theta) = \left\{ \left(1 - \frac{1}{\alpha}\right) [1 - \exp\left(-2\mu t \csc \theta\right)] + \frac{2t \cos \theta \exp\left(-2\mu t \csc \theta\right)}{\omega} \right\} \qquad (2\text{-}38a)$$

for $0 \le t \le \frac{1}{2}\omega \sec \theta$, and

$$A(\theta) = \left\{ 1 - \frac{[1 - \exp\left(-\alpha\right)]}{\alpha} \right\} \qquad (2\text{-}38b)$$

for $t \ge \frac{1}{2}\omega \sec \theta$, where $\alpha = 2\mu\omega \csc 2\theta$, $\omega$ is the width of the beams in the diffraction plane, $2\theta$ is the scattering angle, and $t$ is the thickness of the sample. In addition to the foregoing expressions, Milberg (1958) has derived formulas which apply when the detected beam is wider than the incident beam [a condition which lessens the variation of $A(\theta)$].

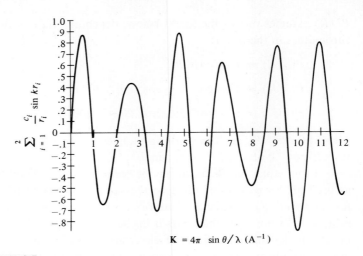

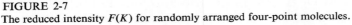

**FIGURE 2-7**
The reduced intensity $F(K)$ for randomly arranged four-point molecules.

or, more generally,

$$F(K) = \int_{\substack{volume \\ irradiated}} 4\pi r \rho(r) \sin Kr \, dr \qquad (2\text{-}40)$$

The summation form is still obviously a special case of the integral with

$$4\pi r^2 \rho(r) = \sum_i c_i \, \delta(r - r_i) \qquad (2\text{-}41)$$

In Fig. 2-7, as an illustration of equation (2-39), the $F(K)$ calculated for the four-point molecules of Fig. 2-4 is plotted assuming random *orientations* of the molecules and including only the two prevalent neighbor distances, as specified below:

| Shell $(i)$ | $r_i(A)$ | $c_i$ |
|---|---|---|
| 1 | 3 | 2 |
| 2 | $3\sqrt{2}$ | 1 |

In Fig. 2-8 the corresponding intensity function, $I/a^2$, is shown. The most important feature in the intensity is at the origin. Since $\sin Kr_i/Kr_i \rightarrow 1$ for all $r_i$ at $K = 0$, the intensity $I/a^2$ approaches $1 + \sum_{i=1} c_i$ as $K$ approaches zero, a result which is a general one. In the plane-wave approximation the amplitudes from all scatterers are expected to add near $K = 0$, regardless of their arrangement in space. Indeed, for that reason, the constant density $\rho_c$ of the ideal molecular gas model pair-distribution function cannot be ignored as it was just now. Since that probability density is a

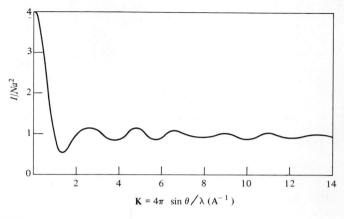

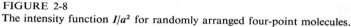

FIGURE 2-8

The intensity function $I/a^2$ for randomly arranged four-point molecules.

continuous function, the integral form

$$F(K) = \int_V 4\pi r \rho(r) \sin Kr \, dr \qquad (2\text{-}42)$$

must be utilized to calculate its contribution to $F(K)$. We continue to assume that the experimental dimensions are such that the plane-wave approximation is valid. For the gaseous sample, then, the complete $F(K)$ is composed of two parts[1]

$$F(K) = \sum_{i=1}^{2} \frac{c_i}{r_i} \sin Kr_i + \int_V 4\pi r \rho_c \sin Kr \, dr \qquad (2\text{-}43)$$

## C.   Finite samples and particle size effects

Since the integrand remains finite as $r \to \infty$, the volume of integration must be considered carefully. Note particularly that in equation (2-42), $\rho(r)$ is the average over that seen by all scatterers, including those close to the boundaries of the irradiated volume of the specimen. For each point $\mathbf{R}$ within the active specimen (as in Fig. 2-9) there is a quantity $f(\mathbf{R}, r)$, which specifies the fraction contained within the active specimen of a spherical shell of radius $r$ centered on a scatterer at $\mathbf{R}$. If $\rho_T(r)$ is the true pair density (in an infinite specimen), then $\rho_T(r)f(r)$ is the quantity that must be used in (2-42), where $f(r)$ is the average over all $\mathbf{R}$ in the active specimen.

As an example, a plot of that function $f(r/D)$, for a spherical active particle of

[1] Note that in taking the integral of $\rho_c$ over the entire volume $V$, one assumes that each molecule excludes *no* volume around it, i.e., that the gas is ideal.

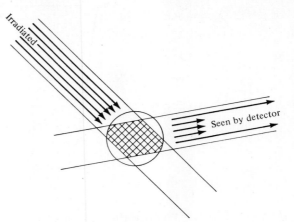

**FIGURE 2-9**
The finite sample (or active volume) geometry.

diameter $D$, is given in Fig. 2-10. The values are given by the analytical expression

$$f\left(\frac{r}{D}\right) = 1 - 1.5\frac{r}{D} + 0.5\left(\frac{r}{D}\right)^3 \qquad (2\text{-}44)$$

The normalized function $r\rho_T(r)f(r/D)/\rho_c D$ is plotted in Fig. 2-11 for $\rho_T(r)$ equal to the constant $\rho_c$.

For a spherical particle, using (2-44) for $f(r)$, the $\rho_c$ term yields

$$\frac{1}{K}\int 4\pi r\rho_c f(r)\sin Kr\, dr = \frac{I_{\rho_c}}{a^2} \qquad (2\text{-}45a)$$

$$\frac{I_{\rho_c}}{a^2} = M\left\{\frac{72}{q^4}\left[1 + \left(1 - \frac{4}{q^2}\right)\cos q\right.\right.$$
$$\left.\left. - 4\frac{\sin q}{q} + \frac{4}{q^2}\right]\right\} \qquad (2\text{-}45b)$$

$$\frac{I_{\rho_c}}{a^2} = MJ(q) \qquad (2\text{-}45c)$$

where $M = \frac{4}{3}\pi\rho_c(D/2)^3$ and $q \equiv KD$. $J(q)$ is plotted in Fig. 2-12, and the universal function $K^3 I_{\rho_c}/a^2\rho_c$ in Fig. 2-13.

Note that $J(q) \to 1$ in the limit $K = 0$ and that $\rho_c = \rho_0(1 - 4/N)$ for the four-point molecule ideal gas. Therefore, $I_{\rho_c}/a^2 \xrightarrow[K=0]{} (N - 4)$. Using (2-43), this yields at $K = 0$ a total relative intensity per scatterer, $I/a^2$, of $[(N - 4) + 3 + 1)] = N$ and, therefore, a total scattered intensity in the forward direction of $N^2 a^2$, as must be the case in general.

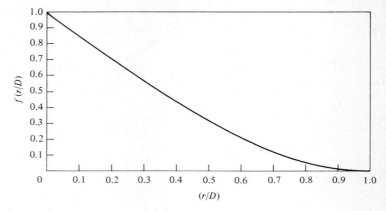

FIGURE 2-10
The particle size function $f(r/D)$ for spherical particles.

## D.   Forward scattering; the contribution of the mean density

It is also noteworthy that the $\rho_c$ contribution to the intensity is confined to the low-$K$ region of the spectrum. This is true not only in the relative sense that is obvious in Figs. 2-12 and 2-13, but also in terms of an absolute comparison with the $\sum_i c_i \sin Kr_i/ Kr_i$ term. That is, although the $\rho_c$ contribution is of the order of $N$ times larger at $K = 0$, it is negligible for $q \gtrsim 40$. Thus, for example, if $D$ is 500 A or more, the $\rho_c$ term can be ignored for $K > 0.08$ A$^{-1}$. The intensity variation for the gas model

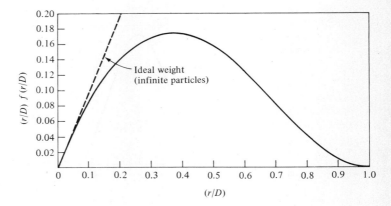

FIGURE 2-11
The reduced density function for a random, finite sample.

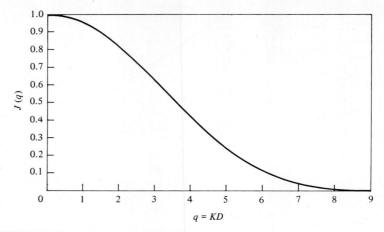

**FIGURE 2-12**
The scattered-intensity pattern for a random, finite sample.

is shown schematically in Fig. 2-14; a value of $3 \times 10^{-5}$ molecule per $A^3$ (and the 500-A size) has been used to obtain the peak value of about 2000.

The foregoing calculation is adequate for the present purpose of showing that the $\rho_0$ term contributes predominantly near $K = 0$. In ordinary experimental situations, however, specimens much larger than 500 A will be used, and a determination of the detailed shape of that peak will require using the full form of (2-6).

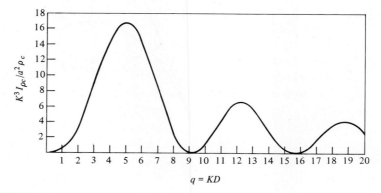

**FIGURE 2-13**
The universal intensity function $K^3 I_{\rho c}/a^2 \rho_c$ for a random, finite sample.

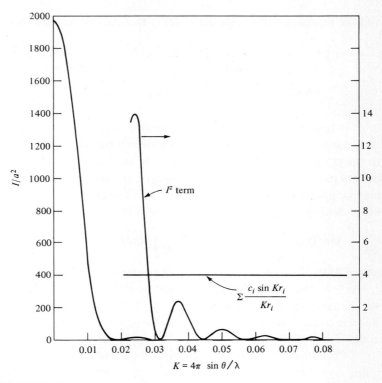

FIGURE 2-14
Small-angle scattering for a finite (500-A) gas-model sample.

In any case, because of the transmitted, unscattered beam, the very-small-angle portion of the intensity pattern will be immeasurable. Ordinarily, the experimental "result" will include an extrapolation to $K = 0$ from a larger value of $K$, as an attempt to exclude the transmitted beam. In the process, the true scattered peak near zero, $I^z$, is also excluded, and it is appropriate to subtract the analogous terms from the distribution function. In fact, for the present ideal-gas model, that is accomplished by leaving out the $\rho_c$ term, as was done originally (see Fig. 2-7).

Considering that example again, but writing a proper expression for the density function including the $\rho_c$ term:

$$4\pi r \rho_T(r) = \sum_{i=1}^{2} \frac{c_i}{r_i} \delta(r - R_i) + 4\pi r \rho_c \qquad (2\text{-}46)$$

The finite-specimen function $f(r)$ must apply to the total density function, and the *modified F(K)*, with the $\rho_c$ term explicitly subtracted, becomes

$$\left(\frac{I}{a^2} - 1\right)K - \frac{I^z}{a^2}K = \int 4\pi r f(r)\rho_T(r) \sin Kr\, dr - \int 4\pi r f(r)\rho_c \sin Kr\, dr \qquad (2\text{-}47)$$

$$F'(K) = \left(\frac{I - I^z}{a^2} - 1\right)K = \int 4\pi r f(r)[\rho_T(r) - \rho_c] \sin Kr\, dr \qquad (2\text{-}48)$$

It is of interest to consider the possible generality of equation (2-48). That is, does the transform of the mean density (which has already been shown to be limited to the small-$K$ region) always simulate the small-$K$ part of the transform of the true density function which is eliminated by an extrapolation from higher $K$? The question may be phrased in a more formal way: Is it true, even when $\rho_T(r)$ is not equal to the mean density over most of the volume as it is in a gas, that

$$\frac{1}{K}\int 4\pi r f(r)\rho_0 \sin Kr\, dr \underset{\text{low } K}{\overset{?}{=}} 1 + \frac{1}{K}\int 4\pi r f(r)\rho_T(r) \sin Kr\, dr + \Delta(K) \qquad (2\text{-}49)$$

where $\Delta(K)$ is small and slowly varying compared to the total intensity in the low-$K$ region? In fact, (2-49) is generally true; to illustrate this, a "worst-case" comparison will be considered: a dense material, in which the $\rho_T(r)$ consists only of a series of discrete $\delta$ functions over the entire volume of the specimen, and a very small specimen size.

Coordination numbers $c_i$ and distances $r_i$, which correspond to a face-centered-cubic *crystalline* material, are used in order to ensure having a physically consistent set. The specimen is to be imagined as being a small sphere of diameter $D$, rotating rapidly in the beam to ensure the spherical symmetry of $\rho_T(r)$.

The lattice parameter chosen is 3.86 A ($r_1 = 2.73$ A), and the specimen size is 19.7 A; 49 neighbor distances are included in a piece that size, and $\rho_0 = 0.06955$ atom per A. The functions $\int 4\pi r f(r)\rho_0 \sin Kr\, dr$ and $\sum (c_i/r_i)f(r_i) \sin Kr_i$ are plotted in Fig. 2-15. Even though the $I^z$ term extends out to relatively large $k$ (because of the small $D$), the extrapolation back to $K = 0$ is feasible. If (2-49) is true, the difference between the two functions should be $\sim -K$ in the low-$k$ region; that this is the case is illustrated in Fig. 2-16. Thus, if $I^z$ is that part of the true scattered intensity which one eliminates experimentally by extrapolating to $K = 0$, and neglecting any difference between $\rho_c$ and $\rho_0$, in general

$$F'(K) = \left(\frac{I - I^z}{a^2} - 1\right)K = \int 4\pi r f(r)[\rho_T(r) - \rho_0] \sin Kr\, dr \qquad (2\text{-}50)$$

In most instances, of course, the specimen will be much larger than 19 or 20 A in diameter. Suppose, in fact, that the small sphere model of the previous section is extended into a polycrystalline specimen, by randomly packing many such pieces of effective diameter $D$ together to form a briquet of random average orientation.

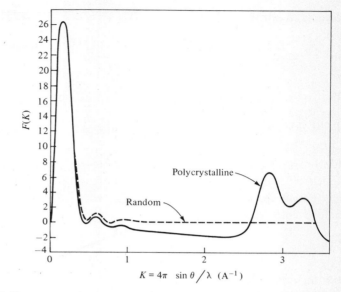

FIGURE 2-15
The small-angle intensity patterns for finite random and polycrystalline samples
of equal density.

Let the small pieces be characterized by a size-shape function $f(r, D)$ and the entire
specimen by another function $q(r, S)$. The expression for the reduced total intensity
function may be constructed as a sum of three terms:

$$\left(\frac{I}{a^2} - 1\right) K = \int_0^D 4\pi r q(r) f(r) \rho_T(r) \sin Kr \, dr$$

$$+ \int_0^D 4\pi r q(r)[1 - f(r)]\rho_0 \sin Kr \, dr$$

$$+ \int_D^S 4\pi r q(r)\rho_0 \sin Kr \, dr \qquad (2\text{-}51)$$

It has been established already that regarding the contribution to $I^z$, the first
term can be replaced by $\int_0^D 4\pi r q(r) f(r)\rho_0 \sin Kr \, dr$; thus the low-$K$ portion of $F(K)$
will be $\int_0^S 4\pi r q(r)\rho_0 \sin Kr \, dr + \Delta(K)$ [see (2-49)], and with $I^z$ subtracted,

$$F'(K) = \int 4\pi r q(r) f(r)[\rho_T(r) - \rho_0] \sin Kr \, dr \qquad (2\text{-}52)$$

Thus, with the composite specimen, although the $I^z$ term is much larger [by a

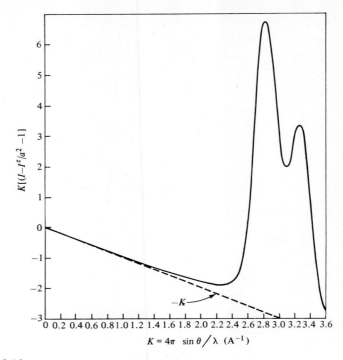

$$K = 4\pi \, \sin \theta / \lambda \;\; (\text{A}^{-1})$$

**FIGURE 2-16**
The small-angle reduced intensity function, with the random scattering subtracted.

ratio of order $(S/D)^3$] and much sharper, the reduced intensity $F'(K)$, formed without the $I^z$, is unchanged in comparison to the isolated particle.[1]

In most instances, particularly with noncrystalline samples, $[\rho_T(r) - \rho_0]$ will approach zero at distances that are very small compared to the active specimen size. The corresponding $f(r)$ is then readily ignored

$$F'(K) = \int 4\pi r [\rho(r) - \rho_0] \sin Kr \, dr \qquad (2\text{-}53)$$

In the remainder of this chapter, it will usually be assumed that $I^z$ is not included in the intensity; $I - I^z$ will generally be referred to as just $I$, and $F'(K)$ as just $F(K)$.

---

[1] In a more precise model for a briquet, however, physical constrictions on the packing would introduce some minor differences. In particular, across-boundary pair distances less than $r_1$, which are allowed by the present model and which show in (2-52) to the extent that $f(r_1) < 1$, would be prohibited.

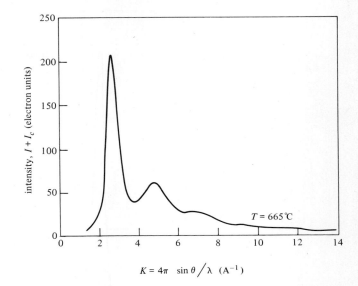

FIGURE 2-17
The experimental intensity pattern for liquid aluminum (including Compton scattering).

Following Fourier methods, it is possible to derive the pair distribution by multiplying both sides by sin $K\varepsilon$ and integrating over $K$.[1]

$$G(r) \equiv 4\pi r[\rho(r) - \rho_0] = \frac{2}{\pi} \int F(K) \sin Kr \, dK \qquad (2\text{-}54)$$

Since $F(K)$ and $G(r)$ are odd functions of $K$ and $r$, respectively [i.e., $F(K) = -F(-K)$ and $G(r) = -G(-r)$], the sine integrals are equivalent to the complete Fourier transforms, which may be indicated by the notation

$$F(K) = \tau[G(r)] \qquad (2\text{-}55a)$$

$$G(r) = \tau^{-1}[F(K)] \qquad (2\text{-}55b)$$

to serve as a reminder that the known properties of Fourier transforms can sometimes provide a rapid insight regarding the behavior of $F(K)$ attendant on changes to $G(r)$, and vice versa.

As an illustration of the sort of scattered x-ray intensity pattern expected for a noncrystalline sample, experimental results are shown in Fig. 2-17 for liquid aluminum

[1] The required integral over $K$ ranges from 0 to $\infty$; the inherent error which arises from the existence of a finite limit $K_{max}$ to experimental data is called the *termination effect*, and is described later.

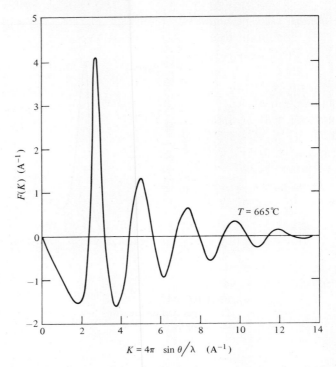

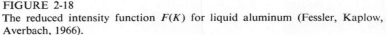

**FIGURE 2-18**
The reduced intensity function $F(K)$ for liquid aluminum (Fessler, Kaplow, Averbach, 1966).

at 665°C. The given function is in electron units, which is to say that

*1*   Extraneous background has been subtracted.
*2*   The known and irrelevant portion of the $K$ dependence of the raw data which is described by the geometrical absorption and polarization functions $A(K)$ and $PL(K)$ has been removed.
*3*   A normalization has been chosen such that at large $K$ values, the intensity modulates $f^2(K) + I_{Compton}$.

The fully reduced intensity function $F(K)$ is depicted in Fig. 2-18, and the corresponding $4\pi r^2[\rho(r) - \rho_0]$ in Fig. 2-19. [Note that the last function is $rG(r)$.] These results for liquid aluminum are taken from the work of Fessler et al. (1966). Their directly comparable results for polycrystalline aluminum are shown later in Section II F, and the contrast is an interesting one.

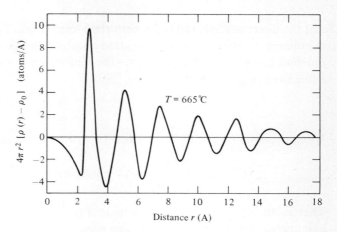

FIGURE 2-19
The atomic pair distribution $rG(r)$ for liquid aluminum (Fessler, Kaplow, and Averbach, 1966).

## E.  The finiteness of data; practical analysis for monatomic samples

In equation (2-54) the integral over $K$ nominally extends from zero to infinity. Actual data always stop below some absolute maximum, however; $K_{max} < 4\pi/\lambda$, for any particular wavelength. With x radiation the upper limit is not readily extended beyond about $18A^{-1}$ by using a shorter wavelength, because the elastic scattering power of the individual atoms decreases so drastically at large $K$.

To be more precise, therefore, the integral in equation (2-54) should be shown as the "limited" transform

$$G'(r) = \frac{2}{\pi} \int_0^{K_{max}} F(K) \sin Kr \, dK \qquad (2-56)$$

$G'(r)$ is clearly not a true representation of the distribution of pairs; because of the termination effect it differs from $G(r)$ by

$$\Delta G(r) = \frac{2}{\pi} \int_{K_{max}}^{\infty} F(K) \sin Kr \, dK \qquad (2-57)$$

How significant the difference is depends not only on the value of $K_{max}$ but also on the relative importance of the high-$K$ $F(K)$. If the transformation from $F(K)$ to $G(r)$ is viewed as a harmonic analysis, it is seen that the termination effect will be more important the sharper the peaks in $G(r)$.

A method or methods are needed to obtain $G(r)$ from $F(K)$, in spite of the fact that the direct analysis yields the perturbed form $G'(r)$. For that purpose it is useful to examine the effect of data termination on particular $G(r)$'s in order to cope with the problem in an actual measurement. The limiting case occurs when $4\pi r\rho(r)$ contains $\delta$ functions; e.g.,

$$4\pi r\rho(r) = \sum_{i=1}^{\infty} \frac{c_i}{r_i} \delta(r - r_i) \qquad (2\text{-}58)$$

Using equation (2-53), $F(K)$ can then be written as a summation of integrals

$$F(K) = \sum_{i=1}^{\infty} \int \frac{c_i}{r_i} \delta(r - r_i) \sin Kr \, dr - \int 4\pi r\rho_0 \sin Kr \, dr \qquad (2\text{-}59)$$

That is, the theoretical contribution to $F(K)$ of one particular (perfectly defined) pair correlation distance is a pure sine wave

$$F_m(K) = \frac{c_m}{r_m} \sin Kr_m \qquad (2\text{-}60)$$

If, as suggested in equation (2-56), a "limited" transform is done on the experimental $F(K)$, the contribution in $r$ space of the $m$th correlation distance is

$$G'_m(r) = \frac{2}{\pi} \int_0^{K_{\max}} F_m(K) \sin Kr \, dK = \frac{2}{\pi} \int_0^{K_{\max}} \frac{c_m}{r_m} \sin Kr_m \sin Kr \, dK \qquad (2\text{-}61)$$

The integral on the right approaches a $\delta$ function as $K_{\max}$ approaches infinity. That is,

$$\frac{2}{\pi} \frac{c_m}{r_m} \int_0^{K_{\max}} \sin Kr_m \sin Kr \, dK \xrightarrow{K_{\max} \to \infty} \frac{c_m}{r_m} \delta(r - r_m) \qquad (2\text{-}62)$$

and equation (2-54) results, for the special conditions of equation (2-58). More generally, the integral, although not a $\delta$ function, is a well-defined function of $K_{\max}$ and $r_m$ which tends to peak at $r_m$ and which may be designated

$$P[K_{\max}, (r_m - r)] = \frac{2}{\pi} \int_0^{K_{\max}} \sin Kr_m \sin Kr \, dK \qquad (2\text{-}63)$$

$P(r)$ is plotted in Fig. 2-20 for $K_{\max} = 16 \text{ A}^{-1}$ and $r_m = 2.5 \text{ A}$. Thus if the true $4\pi r\rho(r)$ were given by equation (2-58), the derived $G'(r)$ would be

$$G'(r) = \sum_{i=1}^{\infty} \frac{c_i}{r_i} P[K_{\max}, (r_i - r)] - 4\pi r\rho_0 \qquad (2\text{-}64)$$

More realistically, note that even when $4\pi r\rho(r)$ can be described as a set of discrete correlation distances (or shells), the corresponding peaks will have finite widths, if only because of vibrations. If the true shape of the $m$th shell is given by the normal-

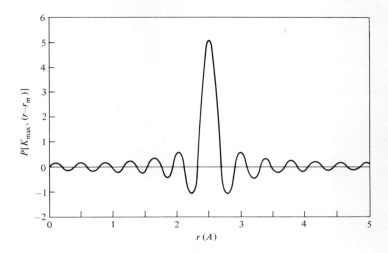

FIGURE 2-20
A sharp correlation peak in $G(r)$, distorted by termination of data.

ized function $B_m(r_m - r)$, the resultant $F(K)$ can be expressed as

$$F(K) = \sum_{i=1}^{\infty} \int \frac{c_i}{r_i} B_i(r_i - r) \sin Kr \, dr - \int 4\pi r \rho_0 \sin Kr \, dr \qquad (2\text{-}65)$$

It is interesting to note that the contribution to $F(K)$ of the $m$th shell is then

$$F(K) = \frac{c_m}{r_m} b_m(K) \sin Kr_m \qquad (2\text{-}66)$$

where $b_m(K)$ is the complete Fourier transform of $B_m(\varepsilon)$. Thus if $B(\varepsilon)$ has a Gaussian shape with $\sigma = 0.1$ A, the damping in $K$ space is $\exp(-0.01K^2/2)$, which is equal to 0.27 at $K = 16$ A$^{-1}$. The derived $G'(r)$ can still be expressed by equation (2-64), but now with

$$P[K_{max}, (r_m - r)] = \frac{2}{\pi} \int_0^{K_{max}} \left[ \int B_m(r_m - q) \sin Kq \, dq \right] \sin Kr \, dK \qquad (2\text{-}67)$$

In Fig. 2-21 the true $B_m(r_m - r)$ is compared to the $P$ function, for a Gaussian-shape correlation peak,

$$B_m(r_m - r) = \frac{1}{\sqrt{2\pi\sigma^2}} \exp - \frac{(r_m - r)^2}{2\sigma^2} \qquad (2\text{-}68)$$

and with $K_{max} = 16$ A$^{-1}$, $r_m = 2.5$ A and $\sigma = 0.1$ A. An asymmetric peak of the same width would show a larger and asymmetric distortion.

It is often true in analysis of x-ray diffraction data that one does not necessarily derive a final result directly. Rather, one shows that a particular model is consistent

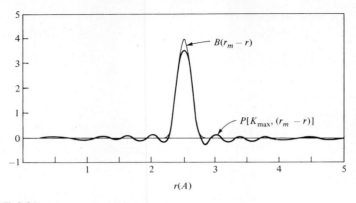

FIGURE 2-21
A Gaussian-shaped peak in $G(r)$, with and without termination distortion.

with the data, and, hopefully, uniquely consistent. That process can be divided into the two steps of (1) finding the model and (2) showing its consistency. Considering the second step first, the important considerations are that (1) the method used should be adequately sensitive to those aspects of the model which may be important physically (such as the width or shape of a peak), and that (2) any deviations between the real experimental data and that predicted by the model should be critically examined with regard to the actual estimates of error in the data.

In principle, the source of the model to be tested is of no importance. In practice, however, it is useful to have a procedure which not only suggests an initial model, but which also provides a mechanism for its refinement. Indeed, it may be desirable that the procedures for refinement and for showing consistency are coupled together, so that one has a systematic and convergent process.

In the present context, the problem is to derive a $G(r)$ from the available and limited $F(K)$, using $G'(r)$ as a basis. Since $G'(r)$ is itself entirely consistent with the available data by definition, it is important to recognize the reasons, if any, for rejecting it in the first place. There are, in fact, two observables—one in $r$ space and one in $K$ space—which may provide evidence that $G'(r)$ is not the true $G(r)$.

*1* The presence of physically implausible oscillations in $G'(r)$, like those in Fig. 2-21, which vary with $K_{max}$, and which extend to small $r$, below the expected nearest-neighbor distance.

*2* The presence of significant oscillations at high $K$ in $F(K)$, below $K_{max}$, which indicate an improbability that $F(K)$ is negligible beyond $K_{max}$.

The problem is to find another $G(r)$ which is also consistent with the data, but which is more satisfying in the above two respects. With the availability of computers, and particularly of time-sharing computers, trial-and-error solutions have become quite feasible. One procedure which seems to work satisfactorily is the following:

*1*   Create a trial function $G^T(r)$ by modifying $G'(r)$, using the results obtained from a few arbitrary terminations (at different "$K_{max}$") as an indication of the trend toward $K_{max} \to \infty$.

*2*   Calculate a trial $K$ function $F^T(K)$ from $G^T(r)$. (Disagreement between $F^T(K)$ and $F(K)$ below $K_{max}$ indicates error in $G^T(r)$, but the comparison does not readily indicate how $G^T(r)$ should be altered.)

*3*   Derive a new function $G^*(r)$:

$$G^*(r) = \frac{2}{\pi} \int_0^{K_{max}} F(K) \sin Kr \, dK + \frac{2}{\pi} \int_{K_{max}}^{\gg K_{max}} F^T(K) \sin Kr \, dK \qquad (2\text{-}69)$$

*4*   Using the difference between $G^T(r)$ and $G^*(r)$ as a guide, create a new $G^T(r)$.

*5*   Iterate steps (2) to (4) until $G^*(r) = G^T(r)$. When that condition holds, it will necessarily be true that $F^T(K) = F(K)$ for $K < K_{max}$, so that the $G^T(r)$ will be entirely consistent with the data, by definition.

Since $F(K)$ will not be known with absolute precision, the equalities can be relaxed to a degree which is consistent with the estimation of possible error $|\varepsilon(K)|$. That relaxation carries with it the implication that a range of physically plausible $G$ functions will be acceptable to the data. The uncertainty will be larger the smaller $K_{max}$ and the larger $|\varepsilon(K)|$, and it may be thoroughly examined, if one desires, by studying $[F^T(K) - F(K)]$ with regard to $|\varepsilon(K)|$.

It may be noted that the procedure outlined above can also be applied if the region of available data is even less complete. For example, if data were to be missing for the regions $0 \to K_1$, $K_2 \to K_3$, and $K_{max} \to \infty$, one would simply use $F^T(K)$ for all those regions in the integral of step (3) to compute $G^*(r)$. From the practical point of view, however, one must recognize that the uncertainty in the result will grow as the extent of data is diminished, for a given level of possible error $|\varepsilon(K)|$ in the available data.

It will also be noted that the above procedure, in which the model $G^T(r)$ is compared to the "corrected experiment" $G^*(r)$, is mathematically equivalent to comparing the "uncorrected experiment" $G'(r)$ to a modified model $G^M(r)$. The modified model $G^M(r)$ is derived from the true model $G^T(r)$, which is transformed to an intensity distribution in $K$ space and there treated exactly as was the experimental intensity. For the simple termination case,

$$G^M(r) = \int_0^{K_{max}} F^T(K) \sin Kr \, dK \qquad (2\text{-}70)$$

The latter procedure may be less desirable, however, because the comparison between $G^M$ and $G'$ provides a less direct indication of how to improve $G^T$ (than does the $G^* - G^T$ comparison) and is also less sensitive to details in $G^T$.

The notion of comparing the modified model to the uncorrected experimental result may be particularly useful when the experimental $G'(r)$ is not readily "unfolded" to a true function, as is the case when atoms with different scattering factor shapes are involved. It also applies to the use of operations (in $K$ space) which are purposely designed to decrease the sensitivity of the model comparison to certain portions of the data. For the latter purpose, $F(K)$ has sometimes been multiplied by an exp $(-\alpha^2 K^2)$ weighting function to decrease the (usually less accurate) high-$K$ oscillations in $F(K)$. The comparison between $F^T(K)$ and $F(K)$, with regard to $|\varepsilon(K)|$, as specified for the original procedure, may provide a more meaningful and a clearer basis for sensing the uncertainty in $G^T(r)$ due to uncertainty in $F(K)$, however. In any case, if the "modified" comparison is used for that purpose, one should use a weighting function more directly related to $|\varepsilon(K)|$, rather than an arbitrary Gaussian form.

In the foregoing discussion, it has been assumed that $G^T(r)$ was specified to sufficiently large $r$ and that the calculation of $F^T(K)$ [or $I^T(K)$] did not itself suffer from termination errors. It is worthwhile to point out, however, that one can forgo, at least initially, the $[F^T(K) - F(K)]/|\varepsilon(K)|$ comparisons (which are useful only for the secondary purpose of studying the uncertainty in the final derived $G^T$). In that case a $G^T(r)$ can be used that is purposely limited in $r$ space. The resultant $F^T(K)$ will be incomplete and not comparable to $F(K)$, but only to the corresponding portion contributed by the same region of the true $G(r)$. Nonetheless, the same procedure holds for the calculation of $G^*(r)$ [or $G^M(r)$]. The comparison of $G^*(r)$ to $G^T(r)$ [or of $G^M(r)$ to $G'(r)$] also holds—within the same limited range in $r$ space—providing only that termination effects due to features outside the considered range are negligible inside the range.

There are two additional effects which, like data termination, are also evidenced by spurious oscillations in the small-$r$ region. These are errors arising from

*1* Incorrect normalization (cf Section II B)
*2* Low-frequency (slowly varying) errors, either additive (as from an inaccurate subtraction of incoherent scattering) or multiplicative [as from use of an inaccurate $f^2(K)$]

Although less fundamental in nature, these effects are so intimately mixed into any careful analysis of pair-distribution data that they require mention here.

The misnormalization effect is nearly equivalent to the addition of a ramp $(\pm \varepsilon K)$ to $F(K)$ and superimposes a high-frequency oscillation on $G(r)$ ($r$ period $\sim 2\pi/K_{max}$), in an envelope which decreases rapidly as $r$ increases from zero. The first maximum in that oscillation, near $r = 0$, is sufficiently distinctive that an error in normalization of much less than 1% can be readily sensed and thereby corrected.

If there is an error of the form $\varepsilon(K)I(K)/f^2(K)$, where $I/f^2$ is the true ratio of intensity to scattering factor squared and $\varepsilon(K)$ is a *smooth* function of $K$, it will be evidenced mainly as a $\Delta G(r)$ deviation in the small-$r$ region. In instances where the nearest-neighbor peaks do not come below (say) 2 to 3 A it may be possible to see clearly and to separate the associated spurious oscillations as a deviation from $-4\pi r\rho_0$. The sine transform of $\Delta G(r)$ yields $K\varepsilon(K)$ approximately, and therefore provides a mechanism for removing or reducing the error (Kaplow, Strong, and Averbach, 1965).

## F. Atomic vibrations and other displacements

In this section, we discuss some of the geometrical considerations relating to the fact that the scattering material (electronic charge) associated with an average atomic scatterer is generally distributed over space. This distribution may be an inherent one, that is, the natural distribution which the electrons assume in the potential of the nucleus and due to their own interactions, as is described by their total spatial wave-function product $\psi\psi^*$. The product $\psi\psi^*$ is a time average in that it represents a statistical probability distribution. Such a smooth function is not what would be seen, in principle, in an *instantaneous* view of an atom in which the discreteness of individual electrons would persist. Rather, it is representative of what one would see as an average (over time) for one atom, or as the average over many different (but identical) atoms.

Because of dynamic motions, an atom, as an entity, can also be considered to be distributed over space when its time-averaged position is considered. In an instantaneous view of a sample, of course, each atom will exhibit not a distribution, but a single specific displacement. Again, the average statistical probability distribution will be seen as the long-term average at one atom, as well as in the sampling of displacements at different (but identical) atoms.

Perturbations which are static in nature (e.g., due to structure defects) may (but are less likely to) also give rise to a statistical probability distribution of small displacements from idealized (central) positions.

These different phenomena have in common the fact that they can be treated—in terms of scattering—as perturbations upon a simpler situation, in which each (atomic) scatterer is viewed as a point at an ideal location. What is shown in part in this section, in the framework of pair-distribution functions, is that, if the instantaneous distribution (or location) associated with each atom is entirely independent of that associated with any other and symmetric with respect to its own center, the spreading of scattering material in $r$ space is entirely accounted for by replacing the $k$-independent scattering probability of point scatterers by $k$-dependent functions. The latter functions turn out to be the Fourier transforms of the spatial distribution functions.

**1.   The r-space broadening functions**   For describing a (pure) material in the pair-distribution sense, consider the function $4\pi r^2 \rho(r)\, dr$, which has been defined previously as the number of scatter centers found in the spherical shell of radius $r$ and thickness $dr$ and centered on an average scatterer. In an idealized situation, one would write

$$4\pi r^2 \rho(r) = \sum_{i=1}^{\infty} c_i\, \delta(r - r_i) \qquad (2\text{-}71)$$

That is, the distribution function would consist of a set of $\delta$ functions, one for each of the allowed interscatterer distances and weighted by the number of scatterers at each such distance $c_i$. As shown earlier, the quantity which is more directly related to the diffracted intensity is $4\pi r \rho(r)$, so that it is preferable to consider the following relation:

$$4\pi r \rho(r) = \sum_{i=1}^{\infty} \frac{c_i}{r_i}\, \delta(r - r_i) \qquad (2\text{-}72)$$

Consider, first of all, the conditions under which it is appropriate to account for spatial spreading of the scatterer by using

$$4\pi r \rho(r) = \sum_{i=1}^{\infty} \frac{c_i}{r_i}\, B(r - r_i) \qquad (2\text{-}73)$$

where $B(\varepsilon)$ is not a $\delta$ function.

Referring to Fig. 2-22a, consider a point scatterer at the origin and an averaged, distributed scatterer centered at point $P$, $r_i$ away. The shell of radius $r$ and thickness $dr$ intersects the distribution about $P$, as shown, and includes contributions from that distribution $c(q)$ over the range $\varepsilon \leq q \leq R$ (where $\varepsilon = |r_i| - |r|$ and $R$ is the maximum radius of the distribution).

If $c(q)$ were constant [i.e., $c(q) = C$], then the distribution as seen from the origin $O$ would vary just as the intersection volume of the sphere at $P$ and the shell centered at $O$. The area of the intersected piece of the shell is $2\pi r h$, where $h = [R^2 - (r_i - r)^2]/2r_i$. One may then write the intersection volume

$$\Delta V = 2\pi r h\, \Delta r = \frac{\pi r}{r_i} (R^2 - \varepsilon^2)\, \Delta r \qquad (2\text{-}74)$$

If one demands that the sphere at $P$ contains $Z$ scatterers, the normalization

$$Z = \int_0^R c(q) 4\pi q^2\, dq \qquad (2\text{-}75)$$

requires that $C = 3Z/4\pi R^3$. If, in addition, the origin point $O$ sees $c_i$ equivalent

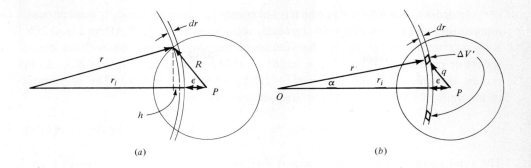

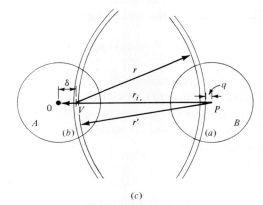

FIGURE 2-22
Geometrical drawings for consideration of the $r$-space broadening functions.

scattering centers at distance $r_i$, the total density contribution to the shell will be

$$4\pi r^2 \rho(r)\, \Delta r = c_i C\, \Delta V \tag{2-76a}$$

$$4\pi r \rho(r)\, \Delta r = \frac{c_i}{r_i}\, \frac{3Z}{4}\left(\frac{R^2 - \varepsilon^2}{R^3}\right) \Delta r \tag{2-76b}$$

$$4\pi r \rho(r) = \frac{c_i}{r_i}\, Q(\varepsilon) \tag{2-76c}$$

Equation (2-76b) does have the form given in equation (2-73); in particular, $Q(\varepsilon) = 3Z(R^2 - \varepsilon^2)/4R^3$ is a symmetric function of $\varepsilon$, and normalized.

$$Z = \int_{-R}^{R} \frac{3Z(R^2 - \varepsilon^2)}{4R^3}\, d\varepsilon \tag{2-77}$$

In order to consider a $c(q)$ which is not constant, but still spherically symmetrical, one requires an expression for the (toroidal) volume element, called $\Delta V'$ in Fig. 2-22b. Referring to the earlier result, the volume of the shell intersection with an entire sphere of radius $q = R$ was $\Delta V = \pi r(R^2 - \varepsilon^2)\,\Delta r/r_i$. If the sphere was $R + \Delta q$ in radius, the volume would be $\pi r(R^2 + 2R\,\Delta q + \Delta q^2 - \varepsilon^2)\,\Delta r/r_i$. The difference, in the limit of small $\Delta q$, is the required volume element:

$$\Delta V' = \frac{2\pi r}{r_i}\, q\,\Delta q\,\Delta r \qquad (2\text{-}78)$$

The total contribution to the spherical shell centered on point $O$ (of radius $r$ and thickness $\Delta r$) may then be derived by multiplying the volume element by the density $[c(q)]$ and integrating to include the entire intersection with the spherical shell.

$$4\pi r^2 \rho(r) = \frac{rc_i}{r_i}\int_\varepsilon^{R_{max}} 2\pi q c(q)\,dq \qquad (2\text{-}79)$$

where, again, a multiplier $c_i$ representing the "coordination number" has been used. The normalization condition for $c(q)$, as given in equation (2-75), is retained.

The integral in equation (2-79) gives the detailed relationship between the distribution centered on point $P$ and the broadening function $Q(\varepsilon)$. That is, if the origin scatterer is still retained as a point,

$$A(\varepsilon) = \int_\varepsilon^{R_{max}} 2\pi q c(q)\,dq \qquad (2\text{-}80)$$

It may be verified that if $c(q)$ is a constant, the earlier result [equation (2-76b)] is obtained directly from equation (2-80).

It is more interesting, however, to consider what particular distribution $c(q)$ is associated with a Gaussian-shaped $Q(\varepsilon)$, a form which seems to have special significance. That is, if $Q(\varepsilon) = \exp(-\varepsilon^2/2\sigma^2)/(2\pi\sigma^2)^{1/2}$, what is $c(q)$? In fact, it may be seen by inspection that

$$\int_\varepsilon^\infty 2\pi q\,\frac{\exp(-q^2/2\sigma^2)}{(2\pi\sigma^2)^{3/2}}\,dq = \frac{\exp(-\varepsilon^2/2\sigma^2)}{(2\pi\sigma^2)^{1/2}} \qquad (2\text{-}81)$$

Therefore, the associated $c(q)$ is a three-dimensional, spherically symmetric, normalized Gaussian!

$$c(q) = \frac{\exp(-q^2/2\sigma^2)}{(2\pi\sigma^2)^{3/2}} \qquad (2\text{-}82)$$

In terms of the separate $x$, $y$, $z$ coordinates,

$$c(x, y, z) = \frac{\exp(-x^2/2\sigma^2)\exp(-y^2/2\sigma^2)\exp(-z^2/2\sigma^2)}{(2\pi\sigma^2)^{3/2}} \qquad (2\text{-}83)$$

Note that the $\sigma$'s which appear in these expressions are the measure of the mean-square displacement along one particular coordinate. The mean-square displacement from the central point $\sigma_T{}^2$ is three times as large

$$\sigma_T{}^2 = 3\sigma^2 \qquad (2\text{-}84)$$

It is beyond the purpose of this discussion to examine the extent to which equations (2-82) and (2-83) may be shown to be a formal representation of thermally activated displacements of individual atoms from their equilibrium positions. Perhaps it is sufficient to note that the envelope of the displacement probability function for a set of quantum-mechanical oscillators *is* a Gaussian function. From the experimental point of view it may be stated, moreover, that the form does account well for measured intensity variations due to thermal vibrations apart from those aspects which arise from anisotropy and/or coupled motions.

In the foregoing, the atom at the origin has been kept as a geometrical point. It is presumably also spread in space on the average, however. The total broadening effect on a particular pair will then be the *convolution* of the two independent displacement distributions, providing that the displacements of the one are not in any way influenced by those of the other atom. Refer to Fig. 2-22c which shows two atoms, labeled $A$ and $B$, centered at points $O$ and $P$, respectively, with equilibrium separation $r_i$. Using the results obtained earlier, note that the contribution to $4\pi r^2 \rho(r)$ of spherical shell segment $a$ in atom $B$ as drawn from point $V$ in atom $A$ is $rQ_B(r' - r) \Delta r/r'$. Moreover, all the points in segment $b$ (i.e., in the shell at $r'$ from point $P$) are equivalent in that regard, since a radius-$r$ sphere centered anywhere in $b$ will have identical intersections with atom $B$. Different points in segment $b$ will have different weights, of course, depending on $c^A(q)$ and their distance from point $O$. However, the total weight of segment $b$ is already known; it is just $r'Q_A(r_i - r') \Delta r'/r_i$. Thus the joint probability for finding a distance $r$ is

$$\text{j.p.} = \int_{-\infty}^{+\infty} \frac{r}{r'} Q_B(r' - r) r \frac{r'}{r_i} Q_A(r_i - r') \, dr' \qquad (2\text{-}85)$$

Letting $r' - r \equiv q$; $r_i - r \equiv \varepsilon$; $r_i - r' = \varepsilon - q$, the joint probability then may be written

$$\text{j.p.} = \frac{r}{r_i} \Delta r \int_{-\infty}^{+\infty} Q_B(q) Q_A(\varepsilon - q) \, dq \qquad (2\text{-}86)$$

It is useful to define the convolution as $B(\varepsilon)$:

$$B_{AB}(\varepsilon) \equiv \int_{-\infty}^{+\infty} Q_B(q) Q_A(\varepsilon - q) \, dq \qquad (2\text{-}87)$$

The subscripts $A$ and $B$ are retained as a reminder that the two atoms in a pair need not be identical. At least one special case of equation (2-87) bears detailing, that for Gaussian-shaped $Q$'s.

$$\frac{\exp\left(-\varepsilon^2/2\sigma^2\right)}{2\pi\sigma^2} = \int_{-\infty}^{+\infty} \frac{\exp\left(-q^2/2\sigma_B^2\right)}{2\pi\sigma_B^2} \frac{\exp\left(-(\varepsilon-q)^2/2\sigma_A^2\right)}{2\pi\sigma_A^2} \, dq \qquad (2\text{-}88)$$

where $\sigma^2 = \sigma_A^2 + \sigma_B^2$. Thus, if the individual displacements give rise to a Gaussian shape [in $4\pi r\rho(r)$], the convolved effect for a pair is also Gaussian, with a $\sigma^2$ which is the sum of the two, provided the individual displacements are independent.

**2. The general effects in K space**  It is now worthwhile to consider the detailed way in which the distribution $B(\varepsilon)$ alters the intensity of scattering from that which would obtain with point scatterers. Reviewing the latter, in terms of the reduced intensity function $F(K)$,

$$F^0(K) = \int_0^\infty 4\pi r\rho(r) \sin Kr \, dr \qquad (2\text{-}89)$$

Note that the (average density) term $\rho_0$ has not been subtracted; i.e., $F^0(K)$ includes the "zero-angle" scattering. For point scatterers one can write

$$F^0(K) = \int_0^\infty \left[ \sum_{i=1}^\infty \frac{c_i}{r_i} \delta(r - r_i) \right] \sin Kr \, dr \qquad (2\text{-}90a)$$

$$F^0(K) = \sum_{i=1}^\infty \frac{c_i}{r_i} \int_0^\infty \delta(r - r_i) \sin Kr \, dr \qquad (2\text{-}90b)$$

$$F^0(K) = \sum_{i=1}^\infty \frac{c_i}{r_i} \sin Kr_i \qquad (2\text{-}90c)$$

Following the same path with "broadened" shells [and allowing the $B(\varepsilon)$'s to be shell dependent]

$$F(K) = \int_0^\infty \left[ \sum_{i=1}^\infty \frac{c_i}{r_i} B_i(r - r_i) \right] \sin Kr \, dr \qquad (2\text{-}91a)$$

$$F(K) = \sum_{i=1}^\infty \frac{c_i}{r_i} \int_0^\infty B_i(r - r_i) \sin Kr \, dr \qquad (2\text{-}91b)$$

Letting $\varepsilon \equiv r - r_i$ and $r = \varepsilon + r_i$, then

$$F(K) = \sum_{i=1}^\infty \frac{c_i}{r_i} \int_{-r_i}^\infty B_i(\varepsilon) \sin (K\varepsilon + Kr_i) \, d\varepsilon \qquad (2\text{-}92)$$

Assuming that $B(\varepsilon)$ is negligible for $\varepsilon > r_i$, the lower limit on the integral may be set at $-\infty$.

$$F(K) = \sum_{i=1}^\infty \frac{c_i}{r_i} \left[ \int_{-\infty}^\infty B_i(\varepsilon) \cos Kr_i \sin K\varepsilon \, d\varepsilon + \int_{-\infty}^\infty B_i(\varepsilon) \sin Kr_i \cos K\varepsilon \, d\varepsilon \right] \qquad (2\text{-}93)$$

If the $B_i(\varepsilon)$ are symmetric, the first integral is identically zero. Then

$$F(K) = \sum_{i=1}^{\infty} \frac{c_i}{r_i} \int_{-\infty}^{+\infty} \sin Kr_i B_i(\varepsilon) \cos K\varepsilon \, d\varepsilon \qquad (2\text{-}94a)$$

$$F(K) = \sum_{i=1}^{\infty} \frac{c_i}{r_i} \sin Kr_i \left[ \int_{-\infty}^{+\infty} B_i(\varepsilon) \cos K\varepsilon \, d\varepsilon \right] \qquad (2\text{-}94b)$$

In general, the integral in (2-94b), which is the Fourier transform of the broadening function, may be defined as follows:

$$V_i(K) \equiv \int_{-\infty}^{+\infty} B_i(\varepsilon) \cos K\varepsilon \, d\varepsilon \qquad (2\text{-}95)$$

Considering the special Gaussian case again, note that the corresponding $V(K)$ has a particularly simple form.

$$e^{-\sigma^2 K^2/2} = \int_{\infty}^{\infty} \frac{\exp(-\varepsilon^2/2\sigma^2)}{(2\pi\sigma^2)^{1/2}} \cos K\varepsilon \, d\varepsilon \qquad (2\text{-}96)$$

Therefore, the reduced intensity function for a sample in which the atoms experienced Gaussian displacements from their central positions might be written as

$$F(K) = \sum_{i=1}^{\infty} \frac{c_i}{r_i} \sin Kr_i \exp\left(\frac{-\sigma_i^2 K^2}{2}\right) \qquad (2\text{-}97)$$

In other words, in the present approximation, for which each type of pair exhibits a specific Gaussian-shaped distribution of distances about its mean, the individual sine-wave contributions to the reduced intensity function are damped; each one, in principle, is damped by its own characteristic damping function. A good deal of generality is retained through keeping the $V(K)$'s within the sum, particularly if the "shell" concept is enlarged so that different subscripts are used for pairs which are not structurally equivalent and/or not of the same species combination, even if the $r_i$ happen to be identical. Thereby, one may include, in principle, effects attributable to

*1* Differences among atom types
*2* Differences among structurally nonequivalent sites
*3* Anisotropy in displacements
*4* Correlated displacements between neighboring atoms

Regarding item (3), above, it is important to note that the local structural relationships are retained, even though the experiment may involve a macroscopic spherical average. For example, the vector between nearest neighbors in a face-centered-cubic lattice is a [110] direction. The displacements seen in a pair distribution which relate to nearest neighbors are displacements along the radial vector between nearest neighbors; i.e., along [110] directions. This means that the fact that the average includes other directions is irrelevant for $r_i \approx r_1$, since there are no other pairs at that distance.

Regarding item (4) above, note that (for identical vibrators) the joint $\sigma^2$ will be twice that for an individual atom examined with respect to its own center [cf equation (2-88)], if they move independently. That value may be called $\sigma_\infty^2$, since independence is certain for infinitely separated atoms. To the extent that the atoms in a pair move together, $\sigma_i^2$ will be smaller. If they are correlated, but with opposite sense, $\sigma_i^2$ will be larger.

**3.  *Thermal diffuse scattering in polycrystalline aggregates*** On the presumption that the $\sigma_i^2$ will approach $\sigma_\infty^2$ at fairly short distances, it is advantageous to introduce a relative quantity $\gamma_i \equiv \sigma_i^2/\sigma_\infty^2$, in terms of which the reduced intensity may be rewritten

$$F(K) = \sum_{i=1}^{\infty} \frac{c_i}{r_i} \sin Kr_i \, e^{-\gamma_i \sigma_\infty^2 K^2/2} \tag{2-98a}$$

$$F(K) = \sum_{i=1}^{\infty} \frac{c_i}{r_i} \sin Kr_i (e^{-\sigma_\infty^2 K^2/2} - e^{-\sigma_\infty^2 K^2/2} + e^{-\gamma_i \sigma_\infty^2 K^2/2}) \tag{2-98b}$$

$$F(K) = e^{-\sigma_\infty^2 K^2/2} \sum_{i=1}^{\infty} \frac{c_i}{r_i} \sin Kr_i + \sum_{i=1}^{\infty} \frac{c_i}{r_i} (e^{-\gamma_i \sigma_\infty^2 K^2/2} - e^{-\sigma_\infty^2 K^2/2}) \sin Kr_i \tag{2-98c}$$

where $0 \le \gamma_i \le 2$.

The first term [cf equation (2-90c)] is just $F^0(K)e^{-\sigma_\infty^2 K^2/2}$, that is, the idealized reduced intensity, but damped by the exponential factor $\exp(-\sigma_\infty^2 K^2/2)$. For a pure crystalline material, the $F^0(K)$ term $\sum_{i=1}^{\infty} c_i/r_i \sin Kr_i$ represents the Bragg peaks; i.e., the sharp diffraction "lines" in a polycrystalline experiment which correspond to the spherically averaged sampling of the intensity maxima at reciprocal-lattice points that would describe a single-crystal diffraction pattern. It should be noticed, however, that multiplying the reduced intensity $F^0(K)$ by $e^{-\sigma_\infty^2 K^2/2}$ implies more than just a damping of the *intensity* peaks for a crystalline sample. In terms of intensity, the first term in equation (2-98c) yields

$$\left(\frac{I}{f^2} - 1\right) K = \left(\frac{I^0}{f^2} - 1\right) Ke^{-\sigma_\infty^2 K^2/2} \tag{2-99a}$$

where $f(K)$ is the $K$-dependent scattering amplitude of a finite-size (but stationary) atom; $f(K)$ has been referred to earlier and will be treated again at a later point in the present section.

$$I = I^0 e^{-\sigma_\infty^2 K^2/2} + f^2(1 - e^{-\sigma_\infty^2 K^2/2}) \tag{2-99b}$$

The first term in (2-99b) represents the damped Bragg peaks for a crystalline sample. In that context, the $e^{-\sigma_\infty^2 K^2/2}$ term is usually called the *Debye-Waller factor*, and written as $e^{-2M}$. The second term is diffuse, and is known as the *Debye thermal diffuse scattering*. As may be seen, it arises from the $-K$ contribution to $F(K)$ (in a pure material), which is due in turn to the self-scattering of an atom with respect to itself, which always contributes just $f^2$ to the total intensity pattern (in electron units). Since an atom cannot be displaced with respect to itself, the second term in (2-99b) can be viewed as the first step in allowing for correlations among the motions; it is the zeroth-order term; i.e., for $r_0 = 0$, $C_0 = 1$, and $\gamma_0 = 0$.

For a polycrystalline sample, it is therefore sometimes convenient to rewrite equation (2-98c) as a relative intensity $I/f^2$ rather than as a reduced intensity $F(K)$; this allows a somewhat natural separation into two summations

$$\frac{I}{f^2} = e^{-2M} \sum_{i=0}^{\infty} \frac{c_i}{Kr_i} \sin Kr_i + \sum_{i=0}^{\infty} \frac{c_i}{Kr_i} (e^{-2\gamma_i M} - e^{-2M}) \sin Kr_i \qquad (2\text{-}100)$$

where $2M \equiv \sigma_\infty^2 K^2/2$.

$$\frac{I}{f^2} = \frac{I^0}{f^2} e^{-2M} + \sum_{i=0}^{\infty} \frac{c_i}{Kr_i} (e^{-2\gamma_i M} - e^{-2M}) \sin Kr_i \qquad (2\text{-}101)$$

The second term in (2-100) is usually referred to as the *thermal diffuse scattering* intensity $I_{TDS}$.

$$\frac{I_{TDS}}{f^2} = \sum_{i=1}^{\infty} \frac{c_i}{Kr_i} (e^{-2\gamma_i M} - e^{-2M}) \sin Kr_i \qquad (2\text{-}102)$$

If all $\gamma$'s are equal to 1.0 except $\gamma_0$, $I_{TDS}$ reduces to the simplified form shown in equation (2-99b). In any case, it is fairly diffuse in $K$ space, because its main contributors are the near neighbors. The $\gamma_i$ can be as small as zero (as is $\gamma_0$) corresponding to fully correlated, in-phase motion, or as large as 2.0, corresponding to out-of-phase motion. The latter situation is not likely to occur, but very small $\gamma$'s are plausible for near neighbors. In particular, a molecular substance with strong covalent intramolecular bonds and weak Van der Waals intermolecular forces would exhibit $\sigma_i^2 \ll \sigma_\infty^2$, for small $i$.

Figure 2-23 shows the experimental diffuse intensity, measured from a polycrystalline copper sample at 900°C [$T$(melting point) = 1080°C] with Cu $K\alpha$ x radiation ($\lambda$ = 1.54 A). The data include the incoherent Compton-modified contribution, but they have not been corrected for polarization. Both of these cause only very smooth variations. These data indicate better than any verbal description might, the degree to which the sharp peak intensities, represented by the first term in equation (2-101), are separable from the second or TDS term. (The separation is less clear-cut at higher $K$ values, where the $e^{-2M}$ factor is larger.) The smooth-drawn curve represents the anticipated diffuse intensity variation if all $\gamma_i$'s beyond the zeroth were equal to 1.0; i.e., if $I_{TDS}/f^2 = (1 - e^{-2M})$. It may be seen that the actual

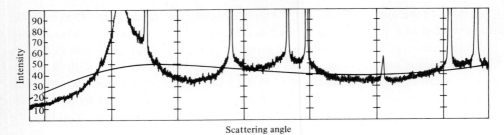

FIGURE 2-23
Experimental diffuse intensity pattern for polycrystalline copper at 900°C.

diffuse scattering modulates that independent vibration approximation. Rewriting (2-102) to show that more explicitly

$$\frac{I_{TDS}}{f^2} = (1 - e^{-2M}) \sum_{i=0}^{\infty} \frac{c_i}{Kr_i} \sin Kr_i \left( \frac{e^{-\gamma_i 2M} - e^{-2M}}{1 - e^{-2M}} \right) \qquad (2\text{-}103)$$

The modulating function may be defined:

$$W(K) \equiv \sum_{i=0}^{\infty} \frac{c_i}{Kr_i} \sin Kr_i \left( \frac{e^{-\gamma_i 2M} - e^{-2M}}{1 - e^{-2M}} \right) \qquad (2\text{-}104)$$

such that

$$\frac{I_{TDS}}{f^2} = W(K)(1 - e^{-2M}) \qquad (2\text{-}105)$$

It may be useful to consider the thermal diffuse scattering alone, in terms of a reduced function $F_{TDS}(K)$, defined analogously to the total reduced intensity.

$$F_{TDS}(K) = \left[ \frac{I_{TDS}}{f^2(1 - e^{-2M})} - 1 \right] K \qquad (2\text{-}106)$$

$$F_{TDS}(K) = [W(K) - 1]K \qquad (2\text{-}107)$$

$$F_{TDS}(K) = \sum_{i=1}^{\infty} \frac{c_i}{r_i} \sin Kr_i Q_i(K) \qquad (2\text{-}108)$$

Here,

$$Q_i(K) = \frac{e^{-\gamma_i 2M} - e^{-2M}}{1 - e^{-2M}} \qquad (2\text{-}109)$$

The extreme similarity in form of equation (2-108) and the more general equation (2-94b) leads one to wonder whether the coupling factors $\gamma_i$ can be retrieved from a direct analysis of $F_{TDS}(K)$, assuming that the experimental $I_{TDS}$ can be separated. To this purpose, note that by taking the sine transform of $F_{TDS}(K)$, one can simply reverse the derivation done earlier in this section [cf equations (2-91)] to show the following:

$$\frac{2}{\pi} \int F_{TDS}(K) \sin Kr \, dK = \sum_{i=1} \frac{c_i}{r_i} \sin Kr_i B_i'(r - r_i) \qquad (2\text{-}110)$$

where $B_i'$ is defined to be the Fourier transform of $Q_i(K)$, which is defined in (2-109). Second, note that a rather good approximation to $Q_i(K)$ is given by

$$Q_i(K) \equiv \frac{e^{-\gamma_i u^2 k^2} - e^{-u^2 k^2}}{1 - e^{-u^2 k^2}} \approx (1 - \gamma_i) e^{-b(\gamma_i) u^2 k^2} \qquad (2\text{-}111)$$

where $b(\gamma_i) = 0.56\gamma_i$, $u^2 \equiv \sigma_\infty^2/2$, $u^2 k^2 = 2M$. Therefore,

$$B_i'(r - r_i) \approx (1 - \gamma_i) \frac{\exp\left[-(r - r_i)^2/2q_i^2\right]}{(2\pi q_i^2)^{1/2}} \qquad (2\text{-}112)$$

where $q_i^2 \equiv 1.12\gamma_i u^2$.

It would appear, therefore, that if the $I_{TDS}$ can be separated from the total scattering, to sufficiently large $K$ values in polycrystalline samples, a sine transform of the reduced TDS intensity [cf equation (2-107)] will yield a sequence of Gaussian-shaped peaks in $r$ space. These will be located at the respective shell distances $r_i$, and their *areas* will be $(1 - \gamma_i)$.

It is necessary to note that $W(K)$ will include a small-angle component, analogous to the $I^z$ term discussed earlier with respect to the total intensity. Again this will be difficult to measure in ordinary (large-angle) experiments, because of interference from the direct beam and because of the low intensities involved (i.e., $I_{TDS}$ approaches zero at $K = 0$). As with the total intensity, it is probably best to ignore the small-angle component altogether [i.e., extrapolate to $W(K) = 1.0$ at $K = 0$] and to subtract the associated contribution in the $r$-space function. By analogy to the discussion which related to the $I^z$ component in small particles, one expects that the appropriate $F_{TDS}'(K)$ (without the zero component) can be expressed, approximately, as follows:

$$F_{TDS}'(K) = \int_0^\infty 4\pi r [\rho(r) - \rho_0][1 - \gamma(r)] \sin Kr \, dr \qquad (2\text{-}113)$$

where $F_{TDS}'(K) \equiv [W'(K) - 1]K$, and $W'(K)$ excludes the small-angle portion by extrapolation, and $4\pi r \rho(r)[1 - \gamma(r)]$ is the summation given in (2-110). From equation (2-113), one would expect that the transformation in $r$ space would exhibit peaks superimposed on a negative, slowly varying $\rho_0$ term $-4\pi r \rho_0[1 - \gamma(r)]$, where $\gamma(r)$ would be a smoothed function drawn through the $\gamma_i$ values.

In the last analysis, there may be no real advantage in utilizing the particular aspects of a polycrystalline sample which led to the foregoing derivations, and the attendant separation of peak and diffuse intensity. It is feasible to analyze the reduced total intensity, including Bragg peaks, and to reduce it to a total pair-distribution

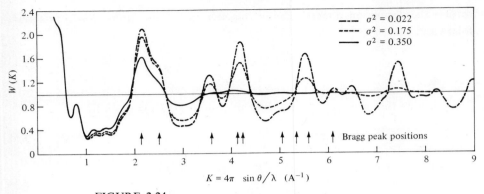

FIGURE 2-24
Calculated thermal diffuse modulating functions $W(K)$ for fcc lead.

function as has been described in general. The resulting $r$-space function, for a poly-crystalline sample, is interpretable as a series of peaks [cf equation (2-73)] located at $r_i$, of area $c_i$, and whose shapes are the respective $B(r - r_i)$. Such an analysis does not require the approximations which led to the simple forms for $W(K)$ in this section; any $B(r - r_i)$ should be reproduced in $4\pi r \rho(r)$. Indeed, use of a polycrystalline sum form for $4\pi r \rho(r)$ is then only a convenience for the interpretation of the $r$-space result.

In Fig. 2-24, three $W(K)$ functions are shown, calculated from empirical values for the $\gamma_i$ values for metallic lead. Only the first twenty-nine shell contributions were included; i.e., all higher-order $\gamma$'s were undetermined and were set equal to 1.0. These values, listed in Table 2-1, were determined by smoothing values obtained from

Table 2-1   SHELL PARAMETERS FOR CRYSTALLINE LEAD USED IN THE CALCULATION OF $W(K)$ AS GIVEN IN FIG. 2-24. (VALUES APPLY TO 655°C)

| $r_i$ | $c_i$ | $\gamma_i$ | $r_i$ | $c_i$ | $\gamma_i$ |
|---|---|---|---|---|---|
| 3.54 | 12 | 0.50 | 14.62 | 48 | 0.93 |
| 5.01 | 6 | 0.63 | 15.04 | 30 | 0.93 |
| 6.14 | 24 | 0.69 | 15.54 | 72 | 0.94 |
| 7.09 | 12 | 0.73 | 15.85 | 24 | 0.95 |
| 7.93 | 24 | 0.75 | 16.24 | 48 | 0.95 |
| 8.68 | 8 | 0.78 | 16.63 | 24 | 0.96 |
| 9.38 | 48 | 0 81 | 17.00 | 48 | 0.96 |
| 10.03 | 6 | 0.83 | 17.36 | 8 | 0.97 |
| 10.63 | 36 | 0.83 | 17.73 | 84 | 0.97 |
| 11.21 | 24 | 0.86 | 18.08 | 24 | 0.98 |
| 11.76 | 24 | 0.87 | 18.42 | 96 | 0.98 |
| 12.28 | 24 | 0.88 | 18.76 | 48 | 0.98 |
| 12.78 | 72 | 0.90 | 19.09 | 24 | 0.99 |
| 13.73 | 48 | 0.92 | 19.74 | 96 | 0.99 |
| 14.18 | 12 | 0.92 | | | |

a fit to the total $4\pi r[\rho(r) - \rho_0]$, allowing the $\gamma$'s to be trial-and-error parameters (Kaplow, Strong, and Averbach, 1964).

The data were obtained at 655°C, just below the melting point, to which temperature the $\gamma$'s therefore apply. The three $W(K)$ correspond to mean-square amplitudes of relative vibration of $\sigma_\infty^2 = 0.022$, $0.175$, and $0.350$ A$^2$. Of these, only the first two are small enough to be physically realistic ($0.175$ A$^2$ is the experimental value at the melting point). It is readily obvious that $W(K)$ is not independent of temperature when $u^2K^2$ is not small, even when the $\gamma_i$'s are all held constant. It is also fairly clear, from a comparison of Figs. 2-23 and 2-24, that terms beyond the twenty-ninth shell must contribute somewhat to the sharpness seen in the experimental modulations.

Since the thermal diffuse scattering does peak up so strongly under Bragg peaks, it is a particular annoyance in applications which do depend critically on being able to measure peak intensities separately. In Chap. 7, therefore, the thermal scattering is discussed in some detail for single-crystal samples. There it is treated as a correction for integrated intensity measurements.

To further illustrate the sort of results that one should expect for polycrystalline samples, experimental reduced intensity functions $F(K)$ are shown schematically in Fig. 2-25 for aluminum at 25 and 655°C (Fessler, Kaplow, and Averbach, 1966). The corresponding atomic-distribution functions $G(r)$ are shown in Fig. 2-26. Calculated fits of the form given in equation (2-73) are also shown, with Gaussian-shaped broadening functions and $\sigma_\infty^2 = 0.022$ and $0.0952$ A$^2$, respectively.

**4. Anharmonic vibrations and asymmetric motions** For crystalline materials, one anticipates slight asymmetries in the peaks, particularly for nearest neighbors and for very high temperatures compared to the bonding energy. This is expected because of the known anharmonicity of the atomic vibrations in crystals (which leads to thermal expansion and other effects). Similar effects are anticipated for amorphous materials, particularly those (if any) which are not constituted of strong molecular units. In liquids, the nearest-neighbor peak is apt to be very asymmetric (again provided the structure is not strongly molecular). In this instance, however, the distortion is probably caused by diffusive rather than vibrational motions. It may be of some value to reconsider the question of the symmetry of $B(r - r_i)$, as a general matter, for noncrystalline as well as crystalline samples. For equation (2-94b), it was assumed that $B_i(\varepsilon)$ was symmetric and that

$$\int_{-\infty}^{+\infty} B_i(\varepsilon) \cos Kr_i \sin K\varepsilon \, d\varepsilon \qquad (2\text{-}114)$$

was therefore identically equal to zero. If that is *not* the case, then each shell with an asymmetric $B_i(\varepsilon)$ contributes an additional term to $F(K)$.

$$\frac{c_i}{r_i} \cos Kr_i \int_{-\infty}^{+\infty} B_i(\varepsilon) \sin K\varepsilon \, d\varepsilon \qquad (2\text{-}115)$$

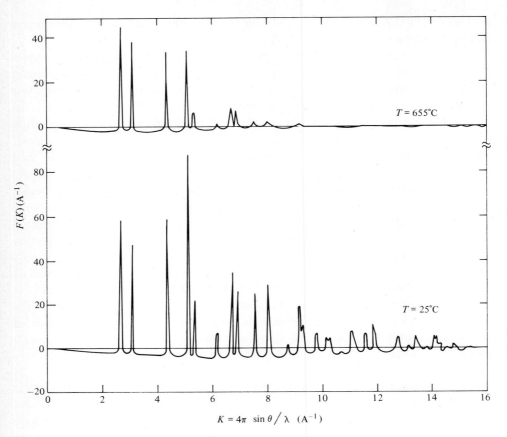

**FIGURE 2-25**
Experimental reduced intensity functions $F(K)$ for polycrystalline aluminum at 25 and 655°C (Fessler, Kaplow, and Averbach, 1966).

That is, in addition to the familiar $\sin Kr_i$ term [damped by the cosine transform of $B_i(\varepsilon)$], there will be a $\cos Kr_i$ [modulated by the sine transform of $B_i(\varepsilon)$]. To get a feeling for the effect, we might assume, for example, that $B_i(\varepsilon)$ can be separated into odd and even functions of $\varepsilon$, and that the odd part looks like two $\delta$ functions of weight $q$ located at $\pm\Delta$ from the center of the peak:

$$^{\text{ODD}}B_i(\varepsilon) = q[-\delta(-\Delta) + \delta(+\Delta)] \qquad (2\text{-}116)$$

This would yield

$$\int_{-\infty}^{\infty} {}^{\text{ODD}}B_i(\varepsilon) \sin K\varepsilon \, d\varepsilon = 2q \sin K\Delta \qquad (2\text{-}117)$$

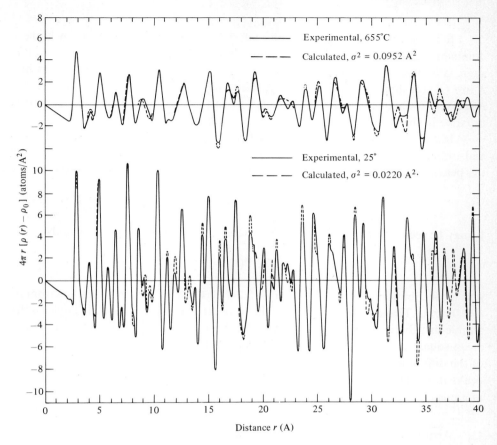

**FIGURE 2-26**
Atomic pair distributions $G(r)$ for polycrystalline aluminum at 25 and 655°C.

More realistically, perhaps, one might assume that the asymmetry could be represented better by broadened $\delta$ functions; using Gaussian shapes, we obtain for an $i$th-shell contribution:

$$^{ODD}B_i(\varepsilon) = \frac{q}{(2\pi\sigma_{0,i}^2)^{1/2}} \left\{ \exp\left[\frac{-(\varepsilon - \Delta)^2}{2\sigma_{0,i}^2}\right] - \exp\left[\frac{-(\varepsilon + \Delta)^2}{2\sigma_{0,i}^2}\right] \right\} \qquad (2\text{-}118)$$

This yields, for the contribution to $F(K)$ of the $i$th term,

$$\frac{c_i}{r_i} \cos Kr_i 2q \sin K\,\Delta \exp\left(\frac{-K^2}{2\sigma_{0,i}^2}\right) \qquad (2\text{-}119)$$

One would expect that $\sigma_{0,i}^2 < \sigma_i^2$, if there were to be an asymmetry, so that the asymmetry contribution would dominate at large $K$.

It is also possible that in some materials the nearest-neighbor term alone is dominant at high $K$ (e.g., for some amorphous solids, liquids, and polycrystalline aggregates at high temperature). Thus, one might seek a cos $Kr_i$ oscillation at high $K$ in $F(K)$ for direct evidence of an asymmetry in $B_i(r - r_i)$. Note in this regard that $\Delta$ is likely to be of the order of 0.2 to 0.4 A, so that the sin $K\,\Delta$ term is likely to be negative in the achievable high-$K$ region, $K \approx 10 - 16$ A.

It is also interesting to consider what is obtained if, having separated the "peak" and diffuse components of the intensity [cf equation (2-101)], one transforms only the peaks (in reduced form) to $r$ space. Analogous to $F(K)$, one could consider

$$F^{\text{peak}}(K) \equiv \frac{Ie^{-2M}}{f^2} K \qquad (2\text{-}120)$$

It may be seen immediately that the associated distribution function is simply

$$G^{\text{peak}}(r) = \sum_{i=1}^{\infty} \frac{c_i}{r_i} \exp\left[\frac{-(r - r_i)^2}{2\sigma^2}\right] \qquad (2\text{-}121)$$

plus a contribution near $r = 0$, $[\exp(-r^2/2\sigma^2) \int K \sin Kr\, dK]$, which arises from the $i = 0$ term. That is, the transform of the peak intensities alone essentially yields that atomic-distribution function which would obtain if the atoms vibrated with a given mean-square amplitude, but entirely independently of one another. Thus, inclusion of the diffuse intensity in the total (ordinary) data reduction *sharpens* the correlation peaks in $r$ space, and the difference is obviously due to the correlated nature of the motions.

## G.  Atomic charge distributions; the scattering factor

Referring to the result shown in equations (2-94b) and (2-95), valid for symmetrical and uncorrelated individual atomic broadening effects,

$$F(K) = \sum_{i=1}^{\infty} \frac{c_i}{r_i} V_i^{AB}(K) \sin Kr_i \qquad (2\text{-}122)$$

where

$$V_i^{AB}(K) = \int_{-\infty}^{+\infty} B^{AB}(\varepsilon) \cos K\varepsilon\, d\varepsilon$$

$$B^{AB}(\varepsilon) = \int_{-\infty}^{+\infty} Q_A(q) Q_B(\varepsilon - q)\, dq$$

$$Q_A = \int_{\varepsilon}^{\infty} 2\pi q c^A(q)\, dq$$

$$Q_B = \int_{\varepsilon}^{\infty} 2\pi q c^B(q)\, dq$$

and $c^A(q)$ and $c^B(q)$ are the distributions of scattering material, with respect to the atom centers of each of the (possibly different) atoms, $A$ or $B$, associated with a particular $i$th $A - B$ type pair, separated by $r_i$.

Avoiding the algebra this time, and utilizing the well-known fact that the transform of the convolution of two functions is the product of the transforms, one can write, at once,

$$V^{AB}(K) = f^A(K)f^B(K) \qquad (2\text{-}123)$$

where

$$f^A(K) = \int_{-\infty}^{+\infty} Q_A(\varepsilon) \cos K\varepsilon \, d\varepsilon$$

$$f^B(K) = \int_{-\infty}^{+\infty} Q_B(\varepsilon) \cos K\varepsilon \, d\varepsilon$$

From the above definitions,

$$f^A(K) = \int_{-\infty}^{+\infty} \cos K\varepsilon \left[ \int_{\varepsilon}^{\infty} 2\pi q c^A(q) \, dq \right] d\varepsilon \qquad (2\text{-}124)$$

Defining $G(q)/2 \equiv 2\pi q c(q)$ and integrating by parts, over $\varepsilon$ [with $u = \int_{\varepsilon}^{\infty} (G(q)/2) \, dq$ and $dv = \cos K\varepsilon \, d\varepsilon$; $du = -G(\varepsilon) \, d\varepsilon/2$ and $v = \sin K\varepsilon/K$], one gets

$$f^A(K) = 2\left[ \int_{\varepsilon}^{\infty} \frac{G(q)}{2} \, dq \, \frac{\sin K\varepsilon}{K} \right]_0^{\infty} + 2 \int_0^{\infty} \frac{G(\varepsilon)}{2} \frac{\sin K\varepsilon}{K} \, d\varepsilon \qquad (2\text{-}125)$$

The first term in equation (2-125) is identically zero at both limits, which leaves

$$f^A(K) = \int_0^{\infty} 4\pi q^2 c^A(q) \frac{\sin Kq}{Kq} \, dq \qquad (2\text{-}126)$$

If $c(q)$ describes the spherically averaged distribution of charge associated with an atom, then $f(K)$ is the atomic scattering factor, a concept which has been utilized in earlier discussions. Thus the contribution to the *intensity* from a shell consisting of $A - B$ pairs at $r_i$ would be

$$f^A(K)f^B(K) \frac{c_i}{Kr_i} \sin Kr_i \qquad (2\text{-}127)$$

if there were no broadening effects other than the inherent charge distribution, and if there were no correlation effects of the sort which arose in the vibration context and which led to the use of the $\gamma_i$ coupling coefficients. Note, in addition, that the properties of the Fourier transform with respect to convolutions allow account to be taken of other simultaneous broadening effects, such as vibrations, for example, by simply multiplying by the appropriate $K$-space function.

Therefore

$$I = \sum_{i=0}^{\infty} f^A(K)f^B(K) \exp\left(\frac{-\gamma_i\sigma_{(A,B)}^2 K^2}{2}\right) \frac{c_i}{Kr_i} \sin Kr_i \quad (2\text{-}128)$$

where $f^A(K)$ and $f^B(K)$ are, respectively, the scattering factors for each of the atoms in an $i$th-type pair, and $\gamma_i\sigma_{(A,B)}^2$ is the mean-square amplitude of relative vibration for such pairs.

The foregoing derivations only apply to the extent that each atom looks spherically symmetrical on the average to its neighbors. If a prominent charge distortion occurs, as might be anticipated in strong covalent or electron transfer bonding, a different approach may be warranted. The aspherical part may be treatable as a perturbation in the pair-distribution framework. Alternatively, it might be preferable to consider larger scattering units, such as molecules, which may present a spherical average appearance to their neighbors, even though they are complex internally. In single-crystal samples, which are discussed in Section III, the unit cell is always an acceptable scattering unit, and its scattering factor (there called "structure factor") can be calculated precisely as the Fourier transform of the electron distribution within the entire cell, if that is known.

## H.   Scattering by two or more elements

In equation (2-9) the relative intensity of scattering for a monatomic specimen was expressed in terms of the interpair vectors among the scatterers.

$$Q(\mathbf{K}) = \left[a(\mathbf{K}) \sum_n e^{i\mathbf{K}\cdot\mathbf{R}_n}\right]\left[a^*(\mathbf{K}) \sum_m e^{-i\mathbf{K}\cdot\mathbf{R}_m}\right] \quad (2\text{-}129)$$

where each sum includes one term for each of the scatterers. If the scattering amplitudes of all the scatterers are not identical, they must be included inside the sums.

$$Q(\mathbf{K}) = \left[\sum_n a_n(\mathbf{K})e^{i\mathbf{K}\cdot\mathbf{R}_n}\right]\left[\sum_m a_m^*(\mathbf{K})e^{-i\mathbf{K}\cdot\mathbf{R}_m}\right] \quad (2\text{-}130)$$

The scattering amplitudes, which are generally complex functions, may be expressed as real functions multiplied by phase factors.

$$a_n(\mathbf{K}) = a_n^{\,0}(\mathbf{K}) \exp\left[-i\theta_n(\mathbf{K})\right] \quad (2\text{-}131)$$

Multiplication of the two sums then yields a cosine form analogous to equations (2-14).

$$Q(\mathbf{K}) = 2 \sum_{n,m} a_n^{\,0}(\mathbf{K})a_m^{\,0}(\mathbf{K}) \cos\left[\theta_{nm}(\mathbf{K}) + \mathbf{K}\cdot\mathbf{R}_{nm}\right] \quad (2\text{-}132)$$

where $\mathbf{R}_{nm} = \mathbf{R}_n - \mathbf{R}_m$, $\theta_{nm}(\mathbf{K}) = \theta_n(\mathbf{K}) - \theta_m(\mathbf{K})$, and the summation includes one term for each scatterer pair. It was indicated earlier (cf Section I C 3) that $Q(K) =$

$Q(-K)$ for pure materials, regardless of the arrangement of scatterers. This is seen to be true from (2-132), provided that the individual scatterers are themselves symmetrical in that respect [i.e., that $a^0(-K) = a^0(K)$]. On the other hand, if the phases $\theta_n(K)$ are not equal for all scatterers, some of the $\theta_{nm}(K)$ will be nonzero, and the corresponding terms in (2-132) will not be symmetrical in $K$. Expansion of the cosine yields

$$Q(K) = 2\left\{\sum_{n,m} a_n{}^0 a_m{}^0 \cos\left[\theta_{nm}(K)\right] \cos K \cdot R_{nm}\right.$$

$$\left. + \sum_{n,m} a_n{}^0 a_m{}^0 \sin\left[\theta_{nm}(K)\right] \sin K \cdot R_{nm}\right\} \quad (2\text{-}133)$$

All terms in the first sum are even, assuming that $\theta_{nm}(K) = \theta_{nm}(-K)$], and if the sample contains a center of inversion the second sum is zero. On the other hand, if there is no inversion center, the difference $Q(K) - Q(-K)$ will be twice the second sum in (2-133). A crystal, for example, which does not have an inversion center, will therefore exhibit a related asymmetry in its diffraction pattern.

In this section, however, the emphasis will be on pair-distribution functions, measured on spherically symmetrical or symmetrized samples, and for which such situations do not arise. For the purposes of that discussion it is useful to be more specific about the scattering units; one generally considers atoms, with scattering amplitudes $f(K)$. It is also more convenient to begin the discussion with the original double summation form.

$$Q(K) = \left[\sum_n f_n(K)e^{iK \cdot R_n}\right]\left[\sum_m f_m^*(K)e^{-iK \cdot R_m}\right] \quad (2\text{-}134)$$

**1.   *Partial distribution functions***   Since there will always be relatively few different types of scatterers in a sample, it is convenient to separate each of the sums according to the type of scatterer involved.

$$Q = \left\{\sum_{j=1}^M f_j(K)\left[\sum_n e^{iK \cdot R_n{}^j}\right]\right\}\left\{\sum_{l=1}^M f_l^*(K)\left[\sum_m e^{-iK \cdot R_m{}^l}\right]\right\} \quad (2\text{-}135)$$

where $R_n{}^j$ is a vector to a $j$-type scatterer whose scattering factor is $f_j$, and there are $M$ different types of scatterers. It is then convenient to rearrange the terms such that the double summation over types of pairs is to be performed last.

$$Q = \sum_{j=1}^M \sum_{l=1}^M \left\{f_j(K)f_l^*(K)\left[\sum_n e^{iK \cdot R_n{}^j} \sum_m e^{-iK \cdot R_m{}^l}\right]\right\} \quad (2\text{-}136)$$

The double summation over sites $n$ and $m$ can be reduced by following the steps traced previously for a single component, with respect to changing the summations to integrals over continuous distributions and with respect to a spherical averaging.

$$\sum_n e^{i\mathbf{K}\cdot\mathbf{R}_n{}^j}\sum e^{-\mathbf{K}\cdot\mathbf{R}_m{}^l} = Nx_j\left[\delta(l-j) + \int_0^\infty 4\pi r^2\rho_{jl}(r)\frac{\sin Kr}{Kr}\,dr\right] \qquad (2\text{-}137)$$

where $4\pi r^2\rho_{jl}(r)\,dr$ is the average number of $l$-type scatterers in a spherical shell of radius $r$ and thickness $dr$, centered on an average $j$-type atom, and $x_j$ is the fraction of $j$-type scatterers. $\delta(l-j)$ arises from the $\mathbf{R}_n{}^j = \mathbf{R}_m{}^l$ terms and is equal to 1 or 0 as $l$ is or is not equal to $j$. The relative scattered intensity per scatterer may then be expressed in terms of the sum over all types of such partial distributions.

$$\frac{Q}{N} = \sum_{j=1}^M x_j f_j f_j^* + \sum_{j=1}^M \sum_{l=1}^M x_j f_j f_l^* \int_0^\infty 4\pi r^2\rho_{jl}(r)\frac{\sin Kr}{Kr}\,dr \qquad (2\text{-}138)$$

The first term is just the average value of $f^2$ (i.e., of $ff^*$) and, as for a pure material, it indicates the scattering that would occur if no interscatterer correlations existed.

As with a pure material, it is necessary to subtract a $\rho_0$ term to account for the "zero-degree" scattering which is ordinarily not included in the experimental intensity. Note that for this purpose it is appropriate to use the average of the scattering strengths for each scatterer. Expressing the intensity $(I^{\text{total}} - I^Z)$ in electron units,

$$I = \sum_{j=1}^M x_j f_j f_j^* + \sum_{j=1}^M \sum_{l=1}^M x_j f_j f_l^* \int_0^\infty 4\pi r^2\rho_{jl}(r)\frac{\sin Kr}{Kr}\,dr$$

$$- \left(\sum_{j=1}^M x_j f_j\right)^2 \int_0^\infty 4\pi r^2\rho_0\frac{\sin Kr}{Kr}\,dr \qquad (2\text{-}139)$$

The double summation in (2-139) contains $M^2$ separate terms, one for each pair combination. For example, for two species, $A$ and $B$, (2-139) takes the special form

$$I = \langle f^2\rangle + x_A f_A f_A^* \int 4\pi r^2\rho_{AA}(r)\frac{\sin Kr}{Kr}\,dr$$

$$+ x_A f_A f_B^* \int 4\pi r^2\rho_{AB}(r)\frac{\sin Kr}{Kr}\,dr + x_B f_B f_A^* \int 4\pi r^2\rho_{BA}(r)\frac{\sin Kr}{Kr}\,dr$$

$$+ x_B f_B f_B^* \int 4\pi r^2\rho_{BB}(r)\frac{\sin Kr}{Kr}\,dr - \langle f\rangle^2 \int 4\pi r^2\rho_0\frac{\sin Kr}{Kr}\,dr \qquad (2\text{-}140)$$

where $\langle f^2\rangle = x_A f_A f_A^* + x_B f_B f_B^*$ and $\langle f\rangle^2 = (x_A f_A + x_B f_B)(x_A f_A^* + x_B f_B^*)$. The number of $j - l$ pairs at a given distance is independent of whether we count from $j$ type or $l$ type, so there must be a redundancy in equations (2-139) and (2-140). In fact, it is necessary that

$$x_j\rho_{jl}(r) = x_l\rho_{lj}(r) \qquad (2\text{-}141)$$

Thus the number of $\rho_{jl}(r)$ functions can be reduced by $M(M-1)/2$ to $(M^2 + M)/2$. On the other hand, if a given type of atom occupies two or more different kinds of positions, the effective number of separable functions is increased. That is, if multiple environments are involved, it may be practical to regard the summations over $j$ and $l$ as summations over $M'(>M)$ types of sites, which happen to have only $M$ different scattering factors.

In any case, it should be immediately obvious that one cannot determine the separate partial $\rho_{jl}(r)$ functions from a single diffraction experiment. As with a pure material, one will ultimately attempt to show consistency between a model and the data, and procedures can be used which are similar to those described for a pure material. In that regard, it is necessary to find an analogy to the transform of equation (2-54), which provided the initial result for a pure material. A similar step, by which one can achieve directly a reasonably accurate and informative representation of the distribution in $r$ space, is essential, unless one has an accurate preconception of the structure.

Subtraction of the independent scattering term $\langle f^2 \rangle$ in equation (2-139), followed by multiplication by $K$ and division by $\langle f \rangle^2$, yields a form which is more like the pure material expression

$$\left[ \frac{I - \langle f^2 \rangle}{\langle f \rangle^2} \right] K = \sum_j \sum_l \omega_{jl}(K) \int 4\pi r \rho_{jl}(r) \sin Kr \, dr - \int 4\pi r \rho_0 \sin Kr \, dr \qquad (2\text{-}142)$$

where $\omega_{jl}(K) = x_j f_l^* f_j / \langle f \rangle^2$. The Fourier inversion, which was used at this point for the pure material, is not as obviously useful here, however. The fact that the $\omega_{jl}$ can be functions of $K$, for x-ray and electron diffraction, complicates the interpretation of the transform in $r$ space. Nonetheless it is worth considering what result is yielded if we define an $F(K)$ for the multielement material

$$F(K) = \frac{I - \langle f^2 \rangle}{\langle f \rangle^2} K \qquad (2\text{-}143)$$

and calculate the sine transform of that experimental function.

$$G(r) = \frac{2}{\pi} \int F(K) \sin Kr \, dK \qquad (2\text{-}144)$$

For a formal view regarding the interpretation of $G(r)$, consider taking the sine integral of the entire right-hand side of (2-142) explicitly. Ignoring, for the moment, the fact that $K_{max}$ is finite

$$G(r) = \frac{2}{\pi} \int \left[ \sum_j \sum_l \omega_{jl}(K) \int 4\pi \xi \rho_{jl}(\xi) \sin k\xi \, d\xi \right] \sin Kr \, dK$$

$$- \frac{2}{\pi} \int \left[ \int 4\pi \xi \rho_0 \sin K\xi \, d\xi \right] \sin Kr \, dK \qquad (2\text{-}145)$$

The second term, like the corresponding term in a pure material, is simply a straight line in $G(r)$ with a slope of $-4\pi\rho_0$. To simplify the first term, one may refer to the inner integral as $Q_{jl}(K)$:

$$Q_{jl}(K) = \int 4\pi r \rho_{jl}(r) \sin Kr \, dr \qquad (2\text{-}146)$$

and then

$$G(r) = \frac{2}{\pi} \sum_j \sum_l \int Q_{jl}(K)\omega_{jl}(K) \sin Kr \, dK - 4\pi r \rho_0 \qquad (2\text{-}147)$$

$Q_{jl}(K)$ is exactly the theoretical contribution to $F(K)$ of unit-weight point scatterers with a distribution given by $\rho_{jl}(r)$. Suppose that in interpreting $G(r)$ it was assumed that the $\omega_{jl}(K)$ in (2-147) were not functions of $K$, but constants, $W_{jl} = \omega_{jl}(0)$, as they would be if all the $f_j f_l$ product functions had the same shape as $\langle f \rangle^2$. In that instance, $G(r)$ would be simply the sum of the different atomic distributions, each with weight $W_{jl}$:

$$G^I(r) = \sum_j \sum_l W_{jl} G_{jl}(r) - 4\pi r \rho_0 \qquad (2\text{-}148)$$

where $G^I$ is the idealized $G(r)$; $W_{jl} = \omega_{jl}(0)$; $G_{jl}(r) = 2/\pi \int Q_{jl}(K) \sin Kr \, dK = 4\pi r \rho_{jl}(r)$.

Considering the convolution theorem in relation to equation (2-147), however, it may be noted that in transforming $F(K)$ one operates on *products* $\omega_{jl}(K)Q_{jl}(K)$. Therefore, each of the partial functions actually appears in $G(r)$ as the idealized form $G_{jl}(r)$, *convolved* with the Fourier transform of $\omega_{jl}(K)$.

Since the $\omega_{jl}(K)$ are even functions of $K$, their transforms are simply the cosine integrals

$$V_{jl}(\varepsilon) = \frac{1}{2\pi} \int_{-\infty}^{+\infty} \omega_{jl}(K) \cos K\varepsilon \, dK \qquad (2\text{-}149)$$

The convolving functions $V_{jl}(\varepsilon)$ are also symmetrical functions. The actual $G(r)$ may then be written as

$$G(r) = \sum_j \sum_l \int_{-\infty}^{+\infty} G_{jl}(\xi) V_{jl}(\xi - r) \, d\xi - 4\pi r \rho_0 \qquad (2\text{-}150)$$

where the integral is the formal specification for the convolution. It is interesting to note that the scale (or weight) of a "broadened" $G_j$ depends on the *area* of the "broadening" function $V_{jl}(\varepsilon)$. That area, in turn, is simply the zeroth coefficient of its Fourier transform $\omega_{jl}(0)$, which has been used for the weight in (2-148):

$$\omega_{jl}(0) = \int V_{jl}(\varepsilon) \cos 0\varepsilon \, d\varepsilon = \int V_{jl}(\varepsilon) \, d\varepsilon \qquad (2\text{-}151)$$

With the idealized form [equation (2-148)] in mind, it is therefore somewhat more convenient to rewrite (2-150)

$$G(r) = \sum_j \sum_l W_{jl} \int_{-\infty}^{+\infty} G_{jl}(\xi) V'_{jl}(\xi - r) \, d\xi - 4\pi r \rho_0 \qquad (2\text{-}152)$$

where $\int V'_{jl}(\varepsilon) \, d\varepsilon = 1$. It is not too difficult to see the physical reasons for the above effects. In formulating $F(K)$ in the pure material case, the division by $f^2$ accomplished two things:

*1* An unfolding for the broadening due to the distribution of electrons associated with each pair of atoms

*2* A scaling down by the number of electron pairs associated with each pair of atoms

In equation (2-142), that unfolding and scaling is simulated through the division by $\langle f \rangle^2$. However, $\langle f \rangle^2$ is not necessarily the proper unfolding or weighting function for any pair type, let alone for all pair types in a multielement material. Indeed, the proper individual functions are those used in the first place to express the intensity in terms of the *atomic* pair distributions, namely, the products $f_j(K) f_l(K)$.

Thus, in equation (2-152), the $W_{jl}$ (to the extent that they deviate from $x_j$) are a measure of the different numbers of electron pairs associated with each type of atom pair. The $V'_{jl}(\varepsilon)$ (to the extent that they deviate from $\delta$ functions) are a measure of the different relative spatial distribution of electrons associated with each pair.

The division by $\langle f \rangle^2$ can be regarded as an approximate way of correcting for the distributions of electrons associated with the atoms. But equations (2-152) and (2-147) are not approximations, except for their neglect of data termination effects. Either or both of those equations can be used to judge the validity of a complete or partial model $G^T(r)$, with termination effects included, using procedures which are directly analogous to those described for a pure material. On the other hand, it is necessary to use an idealized manner of interpreting the experimental $G(r)$, essentially as expressed in equation (2-148), at least until the result is sufficiently well understood to allow formulation of a suitable model. It is useful, therefore, to consider the misinterpretations that might arise in the blind application of equation (2-148).

Assume that $G(r)$ contains one or more peaks, each due to a particular neighbor or bond distance of a particular type of atom pair. One begins by assigning a tentative identification to each peak, probably on the basis of its position. To the extent that a particular peak is properly identified and reasonably well resolved, the peak shape gives the distribution of such bonds, and the area measured on a $4\pi r^2 \rho(r)$ plot and divided by $W_{jl}$ gives the number of them, according to equation (2-148). In fact, if the associated $V'_{jl}(\varepsilon)$ is not narrow compared to the true peak width, both conclusions could be inaccurate. Fortunately, it is possible to examine the $V_{jl}(\varepsilon)$ even before doing any experiments, in order to clarify the analysis. As an example in which distorting effects would be severe, consider the alloy LiH (assuming no charge transfer). The

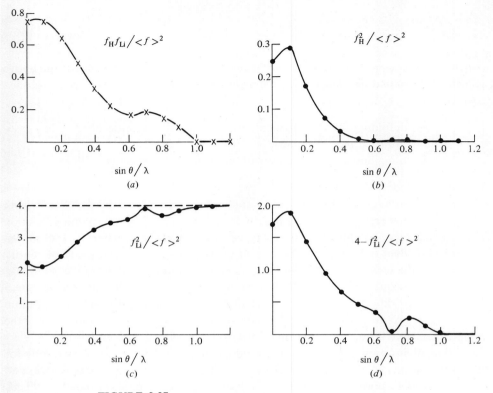

FIGURE 2-27
The atomic scattering factor ratios $f_i f_j / \langle f \rangle^2$ for LiH partial distributions.

atomic form factor ratios $f_H{}^2/\langle f \rangle^2$, $f_H f_{Li}/\langle f \rangle^2$, and $f_{Li}{}^2/\langle f \rangle^2$ are shown in Fig. 2-27, and are seen to vary very strongly with $K$ because the $f_H(K)$ approaches zero so quickly. For a rough estimation of the effects we might approximate the H-Li and H-H functions as simple exponential functions, $\exp(-\sigma^2 K^2/2)$, with half-widths of $\sim 0.36$ and $0.22$ A$^{-1}$, respectively. The $V_{H,Li}^{(\varepsilon)}$ and $V_{H,H}^{(\varepsilon)}$ are then Gaussian functions, $\exp(-\varepsilon^2/2\sigma^2)$, of half-widths $0.31$ and $0.50$ A, respectively ($\sigma_{H,H}^2 \approx 0.18$ A$^2$, $\sigma_{H,Li}^2 \approx 0.068$ A$^2$). These two types of pairs are very much undersharpened. For example, if an H-H pair had a true Gaussian-shaped distribution of distances (due to vibrations) with a $\sigma_{true}^2$ of $0.1$ A$^2$, the peak would appear in $G(r)$ as a Gaussian of $\sigma_{total}^2 = 0.28$ A$^2$. Moreover, although the $\omega_{jl}(0)$ would give the proper weight, it is doubtful that one would include all the area in a measurement since the full $3\sigma$ width would be $\sim 3.4$ A. It may be noted also that since the $\omega_{HH}(K)$ and $\omega_{H,Li}(K)$ both approach zero at low values of $K$, peaks of neither type will exhibit specific termination effects for $K_{max} \geq 12.5$, whatever the true shape.

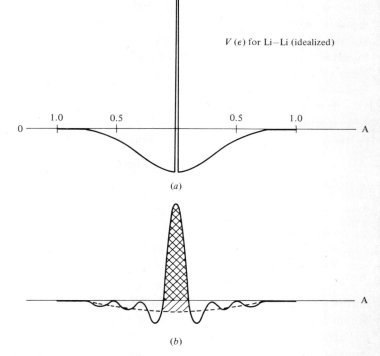

$V(\epsilon)$ for Li−Li (idealized)

(a)

(b)

FIGURE 2-28
Schematic illustrations of a peak distortion due to differences in atomic form factors. (a) Idealized oversharpening; (b) with atomic vibrations and termination error.

On the other hand, a Li-Li peak will be oversharpened and will exhibit clear termination oscillations unless the true peak is sufficiently broad that its contribution to $Q_{Li,Li}(K)$ is negligible beyond $K_{max}$. The ratio $f_{Li}^2/\langle f \rangle^2$ can be written as $4. - Z(K)$ where $Z(K) = 4. - f_{Li}^2/\langle f \rangle^2$ (as shown in Fig. 2.27d). The constant 4.0 yields a $\delta$-function contribution to $V(\varepsilon)$; the $Z(K)$, treated as an exponential form, contributes approximately a negative Gaussian with a half-width of 0.35 A. The resultant $V(\varepsilon)$ is shown schematically in Fig. 2-28a. One may thereby construct the expected shape in $G(r)$ of an actua! Li-Li peak, considering the two components of $V(\varepsilon)$ separately. The $\delta$-function part will yield the exact Li-Li peak, including any oscillations due to terminating the $Q_{Li,Li}(K)$ at $K_{max}$, but scaled by the factor 4.0/2.25. In addition, the broad negative peak of $V(\varepsilon)$ will contribute essentially its own shape, provided that the true peak is not very broad. A schematic drawing of such a peak, including a termination error, is shown in Fig. 2-28b. The total area of the peak [on a $4\pi r^2 \rho(r)$ plot] will correspond to the weight $\omega_{Li,Li}(0)$, the negative portion compensating for the excess in the sharp peak. However, without the calculation one is more likely to measure either the singly or doubly cross-hatched regions (as shown in Fig. 2-28$b$) and overestimate the number of pairs. Since in an actual measurement one is likely to include only the more obvious peak above zero, it may be advantageous to use a mean value of the $\omega_{jl}(K)$ [weighted by the envelope of $F(K)$] rather than the value at $K = 0$.

It may also be advantageous to obtain additional trials in which the division by $\langle f \rangle^2$ is replaced by division by a specific $f_j f_l$. In the current LiH example, if one were to use $f_{Li}{}^2(K)$, it would have the following effects on the derived $G(r)$:

*1* The slope of the $-4\pi r \rho_0$ line is decreased (to an incorrect value) by the ratio $f_{Li}{}^2(0)/\langle f(0) \rangle^2$.

*2* The shape of Li-Li peaks is the natural one, but including termination oscillations; i.e., $V_{Li,Li}(\varepsilon)$ is a $\delta$ function. A weight of 1 should be used for those peaks.

*3* H-H and H-Li peaks will be even broader than before, and the weights will be decreased [by $f_{Li}{}^2(0)/\langle f(0) \rangle^2$, if the $\omega_{jl}(0)$ is used].

It should be mentioned also that the termination effects are usually more severe and more confusing than the distortions caused by the $\omega_{jl}(K)$. Therefore, one may obtain a clearer initial interpretation of $G(r)$, even without specific identifications of peaks, by using the analysis for a pure material (characterized by $\langle f \rangle^2$) which was described earlier. Assuming that a particular model $G^T(r)$ is achieved in which the partial functions $\rho_{jl}(r)$ are individually identified, it may be tested—without approximation—by using equation (2-142) and equation (2-147) or (2-152), modified to account for termination of data. As with a pure material, it is most convenient to test and refine the model via an $r$-space representation. The direct $F^T(K) - F(K)$ comparison is useful again in judging the range of uncertainty in an acceptable $G^T(r)$, with respect to an estimate of error $\varepsilon(K)$. With a pure material, it is possible to compare the model $G^T(r)$ to a corrected experiment $G^*(r)$. With a multielement material, while using x-ray or electron diffraction it is not possible to make that precise comparison of atomic-distribution functions. On the other hand, one would lose too much resolution by working with the actual electron distributions.

**2.** *General analysis techniques* A compromise procedure is feasible, however, which is analogous to the "corrected-experiment-comparison" of the pure material.

*1* Compute $F^T(K)$ from $G^T(r)$, using equation (2-142).

*2* Derive a new function $G^*(r)$:

$$G^*(r) = \int_0^{K_{max}} F(K) \sin Kr \, dK + \int_{K_{max}}^{>>K_{max}} F^T(K) \sin Kr \, dK \qquad (2\text{-}153)$$

*3*  Modify the model to account for $\omega_{jl}(K)$ distortion, by convolving $G^T(r)$ with the appropriate $V_{jl}(\varepsilon)$:

$$G^{TM}(r) = \sum_j \sum_l \int_{-\infty}^{+\infty} G_{jl}{}^T(\xi)V_{jl}(\xi - r)\,d\xi \qquad (2\text{-}154)$$

*4*  Using the difference between $G^{TM}(r)$ and $G^*(r)$ as a guide, create a new $G^T(r)$.

*5*  Iterate steps (1) to (4) until $G^*(r) = G^{TM}(r)$. When that condition holds, it will necessarily be true that $F^T(K) = F(K)$ for $K < K_{max}$.

Alternatively, as with a pure material, one may completely modify the model to simulate the experimental result:

*1*  $G'(r) = \displaystyle\int_0^{K_{max}} F(K) \sin Kr\,dK$

*2*  $G^M(r) = \displaystyle\int_0^{K_{max}} F^T(K) \sin Kr\,dK$

*3*  Modify $G^T(r)$ until $G^M(r) = G'(r)$

As with a pure material, one may optionally multiply both $F(K)$ and $F^T(K)$ by a "damping function" to further lessen the sensitivity of the comparison to particular portions of the data. Also note that both procedures can also be used when data are available over a more limited region of $K$ space, and when the model is formulated over only a limited region of $r$ space.

In the foregoing paragraphs, references to the "shapes of peaks" have been made with the tacit assumption that each particular peak in $G(r)$ could be associated with a particular pair type. In fact, that assumption often holds, particularly in the small-$r$ region for molecular substances, whether in gaseous, liquid, amorphous, or crystalline form. In such instances, moreover, it is often possible to identify the individual peaks by the correlation distance alone, on the basis of a prior knowledge of feasible bond lengths.

In other sorts of materials, however, it is not possible to assign specific, unique identities to specific features in $G(r)$, even though well-defined correlation peaks appear. This is particularly true in metallic alloys, for which even the first (nearest-neighbor) correlation peak is likely to be due to a superposition of pair types. Indeed, as an extreme—but interesting—example of such superposition it is worth considering a material which is random in terms of the specific identities of the atoms in the correlation function. That is, a material for which the following special condition holds at all $r$ for all types of atoms:

$$\rho_{jl}(r) = x_l\rho_j(r) \qquad (2\text{-}155)$$

where $\rho_j(r) = \sum_{k=1}^{M} \rho_{jk}(r)$ is the total distribution about a $j$-type atom. Given condition (2-155) and recalling that $x_j\rho_{jl}(r) = x_l\rho_{lj}(r)$, it follows that each type of atom would—on the average—have an identical total distribution; i.e.,

$$\rho_l(r) = \rho_j(r) = \rho^R(r) \qquad (2\text{-}156)$$

That single distribution function $\rho^R(r)$ describes the entire arrangement, with each partial distribution being a constant portion of it.

$$\rho_{jl}(r) = x_l\rho^R(r) \qquad (2\text{-}157)$$

It is worth reconsidering equation (2-142) for this random case. That expression may be rewritten, using (2-157)

$$F(K) = \int 4\pi r \left( \sum_j \sum_l \frac{x_j f_j f_l^*}{\langle f \rangle^2} x_l \right) \rho^R(r) \sin Kr \, dr - \int 4\pi r \rho_0 \sin Kr \, dr \qquad (2\text{-}158)$$

The quantity in large parentheses in equation (2-158) is identically equal to unity. One thereby obtains the not too surprising result that if the material is actually a random solution, $\langle f \rangle^2$ is the correct "sharpening" function. That is, for the random solution case

$$F(K) = \int 4\pi r [\rho^R(r) - \rho_0] \sin Kr \, dr \qquad (2\text{-}159)$$

and the distribution function may be retrieved in undistorted form by Fourier inversion

$$4\pi r [\rho^R(r) - \rho_0] = \frac{2}{\pi} \int F(K) \sin Kr \, dK \qquad (2\text{-}160)$$

In this instance, the analysis prescribed for a pure material (as regards termination of data) is appropriate.

There are probably not many materials, however, for which the random solution model would be an adequate approximation. One might expect it to be most closely approached in metallic solutions of atoms of nearly the same size, and at temperatures —for the solid crystalline state—which are well above any ordered arrangements. It must be noted, however, that with such materials, and even in an amorphous or liquid state, the deviation from randomness is likely to be the aspect of greatest interest. The measurement of such *short-range order* in crystalline alloys will be considered in a later section, in fact.

It has been suggested [Keating (1963)] that the separate partial distribution functions can be measured directly with multiple diffraction experiments chosen to exploit the fact that different radiations scatter with different relative powers from different atoms. In other words, the relative magnitudes (as well as the shapes) of the $\omega_{jl}(K)$ generally depend on the particular radiation used. Thus, in principle, one may do the diffraction experiment with each of $N$ different radiations, for each one of which there is a particular set of weighting functions $\omega_{jl}{}^n(K)(1 \le n \le N)$. One then achieves a set of $N$ simultaneous equations in $(M^2 + M)/2$ unknown functions, the partial distribution functions. Using the form of equation (2-152), the set of equations may be written

$$G^n(r) = \sum_j \sum_l W_{jl}{}^n \int_{-\infty}^{+\infty} G_{jl}(\xi) V_{jl}'{}^n(\xi - r) \, d\xi - 4\pi r \rho_0 \qquad (2\text{-}161)$$

where all terms have the same meaning defined previously, except that the superscript $n$ designates the $n$th of a set of $N$ equivalent diffraction experiments with different radiations. To the extent that the distortions due to the $V_{jl}'{}^n(\xi)$ are ignorable, one can use the idealized form of equation (2-152)

$$G_n{}^I(r) = \sum_j \sum_l W_{jl}{}^n G_{jl}(r) - 4\pi r \rho_0 \qquad (2\text{-}162)$$

and, if $N$ equals the number of independent partial functions, the set can be solved straightforwardly. On the other hand, note that it would be possible to solve the set of equation (2-161) by trial, even if the $V_{jl}'$ are significant, and even if the termination effects are different for each experiment—as is likely to be the case.

To the present time, such an analysis has not been carried out. It would appear feasible, however, to perform the three experiments necessary for a binary alloy, using x-ray (or electron) diffraction for one and neutron diffraction, with two different isotope combinations in the sample, for the other two. Since the neutron scattering cross sections are unrelated to the atomic number and often very different for different isotopes, some combinations no doubt yield favorable differences in the $\omega_{jl}$ for particular materials. The practical difficulties in achieving accuracy will be significant, nonetheless, since the analysis depends only on the differences among the experiments. In that regard it should be stressed that the three experiments must be equivalent; in particular, the neutron data must include those neutrons which suffer recoil loss (or gain) since the analogous "thermally" scattered photons are necessarily included in the x-ray experiment.

It should be mentioned also that combined neutron and x-ray data may be very valuable toward an interpretation of a complex sample even if the amount of data (i.e., number of experiments) is inadequate for a formally complete separation into all of the partials.

**3.** *Short-range order, particularly in polycrystalline samples* In some multielement materials, liquid and amorphous as well as crystalline, the different partial distributions $\rho_{jl}(r)$ superimpose, at least to a certain extent. That is, different kinds of atom pairs may have effectively the same set of "neighbor distances." Such a situation obtains, for example, in crystalline metallic solid solutions which are disordered with respect to long-range occupancy of specific lattice sites by specific atom types, and yet not significantly distorted locally by atomic size differences. It is of interest in such cases to know the degree of local or "near-neighbor" nonrandomness, as regards the number of pairs of each type at each "neighbor" distance. It is entirely possible, for example, that a particular type of atom has a statistically satisfied preference for unlike (or like) neighbors even though the associated distances are effectively the same.

In the crystalline case, the most detailed information can of course be obtained through the study of single-crystal specimens. In this section, however, the problem is examined in the spherically averaged form, which applies, in principle, to liquids and amorphous materials as well as to polycrystalline solids.

The term "short-range order" and its opposite, "clustering," are ordinarily applied in situations where the basic "randomized" structure is known. In terms of notation previously defined, one is therefore interested in a case where the *total* distribution about each type of atom is nearly the same as a known theoretical average distribution

$$\rho_i(r) = \rho_{ij}(r) + \rho_{ii}(r) \approx \rho^E(r)$$

$$\rho_j(r) = \rho_{ji}(r) + \rho_{jj}(r) \approx \rho^E(r)$$

$$(2\text{-}163)$$

but for which the different pair types do not occur with random frequency; i.e.,

$$\frac{\rho_{ij}}{x_j} \neq \frac{\rho_{ii}}{x_i} \qquad (2\text{-}164)$$

In a previous section, the possibility was discussed of deducing the separate partial distribution functions through a multiple experiment analysis. Such an analysis would provide a general solution, of course, including the present special case. The short-range-order problem in a binary alloy is amenable to a unique solution with a single x-ray experiment, however, at least to the extent that the assumptions of equations (2-163) are satisfied.

In brief, the analysis is centered on examination of the *difference* between the result for the actual sample and that which would be expected if the arrangement were random. In general, that difference has been derived initially in $K$ space, in terms of a difference intensity. One may define an $I_{SRO}$ to be the difference between the actual intensity and that predicted for a random sample. Bypassing, for the moment, the problems inherent in obtaining $I_{SRO}$ experimentally, consider the formal expression for that difference function. Referring to equation (2-139), note that the first and last terms in that expression for the intensity are independent of specific arrangements and disappear in the difference. One may therefore write

$$I_{SRO} \equiv (I - I^R) \qquad (2\text{-}165a)$$

$$I_{SRO} \equiv \frac{1}{K} \sum_{j=1}^{M} \sum_{l=1}^{M} x_j f_j f_l^* \int 4\pi r [\rho_{jl}(r) - \rho_{jl}^R(r)] \sin Kr \, dr \qquad (2\text{-}165b)$$

where $\rho_{jl}{}^R(r)$ is the random distribution function for $jl$ pairs. For a binary alloy which satisfies the "equivalent environment" condition of (2-163), the following special relationships hold:

1  $\rho_{ab} + \rho_{aa} = \rho_{ba} + \rho_{bb} \equiv \rho^E(r)$

2  $x_a\rho_{ab} = x_b\rho_{ba}$

3  $\rho_{ab}{}^R = x_b\rho^E(r);\ \rho_{aa}{}^R = x_a\rho^E(r)$

4  $\langle f^2 \rangle - \langle f \rangle^2 = x_a x_b (f_a - f_b)^2$

5  $M = 2$

Equations (2-165) are thereby readily reduced to

$$I_{\text{SRO}} = \frac{x_a x_b (f_a - f_b)^2}{K} \int 4\pi r \left[ \rho^E(r) - \frac{\rho_{ab}(r)}{x_b} \right] \sin Kr\ dr \qquad (2\text{-}166)$$

Again, it is useful to define a reduced intensity function

$$F_{\text{SRO}}(K) \equiv K \left( \frac{I_{\text{SRO}}}{x_a x_b (f_a - f_b)^2} \right) \qquad (2\text{-}167)$$

and the Fourier sine inversion

$$G_{\text{SRO}}(r) \equiv \frac{2}{\pi} \int F_{\text{SRO}}(K) \sin Kr\ dK \qquad (2\text{-}168)$$

which yields—in undistorted form—

$$G_{\text{SRO}}(r) = 4\pi r \left( \rho^E(r) - \frac{\rho_{ab}(r)}{x_b} \right) \qquad (2\text{-}169)$$

The latter result portrays the deviations from randomness directly. It may be more convenient, however, to express the result as a relative deviation from the average (and known) distribution $\rho^E(r)$

$$G_{\text{SRO}}(r) = 4\pi r \alpha(r) \rho^E(r) \qquad (2\text{-}170)$$

where $\alpha(r) \equiv 1 - \rho_{ab}(r)/x_b\rho^E(r)$. In the particular case of a polycrystalline material, $\rho^E(r)$ can be expressed as a sum of discrete peaks (if not $\delta$ functions), as was done earlier:

$$4\pi r \rho^E(r) \rightarrow \sum_{i=1}^{\infty} \frac{c_i}{r_i} B_i(r - r_i) \qquad (2\text{-}171)$$

where $c_i$ is the number of $i$th neighbors; $r_i$ is the distance to $i$th neighbors; $B_i(r - r_i)$ accounts for the distribution of $i$th neighbors at $r_i$, and is normalized to an area of 1.0. If it is assumed that the short-range order does not vary within a single neighbor shell [i.e., over the range of $B_i(\varepsilon)$], equations (2-170) and (2-166) can be rewritten for the polycrystalline binary alloy case

$$G_{SRO}(r) = \sum_{i=1}^{\infty} \frac{c_i}{r_i} \alpha_i B_i(r - r_i) \tag{2-172}$$

$$I_{SRO}(K) = \frac{x_a x_b (f_a - f_b)^2}{K} \int \sum_{i=1}^{\infty} \frac{c_i}{r_i} \alpha_i B(r - r_i) \sin Kr \, dr \tag{2-173}$$

where $\alpha_i = \alpha(r_i)$. The $\alpha_i$ are called the *Warren short-range-order coefficients* after B. E. Warren, who has been responsible for much of the development of this line of investigation, as well as of many of the other topics described in this chapter.

If the $B(r - r_i)$ are $\delta$ functions, equation (2-173) reduces to

$$I_{SRO}(K) = \frac{x_a x_b (f_a - f_b)^2}{K} \sum_{i=1}^{\infty} \frac{c_i}{r_i} \alpha_i \sin Kr_i \tag{2-174}$$

Otherwise, the effects of finite $B_i(r - r_i)$ can be expressed in terms of their Fourier transforms $b_i(K)$, as is discussed more fully in the section relating to thermal vibrations, and also later in the present section:

$$I_{SRO}(K) = \frac{x_a x_b (f_a - f_b)^2}{K} \sum_{i=1}^{\infty} \frac{c_i}{r_i} \alpha_i b_i(K) \sin Kr_i \tag{2-175}$$

A considerable complication is introduced when it is necessary to consider deviations from the condition of (2-163) associated with the fact that *aa*, *bb*, and *ab* "bond" distances are not identical. For small deviations, however, it has been possible to extend equation (2-173) by introducing "size-effect" coefficients (Roberts). Consider a single shell of neighbors, the $i$th shell, including *aa*, *bb*, and *ab* pairs. The number of pairs of each type (per average atom) may be expressed in terms of the corresponding $\alpha_i$; e.g.,

$$n_{ab} = x_a c_i (1 - \alpha_i) x_b = n_{ba} \tag{2-176a}$$

$$n_{aa} = x_a c_i (x_a + x_b \alpha_i) \tag{2-176b}$$

$$n_{bb} = x_b c_i (x_b + x_a \alpha_i) \tag{2-176c}$$

The specific (mean) neighbor distances for the individual pair types are allowed to differ from $r = r_i$ by small relative amounts, such that $r_{aa} = r(1 + \varepsilon_{aa})$, $r_{ab} = r(1 + \varepsilon_{ab})$, and $r_{bb} = r(1 + \varepsilon_{bb})$. The contribution to the total intensity of that ($i$th) shell then may be written out explicitly

$$KI_i(K) = \frac{c_i}{r} \left\{ \frac{A \sin [Kr(1 + \varepsilon_{aa})]}{1 + \varepsilon_{aa}} + \frac{B \sin [Kr(1 + \varepsilon_{bb})]}{1 + \varepsilon_{bb}} \right.$$

$$\left. + \frac{C \sin [Kr(1 + \varepsilon_{ab})]}{1 + \varepsilon_{ab}} \right\} \tag{2-177}$$

where $A = x_a(x_a + x_b\alpha_i)f_a^2$, $B = x_b(x_b + x_a\alpha_i)f_b^2$, and $C = 2x_ax_b(1 - \alpha_i)f_af_b$. Expanding the sines of the sums yields

$$KI_i(K) = \frac{c_i}{r}\left[\sin Kr\left(\frac{A\cos Kr\varepsilon_{aa}}{1 + \varepsilon_{aa}} + \frac{B\cos Kr\varepsilon_{bb}}{1 + \varepsilon_{bb}} + \frac{C\cos Kr\varepsilon_{ab}}{1 + \varepsilon_{ab}}\right)\right.$$
$$\left. + \cos Kr\left(\frac{A\sin Kr\varepsilon_{aa}}{1 + \varepsilon_{aa}} + \frac{B\sin Kr\varepsilon_{bb}}{1 + \varepsilon_{bb}} + \frac{C\sin Kr\varepsilon_{ab}}{1 + \varepsilon_{ab}}\right)\right] \quad (2\text{-}178)$$

If the $\varepsilon$'s are much less than 1, it is useful to consider the expansions

$$\frac{\cos Kr\varepsilon}{1 + \varepsilon} \approx 1 - \varepsilon + \varepsilon^2\left[1 - \frac{(rK)^2}{2}\right] + \cdots \quad (2\text{-}179a)$$

$$\frac{\sin Kr\varepsilon}{1 + \varepsilon} \approx \varepsilon rK - \varepsilon^2 rK + \cdots \quad (2\text{-}179b)$$

Ignoring all terms beyond those linear in the $\varepsilon$'s, the expansion of equation (2-178) may be written as

$$KI_i(K) = \frac{c_i}{r}\left\{(A + B + C)\sin Kr\right.$$
$$\left. - [\sin Kr - Kr\cos Kr][A\varepsilon_{aa} + B\varepsilon_{bb} + C\varepsilon_{ab}]\right\} \quad (2\text{-}180)$$

Considering the first term

$$A + B + C = \langle f\rangle^2 + \alpha_ix_ax_b(f_a - f_b)^2 \quad (2\text{-}181)$$

Since $\langle f\rangle^2(c_i/Kr_i)\sin Kr_i$ is the total theoretical contribution to the intensity of that shell in the randomized model, the contribution to $I$ (short-range order) $I_i^{SRO}$ is just the rest of equation (2-180)

$$I_i^{SRO}(K) = \frac{x_ax_b(f_a - f_b)^2}{K}\frac{c_i}{r_i}\{\alpha_i\sin Kr_i - \beta_i(K)[\sin Kr_i - Kr_i\cos Kr_i]\} \quad (2\text{-}182)$$

where $\beta_i(K) = (A\varepsilon_{aa}{}^i + B\varepsilon_{bb}{}^i + C\varepsilon_{ab}{}^i)/x_ax_b(f_a - f_b)^2$. Before considering an analysis in terms of the $\beta_i(K)$'s, note that other discernible intensity effects which arise from the "size" differences may be derived with only little more difficulty, if the $\varepsilon^2$ terms of the expansion are included.

$$KI_i(K) = \frac{c_i}{r}\left\{\sin Kr\left[A + B + C + \left(1 - \frac{(rK)^2}{2}\right)(A\varepsilon_{aa}{}^2 + B\varepsilon_{bb}{}^2 + C\varepsilon_{ab}{}^2)\right]\right.$$
$$- (\sin Kr - Kr\cos Kr)(A\varepsilon_{aa} + B\varepsilon_{bb} + C\varepsilon_{ab})$$
$$\left. - Kr\cos Kr(A\varepsilon_{aa}{}^2 + B\varepsilon_{bb}{}^2 + C\varepsilon_{ab}{}^2)\right\} \quad (2\text{-}183)$$

A slight rearrangement then yields

$$KI_i(K) = \frac{c_i}{r} \left\{ \sin Kr \left[ A \left( 1 - \frac{(rK\varepsilon_{aa})^2}{2} \right) + B \left( 1 - \frac{(rK\varepsilon_{bb})^2}{2} \right) + C \left( 1 - \frac{(rK\varepsilon_{ab})^2}{2} \right) \right] \right.$$

$$- (\sin Kr - Kr \cos Kr)[A\varepsilon_{aa}(1 - \varepsilon_{aa})$$

$$\left. + B\varepsilon_{bb}(1 - \varepsilon_{bb}) + C\varepsilon_{ab}(1 - \varepsilon_{ab})] \right\} \qquad (2\text{-}184)$$

Regarding the sin $Kr$ term, and using the definitions $\sigma_{aa}{}^{Di} = r\varepsilon_{aa}{}^i$, one can make the replacements

$$\exp\left[-(\sigma_{aa}{}^{Di})^2 K^2/2\right] \approx 1 - \frac{(r\varepsilon_{aa}{}^i K)^2}{2} \qquad (2\text{-}185)$$

From the definitions of $A$, $B$, and $C$ given in equation (2-177), it is seen that in the approximation of this derivation the effect of the "static" displacements on the sin $Kr$ term is precisely the same as for a "Gaussian" thermal vibration effect; namely, the scattering factor pair products are multiplied by individual and shell-dependent exponential factors. These damp all of the intensity (with increasing $K$), but the effect is most noticeable on the $\langle f \rangle^2$ part of the sin $Kr$ term [see equation (2-181)]. Just as for thermal vibrations, the well-defined Bragg reflections are decreased in intensity, and an additional diffuse component arises. The shape of the latter will differ from the thermal diffuse scattering to the extent that the shell dependence of the relative displacement is different.

Thus, in a binary polycrystalline alloy various scattering effects with different $K$ dependences may arise from the size differences and statistical order preferences of the two species. In summary, these are

*1*  The short-range-order scattering, which is a modulation of $x_a x_b (f_a - f_b)^2$ (the so-called Laue monotonic intensity) to which each shell contributes a sin $Kr_i/Kr_i$ term to the intensity, weighted by its Warren coefficient $\alpha_i$.

*2*  The size-effect scattering, which is also a modulation of $x_a x_b (f_a - f_b)^2$ to which each shell contributes a (cos $Kr_i$ − sin $Kr_i/Kr_i$) term weighted by its "size-effect" coefficient $\beta_i$.

*3*  The static-displacement scattering (often called the *Huang diffuse scattering*) and an associated diminution of the Bragg intensities. Physically, the effect arises from the fact that the atomic size differences cause atoms to be displaced from the geometrically ideal lattice. At sufficiently large separations the displacements can be regarded as independent, but they are clearly coupled for near neighbors. The effect is therefore analogous to the thermal diffuse scattering and has the same general form. In both cases one expects the peak intensities to be diminished by an exp $(-\sigma_\infty{}^2 K^2/2)$ factor, and the diffuse intensity to appear as a modulation on $\langle f \rangle^2 [1 - \exp(-\sigma_\infty^2 K^2/2)]$, where the modulations peak up underneath the Bragg peaks with a form that depends on the effective mean-square displacements for near-neighbor shells, $\sigma_i^2$.

4    The thermal scattering, as reflected both in the diffuse and the peak intensities, certainly depends on the local atomic arrangement, and also because the vibrations themselves must vary somewhat. One would ordinarily (and probably correctly) ignore such effects as being of secondary importance in the already difficult analysis. That point of view is strengthened, of course, by the fact that it is very difficult to accurately predict the thermal diffuse scattering at all, much less its change as a function of the $\alpha$'s and $\beta$'s.

In any case it may be seen that if the aim is to determine the $\alpha$'s and $\beta$'s, one is not really interested in the true (actual minus random) difference intensity, but only in the part which relates to items (1) and (2) in the foregoing list. That is, one would be interested in subtracting from the total intensity at least the following portions: (1) the Bragg peaks $I_B$, (2) the thermal diffuse $I_{TDS}$, (3) the static displacement (Huang) diffuse $I_{SDS}$, as well as (4) the modified (Compton) $I_C$. In this regard, it should be emphasized that the *random* binary alloy will contain another diffuse component, as may be seen below (in the neglect of vibrations):

$$I = \langle f^2 \rangle + \langle f \rangle^2 \sum_{i=1}^{\infty} \frac{c_i}{Kr_i} \sin Kr_i \qquad (2\text{-}186)$$

$$I_B \equiv \langle f \rangle^2 \sum_{i=0}^{\infty} \frac{c_i \sin Kr_i}{Kr_i} \qquad (2\text{-}187)$$

$$I = I_B + (\langle f^2 \rangle - \langle f \rangle^2) \qquad (2\text{-}188a)$$

$$I = I_B + x_a x_b (f_a - f_b)^2 \qquad (2\text{-}188b)$$

Assuming that $I_B$, $I_{TDS}$, $I_{SDS}$, and $I_C$ are subtracted, it is therefore convenient to analyze the remainder $I_{\text{local arrangement}}$, which *includes* the latter term. That is, it may be preferable to work with

$$I_{LA} = I_{SRO} + x_a x_b (f_a - f_b)^2 \qquad (2\text{-}189)$$

$$I_{LA} = x_a x_b (f_a - f_b)^2 \left[ 1 + \sum_{i=1}^{\infty} \frac{c_i \alpha_i \sin Kr_i}{Kr_i} \right.$$

$$\left. + \sum_{i=1}^{\infty} c_i \beta_i \left( \cos Kr_i - \frac{\sin Kr_i}{Kr_i} \right) \right] \qquad (2\text{-}190)$$

Note, once again, that if it is important to take vibrations into account, each shell contribution (2-190) should include a shape "distribution function," such as $\exp[-(r - r_i)^2/2\sigma_i^2]/\sqrt{2\pi\sigma_i^2}$, instead of the (implied) $\delta(r - r_i)$.

Following an analysis technique previously discussed, define an $F_{LA}(K)$

$$F_{LA}(K) = \left[ \frac{I_{LA}}{x_a x_b (f_a - f_b)^2} - 1 \right] K \qquad (2\text{-}191)$$

and its (limited) sine transform

$$G_{LA}(r) = \frac{2}{\pi} \int_0^{K_{max}} F_{LA}(K) \sin Kr \, dK \qquad (2\text{-}192)$$

The explicit integration of the right-hand side of (2-190) yields

$$G_{LA}(r) = \sum_{i=1}^{\infty} \left[ \alpha_k P_i(r) + \beta_i Q_i(r) \right] \qquad (2\text{-}193)$$

where

$$P_i(r) = \int_0^{K_{max}} \frac{c_i}{r_i} \sin Kr_i \, \frac{\exp\left[(r - r_i)^2/2\sigma_i^2\right]}{\sqrt{2\pi\sigma_i^2}} \sin Kr \, dK \qquad (2\text{-}194)$$

$$Q_i(r) = \int_0^{K_{max}} \frac{c_i(K \cos Kr_i - \sin Kr_i) \exp\left(-(r - r_i)^2/2\sigma_i^2\right)}{r_i} \, \frac{}{\sqrt{2\pi\sigma_i^2}} \sin Kr \, dK \qquad (2\text{-}195)$$

The formal representation of equation (2-193) probably appears to be forbiddingly complex. Consider, however, the situation if

*1* The size coefficients are negligible.
*2* The vibrations are negligible.
*3* The $K_{max}$ is very large.

$G_{LA}(r)$ then appears as a set of nearly $\delta$ functions, located at the shell distances $r_i$. Each such sharp peak will have an area equal to $c_i \alpha_i / r_i$ and will be positive or negative depending on whether there are, respectively, more or fewer than the random number of like neighbors at that position. As the $\sigma_i^2$ become more significant the peaks assume the correspondingly broader shapes. The $\beta_i$'s introduce asymmetric distortions in the peak shapes, even to the extent of giving it two positive and one negative regions, corresponding to resolved *aa*, *bb*, and *ab* correlations (provided $\sigma_i^2$ is small enough). As $K_{max}$ is made smaller, the effect of data termination becomes pronounced, which effect appears in the form of a further broadening of the peaks and the introduction of oscillations which may extend to distant shells. The $P_i$ and $Q_i$ are simply the explicit functions which specify the theoretical contribution to $G_{LA}(r)$ of each coefficient. Values of these functions may be calculated by numerical integrations if necessary. $\alpha$'s and $\beta$'s then may be determined using equation (2-193) together with formal fitting procedures or by trial-and-error techniques.

As may have been perceived in the foregoing discussion, the primary difficulty in these analyses lies in the subtraction of the $I_{TDS}$ and, perhaps, $I_{SDS}$. This difficulty is particularly severe at larger $K$ values where the ratio $\langle f \rangle^2 (1 - e^{-\sigma^2 K^2/2})/x_a x_b$ $\times (f - f_b)^2$ is large, even at low temperatures. For this reason, values of $K_{max}$ achieved through this "difference-intensity" approach are often so small that nearly all resolution is lost in $G_{LA}(r)$.

It may be preferable, therefore, to reconsider using an analysis of the entire (unmodified) intensity distribution, even for the crystalline alloys. Just as for the pure polycrystalline sample, this avoids the need to divide the total intensity into those components which appear separately in the theoretical derivation but not in the experiment. It also obviates the need for structural approximations, or at least delays their introduction to the final stage of analysis, at which point their validity is better seen and their effect better estimated. Sufficient work has been done on pure materials to indicate that application of the general transform approach to polycrystalline aggregates is entirely feasible. Therefore consider the general forms

$$F(K) \equiv \left[ \frac{I - \langle f^2 \rangle}{\langle f \rangle^2} \right] K \qquad (2\text{-}196)$$

$$G(r) \equiv \frac{2}{\pi} \int_0^{K_{max}} F(K) \sin Kr \, dK \qquad (2\text{-}197)$$

In general, the interpretation of $G(r)$ would proceed along lines developed in a preceding section. For a short-range-order analysis in a polycrystalline sample, however, one can take good advantage of the presumed knowledge that the total $\rho(r)$ consists of the sum of a set of relatively well-defined peaks, corresponding to the neighbor shells of the crystal structure and of the mean positions of those peaks, $\bar{r}_i$, and the total number of atom pairs associated with each, $c_i$. It is worthwhile to consider an idealized case in which $K_{max}$ is large and the scattering factor shape distortion is small [that is, the differences in the $V_{ij}(\varepsilon)$ are not important to the interpretation]. If, in addition, the vibrations are small compared to the atomic size differences, $G(r)$ simply becomes the sum of $aa$, $bb$, and $ab$ peaks—three peaks for each $r_i$—at least for the first few shells. The desired quantities (i.e., the number and distance of each of the three types of pairs in each shell) then can be immediately and directly obtained. In the more general case, however, the thermal vibrations and the termination effects will wipe out the resolution—within a shell. One may go, nonetheless, to the sort of model-fitting procedures which were discussed for the general case, fitting three separate partial pair-distribution functions, but with the special "polycrystalline" constraints which were listed above and which make such an analysis feasible, at least for the first few shells. In that regard, it is advantageous to compare the actual $G(r)$ to a theoretical, random distribution, and even to develop the difference (actual-random) directly in $r$ space. In the regions where the size differences,

$r_{aa}{}^i - r_{ab}{}^i$ and $r_{bb}{}^i - r_{ab}{}^i$ can be neglected [i.e., $\rho_{aa}(r) + \rho_{ab}(r) \approx \rho_{bb}(r) + \rho_{ba}(r)$],

$$[G(r) - G^R(r)] = \text{convolution} \left\{ V^*(\varepsilon), \left[ \rho^E(r) - \frac{\rho_{ab}(r)}{x_b} \right] \right\} \quad (2\text{-}198)$$

where $\rho^E(r) = \rho_{ii} + \rho_{ij}$ and $V^*(\varepsilon)$ is the transform of $x_a x_b (f_a - f_b)^2/\langle f \rangle^2$. That is, the difference $G(r)$ is just the continuous short-range-order function

$$\alpha(r)\rho^E(r) = \rho^E(r) - \frac{\rho_{ab}(r)}{x_b} \quad (2\text{-}199)$$

albeit scaled by the value of $x_a x_b (f_a - f_b)^2/\langle f \rangle^2$ at $K = 0$, and distorted in a slight and known fashion because of the scattering factor differences. It is interesting to note that the approximation ought to be a good one for more distant shells in the $r$-space function, even if the "size" differences are large for the first few.[1]

It would therefore appear that a fruitful short-range-order analysis of a binary alloy could be achieved through an analysis of the *total* intensity function.

1   One can interpret the short-distance part of $G(r)$ by model fitting, in terms of bond-type probabilities ($\alpha_i$) *and* the distance differences.

2   Beyond the peaks for which the $r_{aa} - r_{bb}$ differences cause visible distortion of a randomized *shape*, the interpretation can be given in terms of the short-range-order function $[\alpha(r)\rho^E(r)]$ derived by subtracting the theoretical $4\pi r(\rho^R(r) - \rho_0)$ from the experimental $G(r)$.

3   In an intermediate region, where the separate shells are reasonably well resolved for the given degree of thermal broadening, the $\alpha(r)$ function can be converted to discrete $\alpha_i$ virtually by observation; in regions of shell overlap some fitting procedure will be necessary.

As mentioned at the outset of the present section, one expects that more detailed information regarding the local arrangements in a binary alloy can be derived from single-crystal data, in which the three-dimensional aspects of $F(\mathbf{K})$ are retained. This is true, in large part, because the non-spherically-averaged data more readily allow effects to be separated from one another, to the extent that the $\mathbf{K}$ dependences of intensity associated with different types of parameters are unique. Specific procedures for achieving such separation (in cubic crystals) and for deriving the short-range-order and size-effect parameters have been described in detail by Borie and Sparks (1966, 1971).

---

[1] It is necessary to point out, however, that the neglect of all size differences beyond the first few would not be valid in calculating a $K$-space function, such as the intensity. Even very small effects which converge to zero in $r$ space can (and usually do) accumulate into major effects at particular points in $K$ space, often in the vicinity of the Bragg peaks.

## I. The use of pair distributions in the determination of crystal structures and the study of structural defects

In equation (2-145), $G(r)$ for a multicomponent system is expressed in terms of the partial distributions $4\pi r \rho_{jl}(r)$ for each of the different pair types. The analysis procedures described in the remainder of Sections H 1 and in H 2 can be applied to polycrystalline as well as to noncrystalline materials. In the former case, if the structure is known, the partial distributions can be written out explicitly in the sum-of-peaks form described for pure elements,

$$4\pi r \rho_{jl}(r) = \sum_{i=1}^{\infty} \frac{c_i^{jl}}{r_i^{jl}} B^{jl}(r - r_i^{jl}) \qquad (2\text{-}200)$$

for which the $c_i^{jl}$ and $r_i^{jl}$ could be readily computed from the structure parameters. Thus, if the $B^{jl}$ functions were expressed as $\exp\left[-(r - r_i^{jl})^2/2\gamma_i^{jl}\sigma_{\infty}^2\right]$, the "unknowns" $\gamma_i^{jl}$ could be determined by fitting the experimental $G(r)$ with calculated functions.

On the other hand, if the structure were not known, it might also be feasible to determine the structure by trial-and-error variation and refinement of all the structure parameters, aimed at matching the experimental $G(r)$. Since the lattice parameters $\mathbf{a}_1$, $\mathbf{a}_2$, and $\mathbf{a}_3$ are usually known from the Bragg intensity peak positions alone, it is "only" necessary to determine the positions of the atoms within the unit cell of the structure. This has been shown to be a practical procedure in at least one instance (Strong and Kaplow, 1968; Strong, Wells, and Kaplow, 1971) where the number of free parameters was not very large. The unknown parameters in the work referred to were determined by a random-walk procedure in which computer-model atoms were allowed to move from plausible (but imprecise) initial arrangements in order to improve the fit of the model to $G(r)$.

Such an analysis, which focuses on $G(r)$, carries the advantage that one is working directly with the near-neighbor arrangements, which are often the main interest. In addition, even if single-crystal data can be obtained for the material in question, the polycrystalline pair distribution may provide useful information preliminary to a full picture of the structure, such as near-neighbor distances and coordination numbers.

As a further example of the application of atomic pair distributions to polycrystalline materials, consider the question of structural defects, particularly as evidenced by perturbations in the near-neighbor correlations. Such defects might be introduced, for example, during solidification (or crystallization), by deformation, or by irradiation. In general, one expects that structural defects might be evidenced in $G(r)$ mainly by (1) different, broader shape functions $B_i(\varepsilon)$ and/or (2) the appearance of additional $r_i$, that is, of metastable pair distances which are characteristic of the specific defects and structure.

It may be recognized that for such an application, proper analysis of data termination is an important and perhaps the most difficult aspect. On the other hand, to the extent that the defects cause the few nearest-neighbor peaks to be less sharp (as is likely), the termination of data effects will be less important than for a "perfect" sample.

The local arrangements in many amorphous materials are known to be reasonably well defined (albeit less well so than in a crystalline material). It is perhaps also meaningful, therefore, to consider studying "structural defects" in the local arrangements in such materials, whether caused by deformation or irradiation or through variation of solidification and subsequent thermal treatment. Relatively little work has been done regarding structural defects (crystalline or amorphous) from the pair-distribution point of view, but a somewhat more detailed discussion than the present one has been given by Kaplow (1972).

More commonly used procedures for studying structural defects in poly-crystalline samples utilize Fourier analysis of the individual Bragg diffraction peaks and consideration of small shifts in the diffraction peak positions. A discussion of those methods, as applied to small average strain and particle size effects and to stacking-fault and twin-type defects, is given by Warren (1969, chap. 13). The $K$-space procedures tend to be complementary to an interpretation of $G(r)$ since they may provide estimates of the densities of defects and of the long-range strain variations, but essentially no direct information regarding the larger, local disturbances. It is also interesting to note that the major difficulty associated with the Bragg peak shape analyses is the separation of the individual peaks, which tend to be broad and to overlap one another. That problem does not arise in the general $r$-space analysis.

As is true for most aspects of crystalline materials, more detailed information can be derived from a complete analysis of single-crystal data. In this chapter, the main discussion of structural defects in crystals is given in that framework, in Sections III E, F, and particularly G.

## III.  SINGLE CRYSTALS

Most solid materials are naturally crystalline, although some are only partly crystalline and others noncrystalline. Quite often the latter can be made crystalline by suitable treatment in the laboratory. One consequence of crystallinity is that the determination of the detailed structure, or atomic arrangement, is considerably simplified for comparably complex compositions. Some ways that crystals interact with x rays are considered in this section while much more extensive discussions are presented in the subsequent chapters.

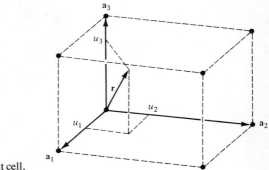

FIGURE 2-29
Parallelepiped unit cell.

## A. The geometry of crystals

In a crystal structure, by definition, a unique arrangement of scatterers is repeated regularly, with three-dimensional translational periodicity extending over large distances compared to the translation period. If the three-dimensional periodicity is defined by a triplet of translations $a_1$, $a_2$, $a_3$, then any scatterer located at point $\mathbf{R}$ in a crystal must be repeated also at

$$\mathbf{R} + \mathbf{T} = \mathbf{R} + m_1 a_1 + m_2 a_2 + m_3 a_3 \quad (2\text{-}201)$$

where $m_1$, $m_2$, and $m_3$ are positive or negative integers. When dealing with a crystalline material, therefore, it is not inherently necessary to refer to a continuous three-dimensional function $p(\mathbf{R})$ extending over the entire sample; rather it suffices to specify the three translational repetitions and the arrangement within the group of scatterers being repeated. In other words, the function describing the distribution in a crystal must itself be periodic so that

$$p(\mathbf{R}) = p(\mathbf{R} + m_1 a_1 + m_2 a_2 + m_3 a_3) \quad (2\text{-}202)$$

**1.** *Unit cells* It follows directly from the above that the structure of an entire crystal is built up from regular repetitions of identical *unit cells* which pack together to fill all space. The three-dimensional periodicity can be represented by a space lattice of points having $a_1$, $a_2$, and $a_3$ as its basic vectors. The same three vectors can be chosen to define a parallelepiped unit cell in this lattice, Fig. 2-29, and any point at $r$ within the cell can be represented in terms of such *unit-cell vectors* by

$$\mathbf{r} = u_1 a_1 + u_2 a_2 + u_3 a_3 \quad (2\text{-}203)$$

where $u_1$, $u_2$, and $u_3$ are fractional coordinates along the three cell edges (Fig. 2-29) and can take on all values from 0 to, but not including, 1.

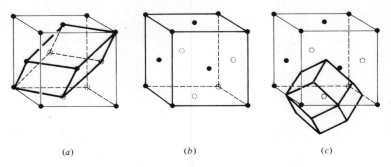

(a)                          (b)                          (c)

FIGURE 2-30
Three possible unit cells in the structure of a "face-centered cubic" monatomic
material. (a) Primitive parallelepiped or reduced cell; (b) face-centered cube;
(c) Wigner-Seitz cell.

There are several possible ways of selecting the unit cell in a lattice. An obvious
one is to choose a cell that contains the minimum amount of scattering matter;
i.e., a cell that includes only one lattice point. Such a cell is called a *primitive cell*.
For parallelepiped cells, if the three unit-cell edges are chosen to be the three shortest,
mutually noncoplanar translations in the lattice, then they are said to define the
*reduced cell* in the lattice. This concept was first proposed by P. Niggli and more
recently propounded by M. J. Buerger, who developed several vector-algebraic
schemes for deriving the reduced cell in any lattice. A tabulation of reduced cells
and the transformation matrices to conventional cells has been compiled by Azároff
and Buerger (1958, chap. 11).

The reason that, conventionally, one may choose a cell that is larger than the
reduced or primitive cell in a lattice is illustrated in Fig. 2-30. The primitive cell in
Fig. 2-30a has the shape of a rhombohedron and does not display the cubic symmetry
of the lattice, whereas the *face-centered* unit cell shown in Fig. 2-30b does. Since the
array of scatterers in the cubic cell also has cubic symmetry, it is clearly advantageous
to select the face-centered cubic cell as the unit cell for the structure. (Note, however,
that this cell contains four lattice points or four times the volume of the primitive
cell.) As first correctly demonstrated by Bravais, one can distinguish a total of 14
unique space lattice types, usually called the *Bravais lattices*.

It is always possible to construct a unit cell in each of the Bravais lattices that is
both primitive and correctly displays the point group symmetry of the lattice. Such
a cell is formed by bisecting the lines joining a lattice point to its nearest neighbors by
planes. The polyhedron formed in this way was called a *Voronoi zone* by Delaunay
(1933), who first tabulated all such possible cells, but is more familiar to solid-state
physicists as a *Wigner-Seitz* cell, originally devised for other purposes. The Voronoi

zone or Wigner-Seitz cell in a face-centered cubic lattice is shown in Fig. 2-30c. Note that these cells also fill all of space by periodic repetition along the translations of the lattice but are not as well suited for depicting the crystal structure as the parallelepiped cells. Voronoi zones constructed in reciprocal or $K$ space, called *Brillouin zones*, however, are very useful for the discussion of certain diffraction phenomena.

**2. Transformations**  It is obvious from the definition of translation **T** in (2-201) that it is a straightforward matter to express the axial relations between the unit-cell vectors of a primitive (unprimed) and a centered (primed) cell

$$\mathbf{a}'_1 = u_1\mathbf{a}_1 + v_1\mathbf{a}_2 + w_1\mathbf{a}_3$$
$$\mathbf{a}'_2 = u_2\mathbf{a}_1 + v_2\mathbf{a}_2 + w_2\mathbf{a}_3$$
$$\mathbf{a}'_3 = u_3\mathbf{a}_1 + v_3\mathbf{a}_2 + w_3\mathbf{a}_3$$

by correctly selecting the *transformation matrix*

$$\begin{Vmatrix} u_1 & v_1 & w_1 \\ u_2 & v_2 & w_3 \\ u_3 & v_3 & w_3 \end{Vmatrix} \tag{2-204}$$

For example, for the face-centered cubic cell in Fig. 2-30b, the transformation matrix expressed in terms of the cell edges of the primitive cell in Fig. 2-30a is

$$\begin{Vmatrix} +1 & +1 & -1 \\ +1 & -1 & +1 \\ -1 & +1 & +1 \end{Vmatrix}$$

It is easy to show (Buerger, 1942, chap. 4) that the same transformation matrix (2-204) also applies to the transformation of the Miller indices $h$, $k$, $l$, from one unit-cell choice to the other.  For the face-centered cubic cell in Fig. 2-30b, therefore, the new indices are

$$h = h_P + k_P - l_P$$
$$k = h_P - k_P + l_P$$
$$l = -h_P + k_P + l_P$$

where the subscript $P$ denotes the indices in the primitive cell. Now, it is well known that the Miller indices $h$, $k$, $l$ can take on all positive or negative integer values in a primitive lattice, but this is no longer the case for a nonprimitive or centered lattice. The reason for this is illustrated by the above transformation relations for the face-centered cubic lattice. Adding any pair of rows yields relations like $h + k = 2h_P$, or, in other words, in a face-centered lattice the three indices $h$, $k$, $l$ must be all odd or

all even. Such *systematic absences*[1] of certain $h$, $k$, $l$ combinations are a direct consequence of the choice of a nonprimitive unit cell and form a very convenient basis for
identifying the appropriate lattice type in diffraction experiments.

## B.   The reciprocal lattice; the diffraction pattern for a single crystal

First, consider a material in which there is one, and only one, point scatterer associated with each lattice point. The intensity distribution diffracted by such an array
will be zero except at *points* in **K** space which themselves form a lattice array, called
the *reciprocal lattice*. This may be seen through the following discussion. The three-
dimensional pair-distribution function of a lattice of point scatterers is the set of all
translation distances

$$\rho(\mathbf{r}) = \sum \delta(\mathbf{r} - \mathbf{D}_{n_1 n_2 n_3}) \qquad (2\text{-}205)$$

where $\mathbf{D}_{n_1 n_2 n_3} = (n_1 - n_1')\mathbf{a}_1 + (n_2 - n_2')\mathbf{a}_2 + (n_3 - n_3')\mathbf{a}_3$ and the sum is to be
taken over all $n_1$, $n_1'$, $n_2$, $n_2'$, $n_3$, and $n_3'$. Ignoring boundary effects, that is, considering
an effectively infinite specimen, the $\mathbf{D}_{n_1 n_2 n_3}$ simply duplicate the set of lattice vectors
$N$ times, where $N$ is the number of lattice points in the sample.

$$\rho(\mathbf{r}) = N \sum_{m_1 m_2 m_3} \delta[\mathbf{r} - (m_1 \mathbf{a}_1 + m_2 \mathbf{a}_2 + m_3 \mathbf{a}_3)] \qquad (2\text{-}206)$$

where the $m$'s can be positive or negative integers. The intensity, which is proportional
to the Fourier transform of $\rho(\mathbf{r})$ [as in equation (2-24)], is given by[2]

$$\frac{I}{Na^2} = \int \rho(\mathbf{r}) \exp(i\mathbf{K} \cdot \mathbf{r}) \, d\mathbf{r} \qquad (2\text{-}207)$$

In this case, using (2-206),

$$\frac{I}{Na^2} = \sum_{m_1 m_2 m_3} \exp[i\mathbf{K} \cdot (m_1 \mathbf{a}_1 + m_2 \mathbf{a}_2 + m_3 \mathbf{a}_3)] \qquad (2\text{-}208)$$

**K** can be specified in terms of any coordinate system that is convenient. Since the
distribution function for the scatterers is not specified in terms of the usual Cartesian
coordinates, there is no inducement to use them for **K**, either. Therefore, let **K** be
specified in a freer format

$$\mathbf{K} \equiv h\mathbf{b}_1 + k\mathbf{b}_2 + l\mathbf{b}_3 \qquad (2\text{-}209)$$

where $h$, $k$, and $l$ are continuous variables and $\mathbf{b}_1$, $\mathbf{b}_2$, and $\mathbf{b}_3$ are three mutually noncoplanar vectors. Noting that the dot product of **K** with vectors constructed from

---

[1] Sometimes called *lattice extinctions*.
[2] In this section it is more convenient to refer to the number of scatterers $N$ explicitly so that the intensity $I$ is defined as a total intensity rather than on a per-scatterer
basis.

$a_1$, $a_2$, and $a_3$ is required (and having some foreknowledge of the result), it is easy to see that it is convenient to define the b's such that they are each orthogonal to two of the a's. Let

$$\mathbf{b}_1 = A(\mathbf{a}_2 \times \mathbf{a}_3) \qquad (2\text{-}210a)$$

$$\mathbf{b}_2 = A(\mathbf{a}_3 \times \mathbf{a}_1) \qquad (2\text{-}210b)$$

$$\mathbf{b}_3 = A(\mathbf{a}_1 \times \mathbf{a}_2) \qquad (2\text{-}210c)$$

where $A$ is a constant. Also, because of the behavior of the circular function $e^{i\alpha}$, it is convenient to scale the b's such that

$$\mathbf{b}_1 \cdot \mathbf{a}_1 = \mathbf{b}_2 \cdot \mathbf{a}_2 = \mathbf{b}_3 \cdot \mathbf{a}_3 = 2\pi \qquad (2\text{-}211)$$

These two conditions can be combined to give

$$\mathbf{b}_1 = \frac{2\pi(\mathbf{a}_2 \times \mathbf{a}_3)}{V} \qquad (2\text{-}212a)$$

$$\mathbf{b}_2 = \frac{2\pi(\mathbf{a}_3 \times \mathbf{a}_1)}{V} \qquad (2\text{-}212b)$$

$$\mathbf{b}_3 = \frac{2\pi(\mathbf{a}_1 \times \mathbf{a}_2)}{V} \qquad (2\text{-}212c)$$

where $V$ is the volume of the lattice unit cell $\mathbf{a}_1 \cdot (\mathbf{a}_2 \times \mathbf{a}_3)$. With that simplifying notation for $\mathbf{K}$, and using (2-208), the intensity becomes

$$\frac{I}{Na^2} = \sum_{m_1 m_2 m_3} \exp\left[2\pi_i(hm_1 + km_2 + lm_3)\right] \qquad (2\text{-}213)$$

Since each combination in the argument of the exponential must have a corresponding negative, the $i \sin \alpha$ term can be dropped in the expansion of exp $(i\alpha)$ [as in equations (2-14)].

$$\frac{I}{Na^2} = \sum_{m_1 m_2 m_3} \cos\left[2\pi(hm_1 + km_2 + lm_3)\right] \qquad (2\text{-}214)$$

If the limits on $m_1 m_2 m_3$ are large, all general values of $h$, $k$, and $l$ will lead to a quite uniform sampling of the angle, and the sum of all negative and positive cosine values will consequently always approach zero. Only if $h$ and $k$ and $l$ are integers, so that the argument is *always* an integer times $2\pi$, will the summation be nonzero. In that instance it approaches $N$. Therefore, the result for the intensity distribution is

$$\frac{I}{N^2 a^2} = \sum_{h,k,l} \delta[\mathbf{K} - (h\mathbf{b}_1 + k\mathbf{b}_2 + l\mathbf{b}_3)] \qquad (2\text{-}215)$$

where $h$, $k$, and $l$ are integers.

The vectors

$$\mathbf{G}_{hkl} \equiv h\mathbf{b}_1 + k\mathbf{b}_2 + l\mathbf{b}_3 \qquad (2\text{-}216)$$

with $h$, $k$, and $l$ integers, are the translation vectors in the *reciprocal lattice* of the real-space lattice defined by $\mathbf{a}_1$, $\mathbf{a}_2$, and $\mathbf{a}_3$. The choice of the particular letters $h$, $k$, and $l$ to designate the reciprocal-lattice points is a purposeful one, since $\mathbf{G}_{hkl}$ bears a particular relationship to the $(hkl)$ planes in the original lattice. In particular, $\mathbf{G}_{hkl}$ is perpendicular to the $(hkl)$ planes and $|\mathbf{G}_{hkl}|$ is equal to $2\pi/d_{hkl}$, where $d_{hkl}$ is the interplanar spacing between $(hkl)$ planes.[1]

In order to extend the foregoing calculation to the more realistic circumstance of a finite specimen, use is made again of the convolution theorem of Fourier transforms. As was discussed in an earlier section, the actual shape of the sample gives rise to a shape function $f(\mathbf{r})$ which yields an effective distribution function

$$\rho(\mathbf{r}) = f(\mathbf{r})\rho_T(\mathbf{r}) \qquad (2\text{-}217)$$

If the intensity distribution that would result from $\rho_T(\mathbf{r})$ is known [see equation (2-215)], the intensity from $\rho(\mathbf{r})$ is that intensity convolved with the Fourier transform of the damping function $f(\mathbf{r})$. Thus if $W(\boldsymbol{\varepsilon}) \equiv \tau[f(\mathbf{r})]$, the intensity resulting from the finite specimen is

$$\frac{I(\mathbf{K})}{N^2 a^2} = \int_{-\infty}^{+\infty} \left[ \sum_{hkl} \delta(\boldsymbol{\kappa} - \mathbf{G}_{hkl}) \right] W(\boldsymbol{\kappa} - \mathbf{K}) \, d\boldsymbol{\kappa} \qquad (2\text{-}218)$$

That is, each $\delta$ function representing the intensity which ideally occurs at each of the reciprocal-lattice points is replaced by a finite peak—having the same area. The shape of the peak is the three-dimensional Fourier transform of the specimen-size shape function.

## C.  The structure factor

The convolution theorem may be used in reverse order to extend the calculation to apply to structures in which there is a general distribution of scatterers $p^0(\mathbf{u})$ associated with every one of the lattice points. In the original equation for $\rho(\mathbf{r})$, equation (2-205), each of the $\delta$ functions represents the correlation between two lattice points. Given a more general $p^0(\mathbf{u})$ associated with each of the lattice points, each $\delta$ function must be replaced by (that is, convolved with) the actual pair distribution associated with the two points, which we will call $\rho^0(\mathbf{q})$. Thus the intensity from the more general structure will be that from the idealized point lattice, *multiplied* by $\tau[\rho^0(\mathbf{q})]$. Further, note that $\rho^0(\mathbf{q})$ is itself the convolution of two identical functions

$$\rho^0(\mathbf{q}) = \int_{\substack{\text{unit} \\ \text{cell}}} p^0(\mathbf{u}) p^{0*}(\mathbf{u} - \mathbf{q}) \, d\mathbf{u} \qquad (2\text{-}219)$$

---

[1] Note that, according to (2-216), it is possible to have reciprocal-lattice vectors whose $h$, $k$, $l$ values contain a common factor $n$. Such vectors correspond to fictitious planes $(nh,nk,nl)$ in the crystal and physically represent the $n$th order of diffraction by the planes $(hkl)$.

Therefore, we define a *structure factor*

$$\mathfrak{F}(\mathbf{K}) \equiv \tau[p^0(\mathbf{u})]$$

$$= \frac{V}{a_1 a_2 a_3} \iiint p^0(u_1 \mathbf{a}_1 + u_2 \mathbf{a}_2 + u_3 \mathbf{a}_3)$$

$$\times \exp\left[i\mathbf{K} \cdot (u_1 \mathbf{a}_1 + u_2 \mathbf{a}_2 + u_3 \mathbf{a}_3)\right] da_1 \, da_2 \, da_3 \qquad (2\text{-}220)$$

and $\mathfrak{F}_{hkl}$ as the value of $\mathfrak{F}(\mathbf{K})$ at $\mathbf{K} = \mathbf{G}_{hkl}$:

$$\mathfrak{F}_{hkl} = \frac{V}{a_1 a_2 a_3} \iiint p^0(u_1 \mathbf{a}_1 + u_2 \mathbf{a}_2 + u_3 \mathbf{a}_3)$$

$$\times \exp\left[2\pi i(h_1 u_1 + h_2 u_2 + h_3 u_3)\right] da_1 \, da_2 \, da_3 \qquad (2\text{-}221)$$

where $V/a_1 a_2 a_3$ is a normalization factor for nonorthogonal cells. A more general expression for the intensity from a crystalline material may then be written

$$\frac{I(\mathbf{K})}{Na^2} = \sum_{hkl} \mathfrak{F}\mathfrak{F}_{hkl}^* W(\mathbf{K} - \mathbf{G}_{hkl}) \qquad (2\text{-}222)$$

where the structure factor $\mathfrak{F}(\mathbf{K})$ is the (possibly complex) Fourier transform of the scatterer distribution associated with each lattice point, and $W(\varepsilon)$ is the Fourier transform of the actual specimen shape function $f(\mathbf{r})$, while $\mathbf{G}_{hkl}$ is a reciprocal-lattice vector.

Equation (2-222) is the fundamental description of the intensity distribution yielded by single-crystal specimens. In experimental application, the mere detection of a significant number of intensity maxima in $\mathbf{K}$ space (that is, the measurement of a number of $\mathbf{G}_{hkl}$ values) is sufficient to uniquely determine the reciprocal and real-space lattices. Beyond that, special conditions of symmetry in $p^0(\mathbf{u})$ can be detected through the systematic *absence* of certain $hkl$ reflections (because $\mathfrak{F}_{hkl}$ is zero) once the correct lattice is firmly established.

In principle, however, the determination of a crystal structure from single-crystal specimens depends on measurement of the total (integrated) intensity at each of the $(hkl)$ reciprocal-lattice points. Those values, corrected for extraneous effects (see Chap. 7), and normalized so that the intensities are converted to the previously discussed electron units, are the squares of the structure factors $\mathfrak{F}(\mathbf{K})\mathfrak{F}^*(\mathbf{K})$ at $\mathbf{K} = \mathbf{G}_{hkl}$, the only values of $\mathbf{K}$ at which they can be measured.

Since the $\mathbf{G}_{hkl}$ constitute a complete set in the framework of Fourier transform theory, however, it is possible to reconstruct the density $p^0(\mathbf{u})$ from the summation

$$p^0(\mathbf{u}) = \frac{1}{V} \sum_{hkl} \sum \sum \mathfrak{F}_{hkl} \exp\left(-i\mathbf{G}_{hkl} \cdot \mathbf{u}\right) \qquad (2\text{-}223)$$

if an adequate number of the *amplitudes* are determined. The fundamental ambiguity of diffraction theory remains, however, since only the magnitude of a generally complex $\mathfrak{F}_{hkl}$ can be deduced from the complex conjugate product $\mathfrak{F}\mathfrak{F}^*$. From the practical point of view, however, there are generally many more data values available than there are unknown (positional) parameters to determine, so that a consistent specification can be achieved by trial-and-error and/or by other more direct methods. A discussion of such specialized procedures lies outside the scope of this book.

## D.  Geometrical constructs in reciprocal space

The incident and scattered x-ray beams were previously defined in terms of unit direction vectors $\mathbf{S}_0$ and $\mathbf{S}$. It is convenient to also define the analogous nonunit vectors

$$\boldsymbol{\kappa}_0 \equiv \frac{2\pi}{\lambda}\,\mathbf{S}_0 \quad (2\text{-}224a)$$

$$\boldsymbol{\kappa} \equiv \frac{2\pi}{\lambda}\,\mathbf{S} \quad (2\text{-}224b)$$

in order to discuss vectors of specific magnitudes in reciprocal space. The previous derivations have been in terms of the *diffraction vector* $\mathbf{K} = \boldsymbol{\kappa} - \boldsymbol{\kappa}_0$, and the condition for ideal crystal diffraction

$$\mathbf{K} = \mathbf{G}_{hkl} \quad (2\text{-}225)$$

is, equivalently,

$$\boldsymbol{\kappa} - \boldsymbol{\kappa}_0 = \mathbf{G}_{hkl} \quad (2\text{-}226)$$

In Fig. 2-31, a beam $\boldsymbol{\kappa}_0$ is shown which is incident on a crystal specimen at an angle $\beta$ to the normal to the $(hkl)$ planes. The normal to the $(hkl)$ planes is designated by $\mathbf{G}_{hkl}$, the associated reciprocal-lattice vector.

Since the magnitudes of $\boldsymbol{\kappa}$ and $\boldsymbol{\kappa}_0$ are identical (for no energy loss upon scattering), note that in order for $\boldsymbol{\kappa} - \boldsymbol{\kappa}_0$ to equal $\mathbf{G}_{hkl}$, $\boldsymbol{\kappa}$ must fall on a sphere (about the origin) of radius $2\pi/\lambda$ *and* that $-\boldsymbol{\kappa}_0$ must end on a sphere of the same radius about the $\mathbf{G}_{hkl}$ terminal point. Thus the intersection of these two spheres defines the only allowable orientations of $\boldsymbol{\kappa}_0$ and $\boldsymbol{\kappa}$ for a given $\lambda$. The intersection is a circle on the plane which is the perpendicular bisector of $\mathbf{G}_{hkl}$.

By symmetry it is obvious that $\boldsymbol{\kappa}$ must also form the angle $\beta$ with $\mathbf{G}_{hkl}$, and that the angles of incidence and diffraction with respect to the $(hkl)$ planes are both equal to $\theta$, one-half the total angle of scattering. By simple geometry,

$$\sin\theta = \frac{|G|_{n(hkl)}}{2|\kappa_0|} \quad (2\text{-}227)$$

and, since

$$|G_{n(hkl)}| = \frac{2\pi n}{d_{hkl}} \quad (2\text{-}228)$$

$$\lambda = \frac{2d_{hkl}\,\sin\theta_{n(hkl)}}{n} \quad (2\text{-}229)$$

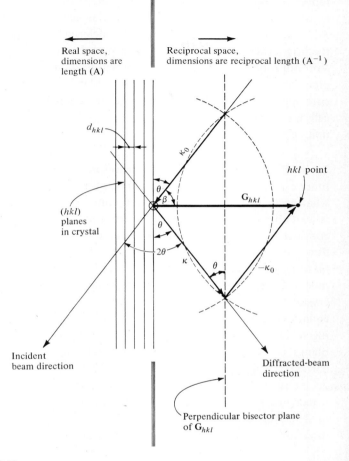

Real space,
dimensions are
length (A)

Reciprocal space,
dimensions are reciprocal length ($A^{-1}$)

$d_{hkl}$

$\kappa_0$

*hkl* point

$\theta$

$\beta$

$G_{hkl}$

(*hkl*)
planes
in crystal

$\theta$

$2\theta$

$\kappa$

$\theta$

$-\kappa_0$

Incident
beam direction

Diffracted-beam
direction

Perpendicular bisector plane
of $G_{hkl}$

**FIGURE 2-31**
Reciprocal-space construction for the crystal diffraction condition.

where $n$ is any positive integer. Equation (2-229) is the well-known Bragg law for the *angles* for diffraction by crystals.

**1.   *Brillouin zones and the Bragg law***   It is clear from examination of Fig. 2-31 that *whenever* a $\kappa_0$ is of such magnitude and orientation that, when it terminates on the origin, its tail is *on* the perpendicular bisecting plane of some **G**, a corresponding beam of radiation will be diffracted. That is, if the conditions are met for $\kappa_0$, there will be a strong scattered beam in the direction specified by $\kappa$. Thus the planes which bisect the **G**'s in the reciprocal lattice are a useful construction. Indeed, because they represent important aspects of the periodicity of crystalline materials they are a useful construct for many aspects of crystalline-state physics. Such planes are called *Brillouin zone boundaries*; the volume enclosed by those planes which are closest to the origin is called the *first Brillouin zone*.[1] It is analogous to the Wigner-Seitz cell or Voronoi zone in the real-space lattice and bears the same relationship to the parallele-piped primitive unit cell of the reciprocal lattice as the Wigner-Seitz cell bears to the corresponding unit cell of the real lattice. Each Brillouin zone or reciprocal-lattice cell has the reciprocal volume $V_{B.Z.} = (2\pi)^3/V$, where $V$ is the volume of a primitive unit cell in the real lattice.

**2.** *Diffraction of a continuum of wavelengths: the Laue pattern* While most quan-titative diffraction work utilizes monochromatic radiation, it is of interest to consider what occurs when the incident beam contains a continuous range of wavelengths, as is available in the Bremsstrahlung from an x-ray tube. As illustrated in Fig. 2-32, consider that x rays having all the available wavelengths are incident in the identical direction, given by $\kappa_0/|\kappa_0|$. The lattice of points shown in Fig. 2-32 represents the $(hk0)$ net of a primitive cubic reciprocal lattice, in which only the 100, 110, and 010 points, **G** vectors, and Brillouin zone boundaries have been shown and labeled. The crystal "selects" out of the wavelengths in the incident beam those which yield $|\kappa_0|$ equal to the distance from the origin to points 1, 2, and 3 (among others) correspond-ing to intersections of the incident beam direction with the perpendicular bisecting planes of $\mathbf{G}_{010}$, $\mathbf{G}_{110}$, and $\mathbf{G}_{100}$, respectively. Diffracted beams then emerge in corresponding directions, as shown. This continuous wavelength analysis is the geometrical basis of the Laue method, in which a film is generally placed so as to detect many of the diffracted beams simultaneously, usually in either the forward- or back-reflection region. Because of their obvious relation to crystallographic directions, Fig. 2-32, Laue patterns are most useful for orienting single crystals.

**3.** *Simultaneous multiple diffraction* As a curiosity, note in Fig. 2-32 that, if the crystal were to be tilted slightly, so that the incident beam passed through point 4, then three diffractions would occur simultaneously, *for a single wavelength*.[2]

In Fig. 2-33, a related phenomenon is illustrated, but one which is less extraor-dinary. The incident, monochromatic beam $\kappa_0$ satisfies the $\mathbf{G}_{100}$ diffraction con-dition and gives a diffracted beam $\kappa$. However, $\kappa$, to the extent that it is still traveling *in* the crystal, must itself be regarded as an incident beam $\kappa_0'$, which, of course, satisfies the diffraction condition to $\mathbf{G}_{\bar{1}00}$, and which gets scattered ($\kappa''$) back into

---

[1] One may speak of the $n$th Brillouin zone in reciprocal space. A point in **K** space lies in the $n$th Brillouin zone if it is necessary to cross $(n-1)$ Brillouin zone boundaries in going to it from the origin.

[2] Of these the 110 would not be detectable because $2\theta_{110} = 180°$ for the corresponding $\lambda$.

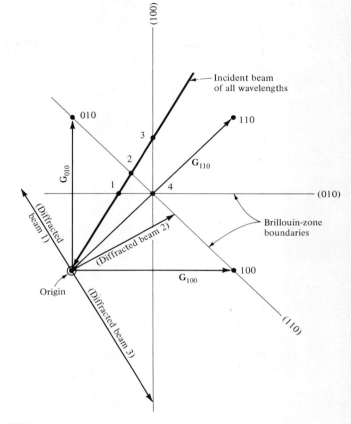

**FIGURE 2-32**
Reciprocal-space construction for single-crystal diffraction of a continuum of wavelengths.

the original beam ($\kappa_0$). Thus, continuing to neglect ordinary (photoelectric) absorption, and assuming that the scattering probability is high, and given a $\kappa_0$ which satisfies the diffraction condition, one may expect that (1) a reflection geometry experiment (as in Fig. 2-34$a$) will transfer all the intensity to $\kappa$, while (2) a transmission geometry experiment will leave an equal division into two beams. The foregoing crude description, however, simply illustrates the need for the more detailed examination of the scattering by perfect crystals that is developed in Chap. 3, in order to predict the absolute intensities for crystals of high perfection.

Another example of multiple diffraction is illustrated in Fig. 2-35. $P_1$ marks the intersection of (1) the intersection of the (100) and (210) Brillouin zone boundaries

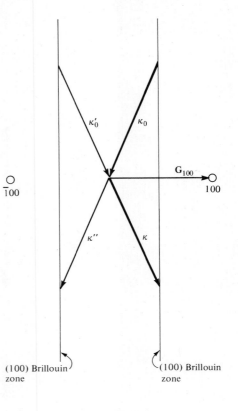

**FIGURE 2-33**
Geometrical mixing of the incident and
strongly diffracted beam.

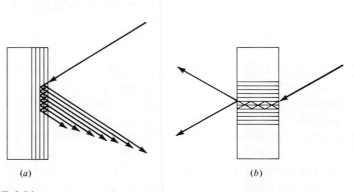

(a)                                        (b)

**FIGURE 2-34**
Schematic behavior under strong Bragg diffraction conditions. (a) Reflection
geometry; (b) transmission geometry.

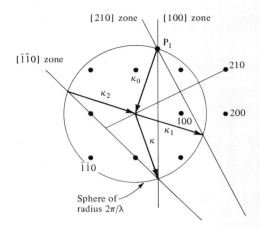

**FIGURE 2-35**
Reciprocal-space construction illustrating simultaneous multiple Bragg diffraction.

(which is a line), and (2) a sphere of radius $2\pi/\lambda$. For Fig. 2-35, $\lambda$ has been chosen such that the sphere is just *tangent* to the intersection line of the two planes so that $P_1$ lies in the $hk0$ net. Otherwise there would be two equivalent $P_1$ points, and a rotation about the [100] axis would be required to show $P_1$ in the plane of the paper, which, by convention, contains the incident and diffracted beams. Thus, given a $\kappa_0$ which emanates from $P_1$, there obtain two diffracted beams, as was illustrated earlier: $\kappa_1$ for the 210 reflection and $\kappa$ for the 100 reflection. Moreover, it can be seen that the 210 reflected beam satisfies the diffraction condition for $G_{\bar{1}\bar{1}0}$ and that the subsequent $\bar{1}10$ reflected beam is equivalent to $\kappa$, the 100 reflected beam! A detector set to measure the intensity of the 100 reflection would then measure more or less than the true intensity, depending (in this instance) on the relative strengths of the 210, $\bar{1}10$, and 100 reflections (and on the magnitude of the photoelectric absorption coefficient). The condition shown would still obtain if the wavelength were different (shorter) with an appropriate rotation about [100], although the $\kappa_1$, $\kappa_2$ beam would not then be coplanar with $\kappa_0$ and $\kappa$. In general, the double-reflection contribution to an $hkl$ reflection from the condition $(h_1 + h_2, k_1 + k_2, l_1 + l_2) = (hkl)$ occurs at two different rotation angles about the [$hkl$] axis. The effect is most obvious when the crystal is fairly perfect and the true $hkl$ reflection is very weak or entirely absent. Figure 2-36 shows a classic example of the apparent 200 silicon reflection intensity as a function of the angle of rotation about its [100] axis. The sharp variations are due to double reflections.

**4.** *Sampling region*  In an actual measurement (with a diffractometer) both the incident and detected beams will include a certain divergence (or angular spread) that will cause the measurement to constitute an average over a small volume element

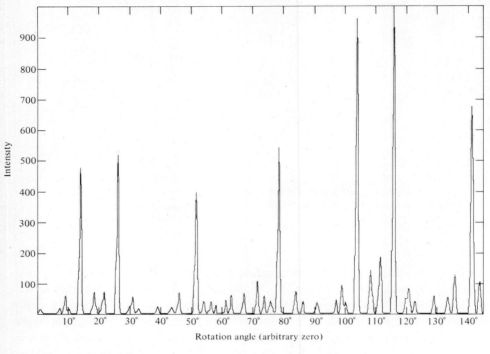

Intensity

Rotation angle (arbitrary zero)

**FIGURE 2-36**
Multiple Bragg diffraction as evidenced in the silicon 200 rotation intensity pattern.

in $\kappa$ space, rather than at a point. (This is discussed in detail in Chap. 7, Section II A.) This is illustrated in Fig. 2-37; the divergence angles $\alpha$ and $\beta$ are usually referred to as the horizontal divergence and $\gamma$ and $\delta$ as the vertical divergence. The volume element is nominally a parallelepiped if the divergence angles are small and if the intensity is uniform within the beam. It can be noted, however, that the shape of the element is not constant; at small $|\mathbf{K}|$ the horizontal cross section tends toward a needlelike shape, oriented parallel to $\mathbf{K}$, while at large $|\mathbf{K}|$ the situation is reversed and the loss of resolution is most pronounced perpendicular to $\mathbf{K}$.

## E.  Particle size effects

One may also use a reciprocal-lattice construction, in which the geometrical points which represent the idealized $\delta$ functions of intensity are each replaced by the more realistic distributions which may result from particular perturbing effects. It has already been shown that a finite-sized specimen yields a peak whose finite shape is the

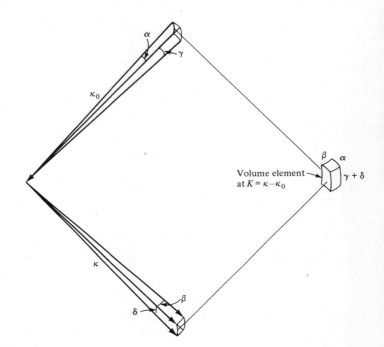

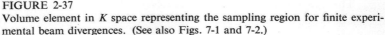

FIGURE 2-37
Volume element in $K$ space representing the sampling region for finite experimental beam divergences. (See also Figs. 7-1 and 7-2.)

transform of the specimen shape. Thus, for example, if a crystal is *exceedingly* thin in one dimension, the diffraction spots are broadened into lines parallel to that dimension, as illustrated for the primitive cubic lattice in Fig. 2-38.[1]  For that sample, if the measurement volume element (with a diffractometer) is displaced along $\mathbf{b}_2$ or $\mathbf{b}_3$ from the center of a diffraction peak, the intensity will drop rapidly—but not along $\mathbf{b}_1$.

This phenomenon is utilized most routinely in high-energy electron transmission diffraction, using film as the detector.  For electron energies of the order of 50 kV, $2\pi/\lambda$ is very large compared to a reciprocal-lattice vector, the scattering angles of interest are correspondingly small, and, because of high absorption, the sample must be thin in the $\boldsymbol{\kappa}_0$ direction.  As Fig. 2-39 illustrates, the locus of $\boldsymbol{\kappa}$ values allowed by the film dimensions is a nearly planar section of a sphere of radius $2\pi/\lambda$, in the vicinity of the origin of $\boldsymbol{\kappa}$ space.  If one of the reciprocal-lattice nets is nearly parallel to that locus, its $\mathbf{G}_{hkl}$ points, elongated into rods along $\boldsymbol{\kappa}_0$, will intersect the allowable

[1] For a general lattice, $\mathbf{a}_1$ and $\mathbf{b}_1$ would not be parallel.

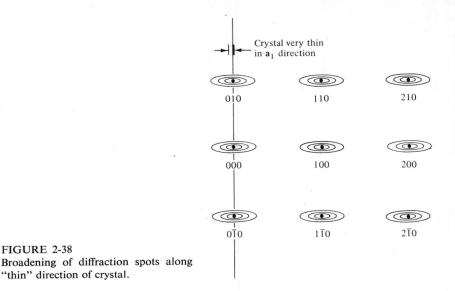

FIGURE 2-38
Broadening of diffraction spots along
"thin" direction of crystal.

$\kappa - \kappa_0$ locus and will be imaged on the film. Such a mapping of an entire section of a reciprocal-lattice net by electron diffraction is illustrated in Fig. 2-40.

The above discussion has illustrated the effect of a small sample dimension on the shape of the diffraction peaks. Similarly, if the sample is broken up into numerous small mosaics (perhaps by deformation) which are incoherently arranged with respect to one another, the shape of the diffraction peaks will be the average of the Fourier transform of the shape functions of the small internally coherent pieces. As a qualitative view, it is important to note that this type of perturbation [a damping of the distribution functions by $f(\mathbf{r})$] leads to a uniform broadening of the peaks (i.e., a convolving of the intensity function with a single $hkl$ independent function

$$\langle V(\varepsilon) \rangle = \langle \tau[f(\mathbf{r})] \rangle \qquad (2\text{-}230)$$

averaged over all pieces).

## F. Lattice parameter variations; the simplest model for K-dependent perturbations

Unlike the above cases, other kinds of perturbations will not affect all peaks identically. Consider, as a simple example, a situation in which the lattice is characterized by precise $\mathbf{a}_1$ and $\mathbf{a}_2$ but which contains a random mixture of three $\mathbf{a}_3$'s: $\mathbf{a}_3{}^0$ and $\mathbf{a}_3{}^0 \pm \delta\mathbf{a}_3{}^0$, where $\delta \ll 1$. Corresponding to each value of $\mathbf{a}_3$, there will be a $\mathbf{b}_3$ in the reciprocal-lattice representation of the diffraction maxima

$$|\mathbf{b}_3| = \frac{2\pi \cos \alpha}{|a_3|} \qquad (2\text{-}231)$$

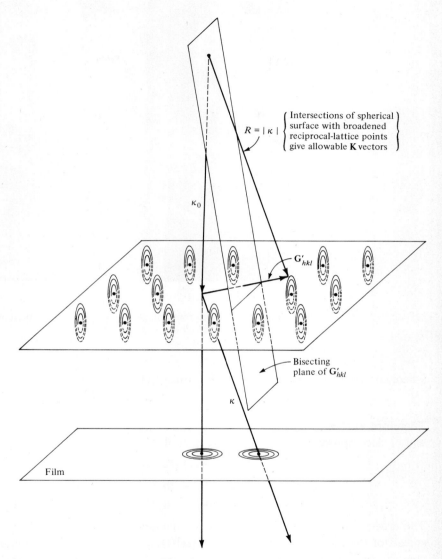

R = |κ|  { Intersections of spherical
          surface with broadened
          reciprocal-lattice points
          give allowable **K** vectors }

$\kappa_0$

$\mathbf{G}'_{hkl}$

Bisecting
plane of $\mathbf{G}'_{hkl}$

$\kappa$

Film

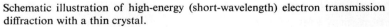

**FIGURE 2-39**
Schematic illustration of high-energy (short-wavelength) electron transmission
diffraction with a thin crystal.

FIGURE 2-40
Electron diffraction pattern from a single crystal of hexagonal titanium, imaging many points in a reciprocal-lattice plane simultaneously.

where $\cos \alpha$ is the $\mathbf{a}_3$, $\mathbf{b}_3$ included angle. Since $1/(1 + \delta) \approx 1 - \delta$, the three $\mathbf{b}_3$'s are

$$\mathbf{b}_3 = \mathbf{b}_3{}^0 \quad \text{and} \quad \mathbf{b}_3{}^0(1 \pm \delta) \quad (2\text{-}232)$$

Therefore, each $\mathbf{G}_{hk}$ in the reciprocal-lattice representation of the diffraction maxima will be decomposed into three, one for each of the $\mathbf{a}_3$:

$$\mathbf{G}_{hkl}^{(1)} = h\mathbf{b}_1 + k\mathbf{b}_2 + l\mathbf{b}_3{}^0 + l\mathbf{b}_3{}^0\delta = \mathbf{G}_{hkl}^0 + l\mathbf{b}_3\delta \quad (2\text{-}233a)$$

$$\mathbf{G}_{hkl}^{(2)} = h\mathbf{b}_1 + k\mathbf{b}_2 + l\mathbf{b}_3{}^0 \qquad\qquad = \mathbf{G}_{hkl}^0 \quad (2\text{-}233b)$$

$$\mathbf{G}_{hkl}^{(3)} = h\mathbf{b}_1 + k\mathbf{b}_2 + l\mathbf{b}_3{}^0 - l\mathbf{b}_3{}^0\delta = \mathbf{G}_{hkl}^0 - l\mathbf{b}_3\delta \quad (2\text{-}233c)$$

The crucial point, as this simple example demonstrates, is the increasingly greater spread of the intensity away from the ideal $\mathbf{G}_{hkl}^0$ as the order of the reflection $l$ increases. In addition, note that each of the three "spots" associated with each $(hkl)$ would not be points, in general. Rather, each would have the shape determined by the transform $V(\varepsilon)$, of the appropriate size function $f(\mathbf{r})$, relevant to the macroscopic specimen size, or to the "coherent particles" discussed in the previous section. That is, the resultant intensity peak is that due to the distribution of $\mathbf{a}_3$'s convolved with $V(\varepsilon)$[1] (or vice versa, if one prefers).

---

[1] This is true within the constraint that no specific correlation exists between the distribution of $\mathbf{a}_3$'s within a particle and the specific shape function.

## G.   Numerical calculation of the diffraction by localized defects

In this section we consider the direct calculation of the diffraction intensities associated with defects in imperfect crystalline materials, that is, calculation by numerical evaluation of the primary diffraction equations for atomic positions selected to model the relevant disturbances. The discussion in this section does not apply therefore to nearly perfect crystals, for which the dynamical scattering theory of Chap. 4 is more appropriate. It must be recognized, moreover, that in certain respects the kinematical and dynamical scattering models (which apply to crystals of different perfection) will yield predictions which are *qualitatively* different. This is particularly true regarding the integrated intensity of a Bragg reflection; i.e., of the total intensity in the sharp peak at a reciprocal-lattice point. For many experimental arrangements, the introduction of defects into a nearly perfect crystal will *increase* the Bragg integrated intensity, while imperfect crystal theory usually predicts a *decrease*. Experimentally, the intensity may increase or decrease in a seemingly magical fashion which is, however, totally understandable qualitatively, if not always precisely quantifiable.

These phenomena are associated with the so-called extinction effects, which are discussed further in Section V B of Chap. 7. Briefly, the situation as regards the integrated Bragg intensities between the two extremes of crystal perfection can be described as follows:

In a perfect crystal the incident beam is Bragg diffracted over an extremely narrow angular range ($\sim 3$ seconds of arc). Generally this angular range is less than the divergence in the incident beam. The crystal therefore selects and reflects by diffraction the small portion of the incident beam which satisfies the precise "dynamical" conditions. As a crystal is rotated, to sweep **K** through the reciprocal-lattice point in order to obtain a measure of the integrated intensity, different but similarly narrow portions of the beam are utilized. The remainder of the beam is either transmitted through the crystal or photoelectrically absorbed. With the introduction of certain gentle imperfections (e.g., elastic strains), coherent diffraction can be maintained over a wider (but still very narrow) range of incident angles, albeit with a smaller maximum intensity per atom layer. In reflection geometry this usually leads to somewhat increased integrated intensity for the Bragg reflection. In transmission it may cause an enhancement or a decrease, depending on the precise atomic displacements incurred and also on the photoelectric absorption coefficient. In transmission through thick samples the situation is further complicated by the fact that the photoelectric absorption essentially disappears for part of those beams which precisely satisfy the diffraction condition throughout the crystal. Therefore, a perfect, properly oriented crystal can yield simultaneously a transmitted and a transmitted-diffracted beam, even though the ordinary absorption factor $\mu t$ is essentially infinite. (See also Section I D of Chap. 4.)

With the introduction of more gross imperfections, the crystal can be thought of as comprising many slightly misoriented pieces ("mosaics"), each reasonably perfect. For any one orientation of the crystal, each diffracts that portion of the incident beam which makes the proper angle for it, so that reflection takes place over a large range of incident angles. Moreover, since only a very thin layer is necessary to diffract essentially all of the beam within that layer's narrow range, the integrated intensity will be much enhanced. Roughly, the increase will be of the order of $\alpha/\alpha$(perfect), where $\alpha$(perfect) is the "natural" angular diffraction width and $\alpha$ is the mosaic angular spread or the beam/detector divergence, whichever is smaller.

It is important to note that the "imperfect-crystal" kinematical theory which is being discussed in this chapter, and which always predicts that imperfections decrease the integrated intensity, requires a significant degree of imperfection to be valid. For example, even a polycrystalline sample of cold-worked powders will generally exhibit larger integrated intensities in the low-order reflections than a comparable annealed specimen.

With the foregoing as a disclaimer, it is nonetheless worthwhile to consider the effects that defects can have on the diffraction by single crystals, in the kinematical approximation. A somewhat different discussion is presented in Section II of the next chapter, while the effects of defects in perfect crystals are considered in Chap. 5.

**1.   *Theory***   In the kinematical theory, the *amplitude* of scattering in electron amplitude units is

$$A(\mathbf{K}) = \sum_{j=1}^{N} f_j(K) \exp\left(-i\mathbf{K}\cdot\mathbf{R}_j\right) \qquad (2\text{-}234)$$

where $N$ is the number of irradiated atoms in the sample, $\mathbf{R}_j$ is the vector to the $j$th atom from an arbitrary origin, and $f_j$ is the atomic scattering factor of the $j$th atom. For a crystal sample, the amplitude can be written

$$A^c(\mathbf{K}) = \mathfrak{F}(\mathbf{K}) \sum_{j=1}^{M} \exp\left(-i\mathbf{K}\cdot\mathbf{R}_j\right) \qquad (2\text{-}235)$$

where $\mathfrak{F}(\mathbf{K})$, the structure factor, is the Fourier transform of the scattering material inside one unit cell of the primitive lattice, $\mathbf{R}_j$ is the vector to the origin of the $j$th-unit cell, and $M$ is the number of unit cells in the irradiated sample.

If the sample is very large, the summation will be equal to zero for all $\mathbf{K}$ except $\mathbf{K} = \mathbf{G}$, where $\mathbf{G}$ is the vector to any one of the $hkl$ reciprocal-lattice points. From previous discussion it is known that $\mathbf{G}_{hkl}\cdot\mathbf{R}_j = 2n\pi$, and $\exp\left(-2n\pi i\right) = 1$. That is, the amplitude of scattering for the single crystal is a sequence of $\delta$ functions in $\mathbf{K}$ space, one at each of the reciprocal-lattice points:

$$A^c(\mathbf{K}) = \mathfrak{F}(\mathbf{G})\,\delta(\mathbf{G})M \qquad (2\text{-}236)$$

(If the sample is not large, $\delta(\mathbf{G})$ is replaced by the Fourier transform of the sample shape in three dimensions, as discussed in an earlier section.)

A defect will be modeled by assuming, first, that its effects are localized. Then, a hole will be made in the crystal big enough to contain the defect, and the atoms removed will be replaced with others in the perturbed positions. The amplitude contributed to the original total by those atoms removed to make the hole is

$$A^H(\mathbf{K}) = \sum_{j=1}^{L} f_j(K) \exp(-i\mathbf{K} \cdot \mathbf{R}_j) \qquad (2\text{-}237)$$

where $L$ is the number of atoms in the hole, $f_j(K)$ is the scattering factor of the $j$th atom, and $\mathbf{R}_j$ is the vector to the $j$th atom. Thus, the amplitude for the crystal with the hole is

$$A^{CH}(\mathbf{K}) = \mathfrak{F}(\mathbf{G}) \, \delta(\mathbf{G})M - A^H(\mathbf{K}) \qquad (2\text{-}238)$$

The amplitude of the defective replacement is defined to be $A^D(\mathbf{K})$, which is due to atoms in perturbed positions:

$$A^D(\mathbf{K}) = \sum_{j=1}^{L'} f'_j(K) \exp(-i\mathbf{K} \cdot \mathbf{R}'_j) \qquad (2\text{-}239)$$

where $\mathbf{R}'_j$ is the vector to the $j$th atom, and $L'$ is the number of atoms. It is not necessary that $L = L'$. The total, new amplitude (with one such defect) is then

$$A^{DT}(\mathbf{K}) = \mathfrak{F}(\mathbf{G}) \, \delta(\mathbf{G})M - A^H(\mathbf{K}) + A^D(\mathbf{K}) \qquad (2\text{-}240)$$

Consider the difference

$$\Delta(\mathbf{K}) \equiv A^H(\mathbf{K}) - A^D(\mathbf{K}) \qquad (2\text{-}241)$$

$$\Delta(\mathbf{K}) = \sum_{j=1}^{L} f_j(K) \exp(-i\mathbf{K} \cdot \mathbf{R}_j) - \sum_{j=1}^{L'} f'_j(K) \exp(-i\mathbf{K} \cdot \mathbf{R}'_j) \qquad (2\text{-}242)$$

If the original crystal and the displacement field both possess inversion symmetry, the complex exponentials can be replaced by cosines. If, in addition, the number of atoms is conserved, and the atoms are all of one type, and if displacements are defined such that $\mathbf{R}'_j = \mathbf{R}_j + \boldsymbol{\delta}_j$,

$$\Delta(\mathbf{K}) = f(K) \sum_{j=1} [\cos \mathbf{K} \cdot \mathbf{R}_j - \cos \mathbf{K} \cdot (\mathbf{R}_j + \boldsymbol{\delta}_j)] \qquad (2\text{-}243)$$

In terms of $\Delta(\mathbf{K})$, one may write the total *intensity*, $A^{DT}(\mathbf{K})A^{DT}(\mathbf{K})^*$:

$$I^{DT}(\mathbf{K}) = \mathfrak{F}\mathfrak{F}^* M^2 \, \delta(\mathbf{G}) - M \, \delta(\mathbf{G})[\mathfrak{F} \, \Delta^*(\mathbf{K}) + \Delta(\mathbf{K})\mathfrak{F}^*] + \Delta(\mathbf{K}) \, \Delta^*(\mathbf{K}) \qquad (2\text{-}244)$$

or, since the second term is nonzero only at $\mathbf{K} = \mathbf{G}$,

$$I^{DT}(\mathbf{K}) = \mathfrak{F}\mathfrak{F}^* M^2 \, \delta(\mathbf{G}) \left\{ 1 - \frac{\mathfrak{F} \, \Delta^*(\mathbf{G}) + \Delta(\mathbf{G})\mathfrak{F}^*}{M\mathfrak{F}\mathfrak{F}^*} \right\} + \Delta(\mathbf{K}) \, \Delta(\mathbf{K})^* \qquad (2\text{-}245)$$

If the crystal contains more than one defect, the amplitude differences must be summed, while account is taken of the relative positions of each. In general, one may write for the *l*th of *n* defects

$$\Delta_l(\mathbf{K}) = \sum_{j=1}^{L} f_j(K) \exp(-i\mathbf{K} \cdot \mathbf{R}_{jl}) - \sum_{j=1}^{L'} f'_j(K) \exp(-i\mathbf{K} \cdot \mathbf{R}'_{jl}) \qquad (2\text{-}246)$$

where $\mathbf{R}_{jl} = \mathbf{R}_j + \mathbf{q}_l$, $\mathbf{q}_l$ is the vector to the center of the *l*th defect (take $\mathbf{q}_l$ to be a lattice vector), $\mathbf{R}'_{jl} = \mathbf{R}_j + \boldsymbol{\delta}_{lj} + \mathbf{q}_l$, and $\boldsymbol{\delta}_{lj}$ is the displacement vector of the *j*th atom in the *l*th defect. All the terms in (2-246) contain the common factor $\exp(-i\mathbf{K} \cdot \mathbf{q}_l)$, which may therefore be extracted,

$$\Delta_l(K) = \exp(-i\mathbf{K} \cdot \mathbf{q}_l) \, \Delta_l^{0}(\mathbf{K}) \qquad (2\text{-}247)$$

where $\Delta_l^{0}(\mathbf{K})$ would be the difference amplitude if the defect were centered at $\mathbf{R} = 0$. The total difference amplitude is

$$\sum_{l=1}^{n} \Delta_l(\mathbf{K}) = \sum_{l=1}^{n} \exp(-i\mathbf{K} \cdot \mathbf{q}_l) \, \Delta_l^{0}(\mathbf{K}) \qquad (2\text{-}248)$$

and the total amplitude [analogous to (2-240)]

$$A^{DT}(\mathbf{K}) = \mathfrak{F}(\mathbf{G}) \, \delta(\mathbf{G})M - \sum_{l=1}^{n} \exp(-i\mathbf{K} \cdot \mathbf{q}_l) \, \Delta_l^{0}(\mathbf{K}) \qquad (2\text{-}249)$$

If all the defects are identical,

$$\sum_{l=1}^{n} \Delta_l(\mathbf{K}) = \Delta^{0}(\mathbf{K}) \sum_{l=1}^{n} \exp(-i\mathbf{K} \cdot \mathbf{q}_l) \qquad (2\text{-}250)$$

The quantity $[\sum_{l=1}^{n} \Delta_l(\mathbf{K})]^2$ which is required for the *intensity* expression can be written as

$$\left[ \sum_{l=1}^{n} \Delta_l(\mathbf{K}) \right]^2 = [\Delta^{0}(\mathbf{K})]^2 \left[ \sum_{l=1}^{n} \exp(-i\mathbf{K} \cdot \mathbf{q}_l) \right]^2 \qquad (2\text{-}251)$$

where the "squaring" is meant to include the use of complex conjugates whenever applicable. Separating out the "self-scattering" of each defect yields

$$\left[ \sum_{l=1}^{n} \Delta_l(\mathbf{K}) \right]^2 = [\Delta^{0}(\mathbf{K})]^2 \left\{ \sum_{l=1}^{n} 1 + \sum_{l,l' \neq l}^{n} \exp[-i\mathbf{K} \cdot (\mathbf{q}_l - \mathbf{q}_{l'})] \right\} \qquad (2\text{-}252)$$

where the second summation includes only cross terms and contains $n(n-1)$ terms. If the positions of the various defects are not correlated, equation (2-252) may be simplified.

$$\left[ \sum_{l=1}^{n} \Delta_l(\mathbf{K}) \right]^2 = [\Delta^{0}(\mathbf{K})]^2 [n + n(n-1)\langle\cos \mathbf{K} \cdot \mathbf{q}_v\rangle] \qquad (2\text{-}253)$$

where $\mathbf{q}_v$ is one of the interdefect vectors $(\mathbf{q}_l - \mathbf{q}_{l'})$. For a random distribution of such vectors, $\langle\cos \mathbf{K} \cdot \mathbf{q}_v\rangle$ will be zero, except at $\mathbf{K} = 0$ or at $\mathbf{K} = \mathbf{G}$ if the $\mathbf{q}_v$ are lattice vectors. At those points $\mathbf{K} \cdot \mathbf{q}_v = 2\pi$ and $\langle\cos \mathbf{K} \cdot \mathbf{q}_v\rangle = 1$.

Returning to equation (2-250), note also that the value of the defect amplitude, $\sum_{l=1}^{n} \Delta_l(\mathbf{K})$, at $\mathbf{K} = \mathbf{G}$ is

$$\sum_{l=1}^{n} \Delta_l(\mathbf{G}) = n \, \Delta^0(\mathbf{G}) \qquad (2\text{-}254)$$

From equation (2-249), it may be seen that the intensity is

$$I^{DT}(\mathbf{K}) = \mathfrak{F}\mathfrak{F}^* \, \delta(\mathbf{G})M - [\mathfrak{F}S^*(\mathbf{G}) + S(\mathbf{G})\mathfrak{F}^*] \, \delta(\mathbf{G})M + [S(\mathbf{K})]^2 \qquad (2\text{-}255)$$

where

$$S(\mathbf{K}) \equiv \sum_{l=1}^{n} \Delta_l(\mathbf{K}) \qquad (2\text{-}256)$$

Using equations (2-253) and (2-254), one obtains

$$I^{DT}(\mathbf{K}) = \delta(\mathbf{G})\{\mathfrak{F}\mathfrak{F}^*M^2 - nM[\mathfrak{F} \, \Delta^0(\mathbf{G})^* + \Delta^0(\mathbf{G})\mathfrak{F}^*]$$
$$+ n(n - 1) \, \Delta^0(\mathbf{G}) \, \Delta^0(\mathbf{G})^*\} + n[\Delta^0(\mathbf{K})]^2 \qquad (2\text{-}257)$$

If the probability of finding a defect per unit cell is small (i.e., $p = n/M$; $p \ll 1$) the $n(n - 1)[\Delta^0(\mathbf{G})]^2$ term will be negligible. Thus the intensity for a crystal which contains a number $(pM)$ of identical defects, each independent of and uncorrelated with the others, is

$$I^{DT}(\mathbf{K}) = \mathfrak{F}\mathfrak{F}^*M^2 \, \delta(\mathbf{G})\left\{1 - \frac{p}{\mathfrak{F}\mathfrak{F}^*} [\mathfrak{F} \, \Delta^*(\mathbf{G}) + \Delta(\mathbf{G})\mathfrak{F}^*]\right\} + pM \, \Delta(\mathbf{K}) \, \Delta^*(\mathbf{K})$$

$$(2\text{-}258)$$

It may be helpful to rewrite (2-258) for the special case of a monatomic crystal, with one atom per unit cell. Then, $f(\mathbf{K}) = \mathfrak{F}(\mathbf{K})$, and (2-258) becomes

$$I^{DT}(\mathbf{K}) = \mathfrak{F}^2M^2 \, \delta(\mathbf{G})[1 - 2p \, D(\mathbf{G})] + pM \, \Delta^2(\mathbf{K}) \qquad (2\text{-}259)$$

where $D(\mathbf{G})$ is the *real part* of $\Delta(\mathbf{K})/f$, evaluated at $\mathbf{K} = \mathbf{G}$, and $\Delta^2(\mathbf{K}) = \Delta(\mathbf{K}) \, \Delta^*(\mathbf{K})$. The first term in equation (2-258) [or (2-259)] represents the original Bragg peaks, but reduced in intensity by the factor in brackets. The reduction factor is sensitive to the crystallographic nature of the defect, since $\Delta(\mathbf{G})$ can vary from reciprocal-lattice point to reciprocal-lattice point, depending on the magnitude of the displacements along particular directions; generally speaking, the larger the component of displacement along a $\mathbf{G}$, the larger the $\Delta(\mathbf{G})$ and the less intense the peak.

The second term represents a more diffuse contribution, the shape of which is quite sensitively related to the displacement field associated with the defects. The latter association is particularly meaningful if the defects maintain a particular crystallographic orientation throughout the crystal. In that instance, the summation over defects is just a constant times the term for one defect, as given in equations

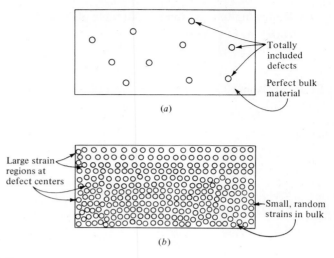

**FIGURE 2-41**
Schematic representation of a localized defect model.

(2-258) and (2-259). If the defects are not all identical, either in form or orientation, it is necessary to consider the actual sum over contributions:

$$I^{DT}(\mathbf{K}) = \mathfrak{F}\mathfrak{F}^* M^2 \, \delta(\mathbf{G}) \left[ 1 - \sum_{l=1}^{n} Q(\mathbf{G}) \right] + \sum_{l=1}^{n} [\Delta_l{}^0(\mathbf{K})]^2 \qquad (2\text{-}260)$$

where $Q_l(\mathbf{G}) = \{\mathfrak{F}[\Delta_l{}^0(\mathbf{G})]^* + \Delta_l{}^0(\mathbf{G})\mathfrak{F}^*\}/\mathfrak{F}\mathfrak{F}^*$. Often, consideration of the summations will be tantamount to averaging over a few equivalent crystallographic orientations. Equation (2-260) is quite general, and should be applicable whenever the defects are uncorrelated in terms of their relative positions, and the displacement fields due to the defects do not overlap. The latter limitation can be relaxed, even for intermediate concentrations of defects which have long-range displacements, by using $\Delta(\mathbf{K})$ to account for the large displacements close to the defects and an approximate model to account for the small and more random displacements in the bulk.

The illustrations in Fig. 2-41a and b portray the nature of the model in a schematic fashion, including the above possibility.

For many sorts of defects, it is quite feasible to evaluate $\Delta(\mathbf{K})$ over interesting areas of $\mathbf{K}$ space directly, using its definition, a set of unperturbed and perturbed positions ($\mathbf{R}_j$ and $\mathbf{R}_j'$) and no further approximations. It should be stressed, moreover, that the schematic description in terms of *spherical* localized defects is purely arbitrary. Whatever shape is appropriate to the displacement field of a particular defect can be used, with little (if any) additional computational complexity. Maps of $\Delta^2(\mathbf{K})$ can be calculated with a very simple computer program, and compared to experimental intensity variations for verification and refinement of defect models. [Of course, $\Delta^2(\mathbf{K})$ should not be calculated directly, since $\Delta(\mathbf{K})$ requires only $1/L$ the number of terms.] A number of examples are presented later in this section.

For numerical calculations, even more than analytic ones, it is important to give careful consideration to deciding what regions of $\mathbf{K}$ space are interesting to examine. Indeed, similar consideration should be given to the experimental measurements, since it will rarely be feasible to measure all of $\mathbf{K}$ space with good resolution. It is worthwhile, therefore, to note some general rules of thumb which will be illustrated by later examples.

    *1*  To the extent that the displacements are correlated within a defect (that is, tend to be in the same direction), the "diffuse" term $[\Delta(\mathbf{K})]^2$ will tend to concentrate near $\mathbf{K} = \mathbf{G}$ (i.e., near reciprocal-lattice points).

    *2*  The more localized a defect, the more spread out its effect will be in $\mathbf{K}$ space. Conversely, if it comprises displacements which are correlated over large distances, the effects will peak up more sharply.

    *3*  The larger the dot products $\mathbf{K} \cdot \boldsymbol{\delta}_j$, the larger the effects. Therefore, if the displacements are not absolutely isotropic, the regions of $\mathbf{K}$ space will be most affected which have $\mathbf{K}$ parallel (or antiparallel) to the predominant $\boldsymbol{\delta}_j$'s. It also follows that the effects will be larger the larger $|\mathbf{K}|$, other factors being equal.

### 2.   *Sample calculations and particular effects*

*a.   Convergence*   From the calculational point of view, it is extremely important that $\Delta(\mathbf{K})$ converges with only that number of atoms which is necessary to describe the defect, not the entire crystal. Convergence must be considered carefully nonetheless, particularly if the displacement field decays slowly as a function of distance from the "center" of the defect. To illustrate the convergence of $\Delta(\mathbf{K})$ for a long-range displacement field, calculations have been done for a simple cubic (one atom per unit cell) structure with $a_0 = 1$ A. The scattering factor $f(K)$ was not included in the calculation, which is equivalent to assuming it to be a constant of unit value. A "defect" has been used in which the atoms are displaced outward radially from the origin (which is the center of the defect) according to the expression

$$\Delta\mathbf{r} = \frac{0.1\mathbf{r}}{r^4} \qquad (2\text{-}261)$$

One might imagine such a displacement field being caused by an overly large substitutional impurity replacing the atom at the origin. It should be stressed, however, that both the structure and the defect model have been selected because they are easily visualized in three dimensions, rather than for physical realism. Because of the symmetry of the defect, the calculations were done for "spherical" pieces; that is, the summations were truncated for three different maximum radii: 7, 12, and 16 times the lattice parameter. These correspond to the inclusion of 1419, 7153, and 17,077 atoms, respectively. It is also of note that the smallest $\delta_j$'s included were respectively 0.0029, 0.0006, and 0.0002 of the ones nearest the origin. The corresponding amplitude $[\Delta(\mathbf{K})]$ and intensity $\{[\Delta(\mathbf{K})]^2\}$ variation along the $\mathbf{K}_x$ axis and in the vicinity of $\mathbf{G}_{100}$ (i.e., $\mathbf{K}_x = 2\pi$; $\mathbf{K}_y = 0$; $\mathbf{K}_z = 0$) are shown in Fig. 2-42. Clearly, the summation has been truncated too soon even for the 7153-atom piece. It is important to note, however, that the fundamental problem is not the additional computer time required to include more terms. Rather, one would be more concerned about *overlap* among defects when the effects do not converge within relatively large pieces. It is worthwhile, therefore, to consider some of the statistical properties of randomly arranged defects. If a defect is centered on a lattice site, the probability that another defect is *not* found within the $n$ nearest sites is $(1 - x)^n$, where $x$ is the probability that a site is a defect center. The radius of the associated volume is

$$R = \left(\frac{3n}{4\pi\rho_0}\right)^{1/3} \qquad (2\text{-}262)$$

For the simple cubic structure used here, $R \approx (0.25n)^{1/3}$. The probability is $x(1 - x)^n$ that the nearest neighboring defect of a given defect occurs at the $n + 1$ site away from the first $[x(1 - x)^n = (1 - x)^n - (1 - x)^{n+1}]$. The average number of sites to the nearest neighboring defect is $\langle n \rangle = 1/x$. More than 60% of the defects will have at least one nearest neighbor closer than $\langle n \rangle$. If $(x \lesssim 0.05)$, the fraction of defects $q$ which have their closest neighboring defect within $n$ sites is given by

$$\ln\left(\frac{1}{1 - q}\right) \approx \frac{n}{\langle n \rangle} \qquad (2\text{-}263)$$

Thus, for example, 20% of the defects have a neighbor within $n = 0.223 \langle n \rangle$ and 50% within $n = 0.693 \langle n \rangle$; nearly 8% have no neighbor within $n = 2.5 \langle n \rangle$, however. From another point of view, it may be seen that lattice sites will see an average of $n/\langle n \rangle$ defects within $n$ sites of themselves.

It would not seem to be meaningful, therefore, to include many more than roughly $1.5 \langle n \rangle$ nearest lattice sites in the calculations for the scattering amplitude associated with one defect. By the same token, the contribution in the real crystal of sites further from the center of defects will tend to be more and more randomized, because of the superposition of displacement fields. Such sites will not contribute to the relatively sharp $[\Delta(\mathbf{K})]^2$ characteristic of the defect, but only to a more diffuse scattering. Thus, when defect overlap is prevalent, the truncation of the sum in the calculation may provide a reasonably good approximation. In other words, the curve in Fig. 2-42 corresponding to 17,077 atoms may be a valid representation of the scattering due to a defect in the $x \to 0$ limit, but the curve for 7153 atoms might be better for $x = 5 \times 10^{-4}$.

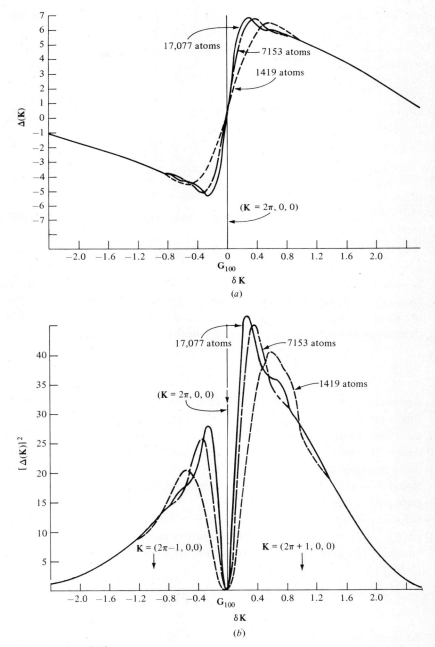

**FIGURE 2-42**
Convergence behavior for a $1/r^3$-type displacement field. (a) $\Delta(\mathbf{K})$; (b) $[\Delta(\mathbf{K})]^2$.

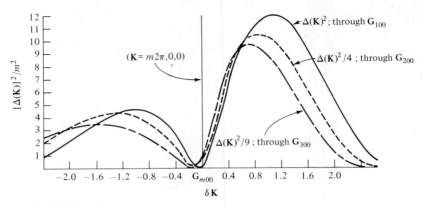

FIGURE 2-43

The variation of $[\Delta(\mathbf{K})]^2$ at $G_{100}$, $G_{200}$, and $G_{300}$ and in the [100] direction for $1/r^3$ displacements.

*b.  Dependence on the magnitude of* **G**   Figure 2-43 illustrates the variation of $[\Delta(\mathbf{K})]^2$ with the magnitude of the scattering vector. Calculations have been done for the same simple cubic structure, and for the same outward, radial displacement form described earlier, but with a different **r** dependence, namely,

$$\Delta\mathbf{r} = \frac{0.1\mathbf{r}}{r^5} \qquad (2\text{-}264)$$

Approximately 2500 atoms were included, which is adequate for convergence with this more localized field. The curves in Fig. 2-43 represent $[\Delta(\mathbf{K})]^2$ along the $\mathbf{K}_{100}$ direction through the $\mathbf{G}_{m(100)}$ reciprocal-lattice points, for $m = 1$, 2, and 3, respectively. It is interesting to note that the shape of $[\Delta(\mathbf{K})]^2$ is nearly independent of the order $m$, while the *magnitude* of $[\Delta(\mathbf{K})]^2$ at its maximum seems to be nearly proportional to $m^2$ (for small $m$). At the same time, the *amplitudes* at $\mathbf{G}_m$, $\Delta(\mathbf{G})$, which are pure real in this instance, are 0.444, 1.63, 3.17, respectively, for the three peaks; this variation is also roughly proportional to $m^2$! This behavior for both $\Delta(\mathbf{G})$ and $[\Delta(\mathbf{K})]^2$ may be seen through consideration of the original definition of $\Delta(\mathbf{K})$ [as in equation (2-242)] with the introduction of appropriate approximations. Equation (2-242) may be written

$$\Delta(\mathbf{K}) = \sum_{j=1}^{L} \exp\left(-i\mathbf{K}\cdot\mathbf{R}_j\right)[1 - \exp\left(-i\mathbf{K}\cdot\boldsymbol{\delta}_j\right)] \qquad (2\text{-}265)$$

Since the interesting variations occur at or near $\mathbf{G}_{m(hkl)}$, it is useful to write

$$\mathbf{K} \equiv m\mathbf{G} + \mathbf{b} \qquad (2\text{-}266)$$

Then, using the fact that $\exp(-im\mathbf{G} \cdot \mathbf{R}_j) = 1$,

$$\Delta(\mathbf{K}) = \sum_{j=1}^{L} \exp(-i\mathbf{b} \cdot \mathbf{R}_j)[1 - \exp(-im\mathbf{G} \cdot \boldsymbol{\delta}) \exp(-i\mathbf{b} \cdot \boldsymbol{\delta})] \qquad (2\text{-}267)$$

At $\mathbf{K} = m\mathbf{G}$, $\mathbf{b} = \mathbf{0}$; whence

$$\Delta(\mathbf{G}_m) = \sum_{j=1}^{L} [1 - \exp(-im\mathbf{G} \cdot \boldsymbol{\delta})] \qquad (2\text{-}268)$$

Expanding the exponential yields

$$\Delta(\mathbf{G}_m) = \sum_{j=1}^{L} \left[ 1 - \left( 1 - im\mathbf{G} \cdot \boldsymbol{\delta} + \frac{(im\mathbf{G} \cdot \boldsymbol{\delta})^2}{2!} - \frac{(im\mathbf{G} \cdot \boldsymbol{\delta})^3}{3!} + \cdots \right) \right] \qquad (2\text{-}269)$$

and

$$\text{Re}\,[\Delta(\mathbf{G}_m)] = \sum_{j=1}^{L} \left[ \frac{(m\mathbf{G} \cdot \boldsymbol{\delta})^2}{2!} - \frac{(m\mathbf{G} \cdot \boldsymbol{\delta})^4}{4!} + \frac{(m\mathbf{G} \cdot \boldsymbol{\delta})^6}{6!} - \cdots \right] \qquad (2\text{-}270)$$

If $m\mathbf{G} \cdot \boldsymbol{\delta}_j$ is small, the terms beyond the first may be ignored, leaving

$$\text{Re}\,[\Delta(\mathbf{G}_m)] \approx m^2 \sum_{j=1}^{L} \frac{(\mathbf{G} \cdot \boldsymbol{\delta}_j)^2}{2} \qquad (2\text{-}271)$$

To approximate the $m$ dependence of $[\Delta(\mathbf{K})]^2$, one may consider $\mathbf{b}$ to be small, compared to $m\mathbf{G}$, but not zero. Equation (2-267) may then be approximated,

$$\Delta(\mathbf{K}) \approx \sum_{j=1}^{L} \exp(-i\mathbf{b} \cdot \mathbf{R}_j)[1 - \exp(-im\mathbf{G} \cdot \boldsymbol{\delta})] \qquad (2\text{-}272)$$

Expanding the exponential [as in (2-269)] and ignoring all terms beyond the first (i.e., $m\mathbf{G} \cdot \boldsymbol{\delta}_j$ is assumed to be small),

$$\Delta(\mathbf{K}) \approx m \sum_{j=1}^{L} i\mathbf{G} \cdot \boldsymbol{\delta}_j \exp(-i\mathbf{b} \cdot \mathbf{R}_j) \qquad (2\text{-}273a)$$

$$[\Delta(\mathbf{K})]^2 \approx m^2 \left[ \sum_{j=1}^{L} i\mathbf{G} \cdot \boldsymbol{\delta}_j \exp(-i\mathbf{b} \cdot \mathbf{R}_j) \right]^2 \qquad (2\text{-}273b)$$

The calculated behavior, for both $\Delta(\mathbf{G})$ and $[\Delta(\mathbf{K})]^2$, is therefore reproduced in the approximate equations (2-271) and (2-273b). Those equations also emphasize the fact that near $\mathbf{G}_{000}$ there is no scattering effect due to defects of this type (that is, due to defects in which the number and scattering power of the atoms are not altered). In Fig. 2-44, however, the intensity variation is plotted along a $\mathbf{K}_{100}$ path from $\mathbf{K} = 0$ to $\mathbf{G}_{100}$. It may be seen that there is some scattering "associated" with $\mathbf{G}_{000}$ but actually occurring at fairly high scattering angles.

*c. The three-dimensional shape of a diffraction spot* The shape of the diffraction spot provides detailed information regarding the nature of the defects when they are

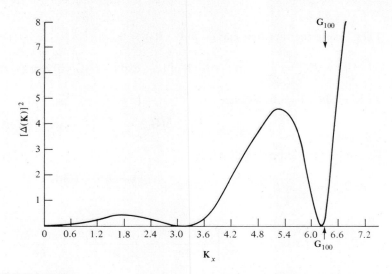

FIGURE 2-44
The variation of $[\Delta(\mathbf{K})]^2$ along the path from $G_{000}$ to $G_{100}$.

coordinated with the crystallographic orientation. Three-dimensional intensity mappings (experimental and/or calculated) are tedious to achieve, however, and difficult to represent pictorially. Fortunately, examination of the variation along a few directions usually suffices to disclose the prominent characteristics. Referring to Fig. 2-45, it may be noted that three paths are easiest to achieve experimentally

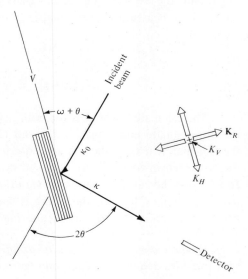

FIGURE 2-45
Easily achieved paths in reciprocal space, for mapping the shape of a diffraction spot.

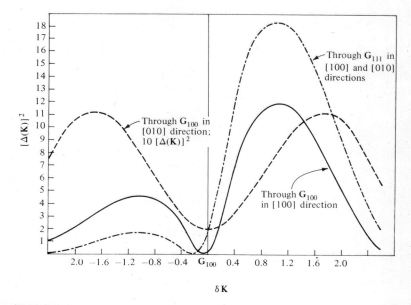

FIGURE 2-46
The asymmetry in $[\Delta(\mathbf{K})]^2$ at $G_{100}$ and $G_{111}$ for an outward, radial displacement field.

with conventional diffractometers:

*1* the radial path through the spot $\mathbf{K} = \mathbf{G}$, which is labeled $\mathbf{K}_R$, and which is obtained by a standard $2\theta/(\omega = 0)$ scan;

*2* the "horizontal" arc, labeled $K_H$, which is achieved by a simple $\omega$ scan while the incident and diffracted beam are both held fixed;

*3* the "vertical" arc, labeled $K_V$ but not shown in the figure because it is perpendicular to the paper, which is achieved through rotation of the sample angle $\chi$ about the horizontal axis $V$. For the present purposes, however, calculations are presented only for $\mathbf{K}_{100}$ and $\mathbf{K}_{010}$ paths.

In Fig. 2-46, the variation of $[\Delta(\mathbf{K})]^2$ is shown for the $\mathbf{K}_{100}$ and $\mathbf{K}_{010}$ paths through $\mathbf{G}_{100}$ and $\mathbf{G}_{111}$, for the previously described outward, radial displacement field with $\Delta\mathbf{r} = 0.1 \ \mathbf{r}/r^5$. The most striking aspect of these results is the marked asymmetry in the shapes and the difference between the $\mathbf{G}_{100}$ and $\mathbf{G}_{111}$ in spite of the *nominal* spherical symmetry of the displacement field. These effects are due only partially to the dependence of the amplitude on the magnitude of $\mathbf{G}$. In addition, it must be noted that the actual displacements (particularly the few largest ones) form a small set of specific crystallographic directions. It is the latter circumstance that

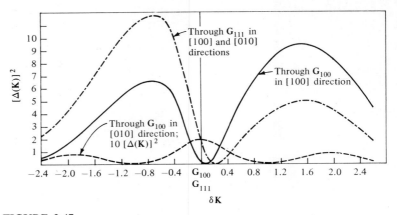

**FIGURE 2-47**
The asymmetry in $[\Delta(\mathbf{K})]^2$ at $G_{100}$ and $G_{111}$ for an outward, radial displacement field.

causes the intensity along the $\mathbf{K}_{010}$ path through $\mathbf{G}_{100}$ to be so much smaller than along the $\mathbf{K}_{100}$ path. That behavior reverses, of course, at $\mathbf{G}_{010}$, which exhibits identical variations but rotated by 90°. In contrast, both paths are identical through $\mathbf{G}_{111}$.

The real part of the amplitude at $\mathbf{G}_{111}$ is 1.332, which is precisely three times the value at $\mathbf{G}_{100}$. This correspondence not only verifies the (approximate) proportionality to $|\mathbf{G}|^2$ but also reflects the fact that, on the average, the components of the $\pmb{\delta}_j$ are the same in the [[100]] and [[111]] directions.

*d. Comparison among different displacement fields*   In order to allow examination of the effect of variations in the displacement directions, two additional defect models similar to those used for the previous examples have been defined. For both of the new ones, the magnitudes of the displacements remain unchanged; i.e., $|\Delta\mathbf{r}| = 0.1/r^4$. For the second model the displacements are reversed (compared to the outward, radial model), so that they point inward to the origin. For the third model, the displacements are all in the [100] direction.

Figure 2-47 represents the intensity variation for the inward, radial defect, near $\mathbf{G}_{100}$ and $\mathbf{G}_{111}$. The paths shown are $\mathbf{K}_{100}$ and $\mathbf{K}_{010}$ again, and the curves are analogous to those in Fig. 2-46 for the outward displacements. The sensitivity to the sense of the strain is striking, particularly as evidenced by the reversal of the symmetry through $\mathbf{G}_{111}$.

As should be expected from equation (2-270), the values for Re $[\Delta(\mathbf{G})]$ are identical for the outward and inward displacement models.

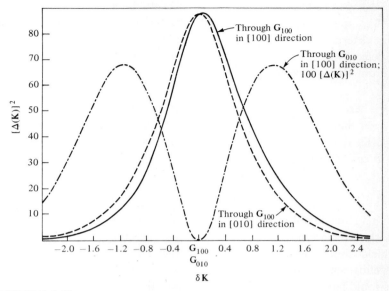

FIGURE 2-48
$[\Delta(\mathbf{K})]^2$ at $G_{100}$ and $G_{010}$ for a unidirectional displacement field; strain in [100] direction.

The third model, with all displacements in the [100] direction, differs from the previous two in a number of important respects.

*1* The displacements do not have inversion symmetry, so complex numbers must be used (which could be avoided for the other computer models).
*2* The displacements are more highly correlated, and so one anticipates a tendency for $[\Delta(\mathbf{K})]^2$ to peak more closely to $\mathbf{K} = \mathbf{G}$.
*3* The components of the $\delta_j$ in different directions vary greatly, and one anticipates a correspondingly large variation at different $\mathbf{G}$'s of both $[\Delta(\mathbf{K})]^2$ and Re $[\Delta(\mathbf{G})]$.

Calculated results are shown in Fig. 2-48, which represents the intensity variation along the $\mathbf{K}_{100}$ and $\mathbf{K}_{010}$ paths through $\mathbf{G}_{100}$ and along the $\mathbf{K}_{010}$ path through $\mathbf{G}_{010}$ (the intensity is zero along the $\mathbf{K}_{100}$ path through $\mathbf{G}_{010}$). It may be seen that the effects near $\mathbf{G}_{010}$ are essentially negligible (but still interesting) compared to those near $\mathbf{G}_{100}$. $[\Delta(\mathbf{K})]^2$ does peak up near $\mathbf{G}_{100}$, but the shift to larger $\mathbf{G}$ (seen in the $\mathbf{K}_{100}$ variation) is not negligible. The Re $[\Delta(\mathbf{G}_{100})]$ is 2.66, while the imaginary part is 18.52. The real part is only about six times as large as for the inward/outward displacements of the same magnitude. At the same time, the Re $[\Delta(\mathbf{G}_{010})]$ is zero.

Perhaps the least expected of these results is the roughly symmetrical shape of the $G_{100}$ spot. That behavior helps to form the rule that the shape of a diffraction spot is determined by the correlations among the interatomic displacements within the defect, while the crystallographic asymmetry is primarily evidenced in the variation from one **G** to another.

*e.*  *Scattering by holes and precipitates*  Returning to equation (2-242), it may be seen that, if the defect consists entirely of the removal of atoms (to make a hole), $\Delta(\mathbf{K})$ is just the amplitude which was due to the removed atoms, and the diffuse intensity term $[\Delta(\mathbf{K})]^2$ becomes

$$[\Delta(\mathbf{K})]^2 = \left[ \sum_{j=1}^{L} f_j(K) \exp\left(-i\mathbf{K}\cdot\mathbf{R}_j\right) \right]^2 \qquad (2\text{-}274)$$

In a realistic hole model, some allowance should be made for the relaxation of atoms which remain surrounding the hole; one expects an inward, radial sort of displacement field similar to that considered earlier. For the present purpose, however, that refinement is ignored. It is interesting to note that $[\Delta(\mathbf{K})]^2$ is just the intensity that would have been diffracted by the atoms that were in the hole, if the rest of the crystal were not there. That is, the summation in (2-274) also represents the intensity diffracted from a small crystallite containing only $L$ atoms.

One may construct a model for a perfectly coherent precipitate, for which atoms are removed from the crystal but replaced with others, in the same positions but with different scattering factors. Equation (2-274) would again apply, albeit with the $f_j$ replaced by $\Delta f_j = f_j - f_j'$. Thus the form of $[\Delta(\mathbf{K})]^2$ is the same for perfectly coherent precipitates as for holes.

The scattering from perfectly incoherent precipitates can be approximated by using equation (2-274) twice, first to obtain the $[\Delta(\mathbf{K})]^2$ associated with the removal of atoms to make room for the precipitate and second to calculate the intensity due to the small, isolated particle. The first contribution will peak near the original reciprocal-lattice points; the latter contribution may come anywhere in **K** space, depending on the crystal structure of the precipitate and its preferred orientation, if any, in the original crystal.

In Fig. 2-49, evaluations of equation (2-274) are presented for a primitive cubic lattice, and $L = 33(R_{max} = 2)$ and $L = 257(R_{max} = 4)$ atoms, respectively. The results are for the $\mathbf{K}_{100}$ and $\mathbf{K}_{110}$ paths in the vicinity of $G_{000}$. For the nominally spherical pieces used, the shape of $[\Delta(\mathbf{K})]^2$ is nearly spherical. Moreover, as anticipated from previous discussion of the finite particle size effect, $[\Delta(\mathbf{K})]^2$ is absolutely identical at every $G_{hkl}$.

Returning to the question of strains which may be associated with holes or precipitates, whether in the surrounding crystal or in the precipitate, one may usefully

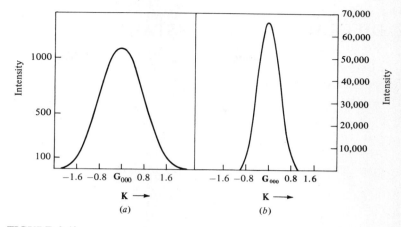

**FIGURE 2-49**
$[\Delta(\mathbf{K})]^2$ for spherical holes or precipitates (through $G_{000}$). (*a*) 33 atoms; (*b*) 257 atoms.

recall the fact that their diffraction effects disappear near $\mathbf{G}_{000}$. Thus, to the extent that the shape of the intensity distribution can be measured near $\mathbf{G}_{000}$, the particle size parameters can be measured independently. It must be stressed, however, that various problems beset such small-angle measurements in crystals, particularly multiple scattering. Alternatively, one might consider extrapolating to $m = 0$ the intensity distributions measured at a series of Bragg spots, $\mathbf{K} = m\mathbf{G} + \mathbf{b}$. Various extrapolation techniques can be applied which may take advantage of the previously derived behavior for the limit of small $\mathbf{G}$.

Once again, however, it should be mentioned that the direct application of a reasonably fundamental equation [e.g., (2-242)] in association with a trial-and-error match to a full set of experimental data is preferred over an approximate analysis which may miss important features.

# 3

# DIFFRACTION BY CRYSTALS[1]

## I. FUNDAMENTALS OF CRYSTAL DIFFRACTION

### A. Maxwell equations

X rays are a form of electromagnetic radiation. When they interact with matter, in general, they display dual properties of waves and particles; however, this chapter is concerned primarily with x-ray diffraction, namely, the wave-optical behavior of x rays inside and outside a crystalline medium. For this reason, the discussions will begin with the classical Maxwell equations as basic laws. Indeed, this approach can be justified by modern quantum electrodynamics, which is a more fundamental theory for the interactions between matter and radiation.

The Maxwell equations are composed of two sets of equations. One set describes the interrelations of four vector fields, the electric vector **E**, the electric displacement

---

[1] The notations employed in this chapter are the same as those in Chaps. 4 and 5 but differ somewhat from the ones used in other chapters of this book in order to facilitate comparison with original citations.

vector $\mathbf{D}$, the magnetic vector $\mathbf{H}$, and the magnetic induction vector $\mathbf{B}$, for a given true charge density $\rho_t$ and current density $\mathbf{J}_t$. The equations are

$$\text{rot } \mathbf{H} = \frac{1}{c} \frac{\partial}{\partial t} \mathbf{D} + \frac{4\pi}{c} \mathbf{J}_t \qquad (3\text{-}1a)$$

$$\text{rot } \mathbf{E} = -\frac{1}{c} \frac{\partial}{\partial t} \mathbf{B} \qquad (3\text{-}1b)$$

$$\text{div } \mathbf{D} = 4\pi\rho_t \qquad (3\text{-}1c)$$

$$\text{div } \mathbf{B} = 0 \qquad (3\text{-}1d)$$

where the constant $c$ is the velocity of light in vacuum, whose numerical value is approximately $3 \times 10^{10}$ cm/sec using the Gaussian system of units. Since the charge must be conserved, the continuity relation

$$\text{div } \mathbf{J}_t + \frac{\partial \rho_t}{\partial t} = 0 \qquad (3\text{-}2)$$

holds in space and time.

The other set of Maxwell equations are called the *material equations*. In the presence of the field vector $\mathbf{E}$, materials are electrically polarized. The electric polarization vector $\mathbf{P}$ of materials is defined by the electric moment per unit volume. The displacement vector $\mathbf{D}$, then, is given by

$$\mathbf{D} = \mathbf{E} + 4\pi\mathbf{P} \qquad (3\text{-}3a)$$

Similarly, the induction vector $\mathbf{B}$ can be expressed as

$$\mathbf{B} = \mathbf{H} + 4\pi\mathbf{M} \qquad (3\text{-}3b)$$

where $\mathbf{M}$ is the magnetic polarization vector which is defined by the magnetic moment per unit volume.

For sufficiently weak fields $\mathbf{E}$ and $\mathbf{H}$, the polarization vectors $\mathbf{P}$ and $\mathbf{M}$ can be assumed to be linear with $\mathbf{E}$ and $\mathbf{H}$, respectively. Then,

$$\mathbf{P} = \eta\mathbf{E} \qquad (3\text{-}4a)$$

$$\mathbf{M} = \kappa\mathbf{H} \qquad (3\text{-}4b)$$

where $\eta$ and $\kappa$ are the dielectric and magnetic susceptibility (polarizability), respectively. In general, they must be tensors, but often they can be assumed to be scalars. Inserting them into equations (3-3a) and (3-3b), one can write

$$\mathbf{D} = \varepsilon\mathbf{E} \qquad (3\text{-}5a)$$

$$\mathbf{B} = \mu\mathbf{H} \qquad (3\text{-}5b)$$

The factors $\varepsilon = 1 + 4\pi\eta$ and $\mu = 1 + 4\pi\kappa$ are called the *dielectric constant* and the *magnetic permeability*, respectively.   Equations (3-5a) and (3-5b) are alternative representations of the material equations (3-3a) and (3-3b).[1]

With the aid of equations (3-3a) and (3-3b), the set of equations (3-1a) to (3-1d) can be written in the forms

$$\text{rot } \mathbf{B} = \frac{1}{c} \frac{\partial}{\partial t} \mathbf{E} + \frac{4\pi}{c} \mathbf{J} \qquad (3\text{-}6a)$$

$$\text{rot } \mathbf{E} = -\frac{1}{c} \frac{\partial}{\partial t} \mathbf{B} \qquad (3\text{-}6b)$$

$$\text{div } \mathbf{E} = 4\pi\rho \qquad (3\text{-}6c)$$

$$\text{div } \mathbf{B} = 0 \qquad (3\text{-}6d)$$

where

$$\mathbf{J} = \mathbf{J}_t + \mathbf{J}_p \qquad (3\text{-}7a)$$

$$\rho = \rho_t + \rho_p \qquad (3\text{-}7b)$$

and $\rho_p$ and $\mathbf{J}_p$ are given by

$$\rho_p = -\text{div } \mathbf{P} \qquad (3\text{-}8a)$$

$$\mathbf{J}_p = \frac{\partial \mathbf{P}}{\partial t} + c \text{ rot } \mathbf{M} \qquad (3\text{-}8b)$$

The quantities $\rho_p$ and $\mathbf{J}_p$ can be interpreted as charge density and current density due to the polarization of the matter.  Indeed, they satisfy the continuity relation (3-2) because the operator (div rot) of any vector is identically null.

Originally, the Maxwell equations (3-1) or (3-6) were established for macroscopic electromagnetic phenomena.  It is useful to postulate, however, that the equations in (3-6) also hold for microscopic phenomena of any scale.  This postulate seems reasonable, but the ultimate justification comes from a comparison between theory and experiments.  From the microscopic viewpoint, $\rho$ in (3-6c) is the charge density of nuclei and electrons which constitute the matter.  Similarly, $\mathbf{J}$ in (3-6a) is the current density owing to the movement of the charged particles.  Obviously they satisfy the continuity relation

$$\text{div } \mathbf{J} + \frac{\partial \rho}{\partial t} = 0 \qquad (3\text{-}9)$$

Usually, electromagnetic waves and light waves can be described by the Maxwell equations in the form given in (3-1), although the material constants $\varepsilon$ and $\mu$ may not be equal to those determined by static experiments.  In fact, they depend on the frequency of the electromagnetic waves involved.  In this case the Maxwell equations

---

[1] The formal generalizations, $\mathbf{D} = \varepsilon\mathbf{E} + \alpha\mathbf{H}$ and $\mathbf{B} = \mu\mathbf{H} + \beta\mathbf{E}$, are easily conceivable.  For simplicity, such general cases are not discussed here.  In crystal optics of visible rays, it is necessary to take into account at least the tensor character of $\varepsilon$.

(3-1) should be interpreted as an average of the microscopic equations of the form (3-6) over a macroscopic or submicroscopic scale of space and a time interval.

For x rays, similarly, relations (3-6) must be used as the basic equations. The equations of the form (3-1) can be used only after the material constants $\varepsilon$ and $\mu$ are justified. As will be shown in Section I C, a reasonable justification can be given by treating classically the movement of the charged particles. In quantum-mechanical treatments, strictly speaking, the justification fails, and nonlocal tensor operators must be substituted for local scalar quantities $\varepsilon$ and $\mu$. Unless the wavelength of x rays is close to the absorption edge of the materials, however, the classical treatment is very nearly correct. Further discussion of these topics lies outside the scope of this book. The interested reader can find the details in Laue's text listed at the end of this chapter.

## B. Energy flux and energy density associated with the electromagnetic field

One can define the electric and magnetic energy density $W_e$ and $W_m$ and the energy-flow vector $\mathbf{S}$ associated with Maxwell equations as follows:

$$W_e = \frac{1}{4\pi} \int_0^{\mathbf{D}} (\mathbf{E} \cdot d\mathbf{D}) \qquad (3\text{-}10a)$$

$$W_m = \frac{1}{4\pi} \int_0^{\mathbf{B}} (\mathbf{H} \cdot d\mathbf{B}) \qquad (3\text{-}10b)$$

$$\mathbf{S} = \frac{c}{4\pi} [\mathbf{E} \times \mathbf{H}] \qquad (3\text{-}10c)$$

When the dielectric constant $\varepsilon$ and the permeability $\mu$ are time independent, $W_e$ and $W_m$ are reduced to

$$W_e = \frac{1}{8\pi} (\mathbf{E} \cdot \mathbf{D}) \qquad (3\text{-}11a)$$

$$W_m = \frac{1}{8\pi} (\mathbf{H} \cdot \mathbf{B}) \qquad (3\text{-}11b)$$

By the use of vector calculus

$$\text{div } \mathbf{S} = \frac{c}{4\pi} [(\mathbf{H} \cdot \text{rot } \mathbf{E}) - (\mathbf{E} \cdot \text{rot } \mathbf{H})]$$

Inserting the Maxwell equations (3-1a) and (3-1b) above, one obtains

$$\text{div } \mathbf{S} = -\frac{1}{4\pi} \left\{ \left( \mathbf{H} \frac{\partial}{\partial t} \mathbf{B} \right) + \left( \mathbf{E} \frac{\partial}{\partial t} \mathbf{D} \right) \right\} - (\mathbf{E} \cdot \mathbf{J}_t)$$

so that

$$\text{div } \mathbf{S} + \frac{\partial}{\partial t} (W_e + W_m) = -(\mathbf{E} \cdot \mathbf{J}_t) \qquad (3\text{-}12)$$

The physical meaning of the right side is the energy gain of the field through the true current $J_t$ per unit volume and unit time. Thus, (3-12) can be interpreted as the continuity relation of the electromagnetic field energy.

## C. The x-ray wave fields in matter

The problem is to find the electromagnetic wave fields for an assemblage of charged particles and a given incident wave field. The general approach to this problem comes from electrodynamics. Here, it is convenient to follow a less general but more tractable approach. First, the dielectric constant $\varepsilon$, adequate for x rays, is derived from a microscopic viewpoint. For the sake of simplicity, a classical treatment is adopted, and the constituent matter is assumed to be an assemblage of electrons. The atomic nuclei are assumed to be fixed. Afterward, the fundamental equations of D are derived both in differential and in integral forms. For doing this, the set of equations in (3-6) are employed as the microscopic physical laws. The integral form is applied to the problems of kinematical scattering by crystals. The differential form is adequate for the problems of wave propagation, which is the subject of the dynamical theory discussed in Chap. 4.

1. *Dielectric constant for x rays*   If a free electron at a position $\mathbf{r}$ is subject to an electromagnetic wave, the electron is moved by the Lorentz force.[1] Then, an additional displacement $\mathbf{x}$, velocity $\dot{\mathbf{x}}$, and acceleration $\ddot{\mathbf{x}}$ are introduced. The classical equation of motion must be

$$m\ddot{\mathbf{x}} = e\left[\mathbf{E} + \frac{1}{c}(\mathbf{v} \times \mathbf{H})\right] \qquad (3\text{-}13)$$

where $\mathbf{v} = \dot{\mathbf{x}} + \dot{\mathbf{r}}$ is the velocity of the electron. Since $v \ll c$ in atomic systems, the magnetic term is neglected. Here one is concerned with the field vectors which have the following time dependence:

$$\mathbf{E} = \mathbf{E}(\mathbf{r})e^{-i\omega t} \qquad \mathbf{B} = \mathbf{B}(\mathbf{r})e^{-i\omega t} \qquad (3\text{-}14)$$

The time dependence of $\mathbf{x}$, then, may be described by the same factor $e^{-i\omega t}$.

The additional velocity and displacement are represented by

$$\dot{\mathbf{x}} = i\frac{e}{m\omega}\mathbf{E} \qquad (3\text{-}15)$$

$$\mathbf{x} = -\frac{e}{m\omega^2}\mathbf{E} \qquad (3\text{-}16)$$

[1] This statement does not mean that the electron is permanently fixed at $\mathbf{r}$ when such fields are absent.

Thus, the oscillating part of the electric polarization is given by

$$\mathbf{P} = eN(\mathbf{r})\mathbf{x} = -\frac{e^2}{m\omega^2} N(\mathbf{r})\mathbf{E} \qquad (3\text{-}17)$$

where $N(\mathbf{r})$ is the equilibrium density of electrons when the fields are absent. The current density, according to equation (3-8b), has the form

$$\mathbf{J} = eN(\mathbf{r})\dot{\mathbf{x}} = i\frac{e^2}{m\omega} N(\mathbf{r})\mathbf{E} \qquad (3\text{-}18)$$

The magnetic part $c$ rot $\mathbf{M}$ is neglected since the magnetization induced by the electromagnetic field is sufficiently small.

Now, consider the Maxwell equations (3-6). The static or stationary parts of $\rho$ and $\mathbf{J}$ produce a static vector field. Since the Maxwell equations are linear with respect to the field vectors and $\rho$ and $\mathbf{J}$, we can subtract the static part from the Maxwell equations. Thus, the time-dependent parts of $\mathbf{E}$ and $\mathbf{B}$ having the form (3-14) also satisfy the Maxwell equations (3-6) in which the current density $\mathbf{J}$ is given by (3-18) and the charge density $\rho$ is given by the divergence of $-\mathbf{P}$. [Note that $\rho(\mathbf{r})$ is not $eN(\mathbf{r})$!]

Under this circumstance, the Maxwell equations are represented by

$$\text{rot } \mathbf{B} = -i\left(\frac{\omega}{c}\right)\mathbf{D} \qquad (3\text{-}19a)$$

$$\text{rot } \mathbf{E} = i\left(\frac{\omega}{c}\right)\mathbf{B} \qquad (3\text{-}19b)$$

$$\text{div } \mathbf{D} = 0 \qquad (3\text{-}19c)$$

$$\text{div } \mathbf{B} = 0 \qquad (3\text{-}19d)$$

where

$$\mathbf{D} = \varepsilon\mathbf{E} \qquad (3\text{-}20)$$

and

$$\varepsilon = 1 - \frac{4\pi e^2}{m\omega^2} N(\mathbf{r}) \qquad (3\text{-}21)$$

Obviously, the quantity $\mathbf{D}$ plays the role of an electric displacement, and $\varepsilon$ can be interpreted as the dielectric constant of matters for x rays. Apparently, no true charge and current density exist. By introducing the quantity $\chi = \varepsilon - 1$ ($4\pi$ times the dielectric polarizability), we obtain

$$\chi(\mathbf{r}) = -\frac{4\pi e^2}{m\omega^2} N(\mathbf{r}) \qquad (3\text{-}22)$$

With reasonable values $\omega = 2\pi c/\lambda \sim 2 \times 10^{19}$ $\text{sec}^{-1}$ and $N \sim 10^{24}$ $\text{cm}^{-3}$, the value of $\chi$ is estimated to have an order of magnitude of $10^{-5}$. The negative sign and the smallness of $\chi$ are significant.

So far, the argument has been based on the free-electron model. It is an easy matter to generalize the argument to an assemblage of bound electrons as has been done in Chap. 1, Section I B. For a bound electron, the equation of motion (3-13) is replaced by

$$m\ddot{\mathbf{x}} = e\mathbf{E} - m\gamma\dot{\mathbf{x}} - m\omega_0^2\mathbf{x} \qquad (3\text{-}23)$$

where $\gamma$ is the damping coefficient and $\omega_0$ is the characteristic frequency of the bound electron. Physically, they must be positive. By the same argument as used for a free electron, it is known that

$$\chi(\mathbf{r}) = -\frac{4\pi e^2}{m\omega^2} gN(\mathbf{r}) \qquad (3\text{-}24)$$

where $g$, the resonance factor, is given by

$$g = \frac{\omega^2}{\omega^2 + i\gamma\omega - \omega_0^2} \qquad (3\text{-}25)$$

It is to be noted that $\chi$ is no longer real. By the use of a complex $\chi$, the absorption of x rays in matter also can be treated phenomenologically. For the assemblage of electrons, (3-24) can be generalized to

$$\chi(\mathbf{r}) = -\frac{4\pi e^2}{m\omega^2} \sum_k g^{(k)} N^{(k)}(\mathbf{r}) \qquad (3\text{-}26)$$

where $g^{(k)}$ and $N^{(k)}$ are the resonance factor and the equilibrium electron density, respectively, of the $k$th electron. It is significant that $\chi$ is a function of position on a microscopic scale. As distinct from the case of visible rays, no space or time average is taken in defining the effective polarizability of the medium for x rays.

Quantum-theoretical considerations of the polarizability are a somewhat cumbersome matter. The main idea here is to obtain the current density $\mathbf{J}$ on the basis of quantum mechanics as a function of the electric field $\mathbf{E}$. The susceptibility is defined by (3-4a) in a formal sense, where the polarization vector $\mathbf{P}$ is now given by $\int_0^t \mathbf{J}\, dt$. For a single atom or ion, the details for calculating $\mathbf{J}$ have been described in Chap. 1. The generalization to a crystalline system has been worked out by G. Molière (1939a, b, c) and more recently by Wagenfeld (1966). The conclusion is that, although $\chi$ must be replaced by a tensorlike operator in general, a scalar quantity $\chi$ can be defined to a good approximation unless the x-ray wavelength is very close to the $K$ or $L$ absorption edges of the constituent atoms. Also, Ohtsuki (1964, 1965), Ohtsuki and Yanagawa (1966), Dederichs (1966), Afanas'ev and Kagan (1968), and

Sano, Ohtaka, and Ohtsuki (1969) presented treatments that take into account thermal vibrations and Compton scattering.

**2.  *Differential form of the field equation***  By eliminating **B** from equations (3-19*a*) and (3-19*b*), one gets

$$\text{rot rot } \mathbf{E} = \left(\frac{\omega}{c}\right)^2 \mathbf{D}$$

Since $\mathbf{E} = \mathbf{D} - \chi\mathbf{E}$ and div $\mathbf{D} = 0$,

$$\Delta\mathbf{D} - \frac{1}{c^2}\frac{\partial^2}{\partial t^2}\mathbf{D} = -\text{rot rot }\frac{\chi}{1 + \chi}\mathbf{D}$$

where $(\omega)^2$ is replaced by the operator $-(\partial^2/\partial t^2)$. In addition, since $\chi$ is of the order of magnitude of $10^{-5}$, $[\chi/(1 + \chi)]$ can be replaced simply by $\chi$. Thus, finally,

$$\Delta\mathbf{D}(\mathbf{r}, t) - \frac{1}{c^2}\frac{\partial^2}{\partial t^2}\mathbf{D}(\mathbf{r}, t) = -\text{rot rot }[\chi(\mathbf{r})\mathbf{D}(\mathbf{r}, t)] \quad (3\text{-}27a)$$

or

$$\Delta\mathbf{D}(\mathbf{r}) + K^2\mathbf{D}(\mathbf{r}) = -\text{rot rot }[\chi(\mathbf{r})\mathbf{D}(\mathbf{r})] \quad (3\text{-}27b)$$

where

$$K = \frac{\omega}{c} \quad (3\text{-}28)$$

These are the fundamental equations of the field vector $\mathbf{D}(\mathbf{r}, t)$ or $\mathbf{D}(\mathbf{r})$ in the differential form. The vector **B** is derived from $\mathbf{E} = \mathbf{D}/\varepsilon$ by the use of (3-19*b*).

**3.  *Integral form of the field equation***  Since (3-27*a*) has the form of an inhomogeneous wave equation, a special solution which vanishes at infinity can be written

$$\mathbf{D}_s(\mathbf{r}, t) = \frac{1}{4\pi}\int\frac{1}{R}\text{rot rot }[\chi(\mathbf{r}')\mathbf{D}(\mathbf{r}', t')]\, dv' \quad (3\text{-}29)$$

where $R = |\mathbf{r} - \mathbf{r}'|$ is the distance from $\mathbf{r}'$ to $\mathbf{r}$, and $t'$ is the retarded time $t - R/c$. A more general solution can be constructed by adding any solution $\mathbf{D}_e(\mathbf{r}, t)$ of the homogeneous wave equation ($\chi = 0$) to the above. In this way, the required boundary condition that the field **D** tends to a field $\mathbf{D}_e$ at infinity is automatically satisfied. This leads to

$$\mathbf{D}(\mathbf{r}, t) = \mathbf{D}_e(\mathbf{r}, t) + \frac{1}{4\pi}\int\frac{1}{R}\text{rot rot }[\chi(\mathbf{r}')\mathbf{D}(\mathbf{r}', t')]\, dv' \quad (3\text{-}30a)$$

This is an integral representation[1] of (3-27*a*) including the boundary condition. Physically, $\mathbf{D}_e$ implies an incident wave, and the second term is the scattered wave

---

[1] Mathematically speaking, (3-30*a*) is an integrodifferential equation.

due to the material. For a special case, in which the field vector has the time dependence $e^{-i\omega t}$, equation (3-30a) reduces to

$$\mathbf{D(r)} = \mathbf{D}_e(\mathbf{r}) + \frac{1}{4\pi} \int \frac{1}{R} \text{ rot rot } [\chi(\mathbf{r'})\mathbf{D(r')}] \exp iKR \, dv' \qquad (3\text{-}30b)$$

where the common factor $e^{-i\omega t}$ is omitted. The equations of the form (3-30a) and (3-30b) are suitable to treat the problem of scattering.

Finally, it is worthwhile to note a similarity of (3-37b) to the time-independent Schrödinger equation for the material wave

$$\Delta\psi + \frac{8\pi^2 m}{h^2} [E - V(\mathbf{r})]\psi = 0 \qquad (3\text{-}31)$$

where $h$ is Planck's constant, $E$ the total energy, and $V(\mathbf{r})$ is the scalar potential. Indeed, by virtue of de Broglie's relation $(8\pi^2 m/h^2)E = K^2 = (2\pi/\lambda)^2$, $\lambda$ being the wavelength in vacuum. $(8\pi^2 m/h^2)V(\mathbf{r})\psi$ corresponds to $-\text{rot rot } \chi\mathbf{D}$ in (3-27b). The integral form of (3-31) is

$$\psi(\mathbf{r}) = \psi_e(\mathbf{r}) - \frac{2\pi m}{h^2} \int \frac{1}{R} V(\mathbf{r'})\psi(\mathbf{r'})e^{iKR} \, dv' \qquad (3\text{-}32)$$

This corresponds to (3-30b). In view of the similarity of the fundamental equations for $\psi$ and $\mathbf{D}$, we can expect that the diffraction theories are essentially similar for material waves and for x rays. Since electromagnetic waves are vector waves, however, a complication is introduced in x-ray diffraction. For this reason, it is often instructive to utilize electron diffraction in order to understand the physical concepts of diffraction theories.

## D.   The propagation of waves

This section will consider simple wavelike fields in materials while the medium is assumed to be extended indefinitely and the dielectric constant $\varepsilon$ or $\chi$ is time independent (nondissipative) and real. They are also assumed to be homogeneous and isotropic. Equation (3-27a) then can be reduced to the homogeneous wave equation[1]

$$\Delta\mathbf{d} - \frac{1}{v^2}\frac{\partial^2}{\partial t^2}\mathbf{d} = 0 \qquad (3\text{-}33)$$

where

$$v = \frac{c}{\sqrt{\varepsilon}} \qquad (3\text{-}34)$$

[1] In the following, the quantities associated with the crystal waves are denoted by small letters, whereas capital letters are used for the quantities of vacuum waves.

The homogeneity and isotropicity of $\chi$ are not adequate to the problems of the diffraction of x rays by crystals. In this sense, the description of this section is a kind of mathematical preparation for the following parts. The results obtained here, however, have a physical significance when the Bragg condition is not satisfied for any crystal plane.

**1.** *Plane waves of complex variables*    The general solution of (3-33) has the form

$$\mathbf{d} = \mathbf{f}[(\mathbf{k} \cdot \mathbf{r}) - \omega t] + \mathbf{g}[(\mathbf{k} \cdot \mathbf{r}) + \omega t] \qquad (3\text{-}35)$$

provided that

$$k^2 = \left(\frac{\omega}{v}\right)^2 \qquad (3\text{-}36)$$

Here, $\mathbf{f}$ and $\mathbf{g}$ are arbitrary analytic vector functions, while $\omega$ and $\mathbf{k}$ are treated as arbitrary constants. The simplest wavelike solution can be obtained by taking

$$\mathbf{f}(\phi) = \mathbf{f}_0 \exp i\phi \qquad \mathbf{g} = 0$$

The explicit expression for $\mathbf{d}$ is written

$$\mathbf{d} = \mathbf{d}_0 \exp i[(\mathbf{k} \cdot \mathbf{r}) - \omega t] \qquad (3\text{-}37a)$$

where $\mathbf{d}_0$ is a constant (not necessarily real) vector.

With this expression

$$\phi = (\mathbf{k} \cdot \mathbf{r}) - \omega t \qquad (3\text{-}38)$$

is the *phase*, and $\omega$ and $\mathbf{k}$ mean *angular frequency* and *wave vector*, respectively. The magnitude of $\mathbf{k}$ is related to the wavelength $\lambda$ by

$$k = |\mathbf{k}| = \frac{2\pi}{\lambda} \qquad (3\text{-}39)$$

and with this physical meaning of $\omega$ and $k$, equation (3-36) is regarded as the dispersion relation. Since the surface of a constant phase is a plane in three-dimensional space, the solution (3-37a) is called a *plane-wave solution*. The surface moves with a constant velocity $v = \omega/k$. Thus, the constant defined by (3-34) is the phase velocity in the material medium. Obviously, the phase velocity is $c$ in vacuum. The refractive index $n$ of the matter is

$$n \equiv \frac{c}{v} = \sqrt{\varepsilon}$$

$$= 1 - \frac{2\pi e^2}{m\omega^2} N \qquad (3\text{-}40)$$

The associated induction vector $\mathbf{b}$ may be written

$$\mathbf{b} = \mathbf{b}_0 \exp i[(\mathbf{k} \cdot \mathbf{r}) - \omega t] \qquad (3\text{-}37b)$$

The amplitudes $\mathbf{d}_0$ and $\mathbf{b}_0$, however, cannot be independent. With the expressions

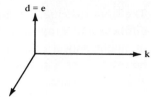

FIGURE 3-1
Transverse relation of the field vectors
$\mathbf{d} = \mathbf{e}$ and $\mathbf{b} = \mathbf{h}$ to the wave vector $\mathbf{k}$.

(3-37a) and (3-37b) for $\mathbf{d}$ and $\mathbf{b}$, the Maxwell equations (3-1) ($\rho_t = 0$ and $\mathbf{J}_t = 0$) are equivalent to

$$[\hat{\mathbf{k}} \times \mathbf{b}_0] = - \sqrt{\frac{\mu}{\varepsilon}} \, \mathbf{d}_0 \quad (3\text{-}41a)^{1,2}$$

$$[\hat{\mathbf{k}} \times \mathbf{d}_0] = \sqrt{\frac{\varepsilon}{\mu}} \, \mathbf{b}_0 \qquad (3\text{-}41b)$$

$$(\hat{\mathbf{k}} \cdot \mathbf{d}_0) = 0 \qquad (3\text{-}41c)$$

$$(\hat{\mathbf{k}} \cdot \mathbf{b}_0) = 0 \qquad (3\text{-}41d)$$

From these relations $\mathbf{k}$, $\mathbf{d}_0$ and $\mathbf{b}_0$ are mutually perpendicular as illustrated in Fig. 3-1. By analogy to elastic waves, electromagnetic waves may be called *transverse waves*. In addition,

$$\frac{\mathbf{d}_0^* \cdot \mathbf{d}_0}{\varepsilon} = \frac{\mathbf{b}_0^* \cdot \mathbf{b}_0}{\mu} \qquad (3\text{-}42)$$

**2.   Plane-wave solution of real variables**   Electromagnetic waves must be real. Consider the following real fields constructed from the complex solutions (3-37a) and (3-37b):

$$\mathbf{d}^r = R_e\{\mathbf{d}_0 \exp i[(\mathbf{k} \cdot \mathbf{r}) - \omega t]\} \qquad (3\text{-}43a)$$

$$\mathbf{b}^r = R_e\{\mathbf{b}_0 \exp i[(\mathbf{k} \cdot \mathbf{r}) - \omega t]\} \qquad (3\text{-}43b)$$

Here, $R_e[a \exp i\phi]$ implies $\frac{1}{2}[a \exp i\phi + a^* \exp(-i\phi)]$. Since the condition (3-36) is quadratic with respect to $\omega$ and $\mathbf{k}$, $\mathbf{d}_0^* \exp(-i\phi)$ also is the solution of the wave equation (3-33). Thus, the vector fields $\mathbf{d}^r$ and $\mathbf{b}^r$ also are solutions. The wave $\mathbf{d}^r$ or $\mathbf{b}^r$ is not to be confused with a standing wave which is composed of complex waves propagating in opposite directions. The conjugate wave $\mathbf{d}^*$ with the phase $\omega t - (\mathbf{k} \cdot \mathbf{r})$ propagates in the same direction as $\mathbf{d}$. The wave propagating in the opposite direction has the phase $\pm[(\mathbf{k} \cdot \mathbf{r}) + \omega t]$.

[1] For generality, $\mu$ is retained. In x-ray problems, $\mu$ can be assumed to be unity.
[2] The symbol ⌃ indicates a unit vector.

In general, if a set of wave fields $\mathbf{d}$, $\mathbf{e}$, etc., satisfies the Maxwell equations of the form (3-1) or (3-6), the complex conjugates $\mathbf{d}^*$, $\mathbf{e}^*$, etc., also are solutions. This fact is derived from the fact that the Maxwell equations are linear with respect to the field vectors and that the dielectric constant for the field vectors having time dependence $e^{i\omega t}$ is the complex conjugate of the dielectric constant of the wave fields which depends on time as $e^{-i\omega t}$. Thus, if the complex field vectors satisfying the Maxwell equations are known, one can construct the real fields just by taking the real part. For this reason and for mathematical convenience, complex fields are often used in what follows.

**3. _Energy flow of the plane wave_**  Energy density and energy flow associated with electromagnetic fields have been defined by equations (3-10) or (3-11). For the special solutions (3-43a) and (3-43b),[1] in the case of real $\varepsilon$ and $\mu$,

$$\mathbf{S} = \frac{1}{16\pi}\left(\frac{c}{\varepsilon\mu}\right)\{(\mathbf{d}_0 \times \mathbf{b}_0^*) + (\mathbf{d}_0^* \times \mathbf{b}_0) + (\mathbf{d}_0 \times \mathbf{b}_0)\exp 2i\phi + (\mathbf{d}_0^* \times \mathbf{b}_0^*)\exp(-2i\phi)\}$$

$$W_e = \frac{1}{32\pi\varepsilon}\{2(\mathbf{d}_0 \cdot \mathbf{d}_0^*) + (\mathbf{d}_0 \cdot \mathbf{d}_0)\exp 2i\phi + (\mathbf{d}_0^* \cdot \mathbf{d}_0^*)\exp(-2i\phi)\}$$

$$W_m = \frac{1}{32\pi\mu}\{2(\mathbf{b}_0 \cdot \mathbf{b}_0^*) + (\mathbf{b}_0 \cdot \mathbf{b}_0)\exp 2i\phi + (\mathbf{b}_0^* \cdot \mathbf{b}_0^*)\exp(-2i\phi)\}$$

Taking an average over a sufficiently long time interval, the terms multiplied by $\exp(\pm 2i\phi)$ drop out. By virtue of the relations (3-41) and the rules of vector calculus

$$\langle\mathbf{S}\rangle = \frac{v}{16\pi}\left\{\frac{1}{\varepsilon}(\mathbf{d}_0 \cdot \mathbf{d}_0^*) + \frac{1}{\mu}(\mathbf{b}_0 \cdot \mathbf{b}_0^*)\right\}\hat{\mathbf{k}} \qquad (3\text{-}44)$$

$$\langle W_e\rangle = \frac{1}{16\pi}\left(\frac{1}{\varepsilon}\right)(\mathbf{d}_0 \cdot \mathbf{d}_0^*) \qquad (3\text{-}45a)$$

$$\langle W_m\rangle = \frac{1}{16\pi}\left(\frac{1}{\mu}\right)(\mathbf{b}_0 \cdot \mathbf{b}_0^*) \qquad (3\text{-}45b)$$

Thus,

$$\langle\mathbf{S}\rangle = v\{\langle W_e\rangle + \langle W_m\rangle\}\hat{\mathbf{k}} \qquad (3\text{-}46)$$

This relation enables one to obtain a hydrodynamical picture of energy flow for the electromagnetic wave. In a nondissipative medium, the phase velocity and the flow velocity are identical.

---

[1] Here again $\mathbf{b} = \mu\mathbf{h}$ and $c/v = \sqrt{\varepsilon\mu}$ are assumed for generality.

More generally, when the complex wave fields have the forms $\mathbf{e}e^{-i\omega t}$ and $\mathbf{h}e^{-i\omega t}$, the time-averaged Poynting vector can be calculated as follows:

$$\langle \mathbf{S} \rangle = \frac{c}{4\pi} \langle [R_e (\mathbf{e}e^{-i\omega t}) \times R_e (\mathbf{h}e^{-i\omega t})] \rangle$$

$$= \frac{c}{16\pi} \{[\mathbf{e} \times \mathbf{h}^*] + [\mathbf{e}^* \times \mathbf{h}]\}$$

$$= \frac{c}{8\pi} \text{Re} [\mathbf{e} \times \mathbf{h}^*] \tag{3-47a}$$

Similarly,

$$\langle W_e \rangle = \frac{1}{16\pi} \text{Re} (\mathbf{e} \cdot \mathbf{d}^*) \tag{3-47b}$$

$$\langle W_m \rangle = \frac{1}{16\pi} \text{Re} (\mathbf{h} \cdot \mathbf{b}^*) \tag{3-47c}$$

**4. *Absorption*** So far, it has been assumed that $\varepsilon$ is a real number unless otherwise stated. As has been seen in Section I C, in general, $\varepsilon$ is complex. The earlier formulas, however, are formally retained. Nevertheless, some remarks are in order regarding the physical implications. It is possible to write

$$\varepsilon = \varepsilon^r + i\varepsilon^i \tag{3-48}$$

where the superscripts $r$ and $i$ represent the real and imaginary parts, respectively. The velocity $v$ defined by (3-34) must be complex. The frequency $\omega$ of the field vectors must be real from its physical meaning. Thus, it is necessary to introduce a complex wave vector

$$\mathbf{k} = \mathbf{k}^r + i\mathbf{k}^i \tag{3-49}$$

From equations (3-34) and (3-36)

$$k^2 = \left(\frac{\omega}{c}\right)^2 \varepsilon$$

$$= (k^r)^2 - (k^i)^2 + 2ik^r k^i (\hat{\mathbf{k}}^r \cdot \hat{\mathbf{k}}^i)$$

Since $\chi^i$ in $\varepsilon = 1 + \chi^r + i\chi^i$ has an order of magnitude of about $10^{-5}$, we can conclude that $k^r \gg k^i$. Thus

$$k^r = K(1 + \tfrac{1}{2}\chi^r) \tag{3-50a}$$

$$k^i = \frac{\tfrac{1}{2}K\chi^i}{\hat{\mathbf{k}}^r \cdot \hat{\mathbf{k}}^i} \tag{3-50b}$$

Inserting (3-49) in equations (3-37a) and (3-37b), it follows that $(\mathbf{k}^r \cdot \mathbf{r}) - \omega t$ has the role of the phase $\phi$ [cf (3-38)]. Therefore the wave front (the surface of a constant phase) is perpendicular to $\mathbf{k}^r$.

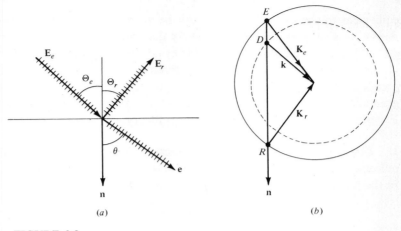

**FIGURE 3-2**
Reflection and refraction. (*a*) Real space; (*b*) reciprocal space.

Since the spatial attenuation of $(\mathbf{d} \cdot \mathbf{d}^*)$ is exp $[-2(\mathbf{k}^i \cdot \mathbf{r})]$, the linear absorption coefficient $\mu_l$ along the direction $\hat{\mathbf{k}}^r$ normal of the wave front is given by

$$\mu_l = K\chi^i \qquad (3\text{-}51a)$$

It is worthwhile to note that expression (3-51a) is true only for the wave of the form (3-37a) and (3-37b) which has the time dependence $e^{-i\omega t}$. If one starts from the expressions of their complex conjugates, one must use

$$\mu_l = -K\chi^i \qquad (3\text{-}51b)$$

because, then, the spatial attenuation of $(\mathbf{d} \cdot \mathbf{d}^*)$ is exp $2(\mathbf{k}^i \cdot \mathbf{r})$. Since, however, the dielectric constant $\varepsilon$ becomes the complex conjugate of the original one for the wave fields having the time dependence $e^{i\omega t}$, the apparent difference between (3-51a) and (3-51b) does not represent a physical difference.

## E. Reflection and refraction

It has been established that x-ray wave fields satisfy equations (3-27a) or (3-27b) inside matter. If the polarizability $\chi$ is constant, a plane wave (3-37a) and (3-37b) is a possible wave whose wave vector $\mathbf{k}$ satisfies the dispersion relation (3-36). Still, the direction of the wave vector and the amplitude are left undetermined. They are determined uniquely by the boundary conditions, namely, the wave fields on the crystal surface. This section considers the wave fields which are excited by a plane wave in the half-bounded medium as shown in Fig. 3-2a. The problem is exactly equivalent to the Fresnel wave-optical theory of reflection and refraction. For this reason, it is sufficient to describe merely the principles and the results.

Although the problem has nothing to do with crystal diffraction, it is one of the interesting subjects of x-ray optics. Moreover, the theory may be considered as a preparation for the ordinary dynamical theory. It is instructive to examine what generalizations and what approximations are required in the transition from the Fresnel theory to the ordinary dynamical theory.

**1.   *Tangential continuity of the wave vectors***   The central concept of the theory of reflection and refraction is the matching of the relevant waves on the crystal boundary. In the present problem, consider the following waves:

$$\text{Incident wave} \qquad \mathbf{E}_e(\mathbf{r}) = \mathbf{E}_e \exp i(\mathbf{K}_e \cdot \mathbf{r}) \qquad (3\text{-}52a)^1$$

$$\text{Reflected wave} \qquad \mathbf{E}_r(\mathbf{r}) = \mathbf{E}_r \exp i(\mathbf{K}_r \cdot \mathbf{r}) \qquad (3\text{-}52b)$$

$$\text{Crystal wave} \qquad \mathbf{e}(\mathbf{r}) \;\; = \mathbf{e} \exp i(\mathbf{k} \cdot \mathbf{r}) \qquad (3\text{-}52c)$$

Here, the time-dependence factor $e^{-i\omega t}$ is omitted. The magnetic vectors can be represented simply by using $\mathbf{H}$ and $\mathbf{h}$ in place of $\mathbf{E}$ and $\mathbf{e}$, respectively. Obviously, they are orthogonal to the corresponding electric fields. Matching occurs whenever the tangential components of the wave vectors $\mathbf{K}_e$, $\mathbf{K}_r$, and $\mathbf{k}$ are identical. Thus, they can be represented by

$$\mathbf{K}_e = \mathbf{T} + \Gamma_e \mathbf{n} \qquad (3\text{-}53a)$$

$$\mathbf{K}_r = \mathbf{T} + \Gamma_r \mathbf{n} \qquad (3\text{-}53b)$$

$$\mathbf{k} = \mathbf{T} + \gamma \mathbf{n} \qquad (3\text{-}53c)$$

where $\mathbf{T}$ is the tangential component common to the three wave vectors and $\Gamma_e$, $\Gamma_r$, and $\gamma$ are their normal components, and $\mathbf{n}$ is the unit vector normal to the boundary. By virtue of the relations $(\mathbf{K}_e)^2 = (\mathbf{K}_r)^2 = \mathbf{K}^2 = (\omega/c)^2$

$$\Gamma_e = -\Gamma_r = \Gamma \qquad (3\text{-}54)$$

From the dispersion relation of the crystal wave

$$\mathbf{k}^2 = \mathbf{T}^2 + \gamma^2$$

$$= K^2(1 + \chi_0{}^r + i\chi_0{}^i) \qquad (3\text{-}55)$$

These analytical relations can be visualized for the cases of real $\gamma$ by the construction in $\mathbf{k}$ space (Fig. 3-2*b*). The dispersion relation of the incident and reflected waves is

---

[1] In order to minimize the number of new symbols, the wave field and the amplitude are denoted in the same fashion. Otherwise stated, the field vector is represented by explicitly stating the argument **r** such as **E(r)**.

represented by the sphere of radius $K$, and that of the crystal wave is represented by the sphere of radius $K(1 + \chi_0')^{1/2}$. The tail points of the wave vectors, $E$, $R$, and $D$, lie on these spheres, respectively, and they must be brought onto a straight line along **n**. When $\Gamma^2 \geq K^2|\chi_0'|$, one can always find a real intersection $D$ of the line **n** and the dispersion surface of the crystal wave. In this case, (3-54) gives the relation of mirror reflection, $\Theta_e = \Theta_r$, in Fig. 3-2a. The tangential continuity of the wave vectors represents Snell's law of refraction; i.e., $K_e \sin \Theta_e = k \sin \theta$.

The normal component $\gamma$ is complex, in general. Writing $\gamma = \gamma' + i\gamma^i$ and substituting it into (3-55),

$$\gamma' = \pm \frac{1}{\sqrt{2}} [(\Gamma^2 + K^2\chi_0') + \sqrt{(\Gamma^2 + K^2\chi_0')^2 + (K^2\chi_0{}^i)^2}\,]^{1/2} \qquad (3\text{-}56a)$$

$$\gamma^i = \pm \frac{1}{\sqrt{2}} [-(\Gamma^2 + K^2\chi_0') + \sqrt{(\Gamma^2 + K^2\chi_0')^2 + (K^2\chi_0{}^i)^2}\,]^{1/2} \qquad (3\text{-}56b)$$

Each of the solutions with different signs is a possible solution. In the half-bounded medium shown in Fig. 3-2a, however, the positive solution must be adopted, because the wave cannot be enhanced as it propagates in the medium. The nonabsorbing case ($\chi_0{}^i = 0$) represents a special case. First, $\gamma^i$ is always zero when $(\Gamma^2 + K^2\chi_0') \geq 0$. On the other hand, if $(\Gamma^2 + K^2\chi_0') \leq 0$, $\gamma'$ is always zero. The normal component $\gamma$ is either purely real or purely imaginary.

**2. The continuity of the amplitudes** Owing to the Maxwell equations, the field vectors must be matched on the boundary surface denoted by a position vector $\mathbf{r}_e$, unless surface currents or charges are present. Thus, one obtains

$$\mathbf{E}_{e,t}(\mathbf{r}_e) + \mathbf{E}_{r,t}(\mathbf{r}_e) = \mathbf{e}_t(\mathbf{r}_e) \qquad (3\text{-}57a)$$

$$\mathbf{H}_{e,t}(\mathbf{r}_e) + \mathbf{H}_{r,t}(\mathbf{r}_e) = \mathbf{h}_t(\mathbf{r}_e) \qquad (3\text{-}57b)$$

$$\mathbf{E}_{e,t'}(\mathbf{r}_e) + \mathbf{E}_{r,t'}(\mathbf{r}_e) = \mathbf{e}_{t'}(\mathbf{r}_e) \qquad (3\text{-}57c)$$

$$\mathbf{H}_{e,t'}(\mathbf{r}_e) + \mathbf{H}_{r,t'}(\mathbf{r}_e) = \mathbf{h}_{t'}(\mathbf{r}_e) \qquad (3\text{-}57d)$$

$$\mathbf{E}_{e,n}(\mathbf{r}_e) + \mathbf{E}_{r,n}(\mathbf{r}_e)' = \mathbf{d}_n(\mathbf{r}_e) \qquad (3\text{-}57e)$$

$$\mathbf{H}_{e,n}(\mathbf{r}_e) + \mathbf{H}_{r,n}(\mathbf{r}_e) = \mathbf{h}_n(\mathbf{r}_e) \qquad (3\text{-}57f)$$

Here, magnetic permeability is assumed to be unity. The subscripts $t$ and $t'$ denote the tangential components normal and parallel to the incident plane, respectively, and the subscript $n$ denotes the normal component to the boundary. The geometrical relations of the field vectors are illustrated in Fig. 3-3.

By the use of Maxwell's equations in the forms (3-41a) and (3-41b), one obtains the various relations between the components of the field vectors. From the relations,

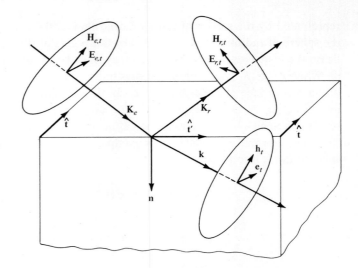

**FIGURE 3-3**
Reflection and refraction of wave fields. One mode is illustrated; in another,
**E** and **H** must be interchanged.

one can conclude that, if equations (3-57a) and (3-57b) are satisfied, equations (3-57e)
and (3-57f) are automatically satisfied, respectively. Also, if (3-57d) is satisfied

$$\Gamma[\mathbf{E}_{e,t}(\mathbf{r}_e) - \mathbf{E}_{r,t}(\mathbf{r}_e)] = \gamma\mathbf{e}_t(\mathbf{r}_e) \qquad (3\text{-}58a)$$

Similarly, from (3-57c) it follows that

$$\varepsilon\Gamma[\mathbf{H}_{e,t}(\mathbf{r}_e) - \mathbf{H}_{r,t}(\mathbf{r}_e)] = \gamma\mathbf{h}_t(\mathbf{r}_e) \qquad (3\text{-}58b)$$

The details of deriving these relations can be found in a textbook on optics. Combining
equations (3-57a) and (3-58a), and equations (3-57b) and (3-58b), one can see that

$$\mathbf{E}_{r,t}(\mathbf{r}_e) = \frac{\Gamma - \gamma}{\Gamma + \gamma} \mathbf{E}_{e,t}(\mathbf{r}_e) \qquad (3\text{-}59a)$$

$$\mathbf{e}_t(\mathbf{r}_e) = \frac{2\Gamma}{\Gamma + \gamma} \mathbf{E}_{e,t}(\mathbf{r}_e) \qquad (3\text{-}59b)$$

$$\mathbf{H}_{r,t}(\mathbf{r}_e) = \frac{\varepsilon\Gamma - \gamma}{\varepsilon\Gamma + \gamma} \mathbf{H}_{e,t}(\mathbf{r}_e) \qquad (3\text{-}59c)$$

$$\mathbf{h}_t(\mathbf{r}_e) = \frac{2\varepsilon\Gamma}{\varepsilon\Gamma + \gamma} \mathbf{H}_{e,t}(\mathbf{r}_e) \qquad (3\text{-}59d)$$

Thus, the amplitudes have the following explicit forms:

$$\mathbf{E}_{r,t} = \frac{\Gamma - \gamma}{\Gamma + \gamma} \mathbf{E}_{e,t} \exp i\{(\mathbf{K}_e - \mathbf{K}_r) \cdot \mathbf{r}_e\} \qquad (3\text{-}60a)$$

$$\mathbf{H}_{r,t} = \frac{\varepsilon\Gamma - \gamma}{\varepsilon\Gamma + \gamma} \mathbf{H}_{e,t} \exp i\{(\mathbf{K}_e - \mathbf{K}_r) \cdot \mathbf{r}_e\} \qquad (3\text{-}60b)$$

$$\mathbf{e}_t = \frac{2\Gamma}{\Gamma + \gamma} \mathbf{E}_{e,t} \exp i\{(\mathbf{K}_e - \mathbf{k}) \cdot \mathbf{r}_e\} \qquad (3\text{-}60c)$$

$$\mathbf{h}_t = \frac{2\varepsilon\Gamma}{\varepsilon\Gamma + \gamma} \mathbf{H}_{e,t} \exp i\{(\mathbf{K}_e - \mathbf{k}) \cdot \mathbf{r}_e\} \qquad (3\text{-}60d)$$

Since the difference vectors $(\mathbf{K}_e - \mathbf{K}_r)$ and $(\mathbf{K}_e - \mathbf{k})$ are normal to the boundary surface, these expressions are constant over the boundary surface.

From the Maxwell equations (3-41a) and (3-41b), one obtains the components of the field vectors within the incident plane determined by the vectors $\mathbf{K}_e$, $\mathbf{K}_r$, and $\mathbf{k}$, the components being denoted by the subscript $s$. The results are given by

$$\mathbf{H}_{e,s}(\mathbf{r}_e) = [\hat{\mathbf{K}}_e \times \mathbf{E}_{e,t}(\mathbf{r}_e)] \qquad (3\text{-}61a)$$

$$\mathbf{H}_{r,s}(\mathbf{r}_e) = [\hat{\mathbf{K}}_r \times \mathbf{E}_{r,t}(\mathbf{r}_e)] \qquad (3\text{-}61b)$$

$$\mathbf{h}_s(\mathbf{r}_e) = \frac{c}{\omega} [(\mathbf{T} + \gamma\mathbf{n}) \times \mathbf{e}_t(\mathbf{r}_e)] \qquad (3\text{-}61c)$$

$$\mathbf{E}_{e,s}(\mathbf{r}_e) = [\mathbf{H}_{e,t}(\mathbf{r}_e) \times \hat{\mathbf{K}}_e] \qquad (3\text{-}62a)$$

$$\mathbf{E}_{r,s}(\mathbf{r}_e) = [\mathbf{H}_{r,t}(\mathbf{r}_e) \times \hat{\mathbf{K}}_r] \qquad (3\text{-}62b)$$

$$\mathbf{e}_s(\mathbf{r}_e) = \frac{c}{\omega} \frac{1}{\varepsilon} [\mathbf{h}_t(\mathbf{r}_e) \times (\mathbf{T} + \gamma\mathbf{n})] \qquad (3\text{-}62c)$$

The time-averaged Poynting vector is given by (3-47a). Using equations (3-60a) and (3-60c) and relations (3-61), one obtains the time-averaged Poynting vectors associated with the plane-polarized wave, whose electric vector is normal to the plane of incidence.

$$\langle \mathbf{S}_e \rangle_\perp = \frac{c}{8\pi} (\mathbf{E}_{e,t} \cdot \mathbf{E}_{e,t}^*) \hat{\mathbf{K}}_e$$

$$\langle \mathbf{S}_r \rangle_\perp = \frac{c}{8\pi} \left(\frac{\Gamma - \gamma}{\Gamma + \gamma}\right) \left(\frac{\Gamma - \gamma}{\Gamma + \gamma}\right)^* (\mathbf{E}_{e,t} \cdot \mathbf{E}_{e,t}^*) \hat{\mathbf{K}}_r$$

$$\langle \mathbf{s} \rangle_\perp = \frac{c}{8\pi} \frac{c}{\omega} \left(\frac{2\Gamma}{\Gamma + \gamma}\right) \left(\frac{2\Gamma}{\Gamma + \gamma}\right)^* (\mathbf{E}_{e,t} \cdot \mathbf{E}_{e,t}^*) R_e(\mathbf{T} + \gamma^*\mathbf{n})$$

For another polarization mode in which the electric field vectors lie within the plane of incidence, the time-averaged Poynting vectors are given according to equations (3-60$b$) and (3-60$d$) and relations (3-62)

$$\langle \mathbf{S}_e \rangle_{\|} = \frac{c}{8\pi} (\mathbf{H}_{e,t} \cdot \mathbf{H}^*_{e,t}) \hat{\mathbf{K}}_e$$

$$\langle \mathbf{S}_r \rangle_{\|} = \frac{c}{8\pi} \left( \frac{\varepsilon\Gamma - \gamma}{\varepsilon\Gamma + \gamma} \right) \left( \frac{\varepsilon\Gamma - \gamma}{\varepsilon\Gamma + \gamma} \right)^* (\mathbf{H}_{e,t} \cdot \mathbf{H}^*_{e,t}) \hat{\mathbf{K}}_r$$

$$\langle \mathbf{s} \rangle_{\|} = \frac{c}{8\pi} \frac{c}{\omega} \left( \frac{2\varepsilon\Gamma}{\varepsilon\Gamma + \gamma} \right) \left( \frac{2\varepsilon\Gamma}{\varepsilon\Gamma + \gamma} \right)^* (\mathbf{H}_{e,t} \cdot \mathbf{H}^*_{e,t}) R_e \left[ \frac{1}{\varepsilon} (\mathbf{T} + \gamma\mathbf{n}) \right]$$

Reflection and refraction (transmission) powers are defined by

$$R = - \frac{(\langle \mathbf{S}_r \rangle \cdot \mathbf{n})}{(\langle \mathbf{S}_e \rangle \cdot \mathbf{n})} \qquad (3\text{-}63a)$$

$$T = \frac{(\langle \mathbf{s} \rangle \cdot \mathbf{n})}{(\langle \mathbf{S}_e \rangle \cdot \mathbf{n})} \qquad (3\text{-}63b)$$

Thus, for each mode of polarization, they have the forms

$$R_\perp = \left( \frac{\Gamma - \gamma}{\Gamma + \gamma} \right) \left( \frac{\Gamma - \gamma}{\Gamma + \gamma} \right)^* \qquad (3\text{-}64a)$$

$$R_\| = \left( \frac{\varepsilon\Gamma - \gamma}{\varepsilon\Gamma + \gamma} \right) \left( \frac{\varepsilon\Gamma - \gamma}{\varepsilon\Gamma + \gamma} \right)^* \qquad (3\text{-}64b)$$

$$T_\perp = \frac{2\Gamma(\gamma + \gamma^*)}{(\Gamma + \gamma)(\Gamma + \gamma^*)} \qquad (3\text{-}64c)$$

$$T_\| = \frac{2\Gamma(\varepsilon^*\gamma + \varepsilon\gamma^*)}{(\varepsilon\Gamma + \gamma)(\varepsilon\Gamma + \gamma)^*} \qquad (3\text{-}64d)$$

Obviously, energy conservation does hold for each mode of polarization; i.e.,

$$R_\perp + T_\perp = 1 \qquad (3\text{-}65a)$$

$$R_\| + T_\| = 1 \qquad (3\text{-}65b)$$

**3.  *Total reflection*** Relations (3-60) and (3-64) are exact even when the wave vectors are complex. When $\chi_0{}^i$ is zero (nonabsorbing case) and $\Gamma^2 < K^2|\chi_0{}^r|^2$, $\gamma$ is purely imaginary. Then, since $\varepsilon$ is real, $\Gamma + \gamma = \Gamma - \gamma^*$ and $\varepsilon\Gamma + \gamma = \varepsilon\Gamma - \gamma^*$. Thus, one can conclude that $R_\perp = R_\| = 1$, whatever value $\chi_0{}^r$ takes on. Total reflection occurs under such conditions. In any case, since $\varepsilon$ is very close to unity for x rays, the difference in $R$ or $T$ is negligibly small between the parallel and perpendicular modes of polarization.

## II. KINEMATICAL THEORY OF CRYSTAL DIFFRACTION

The term "diffraction" is used to describe the modification of the wave front of an incident wave by some physical obstacle. Obviously, the interest here is in a crystal serving as the obstacle. Laue-Bragg reflection occurs at crystal net planes, when the incident wave satisfies a special geometrical condition.

The theories of crystal diffraction are divided into two categories: kinematical and dynamical. In the kinematical theories, it is assumed that the incident wave does not undergo any modification from the presence of the crystalline medium. The idea behind this approximation is similar to that of Kirchhoff's approximation in the diffraction theories of usual optics in which two-dimensional obstacles usually are considered. Also, the approximation is nothing more than the familiar Born approximation in scattering theories. In three-dimensional crystals, however, the validity is less obvious than in two-dimensional cases or in atoms. The incident wave must be modified after propagating a distance in the crystal. Otherwise, no diffracted wave would be produced. Roughly speaking, the approximation is valid only when the interaction between the incident wave and the crystal is sufficiently weak and the crystal size is sufficiently small. Indeed, as will be seen later, the kinematical approximation is not valid for a moderately strong reflection unless the size of the crystal is less than a few microns. Thus, more sophisticated theories are needed in which the modification of the incident wave is taken into account explicitly. Such theories are called collectively the *dynamical theory of diffraction*.

Nevertheless it happens often that, even when the kinematical approximation is not satisfactory, the results derived from it are acceptable in a semiquantitative sense. Moreover, since the framework of the kinematical theory is much simpler than that of the dynamical one, there is no doubt that the kinematical theory is very important for understanding crystal diffraction. Since, however, the details and applications of the kinematical theories are discussed in Chaps. 2 and 7 of this book, only the basics of the theory are described here. In particular, emphasis will be put on the topics which are related to the descriptions of the dynamical theory.

### A. Fraunhofer equation for three-dimensional diffraction

The kinematical theories are a kind of scattering theory. For this reason, the integral form of the fundamental equation (3-29) or (3-30a) is adequate as the starting point. Since, now, one is concerned with the wave field outside the crystal, the displacement vector $\mathbf{D}_s$ can be replaced by the electric vector $\mathbf{E}_s$. One also can assume the monochromatic incident wave to have a time dependence of the form

$$\mathbf{E}_e(\mathbf{r}, t) = \mathbf{E}_e(\mathbf{r}) \exp - i\omega t \qquad (3\text{-}66)$$

Making use of the kinematical approximation, then, the scattered wave can be written

$$E_s\,(\mathbf{r},t) = e^{-i\omega t} \int \frac{1}{R} \mathbf{f}(\mathbf{r}') \exp i\,KR\,dv' \qquad (3\text{-}67)$$

where the vector $\mathbf{f}$ is defined by

$$\mathbf{f}(\mathbf{r}') = \frac{1}{4\pi} \text{rot rot} \left[ \chi(\mathbf{r}')E_e(\mathbf{r}') \right] \qquad (3\text{-}68)$$

and $R = |\mathbf{r} - \mathbf{r}'|$.

In order to obtain a more explicit form of (3-67), use the Fraunhofer approximations

$$R = R_0 \qquad \text{for the amplitude} \qquad (3\text{-}69a)$$

$$R = R_0 - (\hat{\mathbf{R}}_0 \cdot \mathbf{r}') \qquad \text{for the phase} \qquad (3\text{-}69b)$$

The required condition for these approximations is

$$R_0 \gg \text{crystal size} \qquad (3\text{-}70)$$

Then, the scattered wave is given by

$$E_s = \frac{1}{R_0} \exp i(KR_0 - \omega t) \int \mathbf{f}(\mathbf{r}') \exp -i(\mathbf{K}_s \cdot \mathbf{r}')\,dv' \qquad (3\text{-}71)$$

where

$$\mathbf{K}_s = \left( \frac{\omega}{c} \right) \hat{\mathbf{R}}_0 \qquad (3\text{-}72)$$

The operator (rot) in the expression (3-68) can be replaced by the vector multiplication $(i\mathbf{K}_s)$ to the vector involved. By the use of the classical expression (3-26) for the polarizability $\chi$ and the expression of a plane wave for the incident wave,

$$E_e(\mathbf{r}) = \mathbf{E}_e \exp i(\mathbf{K}_e \cdot \mathbf{r}) \qquad (3\text{-}73)$$

one obtains the scattering wave

$$E_s(\mathbf{r},\,t) = \frac{e^2}{mc^2} \cdot \frac{1}{R_0} \exp i(KR_0 - \omega t)[\hat{\mathbf{R}}_0 \times (\hat{\mathbf{R}}_0 \times \mathbf{E}_e)] \cdot \mathfrak{F}(\mathbf{p}) \qquad (3\text{-}74a)$$

where

$$\mathfrak{F}(\mathbf{p}) = \int G(\mathbf{r}) \exp -i(\mathbf{p} \cdot \mathbf{r})\,dv \qquad (3\text{-}75)$$

introducing the generalized electron distribution

$$G(\mathbf{r}) = \sum_k g^{(k)} N^{(k)}(\mathbf{r}) \qquad (3\text{-}76)$$

and the scattering vector

$$\mathbf{p} = \mathbf{K}_s - \mathbf{K}_e \qquad (3\text{-}77)$$

The magnetic field $\mathbf{H} = \mathbf{B}$ is obtained from (3-19b) within the approximation $(KR_0) \gg 1$ as follows:

$$H_s(\mathbf{r}, t) = \frac{e^2}{mc^2} \cdot \frac{1}{R_0} \exp i(KR_0 - \omega t)(\mathbf{E}_e \times \hat{\mathbf{R}}_0) \cdot \mathfrak{F}(\mathbf{p}) \qquad (3\text{-}74b)$$

The field vectors $\mathbf{E}_s$ and $\mathbf{H}_s$ constitute a spherical vector wave field.

Historically, the quantity $\mathfrak{F}$ is called the *form factor*[1] of the charge assemblage. Mathematically, it is the Fourier transform of the generalized electron density $G(\mathbf{r})$. The mathematics of Fourier transforms are well developed. This is the main reason why the kinematical theory is very powerful for determining the structure of matter.

Equations (3-74a) and (3-74b) are the basic equations for the kinematical theory of crystal diffraction which will be described in the following sections. Before going further, however, a few important remarks are required.

**1.  *Energy flow in the spherical wave***  The Poynting vector is defined by (3-10c). The time-averaged one can be calculated with the aid of (3-47a) in terms of the complex representation of the wave fields. In the present case, the wave fields $\mathbf{e}$ and $\mathbf{h}$ are given by the expressions (3-74a) and (3-74b), respectively. Thus

$$\langle \mathbf{S} \rangle = \frac{c}{8\pi} \left( \frac{e^2}{mc^2} \right)^2 \left( \frac{\sin \phi}{R_0} \right)^2 |\mathfrak{F}|^2 (\mathbf{E}_e \cdot \mathbf{E}_e^*) \hat{\mathbf{R}}_0 \qquad (3\text{-}78)$$

where $\phi$ is the angle between $\mathbf{E}_0$ and $\mathbf{R}_0$. Since $\langle \mathbf{S} \rangle$ decreases as $(1/R_0)^2$, the energy $\delta E$ radiated through an infinitesimal solid angle $\delta\Omega$ is constant. For a unit Poynting vector of the incident wave $[(c/8\pi)(\mathbf{E}_e \cdot \mathbf{E}_e^*) = 1]$, $\sigma = \delta E/\delta\Omega$ is called the *differential cross section* or *scattering power*. In the present case, it is

$$\sigma = \left( \frac{e^2}{mc^2} \right)^2 \sin^2 \phi |\mathfrak{F}|^2 \qquad (3\text{-}79)$$

**2.  *Thomson scattering***  If the scatterer is a single free electron at the origin

$$G(\mathbf{r}) = \delta(\mathbf{r})$$

then the form factor turns out to be unity and the expressions (3-74a) and (3-74b) agree with those usually used to describe Thomson scattering.

**3.  *Coherence of scattered waves***  If the electron is located at $\mathbf{r}_0$,

$$G(\mathbf{r}) = \delta(\mathbf{r} - \mathbf{r}_0)$$

---

[1] It is correct practice nowadays to use the term "scattering factor" when speaking of atoms.

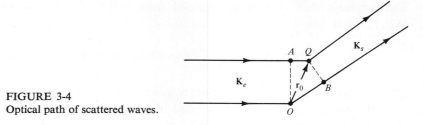

FIGURE 3-4
Optical path of scattered waves.

In this case, the form factor $\mathfrak{F}$ becomes a phase factor

$$\mathfrak{F} = \exp - i(\mathbf{p} \cdot \mathbf{r}_0)$$

The phase can be calculated directly from geometrical considerations of the difference of the path length between two scattered waves from the origin $O$ and the electron position $Q$ in Fig. 3-4. There, $AQ = (\hat{\mathbf{K}}_e \cdot \mathbf{r}_0)$ and $OB = (\hat{\mathbf{K}}_s \cdot \mathbf{r}_0)$. The path difference is simply given by $(AQ - OB)$. The form factor of an assemblage of electrons can be interpreted as the sum of the phase factors associated with the individual electrons. Each scattered wave has a definite phase relation with respect to the incident wave. This situation is called *perfect coherence*.

**4. Polarization**　According to equations (3-74$a$ and $b$), the scattered wave fields $\mathbf{E}_s$ and $\mathbf{H}_s$ bear a definite relation in their vector orientations with respect to the incident electric fields $\mathbf{E}_e$ and the direction $\hat{\mathbf{R}}_0$ (cf Fig. 3-5). This relation is independent of the charge distribution. The mutual relation among $\mathbf{E}_s$, $\mathbf{H}_s$, and $\hat{\mathbf{R}}_0$ is exactly the same as the relation among $\mathbf{d}$, $\mathbf{b}$, and $\mathbf{k}$ illustrated in Fig. 3-1 for a plane wave. From a local viewpoint, therefore, the scattered wave is a transverse one. The magnitude of the vector products $(\hat{\mathbf{R}}_0 \times \mathbf{E}_e)$ and $[\hat{\mathbf{R}}_0 \times (\hat{\mathbf{R}}_0 \times \mathbf{E}_e)]$ is $(\mathbf{E}_e\mathbf{E}_e^*)^{1/2} \sin \phi$, which is the magnitude of the projection of $\mathbf{E}_e$ on a plane normal to the scattering direction $\hat{\mathbf{R}}_0$. The point of observation can see only the projected component. The factor $(\sin \phi)^2$ in equations (3-78) and (3-79) arises from this fact and is called the *polarization factor* for a linearly polarized wave.

　　For natural light, the electric vector can be decomposed into two components which have the same magnitude in a statistical sense; one of these is parallel to the plane determined by $\mathbf{K}_e$ and $\mathbf{K}_s$, and the other is perpendicular to the plane. The polarization factor turns out to be $(\cos 2\Theta)^2$ and unity in these respective cases, where $2\Theta$ is the angle between $\mathbf{K}_e$ and $\mathbf{K}_s$. Thus, the polarization factor for natural light is the arithmetic mean of these two factors. The differential cross section for natural light thus is given by

$$\sigma = \frac{1}{2} \left( \frac{e^2}{mc^2} \right)^2 (1 + \cos^2 2\Theta) |\mathfrak{F}|^2 \qquad (3\text{-}80)$$

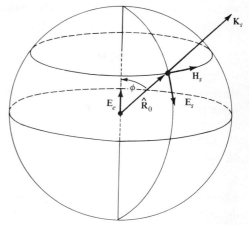

FIGURE 3-5

## B.    The form factor for Laue-Bragg reflection

The form factor for an assembly of electrons is defined by (3-75). From relations (3-74) and (3-79), we know that $\mathbf{E}_s(\mathbf{H}_s)$ and $\sigma$ depend on the structure of crystals only through this form factor. For this reason, the form factor is the central quantity in crystal diffraction. Here, we shall describe three cases briefly: an atom (ion),[1] a unit cell, and the complete crystal.

**1.   Atomic form factor f**   The form factor of an atom is called the *atomic form factor* or *scattering factor* and is denoted by $f$. The explicit form of the scattering factor is

$$f = \sum_k g^{(k)} f^{(k)} \qquad (3\text{-}81)$$

where the superscript $(k)$ specifies the individual electron. Since $g^{(k)}$ is complex in general, $f$ is also complex and can be written

$$f = f_0 + \Delta f' + i\Delta f'' \qquad (3\text{-}82)$$

Here, $f_0$ is the frequency-independent form factor, and the additional term $\Delta f' + i\Delta f''$ is referred to as the *dispersion correction*. The details for calculating $f$ have been described in Chap. 1.

**2.   Crystal-structure factor F**   The crystal-structure factor is defined by the expression

$$F(\mathbf{p}) = \int_{\substack{\text{all}\\\text{space}}} G_1(\mathbf{r}) \exp -i(\mathbf{p} \cdot \mathbf{r})\, dv \qquad (3\text{-}83a)$$

---

[1] Hereafter, ions will be included in the term "atoms."

where $G_1$ is the generalized electron density of atoms belonging to a single unit cell. At this stage, strictly speaking, $F$ is not identical to

$$\bar{F}(\mathbf{p}) = \int_{\substack{\text{unit} \\ \text{cell}}} G(\bar{\mathbf{r}}) \exp -i(\mathbf{p} \cdot \bar{\mathbf{r}}) \, d\bar{v} \qquad (3\text{-}83b)$$

where $\bar{\mathbf{r}}$ is a reduced position vector within the unit cell. As will be seen later, only when the Bragg condition is satisfied exactly, does $F = \bar{F}$.

**3.   *Crystal-shape function S***   Consider now the form of $\mathfrak{F}$ for the complete, perfect crystal.  A general position in the crystal can be written

$$\mathbf{r} = \bar{\mathbf{r}} + \mathbf{r}_{n_1 n_2 n_3} \qquad (3\text{-}84)$$

where $\mathbf{r}_{n_1 n_2 n_3} = n_1 \mathbf{a}_1 + n_2 \mathbf{a}_2 + n_3 \mathbf{a}_3$ represents a Bravais lattice point specified by three integers $n_1$, $n_2$, $n_3$, and $\mathbf{a}_1$, $\mathbf{a}_2$, $\mathbf{a}_3$ are the three translation vectors of the lattice. By the definition of $\mathfrak{F}$ and $\bar{F}$ and by virtue of the translational symmetry of the generalized electron density $G(\mathbf{r})$

$$\mathfrak{F} = \sum_{n_1 n_2 n_3} \int_{\substack{\text{unit} \\ \text{cell}}} G(\bar{\mathbf{r}}) \exp -i[\mathbf{p} \cdot (\bar{\mathbf{r}} + \mathbf{r}_{n_1 n_2 n_3})] \, d\bar{v} \qquad (3\text{-}85)$$

$$= \bar{F}(\mathbf{p}) \cdot S(\mathbf{p})$$

where

$$S(\mathbf{p}) = \sum_{n_1 n_2 n_3} \exp -i[\mathbf{p} \cdot (n_1 \mathbf{a}_1 + n_2 \mathbf{a}_2 + n_3 \mathbf{a}_3)] \qquad (3\text{-}86a)$$

This is called the *crystal-shape function*.  For convenience let

$$\mathbf{p} = \mathbf{K}_s - \mathbf{K}_e = 2\pi \mathbf{g} + \mathbf{\Delta} \qquad (3\text{-}87)$$

where $\mathbf{g}$ and $\mathbf{\Delta}$ are vectors in reciprocal space defined by

$$\mathbf{g} = g_1 \mathbf{b}_1 + g_2 \mathbf{b}_2 + g_3 \mathbf{b}_3 \qquad (3\text{-}88)$$

$$\mathbf{\Delta} = \Delta_1 \mathbf{b}_1 + \Delta_2 \mathbf{b}_2 + \Delta_3 \mathbf{b}_3 \qquad (3\text{-}89)$$

Here, $g_1$, $g_2$, $g_3$ are integers or zero and $\Delta_1$, $\Delta_2$, $\Delta_3$ are arbitrary numbers with the restrictions

$$|\Delta_j| \leq \pi$$

A set of vectors $\mathbf{b}_1$, $\mathbf{b}_2$, $\mathbf{b}_3$, called the *reciprocal-lattice vectors*, are defined by

$$\mathbf{b}_1 = \frac{\mathbf{a}_2 \times \mathbf{a}_3}{v} \qquad (3\text{-}90a)$$

$$\mathbf{b}_2 = \frac{\mathbf{a}_3 \times \mathbf{a}_1}{v} \qquad (3\text{-}90b)$$

$$\mathbf{b}_3 = \frac{\mathbf{a}_1 \times \mathbf{a}_2}{v} \qquad (3\text{-}90c)$$

where $v$ is the volume of the unit cell given by $\mathbf{a}_1 \cdot (\mathbf{a}_2 \times \mathbf{a}_3)$. Obviously, the vector $\mathbf{g}$ denotes a net plane of the index $(g_1, g_2, g_3)$ and $\mathbf{\Delta}$ is the deviation of $(\mathbf{K}_s - \mathbf{K}_e)$ from the reciprocal-lattice point.[1]  By virtue of the orthogonality and normalization of $\mathbf{a}_i$ and $\mathbf{b}_i$, it follows from (3-86a) and (3-88) that

$$S(\Delta) = \sum_{n_1 n_2 n_3} \prod_{j}^{1,2,3} \exp - i\Delta_j n_j \qquad (3\text{-}86b)$$

In order to calculate $S$, it is convenient to introduce a three-dimensional diffraction function

$$D(\Delta) = \int_{\text{crystal}} \exp - i(\Delta \cdot \mathbf{r}) \, dv \qquad (3\text{-}91)$$

Substituting (3-84) and (3-89) in this

$$D(\Delta) = \left[ \int_{\substack{\text{unit} \\ \text{cell}}} \exp - i(\Delta \cdot \bar{\mathbf{r}}) \, d\bar{v} \right] S(\Delta)$$

$$= v \prod_{j=1}^{3} \int_{0}^{1} \exp - i(\Delta_j \, x_j) \, d_j x \, S(\Delta)$$

$$= v \prod_{j=1}^{3} \left\{ \frac{1 - \exp - i\Delta_j}{i\Delta_j} \right\} S(\Delta) \qquad (3\text{-}92)$$

In the following, only the case where $|\Delta_j| \ll \pi$ is considered.  Then, the term multiplying $v$ in (3-92) is practically unity.  Thus, to a good approximation

$$\mathfrak{F}(\Delta) = \bar{F} S(\Delta) = \left( \frac{\bar{F}}{v} \right) D(\Delta) \qquad (3\text{-}93)$$

For any shape of the crystal, when $\Delta = 0$

$$D(0) = Nv = \text{total volume} \qquad (3\text{-}94)$$

where $N$ is the total number of unit cells in the crystal.  Since all waves scattered from the crystal have an identical phase, it is concluded that $D(\Delta)$ has its maximum value when $\Delta = 0$.  On the other hand, if $\Delta$ is large, the scattered wave from different parts of the crystal can have different phases.  In that case $D(\Delta)$ is essentially equal to zero.

In a one-dimensional crystal of length $B$ lying between $B_1$ and $B_2$

$$D_1(\Delta; B) = \int_{B_1}^{B_2} \exp - i\Delta l \, dl = \frac{\sin \frac{1}{2}\Delta B}{\Delta/2} \left[ \exp - \frac{i}{2} \Delta(B_1 + B_2) \right] \qquad (3\text{-}95)$$

---

[1] Strictly speaking, the reciprocal lattice discussed here is one enlarged by the factor $2\pi$.  Hereafter, it also will be called the "reciprocal lattice."

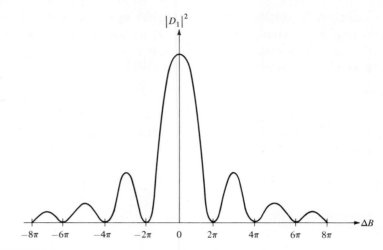

**FIGURE 3-6**
One-dimensional diffraction function, $D_1 (\Delta; B)$.

The appearance of $D_1 \cdot D_1^*$ is shown in Fig. 3-6. It should be noted that this function takes on an appreciable value only when $|\Delta|B \lesssim 2\pi$. The considerable angular spread, therefore, is estimated to be $|\Delta|K \simeq \lambda/B$.

The two-dimensional diffraction function is defined by

$$D_2(\boldsymbol{\sigma}; S) = \iint_s \exp -i(\boldsymbol{\sigma} \cdot \mathbf{s}) \, ds \qquad (3\text{-}96)$$

where $\boldsymbol{\sigma}$ and $\mathbf{s}$ are two-dimensional vectors in the plane of the crystal, and $S$ denotes the two-dimensional crystal.

**4. Laue-Bragg reflection and the Ewald construction** From the above considerations, a large scattering is expected whenever $\boldsymbol{\Delta} = 0$, namely,

$$\mathbf{K}_g = \mathbf{K}_e + 2\pi\mathbf{g} \qquad (3\text{-}97)$$

Here, $\mathbf{K}_g$ is used instead of $\mathbf{K}_s$. This condition is equivalent to the following one which was first proposed by Laue:

$$(\mathbf{g} \cdot \mathbf{a}_j) = g_j \qquad (3\text{-}98)$$

and also to the Bragg condition

$$2d \sin \Theta = n\lambda \qquad (3\text{-}99)$$

where $d$ is the interplanar spacing of the net plane $\mathbf{g}$ and $\Theta$ is the glancing angle formed by $\mathbf{K}_e$ and the net plane.

Equation (3-97) can be interpreted as an expression of momentum conservation.

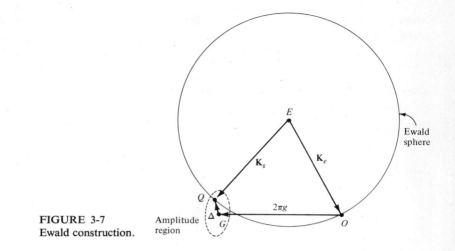

FIGURE 3-7
Ewald construction.

Amplitude region

The wave vectors $\mathbf{K}_e$ and $\mathbf{K}_g$ are proportional to the momentum of the incident and reflected photons, respectively, and $2\pi\mathbf{g}$ is proportional to the lattice momentum. In order to visualize this relation, Ewald proposed an ingenious graphical construction, which is shown in Fig. 3-7. Since $|\mathbf{K}_g| = |\mathbf{K}_e| = (\omega/c)$, a strong reflection is brought about only when the Ewald sphere of radius $K$ and having its origin at $E$ passes through the reciprocal-lattice point $G(\overrightarrow{OG} = 2\pi\mathbf{g})$.

If the crystal size is finite, the diffraction function $D(\Delta)$ may have some value for $\Delta \neq 0$. This implies that a plane wave with the wave vector $\mathbf{K}_e$ is scattered in the various directions $\mathbf{K}_s = \mathbf{K}_e + 2\pi\mathbf{g} + \Delta$. The Ewald construction is extremely helpful in visualizing this situation. Consider a region surrounding the reciprocal-lattice point $G$. Next, assign the diffraction function $D(\Delta)$ to a point $Q$ in this region, where $\Delta = \overrightarrow{GQ}$. The region in which $D(\Delta)$ has a nonvanishing value is called the *amplitude region*.[1] When the Ewald sphere cuts this amplitude region, scattered waves proportional to $D(\Delta)$ are expected in the directions $\overrightarrow{EQ}$. In general, therefore, scattered waves are not a plane wave but have diffuseness with respect to the scattering directions.

**5.  *A remark on the crystal-structure factor***  The quantities $F$ and $\bar{F}$ have been defined previously by (3-83a) and (3-83b), respectively. From this definition, $G_1(\mathbf{r})$ in the integral of (3-83a) must be

$$G_1(\mathbf{r}) = \sum_s G(\mathbf{r} - \mathbf{R}_s) \qquad (3\text{-}100)$$

---

[1] Laue used the term *Intensitätsbereich* (intensity region).  He attributed $|D(\Delta)|^2$ to a point $Q$.  For later descriptions, however, $D(\Delta)$ is more appropriate.

where the subscript $s$ denotes the atoms belonging to the unit cell concerned, and $\mathbf{R}_s$ is the position of the $s$th nucleus. On the other hand, for the reduced vector $\bar{\mathbf{r}}$,

$$G(\bar{\mathbf{r}}) = G_1(\bar{\mathbf{r}}) + \sum_j G(\bar{\mathbf{r}} - \mathbf{R}_j) \qquad (3\text{-}101)$$

where the subscript $j$ specifies the atoms outside the unit cell. The $j$th atom has an equivalent atom specified by $s$ inside the unit cell so that, by a suitable translation, $\mathbf{r}_{n_1 n_2 n_3}$:

$$\int_{\substack{\text{unit} \\ \text{cell}}} \sum_j G(\bar{\mathbf{r}} - \mathbf{R}_j) \exp -i(\mathbf{p} \cdot \bar{\mathbf{r}}) \, d\bar{v}$$

$$= \sum_{n_1 n_2 n_3}' \int_{\substack{\text{unit} \\ \text{cell}}} \sum_s G(\bar{\mathbf{r}} + \mathbf{r}_{n_1 n_2 n_3} - \mathbf{R}_s) \exp -i[\mathbf{p} \cdot (\bar{\mathbf{r}} + \mathbf{r}_{n_1 n_2 n_3})]$$

$$\times \exp i(\mathbf{p} \cdot \mathbf{r}_{n_1 n_2 n_3}) \, d\bar{v} \qquad (3\text{-}102)$$

When the Bragg condition is satisfied exactly for the net plane $\mathbf{g}$, the scattering vector $\mathbf{p}$ is equal to $2\pi\mathbf{g}$. In this particular case, the last phase term is unity. The vector $\mathbf{r} = \bar{\mathbf{r}} + \mathbf{r}_{n_1 n_2 n_3}$ is the position vector outside the unit cell, unless all the $n$ are zero. Thus, the right side of (3-102) equals

$$\int_{\substack{\text{outside} \\ \text{cell}}} G_1(\mathbf{r}) \exp -2\pi i(\mathbf{g} \cdot \mathbf{r}) \, dv$$

and one derives the important consequence that

$$\bar{F}(2\pi\mathbf{g}) = F(2\pi\mathbf{g}) \qquad (3\text{-}103)$$

From the definition of $F$ and the atomic form factor $f$ it follows that

$$F = \left\{ \int \sum_s G(\mathbf{r} - \mathbf{R}_s) \exp -i[\mathbf{p} \cdot (\mathbf{r} - \mathbf{R}_s)] \, d(\mathbf{r} - \mathbf{R}_s) \right\} \exp -i(\mathbf{p} \cdot \mathbf{R}_s)$$

$$= \sum_s f_s \exp -i(\mathbf{p} \cdot \mathbf{R}_s) \qquad (3\text{-}104)$$

where $\mathbf{R}_s$ is the position of the $s$th atom within a unit cell. The crystal-structure factor depends on the atomic position through this relation. It can be seen, therefore, that crystal-structure analysis becomes the science of deducing $\mathbf{R}_s$ from $|F|^2$, in a manner of speaking.

## C.   Imperfect crystals

The charge density of an ideally perfect crystal is characterized by uninterrupted translational symmetry in three dimensions. In nature such perfect crystals are encountered rarely, if ever. It is much more likely that the translational periodicity is interrupted somehow. This, by itself, does not pose a serious problem since it is possible to solve the wave equation for any model of an imperfect crystal. Typically one can only determine a statistical distribution of defects in a crystal, however, so that there is the additional task of averaging statistically the intensities of the scattered waves.

In the present discussion, the latter difficulty will be omitted and the following imperfections will be considered specifically:

*1* Finiteness of the crystal

*2* Isolated plane defects (stacking faults)

*3* Continuously bent lattice plane

The physical origins of these defects will be mentioned below.

All these imperfections give rise to line broadening; i.e., to the spreading of the amplitude region, which is characterized by the scattering factor $\mathfrak{F}$. For finite perfect crystals, as has been discussed already in Section B, $\mathfrak{F}$ reduces to the crystal-shape function $S(\Delta)$, which is proportional to the three-dimensional diffraction function $D(\Delta)$. Consequently, the scattering or form factor is the central quantity to be calculated in this section. The first kind of imperfection above is related literally to the external form of the crystal, while the other two concern the internal form. Such descriptions of imperfect crystals are by no means complete, but the main purpose of the present section is to serve as an introduction to understanding the dynamical theories for distorted crystals.

**1.** *Finiteness of the crystal* A very elegant method was developed by von Laue (1936) for calculating the diffraction function $D(\Delta)$ in three-dimensional cases. He has shown that the intensity (or amplitude) region extends along directions that are normal to the crystal surfaces. Here, a somewhat different approach is followed in order to compare the results with those of the dynamical theory. Considerations are limited to columnar crystals terminated by an entrance surface $S_e$ and an exit surface $S_a$. Since any polyhedral crystal can be divided into several columns (Fig. 3-8), the form factor of the total crystal can be represented by the sum of individual ones. The column direction $v$ may be arbitrary in the kinematical theory. In a special case, later, $v$ will be taken in the direction of the Bragg-reflected beam $\overline{\mathbf{K}}_g$. The position vector $\mathbf{r}$ and the deviation vector $\Delta$ in the expression (3-91) of $D(\Delta)$ are divided into two components as follows:

$$\mathbf{r} = \mathbf{s} + l\mathbf{v} \qquad (3\text{-}105)$$

$$\Delta = \sigma + \zeta\mathbf{v} \qquad (3\text{-}106)$$

where $\mathbf{s}$ and $\sigma$ are two-dimensional component vectors perpendicular to the direction $v$. If $v$ is the direction of $\overline{\mathbf{K}}_g$, $\Delta$ is a vector on the Ewald sphere and $\zeta$ is identical to

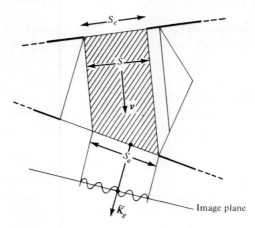

**FIGURE 3-8**
Division of a polyhedral crystal into columns and the crystal image. Here $v$ is the column direction and $\overline{\mathbf{K}}_g$ is the mean direction of the diffracted waves.

Image plane

the normal distance $w$ from $G$ to it. Since $(\boldsymbol{\Delta} \cdot \mathbf{r}) = (\boldsymbol{\sigma} \cdot \mathbf{s}) + \zeta l$ and the volume element $dv = d\mathbf{s}\, dl$, the diffraction function can be written

$$D(\boldsymbol{\Delta}) = \frac{i}{\zeta} \iint_S [(\exp -i\zeta L_a) - (\exp -i\zeta L_e)] \exp -i(\boldsymbol{\sigma} \cdot \mathbf{s})\, d\mathbf{s} \qquad (3\text{-}107)$$

where $L_e$ and $L_a$ denote $l$ values at the entrance and exit surfaces, respectively. The integration covers the cross section $S$ of the column.

Next, equation (3-107) is modified in a way to enable comparison with the dynamical theory. The deviation vector $\boldsymbol{\Delta}$ can be rewritten

$$\boldsymbol{\Delta} = \boldsymbol{\sigma}_e + K\delta_e \mathbf{n}_e \qquad (3\text{-}108a)$$

$$\boldsymbol{\Delta} = \boldsymbol{\sigma}_a + K\delta_a \mathbf{n}_a \qquad (3\text{-}108b)$$

in which $\boldsymbol{\sigma}_e$ and $\boldsymbol{\sigma}_a$ are again the components perpendicular to the direction $v$, but $K\delta_e$ and $K\delta_a$ are the components normal to the surfaces $S_e$ and $S_a$, respectively (see Fig. 3-9). Thus,

$$(\boldsymbol{\sigma} \cdot \mathbf{s}) + \zeta L_e = (\boldsymbol{\Delta} \cdot \mathbf{r}_e) = (\boldsymbol{\sigma}_e \cdot \mathbf{s}) + K\delta_e R_e$$

$$(\boldsymbol{\sigma} \cdot \mathbf{s}) + \zeta L_a = (\boldsymbol{\Delta} \cdot \mathbf{r}_a) = (\boldsymbol{\sigma}_a \cdot \mathbf{s}) + K\delta_a R_a$$

where $\mathbf{r}_e$ and $\mathbf{r}_a$ are the position vectors denoting the surfaces $S_e$ and $S_a$, and $R_e$ and $R_a$ are the normal distances to them from the origin of coordinates (cf Fig. 3-9a). The diffraction function, then, has the form

$$D(\boldsymbol{\Delta}) = \frac{i}{\zeta} \left[ \exp -i(K\delta_a R_a) \iint_s \exp -i(\boldsymbol{\sigma}_a \cdot \mathbf{s})\, d\mathbf{s} \right.$$

$$\left. - \exp -i(K\delta_e R_e) \iint_s \exp -i(\boldsymbol{\sigma}_e \cdot \mathbf{s})\, d\mathbf{s} \right] \qquad (3\text{-}109)$$

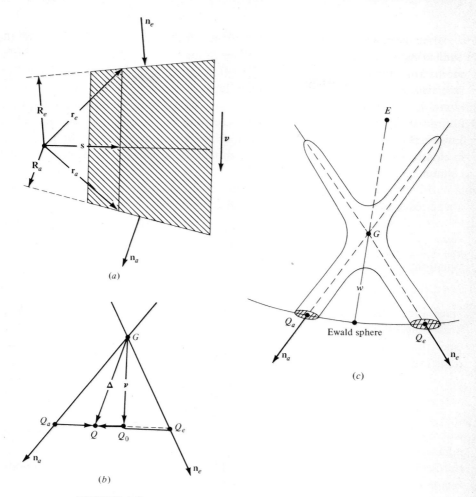

**FIGURE 3-9**

(a) Vectors in real space; (b) vectors in reciprocal space: $\overrightarrow{GQ_0} = \zeta\nu$, $\overrightarrow{Q_eQ} = \sigma_e$, $\overrightarrow{Q_aQ} = \sigma_a$, $\overrightarrow{Q_0Q} = \sigma$, with $GQ_a = K\delta_a$, and $GQ_e = K\delta_e$; (c) amplitude region and Ewald sphere in kinematical theory.

When the cross section $S$ is sufficiently large, the two-dimensional diffraction function $\iint_s d\mathbf{s}$ tends to Dirac's delta function; i.e., the function is appreciable only for small values of $\sigma_e$ and $\sigma_a$, respectively. In this particular case, the amplitude region converges to the lines $\mathbf{n}_e$ and $\mathbf{n}_a$ passing through the reciprocal-lattice point $G$. On the lines $\mathbf{n}_e$ and $\mathbf{n}_a$, the value of the diffraction function is $S$, so that the amplitudes of the two terms in $D(\Delta)$ decrease as $(S/\zeta) = (1/K)(S_e/\delta_e) = (1/K)(S_a/\delta_a)$, respectively.

At the point $G$, the two terms must be considered together. Obviously, then, $D(\Delta)$ takes on the value $V$, the total volume of the column. When the crystal surfaces are narrow, the broadening of the diffraction function $\iint_s ds$ is not negligible. Still, when we are concerned with the region of large $\zeta$, it consists of the two rods elongated along $\mathbf{n}_e$ and $\mathbf{n}_a$ as shown in Fig. 3-9c. Laue called these elongated rods *Stächeln*. Diffraction spots correspond to the cross sections of the amplitude regions with the Ewald sphere. Thus, one can expect two diffraction spots unless the Bragg condition is closely satisfied. The spot size is essentially determined by the thickness of the column.

In the case of parallel-sided crystals, the amplitude regions along $\mathbf{n}_e$ and $\mathbf{n}_a$ always overlap. Equation (3-107), then, has the form

$$D(\Delta) = D_1(\zeta; L) \, D_2(\sigma; S) \qquad (3\text{-}110)$$

where $D_1$ and $D_2$ are the one- and two-dimensional diffraction functions defined by equations (3-95) and (3-96), respectively, and $\sigma = \sigma_e = \sigma_a$.

**2. The direct image and Fraunhofer diffraction**   As pointed out in Section A, the kinematical theory usually is developed in the formalism of Fraunhofer diffraction. This conventional treatment, however, is not essential. If the amplitude density (spectrum) $f(\mathbf{K})$ is known, one can obtain the wave field $E(\mathbf{r})$ in the form of a Fourier transform

$$E(\mathbf{r}) = \frac{r_c}{2\pi} \int\limits_{-\infty}^{+\infty}\!\!\! f(\mathbf{K}) \, \exp \, i(\mathbf{K} \cdot \mathbf{r}) \, d\mathbf{T} \qquad (3\text{-}111a)$$

where $r_c = e^2/mc^2$ is the classical radius of an electron[1] and $\mathbf{T}$ is an arbitrary, two-dimensional component of the vacuum-wave vector $\mathbf{K}$. It is to be noticed that the third component $K_n$ is uniquely determined by $\mathbf{T}$ as $(\mathbf{K}^2 - \mathbf{T}^2)^{1/2}$. When the position vector $\mathbf{r}_s$ on the image plane closely located near the crystal is inserted for $\mathbf{r}$, the equation gives the wave field of the crystal image.

When, on the other hand, we are concerned with the wave field at $\mathbf{R}_0$, far from the crystal, the wave field $E(\mathbf{R}_0)$ can be calculated by means of the stationary-phase method (Appendix 3A), provided that $f(\mathbf{K})$ is a proper function. In that case

$$E(\mathbf{R}_0) = r_c K_n^* f(K^*) \frac{\exp \, iKR_0}{iR_0} \qquad (3\text{-}112)$$

where $\mathbf{K}^*$ is the vacuum-wave vector parallel to $\mathbf{R}_0$ and $K_n^*$ is the normal component. This is nothing more than the Fraunhofer formula (3-74a). According to the definition, $K_n f(\mathbf{K})$ is the scattering factor $\mathfrak{F}(\Delta)$ defined on the Ewald sphere,[2] $\Delta$ being $\mathbf{K} - \mathbf{K}_g$.

---

[1] If necessary, multiplied by the polarization factor $\sin \phi$.

[2] $K_n f(\mathbf{K})$ gives only the amplitude on the Ewald sphere of radius $K$. The whole behavior of the amplitude region around the reciprocal-lattice point $G$ is mapped only when the incident wave covers a certain angular range.

Conversely, given the scattering factor, the direct image of the crystal can be calculated by the Fourier transform of $\mathfrak{F}(\Delta)/K_n$. This relation is very general, not depending upon the kinematical and dynamical treatments (Kato and Uyeda, 1951; Kato, 1952, 1953).

In case that $f(\mathbf{K})$ is appreciable only within a narrow range of $\Delta$, the Ewald sphere is approximated by a plane perpendicular to $\mathbf{K}_g$. Then, it is convenient to write the wave vector $\mathbf{K}$ as

$$\mathbf{K} = \mathbf{K}_g + \hat{\mathbf{K}}_g w + \sigma$$

where $w$ is the distance from the reciprocal-lattice point to the Ewald sphere. Next, consider the wave field on an image plane perpendicular to $\mathbf{K}_g$ through a point $\mathbf{r}_s$. Writing $\mathbf{r} = \mathbf{r}_s + \mathbf{r}_p$, where $\mathbf{r}_p$ is a vector on the image plane, one obtains[1]

$$(\mathbf{K} \cdot \mathbf{r}) = (\mathbf{K} \cdot \mathbf{r}_s) + (\sigma \cdot \mathbf{r}_p)$$

Under these conditions, the wave field $E(\mathbf{r}_p; w)$ has the form

$$E(\mathbf{r}_p; w) = \frac{r_c}{2\pi K} \exp i(\mathbf{K} \cdot \mathbf{r}_s) \int\int_{-\infty}^{+\infty} \mathfrak{F}(\sigma; w) \exp i(\sigma \cdot \mathbf{r}_p)\, d\sigma \quad (3\text{-}111b)$$

Here, $K_n$ is approximated by $K$.

In the specific case of a columnar crystal along the direction of $\mathbf{K}_g$, the Ewald sphere cuts the *Stächeln* as shown in Fig. 3-9c. Thus, the wave field is composed of two plane waves, $\mathbf{K}_g{}^{(e)} = \overrightarrow{EQ_e}$ and $\mathbf{K}_g{}^{(a)} = \overrightarrow{EQ_a}$. Meanwhile, the line broadening is neglected. The difference $\Delta Q = Q_a Q_e$ is easily calculated:

$$\Delta Q = \frac{w}{\gamma}(\tan\varphi_1 + \tan\varphi_2) \quad (3\text{-}113a)$$

where $\gamma$ is the cosine of the angle between $\mathbf{K}_g$ and its projection on the plane containing $\mathbf{n}_e$ and $\mathbf{n}_a$, and $\varphi_1$ and $\varphi_2$ are the angles between the projection of $\mathbf{K}_g$ and $\mathbf{n}_e$ and $\mathbf{n}_a$, respectively. Thus, one may expect a fringe pattern to represent the crystal image, the fringe spacing being given by

$$\Lambda = \frac{2\pi}{|\Delta Q|} = \frac{(2\pi/w)\gamma_g \gamma_g'}{\gamma \sin\varphi} \quad (3\text{-}113b)$$

where $\varphi = \varphi_1 + \varphi_2$ is the wedge angle, and $\gamma_g = \gamma\cos\varphi_1$ and $\gamma_g' = \gamma\cos\varphi_2$ are employed.

In general, the broadening and overlapping of two *Stächeln* must be considered.

---

[1] The approximation is justified when $\Delta^2(K/R) \ll 2\pi$, namely, when optical diffraction is neglected. This situation is called the *geometrical shadow*. When the next term is taken into account, *Fresnel diffraction* is introduced.

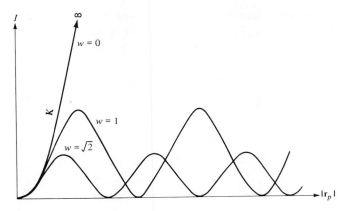

**FIGURE 3-10**
Intensity distribution in the crystal image.

According to (3-93), the Fourier amplitude $\mathfrak{F}(\sigma; w)$ is essentially the diffraction function $D(\sigma; w)$. By using the expression (3-107)[1] for $D(\Delta)$, the wave field is given by

$$E(\mathbf{r}_p; w) = 2\pi \frac{r_c}{K} \frac{\overline{F}}{v} \exp i(\mathbf{K} \cdot \mathbf{r}_s) \int\int_{-\infty}^{+\infty} D_1(w; L)\delta(\mathbf{r}_p - \mathbf{s}) \, ds$$

$$= 2\pi \frac{r_c}{K} \frac{\overline{F}}{v} \exp i(\mathbf{K} \cdot \mathbf{r}_s)D_1(w; L_p) \tag{3-114}$$

where $L_p$ is the thickness of the crystal along the direction $\mathbf{K}_g$ passing through the point $\mathbf{r}_p$ concerned. Changing the position $\mathbf{r}_p$ causes a fringelike pattern to appear, which depends upon the deviation parameter $w$ from the exact Bragg condition. The wave field is appreciable only within the geometrical shadow of the crystal. It is sinusoidal with respect to the length $L_p$. Since $L_p = |\mathbf{r}_p|(\tan \varphi_1 + \tan \varphi_2)/\gamma$ for a suitable choice of the origin for $\mathbf{r}_p$, and since $|D_1|^2$ takes maximal values for $wL_p = 2n\pi(n = \text{integer})$, it can be shown that the fringe spacing is exactly the same as in (3-113b). The intensity decreases as $1/w^2$. Some of the intensity distributions are shown in Fig. 3-10.

**3. Planar defects** We shall consider a crystal including a single stacking fault. The stacking fault can be specified by the normal $\mathbf{n}_f$ of the fault plane and the relative displacement vector $\mathbf{u}_f$ of the divided parts ($A$ and $B$) of the crystal (Fig. 3-11). According to the definition of the scattering factor [cf (3-75)], and recalling that the position $\mathbf{r} = \bar{\mathbf{r}} + \mathbf{r}_{n_1 n_2 n_3}$ in the part $A$ is equivalent to the position

$$\mathbf{r} = \bar{\mathbf{r}} + \mathbf{r}_{m_1 m_2 m_3} + \mathbf{u}_f \tag{3-115}$$

[1] Note that $\zeta$ is replaced by $w$.

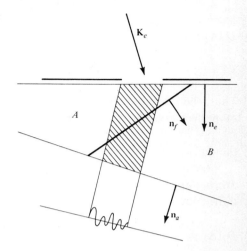

**FIGURE 3-11**
Stacking fault and its crystal image.

in the part $B$, the total scattering factor is given by

$$\mathfrak{F}(\mathbf{\Delta}) = \frac{\bar{F}}{v} [D_A(\mathbf{\Delta}) + \exp{-2\pi i(\mathbf{g} \cdot \mathbf{u}_f)} D_B(\mathbf{\Delta})] \qquad (3\text{-}116)$$

where $\mathbf{\Delta}$ is defined by (3-87), and $(\mathbf{p} \cdot \mathbf{u}_f)$ is approximated by $2\pi(\mathbf{g} \cdot \mathbf{u}_f)$. $D_A$ and $D_B$ are the diffraction functions due to the parts $A$ and $B$, respectively. In addition, $\bar{F}$ is the structure factor of the crystal $A$, and $\bar{F} \exp{-2\pi i(\mathbf{g} \cdot \mathbf{u}_f)}$ is that of the crystal $B$. Again, the amplitude regions will be considered for a few special cases.

*a.  Wedge-shaped crystals ($\mathbf{n}_e$ is not parallel to $\mathbf{n}_a$)*  The functions $D_A$ and $D_B$ contain the terms involving the surface integral over the entrance surface. They are unified as

$$D_e(\mathbf{\Delta}) = -\frac{i}{K\delta_e} \exp{-i(K\delta_e R_e)}$$

$$\times \left[ \iint_{S_{e,A}} \exp{-i(\boldsymbol{\sigma}_e \cdot \mathbf{s})}\, d\mathbf{r}_e \right.$$

$$\left. + \exp{-2\pi i(\mathbf{g} \cdot \mathbf{u}_f)} \iint_{S_{e,B}} \exp{-i(\boldsymbol{\sigma}_e \cdot \mathbf{s})}\, d\mathbf{r}_e \right] \qquad (3\text{-}117a)$$

where $S_{e,A}$ and $S_{e,B}$ are the parts of the entrance surface $S_e$ belonging to the crystal $A$ and $B$, respectively. (Note that $d\mathbf{s}/\zeta = d\mathbf{r}_e/K\delta_e$.) In general, the broadening of

*Stachel* $\mathbf{n}_e$ is modified by the presence of the stacking fault. On the line $\mathbf{n}_e(\sigma_e = 0)$, $D_e(\Delta)$ is reduced to

$$D_e(\delta_e) = -\frac{i}{K\delta_e}\exp-i(K\delta_e R_e)[S_{e,A} + \exp-2\pi i(\mathbf{g}\cdot\mathbf{u}_f)S_{e,B}] \qquad (3\text{-}117b)$$

Thus, the magnitude decreases proportionally to $1/\delta_e$ along the line $\mathbf{n}_e$, as in the case of perfect crystals, but it is always less than the perfect-crystal case by the ratio $|S_{e,A} + \exp[-2\pi i(\mathbf{g}\cdot\mathbf{u}_f)]S_{e,B}|/(S_{e,A} + S_{e,B})$. Similar arguments can be applied to the diffraction function due to the exit surface $S_a$.

A more significant point in the case of faulted crystals is that an additional *Stachel* appears along the direction $\mathbf{n}_f$. It is easily seen that the two-dimensional diffraction function due to the stacking fault can be written

$$D_f(\Delta) = \frac{i}{K\delta_f}\exp-i(K\delta_f R_f)[1 - \exp-2\pi i(\mathbf{g}\cdot\mathbf{u}_f)]\iint_{S_f}\exp-i(\sigma_f\cdot\mathbf{s})\,dr_f \qquad (3\text{-}118)$$

where the subscript $f$ refers to the fault surface. On the line $\mathbf{n}_f(\sigma_f = 0)$, $D_f(\delta_f)$ decreases as $(S_f/K\delta_f)[1 - \exp-2\pi i(\mathbf{g}\cdot\mathbf{u}_f)]$. Obviously, only when $(\mathbf{g}\cdot\mathbf{u}_f)$ is an integer, this *Stachel* disappears. This result is very reasonable, because then the net plane concerned has no irregularity at the fault plane. The total scattering factor can be written as follows

$$\mathfrak{F}(\Delta) = \frac{\bar{F}}{v}[D_e(\Delta) + D_a(\Delta) + D_f(\Delta)] \qquad (3\text{-}119)$$

*b.   Columnar crystals*   Limiting the incident wave by a slit, consider the columnar part along the direction $\mathbf{K}_g$. It is assumed that the fault plane does not appear on the crystal surfaces. With this configuration, $D_e(\Delta)$ is the same as that of the perfect crystal and $D_a(\Delta)$ is multiplied by the phase factor $\exp-2\pi i(\mathbf{g}\cdot\mathbf{u}_f)$. Thus, as far as the amplitude is concerned, the *Stachel* are not altered from the perfect crystal. The additional term due to the fault plane is given by $D_f(\Delta)$ in (3-118).

*c.   Direct image of faulted crystal*   As mentioned above, the amplitude region is composed of three *Stächeln*. Neglecting the broadening, one can see that the crystal image is interpreted by an interference of three plane waves corresponding to the intersections of the three *Stächeln* with the Ewald sphere. Even when the crystal is parallel-sided, a fringe pattern must be observed, because even then two intersections are expected. The difference of the wave vectors is given by the same form as (3-113a); now $\varphi_1$ and $\varphi_2$ are the angles between the projection of $\mathbf{K}_g$ on the $(\mathbf{n}_e, \mathbf{n}_f)$ plane and $\mathbf{n}_e = \mathbf{n}_a$ and $\mathbf{n}_f$, respectively. Thus, the fringe spacing is given by (3-113b). The wave field in general is given by the Fourier transform of (3-116) as follows:

$$E(\mathbf{r}_p; w) = 2\pi \frac{r_c}{K} \frac{\bar{F}}{v} \exp i(\mathbf{K} \cdot \mathbf{r}_s)[D_A(w; L_A) + \exp -2\pi i(\mathbf{g} \cdot \mathbf{u}_f)D_B(w; L_B)] \qquad (3\text{-}120)$$

where $D_A$ and $D_B$ are one-dimensional diffraction functions of the crystal parts $A$ and $B$ along the line in the $\mathbf{K}_g$ direction passing through $\mathbf{r}_p$, respectively, $L_A$ and $L_B$ being the functions of position $\mathbf{r}_p$. Again, the three-dimensional problem can be reduced to one-dimensional problem, as in the case of equation (3-114).

**4. Continuously bent lattice plane** Real crystals always include lattice distortion owing to the presence of dislocations and the inhomogeneous distribution of vacancies, substitutional and/or interstitial impurities. In addition, precipitates also can cause lattice bending in the matrix crystal. Here, any change of the crystal-structure factor $\bar{F}$ corresponding to each Bravais lattice point is neglected. Physically speaking, this assumption can be justified only when the charge distribution is not distorted appreciably in the unit cells or the number of distorted cells is sufficiently small.

An arbitrary position vector $\mathbf{r}$ can be divided into three terms

$$\mathbf{r} = \mathbf{r}_{n_1 n_2 n_3} + \mathbf{u}(\mathbf{r}_{n_1 n_2 n_3}) + \bar{\mathbf{r}} \qquad (3\text{-}121)$$

when $\mathbf{r}_{n_1 n_2 n_3}$ denotes a Bravais lattice point of the reference (perfect) crystal, and $\mathbf{u}$ is its displacement in a distorted state, and $\bar{\mathbf{r}}$ is a reduced position vector within the unit cell. Rigorously speaking, $\mathbf{u}$ is defined only at discrete lattice points, but it can be safely assumed to be a continuous function of position $\mathbf{r}$. As distinct from the case of stacking faults, it is to be noted that the displacement $\mathbf{u}$ may not be smaller than the lattice translational vectors $\{\mathbf{a}_i\}$.

Inserting (3-121) into the definition (3-75) of the scattering factor, one gets an expression similar to (3-85). The shape function $S(\Delta)$ now must be replaced by the generalized one

$$\tilde{S}(\Delta) = \sum_{n_1 n_2 n_3} \exp -i(\mathbf{p} \cdot \mathbf{u}) \exp -i(\Delta \cdot \mathbf{r}_{n_1 n_2 n_3}) \qquad (3\text{-}122)$$

where the difference vector $\Delta$ is redefined by

$$\mathbf{p} = \mathbf{K}_s - \mathbf{K}_e = 2\pi \bar{\mathbf{g}} + \Delta \qquad (3\text{-}123)$$

Here, $\bar{\mathbf{g}}$ is the reciprocal-lattice vector of the reference crystal. For the purpose of evaluating $\tilde{S}(\Delta)$, similarly to the case of perfect crystals, the generalized diffraction function is defined as follows:

$$\tilde{D}(\Delta) = \int_{\text{crystal}} \exp -2\pi i(\bar{\mathbf{g}} \cdot \mathbf{u}) \exp -i(\Delta \cdot \mathbf{r}) \, dv \qquad (3\text{-}124)$$

Again, inserting the expression of the position vector $\mathbf{r}$ in (3-124)

$$\tilde{D}(\Delta) = \int_{\substack{\text{unit} \\ \text{cell}}} \exp -i(\Delta \cdot \bar{\mathbf{r}}) \, d\bar{v} \, \tilde{S}(\Delta)$$

As explained in conjunction with (3-92), the integral can be approximated by $v$, the volume of the unit cell. Thus, the scattering factor has the form

$$\mathfrak{F}(\Delta) = \bar{F}\tilde{S}(\Delta) = \frac{\bar{F}}{v}\,\tilde{D}(\Delta) \qquad (3\text{-}125)$$

Again, the central task in obtaining scattered waves is the calculation of $\tilde{D}(\Delta)$. The diffraction function $D_f(\Delta)$ is easily seen as a particular case of $\tilde{D}(\Delta)$, in which $\mathbf{u}$ is introduced stepwise on the fault plane.

*a.   Reciprocal-lattice vectors in distorted crystals*   Before describing the nature of $\tilde{D}(\Delta)$ it is of interest here to discuss the reciprocal-lattice vectors in distorted crystals. When the crystal is deformed, the translation vectors $\{\mathbf{a}_i\}$ depend upon position. Then, one can write

$$\mathbf{a}_i = \bar{\mathbf{a}}_i + \Delta\mathbf{a}_i \qquad (i = 1, 2, 3) \qquad (3\text{-}126a)$$

The reciprocal-lattice vectors $\{\mathbf{b}_i\}$ can be defined locally from $\{\mathbf{a}_i\}$ by the same rule as those in (3-90). They also are functions of position so that

$$\mathbf{b}_i = \bar{\mathbf{b}}_i + \Delta\mathbf{b}_i \qquad (3\text{-}126b)$$

Because orthogonality and normalization are automatically satisfied, neglecting $(\Delta\mathbf{a}_i \cdot \Delta\mathbf{b}_i)$, there follows

$$(\Delta\mathbf{a}_i \cdot \bar{\mathbf{b}}_i) = -(\bar{\mathbf{a}}_i \cdot \Delta\mathbf{b}_i) \qquad (3\text{-}127)$$

On the other hand,

$$\Delta\mathbf{a}_i = (\bar{\mathbf{a}}_i \cdot \mathrm{grad})\mathbf{u}(\mathbf{r}) \qquad (3\text{-}128)$$

because $\Delta\mathbf{a}_i$ is the increment of $\mathbf{u}$ for changing the position by $\bar{\mathbf{a}}_i$. Combining equations (3-127) and (3-128), one obtains

$$\Delta\mathbf{b}_i = -\mathrm{grad}\,(\bar{\mathbf{b}}_i \cdot \mathbf{u}) \qquad (3\text{-}129)$$

In perfect crystals, any reciprocal-lattice vector is given by

$$\bar{\mathbf{g}} = g_1\bar{\mathbf{b}}_1 + g_2\bar{\mathbf{b}}_2 + g_3\bar{\mathbf{b}}_3 \qquad (3\text{-}130a)$$

When the crystal is deformed, $\bar{\mathbf{g}}$ also is deformed to a vector

$$\mathbf{g} = g_1\mathbf{b}_1 + g_2\mathbf{b}_2 + g_3\mathbf{b}_3 \qquad (3\text{-}130b)$$

By multiplying both sides of (3-129) by $g_i$ and summing over $i$,

$$\Delta\mathbf{g} = -\mathrm{grad}\,(\bar{\mathbf{g}} \cdot \mathbf{u}) \qquad (3\text{-}131a)$$

or

$$\mathbf{g} = \bar{\mathbf{g}} + \Delta\mathbf{g} = \mathrm{grad}\,G \qquad (3\text{-}131b)$$

where

$$G = [\bar{\mathbf{g}} \cdot (\mathbf{r} - \mathbf{u})] \qquad (3\text{-}132a)$$

It should be noted that the vector field $\mathbf{g}$ is derived from the gradient of a scalar $G$. Thus, one obtains an important conclusion

$$\text{rot } \mathbf{g} = 0 \qquad (3\text{-}133)$$

Equation (3-132a) represents a surface in three-dimensional space. In fact, it is a generalization of the equation of a plane

$$G = (\bar{\mathbf{g}} \cdot \mathbf{r}) \qquad (3\text{-}132b)$$

If one assigns integers $\{N\}$ to $G$, (3-132b) represents a set of net planes having Miller indices $(g_1 g_2 g_3)$. In the case of deformed crystals, (3-132a) with $G = \{N\}$ represents a set of curved net planes obtained by the displacement $\mathbf{u}$ from the plane (3-132b). Thus, assuming continuity of $G(\mathbf{r})$,

$$N + 1 = N + |\text{grad } G| \cdot d$$

where $d$ is the interplanar spacing (normal distance) between two neighboring net planes. Thus, the vector $\mathbf{g}$ is normal to the deformed net plane and $|\mathbf{g}|$ is reciprocal to the interplanar spacing $d$.

Return now to the discussion of $\tilde{D}(\boldsymbol{\Delta})$ and the direct image of distorted crystals.

b.  *Calculation of the generalized diffraction function $\tilde{D}(\boldsymbol{\Delta})$*   The integration (3-124) can be carried out approximately by means of the stationary-phase method (Appendix 3A). Although the method is not always satisfactory, the physical nature of $\tilde{D}(\boldsymbol{\Delta})$ is elucidated by consulting the principles used in the approximation. The condition of stationary phase is given by

$$\boldsymbol{\Delta} = -2\pi \text{ grad } (\bar{\mathbf{g}} \cdot \mathbf{u}) = 2\pi \Delta \mathbf{g}(\mathbf{r}_0) \qquad (3\text{-}134)$$

Fixing a vector $\boldsymbol{\Delta}$ in the amplitude region, the amplitude $\tilde{D}(\boldsymbol{\Delta})$ is effectively determined by the crystal part nearby the position $\mathbf{r}_0$ satisfying (3-134). In fact, when the end point of $\boldsymbol{\Delta}$ lies on the Ewald sphere, the Bragg condition is satisfied exactly at the position $\mathbf{r}_0$, where the local $\mathbf{g}$ vector turns out to be

$$\mathbf{g} = \bar{\mathbf{g}} + \Delta \mathbf{g}$$

The size of the effective region for diffraction is given by an ellipsoid associated with the strain gradient tensor at $\mathbf{r}_0$

$$\alpha_0 = \left[ \left( \begin{array}{ccc} \dfrac{\partial^2}{\partial x^2} & \dfrac{\partial^2}{\partial x \, \partial y} & \dfrac{\partial^2}{\partial x \, \partial z} \\[2mm] \dfrac{\partial^2}{\partial y \, \partial x} & \dfrac{\partial^2}{\partial y^2} & \dfrac{\partial^2}{\partial y \, \partial z} \\[2mm] \dfrac{\partial^2}{\partial z \, \partial x} & \dfrac{\partial^2}{\partial z \, \partial y} & \dfrac{\partial^2}{\partial z^2} \end{array} \right) (\bar{\mathbf{g}} \cdot \mathbf{u}) \right]_{\mathbf{r} = \mathbf{r}_0} \qquad (3\text{-}135a)$$

where $x$, $y$, and $z$ are arbitrary rectangular coordinates. If one takes the principal axes $\{s_i\}$, the symmetrical tensor $\boldsymbol{\alpha}_0$ can be transformed to the simple form

$$\boldsymbol{\alpha}_0 = \begin{pmatrix} \lambda_1 & 0 & 0 \\ 0 & \lambda_2 & 0 \\ 0 & 0 & \lambda_3 \end{pmatrix}_{r=r_0} \tag{3-135$b$}$$

where $\lambda_i$ denotes $(\partial/\partial s_i^2)(\bar{\mathbf{g}} \cdot \mathbf{u})$. The volume of the effective region, then, is given by $(2\pi)^{3/2}/|\lambda_1\lambda_2\lambda_3|^{1/2}$. The phase is essentially given by that of the integrand at $\mathbf{r}_0$, namely, $-[2\pi(\bar{\mathbf{g}} \cdot \mathbf{u}_0) + (\boldsymbol{\Delta} \cdot \mathbf{r}_0)]$. An additional phase $(\pi/2)\sigma$ depending upon the curvature of lattice plane at $\mathbf{r}_0$ is required.

Finally, the generalized diffraction function can be written [cf (3A-18)]

$$\tilde{D}(\boldsymbol{\Delta}) = \left[\frac{(2\pi)^{3/2}}{|\lambda_1\lambda_2\lambda_3|^{1/2}}\right]_{r=r_0} \exp - i\left[2\pi(\bar{\mathbf{g}} \cdot \mathbf{u}_0) + (\boldsymbol{\Delta} \cdot \mathbf{r}_0) + \frac{\pi}{2}\sigma\right] \tag{3-136}$$

In this expression, $\{\lambda_i\}$, $\mathbf{u}_0$, and $\mathbf{r}_0$, as well as $\sigma$, are functions of $\boldsymbol{\Delta}$, through $\mathbf{r}_0$ by virtue of (3-134). When the condition of stationary phase (3-134) is satisfied at several points, $\tilde{D}(\boldsymbol{\Delta})$ must be the sum of terms similar to (3-136) at these points. The crystal, under this condition, satisfies the Bragg condition at separate regions.

The approximation used here is not accurate for gently distorted crystals. When one of $\{\lambda_i\}$ is nearly equal to zero, obviously, the stationary-phase method cannot be applied. Even when $\{\lambda_i\}$ are not very small, the effective regions for Bragg reflection may overlap each other, or they are cut by the external crystal surfaces. Under these conditions, the integral (3-124) is rather complicated.

c. *The direct image* By combining equations (3-124) and (3-125) and inserting them into (3-111$b$), the direct image of the distorted crystal is given by the same procedure used in deriving (3-114).

$$E(\mathbf{r}_p \cdot w) = 2\pi\frac{r_c}{K}\frac{\bar{F}}{v} \exp i(\mathbf{K} \cdot \mathbf{r}_s) \int_{L_e}^{L_a} \exp -i\{2\pi[\mathbf{g} \cdot \mathbf{u}(l)] + wl\}\, dl \tag{3-137}$$

where $l$ is the line in the direction $\mathbf{K}_g$ passing through the point concerned, $\mathbf{r}_p$. Again, the problem can be reduced to a one-dimensional one. The stationary-phase method can be easily employed. Arguments paralleling those in three-dimensional cases can be applied. If merely a single region satisfies the Bragg condition along the line $l$, the crystal image (intensity) is proportional to the square of the effective length. When two and more separate regions satisfy the Bragg condition, the intensity distribution is complicated owing to the interference of the waves created at each region.

# APPENDIX 3A. STATIONARY-PHASE METHOD

The stationary-phase method due to Kelvin is not only useful from a mathematical point of view for calculating some types of integrals, as described below, but it is also very helpful in understanding the optical principles. Further details of the mathematical arguments presented can be found in, for example, Jeffreys and Jeffreys (1956), Sneddon (1951), and Born and Wulf (1959).

## I. ONE-DIMENSIONAL CASES

Consider the following function, which can be defined by a superposition of waves specified by a real parameter $w$

$$D(\mathbf{r}) = \int A(w) \exp iT(w) \, dw \qquad (3A\text{-}1)$$

Obviously, the functions $A$ and $T$ have the meanings of amplitude and phase, respectively. Actually, they may include a position vector $\mathbf{r}$ as another parameter,[1] but, to avoid unnecessary complexity, it is omitted here.

If the amplitude is a slowly varying function of $w$, and the phase is a rapidly changing function except near the stationary point $[w]$, the following approximation can be used. The stationary point is defined by the stationary-phase condition

$$\frac{\partial T}{\partial w} = 0 \qquad (3A\text{-}2)$$

The value of a quantity $Q$ at the stationary point is denoted by $[Q]$. The functional forms of $A$ and $T$ are schematically illustrated in Fig. 3A-1a. In principle, the wave field $D(\mathbf{r})$ is the sum of the contributions from all component waves. Since, however, the function $\exp iT$ oscillates very rapidly except near the point $[w]$ (cf Fig. 3A-1b and c), the contributions of waves specified by $w$ far from $[w]$ cancel out. The net contribution comes from a very small region, where $T$ differs from $[T]$ by about $\pi/2$. Therefore, the phase $T$ can be approximated by the quadratic form

$$T = [T] + \frac{1}{2} \left( \frac{\partial^2 T}{\partial w^2} \right) (w - [w])^2 + \cdots \qquad (3A\text{-}3)$$

where the linear term can be omitted by virtue of the definition of $[w]$. The amplitude $A$ is simply approximated by $[A]$. Under these approximations the wave field (3A-1) can be represented by

$$D(\mathbf{r}) = (A \exp iT) \int_{-\infty}^{+\infty} \exp \frac{i}{2} \left( \frac{\partial^2 T}{\partial w^2} \right) (w - [w])^2 \, dw \qquad (3A\text{-}4a)$$

---

[1] In this sense, the problem is three-dimensional in real space. Note that (3A-1) is actually more flexible than a mere superposition of plane waves.

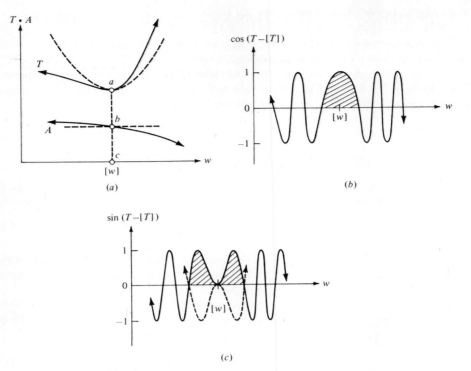

**FIGURE 3A-1**
Mathematical structure of the stationary-phase method. (a) Schematic illustration of the approximation. The solid curves show the behavior of the amplitude $A$ and the phase $T$ and the dashed curves, their approximated behavior. $ac = [T]$ and $bc = [A]$. (b) Real part of $\exp i(T - [T])$. The shaded region is included in the integral. (c) Imaginary part of $\exp i(T - [T])$. The solid curve results when $(\partial^2 T/\partial w^2) > 0$; the dashed curve when $(\partial^2 T/\partial w^2) < 0$. The sign dependence of the phase term in the Fresnel integral (3A-4b) can be seen in this illustration.

The integral is nothing more than the well-known Fresnel integral. Thus, we shall have

$$D(\mathbf{r}) = \left(\frac{2\pi}{|(\partial^2 T/\partial w^2)|}\right)^{1/2} (A \exp iT) \exp \pm \frac{\pi}{4} i \qquad \text{(3A-4b)}$$

The signs must be selected according to the sign of $(\partial^2 T/\partial w^2)$.

Equation (3A-2) and the wave field (3A-4b) include the parameter $[w]$ and a vector variable $\mathbf{r}$. If one considers the left side of (3A-2) as a function of $\mathbf{r}$, one can interpret that equation as representing the trajectory of the beam specified by $[w]$. Equation (3A-4b) is the wavefunction along this trajectory. If one eliminates $[w]$ with the use of (3A-2), $D(\mathbf{r})$ represents the wave field as a function of position.

When the conditions that $A$ changes slowly while $\exp iT$ changes rapidly near the point $[w]$ are not satisfied, the above method cannot be used. In particular, if $(\partial^2 T/\partial w^2) \simeq 0$, the present approximation can no longer be used. For example, consider the spherical wave defined by the integral (4-192). The case is actually two-dimensional with respect to the parameter specifying the component waves; however, one can perform successively the integrations with respect to $K_y$ and $K_x$. For simplicity, but without any loss of generality, one can assume that $y = 0$. The integral to be considered first then is

$$D = \int_{-\infty}^{+\infty} \frac{1}{K_z} \exp i(K_x x + K_z z) \, dK_y$$

where

$$K_z = (K^2 - K_x^2 - K_y^2)^{1/2}$$

The stationary condition is

$$\frac{\partial T}{\partial K_y} = -\frac{K_y}{K_z} z = 0$$

which means $[K_y] = 0$. Furthermore, one can see that

$$\frac{\partial^2 T}{\partial K_y^2} = -\frac{z}{[K_z]}$$

Thus, by virtue of (3A-4b),

$$D = \left(\frac{2\pi}{[K_z]z}\right)^{1/2} \exp i(K_x x + [K_z]z) \exp -\frac{\pi}{4} i \qquad \text{(3A-5)}$$

where $[K_z] = (K^2 - K_x^2)^{1/2}$.

The next step of integration is to consider

$$L = \int_{-\infty}^{+\infty} D \, dK_x$$

where $K_x$ stands for the variable $w$ in (3A-1). The stationary condition is $(\partial T/\partial K_x) = x - [K_x/K_z]z = 0$, namely,

$$\frac{x}{[K_x]} = \frac{z}{[K_z]} = \frac{r}{K} \qquad \text{(3A-6)}$$

which gives $[T] = (K r)$. In addition $(\partial^2 T/\partial K_x^2) = -zK^2/[K_z]^3$. Again, by virtue of (3A-4b)

$$L = \left(\frac{2\pi}{[[K_z]]z}\right)^{1/2} \left(\frac{2\pi[[K_z]]^3}{K^2 z}\right)^{1/2} \exp i(K r) \exp -\frac{\pi}{2} i$$

where $[[K_z]]$ is the value of $[K_z]$ under the condition (3A-6). Thus, finally one obtains

$$\Phi = \frac{i}{8\pi^2} L = \frac{1}{4\pi r} \exp i(K r) \qquad \text{(3A-7)}$$

The coincidence with the original expression (4-191) is rather accidental because it happens that, in the present case, the effects of the approximations in the expressions for phase and amplitude cancel each other out. Nevertheless, the example illustrates the power of the method for calculating this kind of integral.

## II.   TWO-DIMENSIONAL CASES

Next, one considers the following double integral:

$$D(\mathbf{r}) = \int\!\!\!\int_{-\infty}^{+\infty} A(u, v) \exp iT(u, v) \, du \, dv \qquad (3A\text{-}8)$$

In this case, the stationary-phase condition is given by

$$\frac{\partial T}{\partial u} = \frac{\partial T}{\partial v} = 0 \qquad (3A\text{-}9)$$

The phase $T$ is assumed to be approximated by

$$T = [T] + \tfrac{1}{2}\alpha(u - [u])^2 + \tfrac{1}{2}\beta(v - [v])^2 + \gamma(u - [u])(v - [v]) \qquad (3A\text{-}10)$$

and the amplitude is assumed to be $[A]$. With these approximations, one obtains

$$D(\mathbf{r}) = \frac{2\pi}{|\alpha\beta - \gamma^2|} (A \exp iT) \exp i\frac{\pi}{2}\sigma \qquad (3A\text{-}11)$$

where $\sigma$ is defined by

$$\sigma = \begin{cases} 1 & \text{when } \alpha\beta > \gamma^2 \text{ and } \alpha,\beta > 0 \\ 0 & \text{when } \alpha\beta < \gamma^2 \\ -1 & \text{when } \alpha\beta > \gamma^2 \text{ and } \alpha,\beta < 0 \end{cases} \qquad (3A\text{-}12)$$

## III.   THREE-DIMENSIONAL CASES

The following triple integral is considered:

$$D(\mathbf{r}) = \int\!\!\!\int\!\!\!\int_{-\infty}^{+\infty} A(\mathbf{u}) \exp iT(\mathbf{u}) \, d\mathbf{u} \qquad (3A\text{-}13)$$

where $\mathbf{u}$ is a column vector with three components $u_1$, $u_2$, and $u_3$, and $d\mathbf{u}$ represents the volume element $du_1 du_2 du_3$. The amplitude and phase are approximated by

$$A = [A]$$
$$T = [T] + \tfrac{1}{2}(\bar{\mathbf{u}} - [\bar{\mathbf{u}}])\alpha(\mathbf{u} - [\mathbf{u}]) \qquad (3A\text{-}14)$$

where $\bar{\mathbf{u}}$ is a row vector $(u_1, u_2, u_3)$, and $\alpha$ is a three-by-three matrix with components

$$\alpha_{ij} = \frac{\partial^2 T}{\partial u_i \, \partial u_j} \qquad (3A\text{-}15)$$

As before, the square brackets [ ] indicate the value at the point of stationary phase, which is defined by

$$\frac{\partial T}{\partial u_i} = 0 \qquad i = 1, 2, \text{ and } 3 \qquad (3A\text{-}16)$$

Since $T$ is a quadratic form of the real coefficients $\alpha_{ij}$, it can be transformed into the canonical form

$$T = [T] + \lambda_1 u^2 + \lambda_2 v^2 + \lambda_3 w^2 \qquad (3A\text{-}17)$$

where $\lambda_i(i = 1, 2, 3)$ are the eigenvalues of the matrix $\alpha$, and $(u, v, w)$ are the coordinates of the vector $\mathbf{u}$ along the principal axes of the quadratic form of $T$. With this form of $T$, one obtains the approximate value of $D(\mathbf{r})$ as follows:

$$D(\mathbf{r}) = \left[ \frac{(2\pi)^3}{|\lambda_1 \lambda_2 \lambda_3|} \right]^{1/2} (A \exp iT) \exp i\sigma \frac{\pi}{2} \qquad \text{(3A-18)}$$

and $\sigma$ is defined by

$$\sigma = \begin{cases} \frac{3}{2} & \text{when } \lambda_1, \lambda_2, \lambda_3 > 0 \\ \frac{1}{2} & \text{when two of } \lambda_1, \lambda_2, \lambda_3 > 0 \\ -\frac{1}{2} & \text{when two of } \lambda_1, \lambda_2, \lambda_3 < 0 \\ -\frac{3}{2} & \text{when } \lambda_1, \lambda_2, \lambda_3 < 0 \end{cases} \qquad \text{(3A-19)}$$

Incidentally, it is proved that

$$\lambda_1 \lambda_2 \lambda_3 = \text{determinant of } \{\alpha_{ij}\} \qquad \text{(3A-20)}$$

## IV.  FRAUNHOFER DIFFRACTION

The wave field, in general, can be represented by a superposition of plane waves. Equation (3-111a) is an example. The argument used in the example in Section I above can be followed without any essential modification. One can write the wave field (3-111a) for $\mathbf{r}$ in the form

$$E(\mathbf{r}) = \frac{r_c}{2\pi} \frac{2\pi [K_z][f(K)]}{ir} \exp iKr \qquad \text{(3A-21)}$$

because $(i/8\pi^2)(1/K_z)$ can be replaced by $(r_c/2\pi)f(K)$ in the original Fourier representation. If one writes $[K_z]$ and $[f(\mathbf{K})]$ as $K_n^*$ and $f(\mathbf{K}^*)$, respectively, and $r$ is replaced by $R_0$, the expression (3-112) obtains directly.

## V.  REMARKS ON APPLICATION TO RELATIONS

So far, it has been assumed that the physical meanings of the parameters $u$, $v$, or $w$ specifying the component waves are the coordinates in reciprocal space while the implicit parameter $\mathbf{r}$ is the coordinate in real space, particularly in the case when (3A-1) represents the Fourier integral. The mathematical structure, however, is independent of such physical meanings. In equation (3-124) the situation is just the reverse; the parameters specifying the component waves represent the position in real space (the source of the scattered wavelet), and the implicit parameter is the deviation $\boldsymbol{\Delta}$ from the exact Bragg condition; i.e., the vector in reciprocal space.

When the parameter $w$ or $\mathbf{u}$ is complex, the stationary-phase method can be generalized to the steepest descent method. The physical meanings become obscure, but the mathematical results are parallel to those derived above. The details can be found in the references cited at the beginning of this appendix.

# 4

# DYNAMICAL THEORY FOR PERFECT CRYSTALS[1]

## I. PLANE-WAVE THEORY

### A. Historical introduction

The general basis and framework of the dynamical theory were developed soon after the discovery of x-ray diffraction in crystals by three outstanding physicists: Darwin (1914a and b, 1922), Ewald (1916a and b, 1917, 1937), and Laue (1931). In connection with the last named, Bethe's contributions (1928) to electron diffraction also should be cited. The participants in the renaissance of interest in the dynamical theory in the 1950s were surprised to discover just how elegant, complete, and pleasing these theories were. One might say that today's scientists are building small private edifices according to the master plans prepared by the early pioneers. The following discussions in this chapter are, therefore, a description of how subsequent tenants have decorated these edifices to express their individual tastes and interests.

The impetus for such innovations has been provided by advancements in procuring new materials and in the technology of experimentation, namely, the growth

---

[1] The notations employed in this chapter are the same as those in Chaps. 3 and 5 but differ somewhat from the ones used in other chapters of this book in order to facilitate comparison with original citations.

of extremely perfect crystals and the development of topographic x-ray methods, including high-voltage electron microscopy. Specifically, the need for highly perfect germanium and silicon crystals for semiconductor applications spawned a great deal of interest in their characterization. This, in turn, led to the invention of topographic observations of x-ray diffraction images which are complementary to the more conventional analysis of diffraction-line shapes in an optical sense. As is well known, the result of these studies has been the disclosure of a great deal of new information about imperfections present in crystals.

It is interesting to note that the above-mentioned pioneers actually suggested the "life styles" that subsequent investigators have followed. Thus, Darwin's theory is very valuable as a practical method, but it is less general than the other theories. It has been followed largely by electron microscopists. Ewald's and Laue's approaches overlap in many ways. Ewald stressed a more fundamental aspect of the theory by starting explicitly with an elementary scattering process, namely, scattering by a periodic array of dipoles. The current generation, more deeply steeped in modern scattering theories and quantum electrodynamics, therefore, may find his theoretical exposition more to their liking.

The discussion presented below follows more closely the approach of Bethe and Laue, which can be called an "optical" theory. It is the framework within which most of the recent theoretical developments in the dynamical theory have taken place, and it is also most closely related to the development of the kinematical diffraction theory presented above. It should be realized, however, that considerable progress can be expected also from the electron microscopists following in Darwin's footsteps and other physicists pursuing Ewald's development. One also should note the more recent interest in the dynamical theory by neutron and low-energy electron diffractionists. The following discussions, however, are those believed to be best suited for utilization in x-ray diffraction studies.

## B.   Propagation of waves in crystalline media

**1.   *Fundamental equations***   In Section I C, the fundamental equation of the electric displacement vector **d** has been obtained in the differential form (3-27a) and (3-27b).[1] In terms of the generalized electron density $G(\mathbf{r})$, equation (3-76), the microscopic polarizability times $4\pi$ is given by [cf (3-26)][2]

$$\chi(\mathbf{r}) = -\frac{4\pi e^2}{m\omega^2} G(\mathbf{r}) \qquad (4\text{-}1)$$

---

[1] Hereafter, quantities pertaining to matter are denoted by small letters, and the quantities in vacuum are specified by capital letters.

[2] Hereafter, $\chi$ is simply called polarizability.

Since $G(\mathbf{r})$ has a three-dimensional periodicity, $\chi$ can be represented in the form of Fourier series

$$\chi(\mathbf{r}) = \sum_g \chi_g \exp 2\pi i(\mathbf{g} \cdot \mathbf{r}) \qquad (4\text{-}2)$$

where the coefficient is given by

$$\chi_g = -\frac{4\pi e^2}{m\omega^2 v} F_g \qquad (4\text{-}3)$$

Here $v$ is the volume of a unit cell, and the Fourier coefficient $F_g$ is identical to the crystal-structure factor $\bar{F}$ [cf (3-83$b$)].

In this chapter, the wave field is assumed to have the time dependence $e^{-i\omega t}$. Equation (3-27$b$), then, reduces to

$$\Delta \mathbf{d} + K^2 \mathbf{d} + \text{rot rot} \left[ \sum_g \chi_g \exp 2\pi i(\mathbf{g} \cdot \mathbf{r}) \cdot \mathbf{d} \right] = 0 \qquad (4\text{-}4)$$

where

$$K = \frac{\omega}{c} \qquad (4\text{-}5)$$

No longer can a simple plane wave be the solution of (4-4). Noting that many terms having the factor $\exp i[(\mathbf{k}_0 + 2\pi\mathbf{g}) \cdot \mathbf{r}]$ may appear in the square brackets in (4-4), if $\mathbf{d}$ includes the term $\exp i(\mathbf{k}_0 \cdot \mathbf{r})$, assume a special solution having the form

$$\mathbf{d}(\mathbf{r}) = \sum_g \mathbf{d}_g \exp i(\mathbf{k}_g \cdot \mathbf{r}) \qquad (4\text{-}6)$$

where

$$\mathbf{k}_g = \mathbf{k}_0 + 2\pi\mathbf{g} \qquad (4\text{-}7)$$

The waves with subscripts 0 and $g$ are called $O$ and $G$ waves, respectively. Physically, $G$ waves are the Bragg-reflected waves and the $O$ wave is the direct (transmitted) wave. It should be noted, however, that all component waves are equivalent in expression (4-6). The specialization of the $O$ wave is a matter of nomenclature.

An alternative expression for (4-6) is given by

$$\mathbf{d}(\mathbf{r}) = \exp i(\mathbf{k}_0 \cdot \mathbf{r}) \sum_g \mathbf{d}_g \exp 2\pi i(\mathbf{g} \cdot \mathbf{r}) \qquad (4\text{-}8)$$

In this expression, $\mathbf{d}(\mathbf{r})$ is similar to a plane wave, but now the amplitude is modulated by the lattice periodicity. This type of wave is called a *Bloch wave*. The scalar wave of this type is commonly used in solid-state physics to describe electrons in solids. The mathematical implication of taking (4-6) or (4-8) as a special solution of (4-4) is analogous to assuming a plane wave in the case of the homogeneous wave equation (3-33).

Inserting (4-6) into (4-4) and requiring that all the Fourier coefficients must be zero,

$$(K^2 - \mathbf{k}_g^2)\mathbf{d}_g - \sum_h \chi_{g-h}[\mathbf{k}_g \times (\mathbf{k}_g \times \mathbf{d}_h)] = 0 \qquad (4\text{-}9a)$$

for all $g$. The secular equation places a restriction on the wave vectors $\mathbf{k}_0$ (or $\mathbf{k}_g$). In addition, it turns out that a tensorlike ratio between $\mathbf{d}_g$ and $\mathbf{d}_0$ is finite. The details are described for two-beam cases in the next section. Many-beam cases are omitted because they have limited applications in x-ray diffraction. Interested readers are referred to Laue's text and discussion by Ewald and Héno (1968) and Héno and Ewald (1968). (See also Sections I D 5 and II F 3*b*.) Here, a few remarks are in order regarding the Bloch wave.

*a. Transverse component waves* Each component wave is a transverse wave. div $\mathbf{d} = 0$ requires the relation $(\mathbf{k}_g \cdot \mathbf{d}_g) = 0$ for all $g$. When $\mathbf{k}_g{}^r$ and $\mathbf{k}_g{}^i$ have the same direction, this relation implies that $\mathbf{d}_g$ is literally transverse to the wave vector $\mathbf{k}_g$. For general complex vectors of $\mathbf{k}_g$, this is not true. For convenience, however, call the wave satisfying the above relation transverse.

*b. Resonance factor* $R = \mathbf{k}_g{}^2/(k^2 - \mathbf{k}_g{}^2)$ The term for $\chi_0$ in the summation of (4-9*a*) can be included in the first term. Neglecting $\chi^2$, (4-9*a*) can be written

$$(k^2 - \mathbf{k}_g{}^2)\mathbf{d}_g - \sum_h{}' \chi_{g-h}[\mathbf{k}_g \times (\mathbf{k}_g \times \mathbf{d}_h)] = 0 \qquad (4\text{-}9b)$$

where the summation $\sum'$ excludes the term $\mathbf{h} = \mathbf{g}$, and

$$k^2 = K^2(1 + \chi_0) \qquad (4\text{-}10)$$

When $k^2$ is close to $\mathbf{k}_g{}^2$, $\mathbf{d}_g$ must become predominant unless all the amplitudes $\mathbf{d}_h$ are zero. This result is in accord with the results of the kinematical theory, although the Ewald construction must be modified by taking the radius as $k$ instead of $K$. Bragg reflection is brought about by a (spatial) resonance of the wave number $k$ with a characteristic wave number $k_g$. The situation is similar to dispersion, in which a (time) resonance occurs when the frequency $\omega$ of a forced oscillation becomes close to the characteristic frequency $\omega_0$. This is Ewald's central concept in his dynamical theory; the crystal is regarded as an assembly of resonators.

*c. Other field vectors* Electric and magnetic vectors also can be written in the form

$$\mathbf{e}(\mathbf{r}) = \sum_g \mathbf{e}_g \exp i(\mathbf{k}_g \cdot \mathbf{r}) \qquad (4\text{-}11a)$$

$$\mathbf{h}(\mathbf{r}) = \sum_g \mathbf{h}_g \exp i(\mathbf{k}_g \cdot \mathbf{r}) \qquad (4\text{-}11b)$$

Substituting (4-6) and (4-11*a*) and (4-11*b*) into Maxwell's equations and recalling the time-dependence factor $e^{-i\omega t}$, one obtains

$$\mathbf{d}_g = \frac{-(\mathbf{k}_g \times \mathbf{h}_g)}{K} \qquad (4\text{-}12a)$$

$$\mathbf{h}_g = \frac{(\mathbf{k}_g \times \mathbf{e}_g)}{K} \qquad (4\text{-}12b)$$

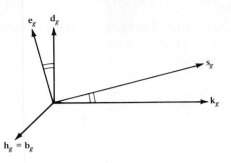

FIGURE 4-1
Relation among the field vectors in the
crystal. $\mathbf{e}_g$, $\mathbf{d}_g$, $\mathbf{s}_g$, and $\mathbf{k}_g$ lie in a plane
perpendicular to $\mathbf{h}_g$; $\mathbf{e}_g \parallel \mathbf{s}_g$ and $\mathbf{d}_g \perp \mathbf{k}_g$.

From these, the orthogonal relations

$$(\mathbf{d}_g \cdot \mathbf{k}_g) = (\mathbf{d}_g \cdot \mathbf{h}_g) = 0 \qquad (4\text{-}13a)$$

$$(\mathbf{h}_g \cdot \mathbf{k}_g) = (\mathbf{e}_g \cdot \mathbf{h}_g) = 0 \qquad (4\text{-}13b)$$

follow directly. In nonabsorbing crystals, the mutual relations of the field vectors are
illustrated in Fig. 4-1. They are similar to Fig. 3-1. In absorbing crystals, however,
they have only a symbolic meaning.

By taking the vector product of $\mathbf{d}_g$ and $\mathbf{k}_g$ and using (4-13b), it turns out that

$$\mathbf{h}_g = \frac{K}{k_g^2} (\mathbf{k}_g \times \mathbf{d}_g) \qquad (4\text{-}14a)$$

In addition, neglecting $\chi^2$, one obtains $\mathbf{e} = (1 - \chi)\mathbf{d}$. Obviously, the Fourier components satisfy

$$\mathbf{e}_g = (1 - \chi_0)\mathbf{d}_g - \sum_h{}' \chi_{g-h}\mathbf{d}_h \qquad (4\text{-}14b)$$

$$\mathbf{e}_g \fallingdotseq \mathbf{d}_g \qquad (4\text{-}14c)$$

Thus, if one obtains the amplitude $\mathbf{d}_g$ and the wave vector $\mathbf{k}_g$, other field vectors
$\mathbf{e}_g$ and $\mathbf{h}_g$ are uniquely determined. For this reason, it is sufficient to consider only the
displacement vector $\mathbf{d}_g$.

**2. Single-beam cases** When a single resonator, say, specified by 0 is excited, only
the equation

$$(k^2 - k_0^2)\mathbf{d}_0 = 0 \qquad (4\text{-}15)$$

is significant. A nontrivial solution of $\mathbf{d}_0$ is obtained only when

$$k^2 - k_0^2 = 0 \qquad (4\text{-}16)$$

This is the dispersion relation for a single plane wave. In this case, the Bloch wave
(4-8) reduces to a single plane wave. Such behavior has been described in Sections
I D and E of Chap. 3.

**3.**  **Two-beam cases**  When two resonance factors specified by 0 and $g$ are large, the set of equations (4-9$b$) is reduced to

$$(k^2 - \mathbf{k}_0{}^2)\mathbf{d}_0 - \chi_{-g}[\mathbf{k}_0(\mathbf{k}_0 \cdot \mathbf{d}_g) - \mathbf{k}_0{}^2 \cdot \mathbf{d}_g] = 0 \qquad (4\text{-}17a)$$

$$(k^2 - \mathbf{k}_g{}^2)\mathbf{d}_g - \chi_g[\mathbf{k}_g(\mathbf{k}_g \cdot \mathbf{d}_0) - \mathbf{k}_g{}^2 \cdot \mathbf{d}_0] = 0 \qquad (4\text{-}17b)$$

This reduction is called the *two-beam approximation*. It is to be noticed that the Fourier coefficient $\chi_{-g}$ appears in the formulation. Since the dynamical theory is concerned with the balance of $O$ and $G$ waves, the reflection of a $G$ wave on the net plane $-\mathbf{g}$ must be taken into account for calculating the amplitude $\mathbf{d}_g$.

Equations (4-17$a$) and (4-17$b$) are somewhat inconvenient because they are vector relations. Multiplying them by $\mathbf{d}_0$ and $\mathbf{d}_g$, respectively, and bearing in mind the relations $(\mathbf{k}_0 \cdot \mathbf{d}_0) = 0$ and $(\mathbf{k}_g \cdot \mathbf{d}_g) = 0$,

$$(k^2 - \mathbf{k}_0{}^2)d_0 + C\chi_{-g}\mathbf{k}_0{}^2 \cdot d_g = 0 \qquad (4\text{-}18a)$$

$$(k^2 - \mathbf{k}_g{}^2)d_g + C\chi_g\mathbf{k}_g{}^2 \cdot d_0 = 0 \qquad (4\text{-}18b)$$

where $d_0$ and $d_g$ are the lengths of $\mathbf{d}_0$ and $\mathbf{d}_g$ defined by $(\mathbf{d}_0 \cdot \mathbf{d}_0)^{1/2}$ and $(\mathbf{d}_g \cdot \mathbf{d}_g)^{1/2}$, respectively, and the factor $C$ stands for $(\mathbf{d}_0 \cdot \mathbf{d}_g)/(\mathbf{d}_0 \cdot \mathbf{d}_0)^{1/2}(\mathbf{d}_g \cdot \mathbf{d}_g)^{1/2}$. In practice, the real and imaginary parts of $\mathbf{d}_0$ or $\mathbf{d}_g$ have the same direction. Then, $d_0$ and $d_g$ represent the complex amplitude. For special cases, $C$ takes the value

$$C = 1: \quad (\mathbf{d}_0 \cdot \mathbf{d}_g \perp R \text{ plane}) \qquad (4\text{-}19a)$$

$$C = |\cos 2\theta_B|: \quad (\mathbf{d}_0 \cdot \mathbf{d}_g \parallel R \text{ plane}) \qquad (4\text{-}19b)[1]$$

where $R$ plane (reflection plane) is the plane which includes the real parts of $\mathbf{k}_0$ and $\mathbf{k}_g$. Equations (4-18$a$) and (4-18$b$) have a nontrivial solution only when the secular equation

$$\begin{vmatrix} k^2 - \mathbf{k}_0{}^2 & C\mathbf{k}_0{}^2\chi_{-g} \\ C\mathbf{k}_g{}^2\chi_g & k^2 - \mathbf{k}_g{}^2 \end{vmatrix} = 0 \qquad (4\text{-}20)$$

is satisfied. This equation is called the *dispersion relation*.[2]

If the wave vectors $\mathbf{k}_0$ and $\mathbf{k}_g$ are known, the amplitude ratio of $O$ and $G$ waves is immediately given by

$$c \equiv \frac{d_g}{d_0} = -\frac{k^2 - \mathbf{k}_0{}^2}{C\mathbf{k}_0{}^2\chi_{-g}} = -\frac{C\mathbf{k}_g{}^2\chi_g}{k^2 - \mathbf{k}_g{}^2} \qquad (4\text{-}21)$$

The Fourier coefficient $\chi_g$ times $C$ stands for the scattering amplitude without dispersion just like the amplitude $-(e^2/mc^2)$ does in the case of Thomson scattering. Owing

---

[1]  In what follows, it is convenient to use the absolute value of $C$ so that the sign of $C$ is transferred either to $d_0$ or to $d_g$.

[2]  By means of (4-5) and (4-10) equation (4-20) gives the relation between $\omega$ and $\mathbf{k}_0$ or $\mathbf{k}_g$. Generally $\omega = \omega(\mathbf{k})$ is called the *dispersion relation*.

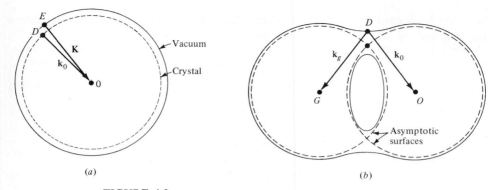

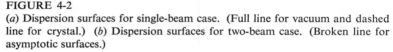

(a)                                                                         (b)

**FIGURE 4-2**
(a) Dispersion surfaces for single-beam case. (Full line for vacuum and dashed
line for crystal.) (b) Dispersion surfaces for two-beam case. (Broken line for
asymptotic surfaces.)

to the dispersion, the amplitude is amplified by the resonance factor $R$. In the
kinematical theory, $k^2$ may coincide with $\mathbf{k}_g^2$ at exact resonance. In the dynamical
theory, however, such exact resonance does not occur by virtue of the coexistence of
$O$ and $G$ waves. This can be anticipated from the dispersion relation (4-20).

**4.   *Dispersion surface.*** For simplicity, assume at first that the wave vectors are real.
Then, any wave vector $\mathbf{k}_0$ can be represented by the end point $D$ in $\mathbf{k}$ space, with the
top point fixed at $O$. The point $D$ is called the *dispersion point*. The surface on which
the dispersion point is located is called the *dispersion surface*. For single-beam cases
(Fig. 4-2a), it turns out that the dispersion surface is a sphere of radius $k$, as in the
case of a homogeneous medium discussed in Section I E.

In the cases of two beams, it is necessary to consider the wave vector $\mathbf{k}_g$ as
well as the vector $\mathbf{k}_0$. They are connected by relation (4-7). If the top point of $\mathbf{k}_g$
is fixed at the reciprocal-lattice point $G$, the end point $D$ is identical to the end point
of $\mathbf{k}_0$. Thus a single point $D$ can represent a pair of wave vectors $\mathbf{k}_0$ and $\mathbf{k}_g$. They are
tied together by the point $D$. This concept is extremely useful for considering many-
beam cases. For this reason, Ewald (1962) proposed the name *tie point* instead of
dispersion point. Ewald's nomenclature will be used only when it is necessary to
emphasize the concept mentioned above.

When the off-diagonal terms of the dispersion relation (4-20) are sufficiently
small, the dispersion surfaces must be spheres of radius $k$ with their centers at $O$
and $G$. These spheres are denoted by $S_0$ and $S_g$, respectively, and the concentric
spheres of radius $K$ are denoted by $\bar{S}_0$ and $\bar{S}_g$, respectively. When the dispersion
point $D$ comes close to the intersection of the spheres $S_0$ and $S_g$, it is necessary to take

into account the off-diagonal terms. Thus, the dispersion surface may deviate slightly from the spheres $S_0$ and $S_g$. The situation is schematically represented in Fig. 4-2b.

When the wave vectors are complex, six-dimensional space is needed in order to represent them, because the real and imaginary parts are generally independent. The dispersion surface is now a hypersurface in the complex reciprocal space. It must be the intersection of the two hypersurfaces of higher rank, which represent the real and imaginary parts of the dispersion relation. In general, the situation is rather complicated, and the ability to visualize the dispersion relation will be lost. If the imaginary parts of $\chi_0$ and $\chi_g\chi_{-g}$ are sufficiently small, however, the surface projected onto the real $\mathbf{k}$ space obtained by letting $k_0{}^i = k_g{}^i = 0$ is approximately the same as the dispersion surface of nonabsorbing crystals ($\chi^i = 0$). Such a surface is often used in order to visualize the relations among the wave vectors.

5. *Approximate forms of the dispersion surface and amplitude ratio*   Equation (4-20) is exact within the two-beam approximation. In practice, since the real parts of $\mathbf{k}_0$ and $\mathbf{k}_g$ are extremely large, the following approximations can be used with sufficient accuracy. First, define the real vectors $\overline{\mathbf{K}}_0$ and $\overline{\mathbf{K}}_g$ of length $K$, satisfying the relation

$$\overline{\mathbf{K}}_g = \overline{\mathbf{K}}_0 + 2\pi\mathbf{g} \qquad (4\text{-}22)$$

where the three vectors $\overline{\mathbf{K}}_0$, $\overline{\mathbf{K}}_g$, and $2\pi\mathbf{g}$ form an isosceles triangle. In other words, $\overline{\mathbf{K}}_0$ satisfies the Bragg condition exactly in a kinematical sense. Next, the vectors $\overline{\mathbf{k}}_0$ and $\overline{\mathbf{k}}_g$ of length $k$ are defined as follows:

$$\overline{\mathbf{k}}_0 = \overline{\mathbf{K}}_0 + \Delta\mathbf{K} \qquad (4\text{-}23a)$$

$$\overline{\mathbf{k}}_g = \overline{\mathbf{K}}_g + \Delta\mathbf{K} \qquad (4\text{-}23b)$$

where the vector $\Delta\mathbf{K}$ is taken along the bisector of $\overline{\mathbf{K}}_0$ and $\overline{\mathbf{K}}_g$. By the definition

$$\overline{\mathbf{k}}_0{}^2 = K^2(1 + \chi_0)$$
$$= K^2 + 2(\overline{\mathbf{K}}_0 \cdot \Delta\mathbf{K}) + (\Delta\mathbf{K})^2$$
$$\overline{\mathbf{k}}_g{}^2 = K^2(1 + \chi_0)$$
$$= K^2 + 2(\overline{\mathbf{K}}_g \cdot \Delta\mathbf{K}) + (\Delta\mathbf{K})^2$$

Assuming $\Delta K \ll K$, therefore, one obtains

$$(\hat{\overline{\mathbf{K}}}_0 \cdot \Delta\mathbf{K}) = (\hat{\overline{\mathbf{K}}}_g \cdot \Delta\mathbf{K}) = \tfrac{1}{2}K\chi_0 \qquad (4\text{-}24)$$

Next, let

$$\mathbf{k}_0 = \overline{\mathbf{k}}_0 + \Delta\mathbf{k} \qquad (4\text{-}25a)$$

$$\mathbf{k}_g = \overline{\mathbf{k}}_g + \Delta\mathbf{k} \qquad (4\text{-}25b)$$

By assuming a common deviation vector $\Delta\mathbf{k}$, one is assured that $\mathbf{k}_0$ and $\mathbf{k}_g$ satisfy relation (4-7).  By the use of the approximation $(\Delta k)^2 \ll (\bar{k}_0)^2$ and $(\bar{k}_g)^2$, it turns out that

$$k^2 - k_0{}^2 \simeq -2(\bar{k}_0 \cdot \Delta k)$$

$$k^2 - k_g{}^2 \simeq -2(\bar{k}_g \cdot \Delta k)$$

Inserting these into (4-20) and approximating $k_0$ and $\bar{k}_0$ by $\bar{K}_0$ and $k_g$, and $\bar{k}_g$ by $\bar{K}_g$, one obtains

$$\Delta\eta_0\Delta\eta_g = \tfrac{1}{4}K^2C^2\chi_g\chi_{-g} \qquad (4\text{-}26)$$

where

$$\Delta\eta_0 = (\hat{\bar{K}}_0 \cdot \Delta k) \qquad (4\text{-}27a)$$

$$\Delta\eta_g = (\hat{\bar{K}}_g \cdot \Delta k) \qquad (4\text{-}27b)$$

They are called *Resonanzfehler* (resonance defects) of the respective waves.  The amplitude ratio (4-21) then can be obtained immediately from

$$c \equiv \frac{d_g}{d_0} = \frac{KC\chi_g}{2\Delta\eta_g} = \frac{2\Delta\eta_0}{KC\chi_{-g}} \qquad (4\text{-}28)$$

In nonabsorbing cases ($\chi_0$ and $\chi_g\chi_{-g}$ real) $\Delta\eta_0$ and $\Delta\eta_g$ are real.  Then, they have the geometrical meaning of distances from the dispersion point $D$ to the surfaces $S_0$ and $S_g$, respectively.  Within the approximations employed in this section, the surfaces $S_0$ and $S_g$ are approximated by the tangential planes at the end point $L$ of $\bar{k}_0$ and $\bar{k}_g$.  Similarly the spheres $\bar{S}_0$ and $\bar{S}_g$ are approximated by the tangential planes at the end point $\bar{L}$ of $\bar{K}_0$ and $\bar{K}_g$.  The points $L$ and $\bar{L}$ are called *Lorentz* and *Laue points*, respectively.  Thus (4-26) states that the dispersion surface is hyperboloid and perpendicular to the $R$ plane.

In absorbing crystals, the construction in (real) $k$ space has only a symbolic meaning.  Nevertheless, the analytical expressions derived in this section can be retained without error.  It should be noticed that the direction of the vector $\Delta K$ is unique, namely, $\overrightarrow{L\bar{L}}$, but the directions of the real and imaginary parts of $\Delta k$ are not necessarily the same.  Their directions are selected depending on the problem with which one is concerned (Section I D).

**6.  *Poynting vector and the wave packet***   The Poynting vector represents the direction and rate of energy flow associated with an electromagnetic field.  Here, consider the Poynting vector of a vectorial Bloch wave, which has the form (4-6) or (4-11a) and (4-11b).  By using the general expression (3-47a), the time-averaged Poynting vector can be calculated straightforwardly:

$$\langle s \rangle = \frac{c}{8\pi} \exp - 2(k^i \cdot r) \cdot \mathrm{Re}\left[ \sum_g (e_g \times h_g^*) + \sum_g \sum_h{}' (e_g \times h_h^*) \exp 2\pi i(g - h) \cdot r \right]$$

$$(4\text{-}29a)$$

where $\mathbf{k}^i$ is the imaginary part of $\mathbf{k}_g$, which is common to all component waves. (Note that $\mathbf{g}$ is real.) The second term in brackets is an interference term which oscillates with the lattice periodicity. The observable Poynting vector must be the vector averaged over a unit cell, so that the interference term drops out. By substituting equations (4-14a) and (4-14c) into the above expression,

$$\langle\!\langle \mathbf{s} \rangle\!\rangle \doteq \frac{c}{8\pi} \exp - 2(\mathbf{k}^i \cdot \mathbf{r}) \sum_g \hat{\overline{\mathbf{K}}}_g (\mathbf{d}_g^* \cdot \mathbf{d}_g) \qquad (4\text{-}29b)^1$$

Here, the approximations $\mathbf{k}_g^* = \mathbf{k}_g$ and $(\mathbf{k}_g^* \cdot \mathbf{d}_g) = 0$ have been made. In addition, $\mathbf{k}_g$ is approximated by $\overline{\mathbf{K}}_g$. The Poynting vector of a Bloch wave is the sum of the Poynting vectors of the component plane waves. It is proportional to a weighted mean of the wave vectors, the weight being the intensity $(\mathbf{d}_g^* \cdot \mathbf{d}_g)$ of the displacement vector.

It is instructive to consider the wave packet of Bloch waves in connection with the Poynting vector. As in the case of a homogeneous medium, the wave packet in the crystal is defined as a superposition of Bloch waves

$$\mathbf{d}(\mathbf{r}, t) = \int f(\mathbf{k}_0{}') \left[ \exp i(\mathbf{k}_0 \cdot \mathbf{r}) \sum_g \mathbf{d}_g(\mathbf{k}_0) \exp 2\pi i (\mathbf{g} \cdot \mathbf{r}) \right] e^{-i\omega t} \, d\mathbf{k}_0{}^r \qquad (4\text{-}30)$$

where $f(\mathbf{k}_0{}')$ is the amplitude density. The expression amounts to a general solution of the fundamental equation (3-27a) since the equation is linear with respect to $\mathbf{d}$. The wave packet represents a wave field which is confined in space. Here, assume that the direction of the imaginary part $\mathbf{k}_0{}^i$ is fixed.[2] The real and imaginary parts of the dispersion relation impose two additional conditions on $\mathbf{k}_0$ and $\omega$. Thus, among seven variables, $\omega$ and the components of $\mathbf{k}_0{}^r$ and $\mathbf{k}_0{}^i$, only three variables can be arbitrarily chosen. In the above expression, three components of the real part of $\mathbf{k}_0{}^r$ are used as independent variables. Since the angular frequency $\omega$ is a function of $\mathbf{k}_0{}^r$, the Bloch waves of various $\omega$ are automatically involved in the above expression of a wave packet.[3]

When the amplitude density is continuous and takes on appreciable values only when $\mathbf{k}_0{}^r$ is close to a vector $\tilde{\mathbf{k}}_0{}^r$, one can write

$$\mathbf{d}(\mathbf{r}, t) = \tilde{\mathbf{d}}(\mathbf{r}, t) \cdot P(\mathbf{r}, t) \qquad (4\text{-}31)$$

---

[1] Later, the notation of double average $\langle\!\langle \; \rangle\!\rangle$ will be omitted.

[2] This assumption is permitted for waves created by refraction on a plane boundary as described in the following sections.

[3] Alternatively, (4-30) can be represented by the less symmetrical form

$$\mathbf{d}(\mathbf{r}) = \int g(\mathbf{t}; \omega) d_B(\mathbf{r}) e^{-i\omega t} \, dt \, d\omega \qquad (4\text{-}30')$$

where $\mathbf{t}$ is a two-dimensional component vector of $\mathbf{k}_0{}^r$ and $d_B(\mathbf{r})$ stands for the expression denoted by the square brackets in equation (4-30).

where

$$\mathbf{d}(\mathbf{r}, t) = f(\tilde{\mathbf{k}}_0{}^r) \exp i(\tilde{\mathbf{k}}_0 \cdot \mathbf{r}) \sum_g d_g(\tilde{\mathbf{k}}_0) \exp 2\pi i(\mathbf{g} \cdot \mathbf{r})e^{-i\tilde{\omega}t} \tag{4-32}$$

$$P(\mathbf{r}, t) = \int P(\delta\mathbf{k}) \exp i(\delta\mathbf{k} \cdot \mathbf{r})e^{-i(\omega - \tilde{\omega})t} \, d(\delta\mathbf{k}) \tag{4-33}$$

Here, the following notations are employed:

$$\delta\mathbf{k} = \mathbf{k}_0{}^r - \tilde{\mathbf{k}}_0{}^r \tag{4-34}$$

and

$$f(\mathbf{k}_0{}^r) = f(\tilde{\mathbf{k}}_0{}^r)P(\delta\mathbf{k}) \tag{4-35}$$

Thus, the wave packet of vectorial Bloch waves is represented by a single Bloch wave modified by a scalar packet function $P(\mathbf{r}, t)$. The intensity field is also represented by the intensity field of a single Bloch wave modified by the intensity packet function $P \cdot P^*$.

Let us take the dispersion point $\tilde{D}$ corresponding to the wave vector $\tilde{\mathbf{k}}_0$ on the dispersion surface of the angular frequency $\omega$. The variable $\delta\mathbf{k}$ can be decomposed into the tangential component $\tau$ and the normal component $\delta\mathbf{k}_\nu$. Then equation (4-33) has the form

$$P(\mathbf{r}, t) = \int P(\tau, \delta\mathbf{k}_\nu) \exp i\{[(\tau + \delta\mathbf{k}_\nu) \cdot \mathbf{r}] - (\omega - \tilde{\omega})t\} \, d\tau \, d(\delta\mathbf{k}_\nu)$$

If the packet function has the distribution $P(\mathbf{r}_0, 0)$ when $t = 0$, the amplitude density is represented by the Fourier inverse transform:

$$P(\tau, \delta\mathbf{k}_\nu) = \left(\frac{1}{2\pi}\right)^3 \int P(\mathbf{r}_0, 0) \exp -i[(\tau + \delta\mathbf{k}_\nu) \cdot \mathbf{r}_0] \, d\mathbf{r}_0 \tag{4-36}$$

Therefore, the packet function at $\mathbf{r}$ and $t$ can be written as follows:

$$P(\mathbf{r}, t) = \left(\frac{1}{2\pi}\right)^3 \iint P(\mathbf{r}_0, 0) \exp i\{[(\tau + \delta\mathbf{k}_\nu) \cdot (\mathbf{r} - \mathbf{r}_0)] - (\omega - \tilde{\omega})t\} \, d\mathbf{r}_0 \, d\tau \, d(\delta\mathbf{k}_\nu)$$

Within a small region of $\delta\mathbf{k}$, $\omega$ can be assumed to be independent of $\tau$ because the dispersion surface is the surface of constant $\omega$ and, approximately, it has the form

$$\omega \doteqdot \tilde{\omega} + \left(\frac{\partial\omega}{\partial k_\nu}\right)_\tau (\delta\mathbf{k}_\nu) \tag{4-37}$$

Thus, one obtains

$$P(\mathbf{r}, t) = \int P(\mathbf{r}_0, 0) \, \delta[(\mathbf{r} - \mathbf{r}_0)_\tau] \, \delta\left[(\mathbf{r} - \mathbf{r}_0)_\nu - \left(\frac{\partial\omega}{\partial k_\nu}\right)_\tau t\right] d\mathbf{r}_0 \tag{4-38}$$

where $\delta$ is the Dirac delta function. From this expression it is concluded that the packet function $P(\mathbf{r}, t)$ is just a translation of $P(\mathbf{r}_0, 0)$ along a direction $\nu$, with the velocity

$$v_g = \left(\frac{\partial\omega}{\partial k_\nu}\right)_\tau \quad v = \mathrm{grad}^{(k)}\omega \tag{4-39}$$

where grad$^{(k)}$ is the gradient operator in $\mathbf{k_0}^r$ space.

The approximation of equation (4-37) or replacement of the dispersion surface by the tangential plane is equivalent to the approximation of geometrical optics. The use of geometrical optics in isotropic media is justified by the Eikonal theory. One can generalize the Eikonal theory also in crystalline media. These topics will be explained in Chap. 5. The above treatments give a justification in the case of perfect crystals.

A question may arise as to whether the expression (4-29$b$) contradicts expression (4-39) or not. One can prove the coexistence of both expressions even in $N$-beam cases. Here, for simplicity, only two-beam cases are described in detail. The dispersion relation, then, is given by equation (4-26).[1] Consider a variation $\tau$ of the wave vector $\mathbf{k_0}$ parallel to the tangential plane at $D$, giving

$$(\hat{\mathbf{K}}_0 \cdot \tau)\Delta\eta_g + (\hat{\mathbf{K}}_g \cdot \tau)\Delta\eta_0 = 0$$

From (4-28), on the other hand, one obtains

$$\left(\frac{d_g}{d_0}\right)\left(\frac{d_g}{d_0}\right)^* = \frac{\Delta\eta_0}{\Delta\eta_g} \qquad (4\text{-}40)$$

employing the approximation $\chi_g^* = \chi_{-g}$. From the above two equations

$$\{\tau \cdot [\hat{\mathbf{K}}_0(d_0 d_0^*) + \hat{\mathbf{K}}_g(d_g d_g^*)]\} = 0 \qquad (4\text{-}41)$$

Thus, it follows that the Poynting vector is perpendicular to the dispersion surface.

Next, consider the variation $\delta\omega$ corresponding to a variation $\delta\mathbf{k}_v$. For this purpose, it is necessary to return to the original dispersion equation (4-20)[2] from which

$$\left(\frac{1}{c}\right)^2 \omega\delta\omega[(k^2 - \mathbf{k_0}^2) + (k^2 - \mathbf{k_g}^2)] = \delta k_v\{[\mathbf{k}_0(k^2 - \mathbf{k_g}^2) + \mathbf{k}_g(k^2 - \mathbf{k_0}^2)] \cdot v\}$$

The quantities $(k^2 - \mathbf{k_0}^2)$ and $(k^2 - \mathbf{k_g}^2)$ can be replaced by $-2K\Delta\eta_0$ and $-2K\Delta\eta_g$, respectively. By virtue of equation (4-40), it is possible to obtain, approximately,

$$\frac{1}{c}\left(\frac{\delta\omega}{\delta k_v}\right)_\tau [(d_0 d_0^*) + (d_g d_g^*)] = \{[\hat{\mathbf{K}}_0(d_0 d_0^*) + \hat{\mathbf{K}}_g(d_g d_g^*)] \cdot v\}$$

On the other hand, from (3-47$b$) and (4-14$c$), with the approximation $\mathbf{k_g}^2 = K^2$, the averaged electric energy density can be written for each polarization mode

$$W_e = \frac{1}{16\pi} [(d_0 \cdot d_0^*) + (d_g \cdot d_g^*)] \exp - 2(\mathbf{k}^i \cdot \mathbf{r}) \qquad (4\text{-}42)$$

---

[1] If one considers expression (4-20), it is possible to prove that equation (4-39) accords with (4-29$a$), omitting the interference term. Here, for simplicity, equations (4-39) and (4-29$b$) have been compared.

[2] Nevertheless, $k_0^2$ and $k_g^2$ in the skew elements are approximated by $K^2 = (\omega/c)^2$ with sufficient accuracy. Thus it turns out that the skew elements are regarded as constant. Note that $\chi_g$ and $\chi_{-g}$ involve $(1/\omega)^2$.

With the same approximation, (4-14$b$) states that the magnetic energy density $W_m$ is identical to $W_e$. Thus, finally,

$$\left(\frac{\delta\omega}{\delta k_v}\right)_\tau (W_e + W_m) \, v = s \qquad (4\text{-}43)$$

where $s$ is the Poynting vector given by (4-29$b$). From the hydrodynamic analogy, one can interpret $(\delta\omega/\delta k_v)v$ as the velocity of electromagnetic energy in accordance with the conclusion (4-39) for the wave packet.

The above theory has been developed by Laue (1952, 1953), Ewald (1958), Kato (1952, 1958), and Wagner (1959), some of whom considered electron cases. As will be seen later, the concepts of the propagation of the wave packet or of the energy flow are extremely important in x-ray diffraction, while not so significant in electron cases except for the problem of channeling through a crystalline material.[1] In diffraction problems in which interference effects cannot be ignored, logically speaking, wave-packet considerations must be regarded as basic compared to the energy-flow considerations. In practice, however, they can be considered to be equivalent. For this reason, at times later on, such expressions as the "phase change along a ray" will be used.

**7.  *Attenuation of the Bloch wave***   Since, from (4-29$b$), one obtains

$$\text{div } s = -2(\mathbf{k}^i \cdot s) \qquad (4\text{-}44)$$

the quantity $2(\mathbf{k}^i \cdot v)$ represents the linear coefficient of the exponential attenuation of the energy flow. The damping of the electromagnetic energy density along the direction $v$ also can be described by the same coefficient. It is reasonable, therefore, to define a linear absorption coefficient of the Bloch wave by

$$\mu_B(v) = 2(\mathbf{k}^i \cdot v) \qquad (4\text{-}45)$$

Both $\mathbf{k}^i$ and $v$ are functions of $\mathbf{k}_0{}^r$. Physically speaking, $\mu_B$ depends on the extent to which the Bloch wave satisfies the Bragg condition. It depends also on the branch to which the Bloch wave belongs. This is the phenomenological derivation of Borrmann absorption which will be explained in Section I D 5.

## C.   Wave fields in two-beam cases

**1.  *Laue and Bragg cases*[2]**   When the Bragg-reflected wave appears in a crystal, it is necessary to distinguish two cases regarding the geometrical relations between the wave vectors and the crystal boundary. First, the entrance surface is considered. In the Bragg case, the Bragg-reflected wave emerges from the same surface as the one

---

[1] The "column approximation" widely used in electron microscopy (Hirsch et al., 1965), however, can be justified by wave-packet considerations.
[2] See Saka, Katagawa, and Kato (1972$a$).

on which the incident wave falls. In the Laue (transmission) case, on the other hand, the Bragg-reflected wave also penetrates through the crystal and emerges from the side opposite that of incidence. For precise descriptions, it is more convenient to use the following analytical definitions:

$$\text{Bragg case:} \qquad \gamma_0 \gamma_g < 0 \qquad (4\text{-}46a)$$

$$\text{Laue case:} \qquad \gamma_0 \gamma_g > 0 \qquad (4\text{-}46b)$$

where

$$\gamma_0 = (\hat{\mathbf{K}}_0 \cdot \mathbf{n}_e) \qquad (4\text{-}47a)$$

and

$$\gamma_g = (\hat{\mathbf{K}}_g \cdot \mathbf{n}_e) \qquad (4\text{-}47b)$$

Here $\hat{\mathbf{K}}_0$ and $\hat{\mathbf{K}}_g$ are the unit vectors of $\overline{\mathbf{K}}_0$ and $\overline{\mathbf{K}}_g$ defined in (4-22), respectively, and $\mathbf{n}_e$ is the inward-directed normal of the entrance surface. Since the incident wave must penetrate into the crystal, $(\mathbf{S}_e \cdot \mathbf{n}_e)$ is always positive, $\mathbf{S}_e$ being the Poynting vector of the incident wave. Consequently, $\gamma_0$ is positive. Laue and Bragg cases are distinguished from each other by the sign $\gamma_g$ at the entrance surface.

Similar definitions can also be established for the exit surface. When the energy flow of a crystal (Bloch) wave reaches a surface, regardless of how it is excited, this surface is called an exit surface. The definitions for the Bragg and Laue cases then are established as follows:

$$\text{Bragg case:} \qquad \gamma_0' \gamma_g' < 0 \qquad (4\text{-}48a)$$

$$\text{Laue case:} \qquad \gamma_0' \gamma_g' > 0 \qquad (4\text{-}48b)$$

where

$$\gamma_0' = (\hat{\mathbf{K}}_0 \cdot \mathbf{n}_a) \qquad (4\text{-}49a)$$

and

$$\gamma_g' = (\hat{\mathbf{K}}_g \cdot \mathbf{n}_a) \qquad (4\text{-}49b)$$

where $\mathbf{n}_a$ is the outward-directed normal of the surface. Since the direction of the energy flow is a weighted mean of $\hat{\mathbf{K}}_0$ and $\hat{\mathbf{K}}_g$ (4-29b), at least one of the $\gamma_0'$ and $\gamma_g'$ values must be positive. In the Laue cases, both $O$ and $G$ waves emerge from the exit surface, whereas in the Bragg case only one of the $O$ and $G$ waves can emerge from the exit surface. The case when the $G$ wave emerges $(\gamma_g' > 0)$ is called type I, whereas the other case $(\gamma_0' > 0)$ is called type II.

Unlike the conventional usage, here the term Laue or Bragg case is used only for a single surface. Since a pair of entrance and exit surfaces must be considered in practice, the geometrical situation can be described by four cases, Laue-Laue, Laue-Bragg, Bragg-Laue, and Bragg-Bragg, as illustrated schematically in Fig. 4-3. As a natural generalization, more complicated cases, such as Laue-(Bragg)$^n$-Laue, are also conceivable and are considered in a later section. The concept described here can be applied also to the boundary between two crystalline media, such as a stacking fault or a twin boundary (see Chap. 5).

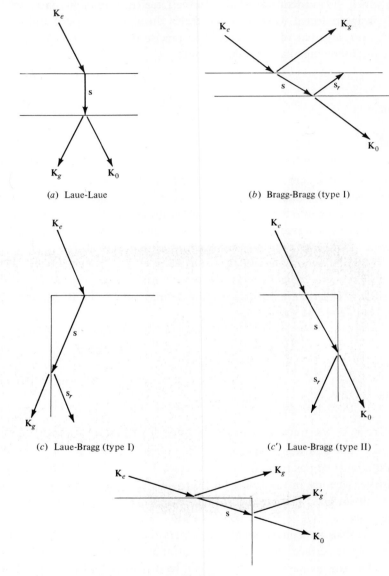

(a) Laue-Laue

(b) Bragg-Bragg (type I)

(c) Laue-Bragg (type I)

(c') Laue-Bragg (type II)

(d) Bragg-Laue

**FIGURE 4-3**
Examples of boundary conditions in which the vacuum waves are denoted by the wave vectors and the crystal waves by Poynting vectors.

**2.  *Wave vectors in crystals*** In this section, some general properties of wave vectors are discussed. In most cases, the distinction between Bragg and Laue cases is not essential from a mathematical point of view. It is advisable, however, to bear the set of waves listed in Table 4-1*a* and *b* in mind for each case. Use is also made of Fig. 4-4*a* and *b* in the subsequent discussion.

*a.*  Anpassung $\delta$ *and* Resonanzfehler $\Delta\eta$  Just as in the single-beam case, the wave vectors of the crystal waves are determined from the wave vector $\mathbf{K}_e$ of the incident wave by the tangential continuity.

$$\mathbf{k}_0 = \mathbf{K}_e - K\delta_e\mathbf{n}_e \qquad (4\text{-}50)[1]$$

The quantity $\delta_e$ may be complex, in general. By introducing a deviation vector $\mathbf{s}_0$, the wave vector $\mathbf{K}_e$ can be written

$$\mathbf{K}_e = \overline{\mathbf{K}}_0 - \mathbf{s}_0 \qquad (4\text{-}51)$$

The vector $\mathbf{s}_0$ must be a real vector on the reflection ($R$) plane and also lie on the surface $\bar{S}_0$. The magnitude is related to the glancing angle $\Theta$ of the wave vector $\mathbf{K}_e$ with respect to the net plane concerned, namely,

$$s_0 = K(\Theta - \theta_B) \qquad (4\text{-}52)$$

Table 4-1  WAVE FIELDS AT THE EN-
TRANCE SURFACE

*a.*  Laue case*

| | $O$ wave | $G$ wave |
|---|---|---|
| Vacuum waves | $(\mathbf{E}_e; \mathbf{K}_e)$ | $0$ |
| | $\downarrow$ | $\downarrow$ |
| Crystal waves | $(\mathbf{d}_0^{(j)}; \mathbf{k}_0^{(j)}) \leftrightarrow (\mathbf{d}_g^{(j)}; \mathbf{k}_g^{(j)})$ | |
| | $(j) = (1)$ and $(2)$ | |

*b.*  Bragg case*

| | $O$ wave | $G$ wave |
|---|---|---|
| Vacuum waves | $(\mathbf{E}_e; \mathbf{K}_e)$ | $(\mathbf{E}_g^{(a)}; \mathbf{K}_g)$ |
| | $\downarrow$ | $\uparrow$ |
| Crystal waves | $(\mathbf{d}_0^{(q)}; \mathbf{k}_0^{(q)}) \rightarrow (\mathbf{d}_g^{(q)}; \mathbf{k}_g^{(q)})$ | |
| | $(q) = (a)$ or $(b)$ | |

\* $(\mathbf{A}; \mathbf{K}) = \mathbf{A} \exp i(\mathbf{K} \cdot \mathbf{r}_e)$ and arrows indicate the sequence in determining the wave fields.

[1] Here, $K\delta_e$ is a small part of the normal component of the wave vector, whereas $\gamma$ in (3-53c) is the total.

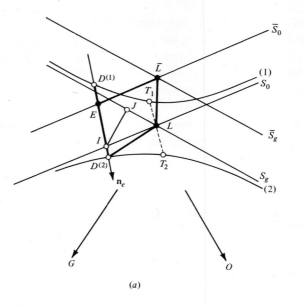

(a)

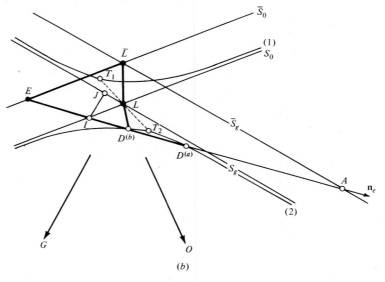

(b)

**FIGURE 4-4**
Construction of the wave vectors at the entrance surface. (a) Laue case.
$(\mathbf{K}_e = \overrightarrow{EO}, \ \mathbf{k}_0^{(J)} = \overrightarrow{D^{(J)}O}, \ \mathbf{k}_g^{(J)} = \overrightarrow{D^{(J)}G}, \ \Delta\mathbf{k}^{(J)} = \overrightarrow{D^{(J)}L}, \ \Delta\mathbf{K} = \overrightarrow{LL}, \ \mathbf{s}_0 = \overrightarrow{LE},$
$\overrightarrow{ED^{(J)}} = K\delta^{(J)}\mathbf{n}_e, \ JI = w, \ LI = s.)$ Solid circles lie on the $R$ plane, while open
circles are the projection on the $R$ plane; (b) Bragg case. $(IJ = w,$ and $T_1$ and $T_2$
are the tangential points of the line $\mathbf{n}_e$ on the dispersion surface.)

Combining (4-50) and (4-51) with (4-23a) and (4-25a)

$$\Delta \mathbf{k} = -(\Delta \mathbf{K} + \mathbf{s}_0 + K\delta_e \mathbf{n}_e) \qquad (4\text{-}53)$$

The vector relations are illustrated in Fig. 4-4a for the Laue case and Fig. 4-4b for the Bragg case. The dispersion point $D$ is the intersection of the dispersion surface and the line $\mathbf{n}_e$ passing through the dispersion point of $\mathbf{K}_e$, labeled $E$.

Substituting from (4-53) into the dispersion relation (4-26), one obtains

$$(\tfrac{1}{2}\chi_0 + \delta_e\gamma_0)\left[\tfrac{1}{2}\chi_0 + \frac{s_0}{K}\sin 2\theta_B + \delta_e\gamma_g\right] = \tfrac{1}{4}C^2\chi_g\chi_{-g} \qquad (4\text{-}54)$$

where $s_0 \sin 2\theta_B = (\mathbf{s}_0 \cdot \hat{\mathbf{K}}_g)$, and $\gamma_0$ and $\gamma_g$ are defined by (4-47a and b). The solution gives

$$K\delta_e = \frac{-\tfrac{1}{2}K\chi_0}{\gamma_0} + \frac{1}{2}\frac{\sin 2\theta_B}{\gamma_g}\left[-s \mp \sqrt{s^2 + \beta^2}\right] \qquad (4\text{-}55a)[1]$$

where $s$ and $\beta$ are defined by

$$s = s_0 + \frac{\tfrac{1}{2}K\chi_0(1 - \gamma_g/\gamma_0)}{\sin 2\theta_B} \qquad (4\text{-}55b)$$

$$\beta = \frac{KC(\chi_g\chi_{-g})^{1/2}(\gamma_g/\gamma_0)^{1/2}}{\sin 2\theta_B} \qquad (4\text{-}55c)$$

In the plane-wave theory, it is convenient to use the following:

$$u = s \sin 2\theta_B = s_0 \sin 2\theta_B + \tfrac{1}{2}K\chi_0\left(1 - \frac{\gamma_g}{\gamma_0}\right) \qquad (4\text{-}56a)$$

$$U = KC(\chi_g\chi_{-g})^{1/2}\left(\frac{\gamma_g}{\gamma_0}\right)^{1/2} \qquad (4\text{-}56b)$$

The transition from $(s, \beta)$ to $(u, U)$ will be utilized often in the following discussions. Then, $K\delta_e$ is written

$$K\delta_e = \frac{-\tfrac{1}{2}K\chi_0}{\gamma_0} + \frac{\tfrac{1}{2}[-u \mp \sqrt{u^2 + U^2}]}{\gamma_g} \qquad (4\text{-}56c)$$

The double signs in equations (4-55a) and (4-56c) are due to the quadratic form of the dispersion relation (4-54). The assignment to the branches (1) and (2) in Fig. 4-4a and b will be explained later.

---

[1] It is necessary here to distinguish the expression $\sqrt{s^2 + \beta^2}$, which refers to a complex quantity having a positive real part, from $(s^2 + \beta^2)^{1/2}$, which refers to the positive Riemann sheet, so that the quantity $-(s^2 + \beta^2)^{1/2}$ denotes the negative sheet. The same applies to $\sqrt{u^2 + U^2}$ and $(u^2 + U^2)^{1/2}$, as further discussed in Section 3b below.

In general, $u$ and $U$ are complex. In what follows, it is convenient to write

$$u = w + iv \qquad (4\text{-}57)$$

and

$$U = W + iV \qquad \text{(Laue case)} \qquad (4\text{-}58a)$$

$$= i(W + iV) \qquad \text{(Bragg case)} \qquad (4\text{-}58b)$$

where

$$W = KC \, R_e[(\chi_g\chi_{-g})^{1/2}] \left(\frac{|\gamma_g|}{|\gamma_0|}\right)^{1/2} \qquad (4\text{-}59a)^1$$

$$V = KC \, J_m[(\chi_g\chi_{-g})^{1/2}] \left(\frac{|\gamma_g|}{|\gamma_0|}\right)^{1/2} \qquad (4\text{-}59b)^1$$

Usually, $R_e[(\chi_g\chi_{-g})^{1/2}]$ is positive, but the sign of $J_m[(\chi_g\chi_{-g})^{1/2}]$ depends on the crystal structure (cf Appendix 4AI).

Next consider the intersection point $I$ of the line $ED$ in Fig. 4-4a and b with the asymptotic surface $S_0$. The component of the vector $\bar{L}I = \mathbf{s}_0 + (-\frac{1}{2}K\chi_0{}^r/\gamma_0)\mathbf{n}_e$ along $\mathbf{K}_g$ is given by $s_0 \sin 2\theta_B - \frac{1}{2}K\chi_0{}^r(\gamma_g/\gamma_0)$. Since the distance between the surface $S_0$ and $\bar{S}_g$ is $(-\frac{1}{2}K\chi_0{}^r)$, the real part $w$ has the geometrical meaning of the normal distance $JI$ in Fig. 4-4a and b. Consequently, $s$ is the length $LI$. Thus, the parameters $w$ and $s$ specify the degree of deviation from the exact Bragg condition in the dynamical sense.

By using the parameter $u$, it is possible, from (4-53) and (4-56), to define a *Resonanzfehler*, or deviation from resonance

$$\Delta\eta_0 = \tfrac{1}{2}[u \pm \sqrt{u^2 + U^2}]\frac{\gamma_0}{\gamma_g} \qquad (4\text{-}60a)$$

$$\Delta\eta_g = \tfrac{1}{2}[-u \pm \sqrt{u^2 + U^2}] \qquad (4\text{-}60b)$$

Again, $\Delta\eta_0$ and $\Delta\eta_g$ are complex in general through $u$ and $U$.

*b.  Nonabsorbing cases*  In nonabsorbing crystals, (4-56a and b) can be reduced to the following forms:

*1*  Laue case ($\gamma_0 > 0$, $\gamma_g > 0$)

$$K\delta_e{}^r = \frac{\tfrac{1}{2}(-K\chi_0{}^r)}{\gamma_0} + \frac{\tfrac{1}{2}[-w \mp (w^2 + W^2)^{1/2}]}{\gamma_g} \qquad (4\text{-}61a)$$

$$K\delta_e{}^i = 0 \qquad (4\text{-}61b)$$

---

[1] At times, hereafter, $R_e[\ ]$ and $J_m[\ ]$ will be denoted by simply affixing the superscripts $r$ and $i$ to the quantity concerned.

The upper and lower signs correspond to branches (1) and (2) of the dispersion surface in Fig. 4-5, respectively. The asymptotic assignment of the wings for $w \leqslant 0$ is illustrated in Fig. 4-5a and b. The points $T_1$ and $T_2$ are the intersections of the line $\mathbf{n}_e$ when it passes through Lorentz point $L$, namely, when $w = 0$. It is easily seen that $T_1 T_2 = KC\, R_e(\chi_g \chi_{-g})^{1/2}/(\gamma_0 \gamma_g)^{1/2}$. The distance between the vertices $V_1 V_2$ is equal to $KC\, R(\chi_g \chi_{-g})^{1/2}/\cos \theta_B$.

2  Bragg case ($\gamma_0 > 0$, $\gamma_g < 0$). For the region of $|w| \geq |W|$.

$$K\delta_e{}^r = \frac{\tfrac{1}{2}(-K\chi_0{}^r)}{\gamma_0} + \frac{\tfrac{1}{2}[w \pm (w^2 - W^2)^{1/2}]}{|\gamma_g|} \qquad (4\text{-}62a)$$

$$K\delta_e{}^i = 0 \qquad (4\text{-}62b)$$

In contrast to the Laue case, the double signs must be assigned to the wings of each branch of the dispersion surface, and $w < -|W|$ and $w > |W|$ correspond to the branches (1) and (2), respectively. The situation is illustrated in Fig. 4-5c. There, as well as in Fig. 4-4b, the points $T_1$ and $T_2$ are the tangential points of the line $\mathbf{n}_e$ on the dispersion surface. In both branches, the right wings of $T_1$ and $T_2$ are called ($a$) and the left wings are called ($b$). The upper and lower signs of (4-62a) can be assigned to ($a$) and ($b$) wings, respectively.

In the range $|w| < |W|$,

$$K\delta_e{}^r = \frac{\tfrac{1}{2}(-K\chi_0{}^r)}{\gamma_0} + \frac{\tfrac{1}{2}w}{|\gamma_g|} \qquad (4\text{-}63a)$$

$$K\delta_e{}^i = \frac{\pm\tfrac{1}{2}(W^2 - w^2)^{1/2}}{|\gamma_g|} \qquad (4\text{-}63b)$$

Again, for convenience, the solutions corresponding to the upper and lower signs are denoted by ($a$) and ($b$), respectively. Like the case of total reflection in the single-beam cases, $\delta_e$ includes the imaginary part even when the crystal is nonabsorbing in a photoelectric sense. In fact, as will be seen later, total reflection occurs within this range. The attentuation of the wave field is purely due to "diffraction absorption." Usually, the thickness to which the wave field is appreciable is less than about 10 $\mu$ below the crystal surface.

c.  *Absorbing cases*   For absorbing crystals, the exact functional form of $\delta_e$ is rather complicated. The details are described in Appendix 3AII, and the discussion in this section is limited to the elementary results.

1  Laue case.  By neglecting the higher terms of $v/(w^2 + W^2)^{1/2}$ and

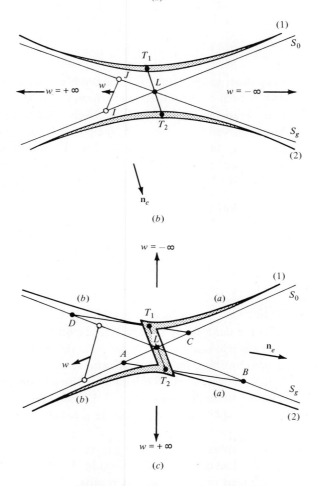

**FIGURE 4-5**
Schematic assignment of the amplitudes to the dispersion surfaces. (*a*) Laue case
(*O* waves). ($T_1$ and $T_2$ mark intersections with $\mathbf{n}_e$.) (*b*) Laue case (*G* waves).
(*c*) Bragg case. ($T_1$ and $T_2$ mark tangential points of the line $\mathbf{n}_e$.)

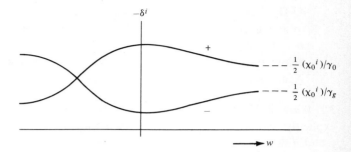

**FIGURE 4-6**
Imaginary part of $\delta$ as a function of the deviation parameter $w$. The plus and minus signs denote the waves of branches (2) and (1), respectively.

$V/(w^2 + W^2)^{1/2}$ in the power series of $\sqrt{u^2 + U^2}$, $\delta_e^r$ has the same form as (4-61a). The imaginary part turns out to take on the form

$$\delta_e^i = \delta_0 \mp (\delta_1 + \delta_2) \qquad (4\text{-}64)$$

where

$$K\delta_0 = \tfrac{1}{4}(-K\chi_0^i)\left(\frac{1}{\gamma_0} + \frac{1}{\gamma_g}\right) \qquad (4\text{-}65a)$$

$$K\delta_1 = \frac{\tfrac{1}{2}[wv/(w^2 + W^2)^{1/2}]}{\gamma_g} \qquad (4\text{-}65b)$$

$$K\delta_2 = \frac{\tfrac{1}{2}[WV/(w^2 + W^2)^{1/2}]}{\gamma_g} \qquad (4\text{-}65c)$$

The behavior of $\delta_e^i$ is illustrated in Fig. 4-6 and the quantities $\delta_e^i$ ($j = 1, 2$) are always negative. The attenuation of the wave field along the direction $\mathbf{n}_e$ depends on the branches (1) and (2) of the dispersion surface and also depends on the parameter $w$. This fact is intimately related to the statement in equation (4-45). Also of significance, $\delta_e^i$ always is negative without using present approximations (cf Appendix 4AVII). This implies that the crystal waves are attenuated in any case, as expected.

*2 Bragg case.* The approximation used in the Laue case is inadequate in the important range $|w| \simeq |W|$. In this case, writing $(u^2 + U^2)^{1/2}$ as $A + iB$, the real and imaginary parts can be represented in the form

$$K\delta_e^r = \frac{\tfrac{1}{2}(-K\chi_0^r)}{\gamma_0} + \frac{\tfrac{1}{2}(w \pm A)}{|\gamma_g|} \qquad (4\text{-}66a)$$

$$K\delta_e^i = \tfrac{1}{4}(-K\chi_0^i)\left(\frac{1}{\gamma_0} + \frac{1}{\gamma_g}\right) \pm \frac{\tfrac{1}{2}B}{|\gamma_g|} \qquad (4\text{-}66b)$$

The expressions for $A$ and $B$ in terms of $w$, $v$, $W$, and $V$ are given in Appendix

4AII. Again, two types of solution exist. They can be distinguished according to the asymptotic behavior of the nonabsorbing cases $A \simeq (w^2 - W^2)^{1/2}$; $B \simeq 0$ for $|w| \geq |W|$; $A \simeq 0$, $B \simeq (W^2 - w^2)^{1/2}$ for $|w| \leq |W|$. As described below, the wave corresponding to one of the double signs is physically significant. Again, it is left to Appendix 4A to prove that $\delta_e{}^i$ of such a wave is negative.

## 3. Wave fields in vacuum and in crystals

Now the wave fields are described explicitly. Solutions $a$ and $b$ deal with the waves excited at the entrance surface, and $c$ and $d$ deal with the waves excited at the exit surface. Afterward, some important cases for crystals bounded by a pair of entrance and exit surfaces are described.

*a. Laue case at the entrance surface*   As explained in the discussion of single-beam cases, strictly speaking, the normal component of the displacement vector and the tangential components of the electric and magnetic vectors must be continuously connected on the boundary surface. Since, however, the polarizability has an order of magnitude of $10^{-5}$, it is experimentally meaningless to distinguish the electric and displacement vectors. The boundary condition, therefore, is approximated by

$$E(\text{vacuum side}) = d(\text{crystal side}) \qquad (4\text{-}67)$$

Here, the incident wave is treated as a linearly polarized wave, the plane of polarization being either perpendicular or parallel to the $R$ plane. Both cases can be treated exactly in the same way, provided that the polarization factor $C$ is introduced [cf equations (4-19$a$ and $b$)].

In Laue cases, since none of $g$-component waves exist in vacuum, the boundary condition (4-67) can be written explicitly (Table 4-1)

$$E_e \exp i(\mathbf{K}_e \cdot \mathbf{r}_e) = d_0{}^{(1)} \exp i(\mathbf{k}_0{}^{(1)} \cdot \mathbf{r}_e) + d_0{}^{(2)} \exp i(\mathbf{k}_0{}^{(2)} \cdot \mathbf{r}_e) \qquad (4\text{-}68a)$$

$$0 = d_g{}^{(1)} \exp i(\mathbf{k}_g{}^{(1)} \cdot \mathbf{r}_e) + d_g{}^{(2)} \exp i(\mathbf{k}_g{}^{(2)} \cdot \mathbf{r}_e) \qquad (4\text{-}68b)$$

where $\mathbf{r}_e$ is a position vector denoting the entrance surface. Writing $d_g{}^{(j)} = c^{(j)} d_0{}^{(j)}$ and recalling that the $\exp 2\pi i (\mathbf{g} \cdot \mathbf{r}_e)$ can be dropped in (4-68$b$), the amplitudes are

$$d_0{}^{(1)} = \left[ \frac{c^{(2)}}{c^{(2)} - c^{(1)}} \right] E_e \exp i[(\mathbf{K}_e - \mathbf{k}_0{}^{(1)}) \cdot \mathbf{r}_e] \qquad (4\text{-}69a)$$

$$d_0{}^{(2)} = \left[ \frac{c^{(1)}}{c^{(1)} - c^{(2)}} \right] E_e \exp i[(\mathbf{K}_e - \mathbf{k}_0{}^{(2)}) \cdot \mathbf{r}_e] \qquad (4\text{-}69b)$$

Consequently,

$$d_g{}^{(1)} = \left[ \frac{c^{(1)} c^{(2)}}{c^{(2)} - c^{(1)}} \right] E_e \exp i[(\mathbf{K}_e - \mathbf{k}_0{}^{(1)}) \cdot \mathbf{r}_e] \qquad (4\text{-}70a)$$

$$d_g{}^{(2)} = \left[ \frac{c^{(1)} c^{(2)}}{c^{(1)} - c^{(2)}} \right] E_e \exp i[(\mathbf{K}_e - \mathbf{k}_0{}^{(2)}) \cdot \mathbf{r}_e] \qquad (4\text{-}70b)$$

By the use of (4-28) and equations (4-60a) and (4-60b), the amplitude ratio of $c^{(j)}$ is given by

$$c^{(j)} = \frac{KC\chi_g[u \pm \sqrt{u^2 + U^2}]}{U^2} \tag{4-71}$$

The upper and lower signs correspond to branches (1) and (2), respectively. The amplitude coefficients, therefore, turn out to be

$$C_0^{(1)} \equiv \frac{c^{(2)}}{c^{(2)} - c^{(1)}} = \frac{\frac{1}{2}[-u + \sqrt{u^2 + U^2}]}{\sqrt{u^2 + U^2}} \tag{4-72a}$$

$$C_0^{(2)} \equiv \frac{c^{(1)}}{c^{(1)} - c^{(2)}} = \frac{\frac{1}{2}[u + \sqrt{u^2 + U^2}]}{\sqrt{u^2 + U^2}} \tag{4-72b}$$

$$C_g^{(1)} \equiv \frac{c^{(1)}c^{(2)}}{c^{(2)} - c^{(1)}} = \frac{\frac{1}{2}(KC\chi_g)}{\sqrt{u^2 + U^2}} \tag{4-72c}$$

$$C_g^{(2)} \equiv \frac{c^{(1)}c^{(2)}}{c^{(1)} - c^{(2)}} = \frac{\frac{1}{2}(-KC\chi_g)}{\sqrt{u^2 + U^2}} \tag{4-72d}$$

By using these coefficients, the wave fields can be written[1]

$$d_0^{(j)}(\mathbf{r}) = C_0^{(j)}E_e \exp i\{(\mathbf{K}_e \cdot \mathbf{r}_e) + [\mathbf{k}_0^{(j)} \cdot (\mathbf{r} - \mathbf{r}_e)]\} \tag{4-73a}$$

$$d_g^{(j)}(\mathbf{r}) = C_g^{(j)} \exp 2\pi i(\mathbf{g} \cdot \mathbf{r}_e) \cdot E_e \exp i\{(\mathbf{K}_e \cdot \mathbf{r}_e) + [\mathbf{k}_g^{(j)} \cdot (\mathbf{r} - \mathbf{r}_e)]\} \tag{4-73b}$$

The physical interpretation of these is obvious. The crystal is an amplitude and phase modulator. The phase terms in braces are the phase changes along the optical path from the source (origin) to the specific observation point. In Fig. 4-5a and b, the magnitudes of $C_{0,g}^{(j)}$ are illustrated schematically by the breadth of the dispersion surface. The asymptotic wings closer to $S_0$ have the effect for the direct wave of increasing the departure from the Bragg condition, whereas only the central part of the dispersion surface is effective for the Bragg-reflected wave.

As explained in connection with expression (3-60), equations (4-73a and b) are independent of the position $\mathbf{r}_e$, provided that it lies on the entrance surface. In fact, alternatively, they can be written

$$d_0^{(j)}(\mathbf{r}) = C_0^{(j)}E_e \exp -iK\delta_e^{(j)}t_e \exp i(\mathbf{K}_e \cdot \mathbf{r}) \tag{4-74a}$$

$$d_g^{(j)}(\mathbf{r}) = C_g^{(j)}E_e \exp -iK\delta_e^{(j)}t_e \exp i[(\mathbf{K}_e + 2\pi\mathbf{g}) \cdot \mathbf{r}] \tag{4-74b}$$

where

$$t_e = [(\mathbf{r} - \mathbf{r}_e) \cdot \mathbf{n}_e] \tag{4-75}$$

---

[1] Hereafter, it is convenient to let $d(\mathbf{r}) = d \exp i(\mathbf{k} \cdot \mathbf{r})$ and to call $d(\mathbf{r})$ and $d$ the wave field and amplitude, respectively.

is the depth of the observation point **r** from the entrance surface. Inserting (4-72) for $C_{0,g}{}^{(j)}$ and the approximate expressions (4-61a) and (4-64) and (4-65) for $K\delta_e{}^{(j)}$, one obtains the crystal wave fields. Usually, the imaginary parts $v$ and $V$ are also neglected in $C_{0,g}{}^{(j)}$. Hereafter, these approximations are called the *small-imaginary-part (SI) approximation*. It follows from relations (4-74a and b) that each crystal wave is gradually decreased in amplitude and modulated in phase along its path and superposed on a carrier wave having an extremely short wavelength.

The intensity of the wave $d_g{}^{(j)}(\mathbf{r})$ is proportional to $|C_g{}^{(j)}|^2$, which turns out to equal $\frac{1}{4}(\gamma_0/\gamma_g)W^2/(w^2 + W^2)$. Roughly speaking, the Bragg reflection takes place only when the deviation parameter $w$ is comparable with $W$. As a measure of the angular range, it is possible to take $\Delta\Theta = \Theta - \Theta_c$, corresponding to $w = W$, where $\Theta$ is the glancing angle and $\Theta_c$ denotes the center of the Bragg reflection. According to equations (4-52), (4-56a and b), and (4-59a), one obtains

$$\Delta\Theta = \frac{CR_e(\chi_g\chi_{-g})^{\frac{1}{2}}\sqrt{\gamma_g/\gamma_0}}{\sin 2\theta_B} \tag{4-76}$$

The center of the Bragg reflection can be defined by $w = 0$. The corresponding Bragg angle is given by

$$\Theta_c = \theta_B - \frac{\frac{1}{2}\chi_0{}^r(1 - \gamma_g/\gamma_0)}{\sin 2\theta_B} \tag{4-77}$$

The deviation from the kinematical Bragg angle thus is caused by a refraction due to the mean polarizability $\chi_0{}^r$.

*b.  Bragg case at the entrance surface*   The crystal is assumed to be semi-infinite and bounded by a single plane on which the incident wave falls. The set of waves that needs to be considered is listed in Table 4-1b. The vacuum wave vector $\mathbf{K}_g$ is given by $AG$ in Fig. 4-4b, so that

$$\mathbf{K}_g = \mathbf{K}_e + 2\pi\mathbf{g} - K\delta_R\mathbf{n}_e \tag{4-78}$$

where

$$K\delta_R = \frac{s_0 \sin 2\theta_B}{|\gamma_g|} \tag{4-79}$$

which, obviously, is real.

The wave vectors $\mathbf{k}_0$ and $\mathbf{k}_g$ of the crystal waves have been discussed in Section 2 above. Here, it is to be noted that only one of the solutions (a) and (b) can really exist. For understanding this, it is necessary to consider the Poynting vector $\mathbf{s} = (c/8\pi)(\hat{\mathbf{K}}_0|d_0| + \hat{\mathbf{K}}_g|d_g|^2)$. From the boundary condition that no wave arrives at the surface from the crystal side, it is neceesary that $(\mathbf{s} \cdot \mathbf{n}_e) \geq 0$. This condition requires that

$$\alpha(\pm) \equiv \left(\frac{|\gamma_g|}{\gamma_0}\right)|c^{(q)}|^2 = \left|\frac{\chi_g}{\chi_{-g}}\right|\left|\frac{u \pm \sqrt{u^2 + U^2}}{U}\right|^2 \leq 1$$

where $(q)$ is used instead of $(j)$ in (4-71) in order to specify the wings $(a)$ and $(b)$. When the crystal is nonabsorbing, it is easily seen that the wing and branch of the acceptable wave is as follows:

$$\text{Wing } (b) \text{ of branch } (2) \qquad \text{for } w > |W|$$
$$\text{Wing } (a) \text{ of branch } (1) \qquad \text{for } w < -|W| \qquad\qquad (4\text{-}80a)[1]$$

In the range $|w| < |W|$, it turns out that $(\mathbf{s} \cdot \mathbf{n}_e) = 0$, namely, the Poynting vector is parallel to the entrance surface. In this case, the acceptable solution cannot be determined simply by the Poynting vector. From (4-63b), the wave associated with the wing $(a)$ increases with increasing depth. For this reason, the acceptable solution must be selected as

$$\text{Solution } (b) \qquad \text{for } |w| < |W| \qquad (4\text{-}80b)$$

For absorbing crystals containing a center of symmetry, again, it is not difficult to select the appropriate branch. In this case, $|\chi_g/\chi_{-g}| = 1$. On the other hand, $|u + \sqrt{u^2 + U^2}| \cdot |u - \sqrt{u^2 + U^2}| = |U|^2$. Then, either one of $\alpha(\pm)$ must be larger than unity and the other is less than unity. Thus, we can select the solution according to the rule that:

The acceptable solution corresponds to the smaller one of

$$|u \pm \sqrt{u^2 + U^2}| \qquad\qquad\qquad \text{(Statement A)}[2]$$

For more general cases, it is necessary to analyze the mathematical structure of $\alpha(\pm)$ more carefully. As shown in Appendix 4AV, however, one can prove that one of the $\alpha(\pm)$ is always larger than unity and the other is less than unity. The selection rule (statement A), therefore, is generally acceptable in any case.

Next, consider some of the mathematical details (Kato, Katagawa, and Saka, 1971). In the theory of complex functions, $z = (\zeta)^{1/2}$ is defined as the solution of $z^2 = \zeta$, which belongs to the positive sheet of the Riemann plane. Similarly $z = -(\zeta)^{1/2}$ is defined as the solution which belongs to the negative sheet. Each of the expressions $[u \pm \sqrt{u^2 + U^2}]$ corresponds to either one of the complex functions $[z \pm (z^2 + U^2)^{1/2}]$ at $z = u$. The latter expression is analytical except at $z = \pm iU$. The double signs in the former specify the branches (Laue case) and the wings (Bragg case) of the dispersion surfaces, whereas the signs in the latter specify one of the Riemann sheets. The solution defined on a single sheet is composed of the solutions attributed to the different wings of the different branches.

Since it can be proved that $|z + (z^2 + U^2)^{1/2}| \geq |z - (z^2 + U^2)^{1/2}|$ for any

---

[1] This rule is derived also from the shape of the dispersion surface and the rule that the Poynting vector is normal to the dispersion surface.

[2] In absorbing crystals, $\alpha(+) = \alpha(-)$ may occur only at a special point satisfying $wA + vB = 0$, where $A + iB = (u^2 + U^2)^{1/2}$, so that this can be overlooked.

complex variable $z$ (Appendix 4AIV), statement A can be expressed in the alternative form:

The acceptable solution corresponds to the negative Riemann sheet.

(Statement B)

This result is based on the theory of complex functions and is particularly important in the spherical wave theory which will be discussed in Section II E.

The continuity of the wave fields at the entrance surface, therefore, can be expressed

$$E_e \exp i(\mathbf{K}_e \cdot \mathbf{r}_e) = d_0{}^{(q)} \exp i(\mathbf{k}_0{}^{(q)} \cdot \mathbf{r}_e) \qquad (4\text{-}81a)$$

and

$$E_g \exp i(\mathbf{K}_g \cdot \mathbf{r}_e) = d_g{}^{(q)} \exp i(\mathbf{k}_g{}^{(q)} \cdot \mathbf{r}_e) \qquad (4\text{-}81b)$$

where the superscript $(q)$ stands for either $(a)$ or $(b)$ as explained above. Given an incident wave, the amplitudes $d_0{}^{(q)}$, $d_g{}^{(q)} = c^{(q)} d_0{}^{(q)}$, and $E_g{}^{(q)}$ are determined in succession (Table 4-1b). Thus, the wave fields are obtained as follows:

$$d_0{}^{(q)}(\mathbf{r}) = E_e \exp i\{(\mathbf{K}_e \cdot \mathbf{r}_e) + [\mathbf{k}_0{}^{(q)} \cdot (\mathbf{r} - \mathbf{r}_e)]\} \qquad (4\text{-}82a)$$

$$d_g{}^{(q)}(\mathbf{r}) = c^{(q)} \exp 2\pi i(\mathbf{g} \cdot \mathbf{r}_e) E_e \exp i\{(\mathbf{K}_e \cdot \mathbf{r}_e) + [\mathbf{k}_g{}^{(q)} \cdot (\mathbf{r} - \mathbf{r}_e)]\} \qquad (4\text{-}82b)$$

$$E_g{}^{(q)}(\mathbf{r}) = c^{(q)} \exp 2\pi i(\mathbf{g} \cdot \mathbf{r}_e) E_e \exp i\{(\mathbf{K}_e \cdot \mathbf{r}_e) + [\mathbf{K}_g \cdot (\mathbf{r} - \mathbf{r}_e)]\} \qquad (4\text{-}82c)$$

Rearranging the phase terms, they can be written alternatively

$$d_0{}^{(q)}(\mathbf{r}) = E_e \exp -iK\delta_e{}^{(q)} t_e \exp i(\mathbf{K}_e \cdot \mathbf{r}) \qquad (4\text{-}83a)$$

$$d_g{}^{(q)}(\mathbf{r}) = c^{(q)} E_e \exp -iK\delta_e{}^{(q)} t_e \exp i[(\mathbf{K}_e + 2\pi\mathbf{g}) \cdot \mathbf{r}] \qquad (4\text{-}83b)$$

$$E_g{}^{(q)}(\mathbf{r}) = c^{(q)} E_e \exp iK\delta_R H \exp i[(\mathbf{K}_e + 2\pi\mathbf{g}) \cdot \mathbf{r}] \qquad (4\text{-}83c)$$

where $t_e$ is the depth defined by (4-75), and $H$ is the height of an observation point $\mathbf{r}$ outside the crystal, the definition being

$$H = [(\mathbf{r}_e - \mathbf{r}) \cdot \mathbf{n}_e] \qquad (4\text{-}83d)$$

$\delta_e{}^{(q)}$, called the *Anpassung*, has been given already by (4-56c) and may be complex. $\delta_R$ is given by (4-79) and is always real. The amplitude ratio $c^{(q)}$ is obtained similarly to (4-71), and it has the same form but $(j)$ must be read as $(q)$.

The reflecting power is defined by

$$P_g \equiv \frac{|(\mathbf{S}_g \cdot \mathbf{n}_e)|}{(\mathbf{S}_0 \cdot \mathbf{n}_e)} = \left(\frac{|\gamma_g|}{\gamma_0}\right) \left|\frac{E_g{}^{(q)}}{E_e}\right|^2 \qquad (4\text{-}84a)$$

where $\mathbf{S}_g$ and $\mathbf{S}_0$ are the Poynting vectors of the reflected and incident waves, respectively. Using equations (4-83c), (4-71), and (4-56b), one obtains

$$P_g = \left|\frac{\chi_g}{\chi_{-g}}\right| \left|\frac{u \pm \sqrt{u^2 + U^2}}{U}\right|^2 \qquad (4\text{-}84b)$$

The plus or minus signs must be selected according to statement A. The functional form in terms of $w$, $v$, $W$, and $V$ is obtained in Appendix 4AII. As proved there, the maximum lies always in the range $0 > w > -|W|$ when $Jm[(\chi_g \chi_{-g})^{1/2}]$ is negative. Sometimes, it is more convenient to introduce the parameter

$$M = \frac{|u^2| + |u^2 + U^2|}{|U^2|} \qquad (4\text{-}85)$$

Then, $P_g$ can be expressed by a single equation in the whole range of $w$ as follows:[1]

$$P_g = \left|\frac{\chi_g}{\chi_{-g}}\right| [M - (M^2 - 1)^{1/2}] \qquad (4\text{-}84c)$$

In nonabsorbing crystals, equation (4-84$b$) is reduced to

$$P_g = [w - (w^2 - W^2)^{1/2}]^2 / W^2 \qquad w > W \qquad (4\text{-}86a)$$

$$= 1 \qquad\qquad\qquad\qquad |w| < W \qquad (4\text{-}86b)$$

$$= [w + (w^2 - W^2)^{1/2}]^2 / W^2 \qquad w < -W \qquad (4\text{-}86c)$$

The range $|w| < W$ gives total reflection. Half of the angular range of the total reflection is given by

$$\Delta\Theta = \frac{CR_e[(\chi_g \chi_{-g})(|\gamma_g|/\gamma_0)]^{1/2}}{\sin 2\theta_B} \qquad (4\text{-}87)$$

as in the Laue case, (4-76). The center of the total reflection corresponds to the angle

$$\Theta_c = \theta_B - \frac{\tfrac{1}{2}\chi_0{}^r(1 + |\gamma_g|/\gamma_0)}{\sin 2\theta_B} \qquad (4\text{-}88)$$

The intensity distribution based on (4-86) is called *Darwin's rocking curve* (1914$b$) and is illustrated in Fig. 4-7. The more general one (4-84$b$ or $c$) is called *Prin's rocking curve* (1930). Some examples are illustrated in Fig. 3.17 of Zachariasen's textbook (1945) and by Bucksh, Ott, and Renninger (1967). The topic of rocking curves will be considered further in Section I D 6.

Here, it is worthwhile to examine the energy relations. At first sight, the solution (4-83) appears strange from the viewpoint of energy conservation, because the intensity of the transmitted wave is identical to that of the incident wave, whereas the reflected vacuum wave certainly exists. If, however, one considers the Poynting vector **s** in the crystal, it turns out that the paradox is superficial. The transmitted power at the crystal surface is given by

$$P_t = \frac{(\mathbf{s} \cdot \mathbf{n}_e)}{(\mathbf{S}_0 \cdot \mathbf{n}_e)} = 1 - \frac{|\gamma_g|}{\gamma_0} |c^{(a)}|^2 \qquad (4\text{-}89)$$

---

[1] See the derivation in Appendix 4A and another one by Cole and Stemple (1962).

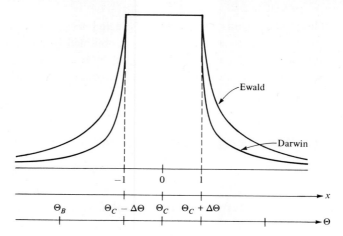

**FIGURE 4-7**
Ewald and Darwin rocking curves (neglecting absorption).

On the other hand, $P_g = (|\gamma_g|/\gamma_0)|c^{(q)}|^2$ from (4-84a). Thus, energy conservation, $P_t + P_g = 1$, is not violated. Also, statement A or B assures that the reflecting power is less than unity.

*c. Laue case at the exit surface* When the crystal is terminated by an exit surface and condition (4-48b) is satisfied, both $O$ and $G$ waves emerge from the surface. The

Table 4-2   WAVE   FIELDS   AT   THE   EXIT
SURFACE

*a.* Laue case

| | O wave | G wave |
|---|---|---|
| Crystal waves | $(\mathbf{d}_0{}^{(j)}; \mathbf{k}_0{}^{(j)})$ | $(\mathbf{d}_g{}^{(j)}; \mathbf{k}_g{}^{(j)})$ |
| | $\downarrow$ | $\downarrow$ |
| Vacuum waves | $(\mathbf{E}_0{}^{(j)}; \mathbf{K}_0{}^{(j)})$ | $(\mathbf{E}_g{}^{(j)}; \mathbf{K}_g{}^{(j)})$ |
| | $(j) = (1), (2), (a),$ and $(b)$ | |

*b.* Bragg case

| | p wave | n wave |
|---|---|---|
| Incident waves | $(\mathbf{d}_p{}^{(j)}; \mathbf{k}_p{}^{(j)})$ | $(\mathbf{d}_n{}^{(j)}; \mathbf{k}_n{}^{(j)})$ |
| | $\downarrow$ | $\downarrow$ |
| Reflected waves | $(\mathbf{d}_{p,r}{}^{(j)}; \mathbf{k}_{p,r}{}^{(j)}) \leftarrow (\mathbf{d}_{n,r}{}^{(j)}; \mathbf{k}_{n,r}{}^{(j)})$ | |
| | $\downarrow$ | |
| Vacuum waves | $(\mathbf{E}_p{}^{(j)}; \mathbf{K}_p{}^{(j)})$ | $0$ |
| | $(j) = (1), (2), (a),$ and $(b)$ | |

surface is not necessarily parallel to the entrance surface, but it is assumed to be sufficiently extended. The waves to be considered are listed in Table 4-2a. How the crystal (or Bloch) waves are excited is not considered in this section. Also, surface reflection is neglected.[1] By virtue of the tangential continuity, the vacuum-wave vector can be written as follows:

$$\mathbf{K}_0 = \mathbf{k}_0 + K\delta_{a,0}\mathbf{n}_a \qquad (4\text{-}90a)$$

$$\mathbf{K}_g = \mathbf{k}_g + K\delta_{a,g}\mathbf{n}_a \qquad (4\text{-}90b)$$

where $\mathbf{k}_0$ and $\mathbf{k}_g$ are the wave vectors of the incident Bloch wave. Meanwhile, the superscript $(j)$ or $(q)$ specifying the type of the Bloch wave is omitted. Since

$$\mathbf{k}_0 = \overline{\mathbf{K}}_0 + \Delta\mathbf{K} + \Delta\mathbf{k} \qquad \text{and} \qquad \mathbf{k}_g = \overline{\mathbf{K}}_g + \Delta\mathbf{K} + \Delta\mathbf{k}$$

it follows that by taking scalar products of $\hat{\mathbf{K}}_0$ and $\hat{\mathbf{K}}_g$ with equations (4-90a and b), respectively (Fig. 4-8a),

$$K\delta_{a,0} = -\frac{\frac{1}{2}K\chi_0}{\gamma_0'} - \frac{\Delta\eta_0}{\gamma_0'} \qquad (4\text{-}91a)$$

$$K\delta_{a,g} = -\frac{\frac{1}{2}K\chi_0}{\gamma_g'} - \frac{\Delta\eta_g}{\gamma_g'} \qquad (4\text{-}91b)$$

At first sight, it is rather strange that the vacuum-wave vectors $\mathbf{K}_0$ and $\mathbf{K}_g$ include the imaginary parts. The imaginary parts $(\Delta\mathbf{K}^i + \Delta\mathbf{k}^i + K\delta_{a,0}{}^i\mathbf{n}_a)$ and $(\Delta\mathbf{K}^i + \Delta\mathbf{k}^i + K\delta_{a,g}{}^i\mathbf{n}_a)$, however, are perpendicular to $\overline{\mathbf{K}}_0$ and $\overline{\mathbf{K}}_g$, respectively, because the dispersion relation $\mathbf{K}_0{}^2 = \mathbf{K}_g{}^2 = K^2$ (real) must be satisfied. Each wave field is essentially a plane wave, and the amplitude attenuates along a direction perpendicular to the wave vector. This situation can be given a more physical interpretation, namely, the Poynting vector is parallel to the wave vector $\mathbf{K}_0$ or $\mathbf{K}_g$. The wave field at $\mathbf{r}$ must be a kind of projection of the wave field at $\bar{\mathbf{r}}_a$, which is the intersection of the exit surface and the line passing through $\mathbf{r}$ in the direction of the wave vector.

By virtue of the continuity of the wave fields at the exit surface, the vacuum-wave fields can be written explicitly

$$E_0{}^{(j)}(\mathbf{r}) = d_0{}^{(j)} \exp i\{(\mathbf{k}_0{}^{(j)} \cdot \mathbf{r}_a) + [\mathbf{K}_0{}^{(j)} \cdot (\mathbf{r} - \mathbf{r}_a)]\} \qquad (4\text{-}92a)$$

$$E_g{}^{(j)}(\mathbf{r}) = d_g{}^{(j)} \exp i\{(\mathbf{k}_g{}^{(j)} \cdot \mathbf{r}_a) + [\mathbf{K}_g{}^{(j)} \cdot (\mathbf{r} - \mathbf{r}_a)]\} \qquad (4\text{-}92b)$$

Obviously, the result is independent of the position $\mathbf{r}_a$ as discussed in connection with relations (4-73) and (4-74). Here the superscripts $(j)$ are written explicitly. If the incident Bloch wave is excited directly from the entrance surface under the conditions of the Bragg case, $(j)$ must be replaced by $(q)$.

[1] The surface reflection examined here is different from the reflection of Bloch waves in Bragg cases which will be described in the next section.

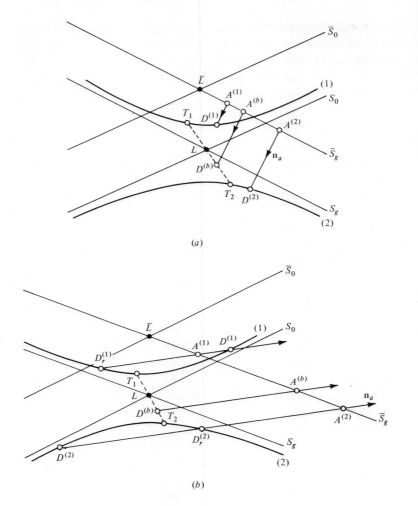

FIGURE 4-8
Construction of the dispersion points at the exit surface. (*a*) Laue case; (*b*) Bragg case.

*d.   Bragg case at the exit surface*   In this section, the reflection of a Bloch wave at the exit surface under the Bragg condition (4-48a) is considered. Like in the Laue case, the origin of the Bloch wave is not considered here. In order to treat $O$ and $G$ waves in a symmetrical way as far as possible, the subscripts 0 and $g$ are replaced by $p$ and $n$, and it should be assumed that $\gamma'_p > 0$ and $\gamma'_n < 0$. Consequently, the $p$ wave is partly reflected and transmitted through the exit surface, whereas the $n$ wave is totally reflected.

The waves to be considered are listed in Table 4-2b. The subscript $r$ is used for denoting the reflected waves. Obviously, the wave vectors of the reflected crystal waves obey the relation

$$\mathbf{k}_{g,r} = \mathbf{k}_{0,r} + 2\pi\mathbf{g} \qquad (4\text{-}93)$$

From the tangential continuity of the wave vectors, they must have the forms

$$\mathbf{k}_{p,r} = \mathbf{k}_p + K\delta_r\mathbf{n}_a \qquad (4\text{-}94a)$$

$$\mathbf{k}_{n,r} = \mathbf{k}_n + K\delta_r\mathbf{n}_a \qquad (4\text{-}94b)$$

To obtain $\delta_r$, it is convenient to introduce *Resonanzfehler* of the reflected waves, which are defined by

$$\Delta\eta_{0,r} = [\hat{\bar{\mathbf{K}}}_0 \cdot (\mathbf{k}_{0,r} - \bar{\mathbf{k}}_0)] \qquad (4\text{-}95a)$$

$$\Delta\eta_{g,r} = [\hat{\bar{\mathbf{K}}}_g \cdot (\mathbf{k}_{g,r} - \bar{\mathbf{k}}_g)] \qquad (4\text{-}95b)$$

By the use of equations (4-94a and b), subtracting $\bar{\mathbf{k}}_0$ or $\bar{\mathbf{k}}_g$ from both sides and multiplying by $\hat{\bar{\mathbf{K}}}_0$ or $\hat{\bar{\mathbf{K}}}_g$ to form a scalar product, one obtains

$$\Delta\eta_{0,r} = \Delta\eta_0 + K\delta_r\gamma_0' \qquad (4\text{-}96a)$$

$$\Delta\eta_{g,r} = \Delta\eta_g + K\delta_r\gamma_g' \qquad (4\text{-}96b)$$

where $\Delta\eta_0$ and $\Delta\eta_g$ are the *Resonanzfehler* of the incident Bloch wave.[1] On the other hand, from the dispersion relation,

$$\Delta\eta_{0,r}\Delta\eta_{g,r} = \Delta\eta_0\Delta\eta_g$$

Thus, $K\delta_r$ is given by

$$K\delta_r = -\left(\frac{\Delta\eta_0}{\gamma_0'} + \frac{\Delta\eta_g}{\gamma_g'}\right) \qquad (4\text{-}97)$$

Inserting this into (4-96a and b),

$$\Delta\eta_{0,r} = -\Delta\eta_g\frac{\gamma_0'}{\gamma_g'} \qquad (4\text{-}98a)$$

$$\Delta\eta_{g,r} = -\Delta\eta_0\frac{\gamma_g'}{\gamma_0'} \qquad (4\text{-}98b)$$

It is worthwhile to note that the quantities $K\delta_r$, $\Delta\eta_{0,r}$, and $\Delta\eta_{g,r}$ specifying the reflected Bloch wave are uniquely determined by the *Resonanzfehler* of the incident Bloch wave.

The vacuum-wave vector $\mathbf{K}_p$ is determined from $\mathbf{k}_p$, as in the Laue case. By using the tangential continuity of the wave vector, we have

$$\mathbf{K}_p = \mathbf{k}_p + K\delta_{a,p}\mathbf{n}_a \qquad (4\text{-}99)$$

and

$$K\delta_{a,p} = -\frac{\frac{1}{2}K\chi_0}{\gamma_p'} - \frac{\Delta\eta_p}{\gamma_p'} \qquad (4\text{-}100)$$

---

[1] They are not necessarily $\Delta\eta_0$ and $\Delta\eta_g$ of the waves excited by the entrance surface.

The geometrical relation is illustrated in Fig. 4-8$b$.

For subsequent purposes, it is desirable also to calculate the amplitude ratio $c_{p,r} = d_{n,r}/d_{p,r}$. Inserting equations (4-98$a$ and $b$) into (4-28) in which $\Delta\eta_{0,g}$ is replaced by $\Delta\eta_{p,r}$, one gets

$$c_{p,r} = -\frac{\frac{1}{2}(KC\chi_{\pm g})(\gamma_p'/\gamma_n')}{\Delta\eta_p} \tag{4-101}$$

where the upper and lower signs in $\chi_{\pm g}$ correspond to $p = 0$ and $p = g$, respectively.

The continuity of the amplitude at the exit surface can be expressed by

$$E_p \exp i(\mathbf{K}_p \cdot \mathbf{r}_a) = d_p \exp i(\mathbf{k}_p \cdot \mathbf{r}_a) + d_{p,r} \exp i(\mathbf{k}_{p,r} \cdot \mathbf{r}_a) \tag{4-102a}$$

$$0 = d_n \exp i(\mathbf{k}_n \cdot \mathbf{r}_a) + d_{n,r} \exp i(\mathbf{k}_{n,r} \cdot \mathbf{r}_a) \tag{4-102b}$$

From these one obtains the wave fields

$$d_{n,r}^{(j)}(\mathbf{r}) = -d_n^{(j)} \exp i\{(\mathbf{k}_n^{(j)} \cdot \mathbf{r}_a) + [\mathbf{k}_{n,r}^{(j)} \cdot (\mathbf{r} - \mathbf{r}_a)]\} \tag{4-103a}$$

$$d_{p,r}^{(j)}(\mathbf{r}) = -\frac{c_p^{(j)}}{c_{p,r}^{(j)}} d_p^{(j)} \exp i\{(\mathbf{k}_p^{(j)} \cdot \mathbf{r}_a) + [\mathbf{k}_{p,r}^{(j)} \cdot (\mathbf{r} - \mathbf{r}_a)]\} \tag{4-103b}$$

where $c_p = d_n/d_p$ and

$$E_p^{(j)}(\mathbf{r}) = \left[1 - \left(\frac{c_p^{(j)}}{c_{p,r}^{(j)}}\right)\right] d_p^{(j)} \exp i\{(\mathbf{k}_p^{(j)} \cdot \mathbf{r}_a) + [\mathbf{K}_p^{(j)} \cdot (\mathbf{r} - \mathbf{r}_a)]\} \tag{4-104}$$

for an incident Bloch wave, whose wave vectors ($\mathbf{k}_p^{(j)}$ and $\mathbf{k}_n^{(j)}$) and amplitudes ($d_p^{(j)}$ and $d_n^{(j)}$) are given. If necessary, the superscripts ($j$) can be read as ($q$).

In the following sections, a few important cases of the combination of Bragg and Laue cases at a pair of entrance and exit surfaces will be considered.

*e.   Laue-Laue case*   By combining (4-73) with $\mathbf{r} = \mathbf{r}_a$ and (4-92), the transmitted wave fields through a wedge-shaped crystal can be obtained straightforwardly as follows:

$$\begin{aligned}
E_0^{(j)}(\mathbf{r}) = C_0^{(j)} E_e \exp i\{(\mathbf{K}_e \cdot \mathbf{r}_e) + [\mathbf{k}_0^{(j)} \cdot (\mathbf{r}_a - \mathbf{r}_e)] \\
+ [\mathbf{K}_0^{(j)} \cdot (\mathbf{r} - \mathbf{r}_a)]\}
\end{aligned} \tag{4-105a}$$

$$\begin{aligned}
E_g^{(j)}(\mathbf{r}) = C_g^{(j)} \exp 2\pi i(\mathbf{g} \cdot \mathbf{r}_e) E_e \exp i\{(\mathbf{K}_e \cdot \mathbf{r}_e) + [\mathbf{k}_g^{(j)} \cdot (\mathbf{r}_a - \mathbf{r}_e)] \\
+ [\mathbf{K}_g^{(j)} \cdot (\mathbf{r} - \mathbf{r}_a)]\}
\end{aligned} \tag{4-105b}$$

Here, by combining (4-50) and (4-90), the vacuum-wave vectors $\mathbf{K}_0^{(j)}$ and $\mathbf{K}_g^{(j)}$ are written in the explicit form

$$\mathbf{K}_0^{(j)} = \mathbf{K}_e - K\delta_e^{(j)} \mathbf{n}_e + K\delta_{a,0}^{(j)} \mathbf{n}_a \tag{4-105c}$$

$$\mathbf{K}_g^{(j)} = (\mathbf{K}_e + 2\pi\mathbf{g}) - K\delta_e^{(j)} \mathbf{n}_e + K\delta_{a,g}^{(j)} \mathbf{n}_a \tag{4-105d}$$

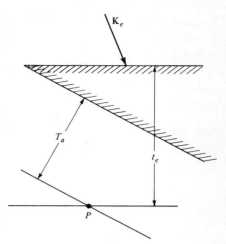

**FIGURE 4-9**
Specification of the observation point $P$.

Each vacuum-wave field of $O$ and $G$ waves is composed of two plane waves specified by $(j) = (1)$ and $(2)$. The sinusoidal distribution of the intensity field is expected if both plane waves overlap in space. The details are described in the discussion of *Pendellösung* phenomena in Section I D 2.

Alternatively, by rearranging the phase terms, equations (4-105a) and (4-105b) can be written

$$E_0^{(j)}(\mathbf{r}) = C_0^{(j)} E_e \exp iK(\delta_{a,0}^{(j)} T_a - \delta_e^{(j)} t_e) \exp i(\mathbf{K}_e \cdot \mathbf{r}) \qquad (4\text{-}106a)$$

$$E_g^{(j)}(\mathbf{r}) = C_g^{(j)} E_e \exp iK(\delta_{a,g}^{(j)} T_a - \delta_e^{(j)} t_e) \exp i((\mathbf{K}_e + 2\pi\mathbf{g}) \cdot \mathbf{r}) \qquad (4\text{-}106b)$$

where

$$t_e = (\mathbf{r} - \mathbf{r}_e) \cdot \mathbf{n}_e \qquad (4\text{-}107a)$$

$$T_a = (\mathbf{r} - \mathbf{r}_a) \cdot \mathbf{n}_a \qquad (4\text{-}107b)$$

They are the normal distances to the observation point measured from the entrance and exit surfaces, respectively. The geometry is illustrated in Fig. 4-9. The *Anpassung* $K\delta_e^{(j)}$ is given by (4-56c). From equations (4-91) and (4-60) it follows that

$$K\delta_{a,0} = -\frac{\frac{1}{2}K\chi_0}{\gamma_0'} - \frac{\frac{1}{2}(\gamma_0/\gamma_g)[u \pm \sqrt{u^2 + U^2}]}{\gamma_0'} \qquad (4\text{-}108a)$$

$$K\delta_{a,g} = -\frac{\frac{1}{2}K\chi_0}{\gamma_g'} - \frac{\frac{1}{2}[-u \pm \sqrt{u^2 + U^2}]}{\gamma_g'} \qquad (4\text{-}108b)$$

The amplitude factors $C_0^{(j)}$ and $C_g^{(j)}$ are given by (4-72). The intensity distribution as a function of $\Theta$, which is given by $w = K(\Theta - \Theta_c) \sin 2\theta_B$, is called the *rocking curve*. The details are described in Section I D 6.

*f.  Laue-Bragg case*  Consider the diffraction problems illustrated in Fig. 4-3c and c′, called type I and II, and described separately.  It is helpful here to consult the preceding solutions a and d.

Type I ($p = g$, $n = 0$): In this case, the O wave is totally reflected on the exit surface, and the G wave penetrates partly through it.  Combining equations (4-73), (4-103), and (4-104), one obtains the wave fields as follows:

$$d_{0,r}^{(j)}(\mathbf{r}) = -C_0^{(j)}E_e \exp i\{(\mathbf{K}_e \cdot \mathbf{r}_e) + [\mathbf{k}_0^{(j)} \cdot (\mathbf{r}_a - \mathbf{r}_e)] + [\mathbf{k}_{0,r}^{(j)} \cdot (\mathbf{r} - \mathbf{r}_a)]\}$$

$$(4\text{-}109a)$$

$$d_{g,r}^{(j)}(\mathbf{r}) = -\frac{c_g^{(j)}}{c_{g,r}^{(j)}} C_g^{(j)} \exp 2\pi i(\mathbf{g} \cdot \mathbf{r}_e)E_e \exp i\{(\mathbf{K}_e \cdot \mathbf{r}_e)$$
$$+ [\mathbf{k}_g^{(j)} \cdot (\mathbf{r}_a - \mathbf{r}_e)] + [\mathbf{k}_{g,r}^{(j)} \cdot (\mathbf{r} - \mathbf{r}_a)]\} \quad (4\text{-}109b)$$

$$E_g^{(j)}(\mathbf{r}) = \left[1 - \frac{c_g^{(j)}}{c_{g,r}^{(j)}}\right] C_g^{(j)} \exp 2\pi i(\mathbf{g} \cdot \mathbf{r}_e)E_e \exp i\{(\mathbf{K}_e \cdot \mathbf{r}_e)$$
$$+ [\mathbf{k}_g^{(j)} \cdot (\mathbf{r}_a - \mathbf{r}_e)] + [\mathbf{K}_g^{(j)} \cdot (\mathbf{r} - \mathbf{r}_a)]\} \quad (4\text{-}109c)$$

Each wave field is essentially a plane wave having the form $AE_e \exp i\phi$.  Table 4-3 lists a set of $(A, \phi)$ for each plane wave.  In constructing the table, the equations for the amplitude ratios $c_g$ and $c_{g,r}$, the amplitude factors $C_0$, $C_g$, and *Anpassungen* $\delta_e$, $\delta_r$, and $\delta_{a,g}$ described in the preceding Sections 2, 3a, and 3d are employed.

Type II ($p = 0$, $n = g$): In this case, the G wave is totally reflected at the exit surface, and the O wave penetrates partly through it.  Similarly to the case of type I, the wave fields can be written straightforwardly

$$d_{0,r}^{(j)}(\mathbf{r}) = -\left(\frac{c_0^{(j)}}{c_{0,r}^{(j)}}\right) C_0^{(j)}E_e \exp i\{(\mathbf{K}_e \cdot \mathbf{r}_e) + [\mathbf{k}_0^{(j)} \cdot (\mathbf{r}_a - \mathbf{r}_e)]$$
$$+ [\mathbf{k}_{0,r}^{(j)} \cdot (\mathbf{r} - \mathbf{r}_a)]\} \quad (4\text{-}110a)$$

$$d_{g,r}^{(j)}(\mathbf{r}) = -C_g^{(j)} \exp 2\pi i(\mathbf{g} \cdot \mathbf{r}_e)E_e \exp i\{(\mathbf{K}_e \cdot \mathbf{r}_e) + [\mathbf{k}_g^{(j)} \cdot (\mathbf{r}_a - \mathbf{r}_e)]$$
$$+ [\mathbf{k}_{g,r}^{(j)} \cdot (\mathbf{r} - \mathbf{r}_a)]\} \quad (4\text{-}110b)$$

$$E_0^{(j)}(\mathbf{r}) = \left[1 - \frac{c_0^{(j)}}{c_{0,r}^{(j)}}\right] C_0^{(j)}E_e \exp i\{(\mathbf{K}_e \cdot \mathbf{r}_e) + [\mathbf{k}_0^{(j)} \cdot (\mathbf{r}_a - \mathbf{r}_e)]$$
$$+ [\mathbf{K}_0^{(j)} \cdot (\mathbf{r} - \mathbf{r}_a)]\} \quad (4\text{-}110c)$$

Again, each wave has the form of $AE_e \exp i\phi$.  The amplitude factors $A$ and the phases $\phi$ are listed in Table 4-3.

In both cases of types I and II, the real crystal waves are a superposition of the incident Bloch waves and the reflected Bloch waves.  For this reason the wave fields are rather complicated.  Nevertheless, the reflected waves are similar to the incident Bloch waves in character, so that no conceptual difficulty is raised.  If the crystal is bounded by several plane surfaces, the reflected Bloch wave may be again reflected by another crystal surface.  Finally, they may emerge from a crystal surface under the conditions of the Laue case. By using the treatment described in the preceding solution $c$, the vacuum waves can be calculated straightforwardly. No crystal wave is newly created at this surface. In the present terminology described in Section 1, the crystal waves are composed of Laue-(Bragg)$^m$ waves, and the vacuum waves are composed of Laue-(Bragg)$^m$ and Laue-(Bragg)$^n$-Laue waves. For details, the reader is referred to the original paper (Saka, Katagawa, and Kato, 1972$b$).

*g.* *(Bragg)$^\infty$ case*   In this section, parallel-sided crystals are considered under the conditions which are conventionally called Bragg cases. In the traditional treatment, the wave fields are determined so as to satisfy the boundary conditions at the entrance and exit surfaces simultaneously.[1] For this purpose, unlike the cases of semi-infinite crystals described in the preceding solution $b$, the crystal waves of types $(a)$ and $(b)$ must be taken into account together. There is no reason to select a particular type of wave because the boundary conditions at infinity are to be imposed on the vacuum waves instead of the crystal waves. Thus, the boundary conditions can be written as follows:

$$E_e \exp i(\mathbf{K}_e \cdot \mathbf{r}_e) = d_0' \exp i(\mathbf{k}_0' \cdot \mathbf{r}_e) + d_0'' \exp i(\mathbf{k}_0'' \cdot \mathbf{r}_e) \quad (4\text{-}111a)$$

$$0 = d_g' \exp i(\mathbf{k}_g' \cdot \mathbf{r}_a) + d_g'' \exp i(\mathbf{k}_g'' \cdot \mathbf{r}_a) \quad (4\text{-}111b)$$

Here, the single prime refers to the type of Bloch wave which is excited in the case of semi-infinite crystals. It stands for either $(a)$ or $(b)$, depending upon the range of the deviation parameter $w$. The double prime denotes the Bloch wave of another type.

Similarly to the Laue case, with the use of $d_g^{(q)} = c^{(q)} d_0^{(q)}$, the amplitudes $d_0^{(q)}$ and $d_g^{(q)}$ are uniquely determined from equations (4-111$a$) and (4-111$b$): Once they are obtained, the vacuum waves are determined by the other boundary conditions

$$E_0 \exp i(\mathbf{K}_0 \cdot \mathbf{r}_a) = d_0' \exp i(\mathbf{k}_0' \cdot \mathbf{r}_a) + d_0'' \exp i(\mathbf{k}_0'' \cdot \mathbf{r}_a) \quad (4\text{-}112a)$$

$$E_g \exp i(\mathbf{K}_g \cdot \mathbf{r}_e) = d_g' \exp i(\mathbf{k}_g' \cdot \mathbf{r}_e) + d_g'' \exp i(\mathbf{k}_g'' \cdot \mathbf{r}_e) \quad (4\text{-}112b)$$

By rearranging the phase terms, the amplitudes $E_0$ and $E_g$ are written in the forms

$$E_0 = \frac{(c'' - c') \exp i\,(\varphi_e - \varphi_a + \phi_a)}{c'' \exp i\,\phi_a - c' \exp i\,\phi_e} E_e \quad (4\text{-}113a)$$

$$E_g = \frac{c'c''(\exp i\,\phi_a - \exp i\,\phi_e)}{c'' \exp i\,\phi_a - c' \exp i\,\phi_e} E_e \exp i\,\psi_e \quad (4\text{-}113b)$$

---

[1] "Entrance" has the literal meaning only with respect to the incident wave. The same surface plays the role of "exit" surface with respect to the crystal waves. For this reason, later on, the entrance and exit surfaces are renamed front and rear surfaces, respectively.

## Table 4-3 SUMMARY OF THE PLANE-WAVE THEORY

### I. The wave fields

(a) Laue case

O: $d_0(\mathbf{r}) = C_0 \exp i\varphi[E_e \exp i(\mathbf{K}_e \cdot \mathbf{r})]$

G: $d_g(\mathbf{r}) = C_g \exp i\varphi\{E_e \exp i[(\mathbf{K}_e + 2\pi\mathbf{g}) \cdot \mathbf{r}]\}$

$C_0, C_g$ and $\varphi$ must be specified by the branch indexes $j = 1$ and 2

(b) Bragg case

O: $d_0(\mathbf{r}) = \exp i\varphi[E_e \exp i(\mathbf{K}_e \cdot \mathbf{r})]$

G: $d_g(\mathbf{r}) = c \exp i\varphi\{E_e \exp i[(\mathbf{K}_e + 2\pi\mathbf{g}) \cdot \mathbf{r}]\}$

$\quad\ E_g(\mathbf{r}) = c \exp i\varphi\{E_e \exp i[(\mathbf{K}_e + 2\pi\mathbf{g}) \cdot \mathbf{r}]\}$

$c$ and $\varphi$ must be specified by the wing indices $q = a$ and $b$; however, only one of the waves ($a$ or $b$) can really exist (cf, statement B)

(c) Laue-Bragg case (specifications: $j = 1$ and 2)

(Type I)

O: $d_{0,r}(\mathbf{r}) = -C_0 \exp i\varphi[E_e \exp i(\mathbf{K}_e \cdot \mathbf{r})]$

G: $d_{g,r}(\mathbf{r}) = C_{g,r} \exp i\varphi\{E_e \exp i[(\mathbf{K}_e + 2\pi\mathbf{g}) \cdot \mathbf{r}]\}$

$\quad\ E_{g,t}(\mathbf{r}) = C_{g,t} \exp i\varphi\{E_e \exp i[(\mathbf{K}_e + 2\pi\mathbf{g}) \cdot \mathbf{r}]\}$

(Type II)

O: $d_{0,r}(\mathbf{r}) = C_{0,r} \exp i\varphi[E_e \exp i(\mathbf{K}_e \cdot \mathbf{r})]$

$\quad\ E_{0,t}(\mathbf{r}) = C_{0,t} \exp i\varphi[E_e \exp i(\mathbf{K}_e \cdot \mathbf{r})]$

G: $d_{g,r}(\mathbf{r}) = -C_g \exp i\varphi\{E_e \exp i[(\mathbf{K}_e + 2\pi\mathbf{g}) \cdot \mathbf{r}]\}$

(d) Laue-Laue case (specifications: $j = 1$ and 2)

O: $E_0(\mathbf{r}) = C_0 \exp i\varphi[E_e \exp i(\mathbf{K}_e \cdot \mathbf{r})]$

G: $E_g(\mathbf{r}) = C_g \exp i\varphi\{E_e \exp i[(\mathbf{K}_e + 2\pi\mathbf{g}) \cdot \mathbf{r}]\}$

(e) Bragg-Laue case (specifications: $q = a$ or $b$)

O: $E_0(\mathbf{r}) = \exp i\varphi[E_e \exp i(\mathbf{K}_e \cdot \mathbf{r})]$

G: $E_g(\mathbf{r}) = c \exp i\varphi\{E_e \exp i[(\mathbf{K}_e + 2\pi\mathbf{g}) \cdot \mathbf{r}]\}$

(f) Bragg-Bragg case (specifications: $q = a$ or $b$)

O: $d_0(\mathbf{r}) = C_{0,r} \exp i\varphi[E_e \exp i(\mathbf{K}_e \cdot \mathbf{r})]$

$\quad\ E_0(\mathbf{r}) = C_{0,t} \exp i\varphi[E_e \exp i(\mathbf{K}_e \cdot \mathbf{r})]$

G: $d_g(\mathbf{r}) = -c \exp i\varphi\{E_e \exp i[(\mathbf{K}_e + 2\pi\mathbf{g}) \cdot \mathbf{r}]\}$

### II. Amplitudes ($C_0, C_g$, etc.) and phases

| Cases | | Amplitude $\{C\}$ | Phase $\varphi$ | $\xi_1 + \xi_2$ | $\xi_1 - \xi_2$ |
|---|---|---|---|---|---|
| Laue | $\begin{cases} d_0 \\ d_g \end{cases}$ | $\frac{1}{2}[-u \pm \sqrt{u^2 + U^2}] \pm \sqrt{u^2 + U^2}$ <br> $\frac{1}{2}KC\chi_g / \pm \sqrt{u^2 + U^2}$ | $-K\delta_e t_e$ | $x_0$ | $(\gamma_0/\gamma_g)x_g$ |
| Bragg | $\begin{cases} d_0 \\ d_g \\ E_g \end{cases}$ | 1 <br> $KC\chi_g[u - (u^2 + U^2)^{1/2}]/U^2$ <br> $KC\chi_g[u - (u^2 + U^2)^{1/2}]/U^2$ | $-K\delta_e t_e$ <br><br> $K\delta_R H$ | $x_0$ <br><br> $\bar{x}_0$ | $-(\gamma_0/|\gamma_g|)x_g$ <br><br> $-(\gamma_0/|\gamma_g|)\bar{x}_g$ |
| Laue-Bragg (I) | $\begin{cases} d_{0,r} \\ d_{g,r} \\ E_{g,t} \end{cases}$ | $\frac{1}{2}[-u \pm \sqrt{u^2 + U^2}] \mp \sqrt{u^2 + U^2}$ <br> $[\frac{1}{2}KC\chi_g \mp \sqrt{u^2 + U^2}] \pm \sqrt{u^2 + U^2}$ <br> $[\frac{1}{2}KC\chi_g] \pm \sqrt{u^2 + U^2}$ | $-K(\delta_e t_e + \delta_{r} t_a)$ <br><br> $K(-\delta_e t_e + \delta_{a,g} T_a)$ | $(|\gamma_0|/\gamma_g)x'_g$ <br><br> $(|\gamma_0|/\gamma_g)\bar{x}'_g$ | $(\gamma'_0/|\gamma'_0|)(\gamma_0/\gamma_g)x'_0$ <br><br> $(\gamma'_g/|\gamma'_0|)(\gamma_0/\gamma_g)\bar{x}'_0$ |

$$\times \left[1 - ((|\gamma'_0|/|\gamma_g|)(\chi_g/\gamma_0)(\gamma_g/\gamma_0)[-u \pm \sqrt{u^2 + U^2}/U](-u \pm \sqrt{u^2 + U^2}/U)^2\right]$$

$$\text{Laue-}\atop\text{Bragg}\;\text{(II)}\left\{\begin{matrix}d_{0,r}\\ d_{g,r}\\ E_{0,t}\end{matrix}\right.\quad\begin{matrix}\frac{1}{4}(|\gamma_g''|/\gamma_0'')(\gamma_0/\gamma_g)[u \mp \sqrt{u^2 + U^2}]\mp\sqrt{u^2 + U^2}\\[4pt] \frac{1}{4}KC\chi_g|\mp\sqrt{u^2 + U^2}\\[4pt] \frac{1}{4}\{[-u \mp \sqrt{u^2 + U^2}]\\ \qquad - (|\gamma_g''|/\gamma_0'')(\gamma_0/\gamma_g)[u \pm \sqrt{u^2 + U^2}]\}/\pm\sqrt{u^2 + U^2}\end{matrix}$$

$$\text{Laue-}\atop\text{Laue}\;\left\{\begin{matrix}E_0\\ E_g\end{matrix}\right.\quad\begin{matrix}\frac{1}{4}[-u \pm \sqrt{u^2 + U^2}]/\pm\sqrt{u^2 + U^2}\\[4pt] \frac{1}{4}KC\chi_g/\pm\sqrt{u^2 + U^2}\end{matrix}$$

$$\left.\phantom{X}\right\}\quad -K(\delta_\ell t_e + \delta_r t_a)\qquad (\gamma_0''/|\gamma_g''|)\chi_g''\quad (|\gamma_g''|/\gamma_0'')(\gamma_0/\gamma_g)\chi_0''$$

$$\begin{matrix}K(-\delta_e t_e + \delta_{a,0}T_a)\\ K(-\delta_e t_e + \delta_{a,0}T_a)\\ K(-\delta_e t_e + \delta_{a,g}T_a)\end{matrix}\qquad\begin{matrix}(\gamma_0''/|\gamma_g''|)\bar\chi_g''\\ \bar\chi_0\\ \bar\chi_0\end{matrix}\quad\begin{matrix}(|\gamma_g''|/\gamma_0'')(\gamma_0/\gamma_g)\bar\chi_0''\\ (\gamma_0/\gamma_g)\bar\chi_g\\ (\gamma_0/\gamma_g)\bar\chi_g\end{matrix}$$

The upper and lower signs indicate branches (1) and (2), respectively. The parameters $(u, U)$ are converted to $(s, \beta)$ by dividing by $\sin 2\theta_B$.

$G \equiv \varphi + K_x\chi_0$ can be represented in the forms

$$G = \tfrac{1}{2}K\chi_0(l_0 + l) + [\pm\xi_1\sqrt{s^2 + \beta^2} + \zeta_2 s]\qquad\text{(for crystal waves)}$$
$$G = \tfrac{1}{2}K\chi_0(l_0 + l) + [\pm\xi_1\sqrt{s^2 + \beta^2} + \zeta_2 s]\qquad\text{(for vacuum waves)}$$

The last two columns list $\{\xi_1 \pm \zeta_2\}$ for the respective cases.

---

*Anpassungen:*

$K\delta_e = -\tfrac{1}{2}K\chi_0/\gamma_0 - \Delta\eta_0/\gamma_0$

$K\delta_R = +s_0 \sin 2\theta_B/|\gamma_g|$

$K\delta_r = -(\Delta\eta_0/\gamma_0' + \Delta\eta_g/\gamma_g')$,

$K\delta_{a,0} = -\tfrac{1}{2}K\chi_0/\gamma_0 - \Delta\eta_0/\gamma_0$

$K\delta_{a,g} = -\tfrac{1}{2}K\chi_0/\gamma_g' - \Delta\eta_g/\gamma_g$

(Type II of Laue-Bragg case: $\gamma_0$ and $\gamma_g'$
must be replaced by $\gamma_0''$ and $\gamma_g''$, respectively)

*Resonanzfehler:*

$\Delta\eta_0 = \tfrac{1}{2}[u \pm \sqrt{u^2 + U^2}](\gamma_0/\gamma_g)$

$\Delta\eta_g = \tfrac{1}{2}[-u \pm \sqrt{u^2 + U^2}]$

Distances:

$t_e = [(\mathbf{r} - \mathbf{r}_e) \cdot \mathbf{n}_e]$

$H = [(\mathbf{r}_e - \mathbf{r}) \cdot \mathbf{n}_e]$

$t_a = [(\mathbf{r}_a - \mathbf{r}) \cdot \mathbf{n}_a]$

$T_a = [(\mathbf{r} - \mathbf{r}_a) \cdot \mathbf{n}_a]$

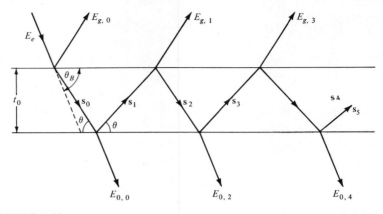

**FIGURE 4-10**

Transmission and reflection of waves in a parallel-sided crystal.

Ewald's solution: $|E_0|^2 = |\Sigma E_{0,2m}|^2$; $|E_g|^2 = |E_{g,0} + \Sigma E_{g,2m+1}|^2$

Darwin's solution: $|E_0|^2 = |E_{0,0}|^2$; $|E_g|^2 = |E_{g,0}|^2$

Real solution: $|E_0|^2 = \Sigma |E_{0,2m}|^2$; $|E_g|^2 = |E_{g,0}|^2 + \Sigma |E_{g,2m+1}|^2$

where the phases[1] are defined by

$$\varphi_e = (\mathbf{K}_e - \mathbf{k}'_0) \cdot \mathbf{r}_e \qquad (4\text{-}114a)$$

$$\varphi_a = (\mathbf{K}_e - \mathbf{k}'_0) \cdot \mathbf{r}_a \qquad (4\text{-}114b)$$

$$\phi_e = (\mathbf{k}''_0 - \mathbf{k}'_0) \cdot \mathbf{r}_e \qquad (4\text{-}114c)$$

$$\phi_a = (\mathbf{k}''_0 - \mathbf{k}'_0) \cdot \mathbf{r}_a \qquad (4\text{-}114d)$$

$$\psi_e = (\mathbf{K}_e + 2\pi\mathbf{g} - \mathbf{K}_g) \cdot \mathbf{r}_e \qquad (4\text{-}114e)$$

The intensity distribution as a function of the deviation parameter $w$ or the glancing angle $\Theta$ is called *Ewald's rocking curve*. The details will be described in Section D 6b and c.

The same problem can be treated by the general viewpoint adopted in this article. The reflected and transmitted waves are created at the front surface in a way described in the preceding solution b. Afterward, as shown in Fig. 4-10, the crystal and vacuum waves are successively created at the rear and front surfaces in a manner described in the preceding solution d. The Bragg-reflected wave, for example, is the total sum of the individual G waves in vacuum created at the front surface. The situation is analogous to the Lummer-Gehrcke interferometer, and may be called (Bragg)$^\infty$ cases according to the presently developed terminology.

---

[1] They include absorption terms as well, if necessary.

From the tangential continuity of the wave vectors, it is easily seen that all the Bloch waves created at the front surface have the same wave vectors $\mathbf{k}_0'$ and $\mathbf{k}_g'$, and the Bloch waves reflected at the rear surface have the same wave vectors $\mathbf{k}_0''$ and $\mathbf{k}_g''$. In the same way, the vacuum waves created by the front surface have the wave vector $\mathbf{K}_g$, and the transmitted waves through the rear surface have the wave vector $\mathbf{K}_e$. Now, denote the amplitudes of the individual crystal waves by $(d_{0,2m}; d_{g,2m})$ and $(d_{0,2m+1}; d_{g,2m+1})$, where the subscripts $2m$ and $2m+1$ mean the number of reflection. Similarly, the amplitudes of the vacuum waves connected with $d_{0,2m}$ and $d_{g,2m+1}$ are denoted by $E_{0,2m}$ and $E_{g,2m+1}$, respectively.

With the use of this notation, the results (4-82) of the preceding solution $b$ can be written in the form

$$d_{0,0} = E_e \exp i\varphi_e \qquad (4\text{-}115a)$$

$$d_{g,0} = c' E_e \exp i\varphi_e \qquad (4\text{-}115b)$$

$$E_{g,0} = c' E_e \exp i\psi_e \qquad (4\text{-}115c)$$

Next, one obtains the recurrence relations among the amplitudes by consulting equations (4-103) and (4-104). At the rear surface, bearing in mind that $p = 0, n = g$; $\mathbf{k}_g^{(j)} = \mathbf{k}_g'$, $\mathbf{k}_{g,r}^{(j)} = \mathbf{k}_g''$, $\mathbf{K}_0^{(j)} = \mathbf{K}_e$; $c_0^{(j)} = c'$, $c_{0,r}^{(j)} = c''$; $d_g^{(j)} = d_{g,2m}$, $d_{g,r}^{(j)} = d_{g,2m+1}$; and $E_0^{(j)} = E_{0,2m}$, one gets

$$d_{g,2m+1} = -d_{g,2m} \exp -i\phi_a \qquad (4\text{-}116a)$$

$$d_{0,2m+1} = -\frac{c'}{c''} d_{0,2m} \exp -i\phi_a \qquad (4\text{-}116b)$$

$$E_{0,2m} = \left(1 - \frac{c'}{c''}\right) d_{0,2m} \exp -i\varphi_a \qquad (m \geq 0) \quad (4\text{-}116c)$$

Similarly, at the front surface $p = g, n = 0$; $\mathbf{k}_g^{(j)} = \mathbf{k}_g''$, $\mathbf{k}_{g,r}^{(j)} = \mathbf{k}_g'$, $\mathbf{K}_g^{(j)} = \mathbf{K}_g$; $c_g^{(j)} = 1/c''$, $c_{g,r}^{(j)} = 1/c'$; $d_g^{(j)} = d_{g,2m+1}$, $d_{g,r}^{(j)} = d_{g,2m+2}$; $E_g^{(j)} = E_{g,2m+1}$; $\mathbf{r}_a = \mathbf{r}_e$, and one obtains the relations

$$d_{0,2m+2} = -d_{0,2m+1} \exp i\phi_e \qquad (4\text{-}117a)$$

$$d_{g,2m+2} = -\frac{c'}{c''} d_{g,2m+1} \exp i\phi_e \qquad (4\text{-}117b)$$

$$E_{g,2m+1} = \left(1 - \frac{c'}{c''}\right) d_{g,2m+1} \exp i(\phi_e - \varphi_e + \psi_e) \qquad (4\text{-}117c)$$

Combining the relations (4-115) to (4-117), one obtains

$$d_{0,2m} = \left[\frac{c'}{c''} \exp i(\phi_e - \phi_a)\right]^m E_e \exp i\varphi_e \qquad (4\text{-}118a)$$

In addition, from (4-116a) and $d_{g,2m} = c'd_{0,2m}$, one gets

$$d_{g,2m+1} = -c' \left[ \frac{c'}{c''} \exp i(\phi_e - \phi_a) \right]^m E_e \exp i(\varphi_e - \phi_a) \quad (4\text{-}118b)$$

Similarly, $d_{g,2m}$ and $d_{0,2m+1}$ are easily written down, if necessary.

From equations (4-116c), (4-117c), and (4-118), the individual vacuum waves are given as follows:

$$E_{0,2m} = \left( 1 - \frac{c'}{c''} \right) \left[ \frac{c'}{c''} \exp i(\phi_e - \phi_a) \right]^m E_e \exp i(\varphi_e - \phi_a) \quad (4\text{-}119a)$$

$$E_{g,2m+1} = (c' - c'') \left[ \frac{c'}{c''} \exp i(\phi_e - \phi_a) \right]^{m+1} E_e \exp i\psi_e \quad (m \geq 0) \quad (4\text{-}119b)$$

The total vacuum waves have the forms

$$E_0 = \sum_{m=0}^{\infty} E_{0,2m} \quad (4\text{-}120a)$$

$$E_g = E_{g,0} + \sum_{m=0}^{\infty} E_{g,2m+1} \quad (4\text{-}120b)$$

where $E_{g,0}$ is given by (4-115c). The summation is merely a geometrical series with the ratio $(c'/c'') \exp i(\phi_e - \phi_a)$. Thus, it is easy to see that $E_0$ and $E_g$ are identical to equations (4-113a and b), respectively. The mathematical complexity of the present method is greater than in the conventional method based on the boundary conditions (4-111) and (4-112). Nevertheless, recalling that the coherent superposition assumed in (4-120) is not adequate for the experimental conditions encountered in practice, the arguments presented here are worthwhile when considering wave fields for the general case in which $E_{0,2m}$ or $E_{g,2m+1}$ do not overlap in space. Also, it is instructive in understanding the optical structure of Darwin and Ewald rocking curves which are further discussed in Section D 6c.

The theoretical developments described in this section follow the analyses of Wagner (1956), Saka, Katagawa, and Kato (1972a and b; 1973), and Kato, Katagawa, and Saka (1971). The recent paper by Fingerland (1971) can be consulted for a discussion of the Bragg case.

## D.   Experimental aspects

The plane-wave theory is fundamental to the dynamical theory of crystal diffraction. In this section, some of the important theoretical results are considered in detail in connection with experimental results. In x-ray cases, however, it is rather difficult to obtain the ideal plane wave required for a critical experiment. The angular divergence of the incident wave must be much less than the angular range $\Delta\Theta$ defined by equation (4-76) or (4-87). The natural width of x-ray emission spectra presents another serious difficulty. Fortunately, some of the theoretical results can be verified by experiments using electron diffraction. It is less difficult to obtain a monodirectional and monochromatic wave by using a point source and a magnetic lens, and the requirement for a plane wave becomes less stringent because of the largeness of $\Delta\Theta$. For this reason, in the following, experimental results from electron diffraction and microscopy will be utilized even though the equations refer to x-ray cases. In fact, comparisons of x-ray, electron, and neutron diffraction are very instructive in understanding diffraction by crystals.

**1.** *Double refraction and amplitude region*    The essence of dynamical phenomena lies in the fact that the crystal wave is represented by a set of Bloch waves. In two-beam cases, two Bloch waves are excited in the crystal. The phenomena may be called double refraction, although the underlying physical principles are different from those of the ordinary double refraction in crystal optics. A direct demonstration is given by the high-resolution electron diffraction experiment for the Laue-Laue case using wedge-shaped crystals.

Figure 4-11 illustrates the experimental conditions for MgO smoke particles which typically have a cubic habit bounded by ((100)) planes. It is assumed that the incident wave is nearly parallel to the face diagonal of a (100) plane and penetrates through two sets of entrance and exit surfaces, $(a, b)$ and $(a', b')$. Each constitutes a wedge-shaped crystal. On each entrance surface, the incident wave splits into two Bloch waves denoted by (1) and (2). The Bloch wave $(j)$ is composed of the direct wave and the Bragg-reflected wave having the wave vectors $\mathbf{k}_0^{(j)}$ and $\mathbf{k}_g^{(j)}$, respectively. They then emerge from the exit surface. Their wave vectors $\mathbf{K}_0^{(1)}$, $\mathbf{K}_0^{(2)}$, $\mathbf{K}_g^{(1)}$, and $\mathbf{K}_g^{(2)}$ are given by (4-105$c$ and $d$). Thus, in the diffraction experiment, a total of two pairs of spots is expected on the photographic plate. In Fig. 4-11, $O$ waves are not indicated because, in practice, they are masked by the undeflected direct wave. Experimentally, the splitting angles for $G$ waves are about $10^{-2}$ of the Bragg angle and are illustrated in Fig. 4-12. This was first observed with electrons by Cowley and Rees (1946, 1947), and a theoretical explanation was given by Sturkey (1948) and Kato (1949), independently, although a theoretical prediction for x rays had been given earlier by Laue (1940).[1] In addition, although the effect of absorption has been neglected in this section, it is expected that the split spots corresponding to waves (1) are weaker than the spots corresponding to waves (2), as a result of Borrmann absorption (Kato, 1952$b$). This was actually observed by Honjo and Mihama (1954). In this connection it should be noted that the roles of branches (1) and (2) are interchanged in electron cases because $J_m(V_g V_{-g})$ is usually positive, whereas $J_m(\chi_g \chi_{-g})$ is usually negative.

In order to visualize the wave vectors of the $G$ waves, it is very useful to consider

---

[1] Full discussions are given by Kato (1952$b$) and Molière and Niehrs (1954).

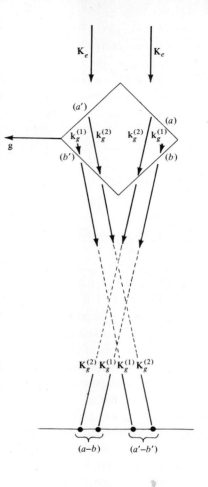

**FIGURE 4-11**
Double refraction in electron diffraction.
(Since the inner potential $V$ is positive,
the refracted rays must cross.)

an *amplitude region* combined with the Ewald construction also in the dynamical theory. To do this, it is sufficient to translate the wave vector $\mathbf{K}_g^{(j)}$ given by (4-105$d$) in such a way that the dispersion point $A^{(j)}$ coincides with the dispersion point $E$ of the wave vector $\mathbf{K}_e$ (Fig. 4-13). Thus, the terminal point of the vector $\mathbf{K}_g^{(j)}$ is translated to a point $Q^{(j)}$. The vector $\overrightarrow{GQ^{(j)}} = \overrightarrow{A_g^{(j)}E} = \mathbf{K}_g^{(j)} - (\mathbf{K}_e + 2\pi\mathbf{g})$, then, is given by

$$\overrightarrow{GQ^{(j)}} = K(\delta_a^{(j)}\mathbf{n}_a - \delta_e^{(j)}\mathbf{n}_e) \quad (4\text{-}121a)^1$$

where $\delta_e^{(j)}$ and $\delta_a^{(j)}$ are given by equations (4-56$c$) and (4-108$b$), respectively. By the

[1] The suffix $g$ is omitted in $\delta_{ag}^{(j)}$.

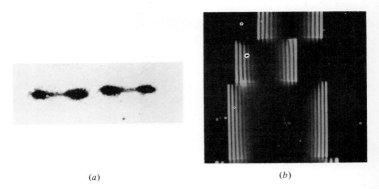

(a) (b)

FIGURE 4-12
Magnesium oxide. (a) Double diffracted spots; (b) *Pendellösung* fringes (electron micrograph).

use of *Resonanzfehler* [(4-60a) and (4-60b)], equation (4-121a) can be rewritten

$$\overrightarrow{GQ^{(j)}} = \tfrac{1}{2}K\chi_0{}^r \left( \frac{\mathbf{n}_e}{\gamma_0} - \frac{\mathbf{n}_a}{\gamma_g'} \right) + \frac{\Delta\eta_0{}^{(j)}}{\gamma_0}\,\mathbf{n}_e - \frac{\Delta\eta_g{}^{(j)}}{\gamma_g'}\,\mathbf{n}_a \qquad (4\text{-}121b)$$

By changing the direction of $\mathbf{K}_e$, that is, the deviation parameter $w$ in $\Delta\eta_0{}^{(j)}$ and $\Delta\eta_g{}^{(j)}$, the point $Q^{(j)}$ moves along a certain trajectory, as discussed further below. In any case, since the magnitude of $\mathbf{K}_g{}^{(j)}$ is $K$, $Q^{(j)}$ automatically falls on the Ewald sphere. If the corresponding amplitude is assigned to each point $Q^{(j)}$, it turns out that the trajectory of $Q^{(j)}$ plays the role of the *amplitude region* as previously defined for the kinematical theory. Because of its size, the Ewald sphere, in practice, is a plane perpendicular to $\overline{\mathbf{K}}_g$.

The first term on the right in equation (4-121b) is denoted by the vector $GG'$ in Fig. 4-13. This term corresponds to the deviation of the wave vectors caused by ordinary refraction at the crystal surfaces. The splitting of the point $G'$ into $Q^{(1)}$ and $Q^{(2)}$, described by the remainder in (4-121b), can be interpreted as a kind of double refraction. The point $Q^{(1)}$ and $Q^{(2)}$ must lie on the plane determined by the lines $\mathbf{n}_e$ and $\mathbf{n}_a$ passing through the point $G'$. Since $(\Delta\eta_0{}^{(j)}/\gamma_0)(\Delta\eta_g{}^{(j)}/\gamma_g')$ is a constant, their trajectories are two branches of the hyperbola whose asymptotes are the lines $\mathbf{n}_e$ and $\mathbf{n}_a$, as shown in Fig. 4-13.

When the last two vectors of equations (4-121), i.e., $\overrightarrow{G'Q^{(j)}}$, are multiplied by $\hat{\mathbf{K}}_g$, the distance from $G'$ to the Ewald sphere is $w$. If $|w|$ is sufficiently large compared to $|W|$, one of $\Delta\eta_0{}^{(j)}$ and $\Delta\eta_g{}^{(j)}$ tends to zero so that the amplitude region is essentially defined by the asymptotic lines $\mathbf{n}_e$ and $\mathbf{n}_a$. This result is in accordance with the amplitude region in the kinematical theory for a crystal having a large lateral size (Chap. 3,

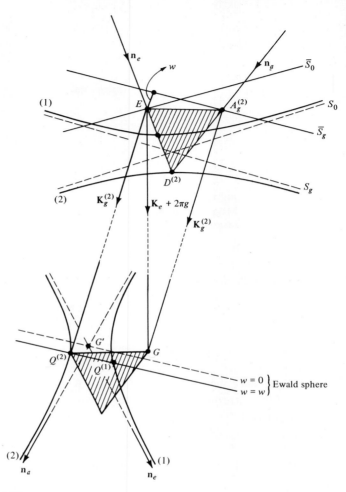

FIGURE 4-13

Relation of dispersion surface to Ewald construction. The stippled triangles are similar and the construction for the dispersion point $D^{(2)}$ only is indicated; an analogous construction can be devised for branch (1).

Section II C 1). Indeed, the condition $|w| \gg |W|$ is one of the conditions required for the validity of the kinematical approximation. In view of this, it should not seem at all strange that splitting of the diffraction spot occurs in the dynamical theory. In the kinematical theory the splitting is interpreted as a consequence of the crystal shape. In the dynamical theory it is interpreted as a refraction effect. In fact, however, this refraction is one of the effects of the shape of the crystal. The correspondence between the kinematical and dynamical amplitude regions will be seen more clearly in Chap. 5, Section I A, where the finiteness of the lateral size of the crystal is examined further.

For later purposes, consider the distance $\Delta Q$ between $Q^{(1)}$ and $Q^{(2)}$. From (4-121a) and the expressions (4-61a) and (4-108b) for $\delta_e^{(j)}$ and $\delta_a^{(j)}$, one can see that

$$\overrightarrow{Q^{(2)}Q^{(1)}} = K[(\delta_a^{(1)} - \delta_a^{(2)})\mathbf{n}_a - (\delta_e^{(1)} - \delta_e^{(2)})\mathbf{n}_e]$$

$$= \frac{(w^2 + W^2)^{1/2}[\hat{\mathbf{K}}_g \times (\mathbf{n}_e \times \mathbf{n}_a)]}{\gamma_g \gamma_g'} \qquad (4\text{-}122)$$

Since the magnitude of the vector in square brackets is $\gamma \sin \varphi$, $\varphi$ being the wedge angle of the crystal and $\gamma$ the cosine of the angle between $\overline{\mathbf{K}}_g$ and the plane of $\mathbf{n}_e$ and $\mathbf{n}_a$ [cf equations (3-113a and b)],

$$\Delta Q = (w^2 + W^2)^{1/2} \frac{\gamma \sin \varphi}{\gamma_g \gamma_g'} \qquad (4\text{-}123)$$

The amplitudes associated with points $Q^{(1)}$ and $Q^{(2)}$ are proportional to $C_g^{(j)}$ as shown by (4-106b). From equations (4-72c and d), they are $\pm\frac{1}{2}KC\chi_g/(w^2 + W^2)^{1/2}$. When $|w| \gg |W|$, again, the dynamical result tends to the kinematical one. The peak value of the amplitude occurs when $w = 0$, where the splitting distance is shortest. Contrary to the kinematical theory, the intensity of each spot is independent of the crystal size, and the splitting of the diffraction spots is expected even when the Bragg condition is satisfied exactly.

**2.  *Pendellösung fringes***    The fringes discussed in this section are those produced by the interference between the two Bloch waves excited at the entrance surface in the Laue case. The fields of the $O$ and $G$ waves are considered separately. The fringes and the associated phenomena were predicted by Ewald as early as 1916 in his general theory of crystal optics. Because he called the wave field in the crystal *Pendellösung* (pendulum solution) of the Maxwell equations, thinking of a mechanical analogy, the fringes are called *Pendellösung fringes*. In the discussion in this section, absorption is neglected. A detailed consideration of absorption is postponed to Sections D 5 and 6.

In considering the origin of these fringes, one first notices that the two Bloch waves have slightly different wave vectors, $\mathbf{k}_0^{(1)}$ and $\mathbf{k}_0^{(2)}$ for $O$ waves, and $\mathbf{k}_g^{(1)}$ and $\mathbf{k}_g^{(2)}$ for $G$ waves. The differences are the same and have the direction $\mathbf{n}_e$. Thus, it is expected that the interference fringes have the spacing

$$\Lambda_p = \frac{2\pi}{[(\mathbf{k}_0^{(1)} - \mathbf{k}_0^{(2)}) \cdot \mathbf{n}_e]} \qquad (4\text{-}124a)$$

The fringes are parallel to the entrance surface. $\Lambda_p$ can be written in the explicit form

$$\Lambda_p = \frac{2\pi\gamma_g}{(w^2 + W^2)^{1/2}} \qquad (4\text{-}124b)$$

$$\Lambda_p = \frac{\lambda\sqrt{\gamma_0\gamma_g}}{C|\chi_g|[1 + (w/W)^2]^{1/2}} \qquad (4\text{-}124c)$$

The crystal wave fields then are given by (4-74a and b) for $O$ and $G$ waves, respectively. By the use of expressions (4-72) and (4-61) for $C_0^{(j)}$, $C_g^{(j)}$, and $\delta_e^{(j)}$, and consulting the descriptions in Appendix B, one obtains the intensity fields

$$i_0 = \frac{1}{2}\left(1 + \frac{w^2}{w^2 + W^2}\right) + \frac{1}{2}\left(\frac{W^2}{w^2 + W^2}\right)\cos\left(\frac{2\pi t_e}{\Lambda_p}\right) \qquad (4\text{-}125a)$$

$$i_g = \frac{1}{2}\frac{\gamma_0}{\gamma_g}\frac{W^2}{w^2 + W^2}\left(1 - \cos\frac{2\pi t_e}{\Lambda_p}\right) \qquad (4\text{-}125b)$$

where $t_e$ is the depth of the observation point defined by (4-75). Remembering the change in width of the wave front of $G$ waves in the crystal, it is readily seen that energy conservation

$$i_0 + \left(\frac{\gamma_g}{\gamma_0}\right)i_g = 1 \qquad (4\text{-}126)$$

is satisfied.

Obviously, it is not possible to insert an instrument to detect the intensity fields inside the crystal. What one actually observes is the intensity field outside the crystal. Here, consider the Laue-Laue case. If the crystal is wedge shaped, each of the $O$ and $G$ wave fields in vacuum is composed of two plane waves having slightly different wave vectors. They are superposed on each other unless the observation point is far from the crystal. The wave vectors $\mathbf{K}_0^{(1)}$ and $\mathbf{K}_0^{(2)}$ and $\mathbf{K}_g^{(1)}$ and $\mathbf{K}_g^{(2)}$ are given by equations (4-105c and d). For $G$ waves, the difference of the wave vectors and the magnitude have been given already by (4-122) and (4-123), respectively. The fringe spacing, therefore, can be obtained directly:

$$\Lambda_g = \frac{2\pi(\gamma_g\gamma_g'/\gamma\,\sin\,\varphi)}{(w^2 + W^2)^{1/2}} \qquad (4\text{-}127a)$$

Since the difference vector is perpendicular to $\overline{\mathbf{K}}_g$ and $(\mathbf{n}_e \times \mathbf{n}_a)$, the fringe plane is determined by these vectors. Incidentally, the vector $(\mathbf{n}_e \times \mathbf{n}_a)$ has the direction of the edge of the wedge-shaped crystal, as can be seen in Fig. 4-14.

Similarly to the $G$ waves, the difference vector $(\mathbf{K}_0^{(1)} - \mathbf{K}_0^{(2)})$ can be obtained from (4-105c). The fringe plane is determined by $\overline{\mathbf{K}}_0$ and the crystal edge, and the spacing is given by

$$\Lambda_0 = \frac{2\pi(\gamma_g\gamma_0'/\overline{\gamma}\,\sin\,\varphi)}{(w^2 + W^2)^{1/2}} \qquad (4\text{-}127b)$$

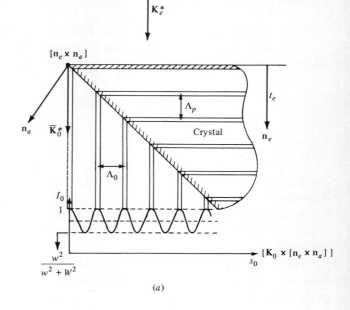

$(a)$

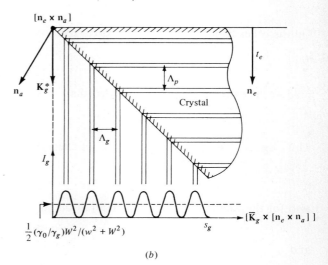

$(b)$

**FIGURE 4-14**
*Pendellösung* fringes in the plane-wave theory. $(a)$ *O* wave. $\mathbf{K}_e^*$ and $\mathbf{K}_0^*$ are projections of $\mathbf{K}_e$ and $\overline{\mathbf{K}}_0$ on the plane of the drawing determined by $\mathbf{n}_e$ and $\mathbf{n}_a$; $(b)$ *G* wave. $\mathbf{K}_e^*$ and $\mathbf{K}_g^*$ are projections of $\mathbf{K}_e$ and $\mathbf{K}_g$ on the plane of the drawing.

269

where $\bar{\gamma}$ is the cosine of the angle between $\bar{\mathbf{K}}_g$ and the plane of $\mathbf{n}_e$ and $\mathbf{n}_a$.

According to Appendix 4B and the expressions (4-105a and b) for the vacuum-wave fields, the intensity fields are given by

$$I_0 = \frac{1}{2}\left(1 + \frac{w^2}{w^2 + W^2}\right) + \frac{1}{2}\left(\frac{W^2}{w^2 + W^2}\right)\cos\frac{2\pi s_0}{\Lambda_0} \qquad (4\text{-}128a)$$

and

$$I_g = \frac{1}{2}\frac{\gamma_0}{\gamma_g}\frac{w^2}{w^2 + W^2}\left(1 - \cos\frac{2\pi s_g}{\Lambda_g}\right) \qquad (4\text{-}128b)$$

where $s_0$ and $s_g$ are the coordinates normal to the fringe planes for $O$ and $G$ waves, respectively, with the origin at a point $\mathbf{r}_s$ on the crystal edge. In deriving these, $\mathbf{r}_a$ and $\mathbf{r}_e$ in (4-105a and b) are replaced by $\mathbf{r}_s$, since the expressions in (4-105) are independent of the position vectors $\mathbf{r}_e$ and $\mathbf{r}_a$ provided that they lie on the entrance and the exit surfaces, respectively.

From equations (4-124b) and (4-127a and b), one obtains relations for the *Pendellösung* fringe spacings which are analogous to Snell's law of refraction in terms of the wavelength

$$\frac{\Lambda_p}{\sin\varphi} = \Lambda_0\frac{\bar{\gamma}}{\gamma_0'} \qquad (4\text{-}129a)$$

$$\frac{\Lambda_p}{\sin\varphi} = \Lambda_g\frac{\gamma}{\gamma_g'} \qquad (4\text{-}129b)$$

On the other hand, since the geometrical relations

$$\frac{t_e}{\sin\varphi} = s_0\frac{\bar{\gamma}}{\gamma_0'} = s_g\frac{\gamma}{\gamma_g'}$$

hold at the exit surface, it is concluded that

$$\frac{t_e}{\Lambda_p} = \frac{s_0}{\Lambda_0} = \frac{s_g}{\Lambda_g} \qquad (4\text{-}129c)$$

Thus, again, one obtains conservation of energy

$$I_0 + \frac{\gamma_g}{\gamma_0}I_g = 1 \qquad (4\text{-}130)$$

also for the vacuum-wave fields. In addition, at the exit surface

$$I_0 = i_0 \qquad (4\text{-}131a)$$

and

$$I_g = i_g \qquad (4\text{-}131b)$$

The intensity fields of $O$ and $G$ waves in vacuum are the projections of the intensity distributions on the exit surface in the directions of $\bar{\mathbf{K}}_0$ and $\bar{\mathbf{K}}_g$, respectively. Since the intensity contours correspond to the portions having equal thickness, sometimes the fringes are called *equal-thickness fringes*. Similarly, when a parallel-sided crystal is slightly bent, the crystal parts having the same orientation give equal-intensity contours called *fringes of equal inclination*.

In electron microscopy, the intensity fields $I_0$ and $I_g$ were actually observed by Heidenreich (1942) and Kinder (1943) independently. Once the diffraction spots are formed in the intermediate stage of an electron microscope, $O$ and $G$ waves can be selected by a suitable hole slit. Subsequently, it is possible to produce the crystal image on the photographic plate from one of the separated waves by means of a lens system. The images produced by using $O$ and $G$ waves are called the *bright-field* image and the *dark-field* image, respectively. A micrograph corresponding to the doubly refracted spots (Fig. 4-12a) is illustrated in Fig. 4-12b. A theoretical explanation of this phenomenon was given by Kossel (1943) and Heidenreich and Sturkey (1945) based on Bethe's dynamical theory for a parallel-sided crystal (Bethe, 1928). The theory in the present form was derived by Kato (1953) and Niehrs (1954).

In x-ray cases, it is not easy to observe fringes of this type because of the lack of ideal plane waves; however, Hart and Milne (1968) recently succeeded in this to a certain extent. The *Pendellösung* oscillations observed in the x-ray case will be explained in Section D 6.

**3. *Double refraction of the rays*** In x-ray cases, direct observation of the splitting of the vacuum-wave vectors is rather difficult, as explained in Section 1 above. The splitting of the ray beam in the crystal, however, can be demonstrated by experiments in the Laue-Laue cases with the use of a sufficiently narrow beam. As discussed in Section 6 above, a narrow beam is a wave packet from the wave-optical view. The narrower the wave front, the wider the region of the dispersion surface that is effective. Thus, various plane waves are included in the packet. For this reason, the phenomenon cannot be described by the plane-wave theory in its strict sense. Nevertheless, the geometrical character of the rays can be described by the Poynting vector associated with a plane wave.

The fundamental concept of ray optics is that the energy flow is normal to the dispersion surface at the corresponding dispersion point. Since two dispersion points become effective in a crystal for a single incident plane wave, the two rays are excited inside the crystal. In general, they have different directions as shown in Fig. 4-15. The analytical expression of the Poynting vector is given by (4-29). Here, the amplitudes of the wave fields (4-73a and b) must be used for $d_0$ and $d_g$. The result can be represented as follows:

$$\mathbf{s} = [\hat{\bar{\mathbf{K}}}_0(C_0 C_0^*) + \hat{\bar{\mathbf{K}}}_g(C_g C_g^*)] \exp(-\mu_B l) S_e \quad (4\text{-}132a)$$

where $S_e$ is the magnitude of the Poynting vector of the incident wave

$$\mathbf{S}_e = \frac{c}{8\pi} (E_e E_e^*) \hat{\mathbf{K}}_e \quad (4\text{-}133)$$

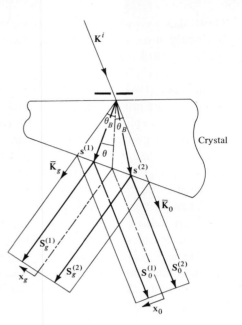

**FIGURE 4-15**
Double refraction in x-ray case. The ray beam is assumed to be infinitely narrow and monochromatic and monodirectional.

The absorption factor $\exp\left(-\mu_B l\right)$ is derived by using (4-45) with the expression $(\mathbf{r} - \mathbf{r}_e) = l\mathbf{v}$, $\mathbf{v}$ being the direction of the ray concerned. The details of this factor will be considered in Section 5 below.

The power[1] transmitted at the entrance surface from the incident ray to one of the doubly refracted rays is given by

$$P_e \equiv \frac{\mathbf{s} \cdot \mathbf{n}_e}{\mathbf{S}_e \cdot \mathbf{n}_e}$$

$$= [(C_0\,C_0^*) + \frac{\gamma_g}{\gamma_0}\,(C_g\,C_g^*)] \qquad (4\text{-}134a)$$

Here, $(\hat{\mathbf{K}}_e \cdot \mathbf{n}_e)$ is approximated by $\gamma_0$. The absorption factor must be omitted, since $l = 0$ at the entrance surface. By the use of (4-72) for $C_0$ and $C_g$, the Poynting vector $\mathbf{s}$ and the transmitted power $P_e$ can be represented in terms of the deviation parameter $w$. Alternatively, introducing a new parameter

$$\bar{w} = \frac{w}{(w^2 + W^2)^{1/2}} \qquad (4\text{-}135)$$

one obtains

$$\mathbf{s} = \frac{1}{4}\left[(1 \mp \bar{w})^2 \hat{\mathbf{K}}_0 + \frac{\gamma_0}{\gamma_g}(1 - \bar{w}^2)\hat{\mathbf{K}}_g\right]\exp\left(-\mu_B l\right)S_e \qquad (4\text{-}132b)$$

[1] The term "power" is often used in place of "power ratio."

and

$$P_e = \tfrac{1}{2}(1 \mp \bar{w}) \quad (4\text{-}134b)$$

where the upper and lower signs specify the branches (1) and (2), respectively.

It is easily seen that the direction of $\mathbf{s}^{(1)}$ points from the direction $\bar{\mathbf{K}}_0$ to that of $\bar{\mathbf{K}}_g$ as the parameter $\bar{w}$ changes from $-1$ to $+1$, and the direction of $\mathbf{s}^{(2)}$ points in the reverse direction. The nearer the ray to the direction $\bar{\mathbf{K}}_0$, the larger the energy transmitted by the ray in the crystal. When the parameter $w$ or $\bar{w}$ has the same magnitude but opposite sign for the rays (1) and (2), the corresponding vectors $\mathbf{s}^{(1)}$ and $\mathbf{s}^{(2)}$ have the same direction. Then, the transmitted powers are identical. Such rays are called conjugate rays, hereafter.

At the exit surface, each ray splits into $O$ and $G$ waves, and they emerge from the crystal. This situation is shown in Fig. 4-15. The Poynting vectors of $O$ and $G$ waves can be obtained from the wave fields $E_0^{(j)}(\mathbf{r})$ and $E_g^{(j)}(\mathbf{r})$ given by equations (4-105a and b), respectively,

$$\mathbf{S}_0 = (C_0 C_0^*) \exp(-\mu_B l) S_e \hat{\bar{\mathbf{K}}}_0 \quad (4\text{-}136a)$$

$$\mathbf{S}_g = (C_g C_g^*) \exp(-\mu_B l) S_e \hat{\bar{\mathbf{K}}}_g \quad (4\text{-}136b)$$

where $l$ is the length of the ray in the crystal. The transmitted powers of the individual rays at the exit surface are given by

$$P_{a,0} \equiv \frac{\mathbf{S}_0 \cdot \mathbf{n}_a}{\mathbf{s} \cdot \mathbf{n}_a}$$

$$= \frac{C_0 C_0^*}{[(C_0 C_0^*) + (\gamma_g'/\gamma_0')(C_g C_g^*)]} \quad (4\text{-}137a)$$

$$P_{a,g} \equiv \frac{\mathbf{S}_g \cdot \mathbf{n}_a}{\mathbf{s} \cdot \mathbf{n}_a}$$

$$= \frac{(\gamma_g'/\gamma_0')(C_g C_g^*)}{[(C_0 C_0^*) + (\gamma_g'/\gamma_0')(C_g C_g^*)]} \quad (4\text{-}137b)$$

One can easily obtain the expressions for $\mathbf{S}_{0,g}$ and $P_{a,0}$ and $P_{a,g}$ in terms of either $w$ or $\bar{w}$.

So far, the Poynting vector and the transmitted powers are represented by a parameter defined in reciprocal space. Since, however, the concept of rays is referred to real space, it is more convenient to represent them in terms of a parameter defined in real space. For this reason, introduce the parameter

$$p = \frac{\tan \theta}{\tan \theta_B} \quad (4\text{-}138)$$

where $\theta$ is the angle formed by $\mathbf{s}$ with the net plane. Taking the ratio $(\hat{\mathbf{x}} \cdot \mathbf{s})/(\hat{\mathbf{z}} \cdot \mathbf{s})$

with the expression (4-132a) for **s** ($\hat{x}$ and $\hat{z}$ are unit vectors normal and parallel to the net plane), one gets the relation

$$p = \frac{(C_g C_g^*) - (C_0 C_0^*)}{(C_g C_g^*) + (C_0 C_0^*)} \quad (4\text{-}139a)$$

The explicit relation between $p$ and $\bar{w}$ is given by

$$p = \frac{(\gamma_0/\gamma_g)(1 \pm \bar{w}) - (1 \mp \bar{w})}{(\gamma_0/\gamma_g)(1 \pm \bar{w}) + (1 \mp \bar{w})} \quad (4\text{-}139b)$$

By the use of this parameter, the transmitted powers are given by

$$P_e = \frac{(\gamma_0/\gamma_g)(1 - p)}{(1 + p) + (\gamma_0/\gamma_g)(1 - p)} \quad (4\text{-}140a)$$

$$P_{a,0} = \frac{1 - p}{(1 - p) + (\gamma_g'/\gamma_0')(1 + p)} \quad (4\text{-}140b)$$

$$P_{a,g} = \frac{(\gamma_g'/\gamma_0')(1 + p)}{(1 - p) + (\gamma_g'/\gamma_0')(1 + p)} \quad (4\text{-}140c)$$

Thus, taking into account the absorption factor, the transmitted powers for the total crystal become

$$P_0 = P_e P_{a,0} \exp - \mu_B l \quad (4\text{-}141a)$$

$$P_g = P_e P_{a,g} \exp - \mu_B l \quad (4\text{-}141b)$$

The split beams were actually observed by Authier (1960a; 1961). By using an ingenious method, he produced (1960b) an incident beam that was sufficiently narrow to detect the splitting and very nearly fits to the requirements for a plane wave. Figure 4-16 illustrates his experimental results.

**4.   *Intensity profiles of O and G beams***   Under the usual experimental conditions, the incident x-ray beams are neither monodirectional nor monochromatic, so that an individual ray associated with a plane wave cannot be detected. The flow of x-ray energy spans the whole angular range between $\bar{K}_0$ and $\bar{K}_g$, and both the transmitted O and G beams have a finite width depending on the crystal thickness and the Bragg angle. Consequently, the examination of the intensity profiles is one of the critical experiments that can be used to test the ray optics based on the dynamical theory.

Since the effective angular range $\Delta\Theta$ of crystal diffraction is extremely small, about $10^{-5}$ in arc, it is very natural to assume that the angular spectrum $I_e(\Theta)$ of the incident beam is constant over this effective range. In this section, the various waves involved in the incident beam are assumed to be optically incoherent and the incident beam width is assumed to be sufficiently narrow, similarly to the assumption in the preceding section.

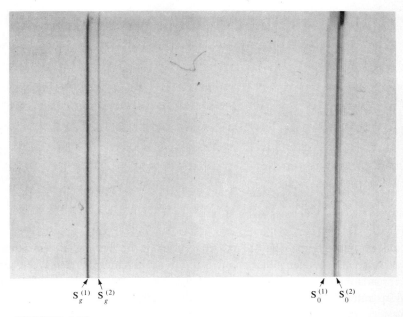

$S_g^{(1)} \quad S_g^{(2)}$ $\qquad\qquad\qquad\qquad S_0^{(1)} \quad S_0^{(2)}$

FIGURE 4-16
Splitting of rays in a crystal (Authier, 1960a and b). The rays $S_g^{(1)}$, $S_g^{(2)}$, $S_0^{(1)}$, and $S_0^{(2)}$ shown in Fig. 4-15 are recorded on a single plate. Multiplet structure of $S_0^{(2)}$ is probably due to an optical diffraction effect of the slit.

In the following, for simplicity, the intensity profiles will be calculated for the symmetrical Laue case of a parallel-sided crystal. In this case ($\gamma_0 = \gamma_g = \gamma_0' = \gamma_g'$), the position parameters $x_0$ and $x_g$ of the $O$ and $G$ beams (Fig. 4-15) are given by $pt_0 \sin \theta_B$, where $p$ is defined by (4-138) and $t_0$ is the crystal thickness. The transmitted powers are simply given by $P_e = P_{a,0} = \frac{1}{2}(1 - p)$ and $P_{a,g} = \frac{1}{2}(1 + p)$. The relation (4-139b) reduces to $p = \bar{w}$ for the (1) wave and $p = -\bar{w}$ for the (2) wave.

Bearing in mind the fact that the energy $I_e \delta\Theta$ of the incident beam falls in the spatial ranges $\delta x_0$ and $\delta x_g$ separately, one obtains

$$I_0(x_0) = I_e[P_0^{(1)}(\bar{w}) + P_0^{(2)}(-\bar{w})] \frac{d\Theta}{dw} \frac{dw}{d\bar{w}} \left|\frac{d\bar{w}}{dp}\right| \frac{dp}{dx_0} \qquad (4\text{-}142a)$$

$$I_g(x_g) = I_e[P_g^{(1)}(\bar{w}) + P_g^{(2)}(-\bar{w})] \frac{d\Theta}{dw} \frac{dw}{d\bar{w}} \left|\frac{d\bar{w}}{dp}\right| \frac{dp}{dx_g} \qquad (4\text{-}142b)$$

Since the relations for $\Theta$, $w$, $\bar{w}$, $p$, and $x_0$ (or $x_g$) already have been given above, one obtains, after straightforward manipulation,

$$\frac{d\Theta}{dw} = (K \sin 2\theta_B)^{-1} \qquad \text{[cf (4-52) and (4-56a)]}$$

$$\frac{dw}{d\bar{w}} = W(1 - p^2)^{-3/2} \qquad \text{[cf (4-135)];} \left|\frac{d\bar{w}}{dp}\right| = 1$$

$$\frac{dp}{dx_0} = \frac{dp}{dx_g} = (t_0 \sin \theta_B)^{-1}$$

From this it follows that

$$I_0(p) = I_e \frac{\Delta\Theta}{2t_0 \sin \theta_B} (1 - p)^{1/2}(1 + p)^{-3/2} \exp -\mu_B l \qquad (4\text{-}143a)$$

$$I_g(p) = I_e \frac{\Delta\Theta}{2t_0 \sin \theta_B} (1 - p^2)^{-1/2} \exp -\mu_B l \qquad (4\text{-}143b)$$

where $\Delta\Theta$ is defined by (4-76). The factor in the first set of parentheses is obvious, because the incident energy in, essentially, an angular range proportional to $\Delta\Theta$ spreads into the beam width $2t_0 \sin \theta_B$. The functional forms of $I_0(p)$ and $I_g(p)$ are illustrated in Figs. 4-17a and b. In this section, only the curves for $\mu_B = 0$ will be discussed.

In the $O$ beam, the waves which do not satisfy the Bragg condition propagate in the direction $\bar{K}_0$, because the dispersion surface $\bar{S}_0$ is assumed to be a plane normal to $\bar{K}_0$. They contribute to the intensity only at $p = -1$. There, therefore, the intensity is singularly increased. More importantly, in the $G$ beam, the intensity increases at the margins. This result seems a little strange. It is known already that the rays propagating near the directions $\bar{K}_0$ and $\bar{K}_g$ are created by the incident waves which deviate considerably from the exact Bragg condition. For this reason, it might be anticipated that the intensity should be decreased at the margins. The proper explanation is as follows: Although the transmitted power, $P_g(\bar{w})$, decreases as $(1 - \bar{w}^2) = (1 - p^2)$, as $p$ approaches to $\pm 1$, most of the crystal rays are concentrated within the angular ranges lying near the margin directions. This situation can be understood easily if one remembers the factor $|dw/dp|$. In fact, the incident beam within the angular range $\pm\Delta\Theta$ contributes to the spatial range $|p| \leq 1/\sqrt{2} \simeq 0.71$. The rest of the rays propagate outside this range. The singularity at the margins is due to the assumption that the asymptotic dispersion surface is replaced by the planes $S_0$ and $S_g$.

The experiments to study the spatial line profiles are called *section experiments*, and the photograph recording the intensity distribution of either the $O$ or $G$ beam is called a *section topograph*.[1] In the section topographs of $O$ and $G$ beams, the margin effects are observed as predicted by the theory, provided that the crystal is not a highly

---

[1] Topographs are further discussed in Section II F 1.

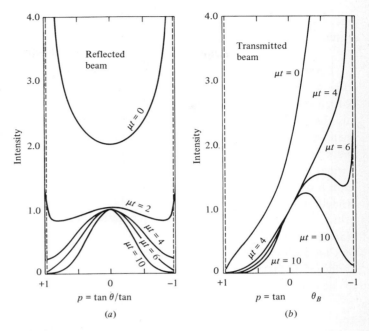

FIGURE 4-17
Spatial intensity profiles in the symmetrical Laue case (Kato, 1960). (a) G beam;
(b) O beam.

absorbing one. The intensity enhancement at the margins of the G beam has been recognized in an early stage of the study of crystal diffraction (Cork, 1932; DuMond and Bollmann, 1936). The phenomenon was observed also in neutron diffraction (Knowles, 1956). Most investigators attributed the enhancement to surface distortions, because they believed that the intensity could increase easily owing to a decrease in the primary or secondary extinction in distorted crystals. However, Elizabeth Wood (Armstrong, 1946) noticed that the margin intensity could never be eliminated by etching the surface. In view of the above, this phenomenon is a consequence of the dynamical diffraction effects and indicates directly the perfectness of the crystal. The important point essential to understanding this phenomenon is the concept of ray propagation, whose theoretical explanation was given by Kato (1960).

One additional observation should be mentioned. The section topographs actually show an intensity modulation superposed on the distribution mentioned above, both in the O and G beams, provided that one uses a very fine slit and a high-resolution photographic plate. This result indicates that there must occur an interference phenomenon whose details will be considered later. Here, it is only noted that observations of this type led to the development of the spherical-wave theory of crystal diffraction.

**5.   Borrmann absorption and Netzebene fringes**   Historically, the anomalous phenomenon which now is called Borrmann absorption had been discovered (Borrmann, 1941; 1950) earlier than the double refraction and *Pendellösung* phenomena. Indeed, this discovery led to the postwar renaissance of interest in dynamical diffraction phenomena. Borrmann absorption occurs when the Bragg condition is nearly satisfied and one of the doubly refracted rays has a smaller absorption coefficient than the normal photoelectric value, whereas the other has the larger coefficient. This phenomenon provides a firm experimental base for the concept of ray optics (e.g., Borrmann, Hildebrandt, and Wagner, 1955; Hildebrandt, 1959a,b). The underlying optical principles were elucidated under the theoretical leadership of von Laue at the Fritz-Haber Institut. Incidentally, the same phenomenon was independently rediscovered by Campbell (1951a,b) who was unaware of the researches of the German group. In addition, Zachariasen had predicted this phenomenon in developing his formulation of the dynamical theory.

The pertinent theory will be developed first on the basis of the plane-wave theory, and next along the line of ray optics. Finally, microscopic physical interpretations will be presented. In the plane-wave theory (Laue, 1949; Zachariasen, 1945, 1952), as described in Section C 2, a Bloch wave is attenuated by an absorption factor exp $2K\delta_e{}^i t_e$, where $\delta_e{}^i$ is given by (4-64) with SI approximations. By the use of (4-65), one obtains

$$2K\delta_e{}^i t_e = -\mu_0 t \mp \left[\frac{(WV + wv)}{(w^2 + W^2)^{1/2}}\right]\left(\frac{t_e}{\gamma_g}\right) \qquad (4\text{-}144)$$

where

$$t = \frac{1}{2}\left(\frac{1}{\gamma_0} + \frac{1}{\gamma_g}\right) t_e \qquad (4\text{-}145)$$

$$\mu_0 = K\chi_0{}^i \qquad (4\text{-}146)$$

Since $t$ is regarded as the mean of the optical paths of $O$ and $G$ waves and $\mu_0$ is the usual linear absorption coefficient, the first term in equation (4-144) is conventionally interpreted as the normal absorption factor. Later, however, a more critical examination of this interpretation will be made. Usually, $VW$ is negative (Appendix 4 A I), so that the wave of branch (1) suffers less attenuation than the wave of branch (2). It should be pointed out, however, that this rule depends upon the sign of $V$, namely $(\chi_g\chi_{-g})^i$, which sometimes is positive for complex crystals. (Notice $W > 0$ always.) The term proportional to $vw$ is asymmetric with respect to the parameter $w$, i.e., the glancing angle $\Theta$. This term disappears in symmetrical Laue cases. As will be seen in the following, this term also must be included in the normal absorption together with the first term. The apparent dependence on the branches is not due to the absorption coefficient itself but the difference between the optical paths of (1) and (2) waves in the crystal. In fact, the conjugate rays ($w^{(1)} = -w^{(2)}$), which propagate in the same direction, suffer the same amount of attenuation.

Under ray-optical considerations (Borrmann and Wagner, 1955; Laue, 1960; Kato, 1964), as mentioned in Section B 7, the attentuation of the ray is represented by $\exp - \mu_B l$, where $\mu_B$ is given by (4-45). Here, for generality, the complex phase changes will be calculated.

$$S_0 = (\mathbf{k}_0 \cdot \mathbf{v})l \quad (4\text{-}147a)$$

$$S_g = (\mathbf{k}_g \cdot \mathbf{v})l \quad (4\text{-}147b)$$

along a ray having the direction $\mathbf{v}$. Twice the imaginary part amounts to $\mu_B l$, which is of primary interest in this section.

Using (4-40) and the definition of the parameter $p$, (4-139a), it follows that

$$p = \frac{\Delta\eta_0 - \Delta\eta_g}{\Delta\eta_0 + \Delta\eta_g} \quad (4\text{-}148)$$

Meanwhile, omit the superscript $r$ indicating the real part. From this result it immediately follows that

$$(\Delta\eta_0 + \Delta\eta_g)^2 = \frac{4\Delta\eta_0\Delta\eta_g}{1 - p^2} \quad (4\text{-}149)$$

and

$$(\Delta\eta_0)^2 + (\Delta\eta_g)^2 = \frac{2\Delta\eta_0\Delta\eta_g(1 + p^2)}{1 - p^2} \quad (4\text{-}150)$$

Next, consider the vector

$$\mathbf{j} = \pm(\Delta\eta_g\hat{\mathbf{K}}_0 + \Delta\eta_0\hat{\mathbf{K}}_g) \quad (4\text{-}151)$$

which obviously has the same direction as $\mathbf{v}$. By the use of (4-150),

$$\mathbf{j}^2 = \frac{2\Delta\eta_0\Delta\eta_g[(1 + p^2) + (1 - p^2)\cos 2\theta_B]}{1 - p^2}$$

$$= \frac{4\Delta\eta_0\Delta\eta_g(\cos\theta_B/\cos\theta)^2}{1 - p^2}$$

From this and equation (4-149), one obtains

$$j = \frac{2(\Delta\eta_0{}^r\Delta\eta_g{}^r)^{1/2}(\cos\theta_B/\cos\theta)}{(1 - p^2)^{1/2}} \quad (4\text{-}152)$$

and

$$(\Delta\eta_0{}^r + \Delta\eta_g{}^r) = \frac{\pm 2(\Delta\eta_0{}^r\Delta\eta_g{}^r)^{1/2}}{(1 - p^2)^{1/2}} \quad (4\text{-}153)$$

where the upper and lower signs correspond to branches (1) and (2). From equations (4-24) and (4-27),

$$(\Delta \mathbf{K} \cdot \mathbf{j}) = \pm \tfrac{1}{2} K \chi_0 (\Delta \eta_0{}^r + \Delta \eta_g{}^r)$$

$$(\Delta \mathbf{k} \cdot \mathbf{j}) = \pm (\Delta \eta_0 \Delta \eta_g{}^r + \Delta \eta_g \Delta \eta_0{}^r)$$

$$= \pm [2 \Delta \eta_0{}^r \Delta \eta_g{}^r + i(\Delta \eta_0{}^i \Delta \eta_g{}^r + \Delta \eta_g{}^i \Delta \eta_0{}^r)]$$

$$\doteq \pm 2 (\Delta \eta_0{}^r \Delta \eta_g{}^r)^{1/2} (\Delta \eta_0 \Delta \eta_g)^{1/2}$$

Here, it is necessary to distinguish the complex $\Delta \eta_0$ and $\Delta \eta_g$ from their real parts and SI approximations are employed. Using equations (4-152) and (4-153), and the dispersion relation (4-26), one gets

$$(\Delta \mathbf{K} \cdot \mathbf{v}) = \tfrac{1}{2} K \chi_0 \frac{\cos \theta}{\cos \theta_B} \tag{4-154a}$$

$$(\Delta \mathbf{k} \cdot \mathbf{v}) = \pm \tfrac{1}{2} K C (\chi_g \chi_{-g})^{1/2} (1 - p^2)^{1/2} \frac{\cos \theta}{\cos \theta_B} \tag{4-154b}$$

After these preliminaries, recalling the relations $\mathbf{k}_0 = \overline{\mathbf{K}}_0 + \Delta \mathbf{K} + \Delta \mathbf{k}$ and $\mathbf{k}_g = \overline{\mathbf{K}}_g + \Delta \mathbf{K} + \Delta \mathbf{k}$, one obtains

$$S_0 = [\overline{\mathbf{K}}_0 \cdot (\mathbf{r} - \mathbf{r}_e)] + \tfrac{1}{2} K \chi_{\mathrm{eff}} \left( \frac{\cos \theta}{\cos \theta_B} \right) l \tag{4-154c}$$

$$S_g = [\overline{\mathbf{K}}_g \cdot (\mathbf{r} - \mathbf{r}_e)] + \tfrac{1}{2} K \chi_{\mathrm{eff}} \left( \frac{\cos \theta}{\cos \theta_B} \right) l \tag{4-154d}$$

where $\chi_{\mathrm{eff}}$ is the effective polarizability and is defined by

$$\chi_{\mathrm{eff}} = \chi_0 \pm C (\chi_g \chi_{-g})^{1/2} (1 - p^2)^{1/2} \tag{4-155}$$

On the other hand, one can write

$$\left( \frac{\cos \theta}{\cos \theta_B} \right) l = l_0 + l_g \tag{4-156}$$

where $l_0$ and $l_g$ are the optical path lengths of $O$ and $G$ waves, respectively, as illustrated in Fig. 4-18. The attenuation of the ray, therefore, is determined by

$$\mu_B l = [\mu_0 \pm \mu_g (1 - p^2)^{1/2}](l_0 + l_g) \tag{4-157}$$

where $\mu_0$ is the usual absorption coefficient defined by (4-146) and $\mu_g$ is given by

$$\mu_g = K C J_m (\chi_g \chi_{-g})^{1/2} \tag{4-158}$$

The second term in brackets in (4-157), viz., $\mu_g (1 - p^2)^{1/2}$, is the anomalous part of the absorption coefficient caused by the $g$th-order Bragg reflection. Using expression (4-157), it is possible to interpret the length $(l_0 + l_g)$ as a zigzag path with the mean direction $\mathbf{v}$ as shown in Fig. 4-18.

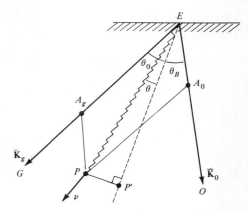

FIGURE 4-18
Optical path of a ray in the direction $\boldsymbol{v}$. The microscopic segments of the ray are composed of two directions, $\bar{\mathbf{K}}_0$ and $\bar{\mathbf{K}}_g$, so that the total lengths of the segments equal $l_0 = EA_0$ and $l_g = EA_g$, respectively ($EP = l$; $EP' = z$).

By inserting (4-157) into equations (4-141$a$ and $b$) or (4-143$a$ and $b$), one can obtain the intensity profiles of $O$ and $G$ waves in the case of absorbing crystals. The results are included in Fig. 4-17. (In practice, it is convenient to use $l_0 + l_g = z/\cos \theta_B$ and $z = l \cos \theta$.) When $\mu_g t_e \gtrsim 10$, it is concluded that x-ray beam of branch (1) penetrates through the crystal mainly along the net plane, and emerges out from the exit surface in the form of narrow beam. This is what Borrmann and Campbell had observed. In the intermediate cases $\mu_g t_e \simeq 1$, the rays of both branches retain sufficient intensity. Their intensity profiles, however, are very different from those expected by taking into account only the normal absorption.

Quantitative measurements of $\mu_g$ have been carried out by several authors (Hunter, 1959; Batterman, 1962); however, the effects of elastic strain were not completely eliminated in these experiments. In Table 4-4, the most reliable results of Okkerse (1962) are shown in comparison with the theoretical values of Wagenfeld (1962).

In the present and previous sections, only the Laue-Laue case has been discussed because the case is of most interest in practice. A similar argument can be applied to the other cases provided that the crystal ray appears from the entrance and exit surfaces. Theoretical and experimental studies have been carried out by Borrmann (1951); Borrmann, Hildebrandt, and Wagner (1955); Wagner (1956); and Authier (1962).

To derive a physical interpretation, first consider the intensity field associated with a single Bloch wave given by the sum of equations (4-73$a$ and $b$)

$$i^{(j)} = [(C_0 C_0^*) + (C_g C_g^*)] \pm 2C|C_0| |C_g| \cos [2\pi(\mathbf{g} \cdot \mathbf{r}) + \phi] \quad (4\text{-}159)^1$$

[1] In absorbing crystals one must multiply through the absorption factor $\exp - \mu_B l$, dependent on ($j$). Note that the polarization factor $C$ is required in the third term.

Table 4-4   QUANTITATIVE STUDY OF
BORRMANN   ABSORPTION
$(1 - \mu_g/\mu_0)$ FOR Ge USING
Cu $K\alpha$ RADIATION

| Reflection | Theory* | Experiment† (at 25°C) |
|---|---|---|
| 220 | 0.046 | 0.0408 |
| 400 | 0.088 | 0.0812 |
| 422 | 0.128 | 0.1216 |

*Wagenfeld (1962).
†Okkerse (1962).

where $\phi$ is the phase of the Fourier coefficient $\chi_g^r$. It is not difficult to obtain the expression in terms of the deviation parameter $w$ by the use of (4-72). Since the difference of the wave vectors for $O$ and $G$ waves is $2\pi g$, the intensity field is sinusoidal, and the spacing is given by

$$\Lambda = \frac{1}{|\mathbf{g}|} = d \qquad (4\text{-}160)$$

where $d$ is the interplanar spacing of the net plane concerned (Appendix 4B). For this reason the fringes of this type may be called *Netzebene fringes*. In electron microscopy, the fringes actually were observed by Menter (1956), first in phthalocyanide. Electron-optics users called them *lattice image*, although, strictly speaking, the term lattice image refers to fringes resulting from $I = |E_0^{(1)}(\mathbf{r}) + E_0^{(2)}(\mathbf{r}) + E_g^{(1)}(\mathbf{r}) + E_g^{(2)}(\mathbf{r})|$, which also has the spacing $d$ in the case of a parallel-sided crystal. Further details can be found in electron microscopy texts (Hirsch et al., 1965).

The first term in (4-159) represents the background and the second term the modulation of the intensity field. The contrast of the fringes is the ratio of the modulation amplitude and the background, namely, $\pm C|C_0| \cdot |C_g|/[(C_0 C_0^*) + (C_g C_g^*)]$. For the symmetrical Laue case, it is $\pm\frac{1}{2}CW/(w^2 + W^2)^{1/2}$. In the more general case, it can be shown with the use of (4-139a) that the contrast can be written $\pm\frac{1}{2}C(1 - p^2)^{1/2}$.

It is very instructive to compare the *Netzebene* fringes (4-159) with the charge distribution $N(\mathbf{r})$, which is related to $\chi^r(\mathbf{r})$ by (4-1). Here, the dispersion is neglected ($G = N$). Then, the modulation of charge density owing to the $\pm$gth coefficient is proportional to $-2|\chi_g^r| \cos [2\pi(\mathbf{g} \cdot \mathbf{r}) + \phi]$. Since $|C_g|$ is proportional to $|\chi_g^r|$, it turns out that the maxima in the *Netzebene* fringes of the branch (1) wave coincide with the minima of the charge distribution, whereas the maxima of the *Netzebene* fringes and the maxima of the charge distribution coincide with each other for branch (2) waves.

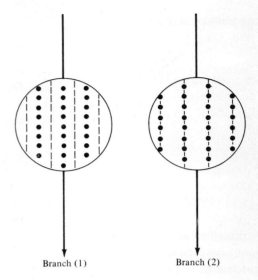

**FIGURE 4-19**
Microscopic view of the intensity modulation of the ray fields. The circles
represent the atomic array while the
dashed lines represent the maximum
intensity contour of the wave field
(*Netzebene* fringes).

Branch (1)    Branch (2)

In x-ray cases, the main part of the absorption is photoelectric absorption due to ejection of $K$ electrons. For low-order reflections, it is reasonable to assume that the atoms are lined up very nearly at the maxima of the charge distribution. For this reason it is expected that the branch (1) wave does not suffer photoelectric absorption while the (2) wave is absorbed more than the mean value (Fig. 4-19). The interpretation of the Borrmann absorption with this model is conventionally accepted, and it is helpful in visualizing the physical phenomenon. The model, however, becomes obscure for the case of higher-order reflections and for crystals having a complex structure, where the atoms or absorption centers do not necessarily lie at the maxima of the Fourier components of the charge distribution.

The interpretation can be developed in a more refined form. It is reasonable to assume that the energy absorbed in a microscopic volume is proportional to the intensity field (4-159) and the distribution of absorbing centers, which is proportional to $\chi^i$, the imaginary part of the polarizability. Thus, the observable attenuation is proportional to attenuation per unit cell and per unit intensity of $I_B = [(C_0 C_0^*) + (C_g C_g^*)]$ is

$$\mu_B l = \left(\frac{\alpha}{v}\right) \int_{\text{cell}} \{1 \pm C(1 - p^2)^{1/2} \cos [2\pi(\mathbf{g} \cdot \mathbf{r}) + \phi]\}$$

$$\times \left\{\chi_0{}^i + 2 \sum_h |\chi_h{}^i| \cos [2\pi(\mathbf{h} \cdot \mathbf{r}) + \varphi_h]\right\} dv$$

where $\alpha$ is a proportionality constant. Dropping the interference terms, and letting $\varphi_g = \phi_2$ and $\phi = \phi_1$,

$$\mu_B l = K[\chi_0{}^i \pm C(1 - p^2)^{1/2}|\chi_g{}^i| \cos(\phi_1 - \phi_2)](l_0 + l_g)$$

where the proportionality constant $K$ can be identified with ordinary absorption. In addition, as shown in Appendix 4AI, $|\chi_g{}^i| \cos(\phi_1 - \phi_2)$ is approximately $J_m(\chi_g \chi_{-g})^{1/2}$. The result is identical to equations (4-157) and (4-158).

With this interpretation, it is possible to develop a theory of the Borrmann absorption which takes into account thermal vibrations. Okkerse (1962) has done this for x rays, while Hall and Hirsch (1965) and Whelan (1965) have done this for electrons to some extent. They concluded that it is sufficient to multiply $J_m(\chi_g \chi_{-g})^{1/2}$ by the Debye-Waller factor $e^{-M}$. The experiments of Okkerse (1962) and Ling and Wagenfeld (1965) confirmed the above result in a wide range of temperatures. It is now possible to justify the factor $e^{-M}$ on the basis of the more fundamental theories described at the end of Section I C 1 of the preceding chapter.

Recently, superanomalous absorption was observed also when simultaneous reflection takes place from **g** and **h** net planes (Borrmann and Hartwig, 1965; Hildebrandt, 1966) in the x-ray case. For this phenomenon, the orthodox theories are given by Henó and Ewald (1968), Penning (1967), and Penning and Polder, (1968$a$,$b$). A theory based on the microscopic model mentioned above also will be developed along similar lines. In electron diffraction, the presence of simultaneous reflections is unavoidable, and this complex subject has been studied extensively. Some of these theories may be applicable also to x-ray cases for understanding simultaneous reflections in general.

**6.  Rocking curves**  So far, the topographic experiments relating to the wave fields and the associated rays have been considered primarily for the Laue-Laue case. In traditional experiments of crystal diffraction, so-called rocking curves and integrated intensities are investigated. In these cases, the transmitted or diffracted beams are measured in their entirety. The relations to the wave fields are illustrated schematically in Fig. 4-20. In this section and the following one, Laue-Laue, Bragg, and (Bragg)$^\infty$ cases are considered.

*a.  Laue-Laue case*  If one has an ideal plane wave, one may obtain the intensity fields of $O$ and $G$ waves similarly to equations (4-128$a$ and $b$) for the case of nonabsorbing crystals. To obtain the total intensity of an incident beam having a finite width, it is necessary to integrate $I_0$ and $I_g$ spatially over the beam width. By virtue of relation (4-129$c$), the integration is equivalent to averaging the intensity field in the crystal over a definite range of crystal thickness. In the following, for simplicity, consider a parallel-sided crystal having a thickness $t_0$. The spatial integration then can be replaced by multiplying by the beam width.

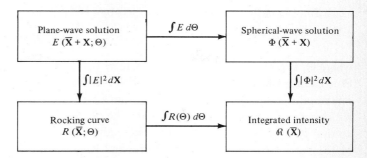

FIGURE 4-20
Interrelation of the fundamental quantities in crystal diffraction studies. ($\overline{\mathbf{X}}$ is the beam position relative to the crystal; $\mathbf{X}$ is the position within the beam; $\Theta$ is the glancing angle of the incident beam.)

To obtain the intensity equations for absorbing crystals, it is necessary to replace $C_{0,g}{}^{(j)}$ in deriving (4-128) by $C_{0,g}{}^{(j)} \exp 2K\delta_e{}^i t_0$, where $2K\delta_e{}^i t_0$ is given by (4-144). Thus, with the SI approximation, the transmitted powers of $O$ and $G$ waves are given as follows:[1]

$$R_0(\Theta) = \frac{I_0}{I_e} \qquad (4\text{-}161a)$$

$$= \exp(-\mu_0 t) \cdot (w^2 + W^2)^{-1}$$
$$\times \left[(w^2 + W^2) - W^2 \sin^2 F + (2w^2 + W^2)\right.$$
$$\left.\times \sinh^2 H - w(w^2 + W^2)^{1/2} \sinh 2H\right]$$

$$R_g(\Theta) = \frac{(\gamma_g/\gamma_0)I_g}{I_e}$$

$$= \exp(-\mu_0 t)\frac{W^2}{w^2 + W^2}(\sin^2 F + \sinh^2 H) \qquad (4\text{-}161b)$$

where

$$F = \frac{1}{2}\frac{t_0}{\gamma_g}(w^2 + W^2)^{1/2} \qquad (4\text{-}162a)$$

$$H = \frac{\frac{1}{2}(t_0/\gamma_g)(wv + WV)}{(w^2 + W^2)^{1/2}} \qquad (4\text{-}162b)$$

The factor $(\gamma_g/\gamma_0)$ in (4-161b) is the ratio between the widths of the incident and Bragg-reflected beams.

[1] These powers are different from those given by equations (4-141a and b). The present ones are referred to a sufficiently wide beam including (1) and (2) beams, whereas the previous ones dealt with a single ray.

Since $w$ is the real part of $u$ defined by (4-56a), one obtains the relation

$$w = K \sin 2\theta_B(\Theta - \Theta_c) \qquad (4\text{-}163)$$

where $\Theta_c$ is given by (4-77). In diffractometer experiments, usually, the glancing angle $\Theta$ is changed by rotating the crystal with respect to the fixed incident beam. Then, one obtains a line profile of the power versus the glancing angle $\Theta$, which is called the *rocking curve*. In ordinary experiments, the oscillating term $\sin^2 F$ is hardly detected, so that it must be replaced by the mean value $(1/2)$. Experimental tests of equations (4-161) have been reported by Schwartz and Rogosa (1954), Yoshioka (1954), and by Brogren and Adell (1954). Recent experiments to detect the oscillating term will be discussed further after describing the Bragg case.

Equations (4-161a and b) can be regarded as the intensity distribution (omitting the factor $\gamma_g/\gamma_0$) as a function of the deviation parameter $w$ and the crystal thickness $t_0$. The equations can be used, therefore, to understand the *Pendellösung* fringes in absorbing crystals.

*b. Bragg case*   The wave fields and the rocking curve already have been discussed in Section 3b for half-bounded crystals. There, Darwin's rocking curve (4-86) and Prins' curve (4-84b or c) were obtained.

The experimental study of the Darwin-Prins rocking curve can be considered a classical subject in x-ray diffraction by crystals. The really critical experiments, however, are made possible by recent technical developments so that efforts to test the theoretical predictions are still going on. The following papers are of interest in this connection: Renninger (1955; 1967; 1968), Fingerland (1962; 1971), Bubáková (1962), Bubáková, Drahokoupil, and Fingerland (1961; 1962a; 1962b), Fingerland and Drahokoupil (1970), Kohra and Kikuta (1968), Kikuta, Kawashima, and Kohra (1970), Kikuta and Kohra (1970), Kikuta (1971), Matsushita, Kikuta, and Kohra (1971).

In very thin crystals, the effect of the rear surface cannot be ignored. If the rear surface is not parallel to the entrance surface, one must define the power ratio for a finite beam width. As in the Laue-Laue cases, the spatial integrals of the intensity fields over the beam width have to be included in the expression. Here, again for simplicity, consider only parallel-sided crystals. Then, the transmission and reflection powers are identical to those associated with the rays and are given by

$$R_0(\Theta) = P_0 = \frac{\mathbf{S}_0 \cdot \mathbf{n}_e}{\mathbf{S}_e \cdot \mathbf{n}_e} \qquad (4\text{-}164a)$$

$$R_g(\Theta) = P_g = \frac{|\mathbf{S}_g \cdot \mathbf{n}_e|}{\mathbf{S}_e \cdot \mathbf{n}_e} \qquad (4\text{-}164b)$$

where $S_e$ is the Poynting vector defined by (4-133) and $S_0$ and $S_g$ are those of transmitted and reflected waves, which are derived from equations (4-113a and b), respectively. After a straightforward manipulation, therefore, one obtains

$$R_0(\Theta) = \left| \frac{(u^2 + U^2)^{1/2}}{u \sin \xi - i(u^2 + U^2)^{1/2} \cos \xi} \right|^2 \exp -\mu_0 t \qquad (4\text{-}165a)$$

$$R_g(\Theta) = \left| \frac{\chi_g}{\chi_{-g}} \right| \left| \frac{U \sin \xi}{u \sin \xi - i(u^2 + U^2)^{1/2} \cos \xi} \right|^2 \qquad (4\text{-}165b)$$

where

$$\xi = \frac{1}{2} \left( \frac{t_0}{|\gamma_g|} \right) (u^2 + U^2)^{1/2} \qquad (4\text{-}166a)$$

$$t = \tfrac{1}{2} t_0 \left( \frac{1}{\gamma_0} - \frac{1}{|\gamma_g|} \right) \qquad (4\text{-}166b)$$

Since $u$ and $U$ are complex in general, they are rather complicated. For numerical calculation, equations (4-57) and (4-58b) and the expression for $(u^2 + U^2)^{1/2}$ in Appendix 4AII should be consulted.

For simplicity, consider in detail the case of nonabsorbing crystals, where $u$ and $U$ are replaced by $w$ and $iW$, respectively, and $|\chi_g|$ is identical to $|\chi_{-g}|$. Then, equations (4-165a and b) are reduced to

$$R_0(\Theta) = \frac{(w^2 - W^2)}{[w^2 \sin^2 |\xi| + (w^2 - W^2) \cos^2 |\xi|]} \qquad |w| \geq W \quad (4\text{-}167a)$$

$$= \frac{(W^2 - w^2)}{[w^2 \sinh^2 |\xi| + (W^2 - w^2) \cosh^2 |\xi|]} \qquad |w| \leq W \quad (4\text{-}167b)$$

$$R_g(\Theta) = \frac{W^2 \sin^2 |\xi|}{[w^2 \sin^2 |\xi| + (w^2 - W^2) \cos^2 |\xi|]} \qquad |w| \geq W \quad (4\text{-}167c)$$

$$= \frac{W^2 \sinh^2 |\xi|}{[w^2 \sinh^2 |\xi| + (W^2 - w^2) \cosh^2 |\xi|]} \qquad |w| \leq W \quad (4\text{-}167d)$$

$$|\xi| = \frac{1}{2} \frac{t_0}{|\gamma_g|} (|w^2 - W^2|)^{1/2}$$

where $w$ has the same form as (4-163), but the expression for $\Theta_c$ must be replaced by (4-88). Obviously the conservation of energy

$$R_0 + R_g = 1 \qquad (4\text{-}168)$$

is always satisfied.

In the range $|w| > W$, the reflected and transmitted powers oscillate as the crystal thickness or the glancing angle changes. This phenomenon is intimately connected with the size effects discussed for the kinematical theory in Chap. 3. In fact, under the kinematical condition $|w| \gg W$, $R_g$ has the approximate form $(W/w)^2 \sin^2 \left[\frac{1}{2}(t_0/\gamma_g)w\right]$, which is identical to the result of the kinematical theory. A similar correspondence can be seen also in the Laue-Laue case (4-161$b$). As distinct from the Laue-Laue case, however, the intensity modulation is not due to the interference between the waves of branches (1) and (2) created at the entrance surface but to the interference between the waves of wings ($a$) and ($b$) created at the entrance and rear surfaces, respectively.[1] These types of oscillation actually have been observed by several authors: For the Laue-Laue case, by Renninger (1968), Kohra and Kikuta (1968), Kikuta and Kohra (1968), and Lefeld-Sosnowska and Malgrange (1968). For the Bragg case, by Batterman and Hildebrandt (1967; 1968) and by Hashizume, Nakayama, Matsushita, and Kohra (1970). Topographic observations have been reported by Nakayama, Hashizume, and Kohra (1971), who devised a supermonodirectional (0.1 ~ 0.01 sec) wave having a wide coherent wave front (~500 $\mu$) in the incident beam, obtained by the use of asymmetric crystal reflection (Renninger, 1961; Kohra, 1962). Other techniques for monodirectional waves were devised also by Authier (1960) and Bonse and Hart (1965). It is important to note that the crystal reflection serves as a kind of condenser lens for producing a parallel coherent wave from a spherical wave emitted by an atom.

The observed rocking curve including the oscillation, however, may not be exactly the same as the one which is expected from equations (4-161$b$) and (4-165$b$), because of the difficulties in obtaining an ideal plane wave. The theoretical curve must be modified by taking into account the finiteness of the wave front, as will be discussed in the next section for Bragg cases. Nevertheless, one must admire the efforts of the investigators cited above, who have verified the correctness of the plane-wave theory, since the experiments are really sophisticated and quite tedious.

In most experimental conditions, the crystal thickness $t_0$ is much larger than $1/W$ and not exactly constant as assumed in the calculation. For this reason, it is more meaningful to consider the averaged powers over one period in $\xi$, in the range $|w| \gtrsim W$. In addition, it is safe to assume that cosh $\simeq$ sinh are very large in the range $|w| \lesssim W$. Under these conditions

---

[1] Inside the crystal one really has an intensity modulation due to the standing wave. Most people call it *Pendellösung* fringes. Here, one faces a difficulty as to terminology. The standing wave is of the Borrmann-Lehmann type and will be discussed in Section II F 4. Complementarity of $i_0$ and $i_g$ is not satisfied in this case, whereas it is satisfied in *Pendellösung* fringes in the Laue case. For this reason, it would be better to call the intensity modulation discussed here "*Pendellösung* fringes in a broad sense." *Pendellösung* fringes of the type observed in Laue cases are expected only in the spherical-wave theory for Bragg cases.

$$\langle R_0(\Theta) \rangle = \frac{(w^2 - W^2)^{1/2}}{|w|} \qquad |w| \geq W \quad (4\text{-}169a)^1$$

$$= 0 \qquad |w| \leq W \quad (4\text{-}169b)$$

$$\langle R_g(\Theta) \rangle = 1 - \frac{(w^2 - W^2)^{1/2}}{|w|} \qquad |w| \geq W \quad (4\text{-}169c)$$

$$= 1 \qquad |w| \leq W \quad (4\text{-}169d)$$

This result is called *Ewald's rocking curve*, which is illustrated in Fig. 4-7.

*c. Remarks on Darwin and Ewald solutions* A difference can be noticed between the tails of the two rocking curves in Fig. 4-7. Conventionally (see, for example, Zachariasen, 1945, p. 141), the difference is attributed to the mathematical processes used in simplifying equations (4-165a and b), which are quite general under the prescribed conditions. Darwin's is derived by taking first a sufficiently large crystal thickness and next neglecting absorption. On the other hand, Ewald's is obtained by the reverse procedure.

In order to see the physical meaning of this difference it is very instructive to consult the second method for deriving the wave fields (4-113a and b). There, it is assumed that all the multireflected waves at the front and rear surfaces contribute coherently to the wave fields $E_0$ and $E_g$. This is true only when (1) the wave front of the incident wave is sufficiently wide, and (2) the parallelism of the crystal surfaces is sufficiently accurate. In addition, the contribution of multiply reflected waves will be noticeable only when (3) the absorption (or thickness) is sufficiently small. In practice, however, conditions (1) and (2) are rather difficult to realize. Considering the rays associated with the reflected waves in crystals (Fig. 4-10), the separation and the path difference of the reflected waves of $n$th and $(n + 2)$th order are $2t_0 \sin \theta_B \cot \theta$ and $2t_0/\sin \theta$, respectively.[2] The former must be much less than $L\Delta\Theta \simeq 5 \ \mu$ under the ordinary experimental conditions for satisfying condition (1), and the latter must be constant with the accuracy of the wavelength. For this reason, the real rocking curve should be

$$R_0(\Theta) = \frac{\sum\limits_{m=0}^{\infty} |E_{0,2m}|^2}{|E_e|^2} \qquad (4\text{-}170a)$$

$$R_g(\Theta) = \frac{|\gamma_g|}{\gamma_0} \frac{|E_{g,0}|^2 + \sum\limits_{m=0}^{\infty} |E_{g,2m+1}|^2}{|E_e|^2} \qquad (4\text{-}170b)$$

---

[1] In deriving these, the integral formula $\int_0^{\pi/2} (a^2 \cos^2 \xi + b^2 \sin^2 \xi)^{-1} \, d\xi = \pi/2ab$ for $ab > 0$ is employed.

[2] For simplicity, the symmetrical Bragg case is considered.

The effect of the crystal thickness appears only through the absorption factor. Again, these are geometrical series so that it is easy to obtain the sum. If absorption is neglected, the results tend to Ewald's result (4-169).[1] It is obvious that Darwin's result is the first term of (4-170b). In practice, perhaps, the first few terms can give the correct result, owing to the presence of x-ray absorption.

Equations (4-170), however, are not applicable to extremely thin crystals. Then, the coherence of the multiply reflected waves has to be taken into account. The thickness beyond which equations (4-170) must be used instead of (4-165) depends on the size of the coherent wave front of the incident beam. In practice, the critical thickness is less than several microns. When the oscillations discussed in the preceding section are included in the rocking curve (intermediate case), the theory must be extended to include the width of the coherent wave front.

One point is worth noting. If one records Berg-Barrett or Lang topographs with the reflected or transmitted beams with the use of a monodirectional but narrow beam, it may be possible to observe a few equally spaced beams corresponding to individual $|E_{g,2m+1}|^2$ or $|E_{0,2m}|^2$. They are not *Pendellösung* fringes, although their appearance may be similar. To the present time, no critical experiment has been reported. In this connection, it is necessary to consult the spherical-wave theory discussed in the next chapter.

**7.** *Integrated power*[2]   The integral of a rocking curve with respect to the glancing angle $\Theta$ is called the *integrated power* or, more precisely, the *power ratio* (Fig. 4-20) or sometimes, less precisely, the *integrated intensity*. In the case of an *O* beam, the transmitted power $R_0$ tends to a constant background $\exp - \mu_0 t_0 / \gamma_0$. To avoid divergence, therefore, this part must be subtracted from $R_0$. Thus, the mathematical expressions for the integrated powers are defined

$$\Re_0^{\Theta} = \int_{-\infty}^{+\infty} \left[ R_0(\Theta) - \exp - \frac{\mu_0 t_0}{\gamma_0} \right] d\Theta \quad (4\text{-}171a)$$

$$\Re_g^{\Theta} = \int_{-\infty}^{+\infty} R_g(\Theta) d\Theta \quad (4\text{-}171b)$$

For convenience, transform the variable to $y = w/W$. Then, the integrated powers

---

[1] Here, again, it must be noticed that the transmitted power (4-89) is zero in the range of total reflection. Then, $|E_{0,2m}|^2$ and $|E_{g,2m+1}|^2$ must be set equal to zero. In any case, *O* and *G* waves in the crystal are coherent.

[2] A good summary is given in the book by Weiss (1966).

can be written as

$$\mathfrak{R}_0^\Theta = \Delta\Theta\mathfrak{R}_0^P \quad (4\text{-}172a)^1$$

$$\mathfrak{R}_g^\Theta = \Delta\Theta\mathfrak{R}_g^P \quad (4\text{-}172b)$$

where $\Delta\Theta$ is the angular range of reflection defined by (4-76) or (4-87).

*a. Laue-Laue case* (Ramachandran, 1954; Kato, 1955.) The integrated powers $\mathfrak{R}_0^P$ and $\mathfrak{R}_g^P$ can be expressed in various forms. Here, only the cases of parallel-sided crystals are considered. The other expressions and the details for calculating them can be found in the papers cited above. By the use of equations (4-161a and b),

$$\mathfrak{R}_0^P = e^{-\mu_0 t}\left[-2\pi \sum_{m=1}^\infty m(\cos m\alpha)\, I_m(h)\right] - \mathfrak{R}_g^P \quad (4\text{-}173a)$$

$$\mathfrak{R}_g^P = e^{-\mu_0 t}\left\{\pi \sum_{m=0}^\infty J_{2m+1}(2A) + \frac{\pi}{2}\left[I_0(h) - 1\right]\right\} \quad (4\text{-}173b)$$

$$= e^{-\mu_0 t}\{[W] + [V]\}$$

where $J_m$ is the Bessel function of the $m$th order and $I_m$ is the modified Bessel function. The arguments $A$, $h$, and $\alpha$ are defined by

$$A = \frac{1}{2}\frac{t_0}{\gamma_g} W \quad (4\text{-}174a)$$

$$h = \frac{t_0}{\gamma_g}(v^2 + V^2)^{1/2} \quad (4\text{-}174b)$$

$$\alpha = \tan^{-1}\frac{V}{v} \quad (4\text{-}174c)$$

and

$$t = \tfrac{1}{2}t_0\left(\frac{1}{\gamma_0} + \frac{1}{\gamma_g}\right)$$

The first term $[w]$ in the braces of (4-173) is the integrated power of the $G$ beam for nonabsorbing crystals ($V = v = 0$). It is originally defined by

$$[W] = \int_{-\infty}^{+\infty}\left[\frac{\sin A(1 + y^2)^{1/2}}{(1 + y^2)^{1/2}}\right]^2 dy \quad (4\text{-}175a)$$

$$= \frac{\pi}{2}\int_0^{2A} J_0(p)\, dp \quad (4\text{-}175b)$$

---

[1] $\mathfrak{R}_g$ and $\mathfrak{R}_g^P$ are essentially the same as $R_g$ and $R_g^y$ of Zachariasen, respectively. The superscript $P$ is used for denoting plane-wave theory in contrast to $\mathfrak{R}^s$ of the spherical-wave theory (Section II D).

The latter expression is called the *Waller integral*. The functional form is discussed later (Fig. 4-32c). The oscillation is due to *Pendellösung* phenomena described in Section D 2. For thin crystals, $[W]$ is given by $\pi A$, so that the integrated power has the form

$$\mathfrak{R}_g^{\ominus} = \pi \Delta\Theta A = \frac{(r_c/v)^2 \lambda^3 C^2 (F_g F_{-g})(t_0/\gamma_0)}{\sin 2\theta_B} \qquad (4\text{-}176a)$$

where $r_c$ is the classical radius of an electron and $v$ is the volume of the unit cell. The result is identical to the kinematical one. For sufficiently thick crystals, $A \to \infty$ and $[W]$ tends to $\pi/2$. Therefore, it turns out that

$$\mathfrak{R}_g^{\ominus} = \frac{\pi}{2} \Delta\Theta = \frac{1}{2}\left(\frac{r_c}{v}\right) \lambda^2 C(F_g F_{-g})^{1/2} \left(\frac{\gamma_g}{\gamma_0}\right)^{1/2} \qquad (4\text{-}176b)$$

which is independent of the crystal thickness.

For absorbing crystals, the terms including $I_m$ must be considered. By the use of a power series and the asymptotic forms of the modified Bessel functions, they have the following special forms:

Thin crystals ($h \ll 1$):

$$\mathfrak{R}_0^P = e^{-\mu_0 t} \pi \left[ -\cos\alpha \cdot h - \tfrac{1}{2}\cos 2\alpha \cdot h^2 + \cdots \right] - \mathfrak{R}_g^P \qquad (4\text{-}177a)$$

$$\mathfrak{R}_g^P = e^{-\mu_0 t} \frac{\pi}{2} \left[ \tfrac{1}{4}h^2 + \tfrac{1}{64}h^4 + \cdots \right] + e^{-\mu_0 t}[W] \qquad (4\text{-}177b)$$

Thick crystals ($h \gg 1$):

$$\mathfrak{R}_0^P = e^{-\mu_0 t} \pi \frac{e^h}{\sqrt{2\pi h}} \left( D_0 + \frac{D_1}{h} + \frac{D_2}{h^2} + \cdots \right) - \mathfrak{R}_g^P \qquad (4\text{-}178a)$$

$$\mathfrak{R}_g^P = e^{-\mu_0 t} \frac{\pi}{2} \frac{e^h}{\sqrt{2\pi h}} \left( 1 + \frac{1}{8}\left(\frac{1}{h}\right) - \frac{9}{128}\left(\frac{1}{h}\right)^2 + \cdots \right) + e^{-\mu_0 t}[W] \qquad (4\text{-}178b)$$

where

$$D_0 = (1 - \cos\alpha)^{-1} \qquad (4\text{-}179a)$$

$$D_1 = \frac{3}{4}\left\{ \frac{1}{4} + \frac{[1 - \tfrac{1}{4}(1 + \cos\alpha)](1 + \cos\alpha)}{(1 - \cos\alpha)^2} \right\} \qquad (4\text{-}179b)$$

and so forth. For the symmetrical Laue case ($\alpha = \pi/2$), the integrated powers $\mathfrak{R}_0^P$ and $\mathfrak{R}_g^P$, omitting $[W]$, are illustrated in Fig. 4-21. In the case of thick crystals, both $\mathfrak{R}_0^P$ and $\mathfrak{R}_g^P$ decrease more slowly than the normal decrease given by $\exp(-\mu_0 t)$. The physical reason for this is the Borrmann absorption. [See the experiments of Hildebrandt (1959a) and Okkerse (1962).]

b. *Bragg case* (Hirsch and Ramachandran, 1950; Afanas'ev and Perstnev, 1969.) For sufficiently thick crystals, Prins' rocking curve (4-84c) is adequate for actual experi-

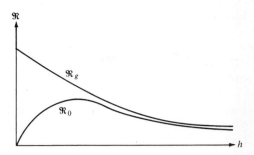

**FIGURE 4-21**
Integrated powers $\Re_0{}^p$ and $\Re_g{}^p$ for the symmetrical Laue case. (Waller's integral is omitted.)

ments.  Recently, Afanas'ev and Perstnev gave an exact solution for the integrated power.  The result is expressed in the form

$$\Re_g{}^\Theta = \tfrac{8}{3}\Delta\Theta P(r, q) \qquad (4\text{-}180a)$$

where $\tfrac{8}{3}\Delta\Theta$ is the well-known Darwin result for nonabsorbing crystals.  The parameters $r$ and $q$ are given by

$$r = \frac{|v|}{(W^2 + V^2)^{1/2}} \qquad (4\text{-}181a)$$

$$q = \left|\frac{V}{v}\right| \qquad (4\text{-}181b)$$

The functional form of $P(r,q)$ is

$$P(r,q) = \frac{(1 - r^2 - 2q^2r^2)E(k)}{k} - \frac{3\pi}{4} r(1 - 2q^2r^2) + kr^2(1 - q^2)$$

$$\times \; [3(1 - q^2r^2)\Pi(-q^2,k) - (2 - r^2 - 2q^2r^2)K(k)] \qquad (4\text{-}182a)$$

$$k = \frac{1}{[1 + r^2(1 - q^2)]^{1/2}} \qquad (4\text{-}183)$$

where $K(k)$, $E(k)$, $\Pi(-q^2, k)$ are the complete elliptic integrals of the first, second, and third type.

Within an accuracy of 1% for $r \le 0.2$ ($q < 1$ always),

$$P(r,q) = 1 - \frac{3\pi}{4} r(1 - 2q^2r^2) + 3r^2$$

$$\times \left[ \tfrac{1}{2}(1 + q^2) \ln \frac{4}{r(1 + q)} - \tfrac{1}{4}(1 + 3q^2) + \tfrac{1}{4}(1 - q^2)^2 \ln \frac{1 + q}{1 - q} \right] \qquad (4\text{-}182b)$$

If one approximates as $P(r, q) = 1 - (3\pi/4)r$, the empirical formula of Hirsch and

Ramachandran,

$$\mathfrak{R}_g{}^{\Theta} = \tfrac{8}{3}\Delta\Theta \left(1 - 2.4\,\frac{v}{W}\right) \quad (4\text{-}180b)$$

can be justified ($\tfrac{3}{4}\pi \simeq 2.36$) approximately.

Equations (4-180$a$ and $b$) cannot be used for extremely thin crystals. Then, Ewald's solutions (4-165$b$) or (4-167$c$ and $d$) must be used for the rocking curves. However, the integration of (4-165$b$) is formidable. Equations (4-167$c$ and $d$) can be integrated

$$\mathfrak{R}_g{}^{\Theta} = \pi\Delta\Theta \tanh A \quad (4\text{-}184)$$

For thin crystals, therefore, the result is identical to the kinematical one and $\mathfrak{R}_g$ of the Laue-Laue case. For thick crystal ($A \gg 1$), equation (4-184) should not be used as discussed in connection with the rocking curve. Nevertheless, (4-184) gives a similar result to Darwin's except for the numerical factor.

*c.* *Extinction*   In order to simplify the problem, photoelectric absorption has been neglected. Both in the Laue-Laue and the Bragg$^{\infty}$ cases, the integrated powers justify the kinematical result for small $A$, namely, $\mathfrak{R}_g$. Upon increasing $A$, however, they show a kind of saturation and are proportional to $\lambda^2|F_g|$. The saturation in the integrated power was observed experimentally quite early and was called *extinction*. Darwin classified it into two categories. For a given crystal having a large size, if the crystal is perfect throughout, the saturation discussed above is expected and is called *primary* extinction. The physical cause is a dynamical or coherent interaction between $O$ and $G$ waves. It is also conceivable that an incoherent interaction, namely, an exchange of the energy between the $O$ and $G$ waves, can occur even when the crystal is not ideally perfect. Since no interference occurs in this case, the energy of the $G$ wave cannot exceed that of the $O$ wave. Statistically, then, a balance is struck when the energies carried by the $O$ and the $G$ beam become identical. Thus, again, a kind of saturation in the integrated power is expected. This phenomenon is called *secondary* extinction (Darwin, 1922). Since these topics are mentioned in other chapters of this book and in standard textbooks on x-ray and neutron diffraction, they are not further discussed here. In general, the extinction effect is minimal in highly distorted crystals.

It is worthwhile to note at this point that the distinction between primary and secondary extinction is rather arbitrary. In principle, any interaction between $O$ and $G$ waves must be coherent for static crystals. The incoherent interaction is merely a consequence of the random distribution of regions in the crystal that contribute to the crystal diffraction. If the crystal is warped or distorted in a specific way, then the problem must be treated from a wave-optical point of view. Some simple lattice distortions are considered in Chap. 5 where the disappearance of extinction in distorted crystals will become apparent.

In connection with these topics, a few comments should be included regarding the recent work of Zachariasen (1967a; 1967b; 1968).[1] He considered the energy balance between $O$ and $G$ beams which penetrate the crystal in the directions $\overline{\mathbf{K}}_0$ and $\overline{\mathbf{K}}_g$. The resulting intensity fields are more flexible than those considered in the original work of Darwin (1922), in which the intensity fields are specified simply by a depth parameter. Although the practical utility of his expression is assured, the theory remains within the phenomenological approach, as is true of Darwin's theory in the sense that the interaction of $O$ and $G$ waves is assumed to be perfectly incoherent. One of the more fundamental approaches to the problem of extinction was recently proposed by Kuriyama and Miyakawa (1970). The theory still remains at a conceptual stage, however, but it is very useful for future developments.

## II. SPHERICAL-WAVE THEORY

Undoubtedly, the x rays emitted by any atomic process constitute a kind of spherical wave. Since no lens system is available, such a wave is used as the incident wave in all experiments of crystal diffraction. Because, however, the x-ray source is usually placed very far from the specimen as compared to the wave front effective in Bragg reflection, it seems reasonable, at a first sight, to approximate the incident wave by a plane wave. For this reason, the plane-wave theory described above has proved to be useful for over 50 years in the development of x-ray diffraction.[2] The observation of *Pendellösung* fringes in x-ray studies (Kato and Lang, 1959), however, casts doubt on this intuitive assumption, at least for the cases of dynamical diffraction. In this section, therefore, the diffraction of a spherical wave emitted from a point lying outside the crystal is considered. This theory is called the *spherical-wave theory*.

First, the need to adopt this new viewpoint is described in terms of elementary optical principles. Next, the theory is presented on the basis of the Fourier transform of the plane-wave theory. Such a theory is essentially correct unless the source is located more than 10 m from the specimen. Third, the details of the connection between the wave-optical theory and ray optics are described by means of Kelvin's stationary-phase method or by the method of steepest descent (Appendix 3A). All these are described for the Laue-Laue case. Recent developments in other cases will be described afterward. Finally, the related experimental effects will be considered.

---

[1] Similar theories have been developed for neutron diffraction by Vineyard (1954), Hamilton (1957), Werner and Arrott (1965), and Werner et al. (1966).

[2] The theories for Kossel figures are an exception since they must treat the diffraction of a spherical wave emitted from a source within the crystal (Laue, 1935a,b). It is very interesting to note that Ewald already had indicated the wave field schematically as early as 1917! Darwin also considered a spherical wave and the effect of a slit but did not include phase relations among the incident waves.

## A.   Necessity for a spherical-wave theory

**1.   *Divergence of the wave vectors***   A plane wave is a special solution of the wave equation (3-27*b*) in vacuum. A more general form of the solution is obtained by a superposition of plane waves in the form

$$E(\mathbf{r}) = \int\!\!\int_{-\infty}^{+\infty} E(\mathbf{K}) \exp i(\mathbf{K} \cdot \mathbf{r}) \, dK_x \, dK_y \qquad (4\text{-}185)$$

Obviously, the plane wave $E_e \exp i(\mathbf{K}_e \cdot \mathbf{r})$ is a special one, for which $E_e \delta(\mathbf{K} - \mathbf{K}_e)$ is used in place of $E(\mathbf{K})$. The component plane waves are characterized by the amplitude density $E(\mathbf{K})$ in $\mathbf{K}$ space. Since only a monochromatic wave is considered here, the magnitude of $\mathbf{K}$ must be $K$ given by (3-28). Thus, the independent variables of the wave vector are $K_x$ and $K_y$, the $z$ component $K_z$ being fixed by

$$K_z = \pm(K^2 - K_x^2 - K_y^2)^{1/2} \qquad (4\text{-}186)$$

Because, here, one is concerned with waves propagating in the positive direction of $z$, it suffices to take the positive sign in equation (4-186).

Under these restrictions, a component plane wave is represented by a dispersion point $E$ on the dispersion surface $\bar{S}_0$. The plane wave excites Bloch waves, each being assignable to a branch of the dispersion surface as discussed in the plane-wave theory. The incident wave, having the general form (4-185), therefore, excites the wave packets consisting of Bloch waves already discussed in Section I B 6. A single wave packet is represented by a certain area of the dispersion surface. The angular width of this area with respect to either $O$ or $G$ is denoted by $\Omega$. When the angle $\Omega$ is comparable with the angular width $\Delta\Theta$ defined by (4-76) or (4-87), the behavior of crystal diffraction is no longer described by a single parameter $w$ as in the plane-wave theory. Thus, a required condition for the applicability of the plane-wave theory must be

$$\Delta\Theta \gg \Omega \qquad (4\text{-}187)$$

In electron diffraction, since $\Delta\Theta \simeq 10^{-2}$ and $\Omega$ can be reduced to $10^{-4}$ by means of a condenser lens, the above condition is attainable with sufficient accuracy. What must be investigated thoroughly is whether (4-187) is satisfied or not for the x rays and neutrons usually employed in experiments.

When the crystal is placed at a distance $L$ from the source and a slit of width $B$ is placed in front of the crystal for collimation, the angular width $\Omega$ is estimated to be $B/L$ according to the principles of geometrical optics. This is legitimate only when the slit width $B$ is sufficiently large. If the width $B$ is extremely narrow, $\Omega$ must be estimated by the principle of optical diffraction, so that it is $\lambda/B$, as already indicated in connection with the diffraction function [cf (3-95)]. Thus, roughly speaking, $\Omega$ may assume a minimum when both principles give the same answer, as shown in Fig.

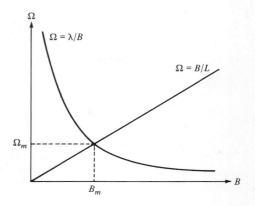

**FIGURE 4-22**
Angular divergence $\Omega$ of the wave vectors caused by limiting the wave front by a slit of width $B$.

4-22. From this argument, the attainable minimum value $\Omega_m$ and the corresponding slit width $B_m$ are given, respectively, by

$$\Omega_m = \left(\frac{\lambda}{L}\right)^{1/2} \quad (4\text{-}188a)$$

and

$$B_m = (\lambda L)^{1/2} \quad (4\text{-}188b)$$

For reasonable magnitudes, $\lambda = 1$ A and $L = 50$ cm, it turns out that $\Omega_m \sim 1.4 \times 10^{-5}$ and $B_m \sim 7 \times 10^{-4}$ cm. On the other hand, $\Delta\Theta \lesssim 10^{-6}$ in most cases. Thus, it is concluded that

$$\Delta\Theta \lesssim \Omega \quad (4\text{-}189)$$

in x-ray and neutron cases. The situation is schematically illustrated in Fig. 4-23. In order to attain the condition (4-187), say $\Omega = 10^{-6}$, a distance $L$ of 100 m is required!! No longer can the condition to the plane-wave approximation be satisfied in practice simply by using a slit system. It should be emphasized that the component Bloch waves excited in the crystal have mutual phase relations. For this reason, the crystal wave field is not merely the superposition of the intensity fields expected from the plane-wave theory. The difficulty here is not merely that of smearing due to the divergence of the incident wave.

**2. *Divergence of the rays in the crystal*** So far, the problem has been considered in reciprocal space. In this and the next sections, some aspects observed in real space owing to the divergence of the wave vectors are considered. As discussed in Section I B 6, the Bloch wave propagates in the direction normal to the dispersion surface at the corresponding dispersion point. As shown there, in general, the doubly refracted beams excited by a single plane wave propagate in different directions, the angle between which is the Bragg angle at maximum, or zero at minimum, depending on the extent to which the Bragg condition is satisfied. On the other hand, the beam width

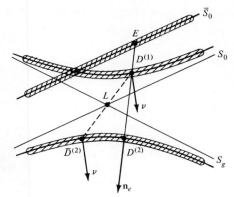

**FIGURE 4-23**
Excitation scheme for the dispersion points. ($\nu$ is normal of dispersion surface; $\mathbf{n}_e$ is normal of entrance surface.)

must be much less than $L\Delta\Theta \sim 5\ \mu$. If the separation angle of the doubly refracted beams is a few tenths of the Bragg angle, the beams are separated in crystal space soon after they pass through a distance of $100\ \mu$ in depth. Thus it is meaningless to consider the interference between the doubly refracted waves, as assumed in the plane-wave theory. It is more reasonable to consider the interference between two conjugate waves specified by $D$ and $\bar{D}$ in Fig. 4-23 because they propagate in the same direction in the crystal.

**3.** *Spherical form of the coherent wave front*  Consider two optical paths in vacuum, one satisfying the Bragg condition exactly and the other deviating by $\Delta\Theta$. Although $\Delta\Theta$ is very small, the real optical path is different from that expected on the basis of a plane wave, and the difference is estimated to be $\frac{1}{2}L(\Delta\Theta)^2$, which amounts to $2.5 \times 10^{-9}$ cm. It is indeed a quarter of the wavelength ($\lambda \sim 1$ A). Thus, if interference is in fact important, this deviation may not be neglected, and it seems that this fact gives another good reason for considering the spherical-wave theory. In the theory which will be developed in the next section (II B), however, partly due to mathematical difficulties, the sphericity of the wave front is neglected. In perfect crystals, fortunately, it turns out that this approximation does not give any serious error, as will be discussed in Section II C.

**B.  The spherical-wave theory for Laue cases**

**1.**  *The incident wave*  Electromagnetic waves emitted from an atomic source can be expressed to the order of $\lambda/r$ as follows (see Schiff, 1949):

$$\mathbf{H} = iK(\mathbf{J} \times \hat{\mathbf{r}})\Phi \qquad (4\text{-}190a)$$

$$\mathbf{E} = iK[(\mathbf{J} \times \hat{\mathbf{r}}) \times \hat{\mathbf{r}}]\Phi \qquad (4\text{-}190b)$$

where $\Phi$ is a scalar spherical wave of the form

$$\Phi = \frac{1}{4\pi r} \exp iKr \qquad (4\text{-}191)$$

and $\mathbf{J}$ is a constant vector proportional to an induced current due to a microscopic process of emission of photons, and $\hat{\mathbf{r}}$ is a unit vector from the source to the observation point. Incidentally, note that $\mathbf{H}$ and $\mathbf{E}$ are similar to their form in elastic scattering ($3\text{-}74a$ and $b$) due to an assemblage of electrons. The scalar field $\Phi$ can be represented by the Fourier transform

$$\Phi = \frac{i}{8\pi^2} \int\limits_{-\infty}^{+\infty}\!\!\int \frac{1}{K_z} \exp i(\mathbf{K} \cdot \mathbf{r})\, dK_x\, dK_y \qquad (4\text{-}192)$$

Since of concern here are the field vectors of x rays which are emitted within a small solid angle limited by a slit system, the polarization factors $(\mathbf{J} \times \hat{\mathbf{r}})$ and $[(\mathbf{J} \times \hat{\mathbf{r}}) \times \hat{\mathbf{r}}]$ are assumed to be constant. Moreover, since two-beam cases are treated in what follows, it is sufficient to consider the wave fields decomposed into two directions which are parallel and perpendicular to the reflection plane ($R$ plane). Each component wave, therefore, can be treated as if it were a scalar wave. Thus, each component of the incident wave can be expressed

$$E = E_e \Phi \qquad (4\text{-}193a)$$

$$H = H_e \Phi \qquad (4\text{-}193b)$$

The amplitudes $E_e$ and $H_e$ include the polarization factors $(\mathbf{J} \times \hat{\mathbf{r}})$ and $[(\mathbf{J} \times \hat{\mathbf{r}}) \times \hat{\mathbf{r}}]$. For ordinary light, they are the same for the two modes of polarization because, in that case, the direction of the current vector $\mathbf{J}$ is statistically random.

**2. Crystal wave fields** As discussed in Section I C 3a, the plane wave $E_e \exp i(\mathbf{K} \cdot \mathbf{r})$ excites the crystal wave fields $d_0^{(J)}(\mathbf{r})$ and $d_g^{(J)}(\mathbf{r})$ given by equations ($4\text{-}73$) or ($4\text{-}74$).[1] For the incident wave having the form ($4\text{-}193a$), therefore, $O$ and $G$ waves can be written

$$\phi_0(\mathbf{r}) = \frac{i}{8\pi^2} E_e \int\limits_{-\infty}^{+\infty}\!\!\int \frac{1}{K_z} d_0(\mathbf{K}; \mathbf{r}) \exp i(\mathbf{K} \cdot \mathbf{r})\, dK_x\, dK_y \qquad (4\text{-}194a)$$

$$\phi_g(\mathbf{r}) = \frac{i}{8\pi^2} E_e \exp 2\pi i(\mathbf{g} \cdot \mathbf{r}) \int\limits_{-\infty}^{+\infty}\!\!\int \frac{1}{K_z} d_g(\mathbf{K}; \mathbf{r}) \exp i(\mathbf{K} \cdot \mathbf{r})\, dK_x\, dK_y \qquad (4\text{-}194b)$$

---

[1] The subscript $e$ in $\mathbf{K}_e$ is omitted.

where, in the present case,

$$d_0(\mathbf{K}; \mathbf{r}) = \sum_{(j)}^{1,2} C_0^{(j)} \exp i[(\mathbf{k}_0^{(j)} - \mathbf{K}) \cdot (\mathbf{r} - \mathbf{r}_e)] \quad (4\text{-}195a)$$

$$d_g(\mathbf{K}; \mathbf{r}) = \sum_{(j)}^{1,2} C_g^{(j)} \exp i[(\mathbf{k}_0^{(j)} - \mathbf{K}) \cdot (\mathbf{r} - \mathbf{r}_e)] \quad (4\text{-}195b)$$

For the purpose of integrating equations (4-194a and b), it is convenient to take the $z$ axis in the direction of $\overline{\mathbf{K}}_0$ and the $x$ and $y$ axes within and perpendicular to the $R$ plane, respectively.[1]  Since, then, $d_0(\mathbf{K}; \mathbf{r})$ and $d_g(\mathbf{K}; \mathbf{r})$ are independent of $K_y$, the integration of $K_y$ can be performed by the use of Kelvin's stationary-phase method [Appendix 3A; cf (3A-5)].  Thus (4-194a and b) can be written

$$\phi_0(\mathbf{r}) = A_0 E_e \int_{-\infty}^{+\infty} d_0(\mathbf{K}; \mathbf{r}) \exp iK_x x_0 \, dK_x \quad (4\text{-}196a)$$

$$\phi_g(\mathbf{r}) = A_g E_e \int_{-\infty}^{+\infty} d_g(\mathbf{K}; \mathbf{r}) \exp iK_x x_0 \, dK_x \quad (4\text{-}196b)$$

where

$$A_0 = \frac{i}{4\pi} (2\pi Kr)^{-1/2} \exp i \left[ (\overline{\mathbf{K}}_0 \cdot \mathbf{r}) - \frac{\pi}{4} \right] \quad (4\text{-}197a)$$

$$A_g = \frac{i}{4\pi} (2\pi Kr)^{-1/2} \exp i \left[ (\overline{\mathbf{K}}_g \cdot \mathbf{r}) - \frac{\pi}{4} \right] \quad (4\text{-}197b)$$

Here, $(\overline{\mathbf{K}}_0 \cdot \mathbf{r}) = K_z$ and $K_z$ is replaced by $K$, approximately.  This is equivalent to neglecting the sphericity of the wave front as mentioned above.  We can see that this approximation is equivalent also to approximating the dispersion surfaces $\overline{S}_0$ and $\overline{S}_g$ by their respective tangential planes at the Laue point.

For integrating (4-196a and b), the phase term

$$G_0 = [(\mathbf{k}_0 - \mathbf{K}) \cdot (\mathbf{r} - \mathbf{r}_e) + K_x x_0] \quad (4\text{-}198)$$

is rewritten.  By introducing an oblique coordinate system $(\hat{\mathbf{K}}_0, \hat{\mathbf{K}}_g)$ having an origin at the entrance point $E$

$$(\mathbf{r} - \mathbf{r}_e) = l_0 \hat{\mathbf{K}}_0 + l_g \hat{\mathbf{K}}_g \quad (4\text{-}199)$$

where $(l_0, l_g)$ are the coordinates (see Fig. 4-24).  According to (4-23a) and (4-25a), it is known that

$$(\mathbf{k}_0 - \mathbf{K}) = \Delta \mathbf{K} + \Delta \mathbf{k} - K_x \hat{\mathbf{x}}_0 \quad (4\text{-}200)$$

Combining the above two,

$$G_0 = \tfrac{1}{2} K \chi_0 (l_0 + l_g) + (\Delta \eta_0 l_0 + \Delta \eta_g l_g) \quad (4\text{-}201a)$$

$$= \tfrac{1}{2} K \chi_0 (l_0 + l_g) + \frac{\Delta \eta_0 x_g + \Delta \eta_g x_0}{\sin 2\theta_B} \quad (4\text{-}201b)$$

---

[1] Later, these coordinate axes will be denoted by the subscript 0.

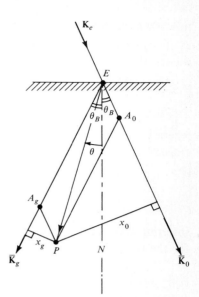

**FIGURE 4-24**
Coordinates used in the spherical-wave theory (Laue case). ($l = EP$, $l_0 = EA_0 + A_gP$, $l_g = EA_g = A_0P$, and $EN$ is the net plane.)

where $x_0$ and $x_g$ are perpendiculars from the observation point $P$ to the two edges $ET$ and $ER$, respectively. The *Resonanzfehler* $\Delta\eta_0$ and $\Delta\eta_g$ are defined by (4-27) and given by (4-60), which are easily transformed to be functions of $s$ and $\beta$. Thus, (4-196a and b) can be written in the explicit form

$$\phi_0(\mathbf{r}) = A_0 E_e \exp\frac{i}{2} K\chi_0(l_0 + l_g) \times \frac{1}{2} \sum_{(j)}^{1,2} \int_{-\infty}^{+\infty} \left[ \frac{-s \pm \sqrt{s^2 + \beta^2}}{\pm\sqrt{s^2 + \beta^2}} \right]$$

$$\times \exp i[\pm\xi_1\sqrt{s^2 + \beta^2} - \xi_2 s]ds^r \qquad (4\text{-}202a)$$

$$\phi_g(\mathbf{r}) = A_g E_e \exp\frac{i}{2} K\chi_0(l_0 + l_g) \times \left(\frac{\chi_g}{\chi_{-g}}\right)^{1/2} \left(\frac{\gamma_0}{\gamma_g}\right)^{1/2} \frac{1}{2} \sum_{(j)}^{1,2}$$

$$\times \int_{-\infty}^{+\infty} \left[ \frac{\beta}{\pm\sqrt{s^2 + \beta^2}} \right] \exp i[\pm\xi_1\sqrt{s^2 + \beta^2} - \xi_2 s]ds^r \qquad (4\text{-}202b)$$

where $\xi_1$ and $\xi_2$ are functions of $x_0$ and $x_g$, and the plus and minus signs correspond to the branch indexes (1) and (2), respectively. The wave fields (4-202a and b) often appear in other cases of spherical-wave theory also. The arguments presented here for deriving them and the following analyses are widely applicable.

As shown in Appendix 4C, the integral sums are essentially a kind of integral definitions of Bessel functions $J_1$ and $J_0$, and it turns out that they are functions of $\xi_1 \pm \xi_2$. For this reason, their explicit forms shall be derived at this time. First, note that the term proportional to $\Delta\eta_0$ in (4-201a and b) does not contribute to $\xi_1 + \xi_2$ since the coefficients of $\pm\sqrt{s^2 + \beta^2}$ and $-s$ have opposite signs. Similarly, the term proportional to $\Delta\eta_g$ is irrelevant to $\xi_1 - \xi_2$. Thus, one can write

$$[G_0]_+ = \Delta\eta_g l_g \qquad (4\text{-}203a)$$

$$[G_0]_- = \Delta\eta_0 l_0 \qquad (4\text{-}203b)$$

where $[\;\;]_\pm$ indicate operators to pick up the relevant terms in obtaining $\xi_1 \pm \xi_2$, respectively. Consequently,

$$\xi_1 + \xi_2 = x_0 \qquad (4\text{-}204a)$$

$$\xi_1 - \xi_2 = \left(\frac{\gamma_0}{\gamma_g}\right) x_g \qquad (4\text{-}204b)$$

Inserting these into equations (4C-7a and b) [$m = 1$ for the $O$ wave and $m = 0$ for the $G$ wave], one obtains the wave fields as follows:

$$\phi_0(\mathbf{r}) = -(\pi\beta_0)A_0 E_e \exp\frac{i}{2} K\chi_0(l_0 + l_g)\sqrt{\frac{x_g}{x_0}} J_1(\beta_0\sqrt{x_0 x_g}) \qquad x_0 x_g > 0 \qquad (4\text{-}205a)$$

$$= 0 \qquad\qquad x_0 x_g < 0$$

and

$$\phi_g(\mathbf{r}) = i(\pi\beta_0)\left(\frac{\chi_g}{\chi_{-g}}\right)^{1/2} A_g E_e \exp\frac{i}{2} K\chi_0(l_0 + l_g)\,\mathrm{sign}\,(x_0)J_0(\beta_0\sqrt{x_0 x_g}) \qquad x_0 x_g > 0$$

$$= 0 \qquad\qquad x_0 x_g < 0$$

$$(4\text{-}205b)$$

where $\beta_0$ is defined by

$$\beta_0 = \sqrt{\frac{\gamma_0}{\gamma_g}}\,\beta = \frac{KC(\chi_g\chi_{-g})^{1/2}}{\sin 2\theta_B} \qquad (4\text{-}206)$$

The crystal wave fields exist only within a triangular region (Borrmann fan). Since the Bessel functions have an oscillating character, one may expect fringe systems in this region. The details will be discussed below and also in Section F 3 in connection with experimental results.

**3.  *Approximate behavior of the wave fields***   Here, an approximate form of (4-205a and b) will be considered. It is well known that the Bessel functions $J_0(\zeta)$ and $J_1(\zeta)$ have the following asymptotic forms for large values of $|\zeta|$, respectively:

$$J_0(\zeta) = \left(\frac{1}{2\pi\zeta}\right)^{1/2}\left[\exp i\left(\zeta - \frac{\pi}{4}\right) + \exp - i\left(\zeta - \frac{\pi}{4}\right)\right] \qquad (4\text{-}207a)$$

$$J_1(\zeta) = \left(\frac{1}{2\pi\zeta}\right)^{1/2}\left[\exp i\left(\zeta - \frac{3\pi}{4}\right) + \exp - i\left(\zeta - \frac{3\pi}{4}\right)\right] \qquad (4\text{-}207b)$$

These are valid for complex $\zeta$. In the present problem,

$$\zeta = \beta_0 \sqrt{x_0 x_g} \qquad (4\text{-}208a)$$

From Fig. 4-24, one can see that

$$x_0 = l \sin (\theta_B + \theta) \qquad x_g = l \sin (\theta_B - \theta)$$

where $l$ is the distance $EP$ which forms an angle $\theta$ with the net plane concerned. Therefore,

$$\zeta = \tfrac{1}{2} KC(\chi_g \chi_{-g})^{1/2} (1 - p^2)^{1/2} \frac{\cos \theta}{\cos \theta_B} l \qquad (4\text{-}208b)$$

$$= \tfrac{1}{2} KC(\chi_g \chi_{-g})^{1/2} (1 - p^2)^{1/2} (l_0 + l_g) \qquad (4\text{-}208c)$$

It is easily seen that the coefficient of $l$ in $\pm \zeta$ is identical to $(\Delta \mathbf{k} \cdot \mathbf{v})$ calculated in (4-154b).

Inserting (4-207a and b) into (4-205a and b), respectively, one obtains the approximate forms of the wave fields

$$\phi_0(\mathbf{r}) = \left( \frac{\pi}{2} \beta_0 \right)^{1/2} A_0 E_e \exp \frac{i}{2} K\chi_0(l_0 + l_g)$$

$$\times \sqrt{\frac{x_g}{x_0}} (x_0 x_g)^{-1/4} \left[ \exp i \left( \zeta + \frac{\pi}{4} \right) + \exp - i \left( \zeta + \frac{\pi}{4} \right) \right] \qquad (4\text{-}209a)$$

$$\phi_g(\mathbf{r}) = \left( \frac{\pi}{2} \beta_0 \right)^{1/2} \left( \frac{\chi_g}{\chi_{-g}} \right)^{1/2} A_g E_e \exp \frac{i}{2} K\chi_0(l_0 + l_g)$$

$$\times (x_0 x_g)^{-1/4} \left[ \exp i \left( \zeta + \frac{\pi}{4} \right) - \exp - i \left( \zeta + \frac{\pi}{4} \right) \right] \qquad (4\text{-}209b)$$

Thus, one can interpret each of the crystal wave fields along a direction $\mathbf{v}$ as being composed of two plane waves. The total phases of the respective waves can be expressed

$$\tilde{S}_0 = (\bar{\mathbf{K}}_0 \cdot \mathbf{r}_e) + S_0 + \{^{\pi/2}_0\} \qquad (4\text{-}210a)$$

$$\tilde{S}_g = (\bar{\mathbf{K}}_g \cdot \mathbf{r}_e) + S_g + \{^{\pi/2}_\pi\} + \delta \qquad (4\text{-}210b)$$

where $\delta$ is the phase of $(\chi_g / \chi_{-g})^{1/2}$ and $S_0$ and $S_g$ are the phase changes of the crystal waves along a direction based on the plane-wave theory. They are defined by (4-147a and b), and the explicit forms are given by (4-154c and d). For this reason, the terms $\exp \pm i(\zeta + \pi/4)$ can be regarded as the waves of branches (1) and (2).

*a.  Pendellösung fringes*  Because the two wave fields overlap in space and their phase relations are definite, interference fringes are expected. The fringe spacing along the ray direction $\mathbf{v}$ is given by

$$\Lambda_s = \frac{2\pi}{(\mathbf{k}_0^{(1)} - \bar{\mathbf{k}}_0^{(2)}) \cdot \mathbf{v}} \qquad (4\text{-}211a)$$

To denote the conjugate relationship a bar is put over $\mathbf{k}_0^{(2)}$. (The wave vectors denote the real parts.) It is worthwhile to compare this with (4-124a) in the plane-wave theory. More explicitly, one can rewrite (4-211a)

$$\Lambda_s = \frac{2\pi}{KCR_e(\chi_g\chi_{-g})^{1/2}(1 - p^2)^{1/2}(\cos\theta/\cos\theta_B)} \quad (4\text{-}211b)$$

In the present approximation, the fringe spacing is constant throughout the crystal, similarly to the case of the plane-wave theory. According to the rigorous solution (4-205a and b), however, the spacing of the first few fringes is shorter than the value given by (4-211b).

The additional phases in braces in (4-210a and b) are intrinsic to the spherical-wave theory. Their origin will be discussed later, after considering the ray trajectories in detail. The presence of these phase terms has nothing to do with the fringe spacing, but it is relevant to their absolute positions. The details will be discussed in Section II F 3.

b.  *Absorption*   The absorption of waves interpreted to be the waves of branches (1) and (2) is determined by the imaginary part of $S_0$ and $S_g$. The results of the plane-wave theory described in Section I D 5 can be used without any modification. In particular, the attenuation of the rays can be determined by $(\mu_B l)$ given by equations (4-157) and (4-158). The spherical-wave theory, however, does not have to assume the SI approximations. In this sense, the present formulation is generally valid even when $(\chi_g\chi_{-g})^i$ is comparable with $(\chi_g\chi_{-g})^r$ in higher order reflections of complex crystals.

c.  *Intensity profiles*   If one neglects interference, the intensity distributions

$$i_0 = \frac{|\beta_0|}{32\pi^2 Kr} |E_e|^2 (x_0 x_g)^{-1/2} \frac{x_g}{x_0} \exp -\mu_B l \quad (4\text{-}212a)$$

$$i_g = \frac{|\beta_0|}{32\pi^2 Kr} \left|\frac{\chi_g}{\chi_{-g}}\right| |E_e|^2 (x_0 x_g)^{-1/2} \exp -\mu_B l \quad (4\text{-}212b)$$

For symmetrical Laue cases, one can show that these results are the same as those given by (4-143a and b) within the SI approximations, if one remembers that the intensity of the incident wave at a distance $r$ per unit glancing angle is $(1/16\pi^2 r)|E_e|^2$, and $\Delta\Theta = |\beta_0|/K$. The rigorous solutions (4-205), however, are different from the relations (4-212a and b) near the margins ($x_0 = 0$ and $x_g = 0$) of the wave fields. For example, $I_g$ tends to infinity at $x_0 = 0$ and $x_g = 0$ in (4-212b), whereas $|\phi_g|^2$ given by (4-205b) is finite, as it should be.

4. **Transmitted wave fields (Laue-Laue case)** Similarly to the crystal wave fields, the transmitted vacuum waves can be represented in the integral forms

$$\Phi_0(\mathbf{r}) = \frac{i}{8\pi^2} E_e \int\!\!\!\int_{-\infty}^{+\infty} \frac{1}{K_z} E_0(\mathbf{K}; \mathbf{r}) \exp i(\mathbf{K} \cdot \mathbf{r}) \, dK_x \, dK_y \qquad (4\text{-}213a)$$

$$\Phi_g(\mathbf{r}) = \frac{i}{8\pi^2} E_e \exp 2\pi i(\mathbf{g} \cdot \mathbf{r}) \int\!\!\!\int_{-\infty}^{+\infty} \frac{1}{K_z} E_g(\mathbf{K}; \mathbf{r}) \exp i(\mathbf{K} \cdot \mathbf{r}) \, dK_x \, dK_y \qquad (4\text{-}213b)$$

where $E_0(\mathbf{K}; \mathbf{r})$ and $E_g(\mathbf{K}; \mathbf{r})$ are obtained, as follows, from the plane-wave solution (4-105).

$$E_0(\mathbf{K}; \mathbf{r}) = \sum_j^{1,2} C_0^{(j)} \exp i[(\mathbf{k}_0^{(j)} - \mathbf{K}) \cdot (\mathbf{r} - \mathbf{r}_e) + (\mathbf{K}_0^{(j)} - \mathbf{k}_0^{(j)}) \cdot (\mathbf{r} - \mathbf{r}_a)] \qquad (4\text{-}214a)$$

$$E_g(\mathbf{K}; \mathbf{r}) = \sum_j^{1,2} C_g^{(j)} \exp i[(\mathbf{k}_0^{(j)} - \mathbf{K}) \cdot (\mathbf{r} - \mathbf{r}_e) + (\mathbf{K}_g^{(j)} - \mathbf{k}_g^{(j)}) \cdot (\mathbf{r} - \mathbf{r}_a)] \qquad (4\text{-}214b)$$

The integration with respect to $K_y$ can be performed in the same way as in the case of the crystal wave. Thus, one gets

$$\Phi_0(\mathbf{r}) = A_0 E_e \int_{-\infty}^{+\infty} E_0(\mathbf{K}; \mathbf{r}) \exp iK_x x_0 \, dK_x \qquad (4\text{-}215a)$$

$$\Phi_g(\mathbf{r}) = A_g E_e \int_{-\infty}^{+\infty} E_g(\mathbf{K}; \mathbf{r}) \exp iK_x x_0 \, dK_x \qquad (4\text{-}215b)$$

where $A_0$ and $A_g$ are defined by (4-197a and b). Here, however, $\mathbf{r}$ is referred to the observation point $P$ in vacuum.

The phase terms in the integrand can be written

$$H_0 = G_0 + (\mathbf{K}_0 - \mathbf{k}_0) \cdot (\mathbf{r} - \mathbf{r}_a) \qquad (4\text{-}216a)$$

$$H_g = G_0 + (\mathbf{K}_g - \mathbf{k}_g) \cdot (\mathbf{r} - \mathbf{r}_a) \qquad (4\text{-}216b)$$

where $G_0$ is defined by (4-198) and is written explicitly in the form (4-201) where the coordinates $(l_0, l_g)$ and $(x_0, x_g)$ are referred to the observation point $P$ in vacuum. From relations (4-108a and b)

$$(\mathbf{K}_0 - \mathbf{k}_0) \cdot (\mathbf{r} - \mathbf{r}_a) = K\delta_{a0} T_a = -(\tfrac{1}{2} K\chi_0 + \Delta\eta_0) \frac{T_a}{\gamma_0'} \qquad (4\text{-}217a)$$

$$(\mathbf{K}_g - \mathbf{k}_g) \cdot (\mathbf{r} - \mathbf{r}_a) = K\delta_{ag} T_a = -(\tfrac{1}{2} K\chi_0 + \Delta\eta_g) \frac{T_a}{\gamma_g'} \qquad (4\text{-}217b)$$

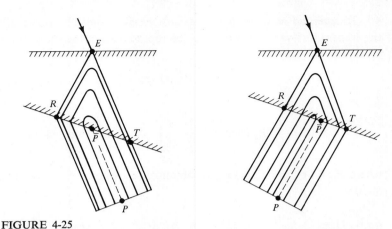

**FIGURE 4-25**
Relation between the crystal and the vacuum-wave fields. $P\bar{P} = T_a/\gamma_0'$ (for the $O$ wave) and $P\bar{P} = T_a/\gamma_g'$ (for the $G$ wave), where $T_a$ is the normal distance from $P$ to the exit surface.

Since $(T_a/\gamma_0')$ and $(T_a/\gamma_g')$ are the distances from the observation point (in vacuum) to the exit surface along $\bar{\mathbf{K}}_0$ and $\bar{\mathbf{K}}_g$, respectively,

$$H_0 \text{ and } H_g = \tfrac{1}{2}K\chi_0(\bar{l}_0 + \bar{l}_g) + \Delta\eta_0\bar{l}_0 + \Delta\eta_g\bar{l}_g \qquad (4\text{-}218)$$

where $(\bar{l}_0, \bar{l}_g)$ are the oblique coordinates of the point $\bar{P}$ on the exit surface corresponding to the observation point $P$ (see Fig. 4-25).

In order to obtain the wave fields $\Phi_0(\mathbf{r})$ and $\Phi_g(\mathbf{r})$, as in (4-202a and b), calculate $\xi_1 \pm \xi_2$, where $\xi_1$ and $\xi_2$ are the coefficients of $\pm\sqrt{s^2 + \beta^2}$ and $-s$ in the expressions for $H_0$ or $H_g$, respectively. By the use of the same arguments as those used in deriving (4-203a and b),

$$[H_0]_+ = [H_g]_+ = \Delta\eta_g\bar{l}_g \qquad (4\text{-}219a)$$

$$[H_0]_- = [H_g]_- = \Delta\eta_0\bar{l}_0 \qquad (4\text{-}219b)$$

From these, finally, one finds that

$$\xi_1 + \xi_2 = \bar{x}_0 \qquad (4\text{-}220)$$

$$\xi_1 - \xi_2 = \frac{\gamma_0}{\gamma_g}\bar{x}_g \qquad (4\text{-}221)$$

where $\bar{x}_0$ and $\bar{x}_g$ are the coordinates of $\bar{P}$, defined above, corresponding to the observation point $P$.

Thus, the expressions for $\Phi_0(\mathbf{r})$ and $\Phi_g(\mathbf{r})$ are very similar to (4-202a and b).

After the integration,

$$\Phi_0(\mathbf{r}) = -(\pi\beta_0)A_0E_e \exp\frac{i}{2}K\chi_0(\overline{l_0} + \overline{l_g})\sqrt{\frac{\bar{x}_g}{\bar{x}_0}} J_1(\beta_0\sqrt{\bar{x}_0\bar{x}_g}) \quad \bar{x}_0\bar{x}_g > 0$$

$$= 0 \qquad\qquad\qquad\qquad\qquad \bar{x}_0\bar{x}_g < 0 \qquad (4\text{-}222a)$$

$$\Phi_g(\mathbf{r}) = i(\pi\beta_0)\left(\frac{\chi_g}{\chi_{-g}}\right)^{1/2} A_gE_e \exp\frac{i}{2}K\chi_0(\overline{l_0} + \overline{l_g})J_0(\beta_0\sqrt{\bar{x}_0\bar{x}_g}) \quad \bar{x}_0\bar{x}_g > 0$$

$$= 0 \qquad\qquad\qquad\qquad\qquad \bar{x}_0\bar{x}_g < 0 \qquad (4\text{-}222b)$$

Except for the phases and amplitudes in $A_0$ and $A_g$, the transmitted vacuum waves are identical to the wave fields at the point $\bar{P}$ corresponding to the observation point $P$. The intensity fields are essentially the projections of the intensity on the exit surface in the directions of $\bar{\mathbf{K}}_0$ and $\bar{\mathbf{K}}_g$. This is what is actually observed in the section topographs of perfect crystals.

## C.  Ray considerations for the crystal wave fields

In the previous section, it was shown that the crystal wave fields can be interpreted by two sets of bundles of rays. In the present section, the ray theory is described in more detail by using the stationary-phase method.

**1.  Rays associated with the wave fields given by (4-202)**  The asymptotic expansions (4-207a and b) are derived from the integral definitions of the Bessel functions by the use of the method of steepest descents, which is applicable to complex functions. Thus, if one is concerned merely with the final expressions for the wave fields, the application of the stationary-phase method is a less general prototype of Section B 3. The method, however, is very powerful even when the exact integration of relations such as (4-202) is not possible. In fact, most cases of crystals containing a fault plane belong to this category. For this reason, and also in order to elucidate the connection between the ray and wave-optical theories, the physical implication of the stationary-phase method are described by application to relations (4-202a and b).

First, assume the SI approximation described in Section I C 2. Then, the real part of the phase of the crystal waves is identical to that in nonabsorbing cases, the imaginary part being included in the amplitude in the form $\exp - \frac{1}{2}\mu_B l$ [cf (4-157)]. The phase of interest to the present problem is

$$T = (\mathbf{K} \cdot \mathbf{r}_e) + [\mathbf{k} \cdot (\mathbf{r} - \mathbf{r}_e)] \qquad K_y = 0 \qquad\qquad (4\text{-}223a)$$

$$\simeq (\bar{\mathbf{K}}_0 \cdot \mathbf{z}) + \tfrac{1}{2}K\chi_0{}^r(l_0 + l_g) \pm \xi_1\sqrt{s^2 + \beta^2} - \xi_2 s \qquad (4\text{-}223b)$$

where $s$ and $\sqrt{s^2 + \beta^2}$ are the real parts. Then, the integrals appearing in (4-202$a$ and $b$) have the form (3A-1).

The stationary condition for the phase gives

$$\frac{\partial T}{\partial s} = \pm \frac{\xi_1 s}{\sqrt{s^2 + \beta^2}} - \xi_2 = 0 \qquad (4\text{-}224)$$

This equation implies a set of straight lines (rays) specified by the parameter $s$, because $\xi_1$ and $\xi_2$ are linear with respect to the coordinates $x_0$ and $x_g$. Obviously, for large $|s|$, the rays tend to either one of the lines ($\xi_1 \pm \xi_2$) = 0; i.e., $x_0 = 0$ and $x_g = 0$, which correspond to $ET$ and $ER$ in Fig. 4-25, respectively. Since, if the conditions $\xi_1 = 0$ and $\xi_2 = 0$ are satisfied simultaneously, (4-224) holds for any value of $s$, and all rays given by (4-224) pass through the entrance point $E(x_0 = x_g = 0)$.

From (4-224), it immediately follows that

$$\frac{s}{\beta} = \pm \frac{(\text{sign } \xi_1)\xi_2}{(\xi_1^2 - \xi_2^2)^{1/2}} \qquad (4\text{-}225a)$$

$$\frac{\sqrt{s^2 + \beta^2}}{\beta} = \frac{(\text{sign } \xi_1)\xi_1}{(\xi_1^2 - \xi_2^2)^{1/2}} \qquad (4\text{-}225b)$$

The plus and minus signs correspond to the waves of branches (1) and (2), respectively. If $\xi_1^2 - \xi_2^2 > 0$; i.e., $x_0 x_g > 0$, (4-225$a$) gives the deviation parameter $s$ for the rays which pass through the point $P(x_0, x_g)$. On the other hand, for $\xi_1^2 - \xi_2^2 < 0$, no real parameter $s$ can be assigned to the position $P$. This indicates that no real ray exists outside the triangular fan $TER$.

Inserting (4-225$a$ and $b$) into (4-223$b$), one obtains the phase as functions of $x_0$ and $x_g$ (for $x_0 x_g \geq 0$),

$$[T] - K_z - \tfrac{1}{2}K\chi_0'(l_0 + l_g) = \pm[\text{sign } \xi_1](\xi_1^2 - \xi_2^2)^{1/2}\beta$$
$$= \pm[\text{sign } x_0](x_0 x_g)^{1/2}\beta_0 \qquad (4\text{-}226)$$

As described in Appendix 3AI, it is necessary to add the constant phase $\pm\pi/4$ to the true value of the phase, the plus or minus signs being selected in accordance with the sign of $\partial T^2/\partial s^2$. In this way one obtains the phases of the two terms in brackets in (4-209$a$ and $b$).

The amplitudes are obtained by multiplying by the expressions for the amplitudes in the plane-wave theory: $(2\pi K/r)^{1/2}$ (owing to the integration of $K_y$) and $(2\pi/|\partial^2 T/\partial s^2|)^{1/2} = (2\pi|\kappa|/l)^{1/2}$ (owing to the integration of $K_x$),[1] where $\kappa$ is $[\partial^2(\mathbf{k} \cdot \mathbf{v})/\partial K_x^2]^{-1}$, the radius of curvature of the dispersion surface. This relation is easily obtained by recalling that $[\mathbf{k} \cdot (\mathbf{r} - \mathbf{r}_e)] = (\mathbf{k} \cdot \mathbf{v})l$ and $ds = -dK_x$.

---

[1] Alternatively, $\partial^2 T/\partial s^2 = \pm[\text{sign } \xi_1](1/\beta)(\xi_1^2 - \xi_2^2)^{3/2}/\xi_1^2$, where the plus and minus signs correspond to the branches (1) and (2). In the present problem [sign $\xi_1$] is positive, since the divergent fan $TER$ is being considered.

The two factors are the effective ranges of the integral variables $K_x$ and $K_y$, respectively. The component waves in this range contribute to the wave fields at $\mathbf{r}$. The similarity of these two factors can be seen from their physical meanings. The radius $\kappa$ plays the role of $K$. The factors $(1/r)^{1/2}$ and $(1/l)^{1/2}$ are reasonable because the x rays emitted from the source spread perpendicularly to the reflection ($R$) plane and propagate in a Borrmann fan from the entrance point $E$. The divergence of the ray direction is amplified by the factor $K/\kappa$ in the crystal. In the vicinity of the exact Bragg condition, $K/\kappa$ is equivalent to $\theta_B/\Delta\Theta \sim 10^4$.

Increasing $r$ and $l$, narrower regions on the dispersion surface contribute to the wave field. Thus, a correspondence is expected between the point on the dispersion surface and the position in real space. This is what is encountered in Fraunhofer diffraction. Owing to the smallness of $\kappa$ in the vicinity of the exact Bragg condition, such a situation characterizing Fraunhofer diffraction is realized for very small distances of wave propagation.

The presence of the factor $|\kappa|^{1/2}$ and its variation on the dispersion surface introduces the intensity enhancement at the margins of $G$-wave section patterns, previously discussed in Section I D 4. The radius $\kappa$ tends to infinity, as the rays come closer to the margins, owing to the plane approximation of the asymptotic surfaces $S_0$ and $S_g$. Obviously, in this case, the stationary-phase method loses its validity.

**2. Effect of the spherical form of the wave front of the incident wave** So far, it has been assumed that the dispersion surface $\bar{S}_0$ is a plane. Here, the exact form is considered, namely, instead of taking $K$ for $[K_z]$ in (4-197a and b),

$$[K_z] = (K^2 - K_x^2)^{1/2} \qquad (4\text{-}227)$$

is assumed. In vacuum, the stationary condition of the phase can be assumed

$$\left(\frac{\partial \mathbf{K}}{\partial K_x} \cdot \mathbf{r}_e\right) = 0 \qquad (4\text{-}228)$$

As discussed in Section I B 6 the ray direction $\mathbf{r}_e$ from the source must be perpendicular to the dispersion surface $\bar{S}_0$, namely, $\mathbf{r}_e$ is parallel to $[\mathbf{K}]$. No longer can $\mathbf{r}_e$ be assumed to be a fixed point. Instead it is a function of $K_x$. The stationary condition of the total phase is given by

$$\frac{\partial T}{\partial K_x} = \left(\frac{\partial \mathbf{K}}{\partial K_x} \cdot \mathbf{r}_e\right) + \left[\frac{\partial \mathbf{k}}{\partial K_x} \cdot (\mathbf{r} - \mathbf{r}_e)\right] + \left[(\mathbf{K} - \mathbf{k}) \cdot \frac{\partial \mathbf{r}_e}{\partial K_x}\right] = 0 \qquad (4\text{-}229a)$$

By virtue of (4-228), the first term is equal to zero. In addition, since $(\mathbf{K} - \mathbf{k})$ is always normal to $\partial \mathbf{r}_e/\partial K_x$ owing to the tangential continuity of the wave vectors,

$$\left[\frac{\partial \mathbf{k}}{\partial K_x} \cdot (\mathbf{r} - \mathbf{r}_e)\right] = 0 \qquad (4\text{-}229b)$$

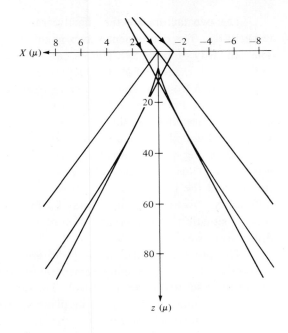

**FIGURE 4-26**
Trajectories of the crystal waves. (After
S. Homma.)

Again, the rays $(\mathbf{r} - \mathbf{r}_e)$ in the crystal form a set of straight lines, which are normal to the dispersion surface. As opposed to the previous treatment, however, the crystal rays start from different positions.

Significant difference between the rays belonging to branches (1) and (2) must be noted. For the latter, $(\partial^2 \mathbf{K}/\partial K_x^2) \cdot \mathbf{r}_e$ and $(\partial^2 \mathbf{k}/\partial K_x^2) \cdot (\mathbf{r} - \mathbf{r}_e)$ have the same sign. For branch (1), in contrast, they have opposite signs. Thus, somewhere in the crystal,

$$\frac{\partial^2 T}{\partial K_x^2} = \left[ \frac{\partial^2 \mathbf{K}}{\partial K_x^2} \cdot \mathbf{r}_e \right] + \left[ \frac{\partial^2 \mathbf{k}}{\partial K_x^2} \cdot (\mathbf{r} - \mathbf{r}_e) \right] \qquad (4\text{-}230)^1$$

changes its sign for branch (1), whereas it increases monotonically for branch (2). Mathematically,

$$\frac{\partial^2 T}{\partial K_x^2} = 0 \qquad (4\text{-}231)$$

is either the envelope or the focus for a bundle of the lines expressed by (4-229b). In optics, they are called *caustics*. Thus, the rays of branch (1) have caustics. The behavior of these rays is illustrated in Fig. 4-26.

---

[1] The additional term $2\{[\partial(\mathbf{K} - \mathbf{k})/\partial K_x] \cdot (\partial \mathbf{r}_e/\partial K_x)\} + [(\mathbf{K} - \mathbf{k}) \cdot (\partial^2 \mathbf{r}_e/\partial K_x^2)]$ equals zero owing to the tangential continuity of $\mathbf{K}$ and $\mathbf{k}$.

As described in Appendix 3A, the additional phase factors $\exp \pm (\pi/4)i$ must be introduced depending on the sign of $\partial^2 T/\partial K_x^2$. Before and after the caustics, therefore, one may expect a phase jump of $\pi/2$. The underlying optical principle is the same as that in the phase jump at the focus of a cylindrical lens. For a further discussion of this topic, the reader is referred to Sommerfeld's textbook on optics (p. 323).

The spherical form of the dispersion surface $\bar{S}_0$ introduces some of the complexities described above, although the meaning of the phase jump of the crystal waves is well understood. The theory described in Section II B is exact only when the x-ray source is located on the entrance surface. Then, the approximation $[K_z] = K$ does not affect the final result. The next step of approximation, obviously, is to take $[K_z] = K - \frac{1}{2}K_x^2/K$. Then, the stationary-phase method described in the preceding section may be used. Such a modification is not serious for the important range $|K_x| < K\Delta\Theta$. Even for $|K_x| > K\Delta\Theta$, the additional correction $(-K_x^2/2K)z$ to $K_z$ in the expression (4-197a and b) must be practically identical for the waves labeled by (1) and (2), because waves (1) and (2), propagating in the same direction $v$, are specified by the conjugate dispersion points, namely, $\pm K_x$, neglecting refraction. Thus, as far as the section patterns of perfect crystals and the phase difference between waves (1) and (2) are concerned, similarly to the case of *Pendellösung* fringes, the phase correction is irrelevant to the final result. For certain problems of distorted crystals and x-ray interferometry, however, the $K_x$ dependence of the phase correction may play a significant role in the interference phenomena. The preceding was developed by Kato (1960; 1961a,b; 1968b) and reviewed by him (Kato, 1963; 1968).

## D. Laue-Bragg case

The problem now is to consider the reflection and the transmission of the crystal wave fields (4-205a and b) at an exit surface. The triangular fan having its origin at $A_1$ in Fig. 4-27 is the region of interest here. Similarly to the Laue and Laue-Laue cases, the spherical-wave solutions can be obtained from the plane-wave solutions. The reflected crystal waves and the transmitted vacuum waves must have the same form as equations (4-196) and (4-213), respectively. After the integration, it turns out that the wave fields are strikingly similar to those for the Laue and Laue-Laue cases. They can be interpreted, therefore, with the ray-optical considerations described in the previous section. Types I and II are treated separately. The analysis presented here and in the next section has been discussed fully in the original paper by Saka, Katagawa, and Kato (1972a,b; 1973). The same problem also has been discussed by Uragami (1969; 1970; 1971) on the basis of Takagi's approach (1969) to the dynamical theory.

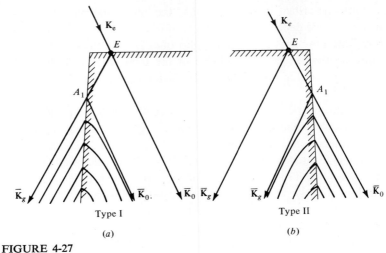

**FIGURE 4-27**
Laue-Bragg cases.

**1.** *Type I* In this case, according to (4-109a and b), the amplitude densities can be written as

$$d_0(\mathbf{K}; \mathbf{r}) = - \sum_{(j)}^{1,2} C_0^{(j)} \exp i[(\mathbf{k}_0^{(j)} - \mathbf{K}) \cdot (\mathbf{r} - \mathbf{r}_e) + (\mathbf{k}_{0,r}^{(j)} - \mathbf{k}_0^{(j)}) \cdot (\mathbf{r} - \mathbf{r}_a)]$$

$$(4\text{-}232a)$$

$$d_g(\mathbf{K}; \mathbf{r}) = - \sum_{(j)}^{1,2} \frac{c_g^{(j)}}{c_{g,r}^{(j)}} C_g^{(j)} \exp i[(\mathbf{k}_0^{(j)} - \mathbf{K}) \cdot (\mathbf{r} - \mathbf{r}_e) + (\mathbf{k}_{0,r}^{(j)} - \mathbf{k}_0^{(j)}) \cdot (\mathbf{r} - \mathbf{r}_a)]$$

$$(4\text{-}232b)$$

Now, the wave fields can be expressed in the same form as (4-202a and b). The phase of each wave in the integrands is given by

$$G_r = G_0 + (\mathbf{k}_{0,r} - \mathbf{k}_0) \cdot (\mathbf{r} - \mathbf{r}_a) \quad (4\text{-}233)[1]$$

where $G_0$ is defined by (4-198) and explicitly given in (4-201a or b). Since $\mathbf{r}_a$ denotes an arbitrary point on the exit surface, either $A_1$ or $A_2$ in Fig. 4-28 is used as this point. The vector $(\mathbf{r} - \mathbf{r}_a)$, therefore, can be replaced by the two alternative vectors

$$\overrightarrow{A_1 P} = l_0 \hat{\mathbf{K}}_0 + l_g' \hat{\mathbf{K}}_g \quad (4\text{-}234a)$$

$$\overrightarrow{A_2 P} = l_0' \hat{\mathbf{K}}_0 + l_g \hat{\mathbf{K}}_g \quad (4\text{-}234b)$$

where $l_0$ and $l_g$ are the coordinates defined in connection with (4-199) and $(l_0', l_g')$ are

---

[1] The superscript $(j)$ is omitted.

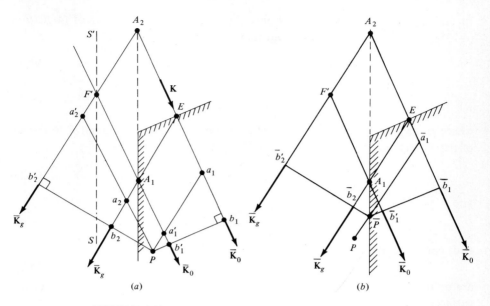

**FIGURE 4-28**
Coordinates in Laue-Bragg cases (type I). (*a*) The observation point $P$ is placed inside the crystal. $[(l_0,l_g) = (Ea_1,Ea_2), (l'_0,l'_g) = (F'a'_1,F'a'_2), (x_0,x_g) = (Pb_1; Pb_2) = \sin 2\theta_B(l_g,l_0), (x'_0,x'_g) = (Pb'_1,Pb'_2) = \sin 2\theta_B(l'_g,l'_0).]$ (*b*) The observation point $P$ is placed in vacuum. $[\bar{l}_0 = l_0,\bar{l}_g) = (E\bar{a}_1,\bar{a}_1\bar{P}), (\bar{x}_0,\bar{x}_g = \bar{x}_g) = (\overline{Pb}_1,\overline{Pb}_2) = \sin 2\theta_B(\bar{l}_g,\bar{l}_0), (\bar{x}'_0,\bar{x}'_g) = (\overline{Pb}'_1,\overline{Pb}'_2).]$

the coordinates in the oblique coordinate system $\hat{\mathbf{K}}_0$ and $\hat{\mathbf{K}}_g$ directions, respectively, with the origin at $F'$. Since $(\mathbf{k}_{0,r} - \mathbf{k}_0)$ is identical to $(\Delta\mathbf{k}_r - \Delta\mathbf{k})$

$$G_r = G_0 + [(\Delta\eta_{0,r} - \Delta\eta_0)l_0 + (\Delta\eta_{g,r} - \Delta\eta_g)l'_g] \quad (4\text{-}235a)$$

$$= G_0 + [(\Delta\eta_{0,r} - \Delta\eta_0)l'_0 + (\Delta\eta_{g,r} - \Delta\eta_g)l_g] \quad (4\text{-}235b)$$

Thus, again, one can express it by

$$G_r = \tfrac{1}{2}K\chi_0(l_0 + l_g) + [\pm\xi_1\sqrt{s^2 + \beta^2} - \xi_2 s] \quad (4\text{-}236)$$

Recalling that $\Delta\eta_{0,r}$ and $\Delta\eta_{g,r}$ are proportional to $\Delta\eta_g$ and $\Delta\eta_0$ (4-98), respectively,

$$[G_r]_+ = \Delta\eta_{0,r}l'_0 = \frac{|\gamma'_0|}{\gamma'_g}\Delta\eta_g l'_0 \quad (4\text{-}237a)$$

$$[G_r]_- = \Delta\eta_{g,r}l'_g = \frac{\gamma'_g}{|\gamma'_0|}\Delta\eta_0 l'_g \quad (4\text{-}237b)$$

In deriving these, either combinations of (4-203a) and (4-235b) or of (4-203b) and (4-235a) are used. Thus, it turns out that

$$\xi_1 + \xi_2 = \frac{|\gamma_0'|}{\gamma_g'} x_g' \qquad (4\text{-}238a)$$

$$\xi_1 - \xi_2 = \frac{\gamma_g'}{|\gamma_0'|} \frac{\gamma_0}{\gamma_g} x_0' \qquad (4\text{-}238b)$$

where $(x_0', x_g')$ are perpendicular distances from the observation point $P$ to the edges of the $F'$-triangular fan. For use later, the geometrical relations

$$\frac{\bar{x}_0}{|\gamma_0'|} = \frac{\bar{x}_g'}{\gamma_g'} \qquad (4\text{-}239a)$$

and

$$\frac{\bar{x}_0'}{|\gamma_0'|} = \frac{\bar{x}_g}{\gamma_g'} \qquad (4\text{-}239b)$$

and, consequently,

$$\bar{x}_0 \bar{x}_g = \bar{x}_0' \bar{x}_g' \qquad (4\text{-}239c)$$

are pointed out at this time, where the overbar is used to indicate a point on the exit surface.

The amplitude factors appearing in (4-232a and b) are easily represented in terms of $(s, \beta)$ with the aid of Table 4-3. Thus, one obtains the wave fields in the following integral forms:

$$\phi_{0,r}(\mathbf{r}) = A_0 E_e \exp \frac{i}{2} K\chi_0(l_0 + l_g) \left(-\frac{1}{2}\right) \sum_{(j)}^{1,2} \int_{-\infty}^{+\infty} \frac{[-s \pm \sqrt{s^2 + \beta^2}]}{\pm\sqrt{s^2 + \beta^2}}$$

$$\times \exp i[\pm\xi_1\sqrt{s^2 + \beta^2} - \xi_2 s] \, ds^r \qquad (4\text{-}240a)$$

$$\phi_{g,r}(\mathbf{r}) = A_g E_e \left(\frac{\chi_g}{\chi_{-g}}\right)^{1/2} \sqrt{\frac{\gamma_g}{\gamma_0}} \frac{|\gamma_0'|}{\gamma_g'} \left(-\frac{1}{2}\right) \sum_{(j)}^{1,2} \int_{-\infty}^{+\infty} \frac{[-s \pm \sqrt{s^2 + \beta^2}]^2}{\pm\beta\sqrt{s^2 + \beta^2}}$$

$$\times \exp i[\pm\xi_1\sqrt{s^2 + \beta^2} - \xi_2 s] \, ds^r \qquad (4\text{-}240b)$$

By the use of the integral calculation shown in Appendix 4CI ($m = 1$ for the $O$ wave, and $m = 2$ for the $G$ wave), one obtains the explicit forms as follows:

$$\phi_{0,r}(\mathbf{r}) = (\pi\beta_0) \frac{\gamma_g'}{|\gamma_0'|} A_0 E_e \exp \frac{i}{2} K\chi_0(l_0 + l_g) \sqrt{\frac{x_0'}{x_g'}} \, J_1(\beta_0\sqrt{x_0' x_g'})$$

$$\qquad\qquad\qquad\qquad\qquad\qquad\qquad\qquad x_0' x_g' > 0$$

$$= 0 \qquad\qquad\qquad\qquad\qquad\qquad\qquad x_0' x_g' < 0 \qquad (4\text{-}241a)$$

$$\phi_{g,r}(\mathbf{r}) = (i\pi\beta_0) \left(\frac{\chi_g}{\chi_{-g}}\right)^{1/2} \frac{\gamma_g'}{|\gamma_0'|} A_g E_e \exp \frac{i}{2} K\chi_0(l_0 + l_g) \, \text{sign}\,(x_0') \frac{x_0'}{x_g'} \, J_2(\beta_0\sqrt{x_0' x_g'})$$

$$\qquad\qquad\qquad\qquad\qquad\qquad\qquad\qquad x_0' x_g' > 0$$

$$= 0 \qquad\qquad\qquad\qquad\qquad\qquad\qquad x_0' x_g' < 0 \qquad (4\text{-}241b)$$

The wave fields are essentially similar to the wave fields (4-205a and b) in Laue cases. They are regarded as a cylindrical wave emitted from the imaginary source $F'$. The asymptotic forms of Bessel-function functions give the approximate expressions for the fields:

$$\phi_{0,r}(\mathbf{r}) = -\left(\frac{\pi}{2}\beta_0\right)^{1/2} \left(\frac{\gamma'_g}{|\gamma'_0|}\right) A_0 E_e \exp\frac{i}{2} K\chi_0(l_0 + l_g)$$

$$\times \sqrt{\frac{x'_0}{x'_g}}\,(x'_0 x'_g)^{-1/4}\left[\exp i\left(\zeta' + \frac{\pi}{4}\right) + \exp -i\left(\zeta' + \frac{\pi}{4}\right)\right]$$

(4-241c)

$$\phi_{g,r}(\mathbf{r}) = -\left(\frac{\pi}{2}\beta_0\right)^{1/2} \left(\frac{\chi_g}{\chi_{-g}}\right)^{1/2}\left(\frac{\gamma'_g}{|\gamma'_0|}\right) A_g E_e \exp\frac{i}{2} K\chi_0(l_0 + l_g)$$

$$\times \frac{x'_0}{x'_g}\,(x'_0 x'_g)^{-1/4}\left[\exp i\left(\zeta' + \frac{\pi}{4}\right) - \exp -i\left(\zeta' + \frac{\pi}{4}\right)\right]$$

(4-241d)

where

$$\zeta' = \beta_0\sqrt{x'_0 x'_g} \quad (4\text{-}241e)$$

The arguments described in Sections B 3 and C 1 can be followed also in the present case. The rays starting from the entrance point $E$ are reflected at the exit surface. They can be interpreted as a bundle of rays starting from the imaginary source $F'$. The situation is similar to mirror reflection in ordinary optics. In the present case, however, the incident and reflected rays have different glancing angles with respect to the exit surface, and $E$ and $F'$ are not combined by the relation of mirror reflection.

In connection with these ray considerations, it is worthwhile to consider the implications of the integral representations (4-240a and b). The Bloch waves propagating between the directions parallel to the surface $S_a$ and $\overline{\mathbf{K}}_g$ are really reflected on the surface $S_a$. The other Bloch waves, however, are reflected hypothetically on the rear side of $S_a$. The above representations include all these waves. Actually, the solution between $A_2 F'b'_2$ and $S'F'S$ in Fig. 4-28 implies the hypothetical part, and the solution between $S'F'S$ and $F'A_1a'_1$ corresponds to the part reflected at the surface $S_a$.

Next, a question arises as to whether the semifiniteness of the crystal surface $S_a$ invalidates the solution since the Bloch waves assumed in the integrand are acceptable only when the crystal extends over all of space. The same question also may be raised as to the wave fields in the Laue case since the crystal surface is limited as shown in Fig. 4-28. The spherical-wave solutions, however, happen to take on appreciable values within the triangular fans mentioned above. Thus, from the physical viewpoint, the behavior of the crystal outside the triangular fans has nothing to do with the spherical-wave solution. For this reason, the solutions obtained above are correct within the real crystal.[1]

The transmitted wave field $\Phi_{g,t}(\mathbf{r})$ can be calculated by consulting the plane-wave solution (4-109c) and the treatment used in the Laue-Laue cases (Section B 4). The phase of interest in the integral representation, similarly to (4-215b), has the same form as (4-218). The amplitude factor includes $C_g^{(j)}$ and $-(c_g^{(j)}/c_{g,r}^{(j)})C_g^{(j)}$ which have been calculated already for the Laue case and in (4-232b). Thus, one has

$$\Phi_{g,t}(\mathbf{r}) = (i\pi\beta_0)\left(\frac{\chi_g}{\chi_{-g}}\right)^{1/2} A_g E_e \exp\frac{i}{2}K\chi_0(\bar{l}_0 + \bar{l}_g)$$

$$\times\left[J_0(\beta_0\sqrt{\bar{x}_0\bar{x}_g}) + \frac{|\gamma_0'|}{\gamma_g'}\frac{\bar{x}_g}{\bar{x}_0}J_2(\beta_0\sqrt{\bar{x}_0\bar{x}_g})\right] \qquad \bar{x}_0\bar{x}_g > 0 \qquad (4\text{-}242a)$$

$$= 0 \qquad\qquad\qquad\qquad\qquad \bar{x}_0\bar{x}_g < 0 \qquad (4\text{-}242b)$$

where $(\bar{l}_0 = l_0, \bar{l}_g)$ and $(\bar{x}_0, \bar{x}_g = x_g)$ are the coordinates of $\bar{P}$ on the exit surface corresponding to the observation point $P$ in vacuum, in a way that $\bar{P}P$ has the direction of $\bar{\mathbf{K}}_g$. By virtue of (4-239), the continuity of the wave fields

$$\phi_0(\mathbf{r}_a) + \phi_{0,r}(\mathbf{r}_a) = 0 \qquad (4\text{-}243a)$$

$$\phi_g(\mathbf{r}_a) + \phi_{g,r}(\mathbf{r}_a) = \Phi_{g,t}(\mathbf{r}_a) \qquad (4\text{-}243b)$$

does hold at the exit surface. The wave field $\Phi_{g,t}(\mathbf{r})$ is a kind of projection of the wave field $\Phi_{g,t}(\mathbf{r}_a)$ along the direction of $\bar{\mathbf{K}}_g$. The fringe system appearing in the $A_1$-triangular fan is schematically illustrated in Fig. 4-27.

**2.  Type II**  The treatments described above can be applied to the wave fields for the case of type II. Here, only the final results are described; mathematical details are presented in Appendix 4CII.

$$\phi_{0,r}(\mathbf{r}) = (\pi\beta_0)A_0 E_e \exp\frac{i}{2}K\chi_0(l_0 + l_g)\sqrt{\frac{x_g''}{x_0''}}J_1(\beta_0\sqrt{x_0''x_g''})$$

$$\qquad\qquad\qquad\qquad\qquad x_0''x_g'' > 0$$

$$= 0 \qquad\qquad\qquad\qquad\qquad x_0''x_g'' < 0 \qquad (4\text{-}244a)$$

$$\phi_{g,r}(\mathbf{r}) = (-i\pi\beta_0)\left(\frac{\chi_g}{\chi_{-g}}\right)^{1/2} A_g E_e \exp\frac{i}{2}K\chi_0(l_0 + l_g)\,\text{sign}\,(x_0'')J_0(\beta_0\sqrt{x_0''x_g''})$$

$$\qquad\qquad\qquad\qquad\qquad x_0''x_g'' > 0$$

$$= 0 \qquad\qquad\qquad\qquad\qquad x_0''x_g'' < 0 \qquad (4\text{-}244b)$$

---

[1] In fact, the solutions satisfy the Maxwell equations in the crystal and the boundary conditions on $S_e$ with the incident wave as well as the conditions on $S_a$ with the transmitted wave, as described below. The uniqueness of the solution in the theory of differential equations justifies their correctness.

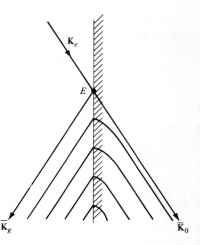

**FIGURE 4-29**
Spherical-wave theory in the Bragg case.

Also,

$$\Phi_{0,t}(\mathbf{r}) = (\pi\beta_0)A_0E_e \exp\frac{i}{2}K\chi_0(\bar{l}_0 + \bar{l}_g)\left(\frac{|\gamma_g''|}{\gamma_0''}\sqrt{\frac{\bar{x}_0}{\bar{x}_g}} - \sqrt{\frac{\bar{x}_g}{\bar{x}_0}}\right)J_1(\beta_0\sqrt{\bar{x}_0\bar{x}_g})$$

$$\bar{x}_0\bar{x}_g > 0$$

$$= 0 \qquad\qquad\qquad \bar{x}_0\bar{x}_g < 0 \qquad (4\text{-}245)$$

As in the case of type I, the crystal wave fields are regarded as a cylindrical wave emitted from an imaginary source $F''$, which can be constructed similarly to $F'$. In this case, the $G$ wave is totally reflected. The vacuum wave is a kind of projection of the wave field $\Phi_{0,t}(\mathbf{r}_a)$ at the exit surface in the direction of $\overline{\mathbf{K}}_0$.

## E.  Bragg case

A half-bounded infinite crystal is considered here. The region in which one is interested is shown in Fig. 4-29. As in the previous two cases, again, the spherical-wave solution in the crystal is given by the expressions in (4-194) and subsequently by (4-196), after performing the integration over $K_y$. The amplitude densities $d_0(\mathbf{K}; \mathbf{r})$ and $d_g(\mathbf{K}; \mathbf{r})$ are obtained by the plane-wave theory which was described in Section I C 3$b$. Although a few significant differences are to be noted between the Laue and Bragg cases, one is able to follow the procedures described for the Laue cases.

The phase of interest has the same form as (4-198) [see (4-82$a$ and $b$)], so that one can write it

$$G_0 = \tfrac{1}{2}K\chi_0(l_0 + l_g) + (\pm\xi_1\sqrt{s^2 + \beta^2} - \xi_2 s) \qquad (4\text{-}246)$$

where $\xi_1$ and $\xi_2$ are given by (4-204). Since $\gamma_g$ is negative in the present problem, they can be written

$$\xi_2 + \xi_1 = x_0 \qquad (4\text{-}247a)$$

$$\xi_2 - \xi_1 = \frac{\gamma_0}{|\gamma_g|} x_g \qquad (4\text{-}247b)$$

The amplitudes of $O$ and $G$ waves are given by unity and $c^{(q)}$, respectively, where $c^{(q)}$ is given by (4-71). The expression can be easily transformed to a function of $(s, \beta)$ defined by (4-55b and c). Thus, the spherical-wave solution can be written

$$\phi_0(\mathbf{r}) = A_0 E_e \exp \frac{i}{2} K\chi_0(l_0 + l_g) \int_{-\infty}^{+\infty} \exp i(\pm \xi_1 \sqrt{s^2 + \beta^2} - \xi_2 s) \, ds^r \qquad (4\text{-}248a)$$

$$\phi_g(\mathbf{r}) = A_g E_e \exp \frac{i}{2} K\chi_0(l_0 + l_g) \left(\frac{1}{i}\right)\left(\frac{\chi_g}{\chi_{-g}}\right)^{1/2} \sqrt{\frac{\gamma_0}{|\gamma_g|}}$$

$$\times \int_{-\infty}^{+\infty} \left(\frac{s \pm \sqrt{s^2 + \beta^2}}{\beta}\right) \exp i(\pm \xi_1 \sqrt{s^2 + \beta^2} - \xi_2 s) \, ds^r \qquad (4\text{-}248b)[1]$$

The plus or minus signs must be selected according to the rule (4-80a and b) for non-absorbing crystals or the statement A (Section I C 3b) in general cases. As described in that section, this complicated situation can be avoided by using a complex variable $z$ instead of real variable $s^r$ in the integrand. The equivalent statement B justifies the alternative representations.

$$\int_{-\infty}^{+\infty} \exp i(\pm \xi_1 \sqrt{s^2 + \beta^2} - \xi_2 s) \, ds^r$$

$$= \int_{-\infty + is^i}^{+\infty + is^i} \exp -i[\xi_1 (z^2 + \beta^2)^{1/2} + \xi_2 z] \, dz \qquad (4\text{-}249a)$$

$$\int_{-\infty}^{+\infty} \left(\frac{s \pm \sqrt{s^2 + \beta^2}}{\beta}\right) \exp i(\pm \xi_1 \sqrt{s^2 + \beta^2} - \xi_2 s) \, ds^r$$

$$= \int_{-\infty + is^i}^{+\infty - is^i} \left[\frac{z - (z^2 + \beta^2)^{1/2}}{\beta}\right] \exp -i[\xi_1 (z^2 + \beta^2)^{1/2} + \xi_2 z] \, dz \qquad (4\text{-}249b)$$

The integral paths are illustrated in Fig. 4C-2. From a physical viewpoint, it is worthwhile to consider the dispersion surface for the case of nonabsorbing crystals. The wings corresponding to the integral path have been drawn in Fig. 4-5c.

---

[1] Note that $\beta = i(|\gamma_g|/\gamma_0)^{1/2}\beta_0$, $\beta_0$ being defined by (4-206).

As shown in Appendix 4C, the integrals (4-249a and b) can be performed by the contour integrals ($m = 0$ for the $O$ wave; $m = 1$ for the $G$ wave). Thus, one can obtain

$$\phi_0(\mathbf{r}) = (\pi\beta_0)A_0E_e \exp\frac{i}{2}K\chi_0(l_0 + l_g)\left[\frac{|\gamma_g|}{\gamma_0}\sqrt{\frac{x_0}{x_g}} - \sqrt{\frac{x_g}{x_0}}\right]$$

$$\times J_1(\beta_0\sqrt{x_0x_g}) \qquad x_0 > 0 \qquad (4\text{-}250a)$$

$$= 0 \qquad\qquad x_0 < 0 \qquad (4\text{-}250b)$$

$$\phi_g(\mathbf{r}) = i(\pi\beta_0)\left(\frac{\chi_g}{\chi_{-g}}\right)^{1/2}A_gE_e \exp\frac{i}{2}K\chi_0(l_0 + l_g)$$

$$\times \left[\frac{|\gamma_g|}{\gamma_0}\frac{x_0}{x_g}J_2(\beta_0\sqrt{x_0x_g}) + J_0(\beta_0\sqrt{x_0x_g})\right] \qquad x_0 > 0 \qquad (4\text{-}250c)$$

$$= 0 \qquad\qquad x_0 < 0 \qquad (4\text{-}250d)$$

In reality, the wave fields have a physical significance only when the observation point lies within the crystal. Each of the relations in (4-250) includes a superfluous part which could exist in space on the vacuum side when the crystal is hypothetically extended throughout all of space. This part is a purely imaginary one. Although it looks different, the underlying idea is similar to the Laue and Laue-Bragg cases. There, also, the superfluous wave fields are imagined to be on the vacuum side of space.

It is interesting to consider the results in (4-250) with the results of Laue and Laue-Bragg case of type I [cf (4-205) and (4-241)]. When the entrance point $E$ happens to be the edge point of the crystal, one can see that

$$\phi_0(\text{Bragg}) = \phi_0(\text{Laue}) + \phi_{0,r} \quad (\text{Laue-Bragg, type I}) \quad (4\text{-}251a)$$

$$\phi_g(\text{Bragg}) = \phi_g(\text{Laue}) + \phi_{g,r} \quad (\text{Laue-Bragg, type I}) \quad (4\text{-}251b)$$

as it should be. (Note that $x_0' = x_0$ and $x_g' = x_g$, and $\gamma_0$ and $|\gamma_g|$ in Bragg cases must be read as $|\gamma_0'|$ and $\gamma_g'$ in Laue-Bragg cases, respectively.)

For vacuum waves, the spherical-wave solution can be represented again by (4-215b) after the integration over $K_y$. The amplitude density $E_g(\mathbf{K}; \mathbf{r})$ is given by the plane-wave solution (4-82c). The phase to be considered, therefore, is given by

$$H_g = G_0 + [(\mathbf{K}_g - \mathbf{k}_g^{(q)}) \cdot (\mathbf{r} - \mathbf{r}_e)] \qquad (4\text{-}252)$$

where $G_0$ is given in the same form as (4-198) or (4-201b). As in the case of Laue-Bragg cases,

$$[(\mathbf{K}_g - \mathbf{k}_g^{(q)}) \cdot (\mathbf{r} - \mathbf{r}_e)] \equiv K\delta_g H = -(\tfrac{1}{2}K\chi_0 + \Delta\eta_g)\frac{H}{|\gamma_g|} \qquad (4\text{-}253)$$

where $H$ is the height of the observation point above the entrance surface, and $H/|\gamma_g|$ is the distance from the observation point $P$ to the corresponding point $\bar{P}$ on the entrance surface in a way that $\bar{P}P$ has the direction of $\bar{\mathbf{K}}_g$. Then, one can see that

$$H_g = \tfrac{1}{2}K\chi_0(\bar{l}_0 + \bar{l}_g) + \Delta\eta_0\bar{l}_0 + \Delta\eta_g\bar{l}_g \qquad (4\text{-}254)$$

where ($\bar{l}_0$ and $\bar{l}_g$) are the oblique coordinates of $\bar{P}$. Thus, if one writes $H_g$ in a similar way to (4-246), it turns out that

$$\xi_2 + \xi_1 = \bar{x}_0 \qquad (4\text{-}255a)$$

$$\xi_2 - \xi_1 = \frac{\gamma_0}{|\gamma_g|}\,\bar{x}_g \qquad (4\text{-}255b)$$

where $\bar{x}_0$ and $\bar{x}_g$ are the perpendicular distances from $\bar{P}$ to the edges of the triangular fan.

Since the amplitude factor in $E_g(\mathbf{K};\mathbf{r})$ is identical to that of $d_g(\mathbf{K};\mathbf{r})$ of the crystal wave [see (4-82b and c)], the spherical-wave solution is given straightforwardly by

$$\Phi_g(\mathbf{r}) = i(\pi\beta_0)\left(\frac{\chi_g}{\chi_{-g}}\right)^{1/2} A_g E_e \exp\frac{i}{2}K\chi_0(\bar{l}_0 + \bar{l}_g)$$

$$\times \{J_0(\beta_0\sqrt{\bar{x}_0\bar{x}_g}) + J_2(\beta_0\sqrt{\bar{x}_0\bar{x}_g})\} \qquad \bar{x}_0 > 0 \qquad (4\text{-}256a)$$

$$= 0 \qquad\qquad\qquad\qquad \bar{x}_0 < 0 \qquad (4\text{-}256b)$$

Here, the relation $\bar{x}_0/\gamma_0 = \bar{x}_g/|\gamma_g|$ has been used at the entrance surface. This result is consistent with the result of Laue-Bragg cases of type I because (4-242a) tends to (4-256a) when the entrance point $E$ is located at the edge of the crystal, $\bar{x}_0/|\gamma'_0|$ then being identical to $\bar{x}_g/\gamma'_g$.

## F.  Experimental verification of the spherical-wave theory

**1.  *Topographic observations***   The spherical-wave theory is particularly useful in explaining x-ray diffraction topographs, and so it is worthwhile here to outline briefly[1] the principles underlying two of the most widely used experimental procedures developed by Lang (1957, 1959) and Berg-Barrett (Barrett, 1945) and typically used in the Laue and the Bragg arrangements, respectively. Although they were originally devised to observe imperfections present in crystals, they are equally valuable for studying x-ray diffraction phenomena.

In the Lang method, the narrow x-ray beam is collimated by a slit system as shown in Fig. 4-30a. The width of the beam conventionally used is about 10 $\mu$ (section

---

[1] Further details can be found, for example, in review articles by Lang (1963, 1970), Azároff (1964), Bonse, Hart, and Newkirk (1967), and Austerman and Newkirk (1967).

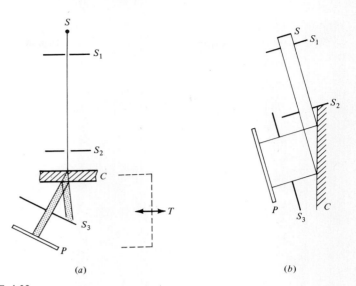

FIGURE 4-30
Topographic methods. (a) Lang arrangement; (b) Berg-Barrett arrangement.
S is the x-ray source; $S_1$, $S_2$, and $S_3$ are slits; C is the crystal; P is a photographic
plate, and T a traversing mechanism.

cases described below) or 100 $\mu$ (traverse cases), and its height is about 10 mm. It is
important that the angular divergence in the reflection plane not exceed approximately
1 minute of arc, which is sufficient to separate $K\alpha_1$ and $K\alpha_2$ in the rocking curve of a
perfect crystal. When such a flat beam pencil satisfies the Bragg condition, the crystal
waves propagate through the crystal in a Borrmann fan resulting in a beam width of
about $2t_0 \tan \theta_B$. For a crystal of the order of 1 mm thick, this amounts to more than
about 200 $\mu$. The photographic plate is placed at right angles to the Bragg-reflected
beam, Fig. 4-30a, so that it records the intensity distribution at the exit surface of the
crystal. By using a nuclear emulsion plate, it is possible to attain positional resolution
of a few microns. Experiments of this type are called *section experiments*, and the
diffraction record produced is called a *section topograph*. By rearranging the slits in
Fig. 4-30, it is possible to record the intensity distribution in the direct beam. As an
example, the section topograph of a perfect silicon crystal is shown in Fig. 4-31a.

It is possible to move the crystal and the photographic plate back and forth
while maintaining a fixed geometrical relation relative to the incident beam and
eliminating the entire direct transmitted beam by a shield placed behind the crystal.
In this case, the section topographs of adjacent portions of the crystal are recorded
continuously on the same plate. This has the effect of permitting the incident beam

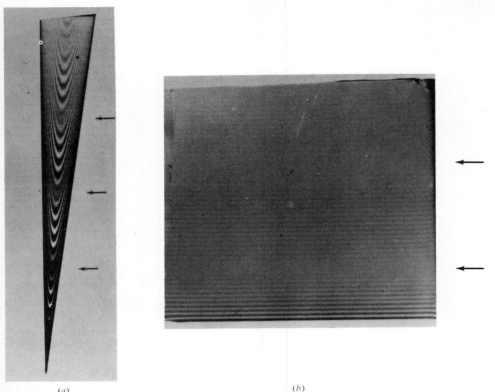

(a)                                                                    (b)

**FIGURE 4-31**
Topographs of perfect silicon crystals; arrows indicate fading due to x-ray
polarization. (*a*) Section topograph; (*b*) traverse topograph. (After Kajimura.)

to scan the entire crystal so that a topograph of the whole crystal is obtained. Such a
topograph is called a *traverse topograph*, and an example is shown in Fig. 4-31*b*.
Although it looks not unlike an electron micrograph, it should be noted that the
underlying optical principles are quite different. An explanation of the origin of the
image in a traverse topograph from the wave-optical point of view and the connection
between the section and traverse topographs are presented below after expressions for
the integrated power in the spherical-wave theory are derived.

The original form of Berg-Barrett arrangement is illustrated in Fig. 4-30*b*. The
incident beam is sufficiently wide to irradiate a crystal area several millimeters wide.
The photographic plate is placed quite closely to the crystal. In effect, the topographs
obtained are of the traverse type although the arrangement is stationary. It is also

possible to prepare topographs of the section type by using an extremely narrow beam. Since the wave field is generated in a triangular region between $\overline{\mathbf{K}}_0$ and $\overline{\mathbf{K}}_g$ (cf Fig. 4-29), the shield $S_3$ must be omitted in order to record the entire wave field.

**2.  *Integrated power in the Laue-Laue case***  From the point of view of the spherical-wave theory, the integrated power is given by the spatial integration of the trans-mitted power at the exit surface.  Here, a wedge-shaped crystal is considered in a general way.  The beam height perpendicular to the $R$ plane is assumed to be in-finitesimal.  Then, one can express the integrated transmission power

$$P_0 = \frac{c}{8\pi}\, \gamma_0' \int_{B_a} |\Phi_0(\mathbf{r}_a)|^2 \, d\mathbf{r}_a \quad (4\text{-}257a)$$

$$P_g = \frac{c}{8\pi}\, \gamma_g' \int_{B_a} |\Phi_g(\mathbf{r}_a)|^2 \, d\mathbf{r}_a \quad (4\text{-}257b)$$

where $B_a$ is the beam area at the exit surface, and the vacuum-wave fields $\Phi_0(\mathbf{r}_a)$ and $\Phi_g(\mathbf{r}_a)$ are given by (4-222).  They are replaced by the crystal wave fields $\phi_0(\mathbf{r}_a)$ and $\phi_g(\mathbf{r}_a)$ given by (4-205).

The integrated power ratios are defined as

$$\mathfrak{R}_0{}^\theta = \frac{P_0}{P_e} \quad (4\text{-}258a)$$

$$\mathfrak{R}_g{}^\theta = \frac{P_g}{P_e} \quad (4\text{-}258b)$$

where $P_e$ is the power of the incident wave at a distance $r$ from the source per unit glancing angle.  For the incident wave obeying the relations in (4-193),

$$P_e = \frac{c}{8\pi} \left(\frac{1}{4\pi r}\right)^2 r|E_e|^2 \quad (4\text{-}259)$$

Similarly to the integrated power ratios of the plane-wave theory, the normalized power ratios are defined as

$$\mathfrak{R}_0{}^s = \frac{\mathfrak{R}_0{}^\theta}{\Delta\Theta'} \quad (4\text{-}260a)$$

$$\mathfrak{R}_g{}^s = \frac{\mathfrak{R}_g{}^\theta}{\Delta\Theta'} \quad (4\text{-}260b)$$

where $\Delta\Theta'$ is $\Delta\Theta$ defined in (4-76) but replacing $\gamma_0$ and $\gamma_g$ by $\gamma_0'$ and $\gamma_g'$, respectively.

For convenience, a position variable $\tau$ is defined in the beam area $B_a$. Then, $d\mathbf{r}_a = d\tau/\gamma'$ where $\gamma'$ is the cosine of the angle between the normal of the exit surface and the $R$ plane. Further, the normalized variable

$$\sigma = \frac{\tau - \frac{1}{2}(a + b)}{\frac{1}{2}(b - a)} \qquad (4\text{-}261)$$

is introduced, where $a$ and $b$ are $\tau$ values of $T$ and $R$ in Fig. 4-25. The width of the wave field $(b - a)$ can be expressed as

$$(b - a) = t_0 \frac{\gamma'}{\gamma'_0 \gamma'_g} \sin 2\theta_B \qquad (4\text{-}262)$$

where $t_0$ is the normal distance from $E$ to the exit surface. The position variables previously used, $l_0$, $l_g$, $\bar{x}_0$, and $\bar{x}_g$, then obey the relations

$$l_0 = \frac{\bar{x}_g}{\sin 2\theta_B} = \frac{1}{2} \frac{t_0}{\gamma'_0} (1 - \sigma) \qquad (4\text{-}263a)$$

and

$$l_g = \frac{\bar{x}_0}{\sin 2\theta_B} = \frac{1}{2} \frac{t_0}{\gamma'_g} (1 + \sigma) \qquad (4\text{-}263b)$$

In addition, the following definitions are useful:

$$t = \frac{1}{2} \left( \frac{1}{\gamma'_0} + \frac{1}{\gamma'_g} \right) t_0 \qquad (4\text{-}264)$$

$$D = \frac{1}{2} \left( \frac{1}{\gamma'_0} - \frac{1}{\gamma'_g} \right) \mu_0 t_0 \qquad (4\text{-}265)$$

$$B = \frac{\frac{1}{2} KC(\chi_g \chi_{-g})^{1/2} t_0}{(\gamma'_0 \gamma'_g)^{1/2}} \qquad (4\text{-}266)$$

$$\kappa = \frac{V}{W} \qquad (4\text{-}267a)$$

$$g = \frac{v}{W} \qquad (4\text{-}267b)$$

Using these definitions

$$\mathfrak{R}_0{}^s = \frac{\pi}{2} (1 + \kappa^2)^{1/2} (\exp -\mu_0 t) W_0 \qquad (4\text{-}268a)$$

$$\mathfrak{R}_g{}^s = \frac{\pi}{2} (1 + \kappa^2)^{1/2} \left| \frac{\chi_g}{\chi_{-g}} \right| (\exp -\mu_0 t) W_g \qquad (4\text{-}268b)$$

where $W_0$ and $W_g$ are the following integrals:

$$W_0 = |B| \int_{-1}^{1} \frac{1 - \sigma}{1 + \sigma} (\exp D\sigma)|J_1[B(1 - \sigma^2)^{1/2}]|^2 \, d\sigma \qquad \text{(4-269a)}$$

$$W_g = |B| \int_{-1}^{1} (\exp D\sigma)|J_0[B(1 - \sigma^2)^{1/2}]|^2 \, d\sigma \qquad \text{(4-269b)}$$

In the following, only the transmission power ratio $\mathfrak{R}_g{}^s$ is considered because $\mathfrak{R}_0{}^s$ is less important in experiments and more difficult to calculate. The mathematical details can be found in the original paper (Kato, 1968c).

Although various expressions for $\mathfrak{R}_g{}^s$ are possible, the following is most convenient for numerical evaluation and for comparison with the integrated power $\mathfrak{R}_g{}^P$ in the plane-wave theory (4-173b):

$$\mathfrak{R}_g{}^s = (1 + \kappa^2) \left| \frac{\chi_g}{\chi_{-g}} \right| (\{W\} + \{V\}) \exp -\mu_0 t \qquad \text{(4-270)}$$

where

$$\{W\} = \frac{\pi}{2} \int_0^{2A} J_0[(1 - g^2)^{1/2}\rho] \, d\rho \qquad \text{(4-271a)}$$

$$\{V\} = \frac{\pi}{2} \sum_{r=1}^{\infty} \frac{1}{r! \, r!} \left(\frac{h}{2}\right)^{2r} g_{2r+1}[2A(1 - g^2)^{1/2}] \qquad \text{(4-271b)}[1]$$

In these expressions, $A$ and $h$ already have been defined by equations (4-174a and b), respectively, after replacing $\gamma_0'$ by $\gamma_0$ and $\gamma_g'$ by $\gamma_g$. The functions $g_{2r+1}$ are related to the multiple integrals of the Bessel function $J_0$; i.e.,

$$g_{2r+1}(z) = \frac{(2r)!}{(2)^{2r}} \int_0^z \cdots \int_0^\rho J(\rho)(d\rho)^{2r+1} \qquad \text{(4-272)}[1]$$

The correspondences between the expressions $\{W\}$ and $[W]$ and between $\{V\}$ and $[V]$ are obvious, where $[W]$ and $[V]$ are obtained in the plane-wave theory with SI approximations. When $\kappa$ and $g$ are less than 0.1, the factors $(1 + \kappa^2)|\chi_g/\chi_{-g}|$ and $(1 - g^2)^{1/2}$ are negligible within an accuracy of 1%. Then, $\{W\}$ can be replaced in practice by $[W]$. According to the numerical calculation $\{V\}/[V]$ is nearly equal to unity, unless one is concerned with the range $2A < 3$ where $[V] = (\pi/2)[I_0(h) - 1]$ itself is almost zero. Thus, it is concluded that the expression $\mathfrak{R}_g{}^P$ is correct as a practical formula. It must be noted, however, that the geometrical factors $\gamma_0'$ and $\gamma_g'$

---

[1] It is possible, therefore, to write concisely

$$\{W\} + \{V\} = \frac{\pi}{2} \int_0^{2A} I_0[(\kappa^2 + g^2)^{1/2}(2A - \rho)]J_0[(1 - g^2)^{1/2}\rho] \, d\rho \qquad \text{(4-271c)}$$

If one assumes $g_{2r+1}$ to be unity, then $\{V\}$ turns out to be a power series of $(\pi/2)[I_0(h) - 1]$, where $I_0$ is the modified Bessel function of zero order.

referred to the exit surface must be used instead of $\gamma_0$ and $\gamma_g$. Moreover, when the experimental accuracy is better than 1% or either one of the parameters $\kappa$ and $g$ is larger than 0.1,[1] the more rigorous expression must be used.

*a. Relation between the integrated intensities in the plane-wave and spherical-wave theories* Both $\Delta\Theta\mathfrak{R}_g^P$ and $\Delta\Theta'\mathfrak{R}_g^s$ are regarded as the total powers of $G$ waves for an incident wave having a unit power per unit beam width and per unit angular range. Nevertheless, their physical meanings are different. In the plane-wave theory, the integration is an angular integration, whereas it is a spatial integration in the spherical-wave theory. In the former, it is assumed that the incident wave is characterized by perfect coherence over a wide wave front and has perfect angular incoherence. In the spherical-wave theory, on the other hand, perfect angular coherence is assumed. When the beam is wide, it is natural to assume perfect spatial incoherence because the parallel waves arriving at different entrance positions arise from different atomic sources. This situation is illustrated schematically in Fig. 4-20.

Using the definitions of $\mathfrak{R}_g^P$ and $\mathfrak{R}_g^s$, one can write

$$\Delta\Theta\mathfrak{R}_g^P = \frac{\gamma_g}{\gamma_0} \int_{-\infty}^{+\infty} |d_g(\mathbf{K};\mathbf{r}_a)|^2 \frac{dK_x}{K} \tag{4-273a}$$

$$\Delta\Theta'\mathfrak{R}_g^s = (4\pi)^2 r \left(\frac{\gamma_g'}{\gamma_0'}\right) \int_{-\infty}^{+\infty} |\phi_g(\mathbf{r}_a)|^2 \, dx_0 \tag{4-273b}$$

where $dK_x/K = d\Theta$ is used and $\mathbf{r}_a$ denotes a position on the exit surface. ($\gamma_0 = \gamma_0'$ and $\gamma_g = \gamma_g'$ for parallel-sided crystals.)

For rigorous definitions, equations (4-194b) and (4-195b) must be used for $\phi_g(\mathbf{r}_a)$ and $d_g(\mathbf{K};\mathbf{r}_a)$, respectively. In practice, however, some approximation are utilized in obtaining the explicit expressions of $\mathfrak{R}_g^P$ and $\mathfrak{R}_g^s$. As explained in conjunction with (4-197b), one has to neglect the curvature of the asymptotic planes $S_0$ and $S_g$, and approximate the wave field by the stationary-phase method in performing the integration over $K_y$. Under these conditions it turns out that $4\pi(Kr)^{1/2}\phi_g(\mathbf{r}_a)$ is the Fourier transform of $d_g(\mathbf{K};\mathbf{r}_a)$ except for the phase factor. Thus, by virtue of Parseval's theorem for the Fourier transform, it is concluded that $\mathfrak{R}_g^P = \mathfrak{R}_g^s$ for parallel sided crystals.[2] The minute differences between the expressions (4-173b) and (4-270) are due to SI approximations for obtaining $\mathfrak{R}_g^P$. For this reason, (4-270) should be used for accurate experiments even in the plane-wave theory.

Here, the integrated power $\mathfrak{R}_0^s$ has not been calculated. As is shown in the original analysis (Kato, 1961b), it turns out that $\mathfrak{R}_0^s + \mathfrak{R}_g^s = \pi A$ in nonabsorbing

---

[1] This case may occur when the structure factor is small.
[2] Rigorously speaking, $\mathfrak{R}_g^P$ can be defined only for parallel-sided crystals. The position variable included in $d_g(\mathbf{K};\mathbf{r}_a)$ in this case is only the thickness $t_0$. Thus, the theorem can be applied to the present problem.

crystals. Recalling that $\pi A$ is the integrated intensity of the kinematical theory, the spherical-wave solutions $\phi_0$ and $\phi_g$ can be interpreted as the transmitted waves of the kinematical waves created along the incident wave. In other words, $\phi_0(\mathbf{r}_a)$ does not include the transmitted wave without an interaction with the crystal.

*b.  Relation between section and traverse topographs*[1]  As shown above, the traverse topograph is a continuous recording of adjacent section topographs obtained with a scanning incident beam. Through ray considerations, it is easily seen that the intensity at a point $P$ in the traverse topograph is the sum of the energy flows arriving at $\bar{P}$ on the exit surface (corresponding to the point $P$). When the crystal is perfect, the lateral position of each ray has nothing to do with the energy carried by the ray, so that one can make the following intuitive assertion:

> The intensity at $P$ (or $\bar{P}$) in the traverse topograph is the integrated intensity of the section topograph for a hypothetical crystal, whose exit surface is parallel to the entrance surface passing through the point $\bar{P}$.　　　(Statement A)

In order to prove this in a general way, it is valuable to consider the reciprocal theorem in optics which was originally worked out by Lorentz (1905) on the basis of Maxwell's equation. Since a proof and a rigorous statement of the theorem are rather lengthy, only the results, in a form suitable to the present problem, are given here:

> The intensity at $P$ in the traverse topograph is the integrated intensity of the section topograph in the reciprocal experiment.　　　(Statement B)

Here, the reciprocal experiment means a stationary experiment, with a source placed at $P$ that emits x rays in reverse but has the same strength of emission as in the real experiment. Obviously, the roles of the entrance and exit surfaces are interchanged in the reciprocal experiment. In addition, as described above, the integrated intensity of a section topograph is relevant only to the exit surface. The integrated intensity relevant to statement B, therefore, is regarded as that referred to a parallel-sided crystal having the same orientation as the entrance surface in the real experiment. Thus, statement A is equivalent to statement B.

The reciprocal theorem is very general. Indeed statement B holds regardless of the optical conditions of the incident beam (polarization and wavelength) and of the medium involved (shape, orientation, polarizability, and absorption). Particularly, it can be used also for distorted crystals which will be discussed in later sections. Statement A is true only for a perfect crystal.

When statement A applies, it turns out immediately that the fringes in the traverse topograph must be contours of equal thickness. Their appearance is very

---

[1] Kato (1968*a* and *c*).

similar to the fringes expected in the plane-wave theory. As has been seen, however, the optical principles involved are very different.

**3.   Pendellösung *fringes in Laue (or Laue-Laue) cases***   The phenomenon of *Pendellösung* fringes involves various aspects directly related to the wave-optical principles of the dynamical theory. In this section, therefore, a quantitative comparison will be made between the theory and experiments, particularly as to their geometrical aspects. For simplicity, nonabsorbing crystals only will be considered, because absorption is less important to the geometrical aspects. In highly absorbing crystals, the fringes lose much of their contrast and the behavior of the wave field tends to the predictions based on the ray-optical considerations. Some of the topics have been described in Sections I D 2 and II B 3 of this chapter.

Pendellösung fringes can be observed in both section and traverse topographs. What one observes in topographic studies is the vacuum wave (4-222). Nevertheless, the section topograph is merely a projection of the intensity field on the exit surface, which is the intersection of the cylindrical intensity field (4-205b) in the crystal with the exit surface. Therefore, if one uses a wedge-shaped crystal as a specimen, the observed fringes are obtained from the fringes defined on the $R$ plane by performing a linear transformation of the coordinates. Similarly, the traverse topograph is a linear transform of a pseudointensity field defined by the integral power ratio $\Re_g^\theta$, (4-258b). Thus, as long as the considerations are limited to the geometrical aspects that are invariant with a linear transformation, it is sufficient to consider the crystal intensity field or pseudointensity field instead of the topograph itself.

The intensity fields in nonabsorbing crystals can be summarized as follows: For section topographs

$$i_g^s = \frac{1}{32\pi} \frac{1}{Kr} |E_e|^2 [\beta_0 J_0(\beta_0 \sqrt{x_0 x_g})]^2 \qquad (274a)$$

cf equation (4-205b).   For traverse topographs

$$i_g^T = \frac{\pi}{2} \Delta\Theta P_e \int_0^{\bar{\beta}_0 t_0} J_0(x)\, dx \qquad (4\text{-}274b)$$

where

$$\beta_0 = \frac{KC|\chi_g|}{\sin 2\theta_B} \qquad (4\text{-}275a)$$

$$\bar{\beta}_0 = \frac{KC|\chi_g|}{(\gamma_0 \gamma_g)^{1/2}} \qquad (4\text{-}275b)$$

$$\Delta\Theta = \frac{C|\chi_g|(\gamma_g/\gamma_0)^{1/2}}{\sin 2\theta_B} \qquad (4\text{-}275c)$$

cf equations (4-172b), (4-173b), and (4-175b). By the use of the asymptotic formulas for $J_0$ and $\int_0^x J_0(x)\,dx$, one can approximate the intensity fields in special cases

$$i_g{}^s \simeq \frac{1}{32\pi^2} \frac{\beta_0}{Kr} \frac{1}{\rho \sin \theta_B}\left[1 - \cos\left(\frac{2\pi\rho}{\Lambda_s} + \frac{\pi}{2}\right)\right]|E_e|^2 \qquad (4\text{-}276a)$$

(along the net plane in the $R$ plane) and

$$i_g{}^T \simeq \frac{\pi}{2}\,\Delta\Theta\left\{1 - \left[\frac{2/\pi}{(\bar\beta_0 t_0)^3}\right]^{1/2}\cos\left(\frac{2\pi t_0}{\Lambda_T} + \frac{\pi}{4}\right)\right\}P_e \qquad (4\text{-}276b)$$

(along the normal to the entrance surface). For comparison, the intensity distribution (4-125b) in the plane-wave theory is presented in the form

$$i_g{}^P = \frac{1}{2}\frac{\gamma_0}{\gamma_g}\left(1 - \cos\frac{2\pi t_0}{\Lambda_P}\right)|E_e|^2 \qquad (4\text{-}276c)$$

(along the normal to the entrance surface; exact Bragg condition). In these expressions,

$$\Lambda_s = \frac{\lambda \cos \theta_B}{C|\chi_g|} \qquad (4\text{-}277a)$$

$$\Lambda_T = \Lambda_P = \frac{\lambda(\gamma_0\gamma_g)^{1/2}}{C|\chi_g|} \qquad (4\text{-}277b)$$

The intensity distributions $i_g{}^s$, $i_g{}^T$, and $i_g{}^P$ are shown in Fig. 4-32.

*a.  Absolute positions of the fringes*[1]   In the present discussion, consider the constant phases $\pi/2$, $\pi/4$, and 0 in the cosine terms of relations (4-276a, b, and c). Neglecting slight modifications in the amplitudes, one may expect that the intensity maxima are located at $(n + \frac{1}{4})\Lambda_s$, $(n + \frac{3}{8})\Lambda_T$, and $(n + \frac{1}{2})\Lambda_p$ from the entrance surface ($\rho = 0$ or $t_0 = 0$) for the respective cases (cf Fig. 4-33). In section topographs, if one plots the fringe positions $R_n(n > 2)$, measured from the tip of the topograph, against the fringe order number $n$, the residual distance $R_0$ at $n = 0$ can be determined experimentally by noting the intercept of the resulting straight line along the ordinate at $n = 0$. Also, the fringe distance $\Lambda_s$ can be obtained from the fringes of higher order. (It is the vertical separation of two adjacent plotted points.) The results are listed in Table 4-5a. It turns out that the residual distance $R_0$ is $\frac{1}{4}\Lambda_s$ as expected theoretically. Similarly, in the traverse topographs, the residual distances $R_0$ are compared with $\frac{3}{8}\Lambda_T$ (Table 4-5b) which is also expected from the theory. Homma and his colleagues also observed a shrinkage of the fringe spacing for the first few fringes.

[1] Homma, Ando, and Kato (1966).

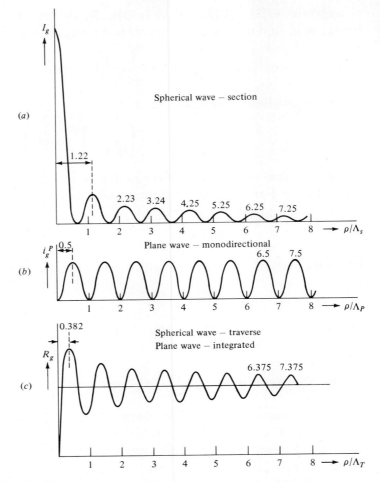

**FIGURE 4-32**
Intensity distributions and fringe positions in the plane-wave and spherical-wave theories.

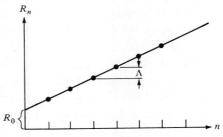

**FIGURE 4-33**
Absolute fringe position.

Subsequent to this, Hart and Milne (1968) performed an interesting experiment. They showed that the absolute positions of fringes are different for the same crystal, depending on whether the incident wave used has a plane-wave character or a spherical-wave character. The results were again in accordance with the theoretical predictions.

The optical meaning of the constant phases involved here has been fully discussed in Section C and Appendix 3A. The phase shift in the pseudointensity fields (traverse cases) is merely a smearing effect of the intensity distribution of the crystal field (4-274a).

b.   *The fringe shape*[1]   Equation (4-274a) implies that the *Pendellösung* fringes observed in section topographs must be a set of hyperbolas, the asymptotes being the edges of the triangular pattern. If this is so, as an elementary property of a hyperbola, it can be seen in Fig. 4-35 that

$$X_h^2 = X_a^2 - A \qquad (4\text{-}278)$$

Table 4-5a   RESIDUAL DISTANCE $R_0$ IN SECTION TOPOGRAPHS

| Plate No. | Exp. | Calculated | |
| | | 0.25Λ | 0.5Λ |
| --- | --- | --- | --- |
| 1 | 139 ($\mu$) | 152 ($\mu$) | 303 ($\mu$) |
| 2 | 142 | 151 | 302 |
| 3 | 170 | 154 | 307 |
| 4 | 135 | 152 | 304 |
| 5 | 167 | 150 | 301 |
| 6 | 156 | 151 | 302 |

Table 4-5b   RESIDUAL DISTANCE $R_0$ IN TRAVERSE TOPOGRAPHS

| Plate No. | Exp. | Calculated | |
| | | 0.375Λ | 0.5Λ |
| --- | --- | --- | --- |
| 1 | 140 ($\mu$) | 168 ($\mu$) | 221 ($\mu$) |
| 2 | 170 | 166 | 218 |
| 3 | 140 | 166 | 218 |
| 4 | 149 | 165 | 217 |

[1] Hattori and Kato (1966).

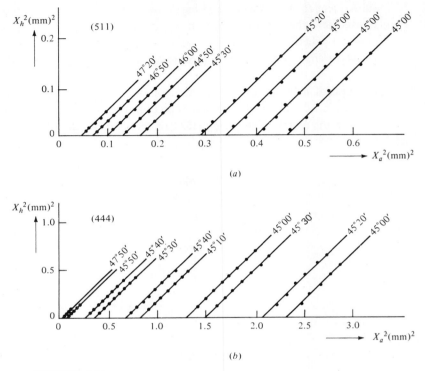

(a)

(b)

FIGURE 4-34
Experimental results for the fringe shape. ($X_a = a - d$ and $X_h = b - d$ in Fig. 4-35.)  After Hattori and Kato (1972).

for any direction of the line $a$–$b$–$c$–$d$, where $X_a = (d - a)$ and $X_h = (c - b)$ and $A$ is a constant.  The relation holds irrespective of the geometry of the wedge crystal and the net plane which is being considered.  Thus, by testing the linearity between $X_h^2$ and $X_a^2$, one can prove whether the fringe shape is really hyperbolic or not.  Figure 4-34 shows an example of plotting $X_h^2$ against $X_a^2$.  As can be seen therein, relation (4-278) is well satisfied.

Next, consider the physical significance of the results mentioned above.  First, assume that the initial phases of the crystal waves are independent of the propagation direction.  Theoretically, this is justified by considering the mathematical representation (4-192) of the spherical wave and the origin of the phase shifts of the crystal waves, so far as the spherical-wave solution (4-205$b$) is valid.[1]  Then, the fringe position is determined simply by the difference of the wave vectors associated with the crystal waves along the direction of propagation.  Thus, the hyperbolic form of the observed

[1] See also Section C 2.

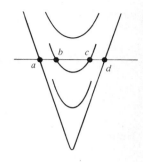

FIGURE 4-35
Geometry for checking the hyperbolic
fringe shape.

fringes is direct evidence of the hyperbolic form of the dispersion surface.

It is also true that most of the dynamical phenomena, for example, the shape of rocking curves, also reflect in some way the shape of the dispersion surface, but the connection is rather indirect. For electron diffraction, many results have been reported on the form of the dispersion surfaces. The experiments of Hoerni (1950) and Kambe (1957a,b) are some examples. More recently, a most significant experiment on this subject has been performed by Lehmpfuhl and Reissland (1968). However, the theoretical correctness of the hyperbolic form is affirmed only in special cases because the two-beam approximation is not adequate for the electron cases.[1] Since the dispersion surface plays a fundamental role in the dynamical theory, it is important to know how accurately the dispersion surface is hyperbolic in the x-ray case, where the two-beam approximation appears to work satisfactorily.

*c. Effects of polarization* So far, only the intensity field for a plane-polarized wave has been considered. Obviously, since x rays emitted from an atomic source are electromagnetic radiation, one must consider the superposition of two intensity fields due to perpendicularly polarized waves, one perpendicular and one parallel to the *R* plane. The polarization factor *C* is given for these respective cases by (4-19). In this section, the polarization effects are considered for the case of section topographs.

The intensity field in the *R* plane is given by

$$
i_g^s = (i_g^s)_\parallel + (i_g^s)_\perp
$$

$$
\simeq \frac{A}{2\rho}(1 + |\cos 2\theta_B|) - \frac{A}{2\rho}(1 + |\cos 2\theta_B|)\cos\left(\frac{2\pi\rho}{\Lambda} + \frac{\pi}{2}\right)\cos(2\pi\Delta\rho)
$$

$$
+ \frac{A}{2\rho}(1 - |\cos 2\theta_B|)\sin\left(\frac{2\pi\rho}{\Lambda} + \frac{\pi}{2}\right)\sin(2\pi\Delta\rho) \tag{4-279}
$$

[1] In fact, most of the experiments mentioned above are concerned with deviations from the hyperbolic form under conditions of simultaneous reflection.

where $A$ is a factor common to the two intensity fields, and

$$\frac{1}{\Lambda} = \frac{1}{2}\left(\frac{1}{\Lambda_\perp} + \frac{1}{\Lambda_\parallel}\right) \qquad (4\text{-}280a)$$

$$\Delta = \frac{1}{2}\left(\frac{1}{\Lambda_\perp} - \frac{1}{\Lambda_\parallel}\right) \qquad (4\text{-}280b)$$

Here, $\Lambda_\perp$ and $\Lambda_\parallel$ stand for $\Lambda_s$ in the cases of $C = 1$ and $C = |\cos 2\theta_B|$, respectively.

Under ordinary experimental conditions for transmission topographs, $\cos 2\theta_B$ is close to unity. Therefore, the third term of (4-279) is a minute correction in the intensity distribution. Keeping this in mind, one can conclude the following:

1   The apparent fringe spacing is given by

$$\Lambda = \frac{2\Lambda_\perp}{(1 + |\cos 2\theta_B|)} \qquad (4\text{-}281)$$

2   The intensity of the fringes is modified by the factor $\cos(2\pi\Delta\rho)$ in the second term. The fringes disappear near the positions $\Delta\rho = \frac{1}{2}(m + \frac{1}{2})$ $(m = 0, 1, 2, \ldots)$. Such regions are called *fading regions*, and the regions where the fringes are observed are called *fringe regions*. In fact, one can see in Fig. 4-31a that the fringes fade out periodically in certain regions indicated by arrows.

3   The number ($N$) of fringes in a single fringe region is given by $(1/2\Delta\Lambda)$

$$N = \frac{\frac{1}{2}(1 + |\cos 2\theta_B|)}{1 - |\cos 2\theta_B|} \qquad (4\text{-}282)$$

except for the first fringe region. There, the number must be $\frac{1}{2}N$.

4   In the fringe regions of odd order,[1] the factor $\cos(2\pi\Delta\rho)$ is positive, whereas it is negative in the fringe regions of even order. Consequently, the positions of maxima and minima of the fringes are interchanged on both sides of the fading region.

5   Owing to the third term of (4-279), a slight displacement of the fringe positions and sometimes a ghost fringe will appear near the midpoint of a fading region.

All these properties of the section topographs are the consequences of the beat effects of two sinusoidal intensity fields. Hattori, Kuriyama, and Kato (1965) confirmed all these experimentally. Similar arguments can be presented also for a traverse topograph, based on (4-276b). In fact, Hart and Lang (1965) have worked

---

[1] This order number should not be confused with the order number of the individual fringes!

out the conditions for this. Particularly, they demonstrated that the fading regions disappear in topographs prepared with plane-polarized x rays.

**4.  *Fringe phenomena in other cases*** As described in Section D, one can expect *Pendellösung* fringes also in Laue-Bragg cases. Actually, Uragami (1971) has demonstrated this experimentally.[1] Since the fringes can be understood without the need for new concepts they shall not be discussed further.

Another interesting observation was performed by Borrmann and Lehmann (1963) and by Lehmann and Borrmann (1967). If the crystal is thin, one may expect rather complicated fringes to appear owing to the interference between Laue-Laue and Laue-Bragg-Laue waves. When the crystal is sufficiently thick, Borrmann absorption eliminates the waves belonging to branch (2). Then, the wave fields are simplified, and the fringe system is due to the interference between Laue-Bragg waves and Laue waves. Since these waves belong to the same branch (1) but different wings of the dispersion surface, the fringes are different from *Pendellösung* fringes from the conceptual viewpoint. For this reason, they should be called *Lehmann-Borrmann fringes*.

By the use of the approximate expressions (4-209) and (4-241c and d) for Laue and Laue-Bragg waves, it can be seen that the intensity distributions both for $O$ and $G$ waves are given essentially by the interference between the waves having the forms $+ \exp i[\zeta + (\pi/4)]$ and $- \exp i[\zeta' + (\pi/4)]$. Thus, the fringe positions can be given correctly by the plane-wave considerations (omitting $\pi/4$ in the phases) as performed by Borrmann and Lehmann. At the first exit surface where $\zeta = \zeta'$ or $\bar{x}_0\bar{x}_g = x'_0 x'_g$, the intensity of the $O$ wave is always zero, whereas the intensity of the $G$ wave is finite. In both cases, however, the intensities are minimum. Thus, the complementarity[2] in the intensity is not satisfied, whereas it is satisfied for *Pendellösung* fringes. All this has been suggested on the basis of plane-wave theory by Lehmann and Borrmann and can be justified by the spherical-wave theory, as described above.

A very interesting phenomenon is that the *Pendellösung* fringes should be observed also in Bragg cases. This was originally predicted by Uragami (1969) based on Takagi's approach and confirmed also experimentally by him. Later, Saka et al. (1973) showed that the same results as those obtained by Uragami can be deduced along the lines of analysis presented here (cf Section E). In nonabsorbing crystals, it is possible to interpret the crystal wave in terms of the interference between two Bloch wave packets associated with the conjugate dispersion points. The ray considerations, however, are rather obscure in Bragg cases as compared to Laue cases. Besides the real rays, there exist a kind of skin waves which are associated with the dispersion points on the complex dispersion surface. These waves and the wave packet propa-

---

[1] Unpublished experiments were also performed by Kato (1960) but perhaps the priority in publication should be given to Uragami.

[2] Here, the term "complementarity" is used for the situation wherein the maxima (or minima) of the $O$ wave coincide with the minima (or maxima) of the $G$ wave.

gating along the surface cannot be treated easily along the lines developed in Section IB6, because there it is assumed that the dispersion surfaces are essentially real. The situation is intimately connected with the failure of the stationary-phase method in mathematics under the conditions involved. One must use the method of steepest descent to obtain the asymptotic forms of (4-250). Although the simple ray interpretation fails, the spherical-wave solution is correct because the wave fields (4-250) and (4-256) satisfy the Maxwell equation under the prescribed boundary conditions.

**5.   *Accurate determination of the structure factor***   One of the central concerns of x-ray crystallography is to obtain accurate values of structure factors on an absolute scale. A symposium, therefore, was convened at Cambridge University in 1968 and a special issue of *Acta Crystallographica*, **A25** (1969) was devoted to the proceedings.[1] The work that has been done can be classified into two main parts: (1) based on the kinematical theory and (2) based on the dynamical theory of x-ray diffraction. Clearly, the first one is universal, but one should not expect an accuracy better than about 1% because extinction effects are unavoidable as "theoretical impurities," if one attempts to push the accuracy further. The second approach has a greater advantage in this respect. So far, however, the crystals available for such studies are very limited. Here, a very brief survey of the second method is presented.

The dynamical method can be classified further into two parts: (1) based on rocking curves and (2) based on fringe phenomena in the case of dynamical x-ray diffraction.

In the first group, the rocking curve is measured with an extremely parallel and monochromatic x-ray beam. Usually, a two-crystal or three-crystal arrangement is used. One of the crystals serves as the specimen, and the Bragg geometry is employed usually, but Laue-Laue cases also can be used. The Darwin-Prins curves (4-84*b* or *c*) and (4-161*b*) provide the theoretical basis for the respective cases. By adjusting the parameters involved so as to fit the observed curve to the theoretical one, one can determine $(\chi_g \chi_{-g})^{1/2}$. Essentially, the half-width $\Delta\Theta$ is measured and $R_e \left[(\chi_g \chi_{-g})^{1/2}\right]$, i.e., $R_e \left[(F_g F_{-g})^{1/2}\right]$, is determined by the use of (4-87).

A less accurate approach is to compare the integrated intensity, which is theoretically given by (4-173*b*) and (4-270) for Laue-Laue cases and (4-180*a*) for Bragg cases. Since the requirements for the plane-wave theory are supposedly less stringent for analyses of the integrated intensity, it is easier to use than the method described above. Both methods were developed from the early days in the history of the x-ray diffraction (e.g., Ewald, 1925). More recent developments can be found in the following articles: The shape of rocking curves is discussed by Kikuta, Matsushita, and Kohra (1970) and in the references cited in Section I D 6*b*; the integrated intensity is treated by DeMarco and Weiss (1964, 1965*a*, 1965*b*) and by Weiss (1966).

---

[1] Also see the review article of Kurki-Suonio (1970).

The fringe method can be classified further into three groups: a method using (1) Borrmann-Lehmann fringes, described in Section F 4, (2) Hart and Milne fringes (1970), and (3) *Pendellösung* fringes in Laue-Laue cases. Although the first method is interesting in principle, its accuracy so far has not been better than a few percent so that is not discussed further here. The second and third ones are accurate, so that they shall be described briefly near the end of this chapter.

The ideas for the *Pendellösung* method can be traced back to Kossel (1943) and Heidenreich and Sturkey (1945), who estimated the spacing of *Pendellösung* fringes in transmission electron microscopy from the structure factor [cf (4-127a and b)]. Its practical utilization in x-ray cases was suggested by Kato and Lang (1959), and this suggestion was followed up by Hattori et al (1965). The theoretical basis is given by (4-211b). The explicit form of the fringe spacing along the net plane can be written

$$\Lambda_s = \frac{(\pi v/\lambda \cos \theta_B)(mc^2/e^2)}{|F_g|} \qquad (4\text{-}283)$$

where the polarization is taken into account by (4-281a) and $v$ is the volume of the unit cell while the other symbols have their usual meanings. The observed fringe spacing $\Lambda_g$ is related to this by

$$\Lambda_g = \Lambda_s \Phi_g \qquad (4\text{-}284)$$

where $\Phi_g$ is a geometrical factor given by the wedge angle of the specimen.

If very perfect crystals are available, it is not difficult to determine $|F_g|$ with an accuracy better than 0.1%. Several practical matters, however, must be considered if it is desired to obtain the true structure factor $|F_g|$ to an accuracy better than 0.5%. The details are described in a review article by Kato (1968) and papers by Hart and Milne (1969, 1970).

Recently, a suggestion made by Kato and Tanemura (1967) to obtain the structure factor on the true absolute scale by a combination of the Bonse-Hart x-ray interferometer fringes and *Pendellösung* fringes was followed up by Tanemura and Kato (1971). According to Bonse and Hart (1965, 1966), the observed spacing of the interferometer fringes is given by (see III A of Chap. 5)

$$\Lambda_0 = \Lambda \Phi_0 \qquad (4\text{-}285)$$

$$\Lambda = \frac{2(\pi v/\lambda)(mc^2/e^2)}{|F_0|} \qquad (4\text{-}286)$$

where $\Phi_0$ again is a geometrical factor. Combining equations (4-284) and (4-285), one can obtain

$$\frac{|F_g|}{|F_0|} = \frac{(\Lambda_0/\Lambda_g)(\Phi_g/\Phi_0)}{2 \cos \theta_B} \qquad (4\text{-}287)$$

Using the same crystal and the same collimated beam, the geometrical factor $(\Phi_g/\Phi_0)$ can be made essentially equal to unity. With this method, $|F_g|$ can be determined on the scale of $|F_0|$ with an accuracy better than 0.1%. Again, several practical difficulties must be overcome to attain an accuracy better than 0.5%. Further details can be found in the original paper.

In order to indicate the present status of the fringe method, the final results of Aldred and Hart and of Tanemura and Kato are listed in Table 4-6. It is significant that the values obtained by independent and different experiments agree with each other with an accuracy of about 0.1%, which would not be expected from the kinematical method. Related to this, recently, Shull and his group have measured the nuclear scattering amplitude for neutrons with an accuracy of $10^{-5}$ in the order of magnitude. As stated before, the dynamical method is still limited to the few crystals available in the form of extremely perfect crystals. Nevertheless, it seems possible that the dynamical investigators may contribute to the kinematical ones more than Greek tutors could do for the sons of Roman statesmen.

# APPENDIX 4A. MATHEMATICAL STRUCTURE OF $Z = z \pm (z^2 + U^2)^{1/2}$

## I.  PROPERTIES OF THE FOURIER COEFFICIENTS $\chi_g$ AND $\chi_{-g}$

If the complex polarizability $\chi = \chi^r + i\chi^i$ is introduced and the gth Fourier coefficients of the real and imaginary parts are written

$$\chi_g{}^r = |\chi_g{}^r| e^{i\phi_1} \qquad (4A\text{-}1a)$$

$$\chi_g{}^i = |\chi_g{}^i| e^{i\phi_2} \qquad (4A\text{-}1b)$$

The $-$gth coefficients must be the complex conjugates, so that

$$\chi_{-g}{}^r = |\chi_g{}^r| e^{-i\phi_1} \qquad (4A\text{-}2a)$$

$$\chi_{-g}{}^i = |\chi_g{}^i| e^{-i\phi_2} \qquad (4A\text{-}2b)$$

Then, one obtains the relations

$$|\chi_g|^2 = |\chi_g{}^r|^2 + |\chi_g{}^i|^2 + 2|\chi_g{}^r| \, |\chi_g{}^i| \sin(\phi_1 - \phi_2) \qquad (4A\text{-}3a)$$

$$|\chi_{-g}|^2 = |\chi_g{}^r|^2 + |\chi_g{}^i|^2 - 2|\chi_g{}^r| \, |\chi_g{}^i| \sin(\phi_1 - \phi_2) \qquad (4A\text{-}3b)$$

$$\chi_g \chi_{-g} = |\chi_g{}^r|^2 - |\chi_g{}^i|^2 + 2i|\chi_g{}^r| \, |\chi_g{}^i| \cos(\phi_1 - \phi_2) \qquad (4A\text{-}3c)$$

From equations (3-25) and (3-26), it turns out that $\chi^r$ is negative and proportional to the charge distribution of the total charge, while $\chi^i$ is positive and proportional to the charge distribution effective in the absorption.  When the crystal has a center of symmetry or the

Table 4-6   ATOMIC SCATTERING FACTORS DETERMINED BY THE FRINGE METHOD FOR SILICON AT 20°C (Ag $K\alpha$)

|       | TK*                | AH†                |
|-------|--------------------|--------------------|
| 111   | $10.66_4$          | $10.66_5$          |
| 220   | $8.46_3$           | $8.43_6$           |
| 333   | $5.84_3$           | $5.83_0$           |
| 440   | $5.40_8$           | $5.38_8$           |
| 444   | $4.17_2$           | $4.17_7$           |

* Tanemura and Kato (1971).
† Aldred and Hart (private communication, 1972).

If nuclear Thomson scattering is corrected, 0.004 must be substracted from all of these figures. The dispersion correction has not been made. According to Cromer (1965) $\Delta f' = 0.06$ (Ag) and 0.09 (Mo).

two kinds of charge distribution are similar, the phase difference $(\phi_1 - \phi_2)$ is essentially $\pi$. Also, in most cases in which one is interested in

$$|\chi_0{}^r| \gg |\chi_0{}^i| \quad \text{(always)} \tag{4A-4a}$$

$$\text{Re } \{\chi_g \chi_{-g}\} \geq 0 \quad \text{(unless } |\chi_g{}^r| \simeq 0) \tag{4A-4b}$$

$$\text{Im } \{\chi_g \chi_{-g}\} < 0 \quad \text{(centrosymmetric crystal)} \tag{4A-4c}$$

However, the sign $\text{Im } \{\chi_g \chi_{-g}\}$ depends on the crystal structure and the reflection vector $\mathbf{g}$, in general cases. Under the condition $|\chi_g{}^r| \gg |\chi_g{}^i|$,

$$(\chi_g \chi_{-g})^{1/2} = |\chi_g{}^r| + i|\chi_g{}^i| \cos (\phi_1 - \phi_2) \tag{4A-3d}$$

## II.   RIGOROUS EXPRESSIONS OF $Z$ FOR $z = u$

For any complex variable $\alpha + i\beta$, it does hold that

$$A \equiv \text{Re } \{(\alpha + i\beta)^{1/2}\} = \frac{\pm 1}{\sqrt{2}} (\sqrt{\alpha^2 + \beta^2} + \alpha)^{1/2} \tag{4A-5a}$$

$$B \equiv \text{Im } \{(\alpha + i\beta)^{1/2}\} = \frac{1}{\sqrt{2}} (\sqrt{\alpha^2 + \beta^2} - \alpha)^{1/2} \tag{4A-5b}$$

The sign $A$ is selected according to the sign of $\beta$, owing to the definition of the Riemann plane.[1] In the present problem $\alpha + i\beta = u^2 + U^2$, and the definitions (4-57) and (4-58) will give

$$\left. \begin{array}{l} \alpha = (w^2 - v^2) + (W^2 - V^2) \\ \beta = 2(wv + WV) \end{array} \right\} \quad \text{Laue case} \tag{4A-6a}$$

$$\left. \begin{array}{l} \alpha = (w^2 - v^2) - (W^2 - V^2) \\ \beta = 2(wv - WV) \end{array} \right\} \quad \text{Bragg case} \tag{4A-6b}$$

Rigorous expressions for $Z$ and *Anpassung* $\delta$ [see (4-56c)] can be obtained in terms of $w$, $v$, $W$, and $V$.

One often needs the expression for $|Z|$. It is given by

$$|Z|^2 = (w^2 + A^2) + (v^2 + B^2) \pm 2|wA + vB|$$

Introducing

$$L \equiv |z|^2 + |z^2 + U^2| = (w^2 + v^2) + (A^2 + B^2) \tag{4A-7}$$

one can see [cf $U^2 = (A + iB)^2 - (w + iv)^2$] that

$$L^2 - |U|^4 = 4(wA + vB)^2$$

---

[1] Note that $\sqrt{\alpha + i\beta} = |A| \pm iB$, where the sign of $iB$ is selected according to the sign of $\beta$.

Then, one obtains the expression

$$|Z|^2 = L \pm (L^2 - |U|^4)^{1/2} \qquad \text{(4A-8)}$$

which is useful for calculating the Prins' rocking curve. In the text, for convenience, $M = L/|U|^2$ has been employed instead of $L$.

## III.  PROOF THAT $v \geq |V|$ IN THE BRAGG CASE

For any positive value of $\gamma_0$ and $|\gamma_g|$,

$$\frac{1}{2}\left(\sqrt{\frac{\gamma_0}{|\gamma_g|}} + \sqrt{\frac{|\gamma_g|}{\gamma_0}}\right) \geq 1 \qquad \text{(4A-9)}$$

From equations (4A-3c) and the relation (4A-5b),

$$\text{Im}\,\{(\chi_g \chi_{-g})^{1/2}\} \leq |\chi_g{}^i| \leq \chi_0{}^i \qquad \text{(4A-10a)}$$

In the Bragg case ($\gamma_g < 0$), inserting the above inequalities and $|C| < 1$ into the definitions of $v$ and $V$, (4-56a), (4-57), and (4-59), one can see that

$$|V| \leq v \qquad \text{(4A-10b)}$$

## IV.  PROOF THAT $|Z(+)| > |U|$ AND $|Z(-)| < |U|$ IN THE BRAGG CASE

For any complex numbers $z$ and $U$,

$$|z + (z^2 + U^2)^{1/2}|^2 - |z - (z^2 + U^2)^{1/2}|^2 = 4|z|^2 R_e \frac{(z^2 + U^2)^{1/2}}{z}$$

Writing $(z + iU) = r_1 e^{i\varphi_1}$ and $(z - iU) = r_2 e^{i\varphi_2}$ and $z = r e^{i\varphi}$ one can see that

$$\frac{(z^2 + U^2)^{1/2}}{z} = \frac{\sqrt{r_1 r_2}}{r} \exp i\left[\tfrac{1}{2}(\varphi_1 + \varphi_2) - \varphi\right]$$

From the geometrical consideration illustrated in Fig. 4A-1, it is easily seen that $|\varphi_1 + \varphi_2 - 2\varphi| = |\theta_1 - \theta_2| < \pi$. Thus, one obtains

$$|Z(+)| > |Z(-)| \qquad \text{(4A-11a)}$$

On the other hand

$$|Z(+)Z(-)| = |U|^2 \qquad \text{(4A-11b)}$$

Combining these, it is concluded that

$$|z + (z^2 + U^2)^{1/2}| > |U| \qquad \text{(4A-12a)}$$

$$|z - (z^2 + U^2)^{1/2}| < |U| \qquad \text{(4A-12b)}$$

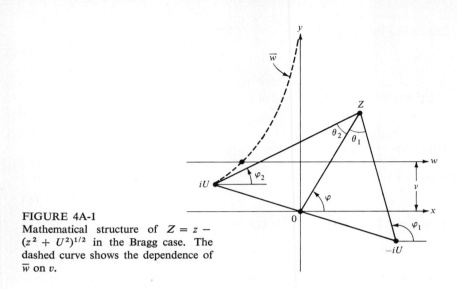

**FIGURE 4A-1**
Mathematical structure of $Z = z -$
$(z^2 + U^2)^{1/2}$ in the Bragg case. The
dashed curve shows the dependence of
$\overline{w}$ on $v$.

## V.   EXTREMAL POINT OF $|Z|$ ON THE LINE $u = w + iv$ IN THE BRAGG CASE

For a fixed value of $v(\geq |V|)$, the extremum is given by the condition

$$\frac{d}{dw} [u \pm (u^2 + U^2)^{1/2}][u \pm (u^2 + U^2)^{1/2}]^* = 0$$

namely,

$$|u \pm (u^2 + U^2)^{1/2}|^2 \operatorname{Re} \frac{1}{(u^2 + U^2)^{1/2}} = 0$$

Since the absolute value never becomes zero, this implies that $A = \operatorname{Re} \{(u^2 + U^2)^{1/2}\}$ must be zero.  More explicitly, from the expression (4A-5a) for $A$ and (4A-6b), it is concluded that

$$(w^2 - v^2) - (W^2 - V^2) \leq 0 \qquad \text{(4A-13a)}$$

$$wv - WV = 0 \qquad \text{(4A-13b)}$$

The extrema are actually the maximum and minimum, respectively, in the cases of $|u - (u^2 + U^2)^{1/2}|$ and $|u + (u^2 + U^2)^{1/2}|$.  The extremum appears only once, in each case at the same value of $w = \overline{w}$.  Because of the relation (4A-10b),

$$|\overline{w}| < |W|$$

The broken line in Fig. 4A-1 shows $\overline{w}$ as a function of $v$.

## VI. PROOF THAT $\alpha = |\chi_g/\chi_{-g}| \, |u - (u^2 + U^2)^{1/2}|^2/|U|^2 < 1$ IN THE BRAGG CASE

As mentioned in the text, the above statement is obvious from equations (4A-12a and b) for the case $|\chi_g| = |\chi_{-g}|$. In general, from (4A-3a, b, and c), one can see that

$$|\chi_g|^2 + |\chi_{-g}|^2 - 2\,\mathrm{Re}\,(\chi_g\chi_{-g}) = 4|\chi_g{}^i|^2 < 4|\chi_0{}^i|^2 \qquad (4A\text{-}14)$$

Multiplying both sides by $(KC)^2(|\gamma_g|/\gamma_0)$ and using relation (4A-9) and $|C| \leq 1$, one obtains

$$|U^2| \left[ \left|\frac{\chi_g}{\chi_{-g}}\right| + \left|\frac{\chi_{-g}}{\chi_g}\right| \right] < 2|u^2| - 2\,\mathrm{Re}\,(u^2 + U^2) \leq 2L$$

When $|\chi_g| > |\chi_{-g}|$, therefore, one has

$$|U^2| \left[ \left|\frac{\chi_g}{\chi_{-g}}\right| - \left|\frac{\chi_{-g}}{\chi_g}\right| \right] < 2(L^2 - |U|^4)^{1/2}$$

Adding up the above, one can conclude that

$$|U^2| \left|\frac{\chi_g}{\chi_{-g}}\right| < |u + (u^2 + U^2)^{1/2}|^2 \qquad (|\chi_g| > |\chi_{-g}|) \qquad (4A\text{-}15a)$$

The reciprocal of this relation gives that

$$|U^2| \left|\frac{\chi_{-g}}{\chi_g}\right| > |u - (u^2 + U^2)^{1/2}|^2 \qquad (|\chi_g| > |\chi_{-g}|) \qquad (4A\text{-}15b)$$

Since relations (4A-12a and b) have been proved, (4A-15a and b) are true also for $|\chi_g| < |\chi_{-g}|$.

## VII. IMAGINARY PART OF ANPASSUNG $\delta_e$ (4-56c)

The Laue and Bragg cases are discussed separately. First, one notices in equation (4A-5b) that

$$
\begin{aligned}
[\mathrm{Im}\,\{(u^2 + U^2)^{1/2}\}]^2 &= \tfrac{1}{2}[|u^2 + U^2| - \mathrm{Re}\,\{u^2 + U^2\}] \\
&\leq \tfrac{1}{2}[|u^2| + |U^2| - \mathrm{Re}\,\{u^2 + U^2\}] \\
&= v^2 + V^2 \leq v^2 + (K\chi_0{}^i)^2 \frac{\gamma_g}{\gamma_0} \\
&= \tfrac{1}{4}(K\chi_0{}^i)^2 \left(1 + \frac{\gamma_g}{\gamma_0}\right)^2 \qquad \text{Laue case} \\
&\geq \tfrac{1}{2}[|u^2| - |U^2| - \mathrm{Re}\,\{u^2 + U^2\}] \\
&= v^2 - V^2 \geq v^2 - (K\chi_0{}^i)^2 \frac{|\gamma_g|}{\gamma_0} \\
&= \tfrac{1}{4}(K\chi_0{}^i)^2 \left(1 - \frac{|\gamma_g|}{\gamma_0}\right)^2 \qquad \text{Bragg case}
\end{aligned}
$$

Recalling that $(K\chi_0{}^i)$ and Im $\{(u^2 + U^2)^{1/2}\}$ are positive, one obtains

$$\tfrac{1}{2}(K\chi_0{}^i)\left(1 + \frac{\gamma_g}{\gamma_0}\right) \pm \text{Im}\,\{(u^2 + U^2)^{1/2}\} \geq 0 \qquad \text{Laue case} \qquad (4\text{A-}16a)$$

$$\text{Im}\,\{(u^2 + U^2)^{1/2}\} - \tfrac{1}{2}(K\chi_0{}^i)\left(1 - \frac{|\gamma_g|}{\gamma_0}\right) \geq 0 \qquad \text{Bragg case} \qquad (4\text{A-}16b)$$

Thus the imaginary parts of *Anpassungen* $\delta_e$ for all Laue cases and for the Bragg case corresponding to the negative Riemann plane must always be negative. This means that the crystal waves actually are attenuated in the cases discussed above.

In this appendix, the analysis has been carried out on the scale of $(u, U)$. All the results obtained can be applied equally to the relations on the scale of $(s, \beta)$.

# APPENDIX 4B. INTERFERENCE

When two plane waves of the forms $A_1 \exp i\,[(\mathbf{k}_1 \cdot \mathbf{r}) + \phi_1]$ and $A_2 \exp i\,[(\mathbf{k}_2 \cdot \mathbf{r}) + \phi_2]$ interfere with each other, the intensity expression becomes

$$I = A_1{}^2 + A_2{}^2 + 2A_1A_2 \cos\,\{[(\mathbf{k}_1 - \mathbf{k}_2)\cdot\mathbf{r}] + (\phi_1 - \phi_2)\} \qquad (4\text{B-}1)$$

The intensity distribution is sinusoidal, and the spatial periodicity of the fringe spacing is given by

$$\Lambda = \frac{2\pi}{|\mathbf{k}_1 - \mathbf{k}_2|} \qquad (4\text{B-}2)$$

The normal of the constant intensity surface has the direction of $(\mathbf{k}_1 - \mathbf{k}_2)$. If one is interested in the fringe spacing along a direction $v$, it is given by

$$\Lambda_v = \frac{2\pi}{(\mathbf{k}_1 - \mathbf{k}_2)\cdot v} \qquad (4\text{B-}3)$$

# APPENDIX 4C. FOURIER INTEGRALS REQUIRED FOR THE SPHERICAL-WAVE THEORY

## I. LAUE-(BRAGG)$^n$ CASES; TYPE I

Consider the sum of the integrals

$$U_m = \tfrac{1}{2}[U_m(+) + U_m(-)] \qquad (4\text{C-}1)$$

$$U_m(\pm) = \left(\frac{1}{\beta}\right)^{m-1} \int_{-\infty}^{+\infty} \frac{[-s \pm \sqrt{s^2 + \beta^2}]^m}{\pm\sqrt{s^2 + \beta^2}}\,\exp i\,[\pm\xi_1\sqrt{s^2 + \beta^2} - \xi_2 s]\,ds^r \qquad (4\text{C-}2)$$

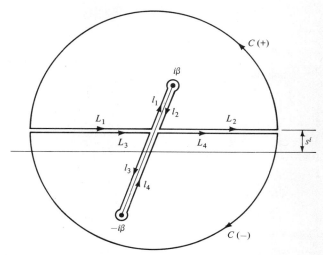

FIGURE 4C-1
Integral contours in Laue-(Bragg)$^n$ cases.

The integration can be performed by the standard method of contour integrals. For this purpose, it is convenient to define the following functions over the whole complex plane $z$:

$$I(\pm) = \left(\frac{1}{\beta}\right)^{m-1} \frac{[-z \pm (z^2 + \beta^2)^{1/2}]^m}{\pm (z^2 + \beta^2)^{1/2}} \exp i[\pm \xi_1 (z^2 + \beta^2)^{1/2} - \xi_2 z] \qquad (4C\text{-}3)$$

Here, Re $(z^2 + \beta^2)^{1/2}$ is the value of $\sqrt{s^2 + \beta^2}$ on the line $z = s^r + is^i$ for $s^r \gg 0$. The functions $I(\pm)$ are singular at $\pm i\beta$ as shown in Fig. 4C-1. Each belongs to one of the Riemann planes which are continuously connected along the Riemann cut between the singular points. The integral sum $U_m$ can be represented as follows:

$$U_m = \frac{1}{2} \left[ \int_{L_1 + L_2} I(+) \, dz + \int_{L_3 + L_4} I(-) \, dz \right] \qquad (4C\text{-}4)$$

The integral paths are shown in Fig. 4C-1.

*1* $\xi_1 \pm \xi_2 > 0$: Under these conditions,

$$\int_{C(+)} I(+) \, dz = 0 \quad \text{and} \quad \int_{C(-)} I(-) \, dz = 0$$

Then, considering the contour integrals avoiding the singular points,

$$U_m = \left(-\frac{1}{2}\right) \left[ \int_{l_1 + l_2} I(+) \, dz + \int_{l_3 + l_4} I(-) \, dz \right] \qquad (4C\text{-}5)$$

Changing the variable $z$ to $\varphi$ as follows,

$$z = i\beta \sin \varphi \qquad \text{along lines } l_1 \text{ and } l_2$$
$$z = -i\beta \sin \varphi \qquad \text{along lines } l_3 \text{ and } l_4,$$

one can see that

$$\int_{l_1} I(+)\, dz = (-i\beta) \int_{\varphi_0}^{\pi/2} (-i \sin \varphi - \cos \varphi)^m \exp \beta(\xi_2 \sin \varphi - i\xi_1 \cos \varphi)\, d\varphi$$

$$\int_{l_2} I(+)\, dz = (i\beta) \int_{\pi/2}^{\varphi_0} (-i \sin \varphi + \cos \varphi)^m \exp \beta(\xi_2 \sin \varphi + i\xi_1 \cos \varphi)\, d\varphi$$

$$\int_{l_3} I(-)\, dz = (-i\beta) \int_{-\varphi_0}^{\pi/2} (i \sin \varphi + \cos \varphi)^m \exp \beta(-\xi_2 \sin \varphi + i\xi_1 \cos \varphi)\, d\varphi$$

$$\int_{l_4} I(-)\, dz = (i\beta) \int_{\pi/2}^{-\varphi_0} (i \sin \varphi - \cos \varphi)^m \exp \beta(-\xi_2 \sin \varphi - i\xi_1 \cos \varphi)\, d\varphi$$

where $\varphi_0$ is defined by $s^i = \beta^r \sin \varphi_0$. Changing the variable $\varphi$ in every exponential function to have the same form as the first one, for example, from $\varphi$ to $\pi - \varphi$ in the second one, the sum of the four integrals becomes

$$U_m = (-)^m \left(\frac{i}{2} \beta\right) \int_0^{2\pi} \exp\left[ im\varphi + \beta(\xi_2 \sin \varphi - i\xi_1 \cos \varphi)\right] d\varphi \qquad (4C\text{-}6)$$

Introducing the complex variable

$$t = \frac{1}{i} \left(\frac{\xi_1 + \xi_2}{\xi_1 - \xi_2}\right)^{1/2} e^{i\varphi}$$

one obtains

$$U_m = (-)^m (i)^{m+1} (\pi\beta) \left(\frac{\xi_1 - \xi_2}{\xi_1 + \xi_2}\right)^{m/2} \frac{1}{2\pi i} \oint t^{m-1} \exp \tfrac{1}{2}\beta(\xi_1{}^2 - \xi_2{}^2)^{1/2} \left(t - \frac{1}{t}\right) dt$$

The contour integral defines the Bessel functions $J_{-m} = (-)^m J_m$, so that

$$U_m = (i)^{m+1} (\pi\beta) \left(\frac{\xi_1 - \xi_2}{\xi_1 + \xi_2}\right)^{m/2} J_m[\beta(\xi_1{}^2 - \xi_2{}^2)^{1/2}] \qquad \xi_1 + \xi_2 > 0,\ \xi_1 - \xi_2 > 0$$

$$(4C\text{-}7a)^1$$

**2**  $\xi_1 \pm \xi_2 < 0$: Under these conditions,

$$\int_{C(-)} I(+)\, dz = 0 \qquad \text{and} \qquad \int_{C(+)} I(-)\, dz = 0$$

Thus, instead of (4C-5), one has

$$U_m = \left(-\frac{1}{2}\right) \left[ \int_{l_1 + l_2} I(-)\, dz + \int_{l_3 + l_4} I(+)\, dz \right] \qquad (4C\text{-}8)$$

---

[1] If one writes $t = e^{i\theta}$, then $(2\pi i)^{-1} \oint dt = (2\pi)^{-1} \int_0^{2\pi} \exp i(m\theta + z \sin \theta)\, d\theta = J_{-m}(z)$.

By changing the variable $z$ to $-z$,

$$U_m = \tfrac{1}{2}(-)^m \left[ \int_{l_1+l_2} I'(+)\, dz + \int_{l_3+l_4} I'(-)\, dz \right]$$

where $I'(\pm)$ are the functions having the same form as $I(\pm)$, but the opposite signs for $\xi_1$ and $\xi_2$. Then, the argument described in 1 can be applied straightforwardly. Thus,

$$U_m = \left(\frac{1}{i}\right)^{m+1} (\pi\beta) \left(\frac{\xi_1 - \xi_2}{\xi_1 + \xi_2}\right)^{m/2} J_m[\beta(\xi_1^2 - \xi_2^2)^{1/2}]$$

$$\xi_1 + \xi_2 < 0, \quad \xi_1 - \xi_2 < 0 \qquad (4C\text{-}7b)$$

3 $\xi_1^2 - \xi_2^2 < 0$: Under these conditions, one of the contours $C(\pm)$ must be used for integrating both $I(\pm)$. Thus, instead of equations (4C-5) and (4C-8),

$$U_m = -\int_{l_1+l_2} [I(+) + I(-)]\, dz$$

or

$$U_m = -\int_{l_3+l_4} [I(+) + I(-)]\, dz$$

On the other hand,

$$\int_{l_1} I(\pm)\, dz = -\int_{l_2} I(\mp)\, dz$$

$$\int_{l_3} I(\pm)\, dz = -\int_{l_4} I(\mp)\, dz$$

Thus, finally,

$$U_m = 0 \qquad \xi_1^2 - \xi_2^2 < 0 \qquad (4C\text{-}7c)$$

## II. LAUE-(BRAGG)ⁿ CASES; TYPE II

In this case, the following integral sum is required.

$$V_m = \tfrac{1}{2}[V_m(+) + V_m(-)] \qquad (4C\text{-}9)$$

where

$$V_m(\pm) = \left(\frac{1}{\beta}\right)^{m-1} \int_{-\infty}^{+\infty} \frac{[s \pm \sqrt{s^2 + \beta^2}]^m}{\pm \sqrt{s^2 + \beta^2}} \exp i[\pm\xi_1 \sqrt{s^2 + \beta^2} - \xi_2 s]\, ds^r \qquad (4C\text{-}10)$$

They are transformed to $U_m(\pm)$ simply by changing the sign of $\xi_2$ and $s$. Thus,

$$V_m = (i)^{m+1}(\pi\beta) \left[\frac{\xi_1 + \xi_2}{\xi_1 - \xi_2}\right]^{m/2} J_m[\beta(\xi_1^2 - \xi_2^2)^{1/2}] \qquad \xi_1 + \xi_2 > 0, \quad \xi_1 - \xi_2 > 0$$

$$(4C\text{-}11a)$$

$$V_m = \left(\frac{1}{i}\right)^{m+1} (\pi\beta) \left(\frac{\xi_1 + \xi_2}{\xi_1 - \xi_2}\right)^{m/2} J_m[\beta(\xi_1^2 - \xi_2^2)^{1/2}] \qquad \xi_1 + \xi_2 < 0, \quad \xi_1 - \xi_2 < 0$$

$$\text{(4C-11}b\text{)}$$

$$V_m = 0 \qquad\qquad\qquad\qquad\qquad\qquad \xi_1^2 - \xi_2^2 < 0 \qquad \text{(4C-11}c\text{)}$$

## III. (BRAGG)" CASES

One needs the following integrals which, as distinct from Laue-(Bragg)" cases, are defined on one of the Riemann planes.

$$W_m = \int_{-\infty + is^i}^{+\infty + is^i} \left[\frac{z - (z^2 + \beta^2)^{1/2}}{\beta}\right]^m \exp - i[\xi_1(z^2 + \beta^2)^{1/2} + \xi_2 z] \, dz \qquad \text{(4C-12)}$$

The integral path, which will be denoted by $L$, and the singular points $\pm i\beta$ are illustrated in Fig. 4C-2. [Note the difference from Laue-(Bragg)" cases shown in Fig. 4C-1.] As already proved in Appendix 3 A III, $v > |V|$; i.e., $s^i > |\beta^r|$. The path $L$, therefore, never crosses the Riemann cut between $\pm i\beta$.

*1*  $\xi_2 \pm \xi_1 > 0$: One immediately obtains the following relations:

$$\int_L dz + \int_{C(-)} dz = \int_{l_1 + l_2 + l_3 + l_4} dz$$

$$\int_{C(-)} dz = 0 \qquad \text{as } |z| \text{ increases} \qquad \text{(4C-13)}$$

Thus, it follows that

$$W_m = \int_{l_1 + l_2 + l_3 + l_4} dz$$

Similarly to the case of (4C-5), by using the variable $\varphi$ defined there, one has that

$$W_m = (i\beta) \int_0^{2\pi} \exp\left[im\varphi + \beta(\xi_2 \sin\varphi + i\xi_1 \cos\varphi)\right] \cos\varphi \, d\varphi$$

Applying the result (4C-7a) to the integral in Section II, one sees that $W_m$ can be represented by $U_{m+1}$ and $U_{m-1}$ with $\xi_1 = -\xi_1$. Thus, finally,

$$W_m = i(\pi\beta) \left\{ \left(\frac{\xi_2 + \xi_1}{\xi_2 - \xi_1}\right)^{(m+1)/2} J_{m+1}[\beta(\xi_1^2 - \xi_2^2)^{1/2}] \right.$$

$$\left. + \left(\frac{\xi_2 + \xi_1}{\xi_2 - \xi_1}\right)^{(m-1)/2} J_{m-1}[\beta(\xi_1^2 - \xi_2^2)^{1/2}] \right\} \qquad \xi_2 \pm \xi_1 > 0 \qquad \text{(4C-14}a\text{)}$$

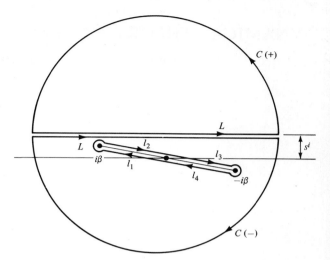

FIGURE 4C-2
Integral contours in (Bragg)$^n$ cases.

When applying this equation to (Bragg)$^n$ cases, where $\beta$ is $i\bar{\beta}$ and $\xi_1 - \xi_2$ is negative, $\beta(\xi_1^2 - \xi_2^2)^{1/2}$ must be replaced by $-\bar{\beta}(\xi_2^2 - \xi_1^2)$.

2  $\xi_2 \pm \xi_1 < 0$: Instead of (4C-13) one has

$$\int_{C(+)} dz = 0 \qquad \text{as } |z| \text{ increases}$$

Since the contours $L$ and $C(+)$ include no singular points, one immediately gets

$$W_m = 0 \qquad \xi_2 \pm \xi_1 < 0 \qquad \text{(4C-14}b\text{)}$$

# 5

# DYNAMICAL THEORY FOR IMPERFECT CRYSTALS[1]

In Section II C of Chap. 3, lattice imperfections were classified into three types. There, statistical problems were not considered, and the same limitations will be imposed in the present discussion. If one asks only what individual imperfections or lattice distortions exist in a nearly perfect crystal, this limitation is not very serious. On the other hand, the applicability of the theory would be broader if one could treat them statistically in a unified theoretical framework. In particular, such theories would be valuable for tackling the problems of extinction.

The dynamical theories described in Chap. 4 partly include the effects of crystal finiteness through the boundary conditions. In Section I of the present chapter the theory is supplemented by taking into account the lateral finiteness of the crystal size so that the connection between the kinematical and dynamical theories can be elucidated in the plane-wave theory. Subsequently, crystals containing a single planar defect, such as a stacking fault, a twin plane, and an idealized misfit boundary, are considered. Both plane-wave and spherical-wave theories are developed for such

---

[1] The notations employed in this chapter are the same as those in Chaps. 3 and 4 but differ somewhat from the ones used in other chapters of this book in order to facilitate comparison with original citations.

cases. The resulting equations are obtained by successive application of the theories for perfect crystals. Continuously deformed crystals will be considered separately in the Section II of this chapter, since some additional concepts are required in the development of the theory.

# I.  FINITENESS OF THE CRYSTAL AND PLANAR DEFECTS

## A.  Finite polyhedral crystals

The present considerations are limited exclusively to a plane-wave theory characterized by a sufficiently small Bragg angle and a fairly large angular range $\Delta\Theta$ of the Bragg reflection. In such a treatment, the Bragg arrangement is not important because the cross sections of the crystal surfaces with respect to both vacuum and crystal waves are extremely small. For this reason, it is sufficient to consider only Laue-Laue cases. It is convenient to regard any polyhedral crystal as a set of columns along a direction $v$ taken in the direction of the energy flow of the crystal wave. With such an arrangement, the waves do not emerge from the sides of the column, and the total Bragg-reflected waves can be represented as the sum of the waves from individual columns. The basic idea is similar to the analysis in the kinematical theory (Fig. 3-8).

  If the cross section is sufficiently wide, or if one considers a hypothetical wedge-shaped crystal denoted by dotted lines in Fig. 3-8, the waves penetrating through the crystal have been given already by equations (4-105$a$ and $b$) or (4-106$a$ and $b$). The only problem remaining, therefore, is to take into account the lateral width of the column. If one imagines two diaphragms on the crystal surfaces, having apertures $S_e$ and $S_a$, respectively, the crystal waves and the penetrating waves must be identical to those in the columnar crystal.[1] The justification for this intuitive treatment can be found in the original papers of Kato (1952$a$,$b$; 1953) and Kato and Uyeda (1951$a$,$b$). The lateral finiteness of the crystal surface merely brings about an optical diffraction of the doubly refracted waves. The broadening, in effect, is determined by the cross section $S$ of the penetrating wave. It should be noted that the column direction depends on the branch of the dispersion surface to which the concerned wave belongs and the degree to which the Bragg condition is satisfied. Owing to the refraction, therefore, the neighboring columns may overlap each other. This does not alter, however, the statement in the previous paragraph.

  Rearranging the phase term, one can rewrite (4-105$b$) also in the form

$$D_g^{(j)}(\mathbf{r}) = C_g^{(j)} E_e \exp iK(\delta_e^{(j)} R_e - \delta_a^{(j)} R_a) \exp i(\mathbf{K}_g^{(j)} \cdot \mathbf{r}) \qquad (5\text{-}1)^2$$

---

[1] This idea was originally suggested by Professor R. Uyeda.
[2] For simplicity, the suffix $g$ in $\delta_{a,g}^{(j)}$ is omitted.

where $R_e$ and $R_a$ are the normal distances from the origin to the entrance and exit surfaces, respectively. The wave is indeed a plane wave having the wave vector $\mathbf{K}_g^{(j)}$. As described in Section II C 2 of Chap. 3, if the plane wave, $\exp i(\mathbf{K}_g^{(j)} \cdot \mathbf{r})$, is modified by an aperture $S$ perpendicular to $\mathbf{K}_g^{(j)}$, the Fraunhofer diffracted wave has the form[1]

$$\frac{K}{2\pi} \frac{\exp iKR_0}{iR_0} \iint_S \exp -i(\sigma^{(j)} \cdot \mathbf{s}) \, d\mathbf{s} \qquad (5\text{-}2)$$

where

$$\sigma^{(j)} = \mathbf{K}_g - \mathbf{K}_g^{(j)}$$

and $\mathbf{K}_g$ is a vacuum-wave vector parallel to a large position vector $\mathbf{R}_0$. Taking up the amplitude (4-72), one gets the Fraunhofer relation for a dynamically diffracted wave:

$$D_g(\mathbf{R}_0) = \frac{\exp iKR_0}{R_0} \frac{r_c C F_g}{v} E_e \left[ \frac{i}{\sqrt{w^2 + W^2}} \right]$$

$$\times \left[ \exp iK(\delta_e^{(1)}R_e - \delta_a^{(1)}R_a) \iint_{S^{(1)}} \exp -i(\sigma^{(1)} \cdot \mathbf{s}) \, d\mathbf{s} \right.$$

$$\left. - \exp iK(\delta_e^{(2)}R_e - \delta_a^{(2)}R_a) \iint_{S^{(2)}} \exp -i(\sigma^{(2)} \cdot \mathbf{s}) \, d\mathbf{s} \right] \qquad (5\text{-}3)$$

where the cross section $S$ must be specified explicitly by branch $(j)$. The notations used in reciprocal space are illustrated in Fig. 5-1a. The corresponding kinematical wave is obtained by the use of equations (3-74a), (3-93), and (3-109):

$$D_g(\mathbf{R}_0) = \frac{\exp iKR_0}{R_0} \frac{r_c C \bar{F}_g}{v} E_e \frac{i}{w} \left[ \exp -iK\delta_a R_a \iint_S \exp -i(\sigma_a \cdot \mathbf{s}) \, d\mathbf{s} \right.$$

$$\left. - \exp -iK\delta_e R_e \iint_S \exp -i(\sigma_e \cdot \mathbf{s}) \, d\mathbf{s} \right] \qquad (5\text{-}4)$$

where the column direction is assumed to be in the direction of $\mathbf{K}_g$, so that $\zeta$ is replaced by $w$. The notations are explained in Fig. 3-9b. The correspondence between (5-3) and (5-4) is straightforward. The dynamical formula (5-3) can be interpreted by means of the Ewald construction in a similar way to the kinematical formula (5-4). The dynamical amplitude region is illustrated in Fig. 5-1b. It is easy to see that $\sigma^{(1)}$ and $\sigma^{(2)}$ tend to either $\sigma_e$ or $\sigma_a$ for large $|w|$. Also, consulting the geometrical meaning of $\delta_e^{(j)}$, $\delta_a^{(j)}$, $\delta_e$, and $\delta_a$, one can see that the phase $K(\delta_e^{(j)}R_e - \delta_a^{(j)}R_a)$ tends to either $K(\delta_e R_e)$ or $K(\delta_a R_a)$, provided that the mean polarizability is ignored. The dynamical

---

[1] Here, $K_n^* = K$; $r_c f(\mathbf{k}) = 1/2\pi \iint_S \exp i[(\mathbf{K}_g^{(j)} - \mathbf{K}_g) \cdot \mathbf{s}] \, d\mathbf{s}$ in equation (3-112)

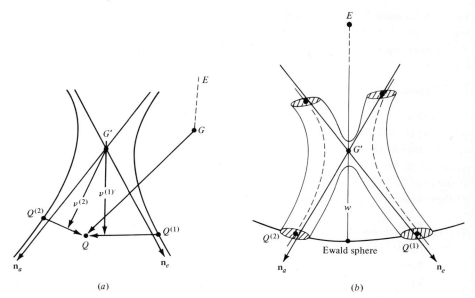

**FIGURE 5-1**

Reciprocal-space constructions. (*a*) Notation used in (5-3). $(\overrightarrow{Q^{(1)}Q} = \sigma^{(1)},$
$\overrightarrow{Q^{(2)}Q} = \sigma^{(2)}, \quad \overrightarrow{GQ} = \Delta = \mathbf{K}_g - (\mathbf{K}_e + 2\pi\mathbf{g}), \quad \overrightarrow{EG} = \mathbf{K}_e + 2\pi\mathbf{g}, \quad \overrightarrow{EQ} = \mathbf{K}_g,$
$\overrightarrow{EQ^{(1)}} = \mathbf{K}_g^{(1)}, \overrightarrow{EQ^{(2)}} = \mathbf{K}_g^{(2)}.)$  When normal refraction is neglected ($\chi_0 = 0$),
$G'Q^{(1)} = K\delta^{(1)}$ and $G'Q^{(2)} = K\delta^{(2)}$. (*b*) Amplitude region and Ewald sphere in
the dynamical theory.

wave can be interpreted in terms of the effects of ordinary and double refraction and
optical diffraction.  In distinction from the kinematical amplitude region, the dynamical one depends on the net plane concerned.

As explained in Section I D 1 of Chap. 4, in high-voltage electron diffraction
experiments, splitting of the diffraction spot is observed for crystals having well-
developed crystal surfaces.  For very tiny crystallites, however, the splitting is masked
by the optical diffraction effects.  For MgO, it is expected that the double spots
should be clearly observable for cubic crystals having edges longer than 200 A.  For
1-A x rays, the corresponding figure is about 800 $\mu$, because the wavelength is 40 times
larger and the reflection range $\Delta\Theta$ is $10^3$ times smaller.  For a crystal having this size,
the plane-wave theory is no longer strictly applicable.  For this reason, the plane-wave
theory described in the present section is not directly relevant to x-ray diffraction
experiments.  Laue (1940) incorrectly expected diffraction spot splitting in x-ray cases.
Nevertheless, x-ray workers may obtain a theoretical satisfaction from its consideration, similarly to a violinist who hears the successful performance of a pianist friend.

In x-ray cases, the lateral finiteness must be treated by the successive reflection and the transmission of the crystal wave fields as discussed in Section II D and F 4 of the preceding chapter.

## B.   Crystal containing a fault plane

In this section, a general approach is presented for solving the problem for the case when the crystal contains a single fault plane, as illustrated in Fig. 5-2. The configuration of the wave vectors with respect to the crystal surfaces and the fault plane is assumed to be a (Laue)[3] case in the nomenclature previously described. The fault plane may be either a stacking fault, a twin plane, or a discrete misfit boundary. Each of these special cases will be considered in the following sections.

**1.   *Plane-wave theory***   In Laue cases, two Bloch waves are excited at the entrance surface of the first crystal $A$ as described in Section I C 2 of Chap. 4. What happens is that there appear four coherent waves which are specified by subscripts $g = 0$ and $g$, and superscripts $(i) = (1)$ and $(2)$. When they strike the fault plane under conditions of the Laue case, each plane wave excites again four plane waves in the second crystal $B$. Consequently, sixteen plane waves or eight Bloch waves must be considered. The waves are specified by four super- and subscripts $\{^{ij}_{gg'}\}$, the first two $\{^{i}_{g}\}$ denoting the state of the wave in crystal $A$, and the last two $\{^{j}_{g'}\}$ denoting the state in crystal $B$. In general, the reflection vector $g'$ in crystal $B$ may be slightly different from the vector $g$ in crystal $A$. This is the reason for introducing the additional indices $0'$ and $g'$.[1]

The wave vectors $k\{^{ij}_{gg'}\}$ are obtained from $k_0{}^{(i)}$ and $k_g{}^{(i)}$ by the tangential continuity on the fault plane. Since a pair of $\{^{ij}_{g0'}\}$ and $\{^{ij}_{gg'}\}$ waves constitutes a Bloch wave, the wave vectors must have the relation

$$k\{^{ij}_{gg'}\} = k\{^{ij}_{g0'}\} + 2\pi g' \qquad (g = 0 \text{ or } g) \qquad (5\text{-}5)$$

Obviously, they must satisfy also the dispersion relation in crystal $B$. In addition the amplitudes must have a definite ratio

$$c^{(ij)} = \frac{d\{^{ij}_{0g'}\}}{d\{^{ij}_{00'}\}} \qquad (5\text{-}6a)$$

and

$$\bar{c}^{(ij)} = \frac{d\{^{ij}_{g0'}\}}{d\{^{ij}_{gg'}\}} \qquad (5\text{-}6b)$$

---

[1] Initially, it might be easier to read through without distinguishing $g$ from $g'$ with reference to Fig. 5-3. Subsequently, it is advisable to consider Figs. 5-8, 5-12, and 5-13.

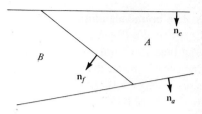

**FIGURE 5-2**
Crystal containing a fault plane.

Like (4-28) for the waves in crystal $A$, they are determined by the *Resonanzfehler* $\Delta\eta_0\{_{g0'}^{ij}\}$ or $\Delta\eta_g\{_{g0'}^{ij}\}$, which are defined by

$$\Delta\eta_0\{_{g0'}^{ij}\} = [\hat{\mathbf{K}}_0 \cdot (\mathbf{k}\{_{g0'}^{ij}\} - \bar{\mathbf{k}}_{0'})] \qquad (5\text{-}7a)^1$$

$$\Delta\eta_g\{_{g0'}^{ij}\} = [\hat{\mathbf{K}}_g \cdot (\mathbf{k}\{_{g0'}^{ij}\} - \bar{\mathbf{k}}_{0'})] \qquad (5\text{-}7b)^1$$

The suffixes 0 and $g$ immediately after $\Delta\eta$ refer to the normals of the asymptotic surfaces $S_0$ and $S_g$. In essence, once the wave vectors $\mathbf{k}\{_{gg'}^{ij}\}$ (or the dispersion points $D_0^{(ij)}$ and $D_g^{(ij)}$) are determined in crystal $B$, merely the amplitudes $d\{_{00'}^{ij}\}$ and $d\{_{gg'}^{ij}\}$ are left as unknowns. They are determined by the boundary conditions at the fault plane, which are equivalent to the boundary conditions (4-68a and b) at the entrance surface.

The wave fields in crystal $A$ [equations (4-73) or (4-74)] can be written

$$d_0^{(i)}(\mathbf{r}) = d_0^{(i)} \exp i(\mathbf{k}_0^{(i)} \cdot \mathbf{r}) \qquad (5\text{-}8a)$$

$$d_g^{(i)}(\mathbf{r}) = d_g^{(i)} \exp i(\mathbf{k}_g^{(i)} \cdot \mathbf{r}) \qquad (5\text{-}8b)$$

where

$$d_0^{(i)} = C_0^{(i)}E_e \exp i[(\mathbf{K}_e - \mathbf{k}_0^{(i)}) \cdot \mathbf{r}_e] \qquad (5\text{-}9a)$$

$$d_g^{(i)} = C_g^{(i)}E_e \exp i[(\mathbf{K}_e - \mathbf{k}_0^{(i)}) \cdot \mathbf{r}_e] \qquad (5\text{-}9b)$$

Similarly, one can write for the waves in crystal $B$

$$d\{_{00'}^{ij}\}(\mathbf{r}) = d\{_{00'}^{ij}\} \exp i(\mathbf{k}\{_{00'}^{ij}\} \cdot \mathbf{r}) \qquad (5\text{-}10a)$$

$$d\{_{0g'}^{ij}\}(\mathbf{r}) = d\{_{0g'}^{ij}\} \exp i(\mathbf{k}\{_{0g'}^{ij}\} \cdot \mathbf{r}) \qquad (5\text{-}10b)$$

$$d\{_{g0'}^{ij}\}(\mathbf{r}) = d\{_{g0'}^{ij}\} \exp i(\mathbf{k}\{_{g0'}^{ij}\} \cdot \mathbf{r}) \qquad (5\text{-}10c)$$

$$d\{_{gg'}^{ij}\}(\mathbf{r}) = d\{_{gg'}^{ij}\} \exp i(\mathbf{k}\{_{gg'}^{ij}\} \cdot \mathbf{r}) \qquad (5\text{-}10d)$$

---

[1] One may define

$$\Delta\eta_0\,\{_{gg'}^{ij}\} = [\hat{\mathbf{K}}_0 \cdot (\mathbf{k}\{_{gg'}^{ij}\} - \bar{\mathbf{k}}_{g'})]$$

and

$$\Delta\eta_g\,\{_{gg'}^{ij}\} = [\hat{\mathbf{K}}_g \cdot (\mathbf{k}\{_{gg'}^{ij}\} - \bar{\mathbf{k}}_{g'})].$$

However, due to the relation (5-5),

$$\Delta\eta_0\,\{_{gg'}^{ij}\} = \Delta\eta_0\{_{g0'}^{ij}\} \qquad \text{and} \qquad \Delta\eta_g\{_{gg'}^{ij}\} = \Delta\eta_g\{_{g0'}^{ij}\}.$$

The boundary conditions, then, can be written

$$d_0^{(i)}(\mathbf{r}_f) = d\{{}^{i1}_{00'}\}(\mathbf{r}_f) + d\{{}^{i2}_{00'}\}(\mathbf{r}_f) \qquad (5\text{-}11a)$$

$$0 = d\{{}^{i1}_{0g'}\}(\mathbf{r}_f) + d\{{}^{i2}_{0g'}\}(\mathbf{r}_f) \qquad (5\text{-}11b)$$

$$0 = d\{{}^{i1}_{g0'}\}(\mathbf{r}_f) + d\{{}^{i2}_{g0'}\}(\mathbf{r}_f) \qquad (5\text{-}11c)$$

$$d_g^{(i)}(\mathbf{r}_f) = d\{{}^{i1}_{gg'}\}(\mathbf{r}_f) + d\{{}^{i2}_{gg'}\}(\mathbf{r}_f) \qquad (5\text{-}11d)$$

By using (5-5) and (5-6a), a pair of equations (5-11a and b) gives the solution $d\{{}^{ij}_{00'}\}$. The other one, $d\{{}^{ij}_{0g'}\}$ is obtained simply by multiplying it by (5-6a). Similarly, from a pair of equations (5-11c and d), $d\{{}^{ij}_{gg'}\}$ and $\{{}^{ij}_{g0'}\}$ are obtained. Here it is convenient to use (5-6b) instead of (5-6a). Thus, it turns out that the amplitudes have the form

$$d\{{}^{ij}_{00'}\} = C_0^{(i)}C\{{}^{ij}_{00'}\}E_e \exp i\{[(\mathbf{K}_e - \mathbf{k}_0^{(i)}) \cdot \mathbf{r}_e] + [(\mathbf{k}_0^{(i)} - \mathbf{k}\{{}^{ij}_{00'}\}) \cdot \mathbf{r}_f]\} \qquad (5\text{-}12a)$$

$$d\{{}^{ij}_{0g'}\} = C_0^{(i)}C\{{}^{ij}_{0g'}\}E_e \exp i\{[(\mathbf{K}_e - \mathbf{k}_0^{(i)}) \cdot \mathbf{r}_e] + [(\mathbf{k}_0^{(i)} - \mathbf{k}\{{}^{ij}_{0g'}\}) \cdot \mathbf{r}_f]\} \qquad (5\text{-}12b)$$

$$d\{{}^{ij}_{g0'}\} = C_g^{(i)}C\{{}^{ij}_{g0'}\}E_e \exp i\{[(\mathbf{K}_e - \mathbf{k}_0^{(i)}) \cdot \mathbf{r}_e] + [(\mathbf{k}_g^{(i)} - \mathbf{k}\{{}^{ij}_{gg'}\}) \cdot \mathbf{r}_f]\} \qquad (5\text{-}12c)$$

$$d\{{}^{ij}_{gg'}\} = C_g^{(i)}C\{{}^{ij}_{gg'}\}E_e \exp i\{[(\mathbf{K}_e - \mathbf{k}_0^{(i)}) \cdot \mathbf{r}_e] + [(\mathbf{k}_g^{(i)} - \mathbf{k}\{{}^{ij}_{gg'}\}) \cdot \mathbf{r}_f]\} \qquad (5\text{-}12d)$$

where the amplitude factors $C$ are given as follows:

$$C\{{}^{i1}_{00'}\} = \frac{c^{(i2)}}{c^{(i2)} - c^{(i1)}} \qquad (5\text{-}13a)$$

$$C\{{}^{i2}_{00'}\} = \frac{c^{(i1)}}{c^{(i1)} - c^{(i2)}} \qquad (5\text{-}13b)$$

$$C\{{}^{i1}_{0g'}\} = \frac{c^{(i1)}c^{(i2)}}{c^{(i2)} - c^{(i1)}} \qquad (5\text{-}13c)$$

$$C\{{}^{i2}_{0g'}\} = \frac{c^{(i1)}c^{(i2)}}{c^{(i1)} - c^{(i2)}} \qquad (5\text{-}13d)$$

and

$$C\{{}^{i1}_{g0'}\} = \frac{\bar{c}^{(i1)}\bar{c}^{(i2)}}{\bar{c}^{(i2)} - \bar{c}^{(i1)}} \qquad (5\text{-}14a)$$

$$C\{{}^{i2}_{g0'}\} = \frac{\bar{c}^{(i1)}\bar{c}^{(i2)}}{\bar{c}^{(i1)} - \bar{c}^{(i2)}} \qquad (5\text{-}14b)$$

$$C\{{}^{i1}_{gg'}\} = \frac{\bar{c}^{(i2)}}{\bar{c}^{(i2)} - \bar{c}^{(i1)}} \qquad (5\text{-}14c)$$

$$C\{{}^{i2}_{gg'}\} = \frac{\bar{c}^{(i1)}}{\bar{c}^{(i1)} - \bar{c}^{(i2)}} \qquad (5\text{-}14d)$$

By inserting the amplitudes (5-12) into equations (5-10), one obtains the wave fields in the crystal $B$.

Sometimes, it is convenient to write the wave fields as follows:

$$d\{^{ij}_{00'}\}(\mathbf{r}) = d\{^{ij}_{00'}\}(\mathbf{K}_e; \mathbf{r})E_e \exp i(\mathbf{K}_e \cdot \mathbf{r}) \tag{5-15a}$$

$$d\{^{ij}_{0g'}\}(\mathbf{r}) = d\{^{ij}_{0g'}\}(\mathbf{K}_e; \mathbf{r})E_e \exp i[(\mathbf{K}_e + 2\pi\mathbf{g}') \cdot \mathbf{r}] \tag{5-15b}$$

$$d\{^{ij}_{g0'}\}(\mathbf{r}) = d\{^{ij}_{g0'}\}(\mathbf{K}_e; \mathbf{r})E_e \exp i[(\mathbf{K}_e + 2\pi\mathbf{g} - 2\pi\mathbf{g}') \cdot \mathbf{r}] \tag{5-15c}$$

$$d\{^{ij}_{gg'}\}(\mathbf{r}) = d\{^{ij}_{gg'}\}(\mathbf{K}_e; \mathbf{r})E_e \exp i[(\mathbf{K}_e + 2\pi\mathbf{g}) \cdot \mathbf{r}] \tag{5-15d}$$

where

$$d\{^{ij}_{00'}\}(\mathbf{K}_e; \mathbf{r}) = C_0^{(i)}C\{^{ij}_{00'}\} \exp i\{[(\mathbf{k}_0^{(i)} - \mathbf{K}_e) \cdot (\mathbf{r} - \mathbf{r}_e)]$$
$$+ [(\mathbf{k}\{^{ij}_{00'}\} - \mathbf{k}_0^{(i)}) \cdot (\mathbf{r} - \mathbf{r}_f)]\} \tag{5-16a}$$

$$d\{^{ij}_{0g'}\}(\mathbf{K}_e; \mathbf{r}) = C_0^{(i)}C\{^{ij}_{0g'}\} \exp i\{[(\mathbf{k}_0^{(i)} - \mathbf{K}_e) \cdot (\mathbf{r} - \mathbf{r}_e)]$$
$$+ [(\mathbf{k}\{^{ij}_{00'}\} - \mathbf{k}_0^{(i)}) \cdot (\mathbf{r} - \mathbf{r}_f)]\} \tag{5-16b}$$

$$d\{^{ij}_{g0'}\}(\mathbf{K}_e; \mathbf{r}) = C_g^{(i)}C\{^{ij}_{g0'}\} \exp i\{[(\mathbf{k}_0^{(i)} - \mathbf{K}_e) \cdot (\mathbf{r} - \mathbf{r}_e)]$$
$$+ [(\mathbf{k}\{^{ij}_{gg'}\} - \mathbf{k}_g^{(i)}) \cdot (\mathbf{r} - \mathbf{r}_f)]\} \tag{5-16c}$$

$$d\{^{ij}_{gg'}\}(\mathbf{K}_e; \mathbf{r}) = C_g^{(i)}C\{^{ij}_{gg'}\} \exp i\{[(\mathbf{k}_0^{(i)} - \mathbf{K}_e) \cdot (\mathbf{r} - \mathbf{r}_e)]$$
$$+ [(\mathbf{k}\{^{ij}_{gg'}\} - \mathbf{k}_g^{(i)}) \cdot (\mathbf{r} - \mathbf{r}_f)]\} \tag{5-16d}$$

**2. Spherical-wave theory** According to the general treatment of the spherical-wave theory in Section II B of Chap. 4, the wave fields produced by the spherical wave expressed in relations (4-191) to (4-193) are given by double integrals of the plane-wave solution, in such forms as equations (4-194a and b). Now, the amplitude densities $d_0(\mathbf{K}_e; \mathbf{r})$ and $d_g(\mathbf{K}_e; \mathbf{r})$ must be replaced by (5-16) in specific cases. Taking the co-ordinate system described there, the integration of $K_y$ can be performed without referring to the crystal diffraction. Thus, the spherical-wave solution is given as follows:[1]

$$\phi\{^{ij}_{00'}\}(\mathbf{r}) = A_0E_e \int_{-\infty}^{+\infty} d\{^{ij}_{00'}\}(\mathbf{K}; \mathbf{r}) \exp i(K_x x_0) \, dK_x \tag{5-17a}$$

$$\phi\{^{ij}_{0g'}\}(\mathbf{r}) = A_{g'}E_e \int_{-\infty}^{+\infty} d\{^{ij}_{0g'}\}(\mathbf{K}; \mathbf{r}) \exp i(K_x x_0) \, dK_x \tag{5-17b}$$

$$\phi\{^{ij}_{g0'}\}(\mathbf{r}) = A_{0'}E_e \int_{-\infty}^{+\infty} d\{^{ij}_{g0'}\}(\mathbf{K}; \mathbf{r}) \exp i(K_x x_0) \, dK_x \tag{5-17c}$$

$$\phi\{^{ij}_{gg'}\}(\mathbf{r}) = A_g E_e \int_{-\infty}^{+\infty} d\{^{ij}_{gg'}\}(\mathbf{K}; \mathbf{r}) \exp i(K_x x_0) \, dK_x \tag{5-17d}$$

---

[1] The subscript $e$ is omitted in $\mathbf{K}_e$, and the coordinate $x$ is written as $x_0$.

where the amplitude factors $A_0$ and $A_g$ are identical to those defined by (4-197a and b). The others are given by

$$A_{0'} = \frac{i}{4\pi} \left( \frac{1}{2\pi Kr} \right)^{1/2} \exp i \left[ (\overline{\mathbf{K}}_g \cdot \mathbf{r}) - 2\pi(\mathbf{g}' \cdot \mathbf{r}) - \frac{\pi}{4} \right] \qquad (5\text{-}18a)$$

$$A_{g'} = \frac{i}{4\pi} \left( \frac{1}{2\pi Kr} \right)^{1/2} \exp i \left[ (\overline{\mathbf{K}}_0 \cdot \mathbf{r}) + 2\pi(\mathbf{g}' \cdot \mathbf{r}) - \frac{\pi}{4} \right] \qquad (5\text{-}18b)$$

It is not easy to perform the integrations in (5-17) exactly. For this reason, the stationary-phase method is used extensively. The details will be described for different cases in the following sections. What results is a triangular wave field, similarly to the case of perfect crystals.

The vacuum-wave field connected with the crystal waves mentioned above is a kind of projection of the wave field on the exit surface along $\overline{\mathbf{K}}_0$ and $\overline{\mathbf{K}}_g$ directions for $O$ and $G$ waves, respectively. It is a straightforward matter to write down the vacuum waves as described in Section II B 4 of Chap. 4.

## C.   Stacking faults

1. *Plane-wave solution*   The problem originally arose in the direct observation (Whelan, Hirsch, Horne, and Bollman, 1957) of fault planes by means of electron microscopy, and the theory was originally developed by Whelan and Hirsch (1957a,b) and subsequently by others. Whelan et al. made a plausible but ad hoc assumption about the shape of the dispersion surface. The assumption, however, is inadequate for x-ray cases because of the large Bragg angle $\theta_B$ and small angular range $\Delta\Theta$. For this reason, one needs to reformulate the theory rigorously (Kato, Usami, and Katagawa, 1967).

A stacking fault is characterized by the normal $\mathbf{n}_f$ and the relative displacement vector $\mathbf{u}_f$ of the Bravais lattice points of crystals $A$ and $B$. The reflection vector $\mathbf{g}'$ is identical to $\mathbf{g}$. In addition $C^2 \chi_g \chi_{-g}$ is common to crystals $A$ and $B$. The dispersion surface of crystal $B$, therefore, is identical to that of crystal $A$.

First, consider the wave vectors and the phases in (5-16). Owing to the identity of the dispersion surfaces for crystals $A$ and $B$, the number of dispersion points excited in crystal $B$ is reduced to four, of which two are identical to the dispersion points excited in crystal $A$. The situation is illustrated in Fig. 5-3. The wave vectors obey the following relations:

$$\mathbf{k}\{{}^{ii}_{00}\} = \mathbf{k}\{{}^{ii}_{g0}\} \equiv \mathbf{k}_0{}^{(ii)} = \mathbf{k}_0{}^{(i)} \qquad (D^{(ii)} = D^{(i)}) \qquad (5\text{-}19a)$$

$$\mathbf{k}\{{}^{ii}_{0g}\} = \mathbf{k}\{{}^{ii}_{gg}\} \equiv \mathbf{k}_g{}^{(ii)} = \mathbf{k}_g{}^{(i)} \qquad (D^{(ii)} = D^{(i)}) \qquad (5\text{-}19b)$$

$$\mathbf{k}\{{}^{ij}_{00}\} = \mathbf{k}\{{}^{ij}_{g0}\} \equiv \mathbf{k}_0{}^{(ij)} \qquad (D^{(ij)}) \qquad (5\text{-}19c)$$

$$\mathbf{k}\{{}^{ij}_{0g}\} = \mathbf{k}\{{}^{ij}_{gg}\} \equiv \mathbf{k}_g{}^{(ij)} \qquad (D^{(ij)}) \qquad (5\text{-}19d)$$

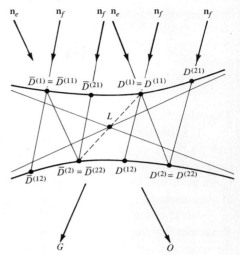

**FIGURE 5-3**
Construction of the dispersion points for a crystal containing a stacking fault. Two groups of dispersion points are indicated. A set of $D^{(ij)}$ corresponds to waves excited by a simple plane wave. Another set of $\bar{D}^{(ij)}$ is conjugate to the set $D^{(ij)}$. The Bloch waves associated with $D^{(ij)}$ and $\bar{D}^{(ji)}$, or $D^{(ii)}$ and $\bar{D}^{(jj)}$, propagate along the same direction in crystal $B$.

The letters in the parentheses denote the relevant dispersion points. In addition, relation (5-5) is reduced to

$$\mathbf{k}_g^{(ij)} = \mathbf{k}_0^{(ij)} + 2\pi\mathbf{g} \qquad (5\text{-}20)$$

Thus, only $\mathbf{k}_0^{(12)}$ and $\mathbf{k}_0^{(21)}$ are independent wave vectors to be introduced in crystal $B$. From the tangential continuity of the wave vectors, they are expressed, as follows, by using *Anpassung* $\delta_f^{(ij)}$:

$$\mathbf{k}_0^{(ij)} = \mathbf{k}_0^{(i)} - K\delta_f^{(ij)}\mathbf{n}_f \qquad (5\text{-}21)$$

Introducing *Resonanzfehler* of the waves in crystal $B$ by

$$\Delta\eta_0^{(ij)} = [\hat{\mathbf{K}}_0 \cdot (\mathbf{k}_0^{(ij)} - \bar{\mathbf{k}}_0)] \qquad (5\text{-}22a)$$

$$\Delta\eta_g^{(ij)} = [\hat{\mathbf{K}}_g \cdot (\mathbf{k}_0^{(ij)} - \bar{\mathbf{k}}_0)] \qquad (5\text{-}22b)$$

which are special forms of (5-7a and b),

$$\Delta\eta_0^{(ij)} = \Delta\eta_0^{(i)} - K\delta_f^{(ij)}\bar{\gamma}_0 \qquad (5\text{-}23a)$$

$$\Delta\eta_g^{(ij)} = \Delta\eta_g^{(i)} - K\delta_f^{(ij)}\bar{\gamma}_g \qquad (5\text{-}23b)$$

where

$$\bar{\gamma}_0 = (\hat{\mathbf{K}}_0 \cdot \mathbf{n}_f) \qquad (5\text{-}24a)$$

$$\bar{\gamma}_g = (\hat{\mathbf{K}}_g \cdot \mathbf{n}_f) \qquad (5\text{-}24b)$$

Since $\Delta\eta_0^{(ij)}$ and $\Delta\eta_g^{(ij)}$ satisfy the same dispersion relations as do $\Delta\eta_0^{(i)}$ and $\Delta\eta_g^{(i)}$,

$$K\delta_f^{(ij)} = \frac{\Delta\eta_0^{(i)}}{\bar{\gamma}_0} + \frac{\Delta\eta_g^{(i)}}{\bar{\gamma}_g} \qquad (5\text{-}25)$$

Inserting this into equations (5-23a and b),

$$\Delta\eta_0^{(ij)} = -\frac{\bar{\gamma}_0}{\bar{\gamma}_g}\Delta\eta_g^{(i)} \qquad (5\text{-}26a)$$

$$\Delta\eta_g^{(ij)} = -\frac{\bar{\gamma}_g}{\bar{\gamma}_0}\Delta\eta_0^{(i)} \qquad (5\text{-}26b)$$

The other *Resonanzfehler* $\Delta\eta_0^{(ii)}$ and $\Delta\eta_g^{(ii)}$ are identical to $\Delta\eta_0^{(i)}$ and $\Delta\eta_g^{(i)}$.

It is important that $\delta_f^{(ij)}$, $\Delta\eta_0^{(ij)}$, and $\Delta\eta_g^{(ij)}$ are represented in terms of $\Delta\eta_0^{(i)}$ and $\Delta\eta_g^{(i)}$ for crystal $A$. The situation is quite analogous to that observed in the surface reflection of Bloch waves (Chap. 4, Section I C 3d), except that, now, $\bar{\gamma}_0$ and $\bar{\gamma}_g$ are positive. Thus, the phases appearing in (5-12) or (5-16) are represented in terms of *Anpassungen*, $\delta_e^{(i)}$ and $\delta_f^{(ij)}$, which are given by equations (4-56c) and (5-25).

Next, consider the amplitudes. The amplitude ratios $c^{(ij)}$ are determined by *Resonanzfehler* $\Delta\eta_0^{(ij)}$ or $\Delta\eta^{(ij)}$ according to the relation (4-28). The other $\bar{c}^{(ij)}$ are the reciprocals of $c^{(ij)}$. The important thing is that the Fourier coefficient $\chi_g$ must be replaced by $\chi_g \exp -2\pi i(\mathbf{g}\cdot\mathbf{u}_f)$. Thus

$$c^{(11)} = c^{(1)}e^{-i\delta} \qquad (5\text{-}27a)$$

$$c^{(22)} = c^{(2)}e^{-i\delta} \qquad (5\text{-}27b)$$

where $\delta$ stands for $2\pi(\mathbf{g}\cdot\mathbf{u}_f)$. Recalling the relations in (5-26) and the relation $\Delta\eta_0^{(i)} = -(\gamma_0/\gamma_g)\Delta\eta_g^{(j)}$ ($i = 1, 2$ and $j = 2, 1$), one gets

$$c^{(12)} = \frac{\bar{\gamma}_0}{\bar{\gamma}_g}\frac{\gamma_g}{\gamma_0} c^{(2)}e^{-i\delta} \qquad (5\text{-}27c)$$

$$c^{(21)} = \frac{\bar{\gamma}_0}{\bar{\gamma}_g}\frac{\gamma_g}{\gamma_0} c^{(1)}e^{-i\delta} \qquad (5\text{-}27d)$$

By using these results, the amplitude coefficients $C\{^{ij}_{00}\}$, etc., are calculated from equations (5-13) and (5-14). As described above, the wave vectors specified by $\{^{ij}_{0g}\}$ and $\{^{ij}_{gg}\}$ are identical. Therefore, the waves specified by them constitute a single plane wave so that there are four crystal waves for each of the $O$ and $G$ waves

$$d_0^{(ij)}(\mathbf{r}) = C_0^{(ij)} \exp i\varphi^{(ij)}E_e \exp i(\mathbf{K}_e\cdot\mathbf{r}) \qquad (5\text{-}28a)$$

$$d_g^{(ij)}(\mathbf{r}) = C_g^{(ij)} \exp i\varphi^{(ij)}E_e \exp i[(\mathbf{K}_e + 2\pi\mathbf{g})\cdot\mathbf{r}] \qquad (5\text{-}28b)$$

where

$$\varphi^{(ij)} = (\mathbf{k}_0^{(i)} - \mathbf{K}_e)\cdot(\mathbf{r} - \mathbf{r}_e) + (\mathbf{k}_0^{(ij)} - \mathbf{k}_0^{(i)})\cdot(\mathbf{r} - \mathbf{r}_f) \qquad (5\text{-}29)$$

and

$$C_0^{(ij)} = C_0^{(i)}C\{^{ij}_{00}\} + C_g^{(i)}C\{^{ij}_{g0}\} \qquad (5\text{-}30a)$$

$$C_g^{(ij)} = C_0^{(i)}C\{^{ij}_{0g}\} + C_g^{(i)}C\{^{ij}_{gg}\} \qquad (5\text{-}30b)$$

The explicit forms of $C_0^{(ij)}$, $C_g^{(ij)}$, and $\varphi^{(ij)}$ are summarized in Table 5-1.

The vacuum waves connected to the crystal waves are obtained by the boundary conditions at the exit surface (Chap. 4, Section I C 3c). Here, only the $G$ wave will be considered in detail. The wave field is composed of four plane waves. The wave vectors connected with $\mathbf{k}_g^{(i)}$ and $\mathbf{k}_g^{(ij)}$ are denoted by $\mathbf{K}_g^{(i)}$ and $\mathbf{K}_g^{(ij)}$. The former has the same form as (4-90b). The latter is given, similarly, by

$$\mathbf{K}_g^{(ij)} = \mathbf{k}_g^{(ij)} + K\delta_a^{(ij)}\mathbf{n}_a \qquad (5\text{-}31)$$

Combining equations (4-50), (5-20), (5-21), and (5-31),

$$\mathbf{K}_g^{(ij)} - (\mathbf{K}_e + 2\pi\mathbf{g}) = K(\delta_a^{(ij)}\mathbf{n}_a - \delta_f^{(ij)}\mathbf{n}_f - \delta_e^{(i)}\mathbf{n}_e) \qquad (5\text{-}32a)$$

**Table 5-1  PLANE-WAVE THEORY FOR A CRYSTAL CONTAINING A FAULT PLANE**

$O$ waves:  $d_0^{(ij)}(\mathbf{r}) = C_0^{(ij)} \exp i\varphi^{(ij)} E_e \exp i(\mathbf{K}_e \cdot \mathbf{r})$

$G$ waves:  $d_g^{(ij)}(\mathbf{r}) = C_g^{(ij)} \exp i\varphi^{(ij)} E_e \exp i[(\mathbf{K}_e + 2\pi\mathbf{g}) \cdot \mathbf{r}]$

| $(i, j)$ | $C_0^{(ij)}$ | $\varphi^{(ij)}$ |
|---|---|---|
| $(1, 1)$: | $\dfrac{1}{2}\dfrac{(-u + \sqrt{u^2 + U^2})}{\sqrt{u^2 + U^2}}\dfrac{\Delta_2 u + \Delta_1\sqrt{u^2 + U^2}}{\Gamma_2 u + \Gamma_1\sqrt{u^2 + U^2}}e^{i\delta}$ | $-K\delta_e^{(1)}t_e$ |
| $(2, 2)$: | $\dfrac{1}{2}\dfrac{(u + \sqrt{u^2 + U^2})}{\sqrt{u^2 + U^2}}\dfrac{\Delta_2 u - \Delta_1\sqrt{u^2 + U^2}}{\Gamma_2 u - \Gamma_1\sqrt{u^2 + U^2}}e^{i\delta}$ | $-K\delta_e^{(2)}t_e$ |
| $(1, 2)$: | $\dfrac{1}{2}\dfrac{U}{\sqrt{u^2 + U^2}}\dfrac{U}{\Gamma_2 u + \Gamma_1\sqrt{u^2 + U^2}}\bar{\Delta}$ | $-(K\delta_e^{(1)}t_e + \delta_f^{(12)}t_f)$ |
| $(2, 1)$: | $\dfrac{1}{2}\dfrac{-U}{\sqrt{u^2 + U^2}}\dfrac{U}{\Gamma_2 u - \Gamma_1\sqrt{u^2 + U^2}}\bar{\Delta}$ | $-K(\delta_e^{(2)}t_e + \delta_f^{(21)}t_f)$ |

$$C_g^{(ij)} = \left(\frac{\gamma_0}{\gamma_g}\right)^{1/2}\left(\frac{\chi_g}{\chi_{-g}}\right)^{1/2} \times \qquad \varphi^{(ij)}$$

| $(i, j)$ | | $\varphi^{(ij)}$ |
|---|---|---|
| $(1, 1)$: | $\dfrac{1}{2}\dfrac{U}{\sqrt{u^2 + U^2}}\dfrac{\Delta_2 u + \Delta_1\sqrt{u^2 + U^2}}{\Gamma_2 u + \Gamma_1\sqrt{u^2 + U^2}}$ | $-K\delta_e^{(1)}t_e$ |
| $(2, 2)$: | $\dfrac{1}{2}\dfrac{-U}{\sqrt{u^2 + U^2}}\dfrac{\Delta_2 u - \Delta_1\sqrt{u^2 + U^2}}{\Gamma_2 u - \Gamma_1\sqrt{u^2 + U^2}}$ | $-K\delta_e^{(2)}t_e$ |
| $(1, 2)$: | $\dfrac{1}{2}\dfrac{U}{\sqrt{u^2 + U^2}}\dfrac{-u + \sqrt{u^2 + U^2}}{\Gamma_2 u + \Gamma_1\sqrt{u^2 + U^2}}\Delta$ | $-K(\delta_e^{(1)}t_e + \delta_f^{(12)}t_f)$ |
| $(2, 1)$: | $\dfrac{1}{2}\dfrac{U}{\sqrt{u^2 + U^2}}\dfrac{u + \sqrt{u^2 + U^2}}{\Gamma_2 u - \Gamma_1\sqrt{u^2 + U^2}}\Delta$ | $-K(\delta_e^{(2)}t_e + \delta_f^{(21)}t_f)$ |

$\Delta = (\gamma_g/\bar{\gamma}_g)(1 - e^{-i\delta})$    $\bar{\Delta} = (\gamma_0/\bar{\gamma}_0)(1 - e^{i\delta})$    $K\delta_e^{(1)} = -\tfrac{1}{2}K\chi_0/\gamma_0 - \tfrac{1}{2}(u + \sqrt{u^2 + U^2})/\gamma_g$

$\Delta_1 = (\gamma_0/\bar{\gamma}_0) + (\gamma_g/\bar{\gamma}_g)e^{-i\delta}$    $\Delta_2 = (\gamma_0/\bar{\gamma}_0) - (\gamma_g/\bar{\gamma}_g)e^{-i\delta}$    $K\delta_e^{(2)} = -\tfrac{1}{2}K\chi_0/\gamma_0 - \tfrac{1}{2}(u - \sqrt{u^2 + U^2})/\gamma_g$

$\Gamma_1 = (\gamma_0/\bar{\gamma}_0) + (\gamma_g/\bar{\gamma}_g)$    $\Gamma_2 = (\gamma_0/\bar{\gamma}_0) - (\gamma_g/\bar{\gamma}_g)$    $K\delta_f^{(12)} = \tfrac{1}{2}(\Gamma_2 u + \Gamma_1\sqrt{u^2 + U^2})/\gamma_g$

$t_f = \mathbf{n}_f \cdot (\mathbf{r} - \mathbf{r}_f)$    $K\delta_f^{(21)} = \tfrac{1}{2}(\Gamma_2 u - \Gamma_1\sqrt{u^2 + U^2})/\gamma_g$

The *Anpassung* $\delta_a^{(ij)}$ is given as follows: Since $\mathbf{k}_g^{(ij)} = \Delta\mathbf{k}_g^{(ij)} + \Delta\mathbf{K} + \overline{\mathbf{K}}_g$, taking the scalar product of (5-31) with $\hat{\mathbf{K}}_g$, one gets

$$K\delta_a^{(ij)} = -\frac{\frac{1}{2}K\chi_0}{\gamma_g'} - \frac{\Delta\eta_g^{(ij)}}{\gamma_g'} \qquad (5\text{-}33a)$$

$$= -\frac{\frac{1}{2}K\chi_0}{\gamma_g'} + \frac{(\bar{\gamma}_g/\bar{\gamma}_0)\Delta\eta_0^{(i)}}{\gamma_g'} \qquad (5\text{-}33b)$$

where (5-26b) is employed. By using the expressions (4-56c), (4-60a), and (5-25) for *Anpassungen* $\delta_e^{(i)}$ and $\delta_f^{(ij)}$, respectively, equation (5-32a) can be rewritten

$$\mathbf{K}_g^{(ij)} - (\mathbf{K}_e + 2\pi\mathbf{g}) = \frac{1}{2}K\chi_0\left(\frac{\mathbf{n}_e}{\gamma_0} - \frac{\mathbf{n}_a}{\gamma_g'}\right) + \left[\left(\frac{\gamma_g}{\gamma_0}\right)\Delta\eta_0^{(i)}\mathbf{n}_f^*/\bar{\gamma}_g^* - \frac{\Delta\eta_g^{(i)}\mathbf{n}_f}{\bar{\gamma}_g}\right] \qquad (5\text{-}32b)$$

where $\mathbf{n}_f^*$ is the unit vector of

$$\mathbf{A} = \frac{\mathbf{n}_e}{\gamma_0} - \frac{\mathbf{n}_f}{\bar{\gamma}_0} + \frac{(\bar{\gamma}_g/\bar{\gamma}_0)\mathbf{n}_a}{\gamma_g'} \qquad (5\text{-}34)^{1,}$$

and

$$\bar{\gamma}_g^* = (\hat{\mathbf{K}}_g \cdot \mathbf{n}_f^*) \qquad (5\text{-}35)^1$$

*a. Amplitude region* In order to understand the optical nature of the wave fields consider the dynamical amplitude regions associated with the wave vectors $\mathbf{K}_g^{(i)}$ and $\mathbf{K}_g^{(ij)}$ (cf Sections I D 1 in Chap. 4 and I A in the present chapter). The shape of the amplitude region of $\mathbf{K}_g^{(i)}$ is identical to that described in Section I D 1 of the preceding chapter although the associated amplitudes are obviously different. It consists of two branches of a hyperbola whose asymptotes are $\mathbf{n}_e$ and $\mathbf{n}_a$ and pass through a point $G'$ as shown previously in Fig. 4-13. The deviation $GG' = \frac{1}{2}K\chi_0'(\mathbf{n}_e/\gamma_0 - \mathbf{n}_a/\gamma_g')$ is due to ordinary refraction.

The amplitude region associated with $\mathbf{K}_g^{(ij)}$ is very similar to $\mathbf{K}_g^{(i)}$. Comparing the expression (4-121b) for $[\mathbf{K}_g^{(i)} - (\mathbf{K}_e + 2\pi\mathbf{g})]$ and (5-32b) for $[\mathbf{K}_g^{(ij)} - (\mathbf{K}_e + 2\pi\mathbf{g})]$ it is seen that the amplitude region of $\mathbf{K}_g^{(ij)}$ also is a hyperbola whose asymptotes are the lines $\mathbf{n}_f$ and $\mathbf{n}_f^*$ passing through the point $G'$.

According to the above consideration, the total amplitude region is composed of a pair of hyperbolas. The situation is illustrated in Fig. 5-4a. It is very interesting to consider the case of large $|w|$. When $w$ is positive, $\Delta\eta_0^{(2)}$ and $\Delta\eta_g^{(1)}$ tend to zero so that the vectors $\overrightarrow{G'Q}$ tend to those listed below, where $Q$ denotes the end point of one of the wave vectors $\mathbf{K}_g^{(ij)}$ (note that $\mathbf{K}_g^{(ij)} = \mathbf{K}_g^{(i)}$). In this case, the amplitudes of $\mathbf{K}_g^{(1)}$, $\mathbf{K}_g^{(2)}$, and $\mathbf{K}_g^{(21)}$ waves decrease proportionately to $1/(w^2 + W^2)^{1/2}$, whereas

---

[1] Letting $\mathbf{n}_f^* = \alpha\mathbf{A}$, it is seen that $\bar{\gamma}_g^* = \alpha(\hat{\mathbf{K}}_g\mathbf{A}) = \alpha(\gamma_g/\gamma_0)$. Thus, it turns out that

$$\left(\frac{\gamma_g}{\gamma_0}\right)\left(\frac{\mathbf{n}_f^*}{\bar{\gamma}_g^*} - \frac{\mathbf{n}_e}{\gamma_g}\right) = \left(\frac{\bar{\gamma}_g}{\gamma_0}\right)\left(\frac{\mathbf{n}_a}{\gamma_g'} - \frac{\mathbf{n}_f}{\bar{\gamma}_g}\right) \qquad (5\text{-}34')$$

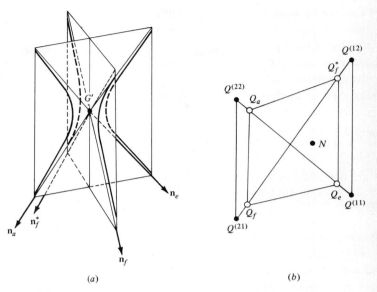

(a)  (b)

FIGURE 5-4

(a) Dynamical amplitude region for a crystal containing a stacking fault. (b) Cross section of the amplitude region with the Ewald sphere ($Q^{(11)}$, $Q^{(22)}$, $Q^{(12)}$, and $Q^{(21)}$ for $w > 0$). $Q_e$, $Q_a$, $Q_f$, and $Q_f^*$ are intersections of $\mathbf{n}_e$, $\mathbf{n}_a$, $\mathbf{n}_f$, and $\mathbf{n}_f^*$ with the Ewald sphere at a distance $w$ from $G'$. [$(\gamma_g/\gamma_0)Q_eQ_f^* = (\gamma_g/\gamma_0)Q_fQ_a$]. $N$ is the projection of $G'$ along the $\overline{\mathbf{K}}_g$ direction.

$\mathbf{K}_g^{(12)}$ waves decrease much more rapidly as $[w - (w^2 + W^2)^{1/2}]/(w^2 + W^2)^{1/2}$. For this reason, the effective amplitude region turns out to be $\mathbf{n}_e$, $\mathbf{n}_a$, and $\mathbf{n}_f$. For negative values of $w$, the same conclusion is reached. The result is in accordance with the results of the kinematical theory (cf Section II C 3c of Chap. 3). When the Bragg condition is nearly satisfied, however, the four rods must be taken into account.

Asymptotic behavior of $G'Q$

|        | $w \to \infty$ | $w \to -\infty$ |
|--------|----------------|-----------------|
| (1, 1) | $(w/\gamma_g')\mathbf{n}_e$ | $(w/\gamma_g')\mathbf{n}_a$ |
| (2, 2) | $(w/\gamma_g')\mathbf{n}_a$ | $(w/\gamma_g)\mathbf{n}_e$ |
| (1, 2) | $(w/\overline{\gamma}_g^*)\mathbf{n}_f^*$ | $(w/\overline{\gamma}_g)\mathbf{n}_f$ |
| (2, 1) | $(w/\overline{\gamma}_g)\mathbf{n}_f$ | $(w/\overline{\gamma}_g^*)\mathbf{n}_f^*$ |

Figure 5-4b illustrates the intersection of the four rods with the Ewald sphere. In electron cases, where the directions of $\overline{\mathbf{K}}_0$ and $\overline{\mathbf{K}}_g$ are practically parallel, relation (5-34') is reduced to

$$\frac{\mathbf{n}_f^*}{\overline{\gamma}_g^*} - \frac{\mathbf{n}_e}{\gamma_g} = \frac{\mathbf{n}_a}{\gamma_g'} - \frac{\mathbf{n}_f}{\overline{\gamma}_g}$$

The intersections of the four asymptotes with the Ewald sphere form a parallelogram.

*b.   Direct image of the stacking fault*   As described previously, the plane-wave solution is realized in electron microscopy. The $O$ wave produces the bright-field image, and the $G$ wave produces the dark-field image. Each is the interference pattern of four plane waves. The image can be interpreted as the projection of the intensity field on the exit surface either in the direction of $\overline{K}_0$ or $\overline{K}_g$, unless the image plane is far from the crystal. Each wave field described above exists only in a limited column along the direction of the energy flow below the stacking fault. In the electron case, the column direction is essentially the direction of $\overline{K}_0$ or $\overline{K}_g$.

Here, the simplest cases [nonabsorbing, parallel slab ($n_e = n_a$)], the exact Bragg condition ($w = 0$), symmetrical Laue, and normal incidence ($\gamma_0 = \gamma_g = 1$) are described in order to give the general idea of the fault image. From Table 5-1, one obtains the intensity field $w \to 0$ as follows (put $\overline{\gamma}_0 = \overline{\gamma}_g = \overline{\gamma}$):

$$I_0 = |d_0^{(11)}(\mathbf{r}_a) + d_0^{(12)}(\mathbf{r}_a) + d_0^{(21)}(\mathbf{r}_a) + d_0^{(22)}(\mathbf{r}_a)|^2$$

$$= \left(\cos\frac{\delta}{2}\cos\frac{1}{2}Wt_e\right)^2 + \left[\sin\frac{\delta}{2}\cos\frac{1}{2}W(t_e - 2t_f/\overline{\gamma})\right]^2 \quad (5\text{-}36a)$$

$$I_g = |d_g^{(11)}(\mathbf{r}_a) + d_g^{(12)}(\mathbf{r}_a) + d_g^{(21)}(\mathbf{r}_a) + d_g^{(22)}(\mathbf{r}_a)|^2$$

$$= \left(\cos\frac{\delta}{2}\sin\frac{1}{2}Wt_e\right)^2 + \left[\sin\frac{\delta}{2}\sin\frac{1}{2}W(t_e - 2t_f/\overline{\gamma})\right]^2 \quad (5\text{-}36b)$$

Obviously, the conservation of energy ($I_0 + I_g = 1$) is satisfied. Since $t' = (t_f/\overline{\gamma})$ is the distance between the fault plane and the exit surface, the intensity field is a sinusoidal fringe pattern parallel to the intersection of the fault plane with the entrance or exit surface. The fringe spacing is half of the *Pendellösung* fringes expected for the single crystal $A$ or $B$. Also, note that $I_0(t' = 0) = I_0(t' = t_e) = \cos^2[(1/2)Wt_e]$ and $I_g(t' = 0) = I_g(t' = t_e) = \sin^2[(1/2)Wt_e]$. This implies that the intensity is continuously connected with the intensity for the perfect part of the crystal, as it should be. The background and the contrast depend on the phase $|\delta|$. The quantitative analysis, therefore, gives the relative displacement $\mathbf{u}_f$ for crystals $A$ and $B$. The sign of $\delta$, however, cannot be determined. The intensity profile is illustrated in Fig. 5-5a.

In actual cases, the Borrmann absorption is very important for analyzing the image. For sufficiently thick crystals, since usually $V = J_m(U)$ is negative, the fringes on the side of crystal $A$ are composed of the waves specified by (1, 1) and (1, 2) whereas only the waves specified by (1, 1) and (2, 1) arrive at the exit surface on the side of crystal $B$. [When $V > 0$, the indices (1) and (2) must be interchanged. This is the case for electrons.] After a straightforward manipulation, one obtains (for $|V|t' \ll 1$, or $|V|t'' \ll 1$ but $|V|t_e \gg 1$)

$$I_0 = I_0^P(1 \mp \sin\delta \sin Wt') \quad (5\text{-}37a)$$

$$= I_0^P(1 \mp \sin\delta \sin Wt'') \quad (5\text{-}37b)$$

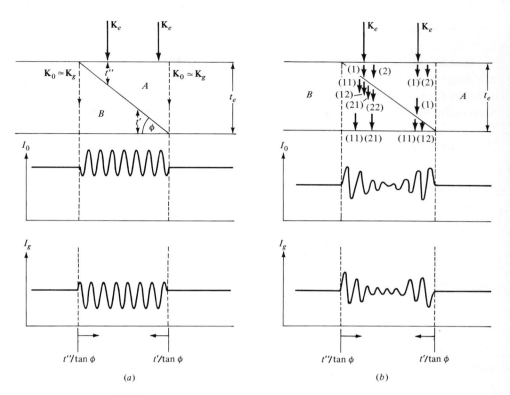

FIGURE 5-5
Stacking fault image in plane-wave theory. (*a*) Nonabsorbing crystal; (*b*) Highly absorbing crystal. ($V < 0, \delta < 0$)

$$I_g = I_g^P(1 \pm \sin \delta \sin Wt') \qquad (5\text{-}37c)$$

$$= I_g^P(1 \mp \sin \delta \sin Wt'') \qquad (5\text{-}37d)$$

where upper and lower signs correspond to the cases $V < 0$ and $V > 0$, respectively, $t''$ is the distance between the entrance surface and the fault plane, and $I_0^P$ and $I_g^P$ imply the intensities of $O$ and $G$ waves for perfect crystals, respectively. The bright-field image is symmetrical and the dark-field image is antisymmetrical with respect to the central line of the fault image. The behavior is illustrated in Fig. 5-5*b*. Since $\sin \delta$ appears in a linear form in (5-37), on the contrary to nonabsorbing crystals, the sign of $\delta$ can be determined. The theory presented here was developed by Hashimoto et al. and experimentally confirmed in electron microscopy images (1960).

For x rays, it may be difficult to obtain the fault image from a single plane wave. Nevertheless, one obtains the fault image in the traverse experiment, which is very similar to that described above. When a fault plane is parallel to the crystal slab, the fault is observed with a different contrast (Kohra and Yoshimatsu, 1962) in the traverse pattern. In fact, in many cases, one can admit a semiempirical rule that the fringe behavior of the traverse pattern is determined essentially by the plane-wave solution for $w = 0$. For this reason, the above-mentioned character of the image is very useful also for understanding the x-ray traverse topographs.

**2. Spherical-wave theory.**   The general expressions (5-17) are reduced to

$$\phi_0{}^{(ij)}(\mathbf{r}) = [\phi\{{}^{ij}_{00}\}(\mathbf{r}) + \phi\{{}^{ij}_{g0}\}(\mathbf{r})]$$

$$= A_0 E_e \int_{-\infty}^{+\infty} d_0{}^{(ij)}(\mathbf{K};\mathbf{r}) \exp i(K_x x_0) \, dK_x \qquad (5\text{-}38a)$$

$$\phi_g{}^{(ij)}(\mathbf{r}) = [\phi\{{}^{ij}_{0g}\}(\mathbf{r}) + \phi\{{}^{ij}_{gg}\}(\mathbf{r})]$$

$$= A_g E_e \int_{-\infty}^{+\infty} d_g{}^{(ij)}(\mathbf{K};\mathbf{r}) \exp i(K_x x_0) \, dK_x \qquad (5\text{-}38b)$$

where

$$d_0{}^{(ij)}(\mathbf{K};\mathbf{r}) = C_0{}^{(ij)} \exp i\varphi^{(ij)} \qquad (5\text{-}39a)$$

$$d_g{}^{(ij)}(\mathbf{K};\mathbf{r}) = C_g{}^{(ij)} \exp i\varphi^{(ij)} \qquad (5\text{-}39b)$$

and the expressions for $\varphi^{(ij)}$, $C_0{}^{(ij)}$, and $C_g{}^{(ij)}$ are given by equations (5-29) and (5-30), and the details in Table 5-1.

First, consider the phase of the integrand in relations (5-38a and b). The treatment is quite analogous to that in Laue-Bragg cases.

$$G^{(ij)} = \varphi^{(ij)} + (K_x x_0)$$

$$= G_0{}^{(i)} + [(\mathbf{k}_0{}^{(ij)} - \mathbf{k}_0{}^{(i)}) \cdot (\mathbf{r} - \mathbf{r}_f)] \qquad (5\text{-}40)$$

where $G_0{}^{(i)}$ is the same as $G_0$ defined by (4-198), the explicit form being given by (4-201). Since $(\mathbf{k}_0{}^{(ij)} - \mathbf{k}_0{}^{(i)})$ is normal to the fault plane, any point on the fault plane can be used for $\mathbf{r}_f$. It is convenient to take either $A_0$ or $A_g$ in Fig. 5-6. Then, $(\mathbf{r} - \mathbf{r}_f)$ can be replaced by either one of the vectors

$$\overrightarrow{A_0 P} = l_0' \hat{\mathbf{K}}_0 + l_g \hat{\mathbf{K}}_g \qquad (5\text{-}41a)$$

$$\overrightarrow{A_g P} = l_0 \hat{\mathbf{K}}_0 + l_g' \hat{\mathbf{K}}_g \qquad (5\text{-}41b)$$

where $(l_0$ and $l_g)$ are the coordinates with respect to the axes $EA_0 T$ and $EA_g R$, and $(l_0'$ and $l_g')$ are the coordinates with respect to the axes $A_g FB_0$ and $A_0 FB_g$. Since $(\mathbf{k}_0{}^{(ij)} - \mathbf{k}_0{}^{(i)}) = (\Delta\mathbf{k}^{(ij)} - \Delta\mathbf{k}^{(i)})$, one can obtain the alternative expressions

$$G^{(ij)} = G_0{}^{(i)} + (\Delta\eta_0{}^{(ij)} - \Delta\eta_0{}^{(i)})l_0' + (\Delta\eta_g{}^{(ij)} - \Delta\eta_g{}^{(i)})l_g \qquad (5\text{-}42a)$$

$$= G_0{}^{(i)} + (\Delta\eta_0{}^{(ij)} - \Delta\eta_0{}^{(i)})l_0 + (\Delta\eta_g{}^{(ij)} - \Delta\eta_g{}^{(i)})l_g' \qquad (5\text{-}42b)$$

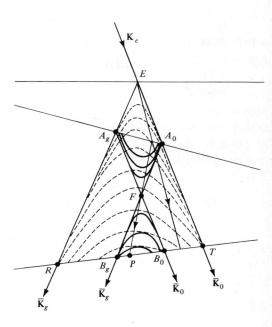

**FIGURE 5-6**
*Pendellösung* fringes and rays in the crystal containing a stacking fault.

Since the *Resonanzfehler* $\{\Delta\eta\}$ are a linear function of $s$ and $\pm\sqrt{s^2 + \beta^2}$, it is expected that

$$G^{(ij)} = \tfrac{1}{2}K\chi_0(l_0 + l_g) \pm \eta_1\sqrt{s^2 + \beta^2} - \eta_2 s \qquad (5\text{-}43)$$

By the same argument as used in connection with (4-236)

$$\eta_1 + \eta_2 = -\frac{\bar{\gamma}_0}{\bar{\gamma}_g} x_g' \qquad (5\text{-}44a)$$

$$\eta_1 - \eta_2 = -\frac{\bar{\gamma}_g \gamma_0}{\bar{\gamma}_0 \gamma_g} x_0' \qquad (5\text{-}44b)$$

where $(x_0', x_g')$ are the perpendicular distances of the observation point $P$ to the edges of the $F$-triangular fan.

The amplitudes $C_0^{(ij)}$ and $C_g^{(ij)}$ are listed in Table 5-1. ($u$ and $U$ are replaced by $s$ and $\beta$.) The integration of the wave fields can be performed readily by the stationary-phase method. The procedures are exactly parallel to those described in Section II C 1 of Chap. 4. Only a few significant points, therefore, will be described.

The crystal wave fields are classified into two types, $P$ and $Q$. The former is composed of the Bloch waves specified by $(i, i)$. They, therefore, pass through the fault plane without changing the wave vector and the ray direction. Accordingly, they are essentially two cylindrical waves starting from the entrance point $E$.

The waves of type $Q$ are specified by $(i, j)$. The associated rays change the direction at the fault plane in general, and all of them pass through the point $F$ in Fig. 5-6. Now, the focal point is inside the crystal as distinct from the Laue-Bragg cases, in which the focal point is outside the crystal.

The amplitudes of the wave fields are essentially the plane-wave solution, but the concentration factor $\sqrt{2\pi}/[|\partial^2 G/\partial s^2|]^{1/2}$ must be multiplied, the square brackets indicating the value under the condition of stationary phase. By consulting the expressions in Table 5-1 and $[\partial^2 G/\partial s^2]$, the component wave fields in the spherical-wave theory can be obtained as summarized in Table 5-2. The total wave fields are given in the form

$$\phi_0(\mathbf{r}) = \sum_{i,j}^{1,2} \phi_0^{(ij)}(\mathbf{r})$$

$$= \sqrt{2\pi}|\beta_0|^{1/2} A_0 E_e[P_0 + \sigma Q_0] \exp \tfrac{i}{2} K\chi_0(l_0 + l_g) \qquad (5\text{-}45a)$$

$$\phi_g(r) = \sum_{i,j}^{1,2} \phi_g^{(ij)}(\mathbf{r})$$

$$= \sqrt{2\pi}|\beta_0|^{1/2} A_g E_e[P_g + \sigma Q_g] \exp \tfrac{i}{2} K\chi_0(l_0 + l_g) \qquad (5\text{-}45b)$$

where

$$\sigma = \begin{cases} \ \ 1 & x_0', x_g' < 0 \quad \text{(inside } A_0FA_g) \\ -1 & x_0', x_g' > 0 \quad \text{(inside } B_0FB_g) \\ \ \ 0 & x_0'x_g' < 0 \quad \text{(outside } A_0FA_g \text{ and } B_0FB_g) \end{cases} \qquad (5\text{-}46)$$

The position variables $(x_0, x_g)$ and $(x_0', x_g')$ appear in $P_{0,g}$ and $Q_{0,g}$ through $(\xi_1, \xi_2)$ and $(\eta_1, \eta_2)$, respectively. The explicit expressions $P_g$ and $Q_g$, for example, are given as follows:

$$P_g = i\left[\frac{x_0/\bar{\gamma}_0 + x_g/\bar{\gamma}_g e^{i\delta}}{x_0/\bar{\gamma}_0 + x_g/\bar{\gamma}_g}\right] \frac{\sin (\beta_0\sqrt{x_0 x_g} + \pi/4)}{(x_0 x_g)^{1/4}} \qquad (5\text{-}47a)$$

$$Q_g = i\left[\frac{(x_0'/\bar{\gamma}_0)(1 - e^{-i\delta})}{x_0'/\bar{\gamma}_0 + x_g'/\bar{\gamma}_g}\right] \frac{\sin [\beta_0\sqrt{x_0' x_g'} + \pi/4]}{(x_0' x_g')^{1/4}} \qquad (5\text{-}47b)$$

The intensity fields are essentially given by $|P_0 + \sigma Q_0|^2$ and $|P_g + \sigma Q_g|^2$ for $O$ and $G$ waves, respectively. Here, the $G$ wave is discussed. For nonabsorbing crystals, it is easily seen that

$$|P_g|^2 = \left[1 - \frac{4(x_0/\bar{\gamma}_0)(x_g/\bar{\gamma}_g)}{(x_0/\bar{\gamma}_0 + x_g/\bar{\gamma}_g)^2} \sin^2 \frac{\delta}{2}\right] \frac{\sin^2 [\beta_0\sqrt{x_0 x_g} + \pi/4]}{(x_0 x_g)^{1/2}} \qquad (5\text{-}48a)$$

$$|Q_g|^2 = \left[\frac{4(x_0'/\bar{\gamma}_0)^2}{(x_0'/\bar{\gamma}_0 + x_g'/\bar{\gamma}_g)^2} \sin^2 \frac{\delta}{2}\right] \frac{\sin^2 [\beta_0\sqrt{x_0' x_g'} + \pi/4]}{(x_0' x_g')^{1/2}} \qquad (5\text{-}48b)$$

The intensity field of the $P_g$ wave is the continuation of the intensity field in crystal $A$, although modified by the factor in the first brackets in (5-48). The intensity field of the $Q_g$ wave is confined within the internal fans $A_0FA_g$ and $B_0FB_g$ in Fig. 5-6. As drawn there, a new type of *Pendellösung* fringes is expected within this region. In this case, again, the constant phase $\pi/4$ must not be ignored. It is worth noting the intensity enhancement along the margin $A_0FB_g(x_g' = 0)$ of the internal fan.

The observed section topograph is the projection of the intensity field at the exit surface. In the special geometry shown in Fig. 5-7c one can expect the characteristic pattern illustrated in Fig. 5-7a and b. In general, the pattern has a trapezoidal shape equivalent to a or b. If the fault plane is inclined to the entrance or exit surface, *Pendellösung* fringes with a hyperbolic form corresponding to $Q$ waves are observed.

**Table 5-2  SPHERICAL-WAVE THEORY FOR A CRYSTAL CONTAINING A FAULT PLANE**

$$O \text{ waves:} \quad \phi_0^{(ij)}(\mathbf{r}) = \sqrt{2\pi}\, A_0 E_e[C_0][|G''|]^{-1/2} \exp i\left\{G + \frac{\pi}{4}\,\text{sign}\,[G'']\right\}$$

$$G \text{ waves:} \quad \phi_g^{(ij)}(\mathbf{r}) = \sqrt{2\pi}\, A_g E_e[C_g][|G''|]^{-1/2} \exp i\left\{G + \frac{\pi}{4}\,\text{sign}\,[G'']\right\}$$

| $(i,j)$ | $[C_0]$ | $[|G''|]^{-1/2}$ | Sign $[G'']$ | $G = \dfrac{\chi_0}{2}(l_0 + l_g) +$ |
|---|---|---|---|---|
| $(1,1)$: | $\dfrac{1}{2}\left(1 - \dfrac{\xi_2}{\xi_1}\right)\dfrac{\Delta_1\xi_1 + \Delta_2\xi_2}{\Gamma_1\xi_1 + \Gamma_2\xi_2}\, e^{i\delta}$ | $\sqrt{|\beta|\xi_1/(\xi_1{}^2 - \xi_2{}^2)^{3/4}}$ | $+$ | $+\beta(\xi_1{}^2 - \xi_2{}^2)^{1/2}$ |
| $(2,2)$: | $\dfrac{1}{2}\left(1 - \dfrac{\xi_2}{\xi_1}\right)\dfrac{\Delta_1\xi_1 + \Delta_2\xi_2}{\Gamma_1\xi_1 + \Gamma_2\xi_2}\, e^{i\delta}$ | | $-$ | $-\beta(\xi_1{}^2 - \xi_2{}^2)^{1/2}$ |
| $(1,2)$: | $\dfrac{1}{2}\dfrac{\overline{\Delta}}{\eta_1}\dfrac{(\eta_1{}^2 - \eta_2{}^2)}{\Gamma_1\eta_1 + \Gamma_2\eta_2}$ | $\sqrt{|\beta|\,|\eta_1|/(\eta_1{}^2 - \eta_2{}^2)^{3/4}}$ | $\pm$ | $\pm\beta(\eta_1{}^2 - \eta_2{}^2)^{1/2}$ |
| $(2,1)$: | $\dfrac{1}{2}\dfrac{\overline{\Delta}}{\eta_1}\dfrac{(\eta_1{}^2 - \eta_2{}^2)}{\Gamma_1\eta_1 + \Gamma_2\eta_2}$ | | $\mp$ | $\mp\beta(\eta_1{}^2 - \eta_2{}^2)^{1/2}$ |

| $[C_g] = \left(\dfrac{\gamma_0}{\gamma_g}\right)^{1/2} \times$ | | $[|G''|]^{-1/2}$ | Sign $[G'']$ | $G = \dfrac{\chi_e}{2}(l_0 + l_g) +$ |
|---|---|---|---|---|
| $(1,1)$: | $\dfrac{1}{2}\dfrac{(\xi_1{}^2 - \xi_2{}^2)^{1/2}}{\xi_1}\dfrac{\Delta_1\xi_1 + \Delta_2\xi_2}{\Gamma_1\xi_1 + \Gamma_2\xi_2}$ | $\sqrt{|\beta|\xi_1/(\xi_1{}^2 - \xi_2{}^2)^{3/4}}$ | $+$ | $+\beta(\xi_1{}^2 - \xi_2{}^2)^{1/2}$ |
| $(2,2)$: | $-\dfrac{1}{2}\dfrac{(\xi_1{}^2 - \xi_2{}^2)^{1/2}}{\xi_1}\dfrac{\Delta_1\xi_1 + \Delta_2\xi_2}{\Gamma_1\xi_1 + \Gamma_2\xi_2}$ | | $-$ | $-\beta(\xi_1{}^2 - \xi_2{}^2)^{1/2}$ |
| $(1,2)$: | $\dfrac{1}{2}\dfrac{(\eta_1{}^2 - \eta_2{}^2)^{1/2}}{|\eta_1|}\dfrac{\eta_1 - \eta_2}{\Gamma_1\eta_1 + \Gamma_2\eta_2}\overline{\Delta}$ | $\sqrt{|\beta|\,|\eta_1|/(\eta_1{}^2 - \eta_2{}^2)^{3/4}}$ | $\pm$ | $\pm\beta(\eta_1{}^2 - \eta_2{}^2)^{1/2}$ |
| $(2,1)$: | $-\dfrac{1}{2}\dfrac{(\eta_1{}^2 - \eta_2{}^2)^{1/2}}{|\eta_1|}\dfrac{\eta_1 - \eta_2}{\Gamma_1\eta_1 + \Gamma_2\eta_2}\overline{\Delta}$ | | $\mp$ | $\mp\beta(\eta_1{}^2 - \eta_2{}^2)^{1/2}$ |

The double sign in the column $G^{(ij)}$ is given by the sign of $\eta_1$. The notations $\Delta_1$, $\Gamma_1$, etc., are defined in Table 5-1. This table can also be used for absorbing crystals within the SI approximations.

$$\xi_1 + \xi_2 = x_0 \qquad \eta_1 + \eta_2 = -(\overline{\gamma}_0/\overline{\gamma}_g)x_g'$$
$$\xi_1 - \xi_2 = (\gamma_0/\gamma_g)x_g \qquad \eta_1 - \eta_2 = -(\gamma_0/\gamma_g)(\overline{\gamma}_g/\overline{\gamma}_0)x_0'$$

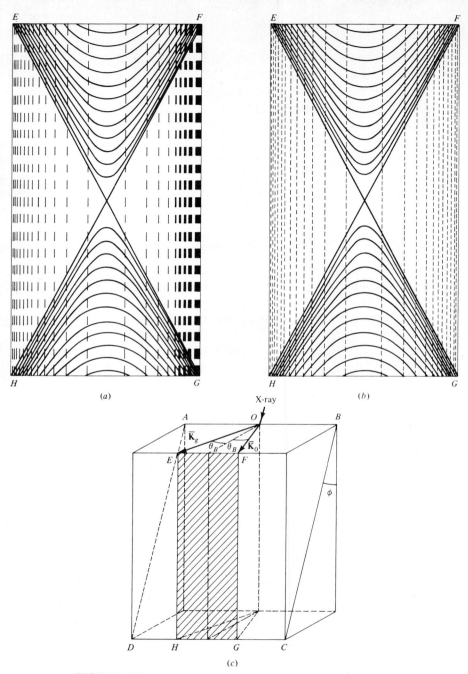

**FIGURE 5-7**
Patterns due to a stacking fault. (Broken straight lines are fringes due to *P* waves, and the hyperbolic lines are fringes due to *Q* waves.) (*a*) Direct wave; (*b*) Bragg-reflected wave; (*c*) geometrical relations among the crystal surfaces, net plane, and fault plane *ABCD*.

Topographs of this type actually have been observed by Yoshimatsu (1965) in $\alpha$ quartz, Authier and Sauvage (1966) in $CaCO_3$, and Chikawa and Austerman (1968) in BeO. The real topographs, however, differ from those predicted by the above theory in certain details. Perhaps the causes are partly due to the fact that the model of the fault plane adopted is oversimplified. A continuous deformation may be overlapped with the discrete misfit owing to the impurity precipitation on the fault plane. In some cases, the crystals $A$ and $B$ may be inclined to each other. In fact, Yoshimatsu's case should be interpreted as a misfit boundary, discussed below. In conjunction with this, Chikawa and Austerman and Authier (1968) pointed out that the hooklike shape of the *Pendellösung* fringes apparently loses sharpness compared to what is expected from the hyperbolic form. They explained this by taking into account the interference between $P$ and $Q$ waves.

Although agreement between the theory and experiments is realized only in a qualitative sense, the appearance of the characteristic patterns mentioned above is intrinsic to the existence of the phase difference $\delta$. Then, one can determine $|\delta|$ as in the case of plane-wave considerations. In fact, Chikawa and Austerman determined the crystallographic nature of the displacement $\mathbf{u}_f$ in the case of BeO, by analyzing the contrast dependence of the intensity field of the $Q$ wave. Since a description of the details of topographic work lies outside the scope of the present discussion, the readers are advised to consult the original papers. The theory described here is based on the paper of Kato, Usami, and Katagawa (1967). Similar considerations have been reported by Authier and Sauvage (1966) and by Authier (1968).

## D.   Twinned crystals

In this section, a crystal is considered that is divided into two parts having different structure factors, both as to magnitude and as to phase. The Bravais lattice is assumed to be common to both parts, as previously. The Dauphiné twin of $\alpha$ quartz is a typical example of such a crystal. The Fourier coefficients of the polarizability of crystal $A$ and $B$ are related by

$$\chi_{\pm g}(B) = M_{\pm g}\chi_{\pm g}(A) \qquad (5\text{-}49)$$

Besides the normal $\mathbf{n}_f$ of the fault plane, the complex coefficient $M_{\pm g}$, characterizes the fault. In absorbing crystals, the $M_{\pm g}$ coefficients may not be complex conjugates of each other. Obviously, the case discussed previously is a special one in which $M_{\pm g} = \exp(\mp i\delta)$.

**1.   *Plane-wave theory*** The construction of the wave vectors in crystal $B$ is straight-forward. Figure 5-8 illustrates the geometrical relations among the dispersion points to be concerned. Now, the wave vectors $\mathbf{k}_0^{(ii)}$ and $\mathbf{k}_g^{(ii)}$ in crystal $B$ are different from $\mathbf{k}_0^{(i)}$ and $\mathbf{k}_g^{(i)}$ in crystal $A$. Nevertheless, the relations up to (5-24) in the previous section need not be altered, if one allows the superscripts $(i)$ and $(j)$ to assume the

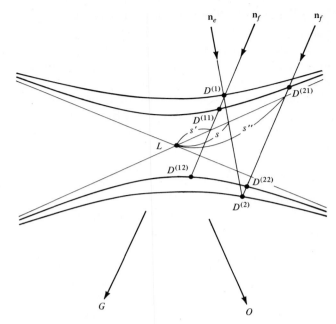

**FIGURE 5-8**
Construction of dispersion points for a crystal containing a twin plane.

values (1) and (2) independently. The amplitude ratio of $O$ and $G$ waves associated with the dispersion point $D^{(ij)}$ is written

$$c^{(ij)} = \frac{\frac{1}{2}KCM_g\chi_g}{\Delta\eta_g^{(ij)}}$$

$$= \frac{\Delta\eta_0^{(ij)}}{\frac{1}{2}KCM_{-g}\chi_{-g}} \qquad (5\text{-}50)[1]$$

The wave fields can be written again similarly to equations (5-28) to (5-30) and it is easy to write down the phase $G^{(ij)}$ and the amplitude factor $C_0^{(ij)}$, etc., in terms of the *Resonanzfehler* $\Delta\eta_0^{(ij)}$ or $\Delta\eta_g^{(ij)}$, and of *Anpassungen* $\delta_e$ and $\delta_f$. The results are summarized in Table 5-3 with the following comments.

Since equation (5-25) is no longer applicable to the present case, it is necessary to find proper representations for the *Resonanzfehler* in terms of the deviation parameter $u$. For this purpose,[2] it is convenient to introduce a new deviation parameter

---

[1] If not otherwise stated, $\chi_{\pm g}$ refers to crystal $A$.
[2] The procedures used in deriving (5-26a and b) can also be used in the present problem. The final results become unwieldy, since the equation for $\delta_f^{(ij)}$ is quadratic.

$u' = s' \sin 2\theta_B$ for the waves specified by $(i) = (1)$ and $u'' = s'' \sin 2\theta_B$ for the waves $(i) = (2)$, $s'$ and $s''$ being defined in Fig. 5-8. Since the mathematical relations between the *Resonanzfehler* $\{\Delta\eta\}$ and the deviation parameters $\{u\}$ or $\{s\}$ have nothing to do with the physical meaning of the directions $\mathbf{n}_e$ and $\mathbf{n}_f$, the *Resonanzfehler* of the waves in crystal $A$ can be written, as follows, by consulting expressions (4-60$a$ and $b$).

$$\Delta\eta_0^{(1)} = \tfrac{1}{2}[u' + \sqrt{u'^2 + \overline{U}^2}]\frac{\overline{\gamma}_0}{\overline{\gamma}_g} \qquad (5\text{-}51a)$$

$$\Delta\eta_g^{(1)} = \tfrac{1}{2}[-u' + \sqrt{u'^2 + \overline{U}^2}] \qquad (5\text{-}51b)$$

$$\Delta\eta_0^{(2)} = \tfrac{1}{2}[u'' - \sqrt{u''^2 + \overline{U}^2}]\frac{\overline{\gamma}_0}{\overline{\gamma}_g} \qquad (5\text{-}51c)$$

$$\Delta\eta_g^{(2)} = \tfrac{1}{2}[-u'' - \sqrt{u''^2 + \overline{U}^2}] \qquad (5\text{-}51d)$$

where

$$\overline{U} = KC(\chi_g\chi_{-g})^{1/2}\sqrt{\frac{\overline{\gamma}_g}{\overline{\gamma}_0}} \qquad (5\text{-}52)$$

**Table 5-3  PLANE-WAVE THEORY FOR A CRYSTAL CONTAINING A TWIN PLANE[1]**

$O$ waves:  $d_0^{(ij)}(\mathbf{r}) = C_0^{(ij)} \exp i\varphi^{(ij)} E_e \exp i(\mathbf{K}_e \cdot \mathbf{r})$
$G$ waves:  $d_g^{(ij)}(\mathbf{r}) = C_g^{(ij)} \exp i\varphi^{(ij)} E_e \exp i[(\mathbf{K}_e + 2\pi g)\cdot\mathbf{r}]$

| $(i, j)$ | $C_0^{(ij)} \times M_g(u^2 + U^2)^{1/2}[(u^i)^2 + M^2\overline{U}^2]^{1/2}$ | $\varphi^{(ij)}$ |
|---|---|---|
| (1, 1): | $\Delta\eta^{(11)}(M_g\Delta\eta^{(1)} - \Delta\eta^{(12)})$ | $-K(\delta_e^{(1)}t_e + \delta_f^{(11)}t_f)$ |
| (2, 2): | $\Delta\eta^{(22)}(M_g\Delta\eta^{(2)} - \Delta\eta^{(21)})$ | $-K(\delta_e^{(2)}t_e + \delta_f^{(22)}t_f)$ |
| (1, 2): | $-\Delta\eta^{(12)}(M_g\Delta\eta^{(1)} - \Delta\eta^{(11)})$ | $-K(\delta_e^{(1)}t_e + \delta_f^{(12)}t_f)$ |
| (2, 1): | $-\Delta\eta^{(21)}(M_g\Delta\eta^{(2)} - \Delta\eta^{(22)})$ | $-K(\delta_e^{(2)}t_e + \delta_f^{(21)}t_f)$ |

| $(i, j)$ | $C_g^{(ij)} \times (u^2 + U^2)^{1/2}[(u^i)^2 + M^2\overline{U}^2]^{1/2}$ | $\varphi^{(ij)}$ |
|---|---|---|
| (1, 1): | $\tfrac{1}{2}(KC\chi_g)(M_g\Delta\eta^{(1)} - \Delta\eta^{(12)})$ | $-K(\delta_e^{(1)}t_e + \delta_f^{(11)}t_f)$ |
| (2, 2): | $\tfrac{1}{2}(KC\chi_g)(M_g\Delta\eta^{(2)} - \Delta\eta^{(21)})$ | $-K(\delta_e^{(2)}t_e + \delta_f^{(22)}t_f)$ |
| (1, 2): | $-\tfrac{1}{2}(KC\chi_g)(M_g\Delta\eta^{(1)} - \Delta\eta^{(11)})$ | $-K(\delta_e^{(1)}t_e + \delta_f^{(12)}t_f)$ |
| (2, 1): | $-\tfrac{1}{2}(KC\chi_g)(M_g\Delta\eta^{(2)} - \Delta\eta^{(22)})$ | $-K(\delta_e^{(2)}t_e + \delta_f^{(21)}t_f)$ |

$\Delta\eta^{(1)} = \tfrac{1}{2}\{-u' + [(u')^2 + \overline{U}^2]^{1/2}\}$
$\Delta\eta^{(11)} = \tfrac{1}{2}\{-u' + [(\overline{u}')^2 + M^2U^2]^{1/2}\}$
$\Delta\eta^{(2)} = \tfrac{1}{2}\{-u'' - [(u'')^2 + \overline{U}^2]^{1/2}\}$
$\Delta\eta^{(12)} = \tfrac{1}{2}\{-u' - [(\overline{u}')^2 + M^2U^2]^{1/2}\}$

$\delta_e^{(i)} = $ see Table 3-5

$K\delta_f^{(11)} = (\Delta\eta^{(1)} - \Delta\eta^{(11)})/\overline{\gamma}_g$
$K\delta_f^{(22)} = (\Delta\eta^{(2)} - \Delta\eta^{(22)})/\overline{\gamma}_g$

$\Delta\eta^{(21)} = \tfrac{1}{2}\{-u'' + [(u'')^2 + M^2\overline{U}^2]^{1/2}\}$
$\Delta\eta^{(22)} = \tfrac{1}{2}\{-u'' - [(u'')^2 + M^2\overline{U}^2]^{1/2}\}$
The suffix $g$ in $\Delta\eta_g^{(i)}$ and $\Delta\eta_g^{(ij)}$ is omitted

$K\delta_f^{(12)} = (\Delta\eta^{(1)} - \Delta\eta^{(12)})/\overline{\gamma}_g$
$K\delta_f^{(21)} = (\Delta\eta^{(2)} - \Delta\eta^{(21)})/\overline{\gamma}_g$

---

[1] The notation $[\ ]^{1/2}$ has the same meaning in this table that $\sqrt{\phantom{x}}$ has elsewhere in this chapter.

Similarly, the *Resonanzfehler* in crystal $B$ can be written

$$\Delta\eta_0^{(1,j)} = \tfrac{1}{2}[u' \pm \sqrt{u'^2 + M^2\overline{U}^2}]\frac{\bar{\gamma}_0}{\bar{\gamma}_g} \qquad (5\text{-}53a)$$

$$\Delta\eta_g^{(1,j)} = \tfrac{1}{2}[-u' \pm \sqrt{u'^2 + M^2\overline{U}^2}] \qquad (5\text{-}53b)$$

$$\Delta\eta_0^{(2,j)} = \tfrac{1}{2}[u'' \pm \sqrt{u''^2 + M^2\overline{U}^2}]\frac{\bar{\gamma}_0}{\bar{\gamma}_g} \qquad (5\text{-}53c)$$

$$\Delta\eta_g^{(2,j)} = \tfrac{1}{2}[-u'' \pm \sqrt{u''^2 + M^2\overline{U}^2}] \qquad (5\text{-}53d)$$

where

$$M^2 = M_g M_{-g} \qquad (5\text{-}54)$$

The plus and minus signs in (5-53) refer to the superscript $(j)$.

By equating the corresponding $\Delta\eta_0^{(i)}$ and $\Delta\eta_g^{(i)}$ in the old form (4-60) and the new form, one can obtain the relations relating the old parameter $u$ and the new ones $u'$ and $u''$.

$$u' = \tfrac{1}{2}[\Gamma_1 u + \Gamma_2\sqrt{u^2 + U^2}]\frac{\bar{\gamma}_g}{\gamma_g} \qquad (5\text{-}55a)$$

$$\sqrt{u'^2 + \overline{U}^2} = \tfrac{1}{2}[\Gamma_2 u + \Gamma_1\sqrt{u^2 + U^2}]\frac{\bar{\gamma}_g}{\gamma_g} \qquad (5\text{-}55b)$$

$$u'' = \tfrac{1}{2}[\Gamma_1 u - \Gamma_2\sqrt{u^2 + U^2}]\frac{\bar{\gamma}_g}{\gamma_g} \qquad (5\text{-}55c)$$

$$\sqrt{u''^2 + \overline{U}^2} = \tfrac{1}{2}[-\Gamma_2 u + \Gamma_1\sqrt{u^2 + U^2}]\frac{\bar{\gamma}_g}{\gamma_g} \qquad (5\text{-}55d)$$

where $\Gamma_1$ and $\Gamma_2$ are already defined in Table 5-1. The inverse transformations also are easily obtained.

As stated previously, the expressions for the wave fields are now ready to be represented in terms of the deviation parameter $u$ with the help of the auxiliary ones $u'$ and $u''$. The explicit equations, however, are rather complicated, but writing them down is a matter of substitution in Table 5-3. The symmetrical nature of the component wave is clearly seen before making the substitution. The amplitude ratio of $G$ and $O$ waves in each $(i, j)$ wave is simply given by $\tfrac{1}{2}(KCM_g\chi_g)/\Delta\eta_g^{(ij)}$, the situation in crystal $A$ being transferable only through the deviation parameter $u'$ or $u''$ in crystal $B$. If one considers a pair of conjugate waves, the amplitudes of $(i, j)$ and $(j, i)$ waves are identical in the case of the $O$ wave and negative with respect to each other in the case of $G$ waves. [Conjugate here is defined: The parameter $w$ for wave (1) is the negative of $w$ for wave (2). By virtue of (5-55a and c), then, it follows that $w'(w) = -w''(-w)$, $\Delta\eta^{(i)} = -\Delta\eta^{(j)}$, and $\Delta\eta^{(ij)} = -\Delta\eta^{(ji)}$.] It is also seen that the phases of the conjugate waves have the same magnitude but opposite signs except for the common phase $\frac{1}{2}K\chi_0(l_0 + l_g)$. These properties have been seen previously in the case of a stacking fault and are important in considering the spherical-wave theory. It is worth noting, however, that the waves in crystals $A$ and $B$ can not satisfy simultaneously the exact Bragg condition in general, because $w' = -w'' = \frac{1}{2}W[(\gamma_0/\bar{\gamma}_0) \cdot (\bar{\gamma}_g/\gamma_g) - 1]$ for $w = 0$.

*a.   Amplitude region*    The wave vectors of the vacuum waves can be written by following the line of consideration used in the case of stacking faults in the same form as (5-32a). The three *Anpassungen* involved there are given by (4-56c), (5-23a), and (5-33a), respectively. Then, one obtains

$$\mathbf{K}_g^{(ij)} - (\mathbf{K}_e + 2\pi\mathbf{g}) = \frac{1}{2}K\chi_0 \left( \frac{\mathbf{n}_e}{\gamma_0} - \frac{\mathbf{n}_a}{\gamma_g'} \right)$$
$$+ \left[ \frac{\Delta\eta_0^{(i)}\mathbf{n}_e}{\gamma_0} + \frac{(\Delta\eta_0^{(ij)} - \Delta\eta_0^{(i)})\mathbf{n}_f}{\bar{\gamma}_0} - \frac{\Delta\eta_g^{(ij)}\mathbf{n}_a}{\gamma_g'} \right] \qquad (5\text{-}56)$$

Since $\Delta\eta_0^{(ij)}$ and $\Delta\eta_g^{(ij)}$ are not reduced to $\Delta\eta_0^{(i)}$ and $\Delta\eta_g^{(i)}$ in the present case, the amplitude region is simply, no longer, a pair of hyperbolas. Nevertheless, the general features are quite similar to the case of stacking faults. Particularly, in the kinematical approximations ($|w| \to \infty$), the amplitude region is reduced asymptotically to the same four spikes listed for the case of stacking faults. Recalling that either of the factors $(M_g\Delta\eta_g^{(1)} - \Delta\eta_g^{(11)})$ in $C_g^{(12)}$ and $(M_g\Delta\eta_g^{(2)} - \Delta\eta_g^{(22)})$ in $C_g^{(21)}$ listed in Table 5-3 tends to zero as $|w|$ increases, one can see again that only three spikes are effective, as expected from the kinematical theory. Obviously, when the Bragg condition is nearly satisfied, four waves must be retained.

*b.   Direct image of the fault plane*    Since the vacuum wave of $O$ and $G$ type is composed of four plane waves, the direct image also is essentially similar to the case of stacking faults (Fig. 5-7). The situation is greatly simplified in the case that a single twin plane is included in a parallel-sided crystal. The fringes are parallel to the edge of the wedge crystal $A$ or $B$.

The fringe spacing is shorter for $|M|^2 \geq 1$ than for $|M|^2 \leq 1$. For the same crystal, however, the image must be the same if the crystal is turned over. (Reciprocity.) For absorbing crystals, the contrasts on the thick side of crystal $A$ and the thick side of crystal $B$ may be different, depending on $J_m(\chi_g\chi_{-g})$ and $J_m(M^2\chi_g\chi_{-g})$.

**2.   Spherical-wave theory**    The phases $G^{(ij)} = (\varphi^{(ij)} + K_x x_0)$ required in the present case can be written in the following two alternative forms:

$$G^{(ij)} = \frac{1}{2}K\chi_0(l_0 + l_g) + \begin{cases} \Delta\eta_0^{(i)}(l_0 - l_0') + \Delta\eta_0^{(ij)}l_0' + \Delta\eta_g^{(ij)}l_g & (5\text{-}57a) \\ \Delta\eta_g^{(i)}(l_g - l_g') + \Delta\eta_0^{(ij)}l_0 + \Delta\eta_g^{(ij)}l_g' & (5\text{-}57b) \end{cases}$$

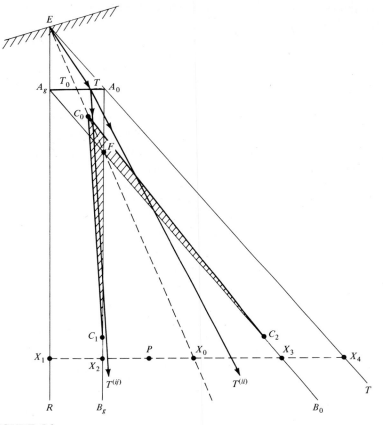

**FIGURE 5-9**
Trajectories $ETT^{(ii)}$ and $ETT^{(ij)}$ and caustics $C_1C_0C_2$ for $M = 0.5$, where $A_0A_g$ is the fault plane, $P$ the observation point, and $F$ the focus in the stacking fault. $(a = 2PX_0, \; b = A_0A_g = X_1X_2 = X_3X_4, \; c = X_1X_3 = X_2X_4.)$

where $(l_0, l_g)$ and $(l'_0, l'_g)$ have the same meanings as in (5-41). The phases also can be written explicitly in terms of the deviation parameter, either $s'$ or $s''$.

$$G^{(ij)} = \tfrac{1}{2}K\chi_0(l_0 + l_g) + (\tfrac{1}{2}\bar{\gamma}_0)$$

$$\times \begin{cases} (1, 1): & as' + b\sqrt{(s')^2 + \bar{\beta}^2} + c\sqrt{(s')^2 + M^2\bar{\beta}^2} & (5\text{-}58a) \\ (2, 2): & as'' - b\sqrt{(s'')^2 + \bar{\beta}^2} - c\sqrt{(s'')^2 + M^2\bar{\beta}^2} & (5\text{-}58b) \\ (1, 2): & as' + b\sqrt{(s')^2 + \bar{\beta}^2} - c\sqrt{(s')^2 + M^2\bar{\beta}^2} & (5\text{-}58c) \\ (2, 1): & as'' - b\sqrt{(s'')^2 + \bar{\beta}^2} + c\sqrt{(s'')^2 + M^2\bar{\beta}^2} & (5\text{-}58d) \end{cases}$$

where

$$\bar{\beta} = \frac{KC(\chi_g \chi_{-g})^{1/2}\sqrt{\bar{\gamma}_g/\bar{\gamma}_0}}{\sin 2\theta_B} \qquad (5\text{-}59)$$

and

$$a = \frac{x_g}{\bar{\gamma}_g} - \frac{x_0}{\bar{\gamma}_0} = \frac{x_g'}{\bar{\gamma}_g} - \frac{x_0'}{\bar{\gamma}_0} \qquad (5\text{-}60a)$$

$$b = \frac{x_0}{\bar{\gamma}_0} - \frac{x_0'}{\bar{\gamma}_0} = \frac{x_g}{\bar{\gamma}_g} - \frac{x_g'}{\bar{\gamma}_g} \qquad (5\text{-}60b)$$

$$c = \frac{x_g}{\bar{\gamma}_g} + \frac{x_0'}{\bar{\gamma}_0} = \frac{x_g'}{\bar{\gamma}_g} + \frac{x_0}{\bar{\gamma}_0} \qquad (5\text{-}60c)$$

The geometrical meanings of the above quantities are explained in the caption of Fig. 5-9. It turns out that $b$ is constant for a given configuration of the fault plane and $c$ is proportional to the distance between the fault plane and the parallel plane on which the observation point $P$ lies, while $a$ specifies the position $P$ on this plane.

The equations for the rays are given by the stationary-phase condition, namely, $\partial G/\partial s = 0$, as follows:[1]

$$(1, 1): \quad a + \frac{bs'}{\sqrt{(s')^2 + \bar{\beta}^2}} + \frac{cs'}{\sqrt{(s')^2 + M^2\bar{\beta}^2}} = 0 \qquad (4\text{-}61a)[1]$$

$$(2, 2): \quad a - \frac{bs''}{\sqrt{(s'')^2 + \bar{\beta}^2}} - \frac{cs''}{\sqrt{(s'')^2 + M^2\bar{\beta}^2}} = 0 \qquad (4\text{-}61b)[1]$$

$$(1, 2): \quad a + \frac{bs'}{\sqrt{(s')^2 + \bar{\beta}^2}} - \frac{cs'}{\sqrt{(s')^2 + M^2\bar{\beta}^2}} = 0 \qquad (4\text{-}61c)[1]$$

$$(2, 1): \quad a - \frac{bs''}{\sqrt{(s'')^2 + \bar{\beta}^2}} + \frac{cs''}{\sqrt{(s'')^2 + M^2\bar{\beta}^2}} = 0 \qquad (4\text{-}61d)[1]$$

The above represent straight lines because they are linear with respect to the position parameters $a$ and $c$. Each set of rays cannot have a single focal point but instead a caustic line given by the condition $\partial^2 G/\partial s^2 = 0$. The results can be represented

$$(1, 1): \quad b\left\{\frac{1}{\sqrt{(s')^2 + \bar{\beta}^2}}\right\}^3 + cM^2\left\{\frac{1}{\sqrt{(s')^2 + M^2\bar{\beta}^2}}\right\}^3 = 0 \qquad (5\text{-}62a)$$

$$(2, 2): \quad -b\left\{\frac{1}{\sqrt{(s'')^2 + \bar{\beta}^2}}\right\}^3 - cM^2\left\{\frac{1}{\sqrt{(s'')^2 + M^2\bar{\beta}^2}}\right\}^3 = 0 \qquad (5\text{-}62b)$$

$$(1, 2): \quad b\left\{\frac{1}{\sqrt{(s')^2 + \bar{\beta}^2}}\right\}^3 - cM^2\left\{\frac{1}{\sqrt{(s')^2 + M^2\bar{\beta}^2}}\right\}^3 = 0 \qquad (5\text{-}62c)$$

$$(2, 1): \quad -b\left\{\frac{1}{\sqrt{(s'')^2 + \bar{\beta}^2}}\right\}^3 + cM^2\left\{\frac{1}{\sqrt{(s')^2 + M^2\bar{\beta}^2}}\right\}^3 = 0 \qquad (5\text{-}62d)$$

[1] For absorbing crystals, the SI approximation must be used. Then, $s'$ and $s''$ and $\bar{\beta}^2$ and $M^2\bar{\beta}^2$ imply their real parts in equations (5-61) and (5-62).

Eliminating $s'$ or $s''$ from (5-61) and (5-62), one obtains the explicit expressions for the caustics.

The wave fields specified by $(i, i)$ and $(i, j)$ have different characters. They are denoted by $P$ and $Q$, respectively, similarly to the case of a stacking fault. Loosely speaking, the wave field of type $P$ is a continuation of the rays in crystal $A$. For instance, the ray specified by $(i)$ in crystal $A$ can be written

$$a \pm (b + c) \left\{ \frac{s'}{\sqrt{(s')^2 + \bar{\beta}^2}} \right\} = 0$$

Comparing this with expression (5-61a), one can see that the ray $(i, i)$ is deflected from the ray $(i)$ by the difference in the coefficient of $c$. In addition, the conjugate rays (1, 1) and (2, 2) specified by $s'$ and $s'' = -s'$, respectively, are identical. This implies that the conjugate rays arriving at $T$ on the fault plane propagate in the same direction also in crystal $B$.

The wave field of type $Q$ is rather complicated. Nevertheless, again, (1, 2) and (2, 1) rays, which are conjugates of each other, propagate in the same direction. As mentioned above, the rays have a caustic $C_1 C_0 C_2$ which is shown in Fig. 5–9. (Incidentally, the rays of type $P$ also have a caustic around the entrance point $E$; i.e. in the imaginary wave field.) For the case of $|M| \geq 1$, the cusp $C_0$ lies on the central line to $F$ but, for the case of $|M| \leq 1$, the caustic is inverted from the previous case and the cusp $C_0$ lies on $FX_0$.

The contours of constant phase are obtained immediately from (5-58) and (5-61) by eliminating the parameter $s'$ or $s''$ in the respective cases. No longer do they have a simple hyperbolic form but, in essence, a hook shape, similarly to the case of a stacking fault. The situation is different for the waves in the shaded region of Figs. 5-9 and 5-10; in the latter, the contours are actually indicated.

In order to obtain the fringe system produced by the interference between the conjugate waves, it is necessary to take into account an additional constant phase of two kinds. This subject has been discussed fully in Chap. 4 in Section II C for the perfect crystal and in Section I C 2 for stacking faults. As to the sign of $\partial^2 G/\partial s^2$, the $P$ waves remain either positive [for (1, 1)] or negative [for (2, 2)]. For $Q$ waves, the role of the focal point $F$ in the case of a stacking fault is now replaced by that of a caustic. The phase of wave (1, 2) changes from $\pi/4$ to $-\pi/4$ and that of (2, 1) changes reversely on the caustic, because the sign of $\partial^2 G/\partial s^2$ changes there. As to the phases of the amplitudes, it is enough to note in Table 5-3 that the $O$ components of the conjugate waves have the same amplitude and the $G$ components have the opposite signs. $C_0^{(ij)} = C_0^{(ji)}$ and $C_g^{(ij)} = - C_g^{(ji)}$.]

Thus, the contours of maximum fringe value are given by

$$P \text{ waves:} \quad G^{(11)} - G^{(22)} + \Phi = 2n\pi \qquad (5\text{-}63a)$$

$$Q \text{ waves:} \quad G^{(12)} - G^{(21)} + \Phi = 2n\pi \qquad (5\text{-}63b)$$

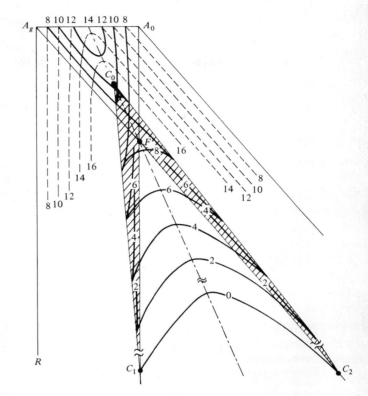

**FIGURE 5-10**
Contours of equal phase difference in $Q$ waves. The numbers indicate the order
of the fringes.

Here, the phases $\Phi$ have the following values in the respective cases:

**Phase $\Phi$ in equations (5-63a and b)**

|  | $P$ | $Q1$ | $Q2$ |
|---|---|---|---|
| $O$ | $\pi/2$ | $\pi/2$ | $-\pi/2$ |
| $G$ | $-\pi/2$ | $-\pi/2$ | $-3\pi/2\ (\pi/2)$ |

In this table, $Q1$ and $Q2$ denote $Q$ waves before and after reaching the caustic. Because of this phase jump, the fringes of the same order in $Q1$ and $Q2$ waves must
dislocate on the caustic by one-half of a fringe.

As in the case of stacking faults, the intensity field can be written symbolically
as $|P + Q1 + Q2|^2$. Besides the interference between $P$ and $Q$ waves, one needs to
consider additionally the interference among $Q1$ and $Q2$ in the shaded region. For
this reason, the intensity fields are rather complicated. Outside the shaded region,

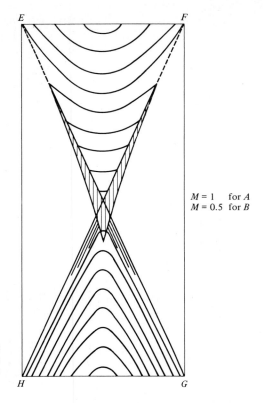

$M = 1$   for $A$
$M = 0.5$  for $B$

**FIGURE 5-11**
Schematic illustration of the fringe system of $Q$ waves for the case of a twin boundary. (The same geometry is assumed as in Fig. 5-7c.)

however, the nature of the intensity field is essentially similar to the case of stacking faults.

The section topograph is given by the intensity distribution on the exit surface. Figure 5-11 gives a typical example, in which the fault plane is oblique but the crystal surfaces are assumed to be parallel. A topograph of this type is expected in $\alpha$ quartz. Although the general characteristics agree with the theoretical results, again, an exact agreement is difficult to obtain. Moreover, the topographs for different positions of a single twinned surface are different in their details. The model assumed here appears to be essentially correct but oversimplified to match reality.

## E. Misfit boundaries

In this section, crystals containing a misfit boundary are discussed. The Bravais lattices of the crystals $A$ and $B$ are supposed to be different so that their reciprocal lattices must be described by different vectors, **g** and **g'**, respectively. Physically

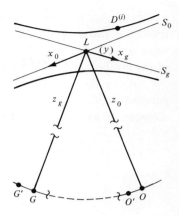

FIGURE 5-12
Dispersion surface, dispersion points $D^{(i)}$, and the reciprocal lattice for a distorted crystal. ($\Delta_0 = OO'$ and $\Delta_g = GG'$.)

speaking, the difference $\Delta \mathbf{g} = \mathbf{g}' - \mathbf{g}$ is due to the misorientation and/or the dilatation of one lattice. The continuous distortion which may accompany the lattice misfit in the crystals $A$ and $B$ is neglected. It is also assumed, for simplicity, that the structure factors of the crystals $A$ and $B$ are identical.

1. **Plane-wave theory** The dispersion surface and the reciprocal-lattice points are shown in Fig. 5-12. Here, the Lorentz point rather than the origin $O$ is fixed. Incidentally, $O$ and $G$ waves can be treated symmetrically by this procedure. (This is the real point of the dynamical theory!) Thus, the end points of $\mathbf{g}$ and $\mathbf{g}'$, i.e., $O$ and $O'$ or $G$ and $G'$, can no longer be identical. They must, however, lie on a sphere of radius $k$ having its center at the Lorentz point according to its definition.[1]

First define the rectangular coordinate systems $(\mathbf{x}_0, \mathbf{y}, \mathbf{z}_0)$ and $(\mathbf{x}_g, \mathbf{y}, \mathbf{z}_g)$ as shown in Fig. 5-12. Next, taking the oblique system $(\mathbf{x}_0, \mathbf{x}_g, \mathbf{y})$

$$2\pi \, \Delta \mathbf{g} = - \, \Delta_0 \mathbf{x}_0 + \Delta_g \mathbf{x}_g + 2\pi \, (\Delta \mathbf{g} \cdot \mathbf{y}) \mathbf{y} \qquad (5\text{-}64)^2$$

where

$$\Delta_0 = -2\pi \, \frac{\Delta \mathbf{g} \cdot \mathbf{z}_g}{\sin 2\theta_B} \qquad (5\text{-}65a)^2$$

$$\Delta_g = 2\pi \, \frac{\Delta \mathbf{g} \cdot \mathbf{z}_0}{\sin 2\theta_B} \qquad (5\text{-}65b)^2$$

If they are sufficiently small compared to the radius $k$, the $k$ sphere can be approximated by the tangential planes at the reciprocal-lattice points $O$ and $G$. This approximation is equivalent to regarding the asymptotic surfaces $S_0$ and $S_g$ as a plane. By

[1] Hereafter, this special sphere is called simply the "$k$ sphere."
[2] Here, $\Delta_0 = \Delta_0 \mathbf{x}_0 = \overline{OO'}$ and $\Delta_g = \Delta_g \mathbf{x}_g = \overline{GG'}$.

virtue of this approximation, the same dispersion surface can be used for both crystals $A$ and $B$. [Note $\chi_g(A) = \chi_g(B)$.] Then, it is straightforward to construct the wave vectors. No longer do the wave vectors $\mathbf{k}\{^{ij}_{0g'}\}$ degenerate to $\mathbf{k}\{^{ij}_{gg'}\}$. The meaning of the symbols in braces has been explained in Section I B of the present chapter.

As in the case of twinned crystals, it is convenient to employ the deviation parameters $u'_0$ and $u''_g$, etc., which are referred to the direction of $\mathbf{n}_f$. Then, the *Resonanzfehler* of the waves in crystal $B$ can be written in a similar way.

$$\Delta\eta_0\{^{1j}_{0g'}\} = \tfrac{1}{2}\{u'_0 \pm \sqrt{(u'_0)^2 + \overline{U}^2}\} \frac{\bar{\gamma}_0}{\bar{\gamma}_g} \qquad (5\text{-}66a)$$

$$\Delta\eta_g\{^{1j}_{0g'}\} = \tfrac{1}{2}\{-u'_0 \pm \sqrt{(u'_0)^2 + \overline{U}^2}\} \qquad (5\text{-}66b)$$

$$\Delta\eta_0\{^{2j}_{0g'}\} = \tfrac{1}{2}\{u''_0 \pm \sqrt{(u''_0)^2 + \overline{U}^2}\} \frac{\bar{\gamma}_0}{\bar{\gamma}_g} \qquad (5\text{-}66c)$$

$$\Delta\eta_g\{^{2j}_{0g'}\} = \tfrac{1}{2}\{-u''_0 \pm \sqrt{(u''_0)^2 + \overline{U}^2}\} \qquad (5\text{-}66d)$$

and

$$\Delta\eta_0\{^{1j}_{gg'}\} = \tfrac{1}{2}\{u'_g \pm \sqrt{(u'_g)^2 + \overline{U}^2}\} \frac{\bar{\gamma}_0}{\bar{\gamma}_g} \qquad (5\text{-}67a)$$

$$\Delta\eta_g\{^{1j}_{gg'}\} = \tfrac{1}{2}\{-u'_g \pm \sqrt{(u'_g)^2 + \overline{U}^2}\} \qquad (5\text{-}67b)$$

$$\Delta\eta_0\{^{2j}_{gg'}\} = \tfrac{1}{2}\{u''_g \pm \sqrt{(u''_g)^2 + \overline{U}^2}\} \frac{\bar{\gamma}_0}{\bar{\gamma}_g} \qquad (5\text{-}67c)$$

$$\Delta\eta_g\{^{2j}_{gg'}\} = \tfrac{1}{2}\{-u''_g \pm \sqrt{(u''_g)^2 + \overline{U}^2}\} \qquad (5\text{-}67d)$$

The subscripts 0 and $g$ of $\Delta\eta$ indicate that the *Resonanzfehler* concerned are referred to either one of the two asymptotic surfaces $S_0$ or $S_g$. The index $(j)$ becomes either (1) or (2), and the upper and lower signs must be selected for their respective cases. The subscript $g'$ can be $0'$ or $g'$. Since the waves $\{^{ij}_{g0'}\}$ and $\{^{ij}_{gg'}\}$ constitute a single Bloch wave, the *Resonanzfehler* are identical in both cases.

The deviation parameters $u'_0$ and $u'_g$ are related to $u'$ by

$$u'_0 = u' + \Delta_0 \sin 2\theta_B \qquad (5\text{-}68a)$$

$$u'_g = u' - \Delta_g \frac{\bar{\gamma}_g}{\bar{\gamma}_0} \sin 2\theta_B \qquad (5\text{-}68b)$$

Similarly, $u''_0$ and $u''_g$ are related to $u''$ by

$$u''_0 = u'' + \Delta_0 \sin 2\theta_B \qquad (5\text{-}69a)$$

$$u''_g = u'' - \Delta_g \frac{\bar{\gamma}_g}{\bar{\gamma}_0} \sin 2\theta_B \qquad (5\text{-}69b)$$

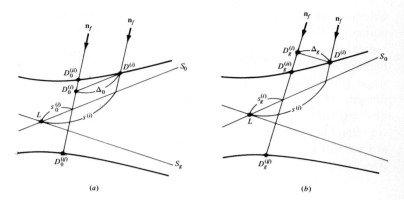

FIGURE 5-13
Construction of dispersion points for the case of misfit boundaries. $\Delta_0$ and $\Delta_g$ are explained in Fig. 5-12 and $s^{(t)} = s'$ or $s''$, in the text.

These relations are easily derived by considering the similar relations for $\{s\} = \{u\}/\sin 2\theta_B$,

$$s'_0 = s' + \Delta_0 \qquad (5\text{-}70a)$$

$$s'_g = s' - \Delta_g \frac{\bar{\gamma}_g}{\bar{\gamma}_0} \qquad (5\text{-}70b)$$

and

$$s''_0 = s'' + \Delta_0 \qquad (5\text{-}71a)$$

$$s''_g = s'' - \Delta_g \frac{\bar{\gamma}_g}{\bar{\gamma}_0} \qquad (5\text{-}71b)$$

which are immediately obtained by consulting Fig. 5-13. By virtue of the tangential continuation of the relevant wave vectors, one can write

$$\mathbf{k}\{^{ij}_{00'}\} = \mathbf{k}\{^{i}_{0}\} - \delta\{^{ij}_{00'}\}\mathbf{n}_f \qquad (5\text{-}72a)$$

$$\mathbf{k}\{^{ij}_{gg'}\} = \mathbf{k}\{^{i}_{g}\} - \delta\{^{ij}_{gg'}\}\mathbf{n}_f \qquad (5\text{-}72b)$$

Since $\mathbf{k}\{^{i}_{0}\} = \overrightarrow{D^{(i)}O} = \overrightarrow{D_0^{(i)}O'}$, equations (5-70a) and (5-71a) are obtained. For the case of equations (5-70b) and (5-71b), one needs to consider the projection of $\Delta_g$ on the asymptotic surface $S_0$ along the direction of $\mathbf{n}_f$. For this reason, the factor $(\bar{\gamma}_g/\bar{\gamma}_0)$ is required. [cf $\mathbf{k}\{^{i}_{g}\} = \overrightarrow{D^{(i)}G} = \overrightarrow{D_g^{(i)}G'}$.]

   The plane-wave theory is formulated according to the general treatment described in Section II B. Amplitude factors $C\{^{ij}_{00'}\}$, etc., and the phases can be written down in terms of *Resonanzfehler* $\{\Delta\eta\}$, which are given by (5-66) and (5-67).[1] Although the formulation is rather complicated, nothing new is required in a conceptual sense. The difference, when compared to the cases of stacking faults and twinned crystals, lies in the fact that the waves created by the $O$ component of a single Bloch wave can no longer be unified with the waves created by the $G$ component. They are classified by denoting them $\{O\}$ and $\{G\}$, respectively. In the practical applications of the theory to electron cases, the formulation can be simplified considerably by the so-called column approximation and matrix formalism, which are not discussed here. The details can be found in textbooks (Hirsch et al., 1965; Amelinckx, 1964).

**2. Spherical-wave theory** A brief sketch of the theory is presented below. One is concerned primarily with the structure of the ray systems and some qualitative aspects of the fringe systems.

By the same arguments as those used in conjunction with equations (5-58) and (5-59), the phases associated with the waves specified by $\{^{ij}_{0g'}\}$ and $\{^{ij}_{gg'}\}$ are written

$$G\{^{ij}_{0g'}\} = \tfrac{1}{2}K\chi_0(l_0 + l_g) + \tfrac{1}{2}c\bar{\gamma}_0\Delta_0 + (\tfrac{1}{2}\bar{\gamma}_0)$$

$$\times \begin{cases} (1,1): & as' + b\sqrt{(s')^2 + \bar{\beta}^2} + c\sqrt{(s'_0)^2 + \bar{\beta}^2} & (5\text{-}73a) \\ (2,2): & as'' - b\sqrt{(s'')^2 + \bar{\beta}^2} - c\sqrt{(s''_0)^2 + \bar{\beta}^2} & (5\text{-}73b) \\ (1,2): & as' + b\sqrt{(s')^2 + \bar{\beta}^2} - c\sqrt{(s'_0)^2 + \bar{\beta}^2} & (5\text{-}73c) \\ (2,1): & as'' - b\sqrt{(s'')^2 + \bar{\beta}^2} + c\sqrt{(s''_0)^2 + \bar{\beta}^2} & (5\text{-}73d) \end{cases}$$

$$G\{^{ij}_{gg'}\} = \tfrac{1}{2}K\chi_0(l_0 + l_g) - \tfrac{1}{2}c\bar{\gamma}_g\Delta_g + (\tfrac{1}{2}\bar{\gamma}_0)$$

$$\times \begin{cases} (1,1): & as' + b\sqrt{(s')^2 + \bar{\beta}^2} + c\sqrt{(s'_g)^2 + \bar{\beta}^2} & (5\text{-}74a) \\ (2,2): & as'' - b\sqrt{(s'')^2 + \bar{\beta}^2} - c\sqrt{(s''_g)^2 + \bar{\beta}^2} & (5\text{-}74b) \\ (1,2): & as' + b\sqrt{(s')^2 + \bar{\beta}^2} - c\sqrt{(s'_g)^2 + \bar{\beta}^2} & (5\text{-}74c) \\ (2,1): & as'' - b\sqrt{(s'')^2 + \bar{\beta}^2} + c\sqrt{(s''_g)^2 + \bar{\beta}^2} & (5\text{-}74d) \end{cases}$$

where $\bar{\beta}$, $a$, $b$, and $c$ have been defined by (5-59) and (5-60). Then, the rays are given immediately by

$$\{^{ij}_{0g'}\}: \quad a \pm bs/\sqrt{s^2 + \bar{\beta}^2} \pm cs_0/\sqrt{(s_0)^2 + \bar{\beta}^2} = 0 \quad (5\text{-}75a)$$

$$\{^{ij}_{gg'}\}: \quad a \pm bs/\sqrt{s^2 + \bar{\beta}^2} \pm cs_g/\sqrt{(s_g)^2 + \bar{\beta}^2} = 0 \quad (5\text{-}75b)$$

[1] If one compares the explicit expressions (4-72) and equations (5-13) and (5-14), one can see that $C\{^{ij}_{00'}\}$ and $C\{^{ij}_{gg'}\}$ are essentially same as $C_0^{(i)}$, and $C\{^{ij}_{0g'}\}$ and $C\{^{ij}_{g0'}\}$ are same as $C_g^{(i)}$. The functional behaviors of $C_0^{(i)}$ and $C_g^{(i)}$ are shown schematically in Fig. 3-5a and b. One thing should be noted, namely, that the role of the lattice points $O'$ and $G'$ must be interchanged in considering $C\{^{ij}_{gg'}\}$ and $C\{^{ij}_{g0'}\}$. In the former case, therefore, the branches close to the asymptotic surfaces $S_g$ have a large amplitude.

The primes (omitted) and the plus and minus signs must be properly selected according to (5-73) and (5-74). The caustics are similarly given by

$$\{^{ij}_{0g'}\}: \quad \pm b/(\sqrt{s^2 + \bar{\beta}^2})^3 \pm c(\sqrt{s_0^2 + \bar{\beta}^2})^3 = 0 \qquad (5\text{-}76a)$$

$$\{^{ij}_{gg'}\}: \quad \pm b/(\sqrt{s^2 + \bar{\beta}^2})^3 \pm c(\sqrt{s_g^2 + \bar{\beta}^2})^3 = 0 \qquad (5\text{-}76b)$$

Again, the explicit expressions for a constant phase and the caustics are obtained by eliminating the deviation parameters $\{s\}$ in the respective equations.

Now, the wave fields or the ray systems can be classified by the following scheme:

$$\{O\} \begin{cases} P \text{ type } [(1, 1) \text{ and } (2, 2)] \\ Q \text{ type } [(1, 2) \text{ and } (2, 1)] \end{cases}$$

$$\{G\} \begin{cases} P \text{ type } [(1, 1) \text{ and } (2, 2)] \\ Q \text{ type } [(1, 2) \text{ and } (2, 1)] \end{cases}$$

$P$-type rays have a caustic only in the imaginary field, whereas the caustic for $Q$-type rays lies in crystal $B$. [Equation (5-76) gives $c < 0$ for $(i, i)$ and $c > 0$ for $(i, j)$.]

The typical form of the caustic is shown for the case of wave $\{O\}$-$(1, 2)$ by the bold full curve in Fig. 5-14. There, $\Delta_0 > 0$ is assumed. The segments $FC_1$, $C_1 C_2$, and $C_2 F$ comprise rays which start from the parts $A_0 T_1$, $T_1 T_2$, and $T_2 A_g$ of the misfit plane in this particular case. The caustic of rays $(2, 1)$ appears in a symmetrical manner on the left side of the central line $EF$ (bold dashed curve). The complete caustic has a butterfly shape. [For $\bar{s} = s' = s''$, the equation of the rays has the form $a = F(\bar{s}; b, c)$ for $(1, 2)$ and $a = -F(\bar{s}; b, c)$ for $(2, 1)$. The condition of caustic $c/b = f(\bar{s})$ is common for $(1, 2)$ and $(2, 1)$, so that the caustic is given by $a = \pm G(c; b)$ in any case.] When $\Delta_0$ is negative, the same caustic and the ray system are obtained simply by interchanging $(1, 2)$ and $(2, 1)$.

The characteristics of $\{G\}$ rays are essentially the same as those of $\{O\}$ rays. [When $\Delta_0 = -\Delta_g(\bar{\gamma}_g/\bar{\gamma}_0)$, the ray systems are identical.] It is worth noting that $\Delta_0$ and $\Delta_g$ have opposite signs for the case of a rotationlike misfit ($|\mathbf{g}| \simeq |\mathbf{g}'|$), whereas they have the same sign for a dilatationlike misfit ($\mathbf{g}$ nearly parallel to $\mathbf{g}'$).

The behavior of the rays depends on the magnitude of $\Delta_0$ and $\Delta_g$. When one of them, say $\Delta_0$, is very small, the ray system $\{O\}$ is essentially the same as in the case of stacking faults   Although the crystal $B$ appears to be perfect for $O$ waves in crystal $A$, $Q$ waves are created in crystal $B$. Only when the crystal is indeed perfect ($\Delta_0 = \Delta_g = 0$) are the waves $\{O\}$ and $\{G\}$ created by a single Bloch wave superposed and canceled out.

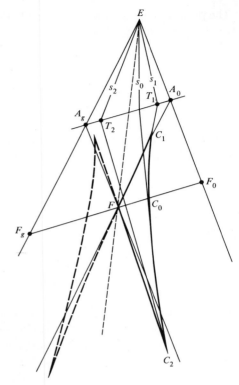

**FIGURE 5-14**
Caustics for a small-angle boundary. $\Delta_0 > 0$ is assumed. The full curve $FC_1C_2F$ corresponds to rays (1,2) and the broken curve to rays (2,1). The parameters $s_1$ and $s_2$ correspond to the tips of $C_1$ and $C_2$, respectively, while $s_0$ corresponds to $C_0$ at the level $F_0FF_g$ which is parallel to the fault $A_0A_g$. $(s_1 = \frac{1}{2}[-\Delta_0 - (\Delta_0{}^2 + 4\beta_0{}^2)^{1/2}],$ $s_2 = \frac{1}{2}[-\Delta_0 + (\Delta_0{}^2 + 4\beta_0{}^2)^{1/2}],$ $s_0 = -\frac{1}{2}\Delta_0.)$

In the other extreme case when $\Delta_0$ and $\Delta_g$ are much larger than $\bar{\beta}$, the situation tends to the case of two independent wedge-shaped crystals. How the ray system $\{O\}$ is changed depending on $\Delta_0$ is illustrated in Fig. 5-15. This sort of diagram is obtained graphically by the construction of the dispersion points (cf Fig. 5-13) and by noting that the ray direction is normal to the dispersion surface at the relevant dispersion points. Analytically, of course, (5-75) must be consulted.

By an examination of the ray system, one can infer the fringe system.

*a.   The case of small $\Delta g$*   The fringe systems $P$ and $Q$ are distinguishable. Since $P$ waves are not very typical, the $Q$-type waves are discussed. Besides the complexity due to the presence of caustics, which was discussed in the case of twinned crystals, the wave fields are rather complicated because, now, each $(i, j)$ wave is composed of $\{O\}$ and $\{G\}$ waves. They are no longer unified as a single wave. Thus, four plane waves specified by either $\{O\}$ and $\{G\}$ and $(i, j)$ or $(j, i)$ have to be considered. In

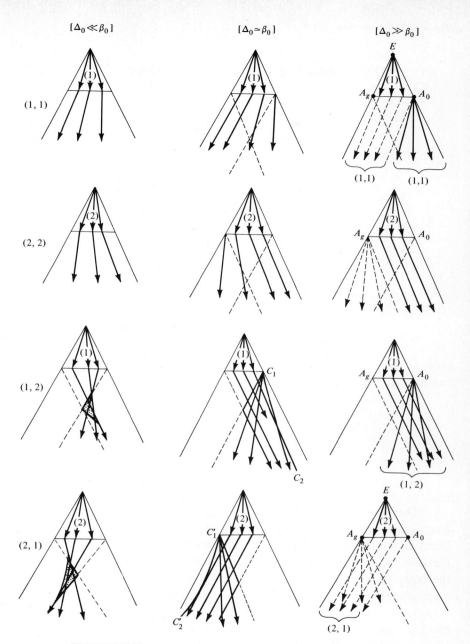

**FIGURE 5-15**

The rays in the case of small-angle boundaries. (The $\{O\}$ rays for $\Delta_0 > 0$ are shown. For $\Delta_0 < 0$, (1) and (2) must be interchanged. For $\{G\}$ rays, $\Delta_g < 0$, the same drawing can be used.) In the case illustrated, (*1, 1*) and (*1, 2*) of the $\{O\}$ rays and (1, 1) and (2, 1) of $\{G\}$ rays are strong.

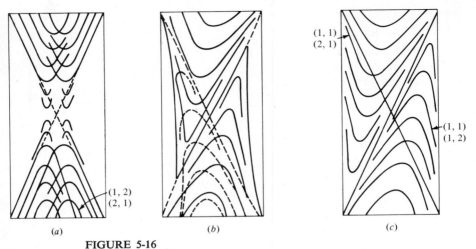

**FIGURE 5-16**
Some of the fringe patterns expected in the case of misfit boundaries.

order to show the connection with the caustics, they are listed separately for rotation and dilatation misfits.

**Rotation-like misfit ($\Delta_o$, $\Delta_g$ have opposite signs)**

| | Left wing | | Right wing |
|---|---|---|---|
| $\{O\}$ | $(ij)$ |  | $(ji)$ |
| $\{G\}$ | $(ij)$ | | $(ji)$ |

**Dilatationlike misfit ($\Delta_o$, $\Delta_g$ have the same signs)**

| | Left wing | | Right wing |
|---|---|---|---|
| $\{O\}$ | $(ij)$ | ⟵⟶ | $(ji)$ |
| $\{G\}$ | $(ji)$ | ⟵⟶ | $(ij)$ |

The important interference is brought about by four pairs of $(i, j)$ and $(j, i)$ waves indicated by arrows above.[1] Each pair forms a set of hook-shaped fringes and forms an hour-glass wave field as in the cases of stacking faults and twinned crystals. Particularly, in the dilatationlike misfit, the waves forming the left or right wings of the butterfly can produce a distinguishable interference system. Then, two fringe systems are separated in the topograph as schematically drawn in Fig. 5-16a.

*b.   The case of large $\Delta g$*   When $\Delta_o$ and $\Delta_g$ are extremely large, the crystals $A$ and $B$ are regarded as independent. In the third column of Fig. 5-15, if one is interested in the $G$ component waves of the rays, as is usually the case in topographic work, only the waves $(1, 1)$ and $(1, 2)$ are strong among the $\{O\}$ waves. On the other hand

---

[1] An interference among $(i,j)$ waves is conceivable, but the spacing must be much larger than those considered here.

(1, 1) and (2, 1) waves among {G} waves are appreciable. Thus, one can expect the topograph to look as illustrated in Fig. 5-16c.

Even when $\Delta_0$ and $\Delta_g$ are comparable to $\bar{\beta}$, the essential characteristics are not much altered. If, then, one considers the rays more precisely, the waves (1, 2) and (2, 1) are limited inside the caustics $C_1 C_2$ and $C'_1 C'_2$ as shown in the middle column of Fig. 5-15. Thus, the fringe patterns will be modified as in Fig. 5-16b.

All these fringe patterns are actually observed in natural and artificially grown crystals (Ikeno, Maruyama, Kato, 1968; see also the papers referred to in Section C 2). Particularly, when a crystal is grown in a solution from a spontaneous nucleus, the crystal is composed of several growth regions, which are divided by a misfit boundary like the one discussed here. Although its real nature is much more complex, one can obtain valuable information about the crystal texture by analyzing the topographs.

## II.   CONTINUOUSLY DEFORMED CRYSTALS

The kinematical theory for diffraction by a continuously bent crystal plane has been given in Section II C 4 of Chap. 3. The distortion is best described by the displacement vector **u(r)**, whose physical meaning has been described at the beginning of that section. The dynamical theory for describing x-ray diffraction by a continuously bent crystal (Kato, 1963c, 1964a,b) is presented below. This theory is quite analogous to the *Eikonal theory*, which was developed for radiation having wavelengths longer than those of visible radiation. Actually, the Eikonal theory provides a foundation for geometrical optics as described in advanced textbooks on optics (Sommerfeld, 1964: Born and Wolf, 1959). The basic concepts of this theory, therefore, are reviewed first and extended to x-ray diffraction by crystals. Next, some special cases and applications of this theory are considered, and, at the end of this chapter, some of the connections with other theories developed during the past 10 years are examined.

### A.   Wave-optical basis of ordinary geometrical optics

**1.   *Eikonal theory***   For the sake of simplicity, consider the case of a scalar wave and isotropic refractive index. As already discussed in Section I of Chap. 3, a plane wave, $A \exp i(\mathbf{kr})$, represents a special solution of the homogeneous wave equation

$$\Delta\psi + (Kn)^2\psi = 0 \qquad (5\text{-}77)$$

provided that $k = Kn$ is constant. An example of this was already encountered in the discussion of the Schrödinger equation in Section I of Chap. 3.

Certainly, one can not reduce Maxwell's equations to (5-77), but it is convenient nevertheless to utilize this mathematical framework. Thus $n$ is called the refractive index in ordinary optics. Even if $n$ changes with position, provided the change is gradual, it is still intuitively reasonable to assume that the plane-wave concept can be retained in a local region. For this reason, the wavefunction can be written in the form

$$\psi = A(\mathbf{r}) \exp iKS(\mathbf{r}) \qquad (5\text{-}78)$$

by introducing the real functions $A(\mathbf{r})$ and $S(\mathbf{r})$.[1] The function $S(\mathbf{r})$ is called an *Eikonal*. Optically speaking, $A(\mathbf{r})$ and $KS(\mathbf{r})$ have the meanings of amplitude and phase, respectively, and $S(\mathbf{r}) = S$ (constant) is the wave surface, by definition. Obviously,

$$\mathbf{k}(\mathbf{r}) = K \text{ grad } S(\mathbf{r}) \qquad (5\text{-}79)$$

has the meaning of the local wave vector.

The problem is to see to what extent (5-78), while satisfying equation (5-77), still retains the properties of the plane wave described in Chap. 3. For this purpose, simply insert (5-78) into equation (5-77). Recalling that $\Delta \equiv \text{div (grad)}$, one can obtain the relations

$$\text{grad } \psi = \left[ i\mathbf{k} + \frac{1}{A} \text{ grad } A \right] \psi \qquad (5\text{-}80a)$$

$$\Delta \psi = \left[ i\mathbf{k} + \frac{1}{A} \text{ grad } A \right]^2 \psi + \left[ i \text{ div } \mathbf{k} + \frac{1}{A} \Delta A - \left( \frac{1}{A} \right)^2 (\text{grad } A)^2 \right] \psi \qquad (5\text{-}80b)$$

If the right of equation (5-80b) is sorted out according to the power order of $K$, the wave equation (5-77) can be represented by

$$K_1 + K_2 + K_3 = 0 \qquad (5\text{-}81)$$

where

$$K_1 = K^2[n^2 - (\text{grad } S)^2]\psi \qquad (5\text{-}82a)$$

$$K_2 = iK \left( \frac{1}{A} \right)^2 \text{div } (A^2 \text{ grad } S)\psi \qquad (5\text{-}82b)$$

$$K_3 = \frac{1}{A} \Delta A \psi \qquad (5\text{-}82c)$$

In the above, $n$ is assumed to be real (nonabsorbing case).

For a large $K$, therefore, $K_1 = 0$ and $K_2 = 0$ will be required asymptotically in order for (5-81) to be satisfied. The first condition

$$(Kn)^2 \equiv k^2 = \mathbf{k}^2 \qquad (5\text{-}83)$$

is the dispersion relation eqiuvalent to (3-36) or (4-16), and the second condition is equivalent to

$$\text{div } (\mathbf{k}A^2) = 0 \qquad (5\text{-}84)$$

---

[1] So far, no approximation has been made. Any complex function $R + iI$ can be written $(R^2 + I^2)^{1/2} \exp i[\tan^{-1} (R/I)]$.

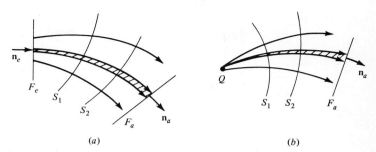

**FIGURE 5-17**
The orthogonal relationship of the Eikonal surfaces to the rays. (a) The boundary condition is specified at the surface $F$; (b) The boundary condition is specified at a source point $Q$. The crosshatched region indicates the ray tube.

If one interprets the expression

$$\mathbf{s} = \mathbf{k}A^2 \qquad (5\text{-}85)$$

as a description of the flow associated with the wave field (5-78), equation (5-84) has the meaning of the conservation law. For this reason, it is possible to retain the concept of a ray, while conserving such physical entities as energy, mass, and particle number. In the present case; i.e., when $n$ is isotropic, the ray direction is identical to the direction of the wave vector. Thus, if the third term $K_3$ in (5-81) is neglected, all properties of a plane wave can be retained in a local sense. Since a plane-wave solution is more easily handled than any other form of waves, it is convenient to adopt the wave (5-78) along with the auxiliary conditions (5-83) and (5-84) as the approximate solution of the wave equation. This is called a *modified plane wave*.

Equation (5-83) can be regarded as the differential equation of $S(\mathbf{r})$ which determines the Eikonal for a given distribution of the refractive index. The amplitude $A(\mathbf{r})$ is determined by (5-84) provided that $\mathbf{k}$ is known from (5-79). The validity of the approximation is assured by the smallness of the term $(1/K)^2(\Delta A/A)$. This implies that the rapid change in grad $A$, not $A$ itself, limits the applicability. Obviously, if $\lambda$ tends to zero, the approximation is correct.

Although the local behavior may appear like a plane wave, the wave field in actual space is not necessarily so. It depends on the functional form of $n$ and on the boundary conditions imposed on $S(\mathbf{r})$ and $A(\mathbf{r})$. If, for example, $n$ has cylindrical symmetry, a cylindrical wave of the form $[A(r)/r] \exp iKS(r)$ is a possible solution.

In any case, it is possible to imagine a set of curved lines which are orthogonal to the Eikonal surface, and the flow vectors assume the directions along these lines. This property is illustrated in Fig. 5-17. Here, a hydrodynamical picture is helpful. An important differential property

$$\text{rot } \mathbf{k} = 0 \qquad (5\text{-}86a)$$

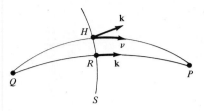

FIGURE 5-18
The real $QRP$ and a hypothetical ray
$QHP$ on which $\kappa = kv$ is not the same as
the wave vector $\mathbf{k} = k$ grad $S$.

is immediately derived from (5-79) because rot grad $\equiv 0$. The equivalent integral
form can be written

$$\oint (\mathbf{k} \cdot \mathbf{v}) \, dl = 0 \qquad (5\text{-}86b)$$

where $v \, dl = d\mathbf{r}$ is the line element of the integral path, which is an arbitrary closed
path.

**2.   Fermat's principle**   Although the Eikonal and the amplitude can be obtained by
the differential equations (5-83) and (5-84), in principle, no practical advantage is
derived from using the Eikonal. The real merit comes from establishing a variational
principle associated with the rays which are orthogonal to the Eikonal surface. For
this, only the existence of an Eikonal function having the above-mentioned properties
is required.

The problem is to find the real trajectory connecting two points $Q$ and $P$. Let
$QRP$ and $QHP$ be the real and hypothetical rays in Fig. 5-18. Applying the integral
property (5-86b) to the closed path $QRPHQ$, one can see that

$$S(P) - S(Q) = \int (\mathbf{k} \cdot \mathbf{v}) \, dl \qquad (Q \to R \to P) \qquad (5\text{-}87a)$$

$$= \int (\mathbf{k} \cdot \mathbf{v}) \, dl \qquad (Q \to H \to P) \qquad (5\text{-}87b)$$

The difference between the Eikonals of two points is calculable independently of the
integral path. Because the vector $\mathbf{k}$ is not known, however, the above equations are
not helpful in obtaining the Eikonal.

Define a hypothetical wave vector $\kappa$ for each hypothetical trajectory. It has the
direction of the trajectory and satisfies the dispersion relation (5-83). (The dispersion
relation itself cannot completely define the wave vector unless one knows the dispersion point. If the direction $v$ is specified, the dispersion point is fixed.) Next, consider
the line integral

$$\int (\kappa \cdot \mathbf{v}) \, dl \qquad (Q \to H \to P)$$

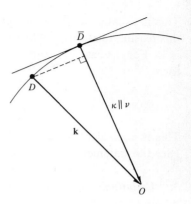

FIGURE 5-19
Geometrical proof for $(\mathbf{k} \cdot \mathbf{v}) \leq (\boldsymbol{\kappa} \cdot \mathbf{v})$
where $\mathbf{k}$ is a real vector and $\boldsymbol{\kappa}$ the wave
vector for a given hypothetical ray.

and compare this with the right side of equation (5-87$b$). From Fig. 5-19, it turns out that always

$$(\boldsymbol{\kappa} \cdot \mathbf{v}) > (\mathbf{k} \cdot \mathbf{v}) \qquad (5\text{-}88)$$

because the dispersion point $D$ of the $\mathbf{k}$ vector is closer to the origin $O$ than the tangential plane of the dispersion surface at the dispersion point $\bar{D}$ of the vector $\boldsymbol{\kappa}$. This results from the fact that the dispersion surface is concave toward the origin $O$. (One does not require the exact form of the dispersion surface.) Thus, one can see with the aid of (5-87$a$ and $b$) that

$$\int_{QHP} (\boldsymbol{\kappa} \cdot \mathbf{v})\, dl > \int_{QRP} (\boldsymbol{\kappa} \cdot \mathbf{v})\, dl \qquad (5\text{-}89)$$

where $\boldsymbol{\kappa}$ is replaced by $\mathbf{k}$ for the real ray $QRP$. (Note that this replacement is only possible for a real ray.)

Equation (5-89) implies that the real ray is determined by the condition

$$\int_Q^P (\boldsymbol{\kappa} \cdot \mathbf{v})\, dl = \text{minimum} \qquad (5\text{-}90a)$$

In the present problem ($n$ is a scalar and isotropic), actually $(\kappa v) = Kn$, because the dispersion surface is a sphere of radius $Kn$. Equation (5-90$a$), therefore, can be written in the form

$$\int_Q^P n\, dl = \text{minimum} \qquad (5\text{-}90b)$$

If one is concerned with light waves, the relation $v = c/n$ holds. Then, equation (5-90$a$) is equivalent also to

$$c \int_Q^P dt = \text{minimum} \qquad (5\text{-}90c)$$

This is usually called the *principle of least time* or *Fermat's principle* in optics. For this reason, relations (5-90a and b) are similarly designated in this section.

For the case of a particle of mass $m_0$, $v = v_0 n$, it is usual to write (5-90b) in the form

$$m_0 \int_Q^P v^2 \, dt = \text{minimum} \qquad (5\text{-}90d)$$

This is equivalent to least action principle provided that $t$ also is fixed at $P$ and $Q$. [See the statement below equation (5-98).]

**3.** *Variational principle and the ray equation*   Frequently, the statements represented by equations (5-90) are called *variational principles*, and, for example, (5-90b) may be written

$$\delta \int_Q^P n \, dl = 0 \qquad (5\text{-}91)$$

including also the case when the integral reaches a maximum. The expression is somewhat analogous to the expression

$$\delta F(x) = 0 \qquad (5\text{-}92a)$$

which might be used for specifying the extremal points of an ordinary function $F(x)$. The integrals of interest here, however, differ from the ordinary functions in that they can have definite values only when a trajectory is specified. In the case of (5-91), for example, the refractive index $n$ is a function of position variables $x$, $y$, and $z$, but it becomes a definite function of $l$ only when the trajectory functions $x(l)$, $y(l)$, and $z(l)$ are given. In other words the integral is "a function of functions." In mathematics such a concept is called a *functional*. The variational principle can be applied to a functional to define the extremal form of the functions involved—the real trajectory in the present problem.

In the case of (5-92a), the extremal points are simply obtained by the differential operation

$$\frac{dF}{dx} = 0 \qquad (5\text{-}92b)$$

From an analytical point of view, equation (5-92b) is more useful than the qualitative or geometrical statement (5-92a). For the same reason, one needs analytical expressions that are equivalent to the statement (5-91). This problem was solved by Euler and by Lagrange. Here, the mathematical results are summarized. The details can be found in any mathematical textbook on variational calculus (cf Jeffreys and Jeffreys, 1956; also, Born and Wolf, 1959, appendixes I and II).

Simplest is the case where the functional depends on a single unknown function $x(\lambda)$, for example,

$$[F] = \int_Q^P F(x, x'; \lambda) \, d\lambda \qquad (5\text{-}93)$$

where $P$ and $Q$ are fixed positions on the $(x, \lambda)$ plane.[1] For generality and for later purposes, $x' = dx/d\lambda$ is also included in the integrand. Indeed $dx/d\lambda$ also depends on the trajectory function. The variational principle can be written

$$\delta[F] = 0 \qquad (5\text{-}94a)$$

The equivalent analytical expression is

$$\frac{\partial F}{\partial x} - \frac{d}{d\lambda}\left(\frac{\partial F}{\partial x'}\right) = 0 \qquad (5\text{-}94b)$$

This is called *Euler's equation*. After the partial differentiation is performed, it is a second-order ordinary differential equation in general because $\partial F/\partial x'$ may include $x'$. For the given initial condition $(x, \lambda) = Q$ and the final condition $(x, \lambda) = P$, the equation can be solved in principle. This is the equation of the real trajectory. By analogy to the relation between (5-92a and b), the left side of equation (5-94b) is sometimes called the *Lagrangian derivative of the functional* $[F]$.

The above results are easily extended to a more general case in which the functional $[F]$ is defined by many unknown functions; e.g.,

$$[F] = \int_Q^P F(x_1, x_2, \ldots; x'_1, x'_2, \ldots; \lambda) \, d\lambda \qquad (5\text{-}95)$$

where $x_1, x_2, \ldots$ are functions of $\lambda$ and $x'_1, x'_2, \ldots$ are their derivatives with respect to $\lambda$, and $P$ and $Q$ are definite points in $(x_1, x_2, \ldots, \lambda)$ space. The problem is somewhat analogous to the problem of finding the extremal points of multivariable function $F(x_1, x_2, \ldots)$ by the partial differentiations $\partial F/\partial x_1 = 0$, $\partial F/\partial x_2 = 0, \ldots$. The analytical expression equivalent to the variational principle applied to the functional (5-95) is given by

$$\frac{\partial F}{\partial x_i} - \frac{d}{d\lambda}\left(\frac{\partial F}{\partial x'_i}\right) = 0 \qquad (i = 1, 2, \ldots) \qquad (5\text{-}96)$$

Now, a set of coupled equations have to be solved, but each one is an ordinary differential equation. The original Eikonal equation was a partial nonlinear equation. The trajectory equation is always an ordinary equation so that one can eliminate the need for a partial. From a mathematical viewpoint, therefore, there is great merit in considering the ray trajectories instead of the Eikonal surfaces.

---

[1] For generality, the independent variable $\lambda$ was used. In optics, it may be replaced by a position variable $z$ and, in mechanics, it may be a time variable. However, the path length $l$ cannot stand for $\lambda$, because $l$ is indefinite at $P$. Such a parameter is called a *floating parameter* in this chapter.

Returning to the optical problem, the above results can be applied to equation (5-91). Since the integral variable $l$ is a floating variable, it is better to choose another variable $\lambda$, which is assumed to have definite values $\lambda_P$ and $\lambda_Q$ at $P$ and $Q$. If one does not mind violating symmetry, $z$ can be used as the variable in place of $\lambda$. Recalling that $dl = [(dx/d\lambda)^2 + (dy/d\lambda)^2 + (dz/d\lambda)^2]^{1/2}\,d\lambda$, equation (5-96) immediately gives

$$\frac{d}{d\lambda}\left\{\frac{n(dX/d\lambda)}{[(dx/d\lambda)^2 + (dy/d\lambda)^2 + (dz/d\lambda)^2]^{1/2}}\right\} = \frac{\partial n}{\partial X}\left[\left(\frac{dx}{d\lambda}\right)^2 + \left(\frac{dy}{d\lambda}\right)^2 + \left(\frac{dz}{d\lambda}\right)^2\right]^{1/2}$$

(5-97a)

where $X = (x, y,$ and $z)$.  Returning to the variable $l$, one can obtain

$$\frac{d}{dl}\left(n\,\frac{dx}{dl}\right) = \frac{\partial n}{\partial x}$$

$$\frac{d}{dl}\left(n\,\frac{dy}{dl}\right) = \frac{\partial n}{\partial y}$$

$$\frac{d}{dl}\left(n\,\frac{dz}{dl}\right) = \frac{\partial n}{\partial z} \qquad (5\text{-}97b)$$

which are the ray equations for an isotropic medium. For a nonrelativistic material wave, where $n \propto m_0 v = [2m_0(E - eV)]^{1/2}$, equation (5-97b) is nothing more than Newton's equation.

It is perhaps worthwhile to say something here about the relation between Newtonian and Einsteinian mechanics. It is well known that the equation of motion for a charged particle also can be derived from the variational principle in the form

$$\delta \int_{t_1}^{t_2} L\,dt = 0 \qquad (5\text{-}98)$$

[Note that the time parameter $t$ is fixed at the end points, whereas $t$ in (5-90c) is not.] The function $L$ is called a *Lagrangian* and the integral is called an *action integral*. The explicit forms are usually written as follows:

$$L = T - N \qquad (5\text{-}99)$$

| | $T$ (kinetic term) | | $N$ (potential term) | |
|---|---|---|---|---|
| Classical | $\frac{1}{2}m_0 v^2$ | (5-100a) | $e\left[V - \dfrac{1}{c}(\mathbf{v}\cdot\mathbf{A})\right]$ | (5-100b) |
| Relativistic | $m_0 c^2\{1 - [1 - (v/c)^2]^{1/2}\}$ | (5-101a) | $e\left[V - \dfrac{1}{c}(\mathbf{v}\cdot\mathbf{A})\right]$ | (5-101b) |

Here, $m_0$ is the rest mass, $e$ is the charge, and $V$ and $\mathbf{A}$ are scalar and vector potentials of electromagnetic fields which must satisfy the Lorentz condition

$$\text{div } \mathbf{A} + \frac{1}{c} \frac{\partial V}{\partial t} = 0 \qquad (5\text{-}102)$$

The field vectors are derived from the definitions of the potentials as follows:

$$\mathbf{E} = -\text{grad } V - \frac{1}{c} \frac{\partial \mathbf{A}}{\partial t} \qquad (5\text{-}103a)$$

$$\mathbf{B} = \text{rot } \mathbf{A} \qquad (5\text{-}103b)$$

In fact, neglecting the vector potential in (5-100b), $L$ has the form $m_0 v^2 - E$, which gives the same variational principle as (5-90d) under the conditions given. Moreover by the orthodox application of (5-96) to the Lagrangian (5-99), one obtains the well-known equations of motion of a charged particle under the Lorentz force

$$\text{Classical:} \quad m_0 \frac{d}{dt} \mathbf{v} = e\left[ E + \frac{1}{c} (\mathbf{v} \times \mathbf{B}) \right] \qquad (5\text{-}104a)$$

$$\text{Relativistic:} \quad m_0 \frac{d}{dt} \left[ \frac{\mathbf{v}}{(1 - (v/c)^2)^{1/2}} \right] = e\left[ E + \frac{1}{c} (\mathbf{v} \times \mathbf{B}) \right] \qquad (5\text{-}104b)$$

The above may seem to be somewhat pedantic; however, the close relationship between ray optics in x-ray diffraction and relativistic mechanics will be shown somewhat later.

**4. Intensity and amplitude** Once the trajectories have been determined, the wave vector is specified for every point along them. Then, equation (5-84) can be integrated on the basis of a hydrodynamic picture; depending on the boundary conditions imposed on the flow, two cases can be considered separately:

*a. Nonsingular cases* In these cases, Fig. 5-17a, it is assumed that the flow vector (5-85) is specified over the entire surface $F_e$ without any singularity. The ray direction is not necessarily perpendicular to this surface. In practice, the surface is the boundary of the medium concerned and the flow vector is given by the incident wave falling on the boundary. Taking an infinitesimal area $\delta F_e$, one can imagine a tube surrounded by such rays passing through the periphery of the area $\delta F_e$. This tube will cut a surface $F_a$ on which it is desired to determine the intensity with a cross section $\delta F_a$ that is also supposedly infinitesimal.

Since the flow never strays outside the wall of the tube, equation (5-84) gives

$$(\mathbf{n}_e \cdot \mathbf{s}_e) \, \delta F_e = (\mathbf{n}_a \cdot \mathbf{s}_a) \, \delta F_a \qquad (5\text{-}105a)$$

where the subscripts $e$ and $a$ refer to the quantities on the surfaces $F_e$ and $F_a$, respectively, and $\mathbf{n}$ denotes the inward and outward normals of $\delta F_e$ and $\delta F_a$, respectively, neglecting true absorption. Absorption will be considered in the next section.

In practice, the source and observation points are placed outside the medium. Then, one must consider the transmission powers at the boundaries as defined by (3-63$b$), (The notation is changed in the present chapter to $P$ from $T$.) For example at the entrance surface, one has the relation

$$(\mathbf{n}_e \cdot \mathbf{S}_e)P_e = \mathbf{n}_e \cdot \mathbf{s}_e \quad (5\text{-}106a)$$

Combining this and the similar relation at the exit boundary with (5-105$a$), one obtains

$$I_a = P_e P_a I_e \frac{(\mathbf{n}_e \cdot \mathbf{K}_e)}{(\mathbf{n}_a \cdot \mathbf{K}_a)} \left(\frac{\delta F_e}{\delta F_a}\right) \quad (5\text{-}107)$$

The last ratio in parentheses represents the change of the cross section of the flow tube and is called the *contraction factor*. The amplitude is given simply by $\sqrt{I_a}$ according to its definition.

*b. Singular cases* When the waves are emitted from either a point or line source, Fig. 5-17$b$, a special analysis is required. It is assumed that on the medium side of the boundary, there is a point source emitting the flow with an angular distribution $k_e i(\omega)$. Again, one can imagine a ray tube which starts with an infinitesimal angular range $\delta\omega$ from the source. The situation at the observation point is identical to the previous case. By the use of the same considerations, one can obtain

$$k_e i(\omega)\, \delta\omega = (\mathbf{n}_a \cdot \mathbf{s}_a)\, \delta F_a \quad (5\text{-}105b)$$

When the source is on the vacuum side with the angular distribution $K_e I(\Omega)$, the connection between $I(\Omega)$ and $i(\omega)$ must be

$$K_e I(\Omega) P_e\, \delta\Omega = k_e i(\omega)\, d\omega \quad (5\text{-}106b)$$

If the observation point also is located on the vacuum side, one can obtain the intensity on the exit surface

$$I_a = P_e P_a I(\Omega) \frac{K_e}{(\mathbf{K}_a \cdot \mathbf{n}_a)} \left(\frac{\delta\Omega}{\delta F_a}\right) \quad (5\text{-}108)$$

The last ratio in parentheses again is called the contraction factor.

The above arguments are valid even when the rays cross each other, unless the crossing point is on or very close to the surface $F_a$. Near the focusing point or the caustic surface, Eikonal theory and geometrical optics fail because the intensity tends to infinity there, and this is not allowed in nature. At this point, it is of interest t note the similarity between how the variational principle treats the phase integral and the stationary-phase method. As has been fully discussed in Chap. 4, the caustic does not at all hinder the wave propagation but merely causes a constant phase jump.

**5.  Absorption**  From the phenomenological viewpoint adopted throughout this chapter, absorption must be described by the imaginary part of the refractive index $n$. Retaining this approach, one may have to assume an imaginary part in the local wave vector $\mathbf{k}$ and, consequently, in the Eikonal $S(\mathbf{r})$ [cf equations (5-79) and (5-83)]. This, in turn, will require some modification of the equations since real functions $A(\mathbf{r})$ and $S(\mathbf{r})$ were assumed at the beginning. Thus it is necessary to transfer some parts of $S(\mathbf{r})$ into $A(\mathbf{r})$, but no ambiguity should arise in the physical concepts or practical procedures, since one can determine the real and imaginary parts of the Eikonal separately. All considerations related to the ray trajectories, variational calculus, and intensity calculation can be performed as if the medium is non-absorbing. It is only necessary to multiply by an attenuation factor

$$\bar{A} = \exp{-2S^i} \qquad (5\text{-}109)$$

where the imaginary part of the Eikonal is given by

$$S^i = K \int_Q^P n^i \, dl \qquad (5\text{-}110)$$

It is interesting to note that the total attenuation depends on the total length of the trajectory. Physically, this result is quite reasonable, because the ray encounters more absorbing materials when the ray is curved.

**6.  Reciprocity**  One of the fundamental theorems in optics is that of reciprocity. Although a complete discussion of its physical meaning is outside the scope of this chapter, it is nevertheless worthwhile to consider some aspects here. What one has to be concerned with is whether the approximations used in the Eikonal theory disturb the reciprocity theorem or not. Fortunately, the answer can be given in the negative. In fact, this is the main reason that the theory works so well, all the way from the radio waves in the ionosphere to x rays in crystals, as will be shown in the next sections. In essence, the reciprocity theorem states that: *The phase and the amplitude must be the same for the real experiment and the corresponding reciprocal experiment.*

The meaning of a reciprocal experiment has been demonstrated previously in conjunction with statement B in Section II F 2b of Chap. 4. First of all, since the ray equation (5-97b) is derived from Euler's equation, which is invariant with respect to the change of the sign of $\lambda$, the trajectory arriving at $P$ in the direction $dX/d\lambda$ (where $X = x, y, z$) can be traced back to the initial point $Q$ by the initial condition $-dX/d\lambda$ at $P$ in the reciprocal experiment. The reciprocity of the phase and the attenuation factor is obvious because they are the line integrals along the same trajectory in opposite directions. To understand the reciprocity of the contraction factors, however, more sophisticated arguments are required.

## B.   Eikonal theory in x-ray diffraction

Once one has understood the basic concepts used in the dynamical theory in perfect crystals and the Eikonal theory in ordinary optics, it is simply a matter of extending the formalism to obtain the Eikonal theory for x-ray diffraction. The following diagram may be valuable for visualizing processes involved.

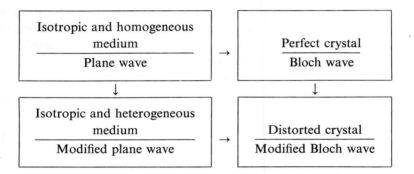

The waves shown in the blocks are the characteristic waves for each type of optical medium.

**1.   *Polarizability in distorted crystals***   As described in Section II C 4 of Chap. 3, it is usual to define the displacement vector $\mathbf{u}(\mathbf{r})$ referring to a perfect crystal, whose reciprocal-lattice vectors then are given by $\bar{\mathbf{g}}$. It is assumed, here, that any distortions are gradual and static in nature. Also, any changes in the charge distribution within a unit cell resulting from such a distortion are neglected.

The polarizability in a perfect crystal can be written [cf (4-2)]

$$\chi^P(\mathbf{r}_p) = \sum_g \chi_g{}^P \exp 2\pi i(\bar{\mathbf{g}} \cdot \mathbf{r}_p) \qquad (5\text{-}111)$$

where $\mathbf{r}_p$ is the position vector referred to the perfect crystal. If each portion of the crystal is displaced by $\mathbf{u}$ without changing the physical state (approximately!) the polarizability $\chi(\mathbf{r})$ in the distorted state must bear a one-to-one correspondence to $\chi^P(\mathbf{r}_p)$. Then, one can say

$$\chi(\mathbf{r}) = \chi^P(\mathbf{r}_p) \qquad (5\text{-}112)$$

where $\mathbf{r} = \mathbf{r}_p + \mathbf{u}$ and $\mathbf{u}$ must be understood to be a function of position $\mathbf{r}$. One obtains, therefore,

$$\chi(\mathbf{r}) = \sum_g \chi_g{}^p \exp 2\pi i[\,\bar{\mathbf{g}} \cdot (\mathbf{r} - \mathbf{u})] \qquad (5\text{-}113)^1$$

This expression for $\chi$ incorporates several important aspects of crystal deformation. First of all, if $\mathbf{u}$ is constant, (5-113) simply gives the polarizability for a perfect crystal whose Fourier coefficients are $\chi_g = \chi_g{}^p \exp -2\pi i(\mathbf{g} \cdot \mathbf{u})$. The situation has been previously encountered in the case of stacking faults. If the displacement $\mathbf{u}$ is linear with respect to the position variable $\mathbf{r}$ ($u_x = \alpha_{xx}x + \alpha_{xy}y + \alpha_{xz}z$; $u_y = \cdots$), then the phase term can be expressed in two alternative ways

$$G \equiv 2\pi[\bar{\mathbf{g}} \cdot (\mathbf{r} - \mathbf{u})] = 2\pi(\mathbf{g} \cdot \mathbf{r}) \qquad (5\text{-}114)$$

where $\mathbf{g}$ must be constant according to the initial assumption above and represents the reciprocal-lattice vectors after deformation. They can be obtained from

$$2\pi\mathbf{g} = \text{grad } G \qquad (5\text{-}115)$$

The deformation postulated, therefore, simply is a rotation and/or a dilatation of the crystal which, itself, remains in a perfect state.

The really interesting cases, therefore, are those for which $\mathbf{u}$ includes nonlinear terms with respect to the position vector $\mathbf{r}$. When such terms are present, the vector introduced by (5-115) turns out to be a function of position. As already proved in Section II C of Chap. 3, the vector $\mathbf{g}$ is normal to the net plane of interest and has a magnitude that is reciprocal to the interplanar spacing. For this reason, it can be interpreted as the local reciprocal-lattice vector.

**2.  Solution of Maxwell's equation**    The problem now is to solve Maxwell's equation or an equivalent one (3-27b) for the polarizability distribution (5-113). Although many-beam cases can be treated by the same line of considerations, here the theory is developed for the two-beam case, because the important applications are limited to this case in x-ray diffraction. For a discussion of the general case, the reader is referred to the original paper (Kato, 1963c).

Analogously to the ordinary Eikonal theory described in a previous section, it is intuitively assumed that a possible solution has the form

$$\mathbf{d}(\mathbf{r}) = \mathbf{d}_0 \exp iKS_0(\mathbf{r}) + \mathbf{d}_g \exp iKS_g(\mathbf{r}) \qquad (5\text{-}116)$$

Now, the amplitudes $\mathbf{d}_0$ and $\mathbf{d}_g$ of the component waves must be regarded as functions

---

[1] Later, the superscript in $\chi_g{}^p$ will be omitted.

of position. They are assumed to be real without any loss of generality. As in the case of the modified plane wave

$$\mathbf{k}_0 = K \operatorname{grad} S_0 \quad (5\text{-}117a)$$

$$\mathbf{k}_g = K \operatorname{grad} S_g \quad (5\text{-}117b)$$

have the meanings of the local wave vectors of the component waves.

For this reason, one writes the relation

$$S_g(\mathbf{r}) = S_0(\mathbf{r}) + G(\mathbf{r}) \quad (5\text{-}118)$$

By doing this, the condition for Bragg reflection

$$\mathbf{k}_g = \mathbf{k}_0 + 2\pi\mathbf{g} \quad (5\text{-}119)$$

can be assured. It is worth noting that

$$\operatorname{rot} \mathbf{k}_0 = 0 \quad (5\text{-}120a)$$

$$\operatorname{rot} \mathbf{k}_g = 0 \quad (5\text{-}120b)$$

which follows from (5-117a and b). The functions $S_0$ and $S_g$ having these properties are called the *Eikonals*.

Now, waves of the form (5-116) and polarizability (5-113) are substituted in equation (3-27b). The following conventions about the notations of differentiation will be employed: $\Delta \equiv (\operatorname{div} \cdot \operatorname{grad}) = (\nabla \cdot \nabla)$, $\operatorname{rot} = (\nabla \times)$, and $\operatorname{div} = (\nabla \cdot)$, but $\Delta$ and $\nabla$ operate only on the vectors and not on the scalars $S_0$ and $S_g$. With this notation the following relations are obtained in the vector calculus:

$$(\operatorname{div} \cdot \operatorname{grad})(\mathbf{d}_0 \exp iKS_0) = (\nabla + i\mathbf{k}_0)^2 \cdot \mathbf{d}_0 \exp iKS_0 \quad (5\text{-}121a)$$

$$(\operatorname{div} \cdot \operatorname{grad})(\mathbf{d}_g \exp iKS_g) = (\nabla + i\mathbf{k}_g)^2 \cdot \mathbf{d}_g \exp iKS_g \quad (5\text{-}121b)$$

$$(\operatorname{rot} \cdot \operatorname{rot})(\chi_g \mathbf{d}_0 \exp iKS_g) = \chi_g\{(\nabla + i\mathbf{k}_g) \times [(\nabla + i\mathbf{k}_g) \times \mathbf{d}_0]\} \exp iKS_g \quad (5\text{-}121c)$$

$$(\operatorname{rot} \cdot \operatorname{rot})(\chi_{-g} \mathbf{d}_g \exp iKS_0) = \chi_{-g}\{(\nabla + i\mathbf{k}_0) \times [(\nabla + i\mathbf{k}_0) \times \mathbf{d}_g]\} \exp iKS_0 \quad (5\text{-}121d)$$

$$(\operatorname{rot} \cdot \operatorname{rot})[\chi_0 \mathbf{d}(\mathbf{r})] = -\chi_0 (\operatorname{div} \cdot \operatorname{grad}) \mathbf{d}(\mathbf{r}) \quad [\operatorname{div} \mathbf{d}(\mathbf{r}) = 0] \quad (5\text{-}121e)$$

where $\chi_0$, $\chi_g$, and $\chi_{-g}$ are regarded as constant. Again, one can sort out the power order of $K$ as follows:

$$\mathbf{K}_1 + \mathbf{K}_2 + \mathbf{K}_3 = 0 \quad (5\text{-}122)$$

where

$$\mathbf{K}_1 = \{(k^2 - \mathbf{k}_0{}^2)\mathbf{d}_0 - \chi_{-g}[\mathbf{k}_0 \times (\mathbf{k}_0 \times \mathbf{d}_g)]\} \exp iKS_0$$
$$+ \{(k^2 - \mathbf{k}_g{}^2)\mathbf{d}_g - \chi_g[\mathbf{k}_g \times (\mathbf{k}_g \times \mathbf{d}_0)]\} \exp iKS_g \quad (5\text{-}123a)$$

$$\mathbf{K}_2 = i\left[\frac{\mathbf{d}_0}{d_0{}^2}\operatorname{div}(\mathbf{k}_0\mathbf{d}_0{}^2)\right] \exp iKS_0 + i\left[\frac{\mathbf{d}_g}{d_g{}^2}\operatorname{div}(\mathbf{k}_g\mathbf{d}_g{}^2)\right] \exp iKS_g \quad (5\text{-}123b)$$

$$\mathbf{K}_3 = \Delta\mathbf{d}_0 \exp iKS_0 + \Delta\mathbf{d}_g \exp iKS_g \quad (5\text{-}123c)$$

In (5-123$b$), the scalars $d_0$ and $d_g$ have the meanings of $(\mathbf{d}_0\mathbf{d}_0)^{1/2}$ and $(\mathbf{d}_g\mathbf{d}_g)^{1/2}$, respectively. In the expressions for $\mathbf{K}_2$ and $\mathbf{K}_3$, the terms proportional to $\chi_0$, $\chi_g$, and $\chi_{-g}$ are neglected. Perhaps they should be included in $\mathbf{K}_3$ but, in the end, the term $\mathbf{K}_3$ is neglected. The mathematics for deriving equations (5-123) are rather difficult, but a kind of correspondence principle to the case of perfect crystals on the one hand and to the case of scalar fields on the other proves to be helpful in the manipulations.

In order to satisfy the condition $\mathbf{K}_1 = 0$ strictly, one must set each of the coefficients of the phase terms equal to zero. They will give the local dispersion surface in the same form as that of the perfect crystal (4-20) as well as the amplitude ratio between $\mathbf{d}_0$ and $\mathbf{d}_g$. In two-beam cases the results are identical to (4-21) in form, so that they need not be repeated here.

Next, consider the condition $\mathbf{K}_2 = 0$. Since the amplitude ratio was fixed by the condition $\mathbf{K}_1 = 0$, it is no longer possible to regard each coefficient of the phase term as equal to zero. If one multiplies the conjugate expression (5-116) for the wave field by $\mathbf{K}_2 = 0$, and neglects the interference terms (*Netzebene* fringes), a more physically meaningful result is obtained, namely,

$$\frac{c}{4\pi} \,\text{div}\,(\mathbf{k}_0{\mathbf{d}_0}^2 + \mathbf{k}_g{\mathbf{d}_g}^2) = 0 \qquad (5\text{-}124)$$

Obviously, $c/4\pi$ times the quantity in parentheses is the expression for the averaged Poynting vector associated with the wave field (5-116). According to its physical representation of energy flow, one can define the ray vector by the unit vector of the Poynting vector. As has been proved in Section I B 6 of Chap. 4, the ray is always normal to the dispersion surface. Thus, again, if one neglects the third term $\mathbf{K}_3$, one can use the expression (5-116) as an approximate solution and keep all the concepts of the Bloch wave in a local sense. For this reason, such wave (5-116) is called a *modified Bloch wave*.

**3. Fermat's principle and the ray equation**   The next step in the analysis is to establish the variational principle. Similarly to the case of modified plane waves in the preceding section, one selects an arbitrary trajectory and defines a hypothetical wave vector $\kappa_0$ at the dispersion point $\bar{D}$ at which the normal to the dispersion surface takes on the ray direction $v$. Since the inequality (5-89) can be derived only from the concaveness of the dispersion surface, one can immediately conclude that

$$\int_Q^P (\kappa_0 \cdot v)\, dl = \begin{cases} \text{max. for branch (1)} & (5\text{-}125a) \\ \text{min. for branch (2)} & (5\text{-}125b) \end{cases}$$

where $\kappa_0$ and $v$ have different directions, in general. Figure 5-20 illustrates the geometry of $\mathbf{k}_0$, $\kappa_0$, and $v$ for obtaining these results.

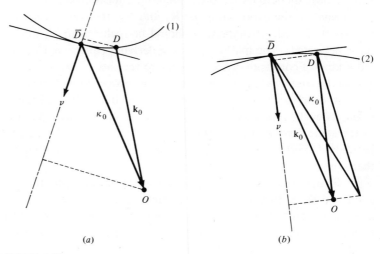

**FIGURE 5-20**
Geometrical proofs that (a) $(\mathbf{k}_0 \cdot \boldsymbol{v}) \geq (\boldsymbol{\kappa}_0 \cdot \boldsymbol{v})$ for wave (1); (b) $(\mathbf{k}_0 \cdot \boldsymbol{v}) < (\boldsymbol{\kappa}_0 \cdot \boldsymbol{v})$ for wave (2).

The relations (5-125a and b) are expressed by a single variational principle

$$\delta \int_Q^P (\boldsymbol{\kappa}_0 \cdot \boldsymbol{v}) \, dl = 0 \quad (5\text{-}126a)$$

Identical arguments can be applied to the hypothetical wave vector $\boldsymbol{\kappa}_g$, which must be $\boldsymbol{\kappa}_0 + 2\pi\mathbf{g}$, and one obtains

$$\delta \int_Q^P (\boldsymbol{\kappa}_g \cdot \boldsymbol{v}) \, dl = 0 \quad (5\text{-}126b)$$

In fact, this result has been expected from the outset because, according to relation (5-115), $\int_Q^P (\mathbf{g} \cdot \boldsymbol{v}) \, dl$ is independent of the trajectory. Both equations (5-126a and b) express Fermat's principle for modified Bloch waves.

Next, the equation of the trajectory is derived. For doing this the coordinate system in real and reciprocal space is fixed as shown in Figs. 5-21 and 5-22. Since $\boldsymbol{\kappa}_0$ and $\boldsymbol{\kappa}_g$ are the wave vectors satisfying the dispersion surface, they can each be divided into three parts as follows:

$$\boldsymbol{\kappa}_0 = (\overline{\mathbf{K}}_0 + \Delta\mathbf{K}) + \Delta\mathbf{k} + \boldsymbol{\Delta}_0 \quad (5\text{-}127a)^1$$

$$\boldsymbol{\kappa}_g = (\overline{\mathbf{K}}_g + \Delta\mathbf{K}) + \Delta\mathbf{k} + \boldsymbol{\Delta}_g \quad (5\text{-}127b)^1$$

---

[1] Although their physical meanings are different, the expressions for $\boldsymbol{\kappa}$ and $\mathbf{k}$ are identical.

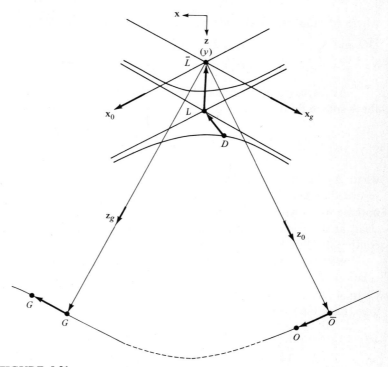

**FIGURE 5-21**

Vector relations in reciprocal space. $\mathbf{k}_0 = \overrightarrow{DO} = DL + L\overline{L} + \overline{LO} + \overline{OO} = \Delta\mathbf{k} + \Delta\mathbf{K} + \overline{\mathbf{K}}_0 + \mathbf{\Delta}_0$ and $\mathbf{k}_g = \overrightarrow{DG} = DL + LL + LG + \overline{G}G = \Delta\mathbf{k} + \Delta\mathbf{K} + \overline{\mathbf{K}}_g + \mathbf{\Delta}_g$.

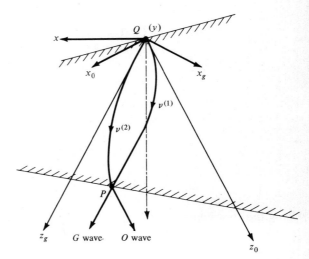

**FIGURE 5-22**
Rays in real space.

where $\Delta \mathbf{K}$ and $\Delta \mathbf{k}$ are defined by equations (4-23) and (4-25), respectively, and $\Delta_0 = \overrightarrow{OO}$ and $\Delta_g = \overrightarrow{GG}$ in Fig. 5-21. Obviously,

$$2\pi \, \Delta \mathbf{g} \equiv 2\pi \, (\mathbf{g} - \bar{\mathbf{g}}) = \Delta_g - \Delta_0 \qquad (5\text{-}128)$$

As discussed in the case of misfit boundaries in Section I E, if one approximates the $k$ sphere by the tangential planes at $\bar{O}$ and $\bar{G}$, one can write

$$\Delta_0 = \Delta_0 x_0 + \Delta_{0y} \mathbf{y} \qquad (5\text{-}129a)$$

$$\Delta_g = \Delta_g x_g + \Delta_{gy} \mathbf{y} \qquad (5\text{-}129b)$$

where $\Delta_0$ and $\Delta_g$ obey the expressions (5-65a and b), respectively. Now they must be regarded as functions of position through $\Delta \mathbf{g}$. (The $y$ components $\Delta_{0y}$ and $\Delta_{gy}$ have nothing to do with x-ray diffraction within the present approximation.) It should be clear, at this point, that the continuously distorted crystal is being represented by a continuous distribution of infinitesimal misfit boundaries.

With the aid of equations (5-127) one can calculate $(\kappa_0 \cdot \mathbf{v})$ and $(\kappa_g \cdot \mathbf{v})$. The calculations for the first two terms already have been performed in Section I D 5 of Chap. 4 [see equations (4-154a and b)]. The functionals appearing in (5-126a and b), then, have the form

$$[S_0] = \{\bar{\mathbf{K}}_0 \cdot (\mathbf{r}_P - \mathbf{r}_Q)\} + \tfrac{1}{2}K\chi_0(l_0 + l_g) + [T] - [N_0] \qquad (5\text{-}130a)$$

$$[S_g] = \{\bar{\mathbf{K}}_g \cdot (\mathbf{r}_P - \mathbf{r}_Q)\} + \tfrac{1}{2}K\chi_0(l_0 + l_g) + [T] - [N_g] \qquad (5\text{-}130b)$$

where

$$[T] = \pm \left[ \frac{\tfrac{1}{2}KC(\chi_g \chi_{-g})^{1/2}}{\cos \theta_B} \right] \int_Q^P (1 - p^2)^{1/2} \, dz \qquad (5\text{-}131)$$

$$[N_0] = - \frac{2\pi}{\sin 2\theta_B} \int_Q^P \frac{\partial}{\partial z_g} (\bar{\mathbf{g}} \cdot \mathbf{u}) \, dx_0 \qquad (5\text{-}132a)^1$$

$$[N_g] = \frac{2\pi}{\sin 2\theta_B} \int_Q^P \frac{\partial}{\partial z_0} (\bar{\mathbf{g}} \cdot \mathbf{u}) \, dx_g \qquad (5\text{-}132b)^1$$

$$[N_0] = -\pi \{\bar{\mathbf{g}} \cdot (\mathbf{u}_P - \mathbf{u}_Q)\} - [\tilde{N}] \qquad (5\text{-}133a)$$

$$[N_g] = \pi \{\bar{\mathbf{g}} \cdot (\mathbf{u}_P - \mathbf{u}_Q)\} - [\tilde{N}] \qquad (5\text{-}133b)$$

---

[1] $(\mathbf{z} \cdot \mathrm{grad} \, A) = \dfrac{\partial A}{\partial z}$

$$\frac{\partial}{\partial z_0} = -\sin \theta_B \frac{\partial}{\partial x} + \cos \theta_B \frac{\partial}{\partial z} \qquad dx_0 = \cos \theta_B \, dx + \sin \theta_B \, dz$$

$$\frac{\partial}{\partial z_g} = \sin \theta_B \frac{\partial}{\partial x} + \cos \theta_B \frac{\partial}{\partial z} \qquad dx_g = -\cos \theta_B \, dx + \sin \theta_B \, dz$$

Using the coordinates $x$ and $z$, one can rewrite the last two relations

where

$$[\tilde{N}] = \pi \int_Q^P \left[ \frac{1}{\tan \theta_B} \frac{\partial}{\partial z} (\bar{\mathbf{g}} \cdot \mathbf{u}) \frac{dx}{dz} + \tan \theta_B \frac{\partial}{\partial x} (\bar{\mathbf{g}} \cdot \mathbf{u}) \right] dz \qquad (5\text{-}134)$$

The change of the independent variable in equations (5-131) and (5-134) from $l$ to $z$ is required because $l$ is a floating parameter.

In variational calculus, only the functionals denoted by the brackets [ ] are important. Thus, one obtains

$$\delta[S] = \delta\{[T] - [N]\} = 0 \qquad (5\text{-}135)$$

There is no need to distinguish between $O$ and $G$ waves since the same trajectory is expected for both. Otherwise it would be necessary to give up the ray concept for the modified Bloch wave. Actually, (5-135) has exactly the same form as the Lagrangian of a relativistic particle (in one dimension) after making the following correspondences [see (5-99) and (5-101)]:

$$z \rightarrow t \text{ (time)} \qquad (5\text{-}136a)$$

$$\tan \theta_B \rightarrow c \text{ (light velocity)} \qquad (5\text{-}136b)$$

$$\frac{dx}{dz} = \tan \theta \rightarrow v \text{ (velocity)} \qquad p = \left( \frac{v}{c} \right) \qquad (5\text{-}136c)$$

$$\frac{\frac{1}{2} K C (\chi_g \chi_{-g})^{1/2}}{\sin \theta_B \tan \theta_B} \rightarrow m_0 \text{ (rest mass)} \qquad (5\text{-}136d)[1]$$

$$\pi \tan \theta_B \frac{\partial}{\partial x} (\mathbf{g} \cdot \mathbf{u}) \rightarrow eV \text{ (scalar potential)} \qquad (5\text{-}136e)[1]$$

$$- \pi \frac{\partial}{\partial z} (\bar{\mathbf{g}} \cdot \mathbf{u}) \rightarrow eA_x \text{ ($x$ component of vector potential)} \qquad (5\text{-}136f)[1]$$

Indeed the last two quantities automatically satisfy the Lorentz conditions (5-102). The double sign may be interpreted as representing the positive and negative charge. By using these relations, one can arrive at the ray equation[2]

$$\pm m_0 \frac{d}{dz} \left[ \frac{v}{(1 - v^2/c^2)^{1/2}} \right] = f \qquad (5\text{-}137)[2]$$

where the plus and minus signs correspond to rays of type (1) and (2), and

$$f = \pi \left[ \cot \theta_B \frac{\partial^2}{\partial z^2} (\bar{\mathbf{g}} \cdot \mathbf{u}) - \tan \theta_B \frac{\partial^2}{\partial x^2} (\bar{\mathbf{g}} \cdot \mathbf{u}) \right] \qquad (5\text{-}138a)$$

$$= \frac{2\pi}{\sin 2\theta_B} \frac{\partial^2}{\partial z_0 \partial z_g} (\bar{\mathbf{g}} \cdot \mathbf{u}) \qquad (5\text{-}138b)$$

---

[1] The correspondences in the original paper (Kato, 1964a) differ by the factor $(1/\tan \theta_B)$.
[2] One retains the variable $z$ in optics.

By analogy to mechanics, one can call $f$ the force. Similarly, in the expressions for the phases (5-130$a$ and $b$), $[T]$ can be called the kinetic term and $[N]$ the potential term. Such an analogy to mechanics frequently is very useful. Note that the ray will never come out from the Borrmann fan, i.e., the light cone.

One can get a physical feeling for (5-137) by considering the case in which the $\bar{\mathbf{g}}$ component of the displacement gives the force zero. Then, obviously, the crystal appears to be perfect, and the ray propagates along a straight line. In other cases, the ray must be bent by the force $f$. By the analogy to mechanics, however, one can expect that the bending becomes more difficult when the ray direction approaches either $\bar{\mathbf{K}}_0$ or $\bar{\mathbf{K}}_g$.

The first experimental demonstration of the concept of the ray bending was presented by Borrmann and Hildebrandt (1959) and by Hildebrandt (1959$a,b$). The theoretical analysis was developed by Penning and Polder (1961). Although based on the postulation of a kind of refractive index, they established the geometrical optics for x-ray diffraction and derived a ray equation that is equivalent to equation (5-137). Together, these works serve as milestones in the development of the x-ray diffraction theory for distorted crystals.

In the force term, the quantity $(\partial^2/\partial z^2)(\bar{\mathbf{g}} \cdot \mathbf{u})$ represents the bending of the net plane $[(\partial/\partial z)(\bar{\mathbf{g}} \cdot \mathbf{u})$ represents inclination], whereas the quantity $(\partial^2/\partial x^2)(\bar{\mathbf{g}} \cdot \mathbf{u})$ is the variation of the lattice expansion or dilatation $[(\partial/\partial x)(\bar{\mathbf{g}} \cdot \mathbf{u})$ represents expansion]. A consequence of the presence of the factors $\cot \theta_B$ and $\tan \theta_B$ is that bending of the rays is caused mainly by the bending of the net planes in ordinary x-ray diffraction experiments because $\tan \theta_B$ is nearly equal to 0.1. If one then neglects the dilatation term, the ray belonging to branch (1) bends in the same sense as the lattice bending, and ray (2) bends in the opposite sense. This point is useful in constructing a picture of the ray behavior in an approximate way. Incidentally, the effect of such lattice bending is more pronounced for small structure factor values.

Sometimes it is helpful to consider the same analysis in terms of the reciprocal-space concept, namely, the movement of the dispersion point. The following rules are then useful:

1. The dispersion points move in the same direction on the branches (1) and (2) for a given force.
2. If the force is positive, they move from the $O$ side to the $G$ side of the dispersion surface, and move in the opposite direction for a negative force.

**4. Phases and Pendellösung fringes** Here, and in the following section, only the spherical-wave theory for the Laue-Laue case is described. For plane-wave considerations, the reader is referred to the original paper (Kato, 1964$a$). The experimental object is a section topograph taken with a point source placed near the entrance surface. The arrangement of the rays is illustrated in Fig. 5-22. The source creates the crystal waves of types (1) and (2), essentially, at a single point $Q$. According to the Eikonal theory, the total wave fields are described by a bundle of rays, each of which is bent owing to the lattice distortion. In ordinary cases, two rays, one of each type, arrive at an exit point $P$. They penetrate through the exit surface and split into $O$ and $G$ waves.

One can assume, in most cases, that the waves close to the entrance point are the same as those in a perfect crystal, because the crystal is always regarded as perfect in a local sense. This assumption, however, would not be acceptable if the crystal is heavily distorted near the entrance point. One can improve this simplified assumption by tracing the ray from its source in vacuum to the observation point and by assigning different **g** vectors at the entrance surface to individual rays. In this way, any spread at the entrance point is taken into account. In practice, however, the actual source size is comparable to or larger than the spread at the entrance point so that the above simplified assumption still suffices for a heavily distorted crystal.

To simplify the analysis, it is assumed that the ray penetration of the exit surface is identical to the case of perfect crystals. Such simplified models of the wave behavior at the crystal surfaces are also used in considering the intensity, which will be discussed in the next section.

Returning to the crystal wave, one has to ask what the phase difference associated with a single ray from $Q$ to $P$ is. The main part, of course, is due to the wave propagation. It is given by

$$S_0 = \int_Q^P (\mathbf{k}_0 \cdot \mathbf{v})\, dl \quad (5\text{-}139a)$$

$$S_g = \int_Q^P (\mathbf{k}_g \cdot \mathbf{v})\, dl \quad (5\text{-}139b)$$

Now, the integration must be carried out along the real trajectory specified by (5-137). For this reason **k** is used instead of $\kappa$. Also, the notation of functional by brackets is removed. $S_0$ and $S_g$ are ordinary functions. In spite of this difference in their mathematical meanings, the expressions themselves are identical to those for $[S_0]$ and $[S_g]$, respectively.

Next, one needs to add a constant phase to the integrals (5-139a and b). Since it has been assumed that the initial state of the crystal waves is the same as that in a perfect crystal, the arguments given in Sections II B 3 and II C of Chap. 4 can be applied as they are [see equations (4-209a and b)]. Therefore, the intrinsic crossing of the rays of branch (1) near the margins of the section topograph must be taken into account. In distorted crystals, it may happen that the rays belonging to the same branch of the dispersion surface cross each other within the Borrmann fan.[1] In that case, the phase

---

[1] The idea was first mentioned by K. Kambe in 1962 in a private discussion, although the possibility of a phase jump was not considered at that time.

must be increased by $\pi/2$ after passing the crossing point, regardless of the type of branch or the mode of the $O$ and $G$ waves. Such arguments have been discussed already in Chap. 4.

Combining the above considerations, the true phase difference between $P$ and $Q$ must be as follows:

$$\{S_0\} = S_0 + \{_0^{\pi/2}\} + m\left(\frac{\pi}{2}\right) \qquad (5\text{-}140a)$$

$$\{S_g\} = S_g + \{_\pi^{\pi/2}\} + m\left(\frac{\pi}{2}\right) + \delta \qquad (5\text{-}140b)$$

where $m$ is the number of the crossing. The phase jump discussed here is not given by the simple Eikonal theory.

Since the two waves arrive at an exit point, it is reasonable to expect that they interfere with each other and give rise to *Pendellösung* fringes. From a practical viewpoint, $G$ waves are considered first, but parallel arguments can be applied also to $O$ waves. According to a general rule of optics, the fringe maxima occur at positions satisfying the condition

$$\{S_g^{(1)}\} - \{S_g^{(2)}\} = 2n\pi \qquad (n = 1, 2, \ldots) \qquad (5\text{-}141)$$

The terms not enclosed in brackets in equations (5-130) and (5-133) are identical in both $S_g^{(1)}$ and $S_g^{(2)}$. They have nothing to do with the interferences. One can see, therefore, that the fringe contour is determined by the phase difference

$$S_g^{(1)} - S_g^{(2)} = m_0 c^2 \int_{L_1+L_2} (1 - p^2)^{1/2}\, dz$$

$$+ \pi \int_{L_1-L_2} \left\{ \frac{1}{c}\left[\frac{\partial(\mathbf{g} \cdot \mathbf{u})}{\partial z}\right] dx + c\left[\frac{\partial(\mathbf{g} \cdot \mathbf{u})}{\partial x}\right] dz \right\} \qquad (5\text{-}142a)$$

where $L_1$ and $L_2$ indicate the trajectories of rays (1) and (2), respectively.

At first sight, the phase difference depends on the strain itself. This, however, is not true. If one applies the Stokes theorem to the potential (second) term, one can easily show that

$$S_g^{(1)} - S_g^{(2)} = m_0 c^2 \int_{L_1+L_2} (1 - p^2)^{1/2}\, dz + \int_{A_{12}} f(x, z)\, dx\, dz \qquad (5\text{-}142b)$$

where the area integral covers the total area bounded by the ray $L_1$ and $L_2$. If the integral path $L_1 - L_2$ is clockwise, a positive sign must be assigned to the area $A_{12}$. In the other case $A_{12}$ is negative. It turns out that the function $f(x, z)$ has exactly the same expression as force in equation (5-138). The kinetic terms depend on the trajectories, which are determined, again, primarily by the force. Thus the fringe phenomena give no information about the strain, only about the strain gradient. This result is quite reasonable, since a crystal having a constant strain is a perfect crystal, as has already been pointed out.

According to (5-140a and b), the phase differences $\{S_0^{(1)}\} - \{S_0^{(2)}\}$ and $\{S_g^{(1)}\} - \{S_g^{(2)}\}$ differ by exactly $\pi$. Therefore, the complementarity of the *Pendellösung* fringes of O and G waves is automatically satisfied.

It is worth noting from equations (5-130) and (5-133) that

$$S_g^{(1)} - S_0^{(1)} = S_g^{(2)} - S_0^{(2)} = 2\pi[\mathbf{g} \cdot (\bar{\mathbf{r}}_p - \bar{\mathbf{r}}_Q)] \qquad (5\text{-}143)$$

where $\bar{\mathbf{r}} = \mathbf{r} - \mathbf{u}$ is the position before the crystal is distorted. Even when a crystal is distorted, the phase difference between the O and G waves, due to the wave propagation within the crystal, is identical to the phase difference in the corresponding perfect crystal. Physically speaking, the phase difference of G and O waves after propagating from P to Q is exactly equal to $2\pi$ times the number of lattice planes which the ray crosses during its propagation. This result is important in considering the problem of moiré fringes.

**5. Intensity and amplitude** As in the case of the modified plane wave, the intensity and the amplitude can be derived from the equation for energy conservation (5-124). Again, only the spherical-wave cases are considered. Within the present approximation, the boundary conditions for singular cases described in Section A 4 are used. (The condition for nonsingular cases can be used for plane waves.) The problem is simplified because all conceivable rays lie in the reflection (R) plane. The solid angle $\Omega$ is replaced by a linear angle $\Theta$, and the surface element $\delta F$ is reinterpreted as a line element.

Following the same arguments that were used in deriving (5-108), one can write down the intensity on the vacuum side of the exit surface as follows. First, one has to specify the mode of O and G waves and the type $(j)$ of ray with which the energy is transported in the crystal.

$$I_0^{(j)} = P_e P_{a0} \frac{K_e}{\overline{\mathbf{K}}_0 \cdot \mathbf{n}_a} \frac{\delta\Theta}{\delta F^{(j)}} I_e \qquad (5\text{-}144a)$$

$$I_g^{(j)} = P_e P_{ag} \frac{K_e}{\overline{\mathbf{K}}_g \cdot \mathbf{n}_a} \frac{\delta\Theta}{\delta F^{(j)}} I_e \qquad (5\text{-}144b)$$

where $\mathbf{K}_a$ is replaced by $\overline{\mathbf{K}}_0$ or $\overline{\mathbf{K}}_g$ in the respective cases, and the angular distribution $I(\Theta)$ is regarded as constant and equal to $I_e$. These approximations are justifiable, as described in the case of perfect crystals (Section I D 4 in Chap. 4). Indeed, the above equations are the extention of equations (4-142), and the transmission powers $P_e$, $P_{a0}$, and $P_{ag}$ are given in suitable forms in that section. They depend on the extent to which the Bragg condition is satisfied by the individual ray at the crystal surfaces, and their expressions, generally, are different for the different kinds of rays. If, however, one uses the parameter $p$ for specifying the ray inclination, it turns out that the expressions do not depend on the ray type. For this reason, the superscript $(j)$ can be deliberately omitted in equations (5-144a and b). Obviously, the contraction factor depends on the ray type.

The contraction factors (omitting the superscripts) can be calculated by the relation

$$\frac{\delta\Theta}{\delta F_a} = \left(\frac{d\Theta}{dw}\frac{dw}{d\overline{w}}\left|\frac{d\overline{w}}{dp}\right|\right)_e \left(\frac{\delta p_e}{\delta F_a}\right) \qquad (5\text{-}145)$$

The factor $(\ )_e$ refers to the entrance point, where $p = p_e$. The calculation is rather involved for general cases, but straightforward, as already shown, for the special case $(\gamma_0 = \gamma_g)$. Putting together all these factors, one finally obtains

$$I_0^{(j)} = \frac{1}{4}\frac{\Delta\Theta_0}{\cos\theta_B}\frac{(1 - p_e)(1 - p_a)}{(1 - p_e^2)^{3/2}}\left(\frac{\delta p_e}{\delta F_a}\right)^{(j)} I_e \qquad (5\text{-}146a)$$

$$I_g^{(j)} = \frac{1}{4}\frac{\Delta\Theta_0}{\cos\theta_B}\frac{(1 - p_e)(1 + p_a)}{(1 - p_e^2)^{3/2}}\left(\frac{\delta p_e}{\delta F_a}\right)^{(j)} I_e \qquad (5\text{-}146b)$$

where

$$\Delta\Theta_0 = \frac{CR_e(\chi_g\chi_{-g})^{1/2}}{\sin 2\theta_B} \qquad (5\text{-}147)$$

is an effective angular width on the $\Theta$ scale. [Compare to $\Delta\Theta$ in (4-76).]

So far, the true (photoelectric) absorption has been neglected. It is a simple matter to take absorption into account by means of the imaginary part of the Eikonal function. In (5-130a and b), the imaginary part is included in $\chi_0$ and $T$ through $\chi_g\chi_{-g}$. The attenuation factors, therefore, turn out to be

$$\{A\} = \exp - 2S^i \qquad (5\text{-}148)$$

$$S^i = \tfrac{1}{2}K\chi_0^i(l_0 + l_g) \pm \tfrac{1}{2}KCJ_m(\chi_g\chi_{-g})^{1/2}\frac{1}{\cos\theta_B}\int_Q^P (1 - p^2)^{1/2}\,dz \qquad (5\text{-}149)$$

Obviously, the first term comes from the ordinary absorption. It is an interesting point that contrary to the case of isotropic media, the ordinary absorption does not depend on the form of the trajectory. This result can be interpreted by means of the zigzag path shown in Fig. 4-18, in a microscopic sense. The second term of (5-149) refers to the Borrmann absorption in a distorted crystal. This term depends on the local inclination of the ray in the same way as in a perfect crystal. Consequently, the total form of the trajectory must be known. The attenuation factor does not depend on the mode of the $O$ and $G$ wave, but it does depend on the ray type $(j)$, even though the superscript is omitted in equations (5-148) and (5-149).

If one uses the notation defined by (4-158), the attenuation[1] due to the Borrmann

---

[1] For branch (1), it is actually an amplification, but the same terminology is retained.

absorption can be written

$$B^{(j)} = \exp \left[ \pm \frac{\mu_g}{\cos \theta_B} \int_Q^P (1 - p^2)^{1/2} \, dz \right] \qquad (5\text{-}150)$$

The plus and minus signs correspond to the rays of branch (1) and (2).

The total intensity fields at the exit surface $F_a$ are given by

O wave: $\quad I_0 = [I_0^{(1)} B^{(1)} + I_0^{(2)} B^{(2)} + 2(I_0^{(1)} I_0^{(2)})^{1/2}$

$$\times \cos (\{S_0^{(1)}\} - \{S_0^{(2)}\})] \exp -\mu_0(l_0 + l_g) \qquad (5\text{-}151a)$$

G wave: $\quad I_g = [I_g^{(1)} B^{(1)} + I_g^{(2)} B^{(2)} + 2(I_g^{(1)} I_g^{(2)})^{1/2}$

$$\times \cos (\{S_g^{(1)}\} - \{S_g^{(2)}\})] \exp -\mu_0(l_0 + l_g) \qquad (5\text{-}151b)$$

The last term in each pair of brackets above represents the *Pendellösung* fringes, and the first two give the background. It is to be noted that the intensity modulation due to the *Pendellösung* fringes is independent of the Borrmann absorption, and the contrast is independent of the normal absorption.

Equations (5-151a and b) include all the information about the intensity at the exit surface. All terms and factors are correlated with the lattice distortion $\mathbf{u(r)}$ in a rather complicated way. Moreover, they include several optical and geometrical parameters. For this reason, it is best to postpone further discussion of the equations and their characteristic features to later sections where actual examples will be presented.

## C.   Case of constant strain gradient

So far, the Eikonal theory has been solved analytically only for the case of a constant force $f$, namely, for the case of a constant strain gradient. This does not mean, however, that only a special case has been worked out. In general, if one obtains a solution for the displacement $\mathbf{u(r)}$, the same solution can be used for another displacement $(\mathbf{u + U})$, provided that $\mathbf{U}$ satisfies a kind of wave equation

$$\frac{\partial^2}{\partial x^2} (\bar{\mathbf{g}} \cdot \mathbf{U}) - \frac{1}{\tan^2 \theta_B} \frac{\partial^2}{\partial z^2} (\bar{\mathbf{g}} \cdot \mathbf{U}) = 0 \qquad (5\text{-}152)$$

because, then, the force is not changed at all. This result is the prototype of what is called gauge invariance in electrodynamics.

When the actual distortion in the diffraction problem is a mixed state involving a bending and a dilatational gradient, one can eliminate either one of them by the use of (5-152). This makes it possible to apply the present theory to a rather wide range of lattice distortions. Moreover, even when the strain gradient changes from point to point, one can often assume that the strain gradient is constant within the Borrmann fan. With such an approximation, the solution which will be presented below can be used to handle long-range distortions in a large crystal.

Here, a rather pessimistic thought may arise. Since the ultimate aim of a diffraction study is to obtain unique information about the crystalline state, breaking the one-to-one correspondence between the actual distortion and the theoretical model may destroy the significance of the experiment. This assertion, however, is not true. One can still gather much information by utilizing reflections from different net planes, and one is then able to combine it to produce a virtually unique picture of the crystal's distortion. Remember that x-ray diffraction is a nondestructive method of analysis.

Returning from philosophy to the mathematical concerns of this chapter, in order to make actual calculations, it is necessary to specify the displacement $\mathbf{u}$ along the $\bar{\mathbf{g}}$ component. Here one considers the case

$$(\bar{\mathbf{g}} \cdot \mathbf{u}) = -\tfrac{1}{2}\alpha x^2 \qquad (5\text{-}153)$$

namely, the case of dilatation is dealt with. The case of bending and the more general case that $(\mathbf{g} \cdot \mathbf{u})$ has a quadratic form in $x$ and $z$ are discussed by Kato (1964b) and by Kato and Ando (1966). For details of some mathematical manipulations the reader is referred to these papers. The central experimental subject is a section topograph, but related topics also will be briefly mentioned.

**1. *Ray trajectory*** The one-dimensional equation of motion (5-137) can be solved for the initial conditions, $v = v_e$ and $(x, z) = (0, 0)$, as follows:

$$\frac{v/c}{(1 - v^2/c^2)^{1/2}} - \left[\frac{v/c}{(1 - v^2/c^2)^{1/2}}\right]_e = \pm \left(\frac{f}{m_0 c}\right) z \qquad (5\text{-}154)$$

$$\frac{f}{m_0 c^2}(x + x_0) = \pm \left[\left(\frac{f}{m_0 c}\right)^2 (z + z_0)^2 + 1\right]^{1/2} \qquad (5\text{-}155a)$$

with the subsidiary condition

$$\left(\frac{f}{m_0 c^2}\right) x_0 = \pm \left[\left(\frac{f}{m_0 c}\right)^2 z_0{}^2 + 1\right]^{1/2} \qquad (5\text{-}156a)$$

where the plus and minus signs correspond to the rays of types (1) and (2), respectively. This will be true throughout this section, unless otherwise specified. In the present case, the force is given by

$$f = \pi \tan \theta_B \alpha \qquad (5\text{-}157)$$

For convenience, the following normalized notations are used:

$$X = \frac{f}{m_0 c}\frac{x}{c} \qquad (5\text{-}158a)$$

$$Z = \frac{f}{m_0 c} z \qquad (5\text{-}158b)$$

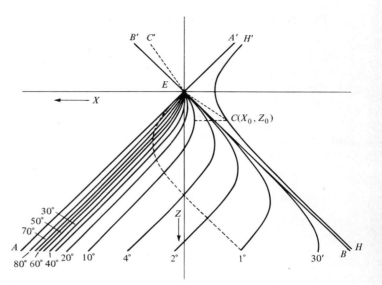

**FIGURE 5-23**
A family of important rays passing through an entrance point $E$. The normalized coordinates in (5-158) are used. Every trajectory is a hyperbola whose asymptotes are parallel to $AA'$ and $BB'$. The centers of these hyperbolas also lie on a hyperbola $HH'$. Given a ray direction $C'E$ for the crystal wave at the entrance surface, the center $C(x_0, z_0)$ of the hyperbola to which this ray belongs is determined by the condition that $\angle C'EB' = \angle CEB$ from equation (5-159c). The hyperbola shown by the dashed line is another type of trajectory.

and

$$p = \frac{dX}{dZ} = \frac{1}{c}\frac{dx}{dz} \qquad (5\text{-}159a)$$

Then, the above equations (5-155a), (5-156a), and (5-159a) can be reduced to

$$(X + X_0)^2 - (Z + Z_0)^2 = 1 \qquad (5\text{-}155b)$$

$$X_0{}^2 - Z_0{}^2 = 1 \qquad (5\text{-}156b)$$

and

$$p = \frac{Z + Z_0}{X + X_0} \qquad (5\text{-}159b)$$

and, at the entrance point,

$$p_e = \frac{Z_0}{X_0} \qquad (5\text{-}159c)$$

The trajectory has the form of a hyperbola while the center $(X_0, Z_0)$ also lies on a hyperbola, as shown in Fig. 5-23.

**2.**  *The Eikonal*  For the purpose of integration, a parameter $h$ is introduced by

$$Z + Z_0 = \sinh h \qquad (5\text{-}160a)$$
$$Z_0 = \sinh h_0 \qquad (5\text{-}160b)$$

By virtue of the ray equation

$$X + X_0 = \pm\cosh h \qquad (5\text{-}161a)$$
$$X_0 = \pm\cosh h_0 \qquad (5\text{-}161b)$$

and

$$p = \pm\tanh h \qquad (5\text{-}162a)$$
$$p_e = \pm\tanh h_0 \qquad (5\text{-}162b)$$

Then, one can calculate the Eikonal as follows:

*Kinetic term* [cf (5-131) and (5-136$d$)]:

$$\frac{T}{m_0 c^2} = \pm\left(\frac{m_0 c}{f}\right)\int_0^Z (1 - p^2)^{1/2}\,dZ = \pm\left(\frac{m_0 c}{f}\right)(h - h_0) \qquad (5\text{-}163)$$

*Potential term* [cf (5-134), (5-153), and (5-157)]:

$$\frac{N}{m_0 c^2} = \left(\frac{m_0 c}{f}\right)\int_0^Z X\,dZ$$
$$= \frac{1}{2}\left(\frac{m_0 c}{f}\right)\{XZ \pm [(h - h_0) - \sinh (h - h_0)]\} \qquad (5\text{-}164)$$

Note that the sign of $X$ in the integral is opposite for rays (1) and (2). Returning to the position variables $X$ and $Z$, one can obtain the phase difference [cf (5-142$a$)]:

$$S^{(1)} - S^{(2)} = 2m_0 c^2 \left(\frac{m_0 c}{|f|}\right)\{\sinh^{-1} \tfrac{1}{2}(Z^2 - X^2)^{1/2}$$
$$+ \tfrac{1}{4}[(Z^2 - X^2)^2 + 4(Z^2 - X^2)]^{1/2}\} \qquad (5\text{-}165a)$$

It is not necessary to specify $O$ and $G$ waves insofar as the phase difference is concerned. [See (5-143).] For small value of $f$, it can be approximated as follows:

$$S^{(1)} - S^{(2)} = 2m_0 c^2 \left(z^2 - \frac{x^2}{c^2}\right)^{1/2}\left[1 + \frac{1}{24}\left(\frac{f}{m_0 c}\right)^2\left(z^2 - \frac{x^2}{c^2}\right) + \cdots\right] \qquad (5\text{-}165b)$$

The behavior of the phase difference is shown in Fig. 5-24. As to the *Pendellösung* fringes, the following conclusions can be reached:

1.  The phase difference is a function of $\rho = (z^2 - x^2/c^2)$. This implies that the fringe shape always is a hyperbola, as in the case of perfect crystals.

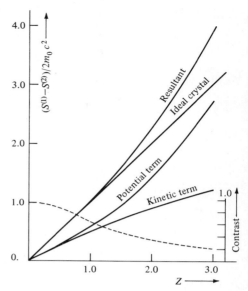

**FIGURE 5-24**
Phase change and fringe contrast along
the net plane.

2. The phase difference increases with $\rho$ more rapidly than in the perfect crystal. This implies that the fringe spacing is shorter than that in perfect crystals. The contraction becomes more predominant with increasing strain and crystal thickness.

These aspects of the *Pendellösung* fringes were confirmed experimentally by Hart (1966). (See Figs. 5-25 and 5-26.) The agreement between theory and experiment turned out to be reasonably close. Ando and Kato (1966) also demonstrated the contraction of the fringes in a qualitative way. They demonstrated that a few new fringes appear in the traverse topograph of a parallel crystal after deliberately introducing an inhomogeneous strain gradient.[1] They proposed the name *equal-strain-gradient fringe* in contrast to the *equal-thickness* or *equal-inclination* fringes mentioned in Chap. 4.

3. *Intensity* The intensity distribution for a fixed $Z$ value will be calculated. The intensity associated with each ray at the exit surface is given by (5-146a and b) in the respective cases.

---

[1] This phenomenon, actually recognized near a dislocation pileup in a traverse photograph (Authier and Lang, 1964), was interpreted theoretically in a special form by Kato (1963b).

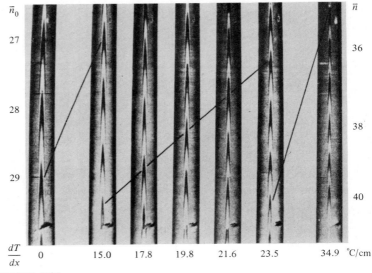

$\bar{n}_0$

27

28

29

$\dfrac{dT}{dx}$   0      15.0    17.8    19.8    21.6    23.5        34.9  °C/cm

$\bar{n}$

36

38

40

**FIGURE 5-25**
Section patterns of silicon deformed by a uniform temperature gradient $dT/dx$ and showing the $\bar{2}20$ reflection obtained with Ag $K\alpha_1$ radiation. $\bar{n}_0$ is the fringe order in perfect crystal, $dT/dx = 0$; $\bar{n}$ is the fringe order in a deformed crystal, $dT/dx = 34.9°C/cm$ (Hart, 1966).

In the present case, the trajectory is given by (5-155b). The line element $\delta x = (m_0c^2/f)\,\delta X$ stands for $\delta F_a$ in the contraction factor, $\delta p_e/\delta F_a$. The parameter $p_e$ is given by (5-159c), and $X_0$ and $Z_0$ are related by (5-156b). Thus the problem of finding $(\delta p_e/\delta F_a) = [(\partial p_e/\partial X_0)_{Z_0} + (\partial p_e/\partial Z_0)_{X_0}(dZ_0/dX_0)]/(\partial X/\partial X_0)_Z$ is a matter of differential calculus. Here, two types of equations will be presented, one in terms of the inclination parameters $p_e$ and $p_a$, and the other in terms of the position parameters $X$ and $Z$:

$$I_0^{(j)} = \frac{1}{4}\frac{\Delta\Theta_0}{\sin\theta_B}\frac{|f|}{m_0c}\frac{(1-p_e)(1-p_a)}{|p_e-p_a|}I_e \tag{5-166a}$$

$$I_g^{(j)} = \frac{1}{4}\frac{\Delta\Theta_0}{\sin\theta_B}\frac{|f|}{m_0c}\frac{(1-p_e)(1+p_a)}{|p_e-p_a|}I_e \tag{5-166b}$$

and

$$I_0^{(j)} = \frac{1}{2}\frac{\Delta\Theta_0}{\sin\theta_B}\frac{|f|}{m_0c}\frac{(Z^2-X^2)^{1/2}}{(X+Z)^2(4+Z^2-X^2)^{1/2}}I_e \tag{5-167a}$$

$$I_g^{(j)} = \frac{1}{2}\frac{\Delta\Theta_0}{\sin\theta_B}\frac{|f|}{m_0c}\left\{\frac{2+Z^2-X^2}{[(Z^2-X^2)(4+Z^2-X^2)]^{1/2}}\pm\frac{f}{|f|}\right\}I_e \tag{5-167b}$$

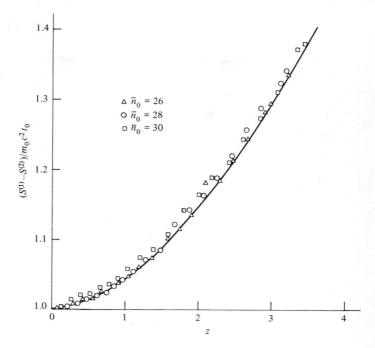

**FIGURE 5-26**
Comparison of equation (5-165a) with experiment. (After Hart, 1966.)

Some of the numerical results are illustrated in Fig. 5-27.

The following points should be noted in the above:

1. The reciprocity theorem is satisfied.

2. If the force is increased, the intensity of $O$ waves decreases, and it tends to a narrow beam having the direction of $\bar{\mathbf{K}}_0$.

3. On the other hand, the intensity of $G$ waves increases as the force is increased. Correctly speaking, ray (1) increases enormously and ray (2) decreases for the case of a positive force, and the roles of rays (1) and (2) are interchanged for a negative force. The intensity at the margins is masked by the increased intensity.

4. The relative changes of these diffraction phenomena are more pronounced in the case of small structure factors.

The conclusions in (2) and (3) above are easily interpreted by ray considerations and by checking the transmission powers of the rays at the crystal surfaces. From the expressions of the powers given by equations (4-140), one can see that the factors $(1 - p_e)$ and $(1 \mp p_a)$ in (5-166a and b) appear owing to the transmission powers.

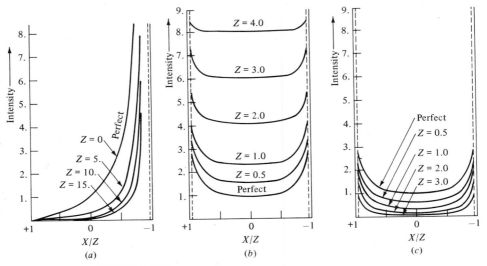

**FIGURE 5-27**
Intensity distributions on the surface perpendicular to the net plane. (*a*) Direct
wave for one branch; no difference between different branches. (*b*) Diffracted
wave corresponding to one branch. (*c*) Diffracted wave corresponding to the
other branch.

Consider the case of a positive force. The energy of the incident beam is mainly
transported by the rays which initially propagate in directions close to the $\overline{K}_0$ direction. Such rays of type (1), however, are bent by the lattice distortion and follow the
direction of $\overline{K}_g$, because they find the place where the Bragg condition is satisfied.
After that, the energy is transported in the state of $G$ waves. At the exit surface,
except in the region close to the incident beam, therefore, the rays pass through the
surface in the $\overline{K}_g$ direction and little is transported in the $\overline{K}_0$ direction. The rays of type
(2) are bent in the opposite direction. Thus, the rays which carry an appreciable
energy are limited to the region close to the incident beam and penetrate through the
exit surface in the $\overline{K}_0$ direction.

The picture presented here is easily extended to more general cases. The key
issue for the intensity distribution on the section topograph is to pay attention to the
rays having appreciable values of the transmission factors $(1 - p_e)$ and $(1 + p_a)$ and
to ask where such rays arrive on the exit surface. When the ray satisfies the Bragg
condition, the ray deflects from $\overline{K}_0$ to $\overline{K}_g$ or from $\overline{K}_g$ to $\overline{K}_0$ in direction. Additionally,
of course, the contraction factor must be considered. In the present case, in most
regions on the exit surface, the spread occurs in a homogeneous way, except along the
margins where the rays accumulate in a narrow region.

The situation described in points (2), (3), and (4) above is in fact what one can expect by the kinematical theory in distorted crystals. (See Section II C 4 of Chap. 3.) For this reason, the present development covers the theory of secondary extinction to some extent. Nevertheless, one must be aware of the limitations in applications to highly distorted crystals (see Subsection 7).

**4.  Absorption**  To understand the absorbing crystal, one needs to calculate the imaginary part of the Eikonal. Since the functional form is exactly the same as the kinetic term, it is straightforward to write the attenuation factor as follows: [cf equations (5-150), (5-163), and (5-165)]:

$$B^{(j)} = \exp \pm \frac{\mu_g}{\cos \theta_B} \frac{2m_0c}{|f|} \sinh^{-1} \tfrac{1}{2}(Z^2 - X^2)^{1/2} \tag{5-168a}$$

$$= \exp \pm \frac{\mu_g}{\cos \theta_B} \left(z^2 - \frac{x^2}{c^2}\right)^{1/2} \left[1 - \frac{1}{24}\left(\frac{f}{m_0c}\right)^2 \left(z^2 - \frac{x^2}{c^2}\right) + \cdots\right] \tag{5-168b}$$

In this particular case the Borrmann effect is suppressed at every point on the exit surface with increasing force because each ray passes through the crystal with a larger inclination on the average than in the case of perfect crystals. Although the general statement cannot be proved, it is very likely that the situation described above does hold in most cases. Thus, for highly distorted crystals, the Borrmann absorption must disappear.

If one inserts all the results from (5-165) to (5-168) into the general expression (5-151), one can obtain the intensity for the section topograph. It is easily expected that the intensity distribution is symmetrical with respect to the central lines ($X = 0$). Also, it is expected that the intensity itself must be different, say, at the fixed point $X = 0$, for the **g** reflection and $-$**g** reflection, when Borrmann absorption is appreciable. This will be discussed further under the heading of Friedel's law, following a discussion of the integrated intensity, and under the heading of symmetry properties, following another example of the strain gradient given in the next section.

**5.  Integrated intensity**  So far, discussions have been limited to section topographs. In a broad application of traverse topography and in the diffractometric study of distorted crystals, the integrated intensity also represents a valuable measure of how a crystal is distorted. As explained in Section II F 1 and 2 of Chap. 4 regarding wedge-shaped crystals, one adopts the viewpoints that only the spatial integrated intensity is definable in distorted crystals, that the traverse topograph corresponds to the distribution of this spatial integrated intensity, and, finally, that the angular integrated intensity in an experimental (operational) sense is proportional to the spatial integrated intensity.

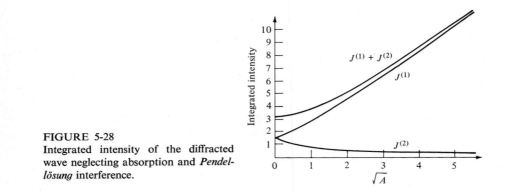

**FIGURE 5-28**
Integrated intensity of the diffracted wave neglecting absorption and *Pendellösung* interference.

For this reason, the intensity distribution obtained from (5-151b) is integrated with respect to the position variable $X$. So far, however, the analytical expression is available in an exact form only for the nonabsorbing case without the interference term. Since the problem is simply a matter of integration, only the final results are presented here. For the integrated intensity:

$$J = \frac{1}{2}\sqrt{\frac{\gamma_g'}{\gamma_0'}}\,\Delta\Theta_0[J^{(1)} + J^{(2)}] \qquad (5\text{-}169)$$

where

$$J^{(j)} = (4 + A)^{1/2}E(k) - 2(4 + A)^{-1/2}K(k) \pm \sqrt{A} \qquad (5\text{-}170a)$$

$$A = \left(\frac{f}{m_0 c}\right)^2 \frac{\cos^2\theta_B}{\gamma_0'\gamma_g'}\,t_0^{\,2} \qquad (5\text{-}170b)$$

$$k = \left(\frac{A}{4 + A}\right)^{1/2} \qquad (5\text{-}170c)$$

Here, $E(k)$ and $K(k)$ are the complete elliptic functions of the first and second kinds, respectively, and $\gamma_0' = (\mathbf{K}_0 \cdot \mathbf{n}_a)$ and $\gamma_g' = (\mathbf{K}_g \cdot \mathbf{n}_a)$, and $t_0$ is the normal distance to the exit surface from the entrance point.

When the force constant is small, $A$ and $k$ tend to zero. Then, by virtue of $E(0) = K(0) = \pi/2$, one can see that $J$ tends to $(\pi/2)\Delta\Theta_0\sqrt{\gamma_g'/\gamma_0'}$, the value for nonabsorbing perfect crystals neglecting oscillation [cf equations (4-260b) and (4-270), noting that $[W] \to \pi/2$]. On the other hand, for a large force, $E(k)$ tends to unity, whereas $K(k)$ tends to log $A$. After all, the integrated intensity increases as $(1/K)(|f|)(t_0/\gamma_0')$, as illustrated in Fig. 5-28.

The importance of absorption is self-evident from an experimental point of view. It is equally important theoretically, however, because the analytical approach is quite formidable, so that it is helpful to work out a numerical calculation (see Table I in

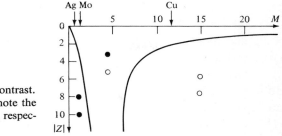

**FIGURE 5-29**
Regions of black-and-white contrast. The upper and lower circles denote the cases of positive and negative $Z$, respectively.

Ando and Kato, 1970). Figure 5-29 illustrates the general features of what one may expect from the theory. If one examines the equation of the integration carefully, it becomes clear that the integrated intensity can be characterized by two parameters:

$$Z = \frac{|f|}{m_0 c} t_0 \qquad (5\text{-}171a)$$

and

$$M = \mu_g t_0 \qquad (5\text{-}171b)$$

Obviously, for a fixed crystal thickness, they indicate the magnitude of the strain gradient and the Borrmann absorption.

**6.    *Departure from Friedel's law***    By the combining of the above two parameters, it is possible to change the integrated intensity considerably. In particular, it is interesting to compare a pair of intensities due to **g** and $-$**g** reflections. As has been previously noticed in the discussion of topographic work, one may encounter black-black, black-white, and white-white contrasts, depending on the combination of these two parameters. Here, black-and-white contrasts mean excess and deficit intensities as compared to the integrated intensities (including absorption) for the perfect crystal ($Z = 0$, not $t_0 = 0$!). In any case, the integrated intensities due to **g** and $-$**g** reflection are different and represent a departure from Friedel's law, which states that the integrated intensity of **g** and $-$**g** reflection must be identical. This law can be easily proved on the basis of the kinematical theory[1] and for any crystal structure and lattice distortion because the integrated intensity must be proportional to $|\chi_g|^2$ within the framework of the kinematical theory. However, it is also easy to show that $|\chi_g|^2$ may not be equal to $|\chi_{-g}|^2$ if there is absorption, namely, when the imaginary part of $\chi$ exists and the crystal structure does not have a center of symmetry. (See Appendix 4AI.) Indeed, this was experimentally demonstrated in the case of ZnS, independently, by Nishikawa and Matsukawa (1928) and Coster, Knol, and Prins (1930).

---

[1] For a nonabsorbing crystal.

This property is utilized in practice to detect the polar direction of noncentrosymmetric crystals, notably in compound semiconductors having the wurtzite structure (Cole and Stemple, 1962), and to distinguish enantiomorphic structures in crystal analysis (Okaya, Saito, and Pepinsky, 1955).

Another example of the departure from Friedel's law was discovered in electron diffraction by Miyake and Uyeda (1950), when simultaneous reflection takes place, and it was theoretically discussed by them (1955) and by Kohra, Uyeda, and Miyake (1950). Since many-beam cases are rare in x-ray cases, this is not further considered here. Readers may find the details in the original papers and in Laue's textbook.

In dynamical two-beam cases, when the crystal has no center of symmetry, the integrated intensity ratio of g and $-$g reflections is, again, $|\chi_g|^2/|\chi_{-g}|^2$ in ideally perfect crystals regardless of absorption. Here one deals with another example of the departure from Friedel's law which should be observed in centrosymmetric crystals.

Changing the sign of g causes an interchange of the rays of type (1) and (2) so that, in general,

$$I_{-g}^{(1)} = I_g^{(2)} \quad \text{and} \quad I_{-g}^{(2)} = I_g^{(1)} \quad (5\text{-}172)$$

On the other hand, the roles of rays (1) and (2) do not interchange as a result of Borrmann absorption; i.e., rays of type (1) are always less absorbed than type (2) rays. But in a perfect crystal, because $I_g^{(1)}$ equals $I_g^{(2)}$ everywhere on the exit surface, the integrated intensity $J_g$ must be the same as $J_{-g}$ even when Borrmann absorption is present. Conversely, the intensities at a particular point on the exit surface, $I_g^{(1)}(X)$ and $I_g^{(2)}(X)$, need not be the same, in general, for a distorted crystal. In the absence of Borrmann absorption, however, changing the sign of g does not alter the total intensities. Thus an anomalous departure from Friedel's law may occur only when both Borrmann absorption is present and the crystal contains distortions. Suppose for such a case that $I_g^{(1)} > I_g^{(2)}$ at some point $X$ on a section topograph. The intensity of the g reflection must be larger than that of the $-$g reflection in that case. Clearly, the reverse will be true when $I_g^{(1)} < I_g^{(2)}$.

Friedel's law is concerned with the integrated intensity, however. The above arguments, therefore, cannot be applied directly to justify the departure from Friedel's law in general cases. Nevertheless, the difference in the intensities $I_g^{(1)}(X)$ and $I_g^{(2)}(X)$ is partly caused by the difference in the transmission factors $(1 - p_e)(1 + p_a)$ and partly by the contraction factor. If the transmission factor is different, on the average, for the ray systems of types (1) and (2), one ray system will transfer more energy than the other to the $G$ wave in the aggregate. This is particularly likely since the two ray systems are different in distorted crystals. Thus, on the average, the above arguments regarding the intensity at a fixed point can be applied to the integrated intensity, as has been seen when a constant strain gradient exists in a crystal.

**7.   Validity of present theory**   It is of obvious interest to assess the validity of the present theory. It is generally accepted that the ordinary Eikonal theory is a three-dimensional case of the Wentzel-Kramers-Brillouin (WKB) approximation for the inhomogeneous wave equation. The theory becomes asymptotically more correct with increasing value of the wave number $K$, and the present theory follows in the same way, broadly speaking.

In the case of x-ray diffraction in crystals, since the dispersion surface is branched, there always exist waves that belong to another branch that are candidates for solutions to the wave equation. The problem becomes clear if one considers a misfit boundary to be an extremely severe distortion. The new rays which have been called type $Q$ in Section I E of the present chapter are always created when a ray passes through the boundary. The present theory neglects this situation, which is usually called *branch jumping*. Also, it is quite possible that the present theory may fail when the strain is rapidly changed perpendicularly to the ray direction. Then, the crystal part which can be regarded as perfect becomes sufficiently narrow for the wave to create diffracted waves in a purely optical sense. This situation may be called *broadening of the dispersion point*, and it is contrary to the present theory, which assumed that the crystal wave in any local region can be represented by a single dispersion point.

Naturally, experimentalists want to know what the consequence is of neglecting these effects. It is not easy to give a definitive answer in general cases. The situation is similar to that existing in the Born approximation or the kinematical theory. Validity depends very much on what measurable quantities one is going to discuss and what accuracy one needs. For example, the Bragg angle can be predicted with sufficient accuracy by the Bragg equation for any size of crystal used in ordinary experiments. On the other hand, the kinematical theory cannot predict the intensity to an accuracy better than 1%, if the crystal size exceeds a few microns.

In what follows, the validity of the present theory is discussed in a semiquantitative way, paying special attention to the intensity. One reason for not discussing the phases is that the crystal distortion must be sufficiently weak so as to maintain the concept of rays (1) and (2) in order for the *Pendellösung* fringes to be observable.

In the following arguments, the relations

$$m_0 c^2 = K \sin \theta_B \Delta \Theta_0 \qquad (5\text{-}173a)$$

$$m_0 c^2 = \frac{\pi}{\Lambda_0} \qquad (5\text{-}173b)$$

are useful, where $\Lambda_0$ is the fringe spacing along the net plane for a perfect crystal. [See equations (5-136d) and (5-147).] Also, note that the bending angle $\delta \Theta$ of the lattice per distance $\delta z$ can be estimated by

$$\delta \Theta = \frac{\partial^2 u}{\partial z^2} \delta z \qquad (5\text{-}174a)$$

Similarly, the lattice expansion can be estimated by

$$\frac{\Delta a}{a} = \frac{\partial^2 u}{\partial x^2} \, \delta x \quad (5\text{-}174b)$$

The corresponding change in the glancing angle from the Bragg condition must be

$$\delta\Theta = \tan^2 \theta_B \frac{\partial^2 u}{\partial x^2} \, \delta z \quad (5\text{-}174c)$$

the directions of the wave propagation being assumed to be either $\overline{K}_0$ or $\overline{K}_g$. In either case, the change in the glancing angle can be written

$$\delta\Theta = \frac{a}{\pi} |f| c \, \delta z \quad (5\text{-}175)$$

where $a$ is the spacing of the net plane [cf (5-138a)].

After these preparations, one can understand why the integrated intensity changes with increased distortion from $\Delta\Theta_0$ to $(1/K)(|f|)(t_0/\gamma_0')$ for the case of homogeneous bending. In a perfect crystal, the crystal picks up the energy of the incident beam within the angular range $\Delta\Theta_0$ and transforms it into $G$ waves. The beams outside this range simply pass through the crystal. If the crystal is distorted, the incident beam within the angular range $\delta\Theta > \Delta\Theta_0$ given by (5-175) with $\delta z = (t_0/\gamma_0') \cos \theta_B$ can find a location where the Bragg condition is satisfied. Setting $(\pi/a \sin \theta_B)$ equal to $K$, one can see that the present picture fits the sophisticated result (5-169) including cases of large $|f|$.

In the above argument for the case of a finite $|f|$, an important assumption was made, namely, that the ray which can locate a point to satisfy the exact Bragg condition can be transformed into the state of a $G$ beam. This is not exactly true for either a very weak distortion or for a very heavy distortion. In the first case, the crystal has a natural reflection breadth $\Delta\Theta_0$, so that the assumption fails and deviations from linearity with $|f|$ and $t_0$ occur.

The objection to the other case arises because a certain crystal thickness is required for the wave to be transformed from an $O$ state to a $G$ state. From the ray equation (5-137), the bending is given by

$$\delta p \approx \langle (1 - p^2)^{3/2} \rangle \frac{|f| c}{\pi} \Lambda_0 \, \delta z \quad (5\text{-}176)$$

where the brackets $\langle \ \rangle$ mean an average over that ray element in which one is interested. To bend the ray appreciably, say $\delta p \approx 1$, one needs a distance

$$\delta z = \left[ \langle (1 - p^2)^{3/2} \rangle \frac{|f| c}{\pi} \Lambda_0 \right]^{-1} \quad (5\text{-}177)$$

which can be shortened to an infinitesimal distance as $|f|$ increases. This, of course, is not true. From an elementary wave consideration, this distance must be larger than one-quarter of the averaged fringe spacing $\Lambda_p$ which can be estimated by $\Lambda_0/$ $\langle(1 - p^2)\rangle^{1/2}$. ($\delta p \approx 1$ implies the transformation of one-half of the energy of the $O$ wave to a $G$ wave. One-half distance of $\Lambda_p$ transforms the entire $O$ wave to a $G$ wave.) The ray considerations must fail below this distance. Thus, the criterion is

$$\tfrac{1}{4}\langle(1 - p^2)\rangle \frac{|f|c}{\pi} \Lambda_0^2 \lesssim 1 \qquad (5\text{-}178)$$

where a crude estimation is made of the average. By virtue of the expression for $|f|$, it turns out that

$$\tfrac{1}{4}\langle(1 - p^2)\rangle \frac{\partial^2 u}{\partial z^2} \Lambda_0^2 \lesssim a \qquad (5\text{-}179a)$$

or

$$\tfrac{1}{4}\langle(1 - p^2)\rangle \frac{\partial^2 u}{\partial x^2} \Lambda_0^2 \tan^2 \theta_B \lesssim a \qquad (5\text{-}179b)$$

For example, (5-179a) states that the displacement of a lattice point due to lattice bending, per fringe distance $\Lambda_0$, should not exceed a few atomic distances. If one takes reasonable figures, $\Lambda_0 \sim 50\,\mu$ and $a = 1\,A\,[p = 0]$, the critical value of $(\partial^2 u/\partial z^2)$ is about $1.6 \times 10^{-3}$ cm, and the corresponding radius of curvature is about 6 m, a rather strong bending, if one remembers that the figure represents an upper limit. By considering also the rays for larger $|p|$, the theory becomes applicable up to larger distortions.

## D. Lattice distortion caused by an oxide film

In most practical applications of the present theory a numerical calculation is inevitably required. As an ad hoc example of such studies, the diffraction topograph of silicon crystals coated with an oxide film is considered.[1] The traverse topograph obtained is shown in Fig. 5-30, in which the square region is covered by the oxide. Figure 5-31a to d is a set of section topographs taken at the position indicated by the arrow in Fig. 5-30. The geometry and the coordinates used below are shown in Fig. 5-31.

At a high temperature (about 1000°C) when the oxide is formed, the base crystal and the oxide may be coherent without any lattice distortion. When such a crystal is brought down to room temperature, an inhomogeneous strain is developed in the base crystal, mainly owing to the differences in thermal expansion; the crystal below the oxide is bent and a high strain is concentrated at the boundary with the oxide. Blech and Meieran (1967) worked out the strain distribution for this problem with some reasonable assumptions; the results are illustrated in Fig. 5-32.

[1] Kato and Patel (1968, 1973) and Patel and Kato (1968, 1973).

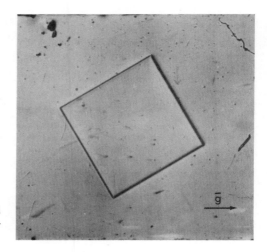

**FIGURE 5-30**
Traverse photograph of Si crystal coated by an oxide film of square shape. (Courtesy of Patel.)

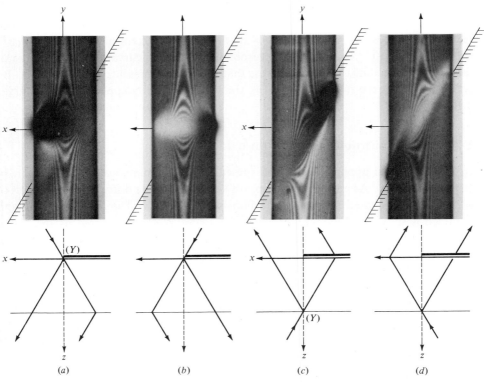

**FIGURE 5-31**
Section topographs at the position indicated by the arrow in Fig. 5-30. (*a*) Standard; (*b*) standard reverse; (*c*) reciprocal; (*d*) reciprocal reverse.

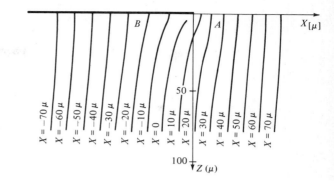

FIGURE 5-32
Distortion of ($2\bar{2}0$) planes due to the edge of an oxide film $1\ \mu$ thick. The horizontal displacements are greatly exaggerated. Region $A$ is compressed while region $B$ is in tension. (After Blech and Meieran, 1967.)

**1.  Analysis of the fringe positions**   It is convenient to rewrite the ray equation (5-137) in the form

$$\frac{d}{dz}\left[\frac{p}{(1 - p^2)^{1/2}}\right] = \pm HB(x, z, Y)  \qquad (5\text{-}180)$$

where

$$H = \frac{2.56}{\pi}\frac{\Lambda_0}{a}\frac{S}{E}\frac{1}{t_0}  \qquad (5\text{-}181)$$

and is called a *force parameter*. ($S$ is the stress per unit length of the oxide boundary, 2.56 is the numerical factor due to Poisson's ratio, $E$ is Young's modulus, $\Lambda_0$ is the *Pendellösung* fringe spacing along the net plane for perfect crystals, $a$ is the lattice spacing, and $t_0$ is the crystal thickness. The function $B$ is given in terms of the strain gradient [see (5-138)]. The normalized coordinate $Y = y/t_0$ is regarded as a parameter specifying the distance from the oxide boundary along the central line of the section topograph. Similarly, later, the capital letter $X = x/t_0 \tan \theta_B$ is used for denoting the $x$ position on the exit surface of the crystal.

The trajectories are calculated numerically, by dividing the crystal into thin slices, and assuming a hyperbolic form for the trajectory (5-155a) in each slice. Some examples are shown in Fig. 5-33. It is important to note that no crossing occurs for the trajectories belonging to either branch (1) or (2). No phase jump discussed in connection with equations (5-140) is present in this particular case.

For any position $(X, Y)$, the phase difference relevant to the *Pendellösung* fringes is given by equation (5-142a). For the numerical calculation, (5-165b) can be

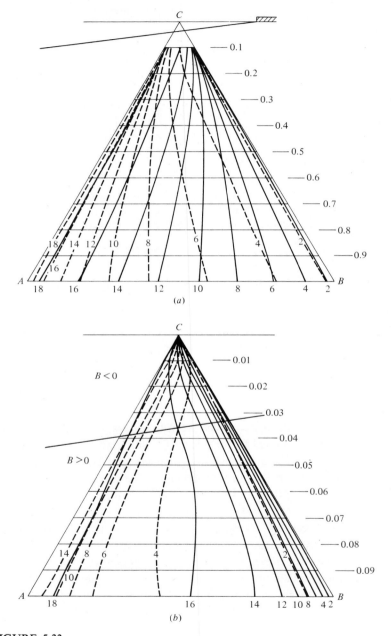

**FIGURE 5-33**
Trajectories in the crystal. (*a*) Full lines for branch (1); dashed lines for branch (2); (*b*) trajectories in top part of Fig. 5-33*a*.

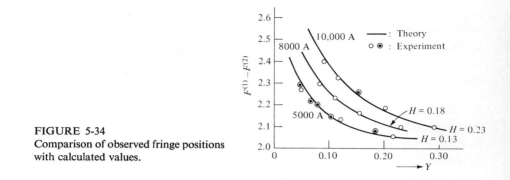

**FIGURE 5-34**
Comparison of observed fringe positions
with calculated values.

used for individual slices. According to the condition (5-141), the fringe positions at
the exit surface are given by

$$\frac{1}{2} \frac{t_0}{\Lambda_0} (F^{(1)} - F^{(2)})_{X,Y,H} - \tfrac{1}{4} = n \qquad \text{(integer)} \qquad \text{(5-182)}$$

where $F^{(j)}$ are the numerical values of $S^{(j)}$, normalized by $m_0 c^2 t_0 = \pi(t_0/\Lambda_0)$.
Obviously, they are functions of positions $(X, Y)$ at the exit surface and the force
parameter $H$, which is essentially the stress $S$ applied at the oxide boundary. If $H$ is
fixed, the above equation gives the contours of the fringes at the exit surface. Alternatively, if $X$ is fixed, the fringe position $Y_n$ is given as a function of $H$ for the $n$th-order
fringes.

   In Fig. 5-34, the calculated $(F^{(1)} - F^{(2)})$ is shown against position $Y$ along the
central line $X = 0$. By normalization, $(F^{(1)} - F^{(2)})$ must equal 2 as $Y$ increases. For
the $i$th extra fringe, therefore, the ordinate must be $2 + 2i(\Lambda_0/t_0)$. The corresponding
position $Y_i$ can be determined by the experiments. They are plotted in the same figure.
The plots lie well on a single theoretical curve in the various cases. Thus, one can
conclude that the diffraction theory and the elastic strain adopted here are reasonably
correct, and the analysis affords a satisfactory nondestructive method for estimating
the stress $S$.

**2.  Symmetrical properties of the section topographs**   From the viewpoint of diffraction theory, it is interesting to study the symmetry of topographs. The set of the
topographs in Fig. 5-31a to d is a good example of this subject. The properties of
symmetry are valuable also for checking the correctness of the computation. The
subjects are discussed further in the papers of Kato (1964b) and Ando and Kato
(1970).

*a.  Reciprocity*   Although the appearances of the standard (S) and the usual reciprocal (R) cases are rather different, they are intimately connected. If one assumes

reciprocity, it is easily concluded that the intensity at $(X, Y)$ of topograph $(S)$ must be identical to that at $(-X, Y + S \tan \theta_B \cot \delta)$ of topograph $(R)$. (Here, $\delta$ is the angle between the $Y$ axis and the normal to the oxide boundary.) Obviously, the same relation must be satisfied between the standard reverse $(SR)$ and the reciprocal reverse $(RR)$ cases. This point is actually confirmed in the relevant topographs.

*b.  Relations between* $(S)$ *and* $(SR)$ *topographs*   As to the intensity, excess and deficit contrasts are expected, depending upon whether $I^{(1)}$ or $I^{(2)}$ is larger in equations (5-151). It is worthwhile, however, to note that the interference terms are identical in both cases because the change of the sign of the **g** vector merely requires interchanging the roles of branches (1) and (2) and changing the sign of each phase $S^{(j)}$. The same holds also for $(R)$ and $(RR)$ cases. Actually, the topographs satisfy this property.

*c.  Symmetry in individual topographs*   In some problems, the strain gradient displays the property of symmetry. In the present case, at any depth $z$, it has the symmetry of inversion. In order to compare the topographs at $\pm Y$, it is convenient to change the $x$ direction for the region of negative $Y$. Then, the strain gradient is identical for any point $x, z$. The problem, therefore, is nothing more than the comparison between $(S)$ and $(SR)$ cases or $(R)$ and $(RR)$ cases. The intensity is different in general, but the interference term must be identical. Thus, one can expect to see inversion symmetry in the fringe shape. The property is actually visible in each topograph, particularly in $(R)$ and $(RR)$ topographs.

**3.   *Other studies***   Intensity enhancement similar to the present problem was perhaps first recognized by Fukushima (1954) at the edge of an abraded region in a quartz plate and studied in detail by Kato (1956). The topographic study was made by Frank, Lawn, Lang, and Wilks (1967). Haruta and Spencer (1966) observed similar effects in the case of metal films on quartz. The problem of oxide films discussed here, which is important for present-day semiconductor technology, has been studied extensively by several authors (Meieran and Blech, 1965, 1968; Howard and Schwuttke, 1967; Blech and Meieran, 1967). In particular, the section topograph presented here has been observed also by Schwuttke and Howard (1968), independently. Recently, Hashizume and Kohra (1971) and Hashizume (1971) studied the ray behavior by the use of an extremely narrow monodirectional beam. Blech and Meieran (1967) calculated the kinematical intensity corresponding to the volume within which the kinematical exact Bragg condition deviates from the perfect region more than the half-width $\Delta\Theta$. It is very interesting to see that the kinematical intensity fits reasonably well to the actually observed enhancement of the intensity. Nevertheless, it is obvious that the problem should be worked out using the dynamical theory, because the actual intensity distribution shows the excess and deficit in both traverse and section topographs. The term "kinematical image" is widely used in topographic work for denoting the region of high intensity in section topographs. In the author's view, however, the excess and deficit observed in a section topograph must be carefully studied by dynamical, not kinematical, theory.

## E.    Related analyses

So far, the Eikonal theory for diffraction by crystals was described in a way that seems most natural as an extension of the ordinary Eikonal and dynamical theories. Several related theories, however, have been developed in the past decade and are briefly reviewed below.

As mentioned above, such a theory follows from the original work of Penning and Polder (1961). Their theory was based on a postulation of the refractive index, limited to the Laue case and formulated, essentially, for a scalar wave field, and concerned itself only with the intensity associated with one type of rays which can survive after penetrating a thick crystal. Nevertheless, the clear-cut presentation of the physical concepts involved encourages one to pursue their approach, namely, the geometrical optics in crystal diffraction. Bonse presented a similar theory which may be used for the Bragg case (1964a,b). The theory can be used to understand defects located on a crystal surface, and he actually worked out the dislocation image for the Berg-Barrett arrangement (Bonse, 1964c). Approximately, if the incident rays fall outside the range $|w| \geq W$, the ray behavior can be treated by the method of Penning and Polder because, in that case, the rays themselves have nothing to do with the crystal surfaces. In the range of total refraction, $|w| \leq W$, the postulation considered by Bonse has not yet been justified by the fundamental Maxwell equations.

The justification of the Penning and Polder theory has been worked out by Kato as presented in this chapter and in a different way by Kambe (1965a, 1968). These articles open the way for considering not only the intensity but also the phase along the ray.

The semiclassical (particle) picture of a single electron in a crystalline medium was well established as an approximation to the wave-mechanical treatment by the use of Wannier wavefunctions (e.g., Ziman, 1964). Kambe applied the theory to electromagnetic waves. The Wannier wave is actually a special form of the wave packet of Bloch waves and localized at each lattice point. The center of the Wannier wave packet moves according to the equation of motion, represented by a Hamiltonian formalism. Moreover, the wave packet can be described by a Lagrangian formalism as in classical mechanics. Once this theoretical framework is established, it is possible to define the Eikonal which is called *action* in mechanics. Although the mathematical machinery is rather sophisticated, the final result is simple and equivalent to the theory described in Section II B.  The theory is particularly useful in considering the criterion of the applicability.

As was explained in Section II C 7, the essence of the approximation of the Eikonal theory lies in neglecting the jumping of the excited dispersion point and a kind of optical diffraction.  Wilkens (1964$a,b$; 1966) improved the theory by calculating the scattered waves caused by the jumping.  His theory, however, is developed for electron cases using the column approximation.  The solution, under this limitation, is a special case of the perturbation solution (Kato, 1963$b$) of the scattering matrix theory (Kato, 1963$a$).  In the theory, the zeroth-order wave is the modified Bloch wave in the Eikonal theory.  The first-order perturbation gives the scattered wave caused by the jumping.  In x-ray cases, however, the column approximation can be used only for a very special case.  For this reason, it is desirable to develop the theory without the use of the column approximation.  A more general theory of x-ray diffraction by crystals, within the framework of optical theory, is presented by Takagi (1962, 1969) and Taupin (1964).  The rays in Section II B are equivalent to the characteristic lines of the partial differential equation of Takagi-Taupin.  The desired development of the present theory, therefore, may be obtained by starting from the Takagi-Taupin equations.

## III.  MULTICRYSTAL SYSTEMS

### A.  X-ray interferometer (three-crystal system)

X-ray interferometry of the Mach-Zehnder type has been developed by Bonse and Hart (1965$a$).  Their original arrangement is shown in Fig. 5-35$a$.  The x-ray beam is split by the Bragg reflection at the splitter crystal $S$ and reflected by the mirror crystal $M$.  The split waves are again reflected at the same place in the analyzer crystal $A$.  Since $E_0$ and $E_g$ waves have a definite phase relation, the superposed waves $E_{0,g,-g}$ and $E_{g,-g,0}$ or $E_{0,g,0}$ and $E_{g,-g,g}$ may also have a definite phase relation.  When the geometry and perfection of the crystals are ideal, the phase difference between the relevant waves must be null.  The intensity observed either on the plate $P_0$ or $P_g$ is, then, homogeneous.  If, however, a wedge-shaped specimen is inserted in one of the beams between the crystals $S$ and $A$, one obtains fringes due to the phase retardation in the specimen.

Subsequently, Bonse and Hart designed various modifications (1965$b$, 1966$b$, 1968).  It was shown that the reflection or Bragg case could also be used instead of the Laue case employed in the original design.  The underlying optical principles are discussed by Bonse and Hart (1965$c$, 1966$a$), Bonse and Te Kaat (1971), and Kuriyama (1971), mainly based on the plane-wave theory.

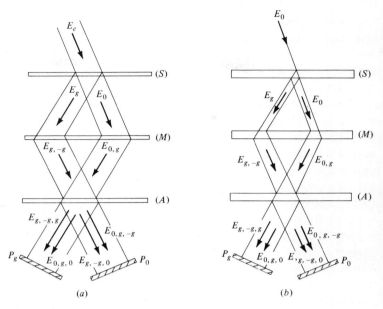

**FIGURE 5-35**
Bonse-Hart interferometer. (*a*) Traverse type; (*b*) section type.

In the arrangement of Fig. 5-35, Bonse and Hart used a wide x-ray beam and thick crystals in the Borrmann absorption sense ($\mu_0 t_0 > 10$). Thus, essentially, only the crystal waves of branch (1) propagating along the net plane are effective in the formation of the interferogram. Obviously, the waves passing through the different parts of crystal $S$ are incoherent. For this reason, the geometry of the crystals as well as the crystal perfection must be rather stringently obeyed in order that the coherent wave fronts, estimated to be a few microns in width, are really superposed in crystal $A$. Nevertheless, modern techniques for growing perfect crystals and cutting and finishing crystal surfaces are precise enough to realize the angstrom interference! In fact, they succeeded in obtaining fringe visibility of about 91% (Bonse and Hart, 1965a).

Besides the success in angstrom interference outside the crystal, their work is really a milestone in the development of x-ray optics. It initiates research on Moiré fringes, which will be discussed in the next section. Technically, the idea of coherent crystal reflectors cut from a large single-crystal block is now widely used in x-ray optical work which requires two- or three-crystal arrangements. The three-crystal technique no longer has been limited to a few ingenious experimentalists. The success of angular resolution of 0.01 second mentioned in Section I D 6 of Chap. 4 owes largely to this experimental technique.

As natural applications of this interferometry, the precise measurement of the refractive index of x rays, and the precise and absolute measurement of x-ray wavelengths, lattice spacing and Avogadro's number are going to be attempted. Also, as a rather special application, the interferometry is used for truly absolute measurement of crystal-structure factors as mentioned in Section II F 5 of Chap. 4.

Since the optical principles have been fully discussed by Bonse and Hart, here only a few ad hoc comments will be presented. They used a wide incident beam and thin crystal slices (in the sense of crystal thickness). This was a clever trick to obtain intensity gain. From an optical point of view, however, it is interesting to see what patterns should be obtained for a fairly thick crystal with a narrow beam compared with the width of the Borrmann fan. This has been done by Kato and Tanemura (1967), and it was shown that nearly parallel interference fringes are recorded over the triply enlarged Borrmann fan on the exit surface of the third crystal $A$. In an analog to the ordinary topograph, the interferogram of Bonse-Hart may be said to be of the "traverse type," whereas the terminology "section type" may be used for Kato-Tanemura's interferograms.

In connection with this, the Borrmann absorption is not essential in x-ray interferometry. Again, the original design of Bonse and Hart was indeed clever, because the wave fields are much simplified by eliminating the waves of branch (2) in every crystal slice. In general, the *Pendellösung* fringes (Angstrom interference inside the crystal) are superimposed on the interference fringes. Also, the margin enhancement of the intensity due to each slice is overlapped in the section-type interferogram.

A wave-optical treatment based on the spherical-wave theory has not been worked out as yet for a three-crystal system. It is worth mentioning, however, that Authier, Milne, and Sauvage (1968) developed a theory of this type for a two-crystal system with a single air gap, for a special case. If the theory is extended to the three-crystal system, the section-type interferograms will be correctly explained. Nevertheless, the theory of Bonse and Te Kaat (1971), which is developed on the basis of plane-wave considerations, is satisfactory in practice for the interferometer having large Borrmann absorption.

## B.   Moiré fringes (two-crystal system)

When a two-crystal system satisfies the Bragg condition and the component crystals are misoriented and/or expanded relative to each other, Moiré fringes are expected. The Moiré fringes associated with diffraction by crystals were first observed by Mitsuishi, Nagasaki, and Uyeda (1951) in an electron micrograph of carbon.

For x rays, perhaps the observation of Chikawa in CdS is the first one (1965). Independently, when the crystals are slightly distorted, Bonse and Hart (1966a) observed intrinsic fringes in the interferogram, which was interpreted correctly as a kind of Moiré fringe. Later, Lang (1968) and Bradler and Lang (1968) succeeded in obtaining Moiré fringes for a two-crystal system, in which the component crystals could be manipulated manually. All these were observed in traverse topographs.

Full understanding must await having a complete theory which would be similar to the theory of planar defects described in Section I E of Chap. 5, but an air gap must be taken into account. In the present section, however, a simple argument regarding some of the fringe characteristics is presented.

According to the theory of perfect crystals, the transmitted $O$ and $G$ waves in vacuum can be represented [cf equations (4-106$a$ and $b$)]

$$E_0(\mathbf{r}) = D_0(\mathbf{K}_e; \mathbf{r}) \exp i(\mathbf{K}_e \cdot \mathbf{r}) \tag{5-183$a$}$$

$$E_g(\mathbf{r}) = \left(\frac{\chi_g}{\chi_{-g}}\right)^{1/2} D_g(\mathbf{K}_e; \mathbf{r}) \exp i[(\mathbf{K}_e + 2\pi\mathbf{g}) \cdot \mathbf{r}] \tag{5-183$b$}[1]$$

where $D_0$ and $D_g$ are gently varying position functions compared to the exponential factors. If they are reflected again from the second crystal, the transmitted $G$ waves would be

$$E_{0,g'}(\mathbf{r}) = \left(\frac{\chi_{g'}}{\chi_{-g'}}\right)^{1/2} D_{0,g'}(\mathbf{K}_e; \mathbf{r}) \exp i[(\mathbf{K}_e + 2\pi\mathbf{g}') \cdot \mathbf{r}] \tag{5-184$a$}$$

$$E_{g,0'}(\mathbf{r}) = \left(\frac{\chi_g}{\chi_{-g}}\right)^{1/2} D_{g,0'}(\mathbf{K}_e; \mathbf{r}) \exp i[(\mathbf{K}_e + 2\pi\mathbf{g}) \cdot \mathbf{r}] \tag{5-184$b$}$$

The nomenclature of the suffixes $(0, g')$ and $(g, 0')$ is analogous to that used in Fig. 5-35. Again, the positional variations of $D_{0,g'}$ and $D_{g,0'}$ are much gentler than the exponential factors, the effects of crystal thickness and the air gap being involved in them. The Moiré fringes are caused by the interference of the exponential factors.

The spacing, therefore, is given by

$$\Lambda_M = \frac{1}{|\mathbf{g} - \mathbf{g}'|} \tag{5-185}$$

The fringes must be perpendicular to the difference vector $\Delta\mathbf{g} = \mathbf{g} - \mathbf{g}'$. (Dilatation Moiré, $\Delta\mathbf{g} \parallel \mathbf{g}$; misorientation Moiré, $\Delta\mathbf{g} \perp \mathbf{g}$.)[2]

If the crystal is simply displaced by $\mathbf{u}$, the intensity observed is homogeneous $[\Lambda_M \to \infty]$. Nevertheless, the effect of interference appears in the intensity owing to the phase factor $e^{-i\delta}[\delta = (\mathbf{g}' \cdot \mathbf{u})]$ in $(\chi_{g'}/\chi_{-g'})^{1/2}$. When it happens that $D_{0,g}$ and $D_{g,0'}$ are nearly equal and $\delta = \pi$, the $G$ wave will disappear. The situation is something like the superposition of two window blinds if one assumes that x rays are prohibited from passing through atomic arrays. Interpretations using this model are useful for neophytes but sometimes lead to a misunderstanding. Optically speaking, the crystal

---

[1] Here, deliberately, the factor $(\chi_g/\chi_{-g})^{1/2}$ is written out explicitly.
[2] No mention is made of the $O$ wave, but the spacing must be identical.

Moiré effect is due purely to the fact that crystals are phase gratings but not absorption gratings. All the fringe characteristics mentioned above can also be derived from a simple consideration of kinematical theory except for the intensity, which is given correctly by the dynamical theory ($D_{0,g'}$ and $D_{g,0'}$).

The spherical-wave theory can be constructed analogously to the theory presented in Section II B. The wave fields, then, must have the forms

$$\Phi_{0,g'}(\mathbf{r}) = \left(\frac{\chi_{g'}}{\chi_{-g'}}\right)^{1/2} B_{0,g'}(\mathbf{r}) \exp i[Kr + 2\pi(\mathbf{g'} \cdot \mathbf{r})] \quad (5\text{-}186a)$$

$$\Phi_{g,0'}(\mathbf{r}) = \left(\frac{\chi_{g}}{\chi_{-g}}\right)^{1/2} B_{g,g'}(\mathbf{r}) \exp i[Kr + 2\pi(\mathbf{g} \cdot \mathbf{r})] \quad (5\text{-}186b)$$

Now, the amplitudes $B_{0,g'}$ and $B_{g,0'}$ include fine fringes owing to the *Pendellösung* phenomena but having nothing to do with the lattice dilatation and misorientation. Thus, the fringe characteristics are analogous to the case of the plane-wave theory but, obviously, fringes appear only in the superposed regions of appreciable $B$'s, which are limited by the Borrmann fans in the component crystals, respectively. In summary, it must be emphasized that the characteristics of Moiré fringes, except the intensity, are relevant neither to the kinematical and dynamical considerations nor to the plane-wave and spherical-wave considerations.

The analysis presented here is easily applied to the interferometer. The relevant reflection vectors in the nonideal interferometer are given by

$$O \text{ beam:} \quad (0, \mathbf{g}_{M1}, -\mathbf{g}_A) \quad \text{and} \quad (\mathbf{g}_S, -\mathbf{g}_{M2}, 0)$$
$$G \text{ beam:} \quad (0, \mathbf{g}_{M1}, 0) \quad \text{and} \quad (\mathbf{g}_S, -\mathbf{g}_{M2}, \mathbf{g}_A)$$

The fringe spacing, therefore, must be

$$\Lambda_M = \frac{1}{|\mathbf{g}_{M1} + \mathbf{g}_{M2} - \mathbf{g}_A - \mathbf{g}_S|} \quad (5\text{-}187)$$

for both $O$ and $G$ beams.

In concluding this chapter, it is again noted that some of the topics requiring statistical treatments have been omitted. Similarly, thermal vibrations and their effect on the dynamical theory have not been considered here. The interested reader is referred to papers by Batterman (1961; 1962a and b), Batterman and Chipman (1962), Kambe (1965b), Khora, Kikuta, and Annaka (1965), Khora et al. (1966), Parathatharasy (1960), and Zener and Bilinsky (1936), listed in the bibliography of this chapter, as well as to papers quoted in Section I C 3 of Chap. 3 and Section I D 5 of Chap. 4.

# POWDER DIFFRACTOMETRY

## I. INTRODUCTION

*Powder diffractometry* is technologically a very important subject, because of its application in phase analysis and quality control. It also has important applications within physics, in the measurement of thermal expansion, in the study of order-disorder and other transitions, and in the study of crystal perfection (stacking faults, elastic strain, particle size). Some of the interesting problems in diffraction theory involved in these investigations are discussed in Chap. 2. This chapter omits further discussion of them, therefore, or of the mechanical side of diffractometer design.

The aim of powder diffractometry is accurate measurement of the positions of diffraction maxima, of the integrated intensities of diffraction maxima, of the significant broadening of diffraction maxima, and, at its most ambitious, of the detailed shape of diffraction maxima. This chapter attempts a brief description of the geometrical principles on which powder diffractometry is based (Section II), a more detailed study of the aberrations arising from the practical impossibility of achieving the ideal arrangement (Sections III and IV), a discussion of the physical aberrations arising from the nonexistence of single-wavelength radiation (Section V), and a description of the methods in use for eliminating as far as practicable the effect of the aberrations on the observed positions, breadths, and profiles of diffraction maxima (Section VI). The final section (VII) discusses sources of experimental error and the minimization of the part of the error arising from statistical effects.

## II.  GEOMETRY OF POWDER DIFFRACTOMETERS

### A.  Bragg-Brentano focusing

A properly oriented single crystal reflects x rays with considerable intensity, a proper orientation being such that the angle of incidence on a set of reflecting planes within the crystal is equal to the Bragg angle for that set of planes. In single-crystal diffractometers the orientation of the crystal is under the control of the experimenter, and the full volume of the specimen, aside from absorption and extinction effects, contributes to the measured intensity of diffraction. In powder diffractometers, on the other hand, the orientation of the crystals making up the specimen is not under the control of the experimenter, and only a small fraction of the volume of the specimen, the fraction consisting of the crystals that happen to be oriented at the proper angle, can diffract x rays from the desired set of reflecting planes. The problem of the geometrical design of a powder diffractometer is, therefore, to arrange a considerable volume of specimen so that rays diverging from a comparatively small source are diffracted so as to converge on a comparatively small detector. The source and detector are required to be comparatively small, as otherwise the angular resolution would be too small to be useful. Existing designs are all based on the theorem of Euclid that all angles in the same arc of a circle are equal.

In Fig. 6-1 the point $A$ represents an x-ray source of negligible dimensions, $O$ represents a negligibly small crystal, and $B$ represents a negligibly small detector. Suppose that it is desired to measure the diffraction from a set of planes of Bragg angle $\theta$. The crystal must then satisfy two conditions:

*1*  The normal to the set of planes must bisect the angle $AOB$.
*2*  The angle $AOB$ must be $\pi - 2\theta$.

The second condition is satisfied if $O$ lies anywhere on a circular arc passing through $A$ and $B$ and having a radius such that the angle $AOB$ has the required value. Since the angle subtended by $AB$ at the center of the circle is twice that subtended at $O$, the required radius is

$$F = K \csc 2\theta \qquad (6\text{-}1)$$

where $2K$ is the length of the chord $FD$. Notice that only the arc $AOB$ is effective; the arc $AO'B$ corresponds to a Bragg angle of $\frac{1}{2}\pi - \theta$. For fixed $A$ and $B$ the circular arc is not confined to the plane of the figure, since the angular condition is satisfied by all points on the surface of the torus obtained by rotating the arc $AOB$ about the chord $AB$.

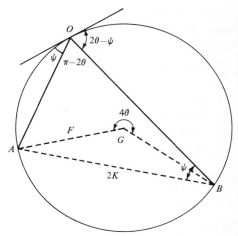

**FIGURE 6-1**

The condition for Bragg-Brentano focusing. The source of radiation is $A$, the
crystal is at $O$, and the detector is at $B$.
The angle between $AO$ produced and
$OB$ is $2\theta$, twice the Bragg angle.

Any particular crystallite in a polycrystalline specimen is unlikely to be oriented
so as to satisfy the first condition even if it is placed so as to satisfy the second. In
order to obtain adequate intensity, therefore, an extended specimen is used, so that
among the numerous crystals that it contains there are enough correctly oriented to
give adequate intensity of diffraction from the desired set of planes. It is fairly obvious
that it will be more efficient to extend it as much as is practicable along the surface
of the torus, since this will not spoil the "focusing" of rays from $A$ to $B$, and as little
as possible perpendicular to the surface of the torus. A displacement normal to the
torus produces a first-order aberration in the "focusing," whereas a displacement
tangential to the torus, thus departing only slightly from its surface, produces a
second-order aberration. Production of a specimen with a curved surface is often
impracticable, and it is then advantageous to make it in the form of a slab tangential
to the torus, rather than, for example, in the form of a solid cylinder (as in the Debye-
Scherrer camera). Ordinarily absorption will limit the effective thickness of the slab,
but if absorption is low the specimen will have to be thin in actuality.

It is obvious from Fig. 6-1 that the rays $AO$ and $OB$ will make equal angles
with the surface of the torus only when $O$ lies on the perpendicular bisector of $AB$,
in which case both angles are $\theta$. In general, if the ray $AO$ makes an angle $\psi$ with the
torus, the ray $OB$ will make an angle of $2\theta - \psi$.

## B.   Practical considerations

It is, of course, necessary to be able to vary the angle $\theta$, so that the intensity of
diffraction can be measured as a function of the Bragg angle. To achieve this at least
one of $A$, $O$, and $B$ must be able to move relative to the other two. (Motion of all

three can be ruled out in a discussion of principles, even in the unlikely event that it were convenient in practice, since the motions could be considered relative to a frame of reference stationary with respect to one of them.) In general it is convenient to keep the source of x rays fixed, since the high-tension cable and the water-cooling connections make it clumsy to move. The detector also needs electrical connections, but they are lighter and more flexible. For measurements on stable materials at room temperature there is no difficulty about moving the specimen, but if a special atmosphere or a controlled temperature is required it may become as difficult to move the specimen as to move the source. One is thus led to consider three design possibilities:

   *1*  The specimen alone moves.
   *2*  The detector alone moves.
   *3*  Both specimen and detector move.

In discussing these there are three practical points to be kept in mind:

   *1*  The simplest motion to achieve mechanically is rotation about an axis.
   *2*  The intensity of the output of an ordinary crystallographic x-ray tube changes rapidly with the angle at which the focal area is viewed (Parrish, quoted by Wilson, 1962*a*, p. 20), and the wavelength distribution is slightly dependent on this angle (Wilson, 1958).
   *3*  The sensitivity of the detector is a function of the path taken by the radiation through the sensitive volume (highly so for Geiger counters, less so for proportional, somewhat for scintillation, possibly negligibly so for solid-state counters).

Motion of the specimen so as to alter $\theta$ could be along almost any path except, of course, the circle $AOB$. The variation of x-ray output with angle of view, in addition to the restriction imposed by the tube window, means that the only practicable motion is along $AO$, thus maintaining a constant angle of view. This would have nothing to recommend it; the angular range is restricted, and position on the line does not bear a simple relation to the Bragg angle. For less directional sources (a radioactive material, for example) linear motion of a cylindrical ring of specimen parallel to and centered on $AB$ might be considered, though the focusing would not be good.

## C.  The Seemann-Bohlin arrangement

Motion of the detector alone is the basis of what is known as the Seemann-Bohlin arrangement (Wassermann and Wiewiorowsky, 1953). Rotation of $B$ about the center of the figure, $G$, can alter $\theta$ from $\frac{1}{2}\psi$ (when the detector would coincide with the specimen) to $\frac{1}{2}\pi$ (when the detector would coincide with the source). Mechanical interference, of course, reduces the range usable in practice. The angle $AGB$ gives $4\theta$ directly. This motion alters neither $\psi$, the angle at which the x rays are incident on the specimen at $O$, nor $\gamma$, the angle at which the specimen views the focal area of the x-ray tube, nor $F$, the radius of the circle $AOB$. It is thus worthwhile to prepare a specimen with this radius of curvature, since it can be used for all reflections.

Although the Seemann-Bohlin arrangement gives, ideally, perfect focusing in the equatorial plane for a single radius of curvature of the specimen, it does not do so in the axial direction. The perpendicular distance of $O$ from $AB$ increases from a very small value when $B$ is near $O$ to a maximum value of $OA$ and then decreases again. It is not practicable to vary the other radius of curvature of the specimen correspondingly, and the allowable axial divergence of the x rays is much less than the allowable equatorial divergence. In practice the focal area and the receiving slit in front of the detector are extended in the axial direction and Soller slits are used to limit the axial divergence.

If the efficiency of detection is a function of the path of the radiation through the counter, it is necessary to rotate the counter about the point $B$ as well as rotating the whole detector assembly about $G$. This subsidiary rotation is done in such a way that the path of the radiation through the counter, $OB$ produced, is always the same, whatever the position of $B$. This can be achieved in various ways; a mechanical linkage suitable for the purpose has been described by Parrish, Mack, and Vajda (1967), but other arrangements can be devised. Pike (1962) has suggested that the necessity for the subsidiary rotation could be avoided by the use of a scintillation counter. The receiving slit can be mounted in two ways: it can either simply rotate about $G$ with the arm carrying the detector assembly, or it can be attached to the counter so that it takes part in the subsidiary motion about $B$. The former arrangement gives a constant angular aperture, and thus reduces the rapid initial decrease of intensity with increasing Bragg angle. To avoid unwanted shadows in the usable range of Bragg angles the slit members must be appropriately staggered (Parrish, Mack, and Vajda) or beveled (Pike).

## D.   The symmetrical Bragg-Brentano arrangement

Motion of both specimen and detector offers several possibilities. That adopted by most manufacturers of diffractometers is to choose a convenient length for $AO$ and to rotate $B$ about $O$. The receiving slit is placed on the rotating arm so that $OB$ is equal to $AO$. The source and the effective position of the detector are thus symmetrically situated with respect to the specimen. As $B$ rotates about $O$ the distances $AO$ and $OB$ remain constant, but the radius of the circle $AOBO'$ contracts, being very large when $\theta$ is small and reducing to $\frac{1}{2}AO$ for $\theta$ approaching $\frac{1}{2}\pi$. Its radius is easily seen to be

$$F = \tfrac{1}{2}AO \csc \theta \qquad (6\text{-}2)$$

since for this arrangement $K$ in equation (6-1) is $AO \cos \theta$. In order to maintain the specimen tangent to the arc $AOB$ it must be rotated about $O$ through half the angle by which $B$ is rotated. This is often achieved by 2:1 reduction gearing, but other mechanical or electrical methods are possible.

The design just described is often called the *Bragg-Brentano arrangement*, since it was that of the first powder diffractometer (Bragg, 1921) and the properties of the focusing surface were discussed by Brentano (1917). Bragg-Brentano focusing, however, is just as basic for the Seemann-Bohlin arrangement. A modification of the symmetrical arrangement, in which the specimen is stationary and both source and detector move in mirror-image fashion with respect to it, has occasionally been used for the study of diffraction by liquids or by specimens too large to manipulate. Though the mechanical details differ, such an arrangement is geometrically the same as that of the usual type, and no special discussion of its aberrations is required.

## E.   Other arrangements without monochromator

A further arrangement involving motion of both the specimen and the detector is to have the specimen in the form of a ring concentric with $AB$, and to move both the ring and the detector along $AB$ in such a way that the ring remains tangent to the focusing circle and $B$ remains on the focusing circle. In the simplest case the ring will be a right circular cylinder halfway between the source and the detector, and the amount that it has to be moved is half the amount that the detector moves. Such a device can cover nearly the whole range of $\theta$ from 0 to $\frac{1}{2}\pi$ and is very close to one of the arrangements suggested by Brentano. For the reason already mentioned, extreme directionality, it is not suitable for x rays from a normal crystallographic tube, but might be useful with a radioactive source or an x-ray tube of the type used for the production of Kikuchi patterns. This arrangement would make effective use of a much greater fraction of the radiation, since, for a given degree of geometrical aberration, a tangent cylinder is capable of intercepting a much greater solid angle than is a tangent plane. Focusing is better and the angular range is greater than with the ring arrangement mentioned at the end of Section B.

## F.   Arrangements with monochromator

The wide range of wavelengths emitted by a crystallographic x-ray tube is a great disadvantage for most purposes, and reflection from a crystal may be used to reduce it. Among other advantages, such a monochromator greatly reduces the number of quanta reaching the detector, with a corresponding reduction of errors due to lost counts. The reflecting crystal may be placed between the x-ray source and the specimen (prereflection) or between the specimen and the detector (postreflection). From the point of view of diffraction geometry there is little to choose between the two arrangements: the situation is geometrically unaltered by reversal of the direction in which the rays travel. The following practical considerations favor postreflection.

*1* Fluorescence radiation from the specimen is prevented from reaching the detector, except for fluorescence from the target element if the specimen contains any of it.[1]

*2* Compton scattering from the specimen is prevented from reaching the detector (Ruland, 1964; Strong and Kaplow, 1966). The change of wavelength on Compton scattering changes with the angle of scattering, however, and at sufficiently low scattering angles will be less than the wavelength resolution of the monochromator. The lower limit for the elimination of Compton scattering can be estimated for any particular apparatus.

*3* It is usually easier to add a postreflection monochromator to an existing piece of equipment.

*4* If the monochromator crystal is one that deteriorates under the action of x rays (urea nitrate, pentaerythritol, mica) its life will be longer in the postreflection position.

The chief advantage of prereflection is the converse of (4):

*1* If the specimen is one that deteriorates under the action of x rays, it will last longer if the monochromator is in the prereflection position. This consideration is usually more important in single-crystal than in powder diffractometry.

The other advantage of prereflection is:

*2* The monochromator does not have to move with the detector arm, so that the overall mechanical design is simplified.

Whether used in the prereflection or postreflection position, the crystal may be plane (undistorted) or curved (bent in such a fashion as to influence the divergence of the beam).

**1.  *Plane-crystal arrangements*** Ideally, a crystal would reflect x rays only when the angle of incidence is exactly that given by the Bragg equation. Rays diverging from a point source and striking an extended plane crystal with reflecting planes parallel to its surface would mostly penetrate into it and ultimately either emerge from the back or be absorbed. Rays lying on the surface of a certain cone, however, would strike the surface of the crystal at the Bragg angle $\theta$ and be reflected along the surface

---

[1] Since the monochromator transmits the characteristic radiation from the target, the same radiation excited in the specimen by the incident continuous radiation is also transmitted by the postreflection crystal.

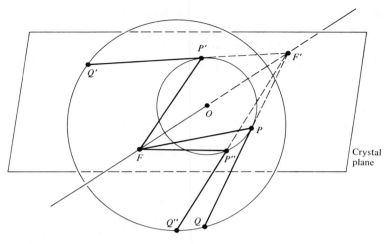

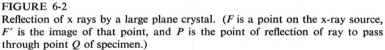

FIGURE 6-2
Reflection of x rays by a large plane crystal. (*F* is a point on the x-ray source, *F'* is the image of that point, and *P* is the point of reflection of ray to pass through point *Q* of specimen.)

of a cone that is the mirror image of the first (Fig. 6-2). It is easily seen that the common axis of the cones lies along a line passing through the source perpendicular to reflecting planes, and that their semiapical angles are $\frac{1}{2}\pi - \theta$. The apex of the first cone is, of course, the point on the source from which the rays diverge, *F* in the figure, and the apex of the second is its mirror image *F'*. A point such as *Q* in the specimen would ideally have only one ray passing through it, appearing to come from the point *F'* but actually traveling along the path *FPQ* with reflection at the point *P*. Actual x-ray sources are extended in space, so that points in the neighborhood of *F* will also be the apices of cones of diverging x rays. Most of these will miss the point *Q*, but it is clear that there will be points on the source displaced a little in the direction *OP* and rather more in the direction perpendicular to *FO* and *OP* that can give rise to rays passing through *Q* in directions slightly inclined to *PQ*. The inclination is mainly but not entirely out of the plane *FOPQ*. It is readily seen that the locus of such points is the intersection with the source of a cone having a semiapical angle of $\frac{1}{2}\pi - \theta$ and apex at the image of *Q* in the crystal surface. This intersection will be a circle only if the source is large, plane, and parallel to the crystal surface. Ordinarily it will be plane, but small in comparison with distances such as *F'Q*, and appreciably inclined to the crystal surface. The locus in question will then be one or two short arcs of a noncircular conic section. If the monochromator crystal is small the locus may be further curtailed by its dimensions. Assuming that the limitations are set by the dimensions of the focus, we may roughly characterize the radiation passing through $Q$ by saying that its axial angular divergence (that is, its angular divergence from the plane $FPQ$) is approximately $Z/L$, where $Z$ is the dimension of the source in the direction perpendicular to $FO$ and $OP$ and $L = F'Q = FP + PQ$ is the total distance traveled by the ray from the source to the specimen, and that its equatorial angular divergence (its divergence in the plane $FPQ$) is approximately $(Z/L)^2/2 \cos \theta$. The equatorial divergence is increased by the rocking curve of the monochromator crystal, but unless this curve is unusually wide or the Bragg angle of the monochromator unusually near $\frac{1}{2}\pi$ the axial divergence will be much larger than the equatorial.

In the preceding discussion any penetration of the x rays into the monochromator crystal has been neglected; the effect of penetration is to extend the point $F'$ into a line extending along $FF'$ for a short distance. This increases somewhat the area of the source contributing to rays through $Q$, but does not alter their angular divergence appreciably.

As explained in Section A, a powder diffractometer acts by concentrating a divergent bundle of x rays to an approximate focus. An arrangement in which the equatorial plane of a plane-crystal monochromator is parallel to the equatorial plane of the diffractometer will be very inefficient; since the equatorial divergence is small the benefit of the focusing is lost, and the large axial divergence must be reduced by Soller slits. The efficiency can be improved considerably by arranging the monochromator so that its equatorial plane is perpendicular to the equatorial plane of the diffractometer (Furnas, 1965). There is then considerable divergence in the equatorial plane of the diffractometer, so that the focusing is fully effective, and very little axial divergence. Soller slits are not required in the incident beam, and the limitation of axial divergence is a great advantage, particularly at low and high Bragg angles, where axial divergence is often the largest geometrical aberration (Section III B 7). The ideal focal spot for this application would be square with one edge parallel to the axial direction of the monochromator, but such foci are not normally available. For the ordinary rectangular spot the long dimension should be in the plane of incidence and the takeoff angle should be as large as possible.

The preceding paragraphs have been written as if for a prereflection monochromator. With minor alterations they apply also to a postreflection monochromator, the word "detector" or "receiving slit" being substituted for "source."

**2. Curved-crystal arrangements** When divergent x rays strike a plane crystal they are reflected only over a small range of angles in any plane such as $FPQ$ (Fig. 6-2), the range being of the order of the mosaic spread of the crystal. If the crystal is bent into a cylinder (not necessarily circular) with its generators perpendicular to this plane, the acceptable range of angles is increased by approximately the angle through which the crystal has been bent. The divergence of the reflected rays is increased or decreased, depending on the direction of bending, and by proper choice of the radius of curvature the rays may be brought to an approximate focus at any convenient point along *PQ*. The divergence in the direction perpendicular to the plane *FPQ* is not greatly altered by the bending; to produce focusing in this direction as well, the crystal would have to have a double curvature. If the crystal is merely bent the focusing is not accurate, the reflected rays forming a caustic and the image of the source having defects similar to those produced by spherical aberration in optics. These defects can be greatly reduced by first grinding the crystal to a certain radius and then bending to another; in the simplest case, in which the source and focus are to be equidistant from the monochromator crystal and focusing is in the plane *FPQ* only, the radius of grinding and the radius of bending are each double that of the circle through source, crystal, and focus, so that the front face of the crystal in its final state lines along this circle.

In the preceding description, reflection from the front face of a crystal has been assumed, and this is the usual situation in practice. For some purposes curved crystals are employed in transmission—probably more frequently than plane crystals in transmission. There are four distinguishable monochromator-diffractometer arrangements, as discussed by Lang (1956*a*). These are sketched in Fig. 6-3. If *F* is regarded as the source and *D* as the detector they represent postreflection, the arrangement envisaged by Lang, and, if *D* is regarded as the source and *F* as the detector, they represent prereflection. The arrangements are drawn and described for the symmetrical Bragg-Brentano diffractometer. There is no reason, however, why they should not be equally applicable for the asymmetrical (Seemann-Bohlin) arrangement.

In the arrangement shown in Fig. 6-3*a* the x rays are "reflected" from both the specimen and the monochromator; Lang calls this the reflection-reflection (RR) arrangement. The figure shows the specimen and the monochromator more or less parallel to one another, and this arrangement is probably convenient, as it makes it easy to avoid mechanical interference between parts of the apparatus when investigating high-angle reflections. There is, however, no reason why either the diffractometer half or the monochromator half of the figure should not be rotated by 180° about the line joining the center of the specimen to the center of the monochromator. These two possibilities correspond with the $(+1, -1)$ and $(+1, +1)$ arrangements in two-crystal spectrometry.

Figure 6-3*b* and *c* correspond to reflection from the specimen and transmission through the monochromator (called the RT arrangement by Lang) and transmission through the specimen and reflection by the monochromator (called the TR arrangement by Lang). Like the arrangement in Fig. 6-3*a*, these arrangements make it easier to reach high Bragg angles without mechanical interference between the x-ray tube and the detector housing.

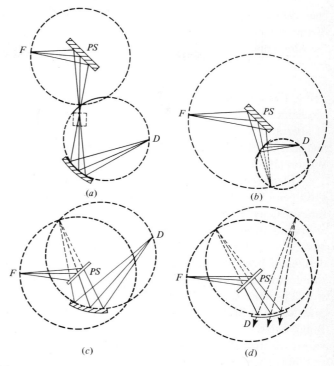

FIGURE 6-3
Four arrangements of powder specimen (*PS*) and curved-crystal diffractometer. For postreflection, *F* represents the x-ray source and *D* the detector; for pre-reflection, the meanings are reversed. (Modified, by permission, from *International tables for x-ray crystallography*, vol. **III**, p. 86.)

In the final arrangement, shown in Fig. 6-3*d*, both the specimen and the mono-chromator are used in transmission. No slits are necessary, and the arrangement can be made very compact. It is, however, limited by the necessity for a reasonably transparent specimen, and the divergence of the ultimate beam necessitates a large and uniformly responsive detector. The reverse arrangement, with a broad source and a slit-limited detector, appears to be advantageous for many purposes.

## III.  GEOMETRICAL ABERRATIONS IN THE GENERAL BRAGG-BRENTANO ARRANGEMENT

There is an extensive literature concerning the geometrical aberrations of the powder diffractometer. Those for the symmetrical Bragg-Brentano arrangement have been collected by Wilson (1963*a*). The theory for the Seemann-Bohlin arrangement has been developed by Kunze (1964*a*,*b*,*c*) and tested by Mack and Parrish (1967). Most of the theory has been developed to the second order of deviations from the ideal focusing condition, but Kunze's first paper (1964*a*) treats higher-order deviations in the equatorial plane. Gale's calculations (1963, 1968) relate primarily to Debye-Scherrer cameras, but are partially applicable to diffractometers. As Kunze has pointed out, the usual Bragg-Brentano arrangement is geometrically (though not mechanically) equivalent to a Seemann-Bohlin arrangement with $\psi = \theta$. Calculations of aberrations carried out for the more general case can thus be applied to the symmetrical case. The following treatment is based on that given in chaps. 2 to 4 of Wilson (1963*a*).[1]

Ideal focusing is possible only with a point source, a point detector, and a specimen of negligible thickness curved to fit the torus described above. Actual x-ray tubes usually have an approximately rectangular focus about 1 by 10 mm. Viewed at grazing incidence ($\sim 6°$) this can appear as a "point" source about 1 by 1 mm, or as a "line" source about 0.1 by 10 mm. The "point" source was used in some early designs of Bragg-Brentano diffractometer, but modern ones almost all use the "line" source with the line perpendicular to the plane of Fig. 6-1. A toroidal specimen would produce something approaching an image of the source, but it is not technically practicable to produce specimens with a double curvature varying appropriately with Bragg angle. The axial divergence is therefore restricted to a few degrees (1 to 4° in various instruments and applications) by Soller slits. These consist of a group of suitably spaced parallel metal foils (Soller, 1924), and intercept all rays diverging by more than $\frac{1}{2}$ to 2° from the equatorial plane. In effect, the source is broken up into a series of approximate point foci. The effective detector, from the geometrical point of view, is a receiving slit placed a short distance in front of the actual detector. Its dimensions are about the same as the effective size of the source, though in step-by-step scanning its width may be increased to equal that of the step. A second set of Soller slits in the diffracted beam may be used to restrict the axial divergence of rays reaching the receiving slit. In the Seemann-Bohlin arrangement it is relatively easy to make specimens curved to the correct radius in the equatorial plane, since this does not alter with Bragg angle. For the symmetrical Bragg-Brentano arrangement it is possible to make a flexible specimen whose curvature in this plane is continuously altered as the Bragg angle changes (Ogilvie, 1963), but ordinarily a flat specimen is used. In neither case, as already pointed out, is it possible to vary the second curvature, so that the specimen is a portion of a plane or cylinder with its generators parallel to the axial direction. We are thus led to consider a ray diverging from a

---

[1] *Note added in proof.* This chapter was written before the author became acquainted with the work of Gillham (1971). The results agree wherever they can readily be compared; Gillham's treatment of axial divergence is much more extensive. It should be noted that his sign convention for $\varepsilon$ is the opposite of that used here and in much other work.

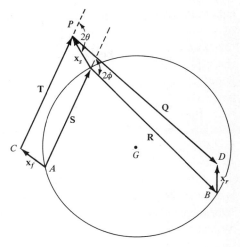

**FIGURE 6-4**
Geometrical aberrations in Bragg-Brentano focusing. The point $C$ on the extended sources is distant $\mathbf{x}_f$ from $A$, the point $P$ in the extended specimen is distant $\mathbf{x}_s$ from $O$, and point $D$ in the extended detector (receiving slit) is distant $\mathbf{x}_r$ from $B$. The diffractometer is set at $2\phi$, and the angle between $CP$ produced and $PD$ is $2\theta$.

point $C$ (Fig. 6-4) situated at a distance $\mathbf{x}_f$ from $A$, the centroid of the source, being diffracted at a point $P$ in the specimen distant $\mathbf{x}_s$ from $O$, the ideal position of the specimen,[1] and passing through the receiving slit at $D$, distant $\mathbf{x}_r$ from $B$, the point at which rays from $A$ would be focused by a specimen of ideal shape. For a particular Bragg angle $\theta$ radiation will be received over quite a range of diffractometer settings (angle between $\mathbf{S}$ and $\mathbf{R}$ in Fig. 6-4) $2\phi$.

## A. General calculation

Unit vectors for a coordinate system may conveniently be chosen with $\mathbf{i}$ and $\mathbf{j}$ in the equatorial plane, $\mathbf{i}$ radially outward and $\mathbf{j}$ tangential, and $\mathbf{k}$ axial. The $X$ direction is thus normal to the torus and $Y$ and $Z$ are tangential to it. Orthogonal components for the small vectors $\mathbf{x}_f$, $\mathbf{x}_s$, $\mathbf{x}_r$ may be chosen so that

$x_s$, $y_s$, $z_s$ are parallel to $\mathbf{i}$, $\mathbf{j}$, $\mathbf{k}$
$x_f$ and $y_f$ are equatorial and respectively parallel and perpendicular to $\mathbf{S}$, and $z_f$ is parallel to $\mathbf{k}$
$x_r$ and $y_r$ are equatorial and respectively parallel and perpendicular to $\mathbf{R}$, and $z_r$ is parallel to $\mathbf{k}$.

[1] In the symmetrical Bragg-Brentano arrangement, $O$ is the intersection of the mechanical axis of rotation with a perpendicular plane passing through the centroid of the source. In the Seemann-Bohlin arrangement with a plane specimen, it is the point at which the specimen is tangent to the focusing torus; with a curved specimen it is somewhat arbitrary, but may conveniently be the center of the tangent arc.

The basic vectors represented in Fig. 6-4 are thus

$$\mathbf{S} = S[\mathbf{i} \sin \psi + \mathbf{j} \cos \psi]$$
$$\mathbf{R} = R[-\mathbf{i} \sin (2\phi - \psi) + \mathbf{j} \cos (2\phi - \psi)]$$
$$\mathbf{x}_f = \mathbf{i}[x_f \sin \psi - y_f \cos \psi] + \mathbf{j}[x_f \cos \psi + y_f \sin \psi] + \mathbf{k}z_f$$
$$\mathbf{x}_s = \mathbf{i}x_s + \mathbf{j}y_s + \mathbf{k}z_s$$
$$\mathbf{x}_r = \mathbf{i}[-x_r \sin (2\phi - \psi) - y_r \cos (2\phi - \psi)]$$
$$\qquad + \mathbf{j}[x_r \cos (2\phi - \psi) - y_r \sin (2\phi - \psi)] + \mathbf{k}z_r$$
$$\mathbf{T} = \mathbf{S} + \mathbf{x}_s - \mathbf{x}_f$$
$$\mathbf{Q} = \mathbf{R} + \mathbf{x}_r - \mathbf{x}_s$$

(6-3)

The angle between $\mathbf{T}$ and $\mathbf{Q}$ is $2\theta$ and between $\mathbf{S}$ and $\mathbf{R}$ is $2\phi$. A slight modification of Wilson's equation (3-8) gives, to the second order of small quantities,

$$\cos 2\phi - \cos 2\theta$$
$$= \left[ \frac{\mathbf{S} \cdot (\mathbf{x}_s - \mathbf{x}_f)}{S^2} + \frac{\mathbf{R} \cdot (\mathbf{x}_r - \mathbf{x}_s)}{R^2} \right] \cos 2\phi - \frac{\mathbf{S} \cdot (\mathbf{x}_r - \mathbf{x}_s)}{RS} - \frac{\mathbf{R} \cdot (\mathbf{x}_s - \mathbf{x}_f)}{RS}$$
$$+ \frac{1}{2} \left[ \frac{|\mathbf{x}_s - \mathbf{x}_f|^2}{S^2} + \frac{|\mathbf{x}_r - \mathbf{x}_s|^2}{R^2} - \frac{3\{\mathbf{S} \cdot (\mathbf{x}_s - \mathbf{x}_f)\}^2}{S^4} \right.$$
$$\left. - \frac{3\{\mathbf{R} \cdot (\mathbf{x}_r - \mathbf{x}_s)\}^2}{R^4} - \frac{2\mathbf{S} \cdot (\mathbf{x}_s - \mathbf{x}_f)\mathbf{R} \cdot (\mathbf{x}_r - \mathbf{x}_s)}{S^2 R^2} \right] \cos 2\phi$$
$$- \frac{(\mathbf{x}_r - \mathbf{x}_s) \cdot (\mathbf{x}_s - \mathbf{x}_f)}{RS} + \left[ \frac{(\mathbf{x}_s - \mathbf{x}_f) \cdot \mathbf{S}}{S^2} + \frac{(\mathbf{x}_r - \mathbf{x}_s) \cdot \mathbf{R}}{R^2} \right]$$
$$\times \left[ \frac{(\mathbf{x}_s - \mathbf{x}_f) \cdot \mathbf{R}}{SR} + \frac{(\mathbf{x}_r - \mathbf{x}_s) \cdot \mathbf{S}}{SR} \right] + \cdots$$

(6-4)

This equation retains terms to the second order in all the small quantities. However, some of these are an order of magnitude smaller than others, and for many purposes the squares of the specimen penetration ($x_s$), of the width of the focal spot ($y_f$) and of the width of the receiving slit ($y_r$), as well as products like $x_s y_f$, can be neglected in comparison with the larger terms. Further, for any particular reflection, the difference between $\phi$ and $\theta$ will not be great, and we may put

$$2\theta = 2\phi + 2\varepsilon \qquad (6\text{-}5)$$

and, correct to the second order in small quantities,

$$2\varepsilon = \frac{\cos 2\phi - \cos 2\theta}{\sin 2\phi} \qquad (6\text{-}6)$$

However, before making these further approximations, we shall put the various constituents of equation (6-4) into scalar form. Unless otherwise noted the equations derived reduce to the corresponding ones in Wilson (1963a) for the symmetrical case $S = R$, $\phi = \psi$. For the sake of brevity we write

$$\delta \equiv \cos 2\phi - \cos 2\theta \qquad (6\text{-}7)$$

### 1 Terms involving the focal spot only

The terms with subscript $f$ only are easily picked out, and reduce to

$$\delta_f = S^{-2} \sin 2\phi [Sy_f + x_f y_f + \tfrac{1}{2}(y_f^2 + z_f^2) \cot 2\phi + \cdots] \qquad (6\text{-}8)$$

### 2 Terms involving the receiving slit only

The terms with subscript $r$ only are

$$\delta_r = R^{-2} \sin 2\phi [Ry_r - x_r y_r + \tfrac{1}{2}(y_r^2 + z_r^2) \cot 2\phi + \cdots] \qquad (6\text{-}9)$$

### 3 Terms involving the specimen only

Extraction of the terms with subscript $s$ only involves a great deal of tedious trigonometric reduction. The result is

$$
\begin{aligned}
\delta_s = {}& (RS)^{-1} x_s \sin 2\phi [R \cos \psi + S \cos (2\phi - \psi)] \\
& + \tfrac{1}{2}(RS)^{-2} x_s^2 [S^2 \cos 2\phi + 2RS \cos 4\phi + R^2 \cos 2\phi] \\
& + (RS)^{-2} x_s y_s \sin (2\psi - 2\phi)[S^2 + 2RS \cos 2\phi + R^2] \\
& + (RS)^{-1} y_s^2 \sin^2 2\phi \\
& + \tfrac{1}{2}(RS)^{-2} z_s^2 [S^2 \cos 2\phi + 2RS + R^2 \cos 2\phi] + \cdots
\end{aligned} \qquad (6\text{-}10)
$$

The quantity in brackets in the term linear in $x_s$ is easily seen to be $2K$ (Fig. 6-1).

For the symmetrical case the coefficient of $x_s^2$ differs from that in equation (3-10) of Wilson (1963a). Since $x_s$ is ordinarily one of the "small" components its square was neglected by Wilson, and the error does not affect any of his applications of the equation.

### 4 Cross terms

The terms involving both focal spot and receiving slit are

$$\delta_{fr} = (RS)^{-1}(y_f y_r \cos 2\phi + z_f z_r) \qquad (6\text{-}11)$$

Those involving both focal spot and specimen are

$$
\begin{aligned}
\delta_{fs} = S^{-2} \Big\{ & x_f x_s \cos \psi \sin 2\phi - y_f y_s \cos \psi \sin 2\phi \\
& - z_f z_s \left( \cos 2\phi + \frac{S}{R} \right) - x_f y_s \sin \psi \sin 2\phi \\
& + y_f x_s \left[ \cos (2\phi + \psi) + \frac{S}{R} \cos (2\phi - \psi) \cos 2\phi \right] \Big\}
\end{aligned} \qquad (6\text{-}12)
$$

The terms involving the receiving slit and the specimen, $\delta_{rs}$, differ from $\delta_{fs}$ in that $\psi$ is replaced by $2\phi - \psi$; $R$ and $S$, $f$ and $r$, are interchanged; and the sign of the first term is reversed. The "small" term in $y_r y_s$ has the sign opposite to that of the corresponding term in equation (3-15) of Wilson (1963a).

## B.   Specific aberrations

It is curious how little extra complexity is introduced by the lack of symmetry of the general Bragg-Brentano arrangement. Most of the equations (6-8) to (6-12) correspond term by term with the corresponding ones for the symmetrical case. The only notable exception is the middle term of equation (6-10), involving $\sin(2\psi - 2\phi)$, and even this term will be small in comparison with others in the same equation.

The displacement of the centroid of the line profile will be given by averaging $2\varepsilon$ (i.e., $\delta/\sin 2\phi$) over the focal spot, the specimen, and the receiving slit, taking into account any correlation between the variables introduced by Soller slits, specimen shape, etc. The additional variance of the line profile, the only simply calculated measure of the broadening, is the mean-square variation of $2\varepsilon$ from its mean

$$W = \langle (2\varepsilon)^2 \rangle - \langle 2\varepsilon \rangle^2 \qquad (6\text{-}13)$$

Its use is discussed in Section VI D 3 below. Calculation of the aberrational line profile arising from slit widths, specimen penetration, and similar purely geometrical effects is possible in principle, and has been carried out for several. The line profile arising from the variation of intensity with takeoff angle and with position on the focal spot can only be determined experimentally, and will depend somewhat on the actual experimental conditions.

Proper adjustment and calibration of a diffractometer aim at reducing the mean values and mean-square values of the various terms entering into equations (6-8) to (6-12). For the mean values this is partly a matter of definition and partly actual adjustment. Since S has been defined as originating at the centroid of the focal spot and R as passing through the centroid of the receiving slit, the mean values

$$\langle x_f \rangle = \langle y_f \rangle = \langle z_f \rangle = \langle y_r \rangle = \langle z_r \rangle = 0 \qquad (6\text{-}14)$$

To make $\langle x_r \rangle$ also equal to zero would involve adjusting the receiving slit along the counter arm, but it is difficult to do this accurately in practice, errors of a few tenths of a millimeter being possible. If the focal spot is symmetrical about the equatorial plane and an axial plane passing through its centroid the mean value of $\langle x_f y_f \rangle$ will also be zero, but if it is tilted with respect to these planes (inaccuracy of manufacture or adjustment of the tube), there will be an additional aberration arising from this product. In discussing aberrations, therefore, the possibility of correlations between the components of $\mathbf{x}_f$, $\mathbf{x}_s$, and $\mathbf{x}_r$ must be kept in mind. Such correlations may be introduced deliberately in order to reduce aberrations—a possibly perverse way of regarding the curvature of a specimen to make it fulfill the focusing condition over its entire length is to say that $x_s$ and $y_s$ are correlated so that

$$Rx_s \cos \psi + y_s{}^2 \sin 2\phi = 0 \qquad (6\text{-}15)$$

For an instrument of given radius the aberrations will tend to be larger for the Seemann-Bohlin than for the symmetrical arrangement. For the latter, $S$ and $R$ are constant and equal to the radius, say $R_0$, whereas for the former,

$$S = 2R_0 \sin \psi \qquad (6\text{-}16)$$

$$R = 2R_0 \sin (2\phi - \psi) \qquad (6\text{-}17)$$

The aberrations are proportional to inverse powers of $S$ and $R$; in the Seemann-Bohlin case $S$ will be constant at a value depending on the choice of $\psi$, but probably less than $R_0$, and $R$ will vary with $2\phi$, going to zero as $\phi$ approaches $\frac{1}{2}\psi$. There will thus be a range of $2\phi$ for which the Seemann-Bohlin aberrations containing $R$ become very large. This behavior has been confirmed in practice by the measurements of Mack and Parrish (1967); the apparent advantage of perfect equatorial focusing is frequently outweighed by the exaggeration of other aberrations.

If one neglects the same squares and products of "small" terms as Wilson did, one obtains for the total geometrical aberration

$$\sin 2\phi (2\varepsilon) \doteqdot \delta$$

$$= \sin 2\phi \left[ x_s [R^{-1} \cos (2\phi - \psi) + S^{-1} \cos \psi] + \frac{y_f}{S} + \frac{y_r}{R} \right.$$

$$+ y_s{}^2 (RS)^{-1} \sin 2\phi - x_f y_s S^{-2} \sin \psi - x_r y_s R^{-2} \sin (2\phi - \psi)$$

$$+ z_s{}^2 [\tfrac{1}{2} R^{-2} \cot 2\phi + (RS)^{-1} \csc 2\phi + \tfrac{1}{2} S^{-2} \cot 2\phi]$$

$$+ \tfrac{1}{2} z_f{}^2 S^{-2} \cot 2\phi + \tfrac{1}{2} z_r{}^2 R^{-2} \cot 2\phi + z_f z_r (RS)^{-1} \csc 2\phi$$

$$- z_f z_s [S^{-2} \cot 2\phi + (SR)^{-1} \csc 2\phi]$$

$$\left. - z_r z_s [R^{-2} \cot 2\phi + (SR)^{-1} \csc 2\phi] + \cdots \right] \qquad (6\text{-}18)$$

The first line contains the first-order terms in the "small" equatorial components, there being no terms that are first order in the "large" equatorial or axial components. The second line contains the second-order terms in the "large" equatorial components, and the remaining lines contain the second-order terms in the "large" axial components. The distinction between "small" and "large" is, however, dependent on the dimensions of the various apertures of the diffractometer and on the transparency of the specimen; with closely spaced Soller slits and a beryllium specimen $x_s$ might become as large as the effective value of the $z$'s. In other cases certain terms contained in equations (6-8) to (6-12) but neglected in equation (6-18) might be as important as some retained. In general, however, the design of commercial diffractometers is such that equation (6-18) is valid.

1.   *Finite width of source*   The width variable $y_f$ appears in three of the four terms of equation (6-8). The second of these, the cross term $x_f y_f$ containing also the extension of the focal spot along the direction of the incident beam, will ordinarily be a small fraction (perhaps 1%) of the first term, $Sy_f$. The third term, $\frac{1}{2}y_f^2 \cot 2\phi$, would be even smaller if it were not for the high value of $\cot 2\phi$ at very large or very small angles (over 89 or under 1° Bragg). In the usual angular range, therefore, the angular aberration arising from the dimension of the focal spot normal to the incident direction is

$$2\varepsilon = \frac{y_f}{S} \qquad (6\text{-}19)$$

The origin of the angle scale has been chosen at the centroid of the focal spot, so that the mean of $y_f$ is zero. Its mean-square value, however, is finite, so that at high and low angles there is a small centroid displacement given by

$$\langle 2\varepsilon \rangle = \tfrac{1}{2}\langle y_f^2 \rangle S^{-2} \cot 2\phi \qquad (6\text{-}20)$$

and at all angles there will be a contribution to the line breadth, amounting to an additional variance of

$$\langle (2\varepsilon)^2 \rangle - \langle 2\varepsilon \rangle^2 = \langle y_f^2 \rangle S^{-2} \qquad (6\text{-}21)$$

the small terms being neglected. The magnitude of $\langle y_f^2 \rangle$ will depend on the way in which the intensity of emission varies across the focal spot; to a first approximation the intensity may be regarded as uniform and

$$W_f = \frac{f_1^2}{12S^2} \qquad (6\text{-}22)$$

where $f_1$ is the projected width of the focal spot.

2.   *Finite width of receiving slit*   In the absence of any other source of line broadening the finite width of the receiving slit would produce a rectangular line profile, the range in $2\varepsilon$ being from $-\tfrac{1}{2}r_1/R$ to $+\tfrac{1}{2}r_1/R$, where $r_1$ is the projected slit width. The variance of the profile is

$$W_r = \frac{r_1^2}{12R^2} \qquad (6\text{-}23)$$

As already mentioned, this will become large for a Seemann-Bohlin arrangement when $R$ becomes small. Whether $r_1$ is constant or varies with $2\phi$ depends on the mechanical details (end of Section II C). If $r_1$ is large, it may be necessary to consider a displacement analogous to equation (6-20).

**3.** *Specimen displacement and transparency* If the specimen is displaced bodily by an amount $x_s$ the diffraction maximum is displaced by

$$2\varepsilon = x_s[S^{-1} \cos \psi + R^{-1} \cos (2\phi - \psi)] \qquad (6\text{-}24)$$

reducing to the familiar form

$$2\varepsilon = 2x_s R^{-1} \cos \phi \qquad (6\text{-}25)$$

in the symmetrical case. The resulting error in lattice spacing is easily dealt with by extrapolation methods, the extrapolation function being $\cos \phi \cot \phi$ for the symmetric case and $\cos^2 \phi / \sin \psi \sin (2\phi - \psi)$ for the Seemann-Bohlin case.

Specimen transparency gives the effect of a continuous series of displacements, weighted in accordance with the absorption experienced by the ray diffracted at a depth $x_s$ within the specimen. In reaching a depth $x_s$ the incident ray has to travel an actual distance $x_s \csc \psi$ and the emergent ray a distance $x_s \csc (2\phi - \psi)$ through the specimen. It is thus reduced in intensity by a factor

$$A = \exp \{- \mu x_s [\csc \psi + \csc (2\phi - \psi)]\} \qquad (6\text{-}26)$$

where $\mu$ is the effective linear absorption coefficient, compared with a ray diffracted at the surface. The additional displacement caused by specimen transparency is thus

$$\langle 2\varepsilon \rangle = \left[ \frac{\cos \psi}{S} + \frac{\cos (2\phi - \psi)}{R} \right] \langle x_s \rangle$$

$$= \left[ \frac{\cos \psi}{S} + \frac{\cos (2\phi - \psi)}{R} \right] \frac{\int_0^t x_s A \, dx_s}{\int_0^t A \, dx_s} \qquad (6\text{-}27)$$

where $t$ is the specimen thickness. The expression may be simplified on remembering that

$$S \sin (2\phi - \psi) = R \sin \psi \qquad (6\text{-}28)$$

so that

$$\langle 2\varepsilon \rangle = \frac{\sin 2\phi}{(R + S)\mu} \left[ 1 - \frac{\mu[(S + R)/R]t \csc \psi}{\exp \{\mu[(S + R)/R]t \csc \psi\} - 1} \right] \qquad (6\text{-}29)$$

This reduces to the familiar form for the symmetric case, from which it differs only by the substitution

$$\mu_{\text{eff}} = \frac{\mu(S + R)}{2R} \qquad (6\text{-}30)$$

for $\mu$ and $\psi$ for $\phi$ [but not $2\psi$ for $2\phi$]. The additional variance may be found similarly, and differs from equation (4-27) of Wilson (1963a) through the same substitutions. In the case, usual in practice, for which $\mu t \gg 1$, the displacement of the centroid reduces to

$$\langle 2\varepsilon \rangle = \frac{\sin 2\phi}{(R + S)\mu} \qquad (6\text{-}31)$$

and the variance to

$$W_\mu = \frac{\sin^2 2\phi}{(R + S)^2 \mu^2} \qquad (6\text{-}32)$$

which is just the square of the displacement of the centroid. For transparent spec-imens, with $\mu t$ not large compared with unity, the calculations become complicated. For the symmetric case they have been carried out by Milberg (1958) and Langford and Wilson (1962). There is no difficulty in principle in extending them to asymmetrical cases, either that of the Seemann-Bohlin arrangement or that, possible with some diffractometers, in which the specimen surface does not pass through the center of the volume potentially effective in diffracting. In the latter case the diffracted intensity can be expressed as a simple exponential factor multiplied by the sum of two functions of the type introduced by Langford and Wilson (Wilson, 1971a).

**4. Specimen tilt**   Specimen tilt, like specimen transparency, gives the effect of a continuous series of displacements. A tilt with respect to the axial direction has no great effect on the position of the centroid, but gives an additional mean-square broadening. If $k$ is the (tangent of the) angle of inclination, the value of $x_s$ corre-sponding to the front face of the specimen becomes

$$x_s = kz_s \qquad (6\text{-}33)$$

instead of zero, giving an aberration of

$$2\varepsilon = kz_s\{R^{-1} \cos (2\phi - \psi) + S^{-1} \cos \psi\} \qquad (6\text{-}34)$$

If the illumination of the specimen is symmetrical about its equator the average value of (6-34) is zero, but its variance is

$$W = \tfrac{1}{3}k^2 h^2\{R^{-1} \cos (2\phi - \psi) + S^{-1} \cos \psi\}^2 \qquad (6\text{-}35)$$

where $2h$ is the axial extension of the specimen.

A tilt of the specimen in the equatorial plane, such as might result from a mis-setting of the 2:1 ratio in the symmetrical arrangement or inexact adjustment in the Seemann-Bohlin arrangement, makes $x_s$ depend linearly on $y_s$, so that

$$x_s = ky_s \qquad (6\text{-}36)$$

where $k$ is the (tangent of the) angle of inclination. The aberration is thus

$$2\varepsilon = ky_s\{R^{-1} \cos (2\phi - \psi) + S^{-1} \cos \psi\} \qquad (6\text{-}37)$$

The mean value of this will be zero only if the illuminated area of the specimen has been arranged so that the centroid of illumination is at $y_s = 0$. This is rather difficult to ensure in practice; though the emission from an x-ray tube is reasonably uniform in the axial direction and symmetric about the equator it varies markedly with takeoff angle and hence with $y_s$, so that the centroid of the illuminated area does not coincide with its geometrical center. The displacement of the peak of the diffraction profile is, however, generally greater than the displacement of the centroid of the diffraction profile from this cause. The mean-square broadening due to tilt in the equatorial plane could only be evaluated if the variation of intensity along the illuminated area were known, but in order of magnitude it will be given by

$$W = \tfrac{1}{3}k^2A^2[R^{-1} \cos (2\phi - \psi) + S^{-1} \cos \psi]^2 \qquad (6\text{-}38)$$

where $2A$ is the illuminated length of the specimen.

A tilt in the equatorial plane alters the angle of incidence from $\psi$ to $\psi - k$, and thus changes the effect of absorption on the integrated intensity. This is not significant in consideration of displacement and broadening, but may be in accurate intensity comparisons (Section VII B).

**5. Equatorial divergence**  The pure equatorial divergence term in equation (6-18) is

$$(2\varepsilon) \sin 2\phi \doteqdot \delta$$
$$= (RS)^{-1}y_s^2 \sin^2 2\phi \qquad (6\text{-}39)$$

As indicated in equation (6-15), in the Seemann-Bohlin arrangement it is not of significance, as the curvature of the specimen is such that it is canceled by the term in $x_s$. For the symmetrical arrangement it has been extensively treated (Wilson, 1950, 1963a), and there seems no point in repeating the calculations here. It may perhaps be worthwhile to mention briefly its effect in the asymmetrical arrangement if, for any reason, a flat instead of a properly curved specimen is used. This differs from the symmetrical case only through the occurrence of the factor $(RS)^{-1}$ instead of $R_0^{-2}$, so that the profiles of individual lines are of the same form, though the broadening and displacement differ in an obvious fashion in their dependence on $2\phi$. The combination of flat-specimen broadening and transparency broadening will also give profiles of the same general form as in the symmetric case, with the additional complication that the effective absorption coefficient is a function of $2\phi$ through equation (6-30).

**6. Equatorial cross terms**  The term involving the equatorial extensions of the focal spot and specimen

$$2\varepsilon = -S^{-2}x_f y_s \sin \psi \qquad (6\text{-}40)$$

can give rise to very little displacement, since there is little correlation between $x_f$ and $y_s$, the axes have been chosen so that $\langle x_f \rangle$ is zero, and the mean value of $\langle y_s \rangle$, though not accurately zero, cannot be large. There will, however, be an additional broadening, of variance

$$W = S^{-4}\langle x_f^2 \rangle \langle y_s^2 \rangle \sin^2 \psi \qquad (6\text{-}41)$$
$$= \tfrac{1}{36}S^{-4}r_1^2 A^2 \sin^2 \psi \qquad (6\text{-}42)$$

where $r_1$ is the actual (not projected) width of the focal spot and $2A$ is the illuminated length of the specimen. The line profile will depend in a complicated fashion on the variation of intensity across the focal spot and on the angle of view.

The term involving the equatorial extension of the specimen and the error in positioning the receiving slit

$$2\varepsilon = -R^{-2}x_r y_s \sin(2\phi - \psi) \qquad (6\text{-}43)$$

will produce an appreciable displacement if (1) there is a significant error $x_r$ in adjusting the receiving slit along the detector arm, and (2) the centroid of the illuminated area is not placed so that $\langle y_s \rangle$ is zero. If $x_r$ is not zero there will be an additional broadening, of mean-square value

$$W = \tfrac{1}{3}R^{-4}x_r^2 A^2 \sin^2(2\phi - \psi) \qquad (6\text{-}44)$$

For the symmetric case both equation (6-42) and equation (6-44) give broadenings varying with Bragg angle, but for the Seemann-Bohlin case only the latter varies.

**7.   *Axial divergence***   The terms in equation (6-18) involving the axial dimensions of the diffractometer can be put into various forms by the use of the identities

$$\cot 2\phi + \csc 2\phi = \cot \phi \qquad (6\text{-}45)$$

and

$$S \sin(2\phi - \psi) = R \sin \psi \qquad (6\text{-}46)$$

In terms of the angles made by the incident and diffracted rays with the equatorial plane,

$$\delta \equiv \frac{z_s - z_f}{S} \qquad (6\text{-}47)$$

and

$$\eta \equiv \frac{z_r - z_s}{R} \qquad (6\text{-}48)$$

it takes the classical form given by Eastabrook (1952)

$$2\varepsilon = \tfrac{1}{2}(\delta^2 + \eta^2) \cot 2\phi - \delta\eta \csc 2\phi \qquad (6\text{-}49)$$

The displacement of the centroid of the diffraction maximum is thus, in a notation similar to that used by previous authors,

$$\langle 2\varepsilon \rangle = Q_1' \cot 2\phi + Q_2' \csc 2\phi \qquad (6\text{-}50)$$

where

$$Q_1' = \tfrac{1}{2}\langle \delta^2 + \eta^2 \rangle$$
$$Q_2' = -\langle \delta\eta \rangle \qquad (6\text{-}51)$$

and the mean-square broadening is

$$W = A' \cot^2 2\phi + B' \cot 2\phi \csc 2\phi + C' \csc^2 2\phi \tag{6-52}$$

where

$$
\begin{aligned}
A' &= \tfrac{1}{4}[\langle(\delta^2 + \eta^2)^2\rangle - \langle\delta^2 + \eta^2\rangle^2] \\
B' &= \langle\delta\eta(\delta^2 + \eta^2)\rangle + \langle\delta\eta\rangle\langle\delta^2 + \eta^2\rangle \\
C' &= \langle\delta^2\eta^2\rangle - \langle\delta\eta\rangle^2
\end{aligned}
\tag{6-53}
$$

The angles $\delta$ and $\eta$ are correlated via the apertures of the Soller slits (if used) and the occurrence of $z_s$ in both, so that the evaluation of the various averages is a matter of rather tedious, though not difficult, integration. In the symmetric case $R$ and $S$ are constant and equal to each other, so that the various coefficients are independent of $\phi$. Eastabrook (1952), Pike (1957, 1959a), and Langford (1962) have examined this case in considerable detail, and tabulate the coefficients in terms of two parameters, the aperture $\Delta$ of the Soller slits,

$$\Delta = \frac{\text{separation of the foils}}{\text{length of the foils}} \tag{6-54}$$

and an unnamed parameter

$$q = \frac{R\Delta}{h} \tag{6-55}$$

where $R$ is the radius of the diffractometer and $2h$ is the axial dimension of the source, specimen, and receiving slit (all taken as equal, as is approximately true for most diffractometers of this type). Their results are summarized by Wilson (1963a, pp. 40–45).

In the Seemann-Bohlin arrangement $R$ is a function of $\phi$, and there seems to be no easy way of applying the results of the calculations for the symmetrical arrangement to the unsymmetrical. A few special cases[1] are readily treated.

### 1 No Soller slits

If no Soller slits are used, there is no correlation between the various $z$'s in equation (6–18); each has the mean and mean-cube value zero, mean-square value $\tfrac{1}{3}h^2$, and mean fourth power $\tfrac{1}{5}h^4$. One has, thus,

$$\langle 2\varepsilon\rangle = \tfrac{1}{3}h^2[(S^{-2} + R^{-2})\cot 2\phi + (RS)^{-1}\csc 2\phi] \tag{6-56}$$

$$
\begin{aligned}
W = \frac{h^4}{45}\big[&(7S^{-4} + 2S^{-2}R^{-2} + 7R^{-4})\cot^2 2\phi \\
&+ 14(RS)^{-1}(S^{-2} + R^{-2})\cot 2\phi \csc 2\phi \\
&+ 19S^{-2}R^{-2}\csc^2 2\phi\big]
\end{aligned}
\tag{6-57}
$$

---

[1] See, however, footnote on p. 445.

which reduce to the familiar form for $S = R$. With a little extra trouble this could be extended to the case of source, specimen, and receiving slits having unequal heights $2h_f$, $2h_s$, $2h_r$.

### 2   One set of Soller slits (narrow, or extended source or receiving slit)

Soller slits intercept entirely any rays for which $|\delta|$ (or $|\eta|$) is greater than $\Delta$, and for rays of lesser inclination they intercept a fraction $(1 - |\delta|/\Delta)$ or $(1 - |\eta|/\Delta)$. In addition, if $h_f$ is not large compared with $h_s$, there are end effects, since a point at the top (or bottom) of the specimen can "see" the focal spot only over a length of approximately $S\Delta$ below (above it), whereas a point near the middle can "see" the focal spot over about twice this length, and is thus illuminated about twice as brightly. If these end effects are neglected, which is legitimate if $\Delta$ is small or if $h_f$ and the axial extent of the Soller slits are sufficiently greater than $h_s$, one can write the axial terms of equation (6-18) as

$$2\varepsilon = \tfrac{1}{2}\delta^2 \cot 2\phi + \tfrac{1}{2}R^{-2}(z_r - z_s)^2 \cot 2\phi - \delta R^{-1}(z_r - z_s) \csc 2\phi \qquad (6\text{-}58)$$

with

$$\langle \delta^2 \rangle = \tfrac{1}{6}\Delta^2$$

$$\langle \delta z_r \rangle = \langle \delta z_s \rangle = \langle z_r z_s \rangle = 0 \qquad (6\text{-}59)$$

so that

$$\langle 2\varepsilon \rangle = \tfrac{1}{12}\Delta^2 \cot 2\phi + \frac{h^2}{3R^2} \cot 2\phi \qquad (6\text{-}60)$$

$$W = \left( \frac{7\Delta^4}{720} + \frac{7h^4}{45R^4} \right) \cot^2 2\phi + \frac{\Delta^2 h^2}{9R^2} \csc^2 2\phi \qquad (6\text{-}61)$$

If the Soller slits are in the diffracted beam instead of in the incident beam, $R$ will be replaced by $S$.

### 3   Two sets of Soller slits (narrow, or extended source and receiving slit)

From equations (6-49) and (6-59),

$$\langle 2\varepsilon \rangle = \tfrac{1}{6}\Delta^2 \cot 2\phi \qquad (6\text{-}62)$$

$$W = \frac{\Delta^4}{360} (10 + 17 \cot^2 2\phi) \qquad (6\text{-}63)$$

just as for the symmetrical case. The asymmetrical case with two sets of Soller slits and no restriction of aperture does not appear to have been examined in detail. The calculations would have to be expressed in terms of two parameters

$$q_f = \frac{S\Delta}{h}$$

$$q_r = \frac{R\Delta}{h} \qquad (6\text{-}64)$$

instead of the single one sufficing for the symmetrical case.

# IV.  GEOMETRICAL ABERRATIONS IN MONOCHROMATOR ARRANGEMENTS

Very little has been written about the detailed geometrical aberrations in arrangements involving a monochromator, the main paper being that of Lang (1956b).

## A.  Plane-crystal monochromators

A plane-crystal monochromator reduces the divergence in its own plane of reflection to little more than its own rocking curve (Section II F 1).  This plane can either coincide with the equatorial plane of the diffractometer, in which case the equatorial divergence (Section III B 5) is severely restricted, or be perpendicular to it, in which case the axial divergence of the incident beam is restricted.  Restriction of equatorial divergence is of no advantage whatever for the Seemann-Bohlin arrangement, and of very little advantage in the symmetrical arrangement.  Restriction of axial divergence, however, is a considerable advantage for both, since this reduces an aberration that is very large at high and low angles.  The detailed effect will depend on the rocking curve of the monochromator crystal, but one might expect it to approximate to that of using a set of very narrow Soller slits in the incident beam.  With no Soller slits in the diffracted beam the displacement of the centroid of the diffraction maximum would then be

$$\langle 2\varepsilon \rangle = \tfrac{1}{2}\langle \delta^2 \rangle \cot 2\phi + \frac{h^2}{3R^2} \cot 2\phi \qquad (6\text{-}65)$$

where $\langle \delta^2 \rangle$ is the mean-square mosaic spread of the monochromator, and the mean-square broadening would be

$$W = \left[ \tfrac{1}{4}(\langle \delta^4 \rangle - \langle \delta^2 \rangle^2) + \frac{7h^4}{45R^4} \right] \cot^2 2\phi + \left( 2\langle \delta^2 \rangle \frac{h^2}{3R^2} \right) \csc^2 2\phi \qquad (6\text{-}66)$$

These expressions would be somewhat modified if the monochromator did not illuminate the full height $2h$ of the specimen.  Narrow Soller slits of aperture $\Delta$ in the diffracted beam would reduce these to

$$\langle 2\varepsilon \rangle = \tfrac{1}{2}[\langle \delta^2 \rangle + \tfrac{1}{6}\Delta^2] \cot 2\phi \qquad (6\text{-}67)$$

and

$$W = \left[ \tfrac{1}{4}(\langle \delta^4 \rangle - \langle \delta^2 \rangle^2) + \frac{7\Delta^4}{720} \right] \cot^2 2\phi + \frac{\langle \delta^2 \rangle \Delta^2}{6} \csc^2 2\phi \qquad (6\text{-}68)$$

## B.  Curved-crystal monochromators in reflection

Lang (1956b) considers three effects: deviation of a Johann-type crystal from the focusing surface, penetration of the beam into the crystal, and axial divergence.  The first two of these correspond closely to flat-specimen error (Section III B 5) and specimen transparency (Section III B 3) in the diffractometer itself.  In order of magnitude the Johann aberration gives an additional broadening of $\frac{1}{8}\alpha^2 \cot \phi$ radians, where $\alpha$ is the total equatorial divergence of the beam falling on the mono-chromator. This aberration can, of course, be avoided by using a Johannson-type monochromator. If the effective absorption coefficient is $\mu_{\text{eff}}$ after bending (which would affect extinction effects, if any), the extra broadening from crystal transparency would be approximately $\sin 2\phi/(R_m + S_m)\mu_{\text{eff}}$, where the subscript refers to the distances $R$ and $S$ for the monochromator. Lang does not consider the axial divergence to be significant; probably its effect is to alter the value of $R$ (if the monochromator is in the diffracted beam) or $S$ (if the monochromator is in the incident beam) in the expressions discussed in Section III B 7.

## C.   Curved-crystal monochromators in transmission

The aberrations are less important for transmission monochromators. The Johann or simple Cauchois aberration gives an additional broadening of order of magnitude $\frac{1}{8}\alpha^2 \tan \phi$, which is a constant fraction of the natural line width, instead of increasing at low angles. Broadening due to the finite thickness of the plate is compensated to a greater or less extent by the change in lattice parameter (and hence Bragg angle) caused by the bending; compensation would be exact for perfectly elastic crystals in symmetric transmission with the neutral surface on the focusing circle. Lang does not mention axial divergence in connection with transmission monochromators, but its effect is probably the same as for reflection monochromators.

## D.   Effect on wavelength distribution

The use of a monochromator produces a change in the distribution of intensity as a function of wavelength—which is, after all, its purpose. In studies of diffraction broadening there may be associated disadvantages: the wavelength distribution must be measured anew after each adjustment of the monochromator, since the effect is not easily predictable and may not be very reproducible. Estimates of the change in line position due to the use of a monochromator are, therefore, hardly worth making.

## V.   PHYSICAL ABERRATIONS IN POWDER DIFFRACTOMETRY

### A.   The effective spectrum

The geometrical aberrations of a diffractometer arise, as has just been discussed, because of the finite dimensions of the apparatus and specimen. The physical aberra-tions arise because x rays are not of a single wavelength. The radiation normally used in crystallographic applications is the $K\alpha$ multiplet of one of the transition metals;

copper has the best physical properties as a target material, but other targets are used to provide other characteristic wavelengths when necessary to avoid excessive absorption (as would arise, for example, if copper radiation were used with a specimen containing a high proportion of iron) or to obtain greater resolution of complex patterns. As is explained in detail in Chap. 1, the $K\alpha$ multiplet consists chiefly of the $\alpha_1\alpha_2$ doublet plus an unresolved satellite group of four or five lines. The satellite group is only about 1% of the intensity of the doublet, but it is sufficient to cause difficulties in line-profile analysis. The $K\beta$ line is sufficiently far removed in wavelength from the $K\alpha$ multiplet to be easily removed by means of a filter or by pre-reflection by a crystal monochromator. Ordinarily no attempt is made to reduce the wavelength spread of the $K\alpha$ multiplet. There are, however, several possibilities that have been or might be used. (1) A monochromator with rather better resolution than those normally used can pick out the $K\alpha_1$ component, suppressing the $K\alpha_2$ and satellite components almost entirely. (2) A monochromator with still better resolution can select a small wavelength range around the peak of the $K\alpha_1$ component. A conventional monochromator of this type has been described by Barth (1960), and the recent development of apparatus cut from large single crystals (Chap. 7, Section IV B 5) offers the possibility of even better ones. (3) Gamma rays in general exhibit a much smaller wavelength spread than x rays, and the wavelength spread of the "recoilless" Mössbauer gamma rays is very small indeed. Gamma-ray sources emitting radiation in the crystallographic wavelength region would thus be ideal for the diffractometric study of line broadening. They have at least two disadvantages: the intensity is small compared with that of an x-ray tube, and there is usually a good deal of short-wavelength radiation that gives rise to shielding difficulties.

## B.  Refraction

The wavelength spread is the source of the aberrations of dispersion and response variation, but even if the incident radiation were strictly monochromatic there would be a change of wavelength on entering the specimen, resulting in a change of direction unless the surface happened to be perpendicular to the incident ray at the point of entry. This refraction of the incident rays leads to a broadening of the diffraction maximum given by a powder specimen (Wilson, 1940, 1962a) and a pseudodisplacement. The pseudodisplacement arises simply because the crystallographer calculates the Bragg angle from the wavelength of the x rays in vacuo, whereas they are actually diffracted inside the specimen, where the wavelength is slightly different. The broadening is, however, real, and its magnitude can be estimated if assumptions are made about the state of the specimen. Wilson (1962a) suggests that the mean-square breadth in $2\theta$ is approximately

$$W_{2\theta} = \frac{\delta^2}{4\mu p} \left[ -6 \log_e (\tfrac{1}{2}\delta) + 25 \right] \qquad (6\text{-}69)$$

where $p$ is an appropriate mean particle size, $\mu$ is the linear absorption coefficient of the specimen, and $\delta$ is the amount by which the refractive index of the material differs from unity. The derivation makes the assumption that the powder particles are separated by voids; in a specimen made with some binder filling the space between the particles $\delta$ would be the difference in refractive index between the material and the binder and the absorption coefficient would need some reinterpretation. For a solid polycrystalline specimen with a smooth surface, such as a polished metal block, the broadening would vanish.

## C.   Dispersion

The origin of the refraction aberration is easy to understand and, though it is logically a physical aberration, it behaves very much like a geometrical one. Dispersion as an aberration arises from the nonlinearity of Bragg's law. Each component of the incident wavelength distribution is diffracted at the appropriate Bragg angle, with the result that, even in the absence of appreciable geometrical aberration and specimen imperfection, radiation is detected over a range of diffractometer settings $2\phi$. If Bragg's law were a linear relationship between wavelength and angle this would be no more of a nuisance than, say, the finite size of the focal spot, since all diffraction maxima would be affected to the same extent. In reality, however, the long-wavelength end of the intensity distribution is stretched over a greater angular range than is the same wavelength range at the short-wavelength end, with the results that (1) the centroid of the angular intensity distribution does not coincide with the Bragg angle corresponding to the centroid of the wavelength distribution, (2) the variance of the angular intensity distribution differs from that calculated in a simple-minded fashion from the variance of the wavelength distribution, the difference depending, to the first approximation, on the third moment of the wavelength intensity distribution, and (3) these effects increase rapidly with increasing mean Bragg angle. First approximations to the difference in centroid position and variance are fairly easily obtained (Wilson, 1965a):

$$\langle 2\theta \rangle - 2\theta_0 = \frac{W}{\lambda_0{}^2} \tan^3 \theta_0 + \cdots \qquad (6\text{-}70)$$

and

$$W_{2\theta} = \frac{4W}{\lambda_0{}^2} \tan^2 \theta_0 + \frac{4\mu_3}{\lambda_0{}^3} \tan^4 \theta_0 + \cdots \qquad (6\text{-}71)$$

where $\lambda_0$ is the centroid wavelength of the distribution, $\theta_0$ is the Bragg angle corresponding to $\lambda_0$, $W$ is the variance of the wavelength distribution, $\mu_3$ is the third moment of the wavelength distribution, $\langle 2\theta \rangle$ is the centroid, and $W_{2\theta}$ is the variance of the angular intensity distribution. Such corrections may be adequate if the Bragg angle is not too large, but they approach infinity very rapidly at high Bragg angles, and in any case the moments of the intensity distributions are not well defined. These difficulties are avoided, at least partially, if the intensity measurements are made or plotted as a function of $\sin \theta$ instead of as a function of $\theta$ (Lang, 1956a; Mitchell and de Wolff, 1967). Routine use of replotting methods would require the reworking of the geometrical aberrations in terms of $\Delta(\sin \theta)$ or of $\Delta(2 \sin \theta/\lambda)$ instead of in terms of $\Delta(2\theta)$, but this involves no difficulty of principle.

## D. Response variations

Dispersion is a distortion of the abscissa of the intensity distribution; response variations are distortions of the ordinate. The most obvious one is perhaps a variation of the sensitivity of the detector with wavelength, so that the quanta with the longer wavelengths, say, are recorded with greater efficiency than those with the shorter wavelengths. This will shift the centroid of the intensity distribution actually recorded to a longer wavelength than that of the distribution incident on the detector. Other distortions of the ordinate are the Lorentz factor; the polarization factor; any other trigonometric factors; absorption in the tube windows, the air path, and the beta filter; asymmetry of the window of the pulse-height analyzer; and any variation with wavelength of the efficiency of diffraction by the specimen. These response variations affect the positions of the peaks as well as of the centroids, but their effects on the peaks are considerably smaller.

It is easy to estimate the effect of response variations on measures derived from the wavelength distribution of intensity. If the incident intensity distribution is $I(\lambda)\, d\lambda$ and the response modifies this to $I(\lambda)R(\lambda)\, d\lambda$, the centroid wavelength, for example, becomes

$$\langle \lambda \rangle = \frac{\int \lambda I(\lambda)R(\lambda)\, d\lambda}{\int I(\lambda)R(\lambda)\, d\lambda} \qquad (6\text{-}72)$$

Near $\lambda_0$, the incident centroid wavelength, $R(\lambda)$ can be expanded as a series in powers of $(\lambda - \lambda_0)$, giving

$$\langle \lambda \rangle = \lambda_0 + \frac{WR'(\lambda_0)}{R(\lambda_0)} + \frac{R''(\lambda_0)}{R(\lambda_0)}\left[\tfrac{1}{2}\mu_3 - \frac{W^2 R'(\lambda_0)}{R(\lambda_0)}\right] + \cdots \qquad (6\text{-}73)$$

where, as before, $W$ is the variance of the wavelength distribution and $\mu_3$ is its third moment. If the response variation is one that expresses itself more naturally in terms of the Bragg angle (such as, for example, the polarization factor), a slight modification of the argument (Wilson, 1963a, p. 60) shows that the angular change in the centroid is

$$\Delta(2\theta) = 2\frac{WR'_\theta}{\lambda^2 R}\tan^2 \theta + \cdots \qquad (6\text{-}74)$$

where $R'_\theta$ indicates the derivative of the response function with respect to $\theta$ at the centroid angle. The effect of response variations on the variance of the intensity distribution, on the peak wavelength, etc., can be evaluated similarly. The change in peak wavelength, for example, is

$$\Delta\lambda_m = -\frac{I(\lambda_m)R'(\lambda_m)}{I''(\lambda_m)R(\lambda_m)} + \cdots \qquad (6\text{-}75)$$

where $\lambda_m$ is the peak wavelength. The question has been treated in some detail by Wilson (1965$a$), whose table is reproduced with some modifications on page 469. It should be noted that in general the effects of physical aberrations on the peak are smaller than the effects on the centroid; this is particularly true of the combined effect of dispersion and Lorentz factor.

Gillham and King (1972) have investigated the effect of dispersion on the centroid and peak positions of the high-angle lines ($2\theta > 163°$) given by silver-zinc alloys of suitable composition with Cu $K\alpha$ radiation. The results are in agreement with the theoretical predictions.

**1.  *Polarization factor*** The intensity of reflection of an unpolarized beam of x rays by an ideally imperfect crystal contains the factor $\frac{1}{2}(1 + \cos^2 2\theta)$. If a mono-chromator is used the relative intensities of the components in and perpendicular to the equatorial plane of the diffractometer are altered, so that the polarization factor becomes

$$R(\lambda) = \frac{1 + B\cos^2 2\theta}{1 + B} \qquad (6\text{-}76)$$

where $\lambda$ and $\theta$ are related by Bragg's law. The value of $B$ depends on whether the monochromator crystal is perfect, ideally imperfect, or intermediate, and on whether the equatorial planes of the monochromator and diffractometer are parallel or perpendicular. For the ideal cases the values of $B$ are:

| Monochromator crystal | Equatorial planes | |
|---|---|---|
| | Parallel | Perpendicular |
| Perfect | $\lvert\cos 2\theta_m\rvert$ | $\lvert\sec 2\theta_m\rvert$ |
| Ideally imperfect | $\cos^2 2\theta_m$ | $\sec^2 2\theta_m$ |

where $\theta_m$ is the Bragg angle for the monochromator crystal. It should be noted that the value of $B$ for crystals of intermediate type does not necessarily lie between those for the ideal types. Substitution of the values of $B$ into Table 6-1 gives the centroid displacement, the peak shift, and the change of mean-square breadth of the diffraction maximum.

**2.** *Lorentz and other trigonometrical factors*    There has been some discussion of the proper form of the Lorentz factor (Lang, 1956; Pike, 1959*b*; and Ladell, 1961), but this appears to have been largely a matter of nomenclature, and the differences have been resolved (Pike and Ladell, 1961). In Table 6-1 the Lorentz factor proper is taken as tan $\theta$, and the other trigonometric factors for the powder method are listed separately.

**3.** *Absorption and extinction*    The absorption coefficient increases with increasing wavelength, except in the immediate neighborhood of absorption edges. Absorption in the beta filter, tube windows, etc., thus shifts the centroid (and to a lesser extent the peak) of the diffraction maximum toward shorter wavelengths (lower Bragg angles), and changes the mean-square breadth by an amount depending on the third moment of the diffraction profile. The effect is proportional to the thickness of the absorber. Absorption in the specimen, if it is thick enough to absorb the beam completely, gives similar changes, as if it had an effective thickness of $1/\mu$. [With a different definition of "Lorentz factor" and $\mu$ proportional to $\lambda^3$, absorption in the specimen would have no effect (Wilson, 1958; Ladell, Mack, Parrish, and Taylor, 1959).]

Table 6-1    CENTROID, PEAK, AND VARIANCE CHANGES RESULTING FROM CERTAIN PHYSICAL ABERRATIONS. FOR THE PEAK DISPLACEMENT RESULTING FROM DISPERSION, SEE WILSON (1965)

| Aberration | $R(\lambda)$ | $\dfrac{R'(\lambda_0)}{R(\lambda_0)} \equiv R_1(\lambda_0)$ | Centroid displacement $\Delta(2\theta)$ | Peak displacement $\Delta(2\theta)$ | Variance change $\Delta W_{2\theta}$ |
|---|---|---|---|---|---|
| Dispersion | — | — | $W\dfrac{\tan^3 \theta_0}{\lambda_0^2}$ | — | $\dfrac{4\mu_3 \tan^4 \theta_0}{\lambda_0^3}$ |
|  | "cos $\theta$" | $\dfrac{-\tan^2 \theta_0}{\lambda_0}$ | — |  | — |
| Polarization factor | $\dfrac{1 + B\cos^2 2\theta}{1 + B}$ | $-\dfrac{2B\sin 4\theta_0 \tan \theta_0}{\lambda_0(1 + B\cos^2 2\theta_0)}$ |  |  |  |
| Lorentz factor | $\tan \theta$ | $\dfrac{\sec^2 \theta_0}{\lambda_0}$ |  |  |  |
| Absorption in specimen | $\mu^{-1}$ | $-\left(\dfrac{1}{\mu}\dfrac{\partial\mu}{\partial\lambda}\right)_{\lambda=\lambda_0}$ | $R_1 \times 2W\tan\theta_0/\lambda_0$ | $R_1 \times -2I(\lambda_0)\tan\theta_0/\lambda_0 I'(\lambda_0)$ | $R_1 \times 4\mu_3\tan^2\theta_0/\lambda_0^2$ |
|  | $\mu \propto \lambda^3$ | $-3\lambda_0^{-1}$ |  |  |  |
| Absorption in filter, etc. | $\exp(-\mu t)$ | $-t\left(\dfrac{\partial\mu}{\partial\lambda}\right)_{\lambda=\lambda_0}$ |  |  |  |
|  | $\mu \propto \lambda^3$ | $\dfrac{-3\mu_0 t}{\lambda_0}$ |  |  |  |
| Other response variations | $f(\lambda)$ | $\dfrac{f'(\lambda_0)}{f(\lambda_0)}$ |  |  |  |

It ought to be mentioned that different effective absorption coefficients are required for different purposes if the specimen is not homogeneous. The inhomogeneity may arise because the specimen consists of two significant components (as in a two-phase alloy) or be merely incidental (powder grains with air or binder in the interstices). For the calculation of the transparency aberrations (Section III B 3) it is the weighted mean absorption coefficient of the composite specimen that is required; for the physical aberration it is the individual absorption coefficient of the component in question. Unless the material is very finely divided it may be necessary to allow for statistical effects in the averaging (Section VII). The effective absorption coefficient may also be increased by extinction. Primary extinction is only rarely a problem in powder diffractometry, but secondary extinction effects, amounting to some percent in the diffracted intensity, are more frequently reported.

**4.   *Quantum-counting efficiency*** The efficiency of the various types of counter as a function of the wavelength of the radiation detected has been discussed by Parrish (1962) and others. If the efficiency varies rapidly with wavelength it will affect the mean wavelength recorded; for various wavelengths and detectors in common use the change ranges from $-0.5$ to $+2.8$ parts in $10^5$, which is just about significant in some diffractometric applications (Wilson and Delf, 1961). It should perhaps be remarked that the count centroid of a distribution differs slightly from the wavelength centroid and the energy centroid.

**5.   *Pulse-size discrimination*** Scintillation and proportional counters are ordinarily used with a pulse-discriminating circuit that passes only pulses of the size that ought to be produced by the characteristic wavelength of interest. The discrimination can only be very rough, because of statistical variations in the size of the pulses arising from quanta of the same energy. If the discriminator "window" is asymmetric, or not correctly centered, it will produce a change in effective mean wavelength, just as the more fundamental response variations do. Even with a symmetric and correctly centered "window," the asymmetry of the $K\alpha$ doublet will produce some small change, depending on the third moment of the wavelength distribution. For the broad "windows" used with scintillation and proportional counters no significant effect is to be expected, but for the narrow ones that could be used with solid-state detectors it may be necessary to use some care in adjustment in order to avoid appreciable distortion of the diffraction maximum.

By the use of a solid-state detector and pulse-height analysis, it is possible to produce new types of diffractometer in which the diffraction maxima are distinguished by the different energies of the scattered photons rather than by the different angles of diffraction of photons of the same energy. Various arrangements and applications of this principle have been described by Giessen and Gordon (1968), Cole (1970),

Lauriat and Pério (1972), and Fukamachi, Hosoya, and Terasaki (1972, 1973). The angle between the incident and diffracted rays is fixed, leading to great simplification in the mechanical design of the apparatus, and the energy distribution of the diffracted photons is displayed by a multichannel analyzer. Intensity measurements are complicated by the necessity of knowing the intensity of the continuous radiation as a function of frequency. So far there do not seem to have been any applications to line broadening, but Fukamachi, Hosoya, and Terasaki have shown that spacing measurements can be made with an accuracy of about 0.01%. Wilson (1972, 1973) has given a preliminary discussion of the geometrical and physical aberrations.

## VI. EXPERIMENTAL DETERMINATION OF INTRINSIC LINE PROFILES

### A. Origin of the observed profile

The line profile observed in powder diffractometry is basically the emission profile of the x-ray tube target, distorted by the instrumental aberrations, the physical aberrations, and the diffraction profile of the crystals making up the specimen. The emission profile could be trimmed by special monochromators (Section V A), so that the wavelength spread becomes less, but this is not done as a routine. In principle the trimming could be done by pulse-height analysis, but for the usual detectors the discrimination is insufficient for this to be effective. In normal practice the emission profile is "stretched" by the physical aberrations, and in the absence of other effects this distortion could be dealt with by replotting the profile (Lang, 1956; Mitchell and de Wolff, 1967). The instrumental aberrations and the diffraction profile of the specimen, on the other hand, affect the profile more fundamentally, by taking the intensity that ought to appear at a particular angle and smearing it out over an angular range in the neighborhood of the position it should have. In mathematical terms the emission profile is "folded" with the aberration or diffraction profile. The process is illustrated in Fig. 6-5. If $f(x)$ and $g(x)$ are the functions folded together, the resultant $h(x)$ is given by

$$h(x) = \int f(y)g(x - y)\, dy \qquad (6\text{-}77)$$

where the range of integration extends over all values of $y$ for which $f$ and $g$ are nonzero. It is fairly obvious from the figure that the breadth of $h(x)$, however breadth is defined, cannot be less than that of the broader of $f(x)$ and $g(x)$. If $f$ and $g$ have finite ranges of existence (as is the case if both are geometrical aberrations), the total range of $h$ is the sum of the ranges of $f$ and $g$. The ranges for the emission profile and for the diffraction profile are, however, "infinite." Most emission and diffraction

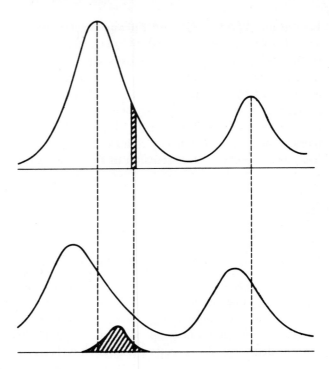

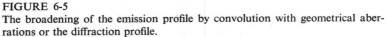

FIGURE 6-5
The broadening of the emission profile by convolution with geometrical aberrations or the diffraction profile.

profiles have inverse-square tails, so that the area under the profile is finite. It is
convenient, and involves no loss of generality, to assume that the profiles have been
normalized by dividing by the area under them, so that, for example,

$$\int f(x)\, dx = 1 \qquad (6\text{-}78)$$

## B.   Properties of convoluted functions

Functions such as $h(x)$ in equation (6-77) are known as the fold, *Faltung*, convolution,
or resultant of the component functions $f(x)$ and $g(x)$, and there are many relations
between the properties of the resultant and the properties of its components. The $n$th
moment of $h(x)$

$$h_n \equiv \int x^n h(x)\, dx \qquad (6\text{-}79)$$

is given by a sort of "binomial" expansion in terms of the moments up to the $n$th of the components, provided that these moments exist. From equations (6-77) and (6-79),

$$h_n = \iint x^n f(y) h(x - y)\, dx\, dy \qquad (6\text{-}80)$$

Putting $z = x - y$ and using $y$ and $z$ as the variables of integration gives

$$h_n = \iint (y + x)^n f(y) g(z)\, dy\, dz \qquad (6\text{-}81)$$

$$= f_n + n f_{n-1} g_1 + \frac{n(n - 1)}{2!} f_{n-2} g_2 + \cdots + n f_1 g_{n-1} + g_n \qquad (6\text{-}82)$$

where

$$f_m = \int x^m f(x)\, dx \qquad (6\text{-}83)$$

and similarly for $g_m$. The simplest application of this theorem, for $n = 0$, shows that if $f$ and $g$ are normalized their resultant $h$ is also normalized. For inverse-square functions only the zeroth moment exists in the strict sense of the word, but in the first moment (centroid) can be given a meaning, and forms the basis of one method of determining lattice parameters corrected for aberrational errors (see, for example, Parrish and Wilson, 1959, or Wilson, 1963a, and the references cited by them). Line profiles are in practice observable only over a finite range, and for finite ranges the second moment increases linearly with range, a property used by Tournarie (1956a,b) and Wilson (1962b) in the study of line broadening.

There is also a simple relation between the Fourier transform of a resultant and the Fourier transforms of its components, provided that these transforms exist. From equation (6-77) the Fourier transform $H(t)$ of $h(x)$ is

$$H(t) = \int \exp(2\pi i t x) h(x)\, dx \qquad (6\text{-}84)$$

$$= \iint \exp(2\pi i t x) f(y) g(x - y)\, dy\, dx \qquad (6\text{-}85)$$

On making use again of the change of variables $z = x - y$ the double integral decomposes into the product of two single integrals, so that

$$H(t) = \int \exp(2\pi i t y) f(y)\, dy \times \int \exp(2\pi i t z) g(z)\, dz$$

$$= F(t) G(t) \qquad (6\text{-}86)$$

The Fourier transform of a resultant is thus just the product of the transforms of its components. This result is the basis of one of the most useful methods of profile analysis and profile synthesis.

In principle a profile would be observable over a range extending in reciprocal space from halfway between the reflection in question and the next lower order to halfway between the reflection and the next higher order, but unrelated reflections always interfere and the practical ranges in powder diffractometry are of the order of one-tenth of this. This limitation of the range of observation introduces difficulties into the analysis of line profiles, and in particular into the use of measures of line broadening. The limitation is less serious in the comparison of synthesized line profiles with those observed.

## C.   Synthesis of line profiles

Synthesis of line profiles from known components presents no particular difficulties. The components to be considered are the emission profile, the geometrical aberrations of the instrument, and the diffraction profile. If these are known they can be convoluted by computer, either directly by the repeated use of equation (6-77), or perhaps more readily by multiplying their Fourier transforms and inverting the product. The emission profiles of the usual crystallographic radiations are not known as accurately as is really desirable, nor over as wide a range. Some geometrical aberrations are readily determinable by measurement of slit dimensions, but some (intensity variation with takeoff angle, for example) have to be measured for the particular tube and applied voltage, and others (specimen transparency, for example) have to be measured or estimated for the particular specimen. The diffraction profile is in most cases derived theoretically from the postulated state of the specimen, and is available as accurately as the model allows. Synthesis of line profiles is thus a rather tedious procedure, and little routine use has been made of it. The work of Boom (1966) is an exception; he used synthetic profiles in the determination of lattice parameters by what he named the *LPC* (= line-profile comparison) method. Neutron diffraction profiles have a simpler shape, and it is thus easier to perform routine syntheses (Rietveld, 1969).

## D.   Analysis of line profiles

**1.   *The available methods***   Analysis of line profiles would normally be undertaken to separate the diffraction profile, or some measure of its breadth, from the observed profile, the emission profile and the geometrical aberrations being taken as known. As Stokes (1948) pointed out, the convolution of the emission profile and the geometrical aberrations can be obtained experimentally by substituting a specimen with the same characteristics (such as size and absorption coefficient) as that under investigation, but not exhibiting diffraction broadening. The diffraction profile is then obtained by deconvoluting the observed broadened profile with the observed unbroadened one. Three methods have been proposed:

*1*  Fourier-transforming the two observed profiles, dividing one transform by the other, and transforming the ratio (Stokes, 1948).

*2*  A relaxation method (Paterson, 1950).

*3*  An iterative method based on repeated convolution (Ergun, 1968).

The Fourier method is perhaps best adapted to reasonably well resolved lines even in the broadened pattern, and the other two to patterns with greater diffraction broadening (carbon black, for example).  The principle of the iterative method was used earlier by Burger and van Cittert (1932) and Wooster (1962, pp. 77–78).  The idea is to broaden the observed profile still further by convoluting it with the instrumental plus emission profile, and then adding to it the differences between it and the doubly broadened profile.  The result is a sharper profile which is a closer approximation to the diffraction profile alone.  The systematization and iteration of this process are developed by Ergun.

**2.  *Fourier methods***  Suppose that the observed profile is $h(s)$, and that it has been measured over the range $-\sigma_1$ to $+\sigma_2$.  The $n$th Fourier coefficient in the series expansion for the observed profile is then

$$H_n = \frac{1}{\sigma_1 + \sigma_2} \int_{-\sigma_1}^{+\sigma_2} h(s) \exp\left(\frac{2\pi i n s}{\sigma_1 + \sigma_2}\right) ds \qquad (6\text{-}87)$$

whereas the Fourier transform of the profile is

$$H(t) = \int_{-\infty}^{\infty} h(s) \exp\left(2\pi i t s\right) ds \qquad (6\text{-}88)$$

If the profile had fallen to zero values at the limits of integration in equation (6-87) there would have been the relationship

$$H_n = (\sigma_1 + \sigma_2)^{-1} H\left(\frac{n}{\sigma_1 + \sigma_2}\right) \qquad (6\text{-}89)$$

but in general $h(\sigma)$, though small, is still finite, and the relationship between the Fourier transform and the Fourier coefficients becomes complicated.  In one way of looking at the problem

$$H_n = (\sigma_1 + \sigma_2)^{-1} H_1\left(\frac{n}{\sigma_1 + \sigma_2}\right) \qquad (6\text{-}90)$$

where $H_1(t)$ is the Fourier transform of a function $h_1(s)$ such that

$$h(s) = \sum_{j=-\infty}^{\infty} h_1[s + j(\sigma_1 + \sigma_2)] \qquad (6\text{-}91)$$

a relation that has been expressed in slightly differing forms by various workers with line profiles (see, for example, Eastabrook and Wilson, 1952; Doi, 1961; Wilson, 1963a). It is not clear that it is possible to construct and interpret $h_1(s)$ when it differs greatly from $h(s)$. Alternatively, from equation (6-88),

$$H(t) = \int_{-\sigma_1}^{\sigma_2} h(s) \exp{(2\pi its)}\, ds + \int_{-\infty}^{-\sigma_1} h(s) \exp{(2\pi its)}\, ds + \int_{\sigma_2}^{\infty} h(s) \exp{(2\pi its)}\, ds$$

$$(6\text{-}92)$$

The first integral is $H_1(t)$ or $(\sigma_1 + \sigma_2)H_n$ for $t = n/(\sigma_1 + \sigma_2)$. The other two are in the nature of correction terms if $h(s)$ decreases monotonely outside the range $-\sigma_1$ to $+\sigma_2$. (They could be large if $h(s)$ has any periodic components, as might arise from a specimen containing crystallites of essentially uniform size and shape.) One can get an upper limit to their size, since for $h(s)$ decreasing monotonely both the real and the imaginary parts consist of alternately positive and negative loops whose sum will lie between $h(\sigma) \times$ |area of one loop| and minus the same quantity. Since the area of one loop is $1/\pi t$,

$$H(t) = H_1(t) \pm \frac{(1 \pm i)[h(\sigma_1) + h(\sigma_2)]}{\pi t} \qquad (6\text{-}93)$$

$$H\left(\frac{n}{\sigma_1 + \sigma_2}\right) = (\sigma_1 + \sigma_2)\left\{H_n \pm \frac{(1 \pm i)[h(\sigma_1) + h(\sigma_2)]}{\pi n}\right\} \qquad (6\text{-}94)$$

The error estimates can be improved if it is reasonable to make any assumptions about the behavior of $h(s)$ outside the observed range. For example, it would often be reasonable to assume that the tails of $h(s)$ are inverse square, $h(s) \sim As^{-2}$ say, giving the correction in terms of sine and cosine integrals (Abramowitz and Stegun, 1968, p. 231). However, an asymptotic expansion can be obtained directly from equation (6-92) by integration by parts. For example,

$$\int_{\sigma_2}^{\infty} h(s) \exp{(2\pi its)}\, ds$$

$$= \left[\frac{h(s) \exp{(2\pi its)}}{2\pi it} + \frac{h'(s) \exp{(2\pi its)}}{(2\pi t)^2} - \frac{h''(s) \exp{(2\pi its)}}{i(2\pi t)^3} - \cdots\right]_{\sigma_2}^{\infty} \qquad (6\text{-}95)$$

$$\int_{\sigma_2}^{\infty} = -\frac{\sin 2\pi\sigma_2 t}{2\pi t} h(\sigma_2) - \frac{\cos 2\pi\sigma_2 t}{(2\pi t)^2} h'(\sigma_2) + \frac{\sin 2\pi\sigma_2 t}{(2\pi t)^3} h''(\sigma_2) + \cdots$$

$$+ i\left[\frac{\cos 2\pi\sigma_2 t}{2\pi t} h(\sigma_2) - \frac{\sin 2\pi\sigma_2 t}{(2\pi t)^2} h'(\sigma_2) - \frac{\cos 2\pi\sigma_2 t}{(2\pi t)^3} h''(\sigma_2) + \cdots\right] \qquad (6\text{-}96)$$

or, for inverse-square variation with $h(s) = As^{-2}$,

$$\frac{A}{\sigma_2}\left\{-\frac{\sin 2\pi\sigma_2 t}{2\pi\sigma_2 t} + 2\frac{\cos 2\pi\sigma_2 t}{(2\pi\sigma_2 t)^2} + 6\frac{\sin 2\pi\sigma_2 t}{(2\pi\sigma_2 t)^3} + \cdots\right.$$

$$\left. + i\left[\frac{\cos 2\pi\sigma_2 t}{2\pi\sigma_2 t} + 2\frac{\sin 2\pi\sigma_2 t}{(2\pi\sigma_2 t)^2} - 6\frac{\sin 2\pi\sigma_2 t}{(2\pi\sigma_2 t)^3} + \cdots\right]\right\} \qquad (6\text{-}97)$$

The terms arising from the negative tail have $\sigma_1$ instead of $\sigma_2$ and $-i$ instead of $i$. For large $t$, then, the correction is inversely proportional to $t$ and oscillates with period $1/\sigma$ as $t$ increases for fixed $\sigma$, or with period $1/t$ as $\sigma$ increases for fixed $t$. It may therefore be investigated experimentally by deliberately truncating at different $\sigma$'s. For $\sigma_1 = \sigma_2$ the imaginary part of the error vanishes for symmetrical tails. The magnitude of the oscillation of the real part is $A/\pi\sigma^2 t$, as previously estimated. For $t = 0$ the error is just the area of the unobserved tails, which is $A(\sigma_1 + \sigma_2)/\sigma_1\sigma_2$. In terms of the Fourier coefficients

$$H(0) = (\sigma_1 + \sigma_2)\left[H_0 + \frac{A}{\sigma_1\sigma_2}\right] \qquad (6\text{-}98)$$

$$H\left(\frac{n}{\sigma_1 + \sigma_2}\right) = (\sigma_1 + \sigma_2)\left[H_n \pm \frac{A(\sigma_1^4 + \sigma_2^4)^{1/2}}{2\pi n \sigma_1^2 \sigma_2^2}\right] \qquad (6\text{-}99)$$

where the limits in equation (6-99) indicate the root mean square of the amplitude of oscillation. The line profile and its derivative are continuous functions of $s$, and hence $H(t)$ and $H_n$ decrease in magnitude at least as fast as $t^{-2}$ or $n^{-2}$ (Hardy and Rogosinski, 1944, p. 26). (For a Cauchy profile they decrease exponentially with $t$ or $n$.) The importance of the error term thus increases rapidly with increasing $n$ or $t$.

In the preceding discussion of the difficulties resulting from truncation of the range of observation it has been tacitly assumed that the background level above which $h(s)$ is measured is known. In practice the background level tends to be overestimated by an amount depending on the relation between the line width and the range of observation. (Experimenters familiar with this effect may overcompensate for it, thus producing an underestimate.) As is discussed by Young, Gerdes, and Wilson (1967), the effect of overestimating the background by a constant amount $C$ is to subtract an amount

$$C\frac{\sin \pi n}{\pi n}\exp\left[\frac{\pi i n(\sigma_2 - \sigma_1)}{(\sigma_1 + \sigma_2)}\right] \qquad (6\text{-}100)$$

from $H_n$ or

$$C\frac{\sin \pi(\sigma_1 + \sigma_2)t}{\pi t}\exp\left[\pi i(\sigma_1 - \sigma_2)t\right] \qquad (6\text{-}101)$$

from $H_1(t)$. The Fourier coefficient $H_0$ is thus the only one affected; background overestimation contributes to the "hook" effect. The Fourier transform, however, has a spurious ripple introduced into it, in addition to that resulting from limitation of range. Background overestimation is sometimes referred to as *horizontal truncation* and limitation of range as *vertical truncation*. Linearization of the variance-range curve (Tournarie, 1956; Langford and Wilson, 1963) is a conscious and explicit attempt to correct the background level; extrapolation and renormalization, as is sometimes done with Fourier coefficients, is a less explicit attempt at the same thing and may alter other parameters in addition to the background level.

Truncation errors are important because the theorem that convolution corresponds to multiplication of Fourier transforms holds strictly only for the "infinite" ranges considered in equations (6-84) and (6-86), and these transforms are not obtainable from the limited range $-\sigma_1$ to $\sigma_2$. To apply Stokes' method we need to know $H(t)$ for the broadened profile and (say) $G(t)$ for the unbroadened profile, so that the transform of the diffraction profile that we seek would be given by

$$F(t) = \frac{H(t)}{G(t)} \qquad (6\text{-}102)$$

In practice we have only $H_n$ and $G_n$ or $H_1(t)$ and $G_1(t)$, and $F_n$ or $F(t)$ is not given by their ratio unless both $h(s)$ and $g(s)$ have fallen to zero before $s$ reaches the limits $-\sigma_1$ and $+\sigma_2$ of the measured range. This may often be a reasonable assumption for the unbroadened profile $g(s)$, and was in fact assumed by Young, Gerdes, and Wilson, but it will rarely be reasonable for the broadened profile $h(s)$.

**3.  *Variance methods***  As already mentioned, the second moment of an inverse-square profile tends to infinity as the range over which it is evaluated increases. In terms of the Fourier transform the second moment of a line profile is given asymptotically by

$$M_2 = -\frac{1}{4\pi^2} \left[ 2A'(0)(\sigma_1 + \sigma_2) + A''(0) - \frac{1}{\pi} B''(0) \log_e \frac{\sigma_1}{\sigma_2} \right.$$
$$\left. + \frac{1}{2\pi^2} A'''(0)(\sigma_1^{-1} + \sigma_2^{-1}) + \cdots \right] \qquad (6\text{-}103)$$

where $A(t) - iB(t)$ is the decomposition of $H(t)$ into its real and imaginary parts (Wilson, 1969a, 1970) and $A'(0)$, $B''(0)$, etc., represent the limiting values of the derivatives with respect to $t$ as $t$ approaches zero through positive values. Calculation of the second moment of the line profile for a sequence of ranges increasing up to the maximum observable will thus give values of the initial derivatives of the Fourier transform, and it is these that are necessary for the determination of mean particle size, mean-square strain, mean distance between stacking faults, etc. (Eastabrook and Wilson, 1952). Other initial derivatives, such as $B'(0)$, are involved in the first moment (centroid), third moment, etc. If the background has not been estimated correctly, terms in $\sigma^3$ are introduced. One can thus determine the initial derivatives by fitting a polynomial (with a logarithmic term in addition if $\sigma_1 \neq \sigma_2$) to the observed variation of the second moment with range, and vary the background level to reduce the component in $\sigma^3$. This process has been carried out in practice by Langford (1968a,b) and Edwards and Toman (1969, 1970a). It is found that with the degree of accuracy attainable at present the background level can be satisfactorily adjusted, the value of $A'(0)/A(0)$ obtained with reasonable accuracy, the value of $A''(0)/A(0)$

with rather less accuracy, but that other adjustments (for example, of background slope) and determinations [for example, of $A'''(0)/A(0)$] cannot be carried out with any degree of confidence. The reason for the lack of confidence is that successive terms in a power series are not orthogonal, and a large percentage change in the coefficient of one of the later terms can be compensated by a comparatively small percentage change in the coefficients of the earlier ones. The work quoted, however, was all done with $\sigma_1 = \sigma_2$, and it is possible that independent variation of $\sigma_1$ and $\sigma_2$ would make possible the determination of one or two more parameters, even without an increase in the accuracy of measurement. This particular correlation difficulty does not arise in the determination of Fourier coefficients, since the sines and cosines involved are orthogonal functions; Fourier coefficients are, however, correlated through statistical fluctuations (Wilson, 1967, 1969b).

It should also be noticed that the moments of a line profile do not add exactly in accordance with equation (6-82) above, since this applies in the form given only to moments that are finite even when the range of integration is infinite. Equation (6-82) thus holds for convolution of the instrumental aberrations (Spencer, 1931, 1949), but not for convolution of emission and diffraction profiles. The necessary corrections have been discussed by Edwards and Toman (1970b) and Wilson (1964, 1970). It is found that the leading terms in equation (6-103) and similar equations for other moments are additive in the same way as for finite moments, but the higher terms involve also cross products of lower derivatives. The simplest example is

$$\frac{H''(0)}{H(0)} = \frac{F''(0)}{F(0)} + \frac{G''(0)}{G(0)} + 2\frac{F'(0)G'(0)}{F(0)G(0)} \qquad (6\text{-}104)$$

where $F$, $G$, $H$ have the same meanings as in equation (6-86). For finite moments only the first terms would appear.

The satellite group in the $K\alpha$ spectrum is a complication, especially with the wider lines. It produces a "step" in the otherwise linear part of the variance-range curve, and unless this is either avoided by an appropriate choice of range or corrected for (Langford, 1968a; Edwards and Toman, 1970a; Edwards and Langford, 1971), the slope of the variance-range curve will be overestimated and its intercept will be underestimated. The latter, in particular, may lead to large errors in particle-size and strain determinations. The effect of the satellite group on the Fourier coefficients has not been closely studied, but it will presumably introduce a ripple of period dependent on the separation of the satellite group from the centroid of the $K\alpha$ doublet. This ripple would be largely but not entirely eliminated if Stokes' method is used for deconvolution.

The interpretation of the Fourier coefficients and the moments of the line profile does not fall within the allotted scope of this chapter. Various aspects of it are discussed in the papers already referred to, especially Stokes (1948), Tournarie (1956a,b), and Eastabrook and Wilson (1952). The standard reference for the Fourier method is the review by Warren (1959); other aspects have been discussed by Waser and Schomaker (1953). For the interpretation of the moments, see Fingerland (1960), Mitra (1964), and Wilson (1962b, 1963b, 1963c, 1965, 1970, 1971b,c).

**4.  *Other methods***  Diffraction broadening of line profiles was first investigated through measurement of the width at half the peak height. It will be obvious that it is difficult to correct the experimental values of this for aberrations and the emission profile, and it is not always easy to calculate the theoretical value corresponding to a postulated model. The integral breadth, the ratio of the total intensity to the peak height, is somewhat more satisfactory on the latter ground, but corrections for other sources of broadening are no easier. These measures of breadth are now only of historical importance or for rough and quick estimates. No serious reliance can be placed on results obtained by the assumption of Cauchy or Gaussian forms for the composite profile.

  Szántó and Varga (1969) have proposed fitting an inverse-square curve plus a linear background variation to the tail of the diffraction profile. The coefficient of the inverse-square term is then equivalent to the slope of the variance-range curve or the initial slope of the Fourier coefficients. This approach does not give the opportunity of seeking useful information also from the intercept of the variance-range curve or the initial curvature of the Fourier coefficients, but it can be used when overlapping is serious on one side of the profile, and it is claimed that the computations are easier.

## VII.  INTENSITY ANALYSIS

There are certain general problems inherent in all experimental measurements of the intensity of x-ray diffraction maxima. These include various statistical fluctuations, correction for thermal and other diffuse scattering, non-x-ray and noncharacteristic background, effective integration over the entire source and specimen, and elimination of "trivial" but important effects such as absorption and extinction. Some of these must be discussed specifically for each experimental setup, but a good many are common to all or nearly all measurements. All specific applications in the following paragraphs are to powder diffractometry, but the general ideas are applicable to single-crystal diffractometry as well.

  Confidence in any measurement depends on a critical consideration of the sources of error. These sources are broadly of three types:

  *1*  Blunders (limitations of the experimenter).
  *2*  Systematic errors (limitations of the apparatus).
  *3*  Statistical fluctuations (limits set by nature).

Blunders can be avoided by sufficient care and detected by repetition, preferably by a different experimenter in a different laboratory. Systematic errors can usually be predicted and corrections calculated after a sufficiently close analysis of the equipment —examples have been presented in Sections III to V above. Some statistical fluctuations are unavoidable in the nature of things, such as the random emission of x-ray photons. Others are unavoidable in practice, such as the random variation in the number of small crystals in a position to reflect in a powder sample. One may include in this class irregular instrumental phenomena, such as variable backlash in gears or stepping mechanisms. Fortunately there is a considerable body of theory concerned with statistical fluctuations, so that the purely statistical uncertainty of an intensity measurement can usually be assessed with greater confidence than can any other uncertainty.

## A. Statistical matters

The quantity of interest is only rarely an actual single count of the number of photons detected during a known time. One is usually concerned with some function of several counts. This may be as simple as the difference between the counting rate at the peak of a diffraction profile and the counting rate in the background, or as complicated as the distribution of particle dimensions (to take an example from powder diffractometry) or a bond length in a crystal structure (to take an example from single-crystal diffractometry). The accuracy of the measurement is most conveniently expressed in terms of its variance, the mean-square deviation of the measurement from its average value. If the simple measurements to be combined are $x_1, x_2, \ldots, x_n$, and the desired compound measurement is $F(x_1, x_2, \ldots, x_n)$, the variance of $F$ is given by

$$\sigma^2(F) = \sum_j \left(\frac{\partial F}{\partial x_j}\right)^2 \sigma^2(x_j) \qquad (6\text{-}105)$$

where $\sigma^2(x_j)$ is the variance of $x_j$, provided that *either* (1) $F$ is a linear function of the $x$'s, *or* (2) the variances of the $x$'s are not too big, *and* (3) the variation of no one $x$ is correlated with the variation of any other $x$.

The third condition is usually satisfied for photon counts made consecutively, as in most diffractometers, but it would not be true for measurements made simultaneously with several detectors—all would vary in the same sense with fluctuations in the tube output or barometric pressure. Nonindependence of the simple measurements can be allowed for through their covariances. The covariance of $x_i$ with $x_j$ is defined as the mean value of the product of (deviation of $x_i$ from its mean) and (deviation of $x_j$ from its mean), and is denoted by $\text{cov}(x_i, x_j)$. Obviously

$$\text{cov}(x_i, x_j) = \text{cov}(x_j, x_i) \qquad (6\text{-}106)$$

and

$$\text{cov}(x_i, x_i) = \sigma^2(x_i) \qquad (6\text{-}107)$$

When it is necessary to allow for correlations, for reasons such as that mentioned above, or because the $x$'s are already composite functions of the photon counts (an example is the Fourier coefficients used in line-profile analysis), equation (6-105) for the variance of $F$ becomes

$$\sigma^2(F) = \sum_{i,j} \left(\frac{\partial F}{\partial x_i}\right)\left(\frac{\partial F}{\partial x_j}\right) \text{cov}\,(x_i, x_j) \qquad (6\text{-}108)$$

The same formulas apply for the combination of nonstatistical experimental errors, subject to conditions (1) and (2) above.

**1.   Counting statistics**   There are two common methods of determining the counting rate. The first is to observe the number of counts $n_j$ accumulated at the $j$th diffractometer setting during a fixed time $\tau$, so that the counting rate at this setting is given by

$$I_j = \frac{n_j}{\tau} \qquad (6\text{-}109)$$

A less frequent but still common method is to measure the time $t_j$ required to accumulate a fixed number of counts $m$, so that the counting rate is

$$I_j = \frac{m}{t_j} \qquad (6\text{-}110)$$

Because of the statistical fluctuations in the emission of x-ray quanta, a repeat measurement at the same setting will only rarely give the same counting rate. It can be shown that the probability of collecting $n$ counts in a time $\tau$ by the first method is

$$p_\tau(n) = \frac{(\lambda\tau)^n \exp\,(-\lambda\tau)}{n!} \qquad (6\text{-}111)$$

where $\lambda$ is a parameter depending on the mean counting rate. This is called the *Poisson distribution function*, and it is easy to show that the average value of $n$ is $\lambda\tau$ and that the variance of $n$ has the same value. In terms of the counting rate

$$\langle I \rangle = \lambda \qquad (6\text{-}112)$$

and

$$\sigma^2(I) = \frac{\lambda}{\tau} \qquad (6\text{-}113)$$

Ordinarily, of course, the true value of the parameter $\lambda$ is unknown, and the only estimate of it available is the counting rate actually observed. We must thus take as the best estimate of the variance of the counting rate

$$\sigma^2(I_j) = \frac{I_j}{\tau} \qquad (6\text{-}114)$$

In the second method, fixed-count timing, the time is a continuous variable, and the probability of requiring a time between $t$ and $t + dt$ to record $m$ counts is found to be

$$p_m(t)\, dt = \frac{(\lambda t)^{m-1}\, \exp\,(-\lambda t)}{(m-1)!}\, d(\lambda t) \qquad (6\text{-}115)$$

The mean counting rate becomes

$$\langle I \rangle = \left\langle \frac{m}{t} \right\rangle = \frac{m\lambda}{(m-1)} \qquad (6\text{-}116)$$

and its variance becomes

$$\sigma^2(I) = \frac{m^2\lambda^2}{(m-1)^2(m-2)} \qquad (6\text{-}117)$$

$$\sigma^2(I) = \frac{\langle I \rangle^2}{(m-2)} \qquad (6\text{-}118)$$

In most practical applications $m$ would be a large number and the mean and variance could be written

$$\langle I \rangle = \lambda \qquad (6\text{-}119)$$

and

$$\sigma^2(I) = \frac{\lambda^2}{m} \qquad (6\text{-}120)$$

However, Killean (1967) has shown that structures can be determined with $R$ as low as 10% with $m$ as low as 25 to 30, for which the two expressions for the variance would differ by 15% or so. As in fixed-time counting, one must take the observed $I_j$ as the only available estimate of $\lambda$.

These results for the variances of counting rates find applications in the estimates of the reliability of practically all quantities depending on intensity measurements, and are particularly important in connection with the optimization and minimization problems discussed in Section VII C below.

**2.  *Random setting errors***   The setting mechanism of a counter diffractometer (or of a densitometer used in connection with x-ray photographs) suffers from two types of error. There may be a systematically varying difference between the angle indicated by the apparatus and the actual angle of diffraction. This difference can be measured by an appropriate calibration procedure, and any necessary correction applied to the indicated angle. In addition, settings will not be perfectly reproducible, because of backlash in gearing, lost counts in moiré-fringe techniques, variability in stepping mechanisms, and similar phenomena. Random setting errors of this kind may affect derived measurements directly, because the setting appears explicitly in the function to be evaluated, or indirectly, because the counting rate at the actual setting differs from the counting rate at the supposed setting. Measurements of integrated intensity, for example, suffer only from the indirect effect, whereas measurements of position or breadth of diffraction maxima are affected both directly and indirectly. The magnitude of the systematic and random errors depends on the care with which the mechanism has been constructed and maintained; in order of magnitude one may expect the systematic errors to be about 0.001° and the setting errors to be about 0.0001°, but they may be much larger in individual cases (Beu, 1962; Wilson, 1965$b$).

The indirect effect of random setting errors is formally included in equation (6-108) by regarding $x_i$, say, as the counting rate at the position $x_j$ and inserting the appropriate covariance. It is more easily visualized by writing the measurement to be made as an explicit function of both the settings and the counting rates:

$$F(x_1, x_2, \ldots, x_n; I_1, I_2, \ldots, I_n)$$

Equation (6-105) then becomes

$$\sigma^2(F) = \sum_j \left[ \frac{\partial F}{\partial x_j} + \frac{\partial F}{\partial I_j} \frac{\partial I_j}{\partial x_j} \right]^2 \sigma^2(x_j) \qquad (6\text{-}121)$$

Some applications of this equation are made in Section VII B 3.

There are some other sources of error that affect derived measurements in an analogous fashion. These include short-term variations in the voltages applied to the x-ray tube, proportional counter or photomultiplier, and possibly in the atmospheric pressure and ambient temperature. As a statistical exercise they could be introduced as additional variables in equation (6-108), but if the counting rates recorded are found to be sensitive to such variables it is preferable to control them experimentally, so that their effect becomes negligible.

**3.   *Statistical effects in particle orientation***   The theory of powder diffractometry assumes that the specimen contains a very large number of small crystals in random orientation, so that the number of crystals in an orientation to reflect is insensitive to changes in the position and orientation of the specimen. This situation is effectively achieved in practice if the crystals are of the order of $10^{-6}$ m in linear dimensions and of roughly spherical shape (and in particular not needlelike and not platelike). Such specimens are not always easy to prepare, and actual specimens may depart from the ideal in two ways:

*1*   The crystals may be so large that only a few are in an orientation to reflect, with resultant statistical fluctuations in the number reflecting.

*2*   The crystals, although small, have a tendency to align themselves in some systematic fashion relative to the specimen surface.

Any particular specimen, of course, may suffer from both departures from the ideal. Neither departure is easy to treat in a general fashion.

Fluctuations in the number of crystals in a position to reflect may be expected to be of the order of the square root of the number, so that any increase in the effective number will reduce the effect of the fluctuation. For a given crystal size the effective number may be increased by moving the specimen during the accumulation of counts, so that crystals in other orientations are brought into the beam (translation of the specimen in its own plane) or other orientations are given to crystals already in the beam (oscillation of the specimen through a small angle). Rotation of the specimen about an axis perpendicular to its own plane produces both effects, and is the method usually adopted in practice; sufficient translation is difficult to arrange, and oscillation produces line broadening (it amounts to a variable specimen tilt of the type considered in Section III B 4). The beneficial effect of specimen rotation has been treated both experimentally and theoretically by de Wolff, Taylor, and Parrish (1959). The need for specimen motion may be assessed by setting the diffractometer on a diffraction maximum and noting whether the counting rate is sensitive to a change in the specimen position. If it is, and if no motion device is available, the crystal size must be reduced before accurate measurements can be made. [Although these statistical effects are a nuisance in the present context, Warren (1960) has been able to make use of them as a method of crystal-size measurement.]

The second departure from the ideal, preferred orientation of the crystals in the specimen, leads to great difficulty in achieving reliable intensity measurements, and there is no certain method of avoiding it. It may sometimes be detected by using several specimens prepared in different ways (packed from the front, packed from the back, "drifted" from the side), or by using the same specimen in both a reflection arrangement and a transmission arrangement (Section II F 2).

**4.** *Statistical absorption effects* Specimens used in powder diffractometry are not usually homogeneous; they may consist of particles of a single substance plus voids, or particles of a single substance plus binder, or a mixture of particles of several substances plus voids or binder, or a compact block containing several substances but no voids or binder. If the particle size is small compared with the x-ray path length the absorption coefficient of a composite specimen is given by

$$\mu = \sum_j p_j \mu_j \qquad (6\text{-}122)$$

$$1 = \sum_j p_j \qquad (6\text{-}123)$$

where the $\mu_j$ are the linear absorption coefficients of the various components and the $p_j$ are their proportions. If the particle size is not small in comparison with the path length there will be statistical fluctuations in the number of particles of each kind encountered along the path, and the actual absorption will be greater or less than that calculated from equation (6-122), depending on whether particles of high absorption or particles of low absorption are in excess of expectation. In this connection voids are to be regarded as particles of zero absorption coefficient. In simple transmission it is clear that the effective absorption coefficient is less than the average given by equation (6-122); paths with an excess of particles of low absorption let through more than enough extra x rays to compensate for those stopped in paths with an excess of particles of high absorption (Mitra and Wilson, 1960). The exponential character of the absorption function $\exp(-\mu t)$ acts somewhat like a rectifier. In powder diffractometry, however, the situation is not that of simple transmission, and the statistical situation is complicated in that the character of one particle in the path, that actually diffracting, is known. There have been many attempts to treat the diffractometer problem; that of Harrison and Paskin (1964) is rigorous for the case of one component plus voids, but it is difficult both to understand and to apply in practice. The work of de Wolff (1947) and Wilchinsky (1951) is simpler but less rigorous. An empirical method for correcting the diffracted intensity, based on the use of specimens with different proportions of voids, has been described by DeMarco and Weiss (1962). It may also be possible to base corrections on measurements of fluorescence (de Wolff, 1956) or Compton scattering (Weiss, 1966, pp. 97–98). In the absence of statistical effects, the intensity diffracted by a specimen consisting of a single component plus voids would be the same as that given by a compact specimen, unless the specimen were unusually transparent (Milberg, 1958; Langford and Wilson, 1962; see also Section III B 3). The geometrical transparency aberration, however, depends on the effective absorption coefficient of the specimen.

## B.   Measurement of Bragg intensity

Powder diffractometers are usually operated in a step-scanning mode when more than very modest accuracy in the measurement of line intensity or line profile is required. The diffractometer is set at a predetermined angle, and counts are accumulated for a predetermined time (or timed till a predetermined count is accumulated). The setting is then increased by a small step, of the order of $0.01°$, and counts are again accumulated. This process of determining the counting rate at a series of equally spaced steps is repeated until the whole region of interest has been scanned. (There would be some advantage in using steps equal in $\sin\theta$ instead of equal in $\theta$, but few instruments are capable of doing this as a routine.) The energy diffracted per unit time into a particular angular range is then proportional to the sum of the counting rates observed at all the steps falling within that range. If the width of the receiving slit is chosen equal to that of the step (or the step chosen equal to the width of the receiving slit, if this is easier to achieve in practice) the statement just made is correct without any assumption about the propriety of replacing an integral by a sum over the steps; the receiving slit does in fact integrate the line profile over its width. If the receiving slit has any other width there is some approximation involved, as some parts of the line profile are omitted entirely if the slit is narrower, or included twice over if the slit is wider. Unless the line is unusually narrow in comparison with the slit width this reservation is unimportant.

The diffracted intensity (energy per unit time per unit solid angle) is generally regarded as consisting of two parts: "line" (or "peak") and "background," the line being the intensity corresponding to a Bragg reflection from the specimen, and the background corresponding to all other sources of counts. In most applications it is the line that is of interest.

**1.  *Correction for background***  The "true" counting rate at each setting is obtained from the observed counting rate $I_j$ by subtracting from it a background counting rate $G_j$. Ideally this would be obtained by measuring or calculating the counting rate corresponding to all nonline sources of counts, but the measurements and calculations are often very difficult to carry out. Among the recognized sources of nonline counts are the following:

1   Air scattering.
2   Specimen fluorescence.
3   Compton scattering.
4   Non-x-ray background.
5   Beam obstructions.
6   Electronic troubles.
7   White radiation.
8   Thermal and other diffuse scattering.

The fourth of these is, of course, easily measured simply by closing the window of the x-ray tube—shutting the tube off is not quite equivalent, as the tube circuits may affect the counting circuits (source 6).

1   The air path between the tube window and the detector affects the counting rate in two ways. The first is a simple reduction of the intensity reaching the detector by absorption. This effect hardly changes with angle, and can often be ignored. The other is x-ray scattering from the air in the effective volume (Section III B 3); ordinarily this is about half filled with air and half with specimen. The air scattering can thus be estimated by removing the specimen. Provided that indirect scattering is guarded against, the counting rate without a specimen will be about double that of the air-scattering rate when the specimen is in position. A more exact factor could be obtained from measurements of the effective volume and the specimen position if necessary. The air scattering varies smoothly with diffractometer setting, becoming large at small and large angles.

Air scattering can be eliminated by use of an evacuated or helium-filled enclosure, but the experimental inconvenience of such arrangements is often more troublesome than the air scattering. They are worthwhile, however, for the longer wavelengths, as the loss of intensity through absorption is serious for these.  Air scattering is more of a problem with the Seemann-Bohlin arrangement, as the length of the air path varies with diffractometer setting.

*2*   The short-wavelength components of the incident x rays excite fluorescent radiation in the specimen.  It can be avoided almost entirely by the use of a premonochromator, unless an important constituent of the specimen has an absorption edge near the characteristic wavelength of the target material (as, for example, a specimen containing much iron used with a copper target).  A postmonochromator is effective unless the target material is an important con-stituent of the specimen.  In most other cases pulse-height discrimination is sufficient to reduce the counting rate due to fluorescent radiation to a very low level.  The fluorescent counting rate changes only slowly with diffractometer setting.

*3*   Compton scattering is an inevitable concomitant of x-ray scattering, but the increased wavelength of the Compton-scattered x radiation makes it possible to eliminate most of it by the use of a postmonochromator.  There is, however, a lower limit of angle below which this elimination is not practicable.  Compton scattering varies rather slowly with diffractometer setting, and it is, in principle, possible to calculate its intensity.

*4 and 5*   There is a small non-x-ray background arising from cosmic rays and radioactivity.  This makes an appreciable contribution to the counting rate when Geiger counters are used as detectors, but the effect is very small for proportional and scintillation counters with pulse-height discrimination—particularly for the latter, as the active volume is so small.  False counts arising from x rays scattered by parts of the apparatus obstructing the beam ought not to occur if the diffractometer is properly designed and adjusted.  Beam obstruction would probably reveal itself through anomalously high background or anomalously low line intensity.

*6*   Electronic troubles are mentioned only for the sake of completeness, as they lie quite outside the field of powder diffractometry.  Unless the circuits are well designed, shielded, and stabilized, false counts and lost counts can arise from line-voltage surges, high-frequency transients, faulty switchgear on neighboring apparatus, and other unclassifiable phenomena.

*7*   The intensity of noncharacteristic radiation reaching the detector is greatly reduced by pulse-height discrimination and the use of a monochromator.  There is always some, however, with a wavelength so close to that of the characteristic that it is not resolved by the monochromator.  For many purposes unresolved white radiation is of negligible importance.  When it is important the wavelength distribution of the detected radiation can be analyzed by replacing the specimen by a single-crystal slip of "intrinsic" silicon or other approximately perfect crystal.

*8*   The preceding sources of background have been incidental, and can be reduced or eliminated by appropriate techniques. Diffuse scattering by the specimen, however, is an essential concomitant of the diffraction process, and should perhaps be looked on as part of the diffraction pattern, rather than as background. In general thermal and other diffuse scattering (short-range order, Huang, cf Chap. 2) has a broad peak at the same angle as the Bragg reflection; its maximum intensity is low in comparison, and its falloff slow (inverse square for the Bragg reflection, logarithmic for thermal diffuse scattering). The intensity of thermal diffuse scattering can be calculated with reasonable accuracy for simple symmetrical substances, as discussed in Chap. 2, Section II F 3. If the thermal diffuse scattering is not the phenomenon under investigation, it can be reduced greatly in intensity by making the measurements at liquid-nitrogen temperature or lower.

Generally, most of the sources of background show only a slow variation with diffractometer setting; the exceptions are the diffuse scattering and the residual white radiation, which are concentrated near the Bragg peak. For many purposes it suffices to measure the counting rates in comparatively level regions on either side of the line and interpolate between them, either linearly or with a smooth curve, to obtain the background counting rate $G_j$. The line profile as a function of the diffractometer setting $x_j$ is then given by $I_j - G_j$, and the integrated intensity of the reflection is proportional to

$$L_0 = \sum_j (I_j - G_j) \qquad (6\text{-}124)$$

It should perhaps be mentioned that some detectors and some counting circuits become overloaded and lose counts at high counting rates. This affects chiefly the peak of the line, producing a sort of "antibackground," compensating to some extent for the peaking up of the thermal diffuse scattering. The effect was important for Geiger counters, but may usually be ignored for scintillation and proportional counters.

Equation (6-124), with calculated or interpolated $G_j$, suffices to give the relative intensities of the reflections appearing in a powder pattern. When it is desired to obtain "absolute" intensities (for example, in verifying that the atomic scattering factors are properly accounted for by quantum mechanics), it is necessary to measure also the intensity in the incident beam of the radiation in the same wavelength range as is included in the summation. This is a matter of considerable experimental difficulty, and is ordinarily practicable only if a high-quality monochromator is used. It is also necessary to know the linear absorption coefficient with considerable accuracy and to measure the dimensions of the receiving slit with greater accuracy than usual. When all precautions are taken the absolute accuracy attainable is of the order of 0.5%. The subject has recently been reviewed by Weiss (1966), and will not be considered further here.

**2. *Correction for experimental effects***   The geometrical aberrations broaden and displace the diffraction maxima, but most of them do not affect the integrated intensity significantly. The exception is the transparency of the specimen (Section III B 3). For a highly absorbing specimen the integrated intensity is proportional to $\mu^{-1}$, where $\mu$ is the true linear absorption coefficient, for a one-component specimen (no binder, but possibly some voids). This proportionality is independent of diffractometer setting for the symmetrical Bragg-Brentano arrangement, but not for the Seemann-Bohlin arrangement. For a multicomponent specimen the factor becomes

$$\frac{\alpha}{\mu''} = \frac{\alpha}{\alpha\mu + (1 - \alpha)\mu'} \qquad (6\text{-}125)$$

where $\alpha$ is the volume fraction of the specimen occupied by the material of interest, $\mu''$ is the overall linear absorption coefficient, $\mu$ is that of the material of interest, and $\mu'$ that of the rest of the specimen. In equation (6-125) it is, of course, assumed that statistical effects of the type discussed in Section VII A 4 are negligible. If they are not, it becomes very difficult to determine correctly the relative intensities of the lines from two or more components of the specimen. Probably the best solution experimentally is to reduce the particle size of the specimen by grinding, but this may not always be possible, and an approximate correction based on one of the methods referred to in Section VII A 4 has to be used. This problem has also been discussed by Schäfer (1933), Taylor (1944), Brindley (1945), and others.

Missetting of the 2:1 ratio mechanism makes the Bragg-Brentano arrangement slightly unsymmetrical, and hence affects the integrated intensity through a change in the absorption factor. It is thus worthwhile to avoid any appreciable missetting, even when the additional broadening (Section III B 4) is not objectionable. Corrections for the effect of this and other geometrical aberrations have been derived by Suortti and Jennings (1971).

When the absorption coefficient is not large, the variation of integrated intensity with diffractometer setting and specimen adjustment becomes complex; references to some papers and particular cases have been given in Section III B 3. With a greater volume of specimen (greater number of particles) effective statistical effects become less troublesome.

**3. *Propagation of errors***   Measurement of a single counting rate is rarely of any interest in itself; the number of individual measurements that have to be combined to produce a result of interest may range from several to several hundred. If blunders have been avoided and systematic errors assessed, the standard deviation of any result can be estimated in principle by the application of equation (6-108) or its special cases (6-105) or (6-121). Carrying through the calculation becomes more difficult as the function $F$ becomes more complex. Typical functions are the integrated intensity, the line centroid or peak, the line variance or other measure of its breadth, and the Fourier coefficients. The simplest of these is the integrated intensity, but it is too simple to illustrate all the features of the problem. The line centroid is given by

$$\langle x \rangle = \frac{\sum_j x_j(I_j - G_j)}{\sum_j (I_j - G_j)} \tag{6-126}$$

and is a better illustration, as it depends on both the counting rates $I_j$ and the diffractometer settings $x_j$. Fixed-time counting will be assumed, so that the variance of $I_j$ is $I_j/\tau$, where $\tau$ is the time spent at each step. There are five sources of statistical fluctuation in $\langle x \rangle$ to be considered:

*1*  Statistical fluctuations in the counting rates $I_j$.
*2*  The direct effect of statistical fluctuations in the settings $x_j$.
*3*  The indirect effect of fluctuations in the settings through the counting rates.
*4*  Statistical fluctuations in the interpolated background $G_j$.
*5*  Statistical fluctuations in the limits of the summations, if these depend on the measured centroid $\langle x \rangle$.

It will be simplest to consider these in turn.

*1*  The counting rates occur in both the numerator and the denominator of equation (6-126), so that

$$\frac{\partial \langle x \rangle}{\partial I_j} = \frac{x_j}{L_0} - \frac{\langle x \rangle}{L_0} \tag{6-127}$$

$$\sigma^2(\langle x \rangle) = \sum_j \left( \frac{\partial \langle x \rangle}{\partial I_j} \right)^2 \sigma^2(I_j)$$

$$= \frac{1}{L_0^2 \tau} \sum_j (x_j - \langle x \rangle)^2 I_j \tag{6-128}$$

where $L_0$ is the integrated intensity [equation (6-124)]. The statistical variance of the centroid arising from fluctuations in the counting rates only is thus equal to $(L_0 \tau)^{-1}$ times the geometrical variance (square of the radius of gyration) of the line plus background (Section VI D 3).

*2*  For the diffractometer settings,

$$\frac{\partial \langle x \rangle}{\partial x_j} = \frac{I_j - G_j}{L_0} \tag{6-129}$$

$$\sigma^2(\langle x \rangle) = \frac{1}{L_0^2} \sum_j (I_j - G_j)^2 \sigma^2(x_j) \tag{6-130}$$

The variance of the diffractometer setting $x_j$ is probably much the same for all $j$, so that the term on the right is (variance of a diffractometer setting) times (integrated intensity of the square of the line profile) divided by (integrated intensity of line profile)$^2$.

3   Equation (6-121) leads to two terms in addition to that considered in equation (6-130). These are

$$\sigma^2(\langle x \rangle) = \frac{2}{L_0{}^2} \sum_j (I_j - G_j)(x_j - \langle x \rangle) I_j' \sigma^2(x_j) + \frac{1}{L_0{}^2} \sum_j (x_j - \langle x \rangle)^2 (I_j')^2 \sigma^2(x_j)$$

(6-131)

where $I_j'$ is the slope of the line profile at the setting $x_j$. It is not easy to see simple geometrical interpretations of these terms; the second is related to the variance of a profile given by the square of the slope of the actual profile.

4   The background $G_j$ is usually interpolated linearly between a value $G_1$ found by counting for a time $p\tau$ at a setting just to the left of range used for the line profile and a value $G_2$ similarly found at a setting just to the right of the profile. Alternatively the $G$'s may be found by averaging $p$ values found by counting at $p$ points just to the left and the right of the line for a time $\tau$ each. By either method the variance of $G_1$ is essentially $G_1/p\tau$, and $G_2/p\tau$ for $G_2$. The inter-polated background in equations like (6-126) is then essentially

$$G_j = \tfrac{1}{2}(G_1 + G_2) + \frac{j(G_2 - G_1)}{R}$$

(6-132)

where $R$ is the number of steps in the range and $j$ is counted from the center of the range. (There are trivial adjustments depending on whether $R$ is odd or even and how far outside the range the measurements of $G_1$ and $G_2$ are made.) With this choice of background, equation (6-126) becomes

$$\langle x \rangle = \frac{\sum_j x_j [I_j - j(G_2 - G_1)/R]}{\sum_j [I_j - \tfrac{1}{2}(G_1 + G_2)]}$$

(6-133)

so that

$$\frac{\partial \langle x \rangle}{\partial G_1} = \frac{\sum_j j x_j}{L_0 R} + \frac{\langle x \rangle R}{2 L_0}$$

(6-134)

$$\frac{\partial \langle x \rangle}{\partial G_2} = - \frac{\sum_j j x_j}{L_0 R} + \frac{\langle x \rangle R}{2 L_0}$$

(6-135)

The sum in the first term of these equations is nearly enough $R^3 \delta/12$, where $\delta$ is the step length, and $\langle x \rangle$ is of the order of $\delta$. The second term is thus negligible in comparison with the first, and the variance of $\langle x \rangle$ arising from the background is

$$\sigma^2(\langle x \rangle) = \frac{R^4 \delta^2 (G_1 + G_2)}{144 L_0^2 p\tau} \qquad (6\text{-}136)$$

5   In the practical application of centroids and variances in powder diffractometry it is usual to adjust the range so that it is symmetrical about the centroid. This is done by dropping a few observations at one end or the other, and statistical fluctuations in the centroid position will cause statistical fluctuations in the number dropped. The effect of this has been considered by Thomsen and Yap (1968*a,b*), who conclude that the variances derived above should be multiplied by $(1 - r)^{-1}$, where

$$r = \frac{(\text{mean counting rate at ends of range}) \times (\text{range})}{(\text{counting rate integrated over range})} \qquad (6\text{-}137)$$

Obviously $r$ is less than unity; Thomsen and Yap estimate that in normal crystallographic practice it amounts to about 5%.

The variances of many other parameters derived from line profiles have been treated by Mack and Spielberg (1958); Pike and Wilson (1959); Zevin, Umanskij, Khejker, and Pančenko (1961); Wilson (1965*b*, 1967, 1968, 1969*b*); Thomsen and Yap (1968*a*) and others; many of the results are collected in the paper by Wilson (1967). It must be emphasized that such results refer only to the statistical fluctuations in the numerical results obtained by the method of calculation adopted, and give no indication of any systematic errors introduced by the adopted method. In the determination of the centroid, used as an example above, there are two sources of systematic error immediately obvious: equation (6-126) is written with summations instead of integrations, and the background has been linearly interpolated. Estimation of the systematic error introduced by these procedures involves analysis of quite another kind. Another example arises in interpretation of line profiles through the Fourier coefficients. It is not too difficult to estimate the variance of the Fourier coefficients, but the method of analysis itself introduces the so-called hook effect, making the interpretation of the plots of Fourier coefficient vs. order very difficult (Eastabrook and Wilson, 1952; Young, Gerdes, and Wilson, 1967; Kukol', 1962).

## C.   Optimization of effort

*Ars longa, vita brevis*, says the Latin proverb. Within limits, the accuracy attainable in a physical measurement is limited by the time that the experimenter is willing to put into it. This time dependence is obvious in the expressions for variance just discussed; the standard deviation of a measurement goes down as the inverse square root of the time.  The time dependence of the reduction of systematic errors is less (or even not) predictable, but is often very tedious.  For most people the tolerable time is measured in months rather than years, and for most of the rest in years rather than decades.  It is, therefore, of some importance to divide the time devoted to the different parts of the measurement in such a way that the best possible return on the investment is achieved.  In recent years the problem of the efficient use of time has been discussed from three related but not identical standpoints:

*1*  The aim is to obtain the best possible value of a single physical parameter in the available time.

*2*  The aim is to obtain an acceptable value of some overall measure of agreement in the minimum possible time.

*3*  The aim is to obtain the best possible values of a set of physical parameters making conflicting claims on the division of the available time.

A typical example of the first kind is the measurement of (peak minus background) intensity, or measurement of the centroid of a powder diffraction maximum.  An example of the second kind has been discussed by Killean (1967*a*,*b*): how to divide counting time between the various reflections in order to achieve an acceptable value of the agreement index $R$ (say 10%) in the shortest time.  A simple example of the third type is the division of counting time among the various steps along a powder diffraction profile so that a best least-squares equation can be fitted to it (Thomsen and Yap, 1968), a more complex one is the reconciliation of the competing claims of, say, accuracy of bond lengths and accuracy of temperature parameters in crystal-structure determination (Shoemaker, 1968).  Problems of the third type are mathematically more complex than those of the first type, not only because more parameters are involved but because there are likely to be weighting factors that vary with the counting times.  Variable weights may also be encountered in some problems of type (2).

All the above examples involve optimization of random errors, probably because an acceptable theory exists for these.  There is no general theory of systematic errors, and reduction of them often has to wait upon a bright idea or a flash of genius, resulting in a new experimental method or a better process of interpretation.

**1.**   *Minimization of the variance of a single parameter*    Particular instances of the minimization of a single parameter have been discussed by Mack and Spielberg (1958) (net peak height and various combinations of net peak heights) and by Zevin. Umanskij, Khejker, and Pančenko (1961) (centroid of a powder diffraction maximum). The process was generalized by Wilson, Thomsen, and Yap (1965).  Suppose that the desired parameter $F$ is a function of $n$ counting rates $I_j$, each determined by counting for a time $t_j$, and that these counting times enter into $F$ only through the corresponding $I_j$. A total time $T$ is allowed for the $n$ observations, and this is to be divided among them so that the variance of $F$ is a minimum. The relevant equations are

$$\sigma^2(F) = \sum_j \left(\frac{\partial F}{\partial I_j}\right)^2 \sigma^2(I_j) \qquad (6\text{-}138)$$

$$\sigma^2(F) = \sum_j \left(\frac{\partial F}{\partial I_j}\right)^2 \frac{I_j}{t_j} \qquad (6\text{-}139)$$

and

$$T = \sum_j t_j \qquad (6\text{-}140)$$

The expression (6-139) will be a minimum (or maximum) if for any small variations $dt_j$ in the $t_j$

$$0 = \sum_j \left(\frac{\partial F}{\partial I_j}\right)^2 \frac{I_j}{t_j^2} \, dt_j \qquad (6\text{-}141)$$

and

$$0 = \sum_j dt_j \qquad (6\text{-}142)$$

These two equations are compatible for all $dt_j$ only if

$$\left(\frac{\partial F}{\partial I_j}\right)^2 \frac{I_j}{t_j^2} = k^2 \qquad (6\text{-}143)$$

where $k$ is a constant determined by equation (6-140). The minimum variance is thus

$$\sigma^2_{\min}(F) = k^2 T \qquad (6\text{-}144)$$

$$\sigma^2_{\min}(F) = \frac{1}{T} \left\{ \sum_j \left|\frac{\partial F}{\partial I_j}\right| \sqrt{I_j} \right\}^2 \qquad (6\text{-}145)$$

and

$$k = \frac{1}{T} \sum_j \left|\frac{\partial F}{\partial I_j}\right| \sqrt{I_j} \qquad (6\text{-}146)$$

$$t_j = \frac{T \left|\frac{\partial F}{\partial I_j}\right| \sqrt{I_j}}{\sum_j \left|\frac{\partial F}{\partial I_j}\right| \sqrt{I_j}} \qquad (6\text{-}147)$$

These equations do in fact give a minimum, not a maximum. As equation (6-145) shows, the minimum variance is a perfect square, aside from the factor $T^{-1}$, so that the minimum standard deviation achieves a simple form.

It is easily deduced from the above calculation that there is no single "best" way of measuring, for example, a line profile, since the recipe for the choice of times, equation (6-147), depends on the function $F$. For determination of the integrated intensity [equation (6-124)] $\partial L_0/\partial I_j$ is unity for all $j$, and the optimum times are proportional to the square root of the counting rates. Optimization of centroid determination gives times proportional to $|x_j - \langle x \rangle|/\sqrt{I_j}$, which in the tails of the lines would approximate to fixed-time counting. Optimization of the determination of the geometrical variance of the line gives zero counting times near the radius of gyration of the profile, and something between fixed time and fixed count in the tails.

Minimization of the variance of other measures of intensity, location, and breadth have been discussed by Wilson (1967, 1969b), and measures of location have been treated in particular detail by Thomsen and Yap (1968). It will have been noticed that the counting times are in general functions of the counting rates, and it is thus logically necessary to know the $I_j$ before starting to measure them. This difficulty can be overcome by what Shoemaker (1968) calls a two-pass method: a first set of measurements is made rapidly, so that an approximate set of counting rates is obtained, and then after a rapid calculation (ideally done by an on-line computer) the set is repeated with counting times adjusted so that the quantities of interest are optimally determined.

**2.  *Minimization of an agreement index***   Minimization of an agreement index with respect to changes in adjustable parameters is usual in structure determination; the agreement index is generally of the form

$$R = \sum_j w_j(F_j - C_j)^2 \qquad (6\text{-}148)$$

where $F_j$ is some function of the counting rate $I_j$ (for example, the positive square root of the counting rate adjusted for the Lorentz-polarization factor and absorption) and $C_j$ is a corresponding model function (the modulus of the structure factor in this example) of several or many parameters (atomic positions, thermal parameters, scaling factor, atomic scattering factors). However, such problems do arise in powder diffractometry, as in the fitting of a polynomial or other function to measurements of the counting rates near the peak of a line with a view to finding the peak position by differentiation. The $w_j$ in equation (6-148) are weights, conventionally taken as the reciprocals of the variances of $F_j - C_j$. These variances may reasonably be taken as having the form

$$\sigma^2 = k_j{}^2 + \frac{\lambda_j}{t_j} \qquad (6\text{-}149)$$

where $k_j{}^2$ represents the variance arising from sources other than counting statistics and $\lambda_j/t_j$ is the counting variance. For each observed $I_j$ the values of $k_j{}^2$ and $\lambda_j$ are known or determinable or estimated from experience; they cannot readily be reduced, so that no choice of counting time can make the variance less than $k_j{}^2$. The usual practice is to minimize $R$ with respect to the parameters of the model function $C_j$, the values of $t_j$ being whatever the experimenter regards as conventional or convenient. The question now arises: can a deliberate reallocation of the total allowed time $T$ among the $t_j$ lead to a lower value of $R$? For $R$ having the form of (6-148) with weights equal to the reciprocal of (6-149) there is no true minimum, but by giving $t_j$ a large value when $|F_j - C_j|$ is small and a small or zero value when $|F_j - C_j|$ is large it is obviously possible to arrive at a very small value of $R$. Anyone adopting such a procedure would rightly be accused of cooking his results.

**3. Optimization of many parameters** The difficulty of optimizing the counting times when there are many parameters has been treated in a different fashion by Shoemaker (1968). His approach was specifically aimed at structure determination by single-crystal diffractometry, but there is no reason why it should not be applied in powder diffractometry, so it will be outlined briefly here. The method is to optimize $R$ with respect to the parameters of interest, and then optimize some function of the variances of these parameters with respect to the counting times. The function that he chooses to optimize (actually maximize) is

$$\Omega = \sum_{i=1}^{n} \frac{W_i}{\sigma^2(\xi_i)} \qquad (6\text{-}150)$$

where the $\xi$'s are the $n$ parameters of interest and $W_i$ is a subjective weight representing the importance to the experimenter of the $i$th parameter. He could, for example, put $W_i$ equal to unity for all positional parameters and zero for all thermal ones, or give higher weights to the parameters of some atoms than to those of the rest. Obviously $\Omega$ is a very messy function of the counting times, via the variances of the parameters, which depend on the weights $w_j$ and hence on $k_j^2$ and $\lambda_j/t_j$ [equation (6-149)]. The condition for $\Omega$ to be a maximum is

$$\sum_j \frac{\partial \Omega}{\partial w_j} \frac{dw_j}{dt_j} dt_j = 0 \qquad (6\text{-}151)$$

subject to

$$\sum_j dt_j = 0 \qquad (6\text{-}152)$$

since the total counting time is taken as fixed. In structure determination the values of $\partial \Omega / \partial w_j$ are mostly evaluated by the computer in the course of refinement, so that they would be available without great extra trouble. If the covariances can be neglected (a matter that would need some consideration in each different type of problem) they take a fairly simple form:

$$\frac{\partial \Omega}{\partial w_j} = \sum_i W_i \left[ \frac{\partial |F_j - C_j|}{\partial \xi_i} \right]^2 \qquad (6\text{-}153)$$

but it is not necessary to assume this in the following. Equations (6-151) and (6-152) are compatible only if for all $j$,

$$\frac{\partial \Omega}{\partial w_j} \frac{dw_j}{dt_j} = \alpha^2 \qquad (6\text{-}154)$$

where $\alpha$ is a constant determined from the constancy of the total available time $T$. After a little manipulation one obtains

$$t_j = \frac{1}{k_j^2} \left[ \frac{(\Omega_j \lambda_j)^{1/2}}{\alpha} - \lambda_j \right] \qquad (6\text{-}155)$$

$$\alpha = \frac{\sum\limits_j (\Omega_j \lambda_j)^{1/2} k_j^{-2}}{T + \sum\limits_j \lambda_j k_j^{-2}} \qquad (6\text{-}156)$$

where $\Omega_j$ has been written for $\partial \Omega / \partial w_j$ for the sake of brevity. The calculation is obviously closely parallel to that of Section VII C 1. The second term in the denominator of $\alpha$ has the dimensions of a time, say $T_0$, and indicates the order of magnitude of the time that it is worth spending on collecting data. Shoemaker considers that $T$ should not exceed $T_0$; others might think it worthwhile to make $T$ a small multiple of $T_0$. Substitution in equation (6-148) with weights given by equation (6-149) shows that the minimum agreement index is

$$R_{\min} = \sum_j \left( 1 - \frac{\alpha \sqrt{\lambda_j}}{\sqrt{\Omega_j}} \right) \frac{(F_j - C_j)^2}{k_j^2} \qquad (6\text{-}157)$$

Whether this is a significant reduction below the values obtained for equal counting times or equal counts per reflection depends on the relative magnitudes of $k_j^2$ and $\lambda_j$. Killean (1969) has expressed some doubt about whether the improvement would be worth the extra complication.

If $\partial \Omega / \partial w_j$ is small for some reflections (in nonmathematical terms, if these reflections are insensitive to the parameters of interest), equation (6-155) will indicate zero or negative counting times $t_j$. This means that any time spent on these reflections is wasted, since it could have been used for reflections more sensitive to the parameters. If this occurs, the remaining $t_j$ should be recalculated with those that were previously indicated as negative or zero constrained to be zero. This process may have to be repeated, since the new calculation may indicate more zero or negative times. Shoemaker conjectures that in the limit when statistical errors are the main source of variance ($k_j^2$ all small) this process would lead to concentrating all the counting time on the $n$ reflections most sensitive to the $n$ chosen parameters. Thomsen and Yap encountered an analogous phenomenon in optimizing the determination of the peak of a diffraction maximum by curve fitting; for a parabolic curve, for example, most of the time should be spent at the two end points of the range and a little at the middle point, no other measurements being made. Such a procedure leaves open the question of whether a parabola is actually a good approximation to the peak shape over the range utilized, and one may wonder if statistical efficiency corresponds with sound scientific practice in such extreme cases.

# 7

## SINGLE-CRYSTAL INTENSITIES

## I. INTRODUCTION

Whatever the ultimate objectives, in every x-ray diffraction experiment what one does is record the intensities of diffracted beams. A photographic film provides a most convenient means for doing this because it is a large-area detector having a uniform response for each particular wavelength. It has the further advantage of forming a conveniently sized permanent record of the spatial disposition of the diffracted beams that is easy to store and to examine. The relationship between the blackening on a film, when viewed with visible light, and the intensity of the x-ray beam that produced it, however, generally is not linear. This greatly complicates quantitative evaluation of the x-ray intensities and, even under the best of circumstances, limits the ultimate accuracy attainable. Nevertheless, there are many occasions when photographic films provide the most convenient and rapid means for recording the diffraction intensities of a large number of diffracted beams, for example, in the case of organic compounds having extremely large unit cells. Their quantitative interpretation requires the use of specialized scanning densitometers whose optical characteristics are unrelated to x-ray diffraction, however, and are not discussed further in this book for that reason.

For most purposes, whenever quantitative assessment of the diffraction intensities is desired, the most convenient instrument to use is an x-ray diffractometer. If one desires to know the intensities on an absolute scale, then it is necessary to use large crystals that intercept the entire incident beam so that its total intensity can be evaluated along with that of the diffracted beams. This requires that the crystal have a large surface parallel to the diffracting planes and generally limits the number of planes whose reflections can be measured conveniently. Fortunately, for most purposes, it is quite adequate to measure the diffracted intensities on a relative scale, so that one can use small single crystals that are entirely bathed within the incident x-ray beam. The measurement of intensities diffracted by such small single crystals is the primary subject of this chapter. First, the principles underlying the operation of single-crystal diffractometers are examined in Section II. This is followed by a consideration of different schemes for achieving beam monochromatization (Section III) and proper corrections for background intensities (Section IV). Special attention is paid to the problems inherent in recording the total diffracted intensity, commonly called the *integrated intensity*. The concluding section of this chapter points up some factors handicapping these measurements that are sometimes overlooked.

## II.  SINGLE-CRYSTAL DIFFRACTOMETRY

### A.  Reciprocal-space sampling region

Consider a strictly monochromatic, perfectly parallel bundle of x rays incident on an infinitesimally small crystal perfectly centered in the beam. The region of reciprocal space that can be sampled in this case is wholly determined by the area on the Ewald sphere corresponding to the solid angle subtended at the center of the sphere by the aperture of the detector employed (Fig. 7-1). Even slight variations in wavelength or ray directions then create a nest or continuum of Ewald spheres, and the successively displaced areas sweep out a volume in reciprocal space. This is illustrated for a beam of converging rays in Fig. 7-2, which shows a cross-sectional plane of the Ewald sphere containing the incident and diffracted rays called the *plane of incidence*. The central ray, 2, gives rise to the diffraction cone shown. The nonparallel rays 1 and 3 all pass through the same reciprocal-space origin $O$ and have the effect of displacing their Ewald spheres in the manner indicated. The corresponding sampling regions are similarly displaced and give rise to the shaded sampling region indicated. Remember that this region is for incident rays lying in the plane of incidence so that, in general, a sampling volume is produced in three dimensions.

To understand something of the effect that finite specimen size has on the geometry of the sampling region, consider first a broad beam of strictly parallel monochromatic radiation incident on a stationary specimen. In this case each point of the

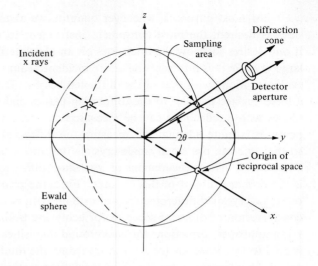

**FIGURE 7-1**
Reciprocal-space sampling region determined by the intersection of the diffraction cone entering the detector aperture and the Ewald sphere. $x$, $y$ and $z$ are orthogonal axes.

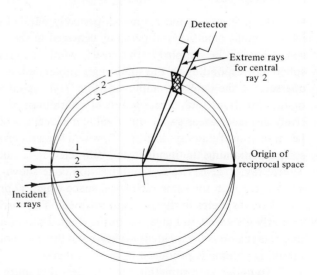

**FIGURE 7-2**
Two-dimensional view of effect of convergence in the incident beam.

specimen has incident on it a ray that is geometrically equivalent to those incident on every other point. The range of scattered rays which can enter the detector differs for each point, however, because, in laboratory space, each point bears a slightly different geometric relationship to the receiving aperture. Two effects can be distinguished.

First, the cone of scattered rays acceptable by the receiving aperture is larger if it emanates from a nearer point in the specimen. Figure 7-3a illustrates this in exaggerated schematic style. With a 0.5-mm-diameter spherical specimen 20 cm from the detector, this effect will produce an extreme difference of approximately 2.5% in the cone half-angle and, hence, approximately 5% in the size of the area "seen" by the detector on the Ewald sphere. In reciprocal space, this means that specimen points lying on a line passing through the center of the receiving aperture will generate sampling areas of differing sizes, but all these areas will be centered at the same point on the particular Ewald sphere corresponding to a given incident-ray direction (Fig. 7-3b).

Second, the diffraction angles $2\theta$ corresponding to the edges of the cone will differ according to the perpendicular distance of the specimen point from the nominal central scattered x-ray line (i.e., the line drawn from the center of the receiving aperture to the center of the specimen) even though $\Delta 2\theta$ (the full cone angle) will be essentially unaffected by this displacement. This means that specimen points lying on a line perpendicular to the nominal central-scattered-ray line will generate sampling areas of the same size but different locations on the same Ewald sphere. Figure 7-3c and d represents this situation in laboratory space and reciprocal space, respectively.

In the practice of single-crystal diffractometry, of course, one expects rarely, if ever, to encounter even good approximations to strictly parallel incident beams except in special-purpose arrangements employing highly perfect crystals or monochromators, such as those involving the Borrmann beam (Cole, Chambers, and Wood, 1961) or other features discussed in Section IV B 5.

Let consideration of monochromators which do not thoroughly parallelize the beam be set aside for the present. Then the nonparallelism, or cross fire, in the incident beam depends only on the sizes of the source and specimen in the simple fashion indicated by Fig. 7-4. Cross fire is seen to be the sum of *divergence*, the angle subtended at a source point by the specimen, and *convergence*, the angle subtended at a specimen point by the source. (Further use of this definition will be made in later sections.) On any one point in the specimen, rays converge from all points on the source unless some interfering object is introduced. This, a convergent beam incident on a point specimen, is the situation assumed for Fig. 7-2. In general, however, the convergence differs in detail for every point in the specimen. The differences occur either in the range of convergence angle or in the particular angular limits, measured

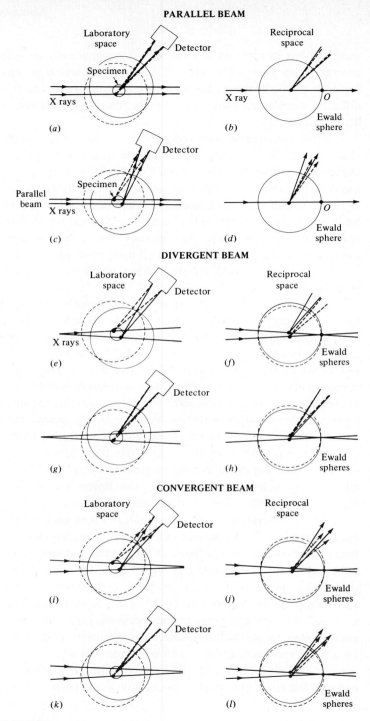

**FIGURE 7-3**

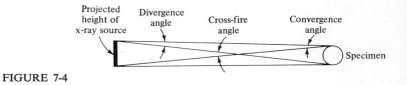

FIGURE 7-4

relative to a fixed direction in the specimen, between which the convergence rays are incident, or both. For example, take the specimen points on an arbitrarily chosen plane perpendicular to the incident beam. Since the sizes of source and specimen are ordinarily small compared to the distances between them, the convergence range is essentially the same for all points on the plane but the convergence limits are not. (If the problem were confined to the two dimensions of the plane of incidence, the points lying on a common line perpendicular to it would have essentially the same convergence limits.) For specimen points lying on an arbitrarily chosen line parallel to the (average) incident beam direction, i.e., from left to right in Fig. 7-4, the convergence range steadily diminishes as points farther to the right are selected.

The consequence is that, to each point in the specimen, there corresponds a slightly different version of Fig. 7-2 which is unique to that point when the third dimension is considered. This conclusion is a corollary of a fact that is now seen to be implicit in the preceding development: *The region of reciprocal space sampled differs for each combination of source point, specimen point* and, of course, *wavelength.*

A useful mental picture of the reciprocal-space region being sampled in the usual case can now be built up rather readily by proper combination of Figs. 7-2 and 7-3*b* and *d* with appropriate modification for extension of source and specimen perpendicular to the diffraction plane. Finally, the separate Ewald spheres can be given differing weights in accordance with the differing intensities in the rays to which they correspond.

Figure 7-3*b* and *d* shows the effect of different choices of specimen point in the plane of incidence. Selection of a specimen point out of the plane simply moves the sampling region perpendicular to the plane in a mutual relationship similar to that shown in Fig. 7-3*c* and *d*.

Figure 7-4 shows the effect of source extension in the plane of incidence for a given specimen point. Selection of a source point out of the plane causes the center of the associated Ewald sphere to be displaced perpendicular to the plane, just as centers are shown displaced within the plane in Figs. 7-2 and 7-3. In the usual case, however, the Ewald-sphere radius is expected to be large compared to this displacement, and the region on the Ewald sphere viewed by the detector is very small compared to the total surface area of the sphere. Therefore, this small lateral displacement

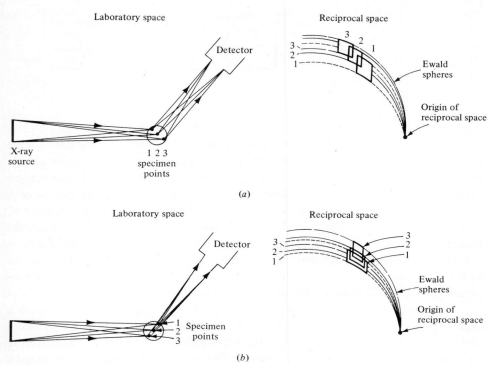

(a)

(b)

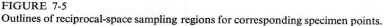

**FIGURE 7-5**
Outlines of reciprocal-space sampling regions for corresponding specimen points.

of a sphere center effects no significant change in the region of reciprocal space co-incident with the viewed area of the Ewald sphere.

For each point in the specimen one constructs the version of Fig. 7-2 appropriate to the range and limits of convergence and to the relative size (Fig. 7-3b) and location (Fig. 7-3c extended) of the single-ray sampling region pertaining to that point. The next step is integration over all specimen points; i.e., superposition of all the shaded regions (remembering that the shaded region in Fig. 7-2 represents a three-dimensional region). The accumulated shading density at any point in reciprocal space then represents the relative number of times the sampling region for one or another specimen point included that reciprocal-space point.

Figure 7-5 shows schematically something of the relationships among the sizes, shapes, and locations of the reciprocal-space sampling regions of different specimen points in an incident beam exhibiting cross fire. Figure 7-5a is, in effect, an extension of Fig. 7-3c and d made to include incident beam cross fire and a third point. Figure 7-5b is similarly related to Fig. 7-3a and b. Differing specimen-point-to-source distances produce the convergence-range differences which result in differing height (as shown) for the sampling regions. The vertical (in these figures) displacements of the sampling regions are the consequence of differing distances of the specimen points from the central specimen-to-detector line; the "lateral" size differences (Fig. 7-5*b*) are the consequence of differing specimen-point-to-detector-aperture distances. One other relationship which should be included in this visualization is apparent, that of a kind of reciprocity between the two spaces: shifting of a specimen point to the right causes its sampling region to move to the left in the figure; shifting of the point up moves the region down, and so forth.

The geometric result of superposition of all sampling regions for all specimen points will, of course, depend on the specimen's shape and, unless it is spherical, the specimen's orientation. For example, consider a spherical specimen smaller than the projected source size. In this case the resultant composite sampling region will be a roughly cylindrical object, with end faces cut at an angle to the cylinder axis. The shading density will diminish to zero as the cylinder walls and ends are approached, but there will ordinarily be a cylindrical core region of nearly uniform density. The overall diameter will be approximately that of the largest circular sampling region, Fig. 7-3*b*, plus the maximum displacement associated with specimen point location, Fig. 7-3*d*. The diameter of the uniformly dense core region, if one exists, will be smaller than the diameter of the smallest individual sampling region, Fig. 7-3*b*. If two or more wavelengths are active, the composite reciprocal-space sampling region is, obviously, the sum of the separate regions, one for each wavelength, constructed as just described.

Finally, account should also be taken of the fact that, in general, it is possible, and even probable, that a different relative intensity is associated with each incident ray direction and wavelength. Graphically, this relative intensity may be represented by the shading density assigned to the particular Ewald sphere (Fig. 7-1) associated with that incident ray. Then the shaded area represented in Fig. 7-2 is, in general, nonuniformly dense. The shading-density gradient at any point near the center of it is essentially perpendicular to the surfaces of the Ewald sphere passing through that point. The subsequent superposition (i.e., integration over all specimen points) of these shaded regions produces a composite sampling region wherein there may be considerable variation in the shading representing the relative weight with which that portion of reciprocal space is sampled.

There is probably little point in defining the mental picture, or one actually drawn out, more precisely. For any quantitative work a mathematical description is needed in any event, and the preceding discussion and geometric visualization developed are sufficient to guide and to provide understanding for the mathematical approach. Reference will be made to this "picture" at various points throughout this chapter. It will perhaps be obvious to the reader at this point that the appropriate mathematical approach is that of convolutions. That approach has been considered in detail by Ladell and Spielberg (1966), in particular.

## B.   Task of the diffractometer

The single-crystal diffractometer may be described as a device for providing controlled motion of selected reciprocal-space points to and through the reciprocal-space sampling region with simultaneous registration of the diffracted intensity. This can be accomplished either by detector motion or wavelength variation, or both. It is also required at times that the reciprocal-space sampling region be moved in controlled fashion during the intensity-data-collection, or scanning, process. Scanning may also be defined as production of controlled relative motion, either continuous or stepwise, of the sampling region and the reciprocal space of the specimen crystal while the reciprocal-space point of interest is in or near the sampling region and the scattered intensity is being measured.

The points selected in reciprocal space need not necessarily be reciprocal-lattice points, of course; they may constitute simply a continuous path along which sampling is desired, as it is for diffuse scattering studies (Chap. 2). Similarly, frequently it is desired that the traverse path of a reciprocal-lattice point through the sampling region be coincident with a specified reciprocal-lattice direction, e.g., that the orientation of the plane of incidence in the reciprocal lattice be under the operator's control.

The location of the specimen must remain fixed while its orientation is varied in order that the sampling region not be moved or changed by variation in the convergence limits, convergence range, and scattered-ray-cone size consequent from specimen displacements and so that the specimen, if small, remain fully bathed by the x-ray beam or, if large enough to intercept the beam, maintain the same portion of itself in the beam.

It is also the task of the diffractometer, or at least of that part of the diffractometer system between and including the x-ray source and the detector, not to contribute unnecessarily to an unfavorable character of the reciprocal-space sampling region, such as unnecessarily large size and nonuniformity of weighting of the sampling; i.e., "shading density" in Section II. The principal instrumental factors involved are: (1) illumination gradients and other characteristics of the source, (2) selection and placement of objects such as windows, filters, apertures, and crystal monochromators interposed between the source and the specimen, (3) the selection and placement of the detector aperture and any other objects in the scattered-beam path, and (4) wavelength discrimination. For example, an object placed in the incident beam path which had the effect of obscuring a portion of the source from a portion, only, of the specimen could have the effect of reducing the size of the nearly uniformly dense region of the sampling region without significantly reducing its overall size. One reason for this concern with the sampling region is that a tightly defined region of the minimum size required, for the particular sampling task, both improves signal-to-noise ratio and makes easier a subsequent sorting out of different intensity contributions on the basis of their differing spatial variations.

Control of the background scattering, such that the signal-to-noise ratio is not unnecessarily impaired, is also part of the task of the diffractometer. Generally, such control involves choices of beam tunnels, apertures, ambient gases, collimators, and beam traps. These matters are considered further in Section IV.

The detailed task of the diffractometer differs with the intended use and, hence, so does the optimum choice of instrument parameters. The protein crystallographer working with very large unit cells and crystals undergoing change during examination, for the purpose of making the first determination of the positions of the atoms in that unit cell, assigns a certain relative importance to features such as resolution and precision of intensity measurement. But a very different relative importance is assigned by a physical crystallographer, working with a small unit cell and a stable crystal of known structure, who is seeking to determine details about thermal vibrations and precise atomic locations for the purpose of determining the mechanism of a ferroelectric phase transition. For precision determination of lattice parameters by Bond's (1960) method, angular precision, beam homogeneity, and a minimum size for the reciprocal-lattice sampling region are emphasized. For diffuse scattering studies and radial distribution studies the sampling region normally is chosen large deliberately to increase the detected intensity; incident beam inhomogeneity will be relatively unimportant, and emphasis will be placed on low background, incident beam mono-chromaticity, and wavelength discrimination against Compton-scattered x rays in the detected beam.

For some types of crystallographic work other instruments may be as good as, or more suitable than, the diffractometer. For ordinary chemical structure determination with stable crystals, for example, film methods incorporating automated micro-densitometry are preferred by some workers (Jeffrey and Whitaker, 1965; Wooster, 1967; Xuong, Kraut, Seely, Freer, and Wright, 1968; Arndt, Crowther, and Mallett, 1968; Milledge, 1969; Jeffrey, 1969). Certainly the two-dimensional sampling of reciprocal space provided by a film has distinct advantages for survey purposes. But for work which requires precision measurement of intensities, particularly when something about the crystal is changing, e.g., temperature, defect content, etc., the diffractometer offers unique advantages.

## C.  Diffractometer motions

*1.  The two general types of instruments*  The present chapter is primarily concerned with the reciprocal-space sampling region and scanning paths associated with the various diffractometers and not with the mechanical designs used to achieve them. Descriptions and comparisons of such design aspects of various diffractometers have been given by Arndt and Willis (1966) in their book *Single-Crystal Diffractometry* as well as by various authors of journal articles (e.g., Davis, Groter, and Kay, 1968; Binns, 1964; Arndt, 1964; Arndt and Phillips, 1963; Abrahams, 1964; Ladell and

Lowitzsch, 1960; Baker et al., 1968, 1969, among many others). A brief discussion leading to a very general description of two of the principal types of diffractometers in use today, however, will help add dimension to the succeeding discussions of scanning motions in reciprocal space. W. H. Bragg's ionization-chamber "spectrometer" (see Compton and Allison) and copies or minor modifications of it must properly be accounted the first single-crystal diffractometers. However, the chain of development leading directly to present-day single-crystal diffractometers appears to have been set off by adaptation of Weissenberg cameras for intensity measurement with Geiger counters (Clifton, Fuller, and McLachlan, 1951; Evans, 1953). This geometry has persisted in some instruments while others have utilized the geometry of other kinds of moving-film cameras. The design of still others incorporated independent ideas of what geometry was most appropriate, but irrespective of their genesis, single-crystal diffractometers may be considered conveniently in two categories, according to whether or not their axial motions are mechanically coupled.

**2.   Mechanical analog instruments**   The purpose of coupling is to provide an instrument that can operate semiautomatically, very simply, without need for complex calculations of angular settings and without need for external (e.g., electronic) co-ordination of axial motions during a scan. Even without coupled motions, the counter-adapted Weissenberg instruments of the 1950s did offer relatively convenient exploration of reciprocal space in a simply comprehended geometry, i.e., along concentric circles, or arcs of circles, within one of a set of parallel planes in reciprocal space. The plane was selected by the Buerger equi-inclination setting. The circle was selected by the counter-elevation angle $\Upsilon$ (Buerger, 1942) and scanning along the circular arc was produced by motor-operated rotation of the spindle axis. However, a better and less restrictive selection of scanning path was needed. Ladell and Lowitzsch (1960) credit Hirshfeld and Schmidt (1953) with recognizing that "the feasibility of single-crystal counter techniques would be enhanced by adopting a general scanning procedure" and with being the first to advocate exploration of reciprocal space along parallel lines under the guidance of a mechanical analog of the reciprocal lattice, provided by appropriate mechanical coupling of axes.

**a.   Linear-tracking diffractometers**   The two most widely used instruments based on the mechanical analog design are the Linear Diffractometer (Arndt and Phillips, 1959, 1961) and the PAILRED II (Ladell, 1962; Ladell and Cath, 1963; Ladell and Lowitzsch, 1960). Both utilize a system of bars and cross slides. These provide a mechanical analog of the reciprocal lattice in which each of the bars is maintained parallel to one or another of three principal reciprocal-lattice directions, usually $a^*$, $b^*$, or $c^*$. Figure 7-6, due to Arndt and Willis (1966), shows schematically the arrangement for zero-level measurements. The bars and cross slides are used only

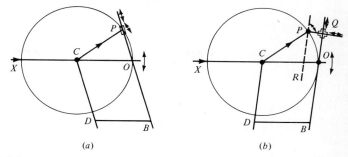

FIGURE 7-6
Linkage system in a mechanical analog linear diffractometer. (*a*) Scan of the
central reciprocal-lattice row $h00$ ($POB$): a reflection occurs whenever $OP = ha*$.
The carriage $P$ moves along $POB$; the counter arm $CP$, assumed to be of unit
length, is pivoted at $C$ and on the carriage $P$; $POB$ is pivoted at $O$; the crystal
arm $CD$ is kept parallel to $POB$. (*b*) Scan of a noncentral reciprocal-lattice
row $hk0$ ($PR$); $PQ = kb* = $ constant for the row as $Q$ moves along $QOB$.

for bringing the desired reciprocal-lattice row or point into scanning position; the
scanning of a reciprocal-lattice point is done with an additionally provided oscillatory
motion of the crystal which, unlike the linear-tracking assembly, provides a constant
speed of traverse in reciprocal space with the use of a constant-speed motor.
PAILRED II, in particular, permits considerable adjustment of the crystal orientation
and, hence, selection of the three principal directions by means of relatively large-angle
mechanical arcs mounted on the spindle parallel to $CD$ in Fig. 7-6.

The object of the linear-tracking device is to produce exploration of reciprocal
space along a chosen straight-line path which does not necessarily pass through the
origin. (The term *scanning* will be reserved for the motions involved in a deliberate
systematic determination of the diffracted intensities as a function of position along a
relatively restricted path, e.g., in the neighborhood of a reciprocal-lattice point or of
another point or region of special interest. In contrast, *exploring* need not involve
recording of any data and will generally be carried out with much higher speed
motions.) The problem of coupling the counter detector to the crystal rotations to
produce exploration of a straight line is shown in Fig. 7-7 to be that of causing the
point $u_1$ (or $u_2$), the intersection of the detected-ray line with the Ewald sphere, to
move along a straight line $PP'$ in the reciprocal space of the specimen.

For equi-inclination geometry with equi-inclination angle $v$, unit radius assigned
to the Ewald sphere, and $v$ the distance from $PP'$ to the axis of rotation of the reciprocal
lattice [from $Q$ to $(O, g)$ in Fig. 7-7*b*], define

$$g = \frac{v}{\cos v} \qquad (7\text{-}1)$$

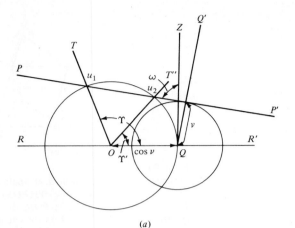

(a)

**FIGURE 7-7**
Geometry of PAILRED diffractometers.
(Ladell and Lowitzsch, 1960.) (*a*) Geometric construction for derivation of
tracking equation (7-2). *TO* and *T"O* are
the directions along which two counter
detectors can be placed to detect diffraction effects occurring at $u_1$ and $u_2$ on the
circle of reflection. (*b*) Geometric construction for one counter arrangement.
The scale of the drawing is such that the
intercept of the sphere of reflection with
the *n*th level, i.e., the circle of reflection
for the *n*th level in the equi-inclination
scheme, has unit radius. $\overline{SS'}$ and $\overline{QQ'}$
are coordinate axes with reference to
which the point $(u, g)$ can be located in
the reciprocal-lattice level.

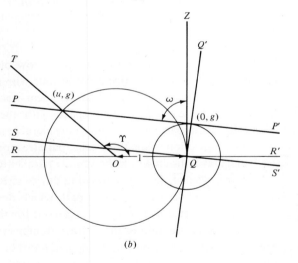

(b)

and the tracking equation becomes

$$\Upsilon = \omega \pm \cos^{-1}(\cos \omega + g) \qquad (7\text{-}2)$$

where $\Upsilon$ is the counter-detector elevation angle, and the crystal angle $\omega$ is the angle
between *PP'* and *ZQ*, the normal to *QQ*.

An interesting aspect of stating the tracking problem this way is that a direct
mechanical analog of Fig. 7-7 can be constructed. This was done by Ladell and
Lowitzsch (1960), who called it a point-drive linkage mechanism. They arranged to
move the point $u_2$, Fig. 7-7a, at constant velocity along any selected line *PP'* by
using a worm as the mechanical analog of the exploration or scanning path *PP'*.

Using $u$ as the positional coordinate of $u_1$ measured along $PP'$ from $ZQ$, one finds that the range in $u$ is from $-(4 - g^2)^{1/2}$ to $+(4 - g^2)^{1/2}$ and that the velocities of the detector ($\dot{\Upsilon}$), crystal ($\dot{\omega}$), and $u_1$ ($\dot{u}$) are related as follows:

$$\dot{u} = \text{constant}$$

$$\dot{\omega} = \frac{g + u\beta^{-1}}{(u^2 + g^2)}\dot{u}$$

$$\dot{\Upsilon} = \frac{2u}{(u^2 + g^2)\beta} \tag{7-3}$$

where $\beta = [4(u^2 + g^2)^{-1} - 1]^{1/2}$ and the linear measures are in units of Ewald sphere radii.

*b. Lorentz factor* For an instrument using this geometry, obviously, the Lorentz factor differs from that for the many instruments employing a constant crystal rotation velocity $\dot{\omega}$. Since the Lorentz factor $L$ may be regarded as an expression of the relative time required for a reciprocal-lattice point to pass through the Ewald sphere surface during a scan, the component of the point's velocity along the sphere radius is wanted. Let $V_r$ be the reciprocal-lattice point velocity along the radius, $r$, of the reflection circle. Then

$$V_r = \dot{\omega}(g^2 + u^2)^{1/2}\frac{[r^2 - (g^2 + u^2)/4]^{1/2}}{r} \tag{7-4}$$

The component of velocity $V_R$ along the radius $R$ of the Ewald sphere is then found with the relation $r = R \cos v$, leading to $V_R = V_r \cos v$, to be

$$V_R = \dot{\omega} \cos v(g^2 + u^2)^{1/2}\left(1 - \frac{g^2 + u^2}{4r^2}\right)^{1/2} \tag{7-5}$$

With the definition in Fig. 7-7 of the counter angle $\Upsilon$, it follows that

$$(g^2 + u^2)^{1/2}/2r = \sin(\Upsilon/2)$$

and, hence,

$$V_R = \dot{\omega}R \cos^2 v \sin \Upsilon \propto L^{-1} \tag{7-6}$$

This relation also can be written

$$L^{-1} \propto \dot{\omega} \sin 2\theta \frac{(\sin^2 \theta - \sin^2 v)^{1/2}}{\sin \theta} \tag{7-7}$$

or

$$L^{-1} \propto \dot{\omega}(\sin^2 2\theta - \cos^2 \theta \sin^2 v) \tag{7-8}$$

Equations (7-7) and (7-8) are the standard relations given, for example, in the *International tables for x-ray crystallography*, but they are developed here to show the explicit dependence of the Lorentz factor on the crystal rotation velocity $\dot{\omega}$. Using relation (7-3), one recognizes that, for the single-crystal diffractometer using the point-drive mechanism and keeping $\dot{u}$ constant,

$$L^{-1} = \frac{g + [4(u^2 + g^2)^{-1} - 1]^{1/2}}{u^2 + g^2} (\sin^2 2\theta - \cos^2 \theta \sin v) \qquad (7\text{-}9)$$

which, using (7-6) rather than (7-8), can be written

$$L^{-1} = \frac{\sin \Upsilon}{4R \sin^2 \Upsilon/2} \left[ g + \frac{u}{2(R^2 \sin^2 \Upsilon/2 - 1)^{-1/2}} \right] \qquad (7\text{-}10)$$

or

$$L^{-1} = \left[ \frac{g}{\delta} + \frac{u}{(4 - \delta^2)^{1/2}} \right] (4R^2 \cos^2 v - \delta^2)^{1/2} \qquad (7\text{-}11)$$

where $\delta = u^2 + g^2$ and constant factors have been omitted. Since the instrument settings are in terms of $g$ and $u$ (the mechanical analogs of reciprocal-space distances which often can be expressed simply in terms of reflection indices and reciprocal-lattice repeat distances) it is clear that (7-11) offers the more convenient form. It also follows from a comparison of (7-6) for constant $\dot{\omega}$ with (7-10) that the Lorentz factor for this mechanical-analog device differs considerably from that for a Weissenberg camera, whose geometry the analog device might, at first sight, seem to resemble.

**3.** *Independent-axes diffractometers*   *a.*   *The axes and their functions*   Automated single-crystal diffractometers of the types that do not involve mechanically coupled motions now outnumber those that do and appear to be growing in popularity. The reason is, apparently, their flexibility when operated from a control unit whose operating program can be changed rather readily, e.g., from punched-tape or card readers operating through appropriate interfaces or from on-line computers: The most widely used of these instruments are based on independent axes related as the axial system for definition of Euler's angles. The first such device to become widely used was the "goniostat" designed by Furnas (1957) and also referred to by him as an Eulerian cradle.

A minimum of three independently controllable axes are required, one for the coupled rotation in 2:1 relation ($2\theta$, $\theta$) of the counter detector and goniostat base, respectively, and two others, $\phi$ and $\chi$, for positioning of the selected reciprocal-lattice point in the diffracting position. Figure 7-8a and b shows schematically a diffractometer and, in relation to each other, the three axes and the additional axis $\omega$, which is incorporated in the usual four-circle instrument. In the most common form of these instruments the counter detector is confined to the equatorial plane. For $\omega \neq 90°$, $\phi$ and $\chi$ motions suffice to bring a chosen reciprocal-lattice point into the equatorial plane on a line coincident with $(S - S_0)$, the diffraction vector. (For $\omega = 0$, this line is also coincident with the mutual intersection of the $\chi$ and equatorial

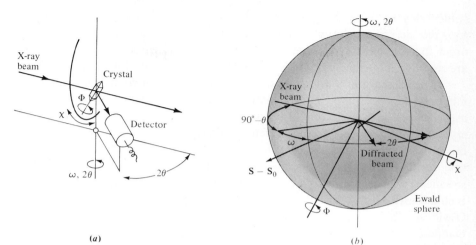

*(a)*                             *(b)*

**FIGURE 7-8**

Axial designations and geometry for three-circle and four-circle diffractometry employing Eulerian angle geometry. (*a*) Relative disposition of the basic components. (After Azároff, 1968.) (*b*) The crystal specimen is at the center of the sphere. The mechanical embodiment of the $\phi$ axis includes the goniometer head (or other crystal support) and the $\phi$ rotation mechanism. It keeps the $\phi$ axis aligned through the center of the sphere (intersection point of all axes) as it is moved around the $\chi$ circle. For the usual three-circle instrument, $\omega$ is not a variable and is fixed at zero, thus maintaining the plane of the $\chi$ circle coincident with the plane containing $S - S_0$ and the $2\theta$ axis, which plane is the symmetric bisector of the angle between the incident and diffracted rays.

planes.) Coupled $\theta - 2\theta$ motion of crystal and counter detector then brings the reciprocal-lattice point into contact with the Ewald sphere at a position within the reciprocal-space sampling region "viewed" by the counter as shown in Fig. 7-1.

The normally used scanning motions consist of crystal rotation about either $2\theta$ or $\omega$, which axes are coincident. Rotation about $2\theta$ involves a coupled $\theta - 2\theta$ motion of crystal specimen and detector; rotation about $\omega$ involves a stationary detector. The reciprocal space sampling region remains fixed in laboratory space (as presented in Figs. 7-3 through 7-5, for example) in the $\omega$-scan case but moves in the $2\theta$-scan case. Insofar as measurement of a Bragg reflection is concerned, the two scans are equivalent, even though the paths followed within the framework of the reciprocal lattice are mutually perpendicular. This point is developed further in Section IV D. In consequence (as an example), the two scans will sample most favorably different portions of the surroundings of a reciprocal-lattice point. In both scans, of course, specific consideration must be given to the size and uniformity of the reciprocal-space sampling region and to the extent of the reciprocal-lattice "point" (which may be thought of as extended to embody several specimen characteristics, e.g., mosaic spread, intrinsic diffraction profile, absorption effects, and others) in order that the most appropriate type and length of scan be selected. These aspects are considered briefly in Section V C in relation to integrated intensity measurements.[1]

The presence of a fourth axis, additional to and not coincident with any of the minimum three required by kinematic design, permits some selection of specimen aspect as well as the reflection. In crystal space this selection consists of rotation of the crystal about the normal to the chosen set of Bragg planes, $(hkl)$. In reciprocal space, this selection constitutes choice of rotation of the reciprocal lattice about the diffraction vector $\mathbf{S}$, and hence about $\mathbf{G}_{hkl}$, in addition to the required positioning of the specified reciprocal-lattice point. The usual four-circle diffractometer incorporates such a fourth axis as an $\omega$ axis defined and related to the others as shown in Fig. 7-8. Obviously, this rotation about $\mathbf{S}$ could also be supplied directly by a mechanical axis maintained parallel to $\mathbf{S}$. This actually has been done by Young, Goodman, and Kay (1964) and by Young (1969) by adding a fifth rotation to the usual four. In their design, the additional axis is kept parallel to $(\mathbf{S} - \mathbf{S}_0)$ only when $\omega = 0$. The desirability of rotating about $(\mathbf{S} - \mathbf{S}_0)$ is considered in Section V A, in connection with the detection of simultaneous diffraction effects.

*b.   Angle-setting procedures*   Several authors have devised routines for calculating the angular settings required to bring an arbitrarily selected reciprocal-lattice point into diffracting position on three- and four-circle instruments. For three-circle instruments and an initially simple orientation of a highly symmetric crystal, e.g., $\mathbf{c}^*$ of an hexagonal crystal parallel to $\Phi$ and $\mathbf{a}^*$ perpendicular to the $\chi$ plane when $\phi = 0$, the angle-setting calculations can be made rather simply. In this case, the $\phi$ shaft first may be rotated by the easily calculated amount needed to bring the $\mathbf{G}_{hkl}$ vector of interest into the $\chi$ plane on the near side of the sphere shown in Fig. 7-8. The $\mathbf{G}$ vector then can be brought into the equatorial plane by rotation of $\chi$ through $\pm(90° - \alpha)$ where $\alpha$ is the angle between $\mathbf{G}$ and $\Phi$ and where the choice of sign depends on whether $\mathbf{G}$ lies above or below the equatorial plane at the completion of the $\phi$ rotation.

For a less symmetric crystal and, particularly, for a nonspecial orientation of the crystal (to be preferred for intensity measurements in order that simultaneous diffraction effects be minimized—Section IV A) the calculations are sufficiently cumbersome so that it is worthwhile to work out a systematic general procedure such as have, e.g., Busing and Levy (1967). By the use of matrix arithmetic they have been able

---

[1] For the usual case of constant angular velocity during the scan, the Lorentz factor is

$$L^{-1} = \sin 2\theta \qquad (7\text{-}12)$$

This simple expression may be contrasted to the more complicated ones like (7-9), for example.

to keep the expressions compact, few in number, and reasonably understandable. This and closely related schemes have been incorporated directly in the computer programs for single-crystal diffractometer control. A central element of the Busing and Levy procedure is the crystal orientation matrix $\mathbf{U}$ which relates a vector $\mathbf{V}_C$, specified in a cartesian coordinate system rigidly attached in convenient orientation to the crystal specimen, to the same vector $\mathbf{V}_\Phi$, specified in a cartesian coordinate system similarly attached to the $\phi$ axis, by

$$\mathbf{V}_\Phi = \mathbf{U}\mathbf{V}_c \qquad (7\text{-}13)$$

The initial vector $\mathbf{V}$ is defined in terms of the reciprocal-lattice vectors $\mathbf{b}_i$ (where $\mathbf{b}_1 = \mathbf{a}^*$, $\mathbf{b}_2 = \mathbf{b}^*$, and $\mathbf{b}_3 = \mathbf{c}^*$) as

$$\mathbf{V} = \sum_{i=1}^{3} V_i \mathbf{b}_i \qquad (7\text{-}14)$$

and a transformation matrix $\mathbf{B}$ is then required to express $\mathbf{V}$ in terms of the cartesian system; i.e.,

$$\mathbf{V}_c = \mathbf{B}\mathbf{V} \qquad (7\text{-}15)$$

where

$$\mathbf{B} = \begin{pmatrix} b_1 & b_2 \cos \beta_3 & b_3 \cos \beta_2 \\ 0 & b_2 \sin \beta_3 & -b_3 \sin \beta_2 \cos \alpha_1 \\ 0 & 0 & 1/a_3 \end{pmatrix} \qquad (7\text{-}16)$$

and the $a$'s and $\alpha$'s and the $b$'s and $\beta$'s refer to crystal and reciprocal-lattice parameters, respectively. Since $\omega$, $\chi$, and $2\theta$ are also angles needing to be properly set, Busing and Levy find it expedient to define additional cartesian systems rigidly attached to the $\omega$, $\chi$, and $\theta$ axes within which the initial vector $\mathbf{V}$ is described as $\mathbf{V}_\omega$, $\mathbf{V}_\chi$, and $\mathbf{V}_\theta$, respectively. The transformations are

$$\mathbf{V}_\chi = \boldsymbol{\Phi}\mathbf{V}_\Phi \qquad (7\text{-}17)$$

$$\mathbf{V}_\omega = \mathbf{X}\mathbf{V}_\chi \qquad (7\text{-}18)$$

and

$$\mathbf{V}_\theta = \boldsymbol{\Omega}\mathbf{V}_\omega \qquad (7\text{-}19)$$

where

$$\boldsymbol{\Phi} = \begin{pmatrix} \cos \phi & \sin \phi & 0 \\ -\sin \phi & \cos \phi & 0 \\ 0 & 0 & 1 \end{pmatrix} \qquad (7\text{-}20)$$

$$\mathbf{X} = \begin{pmatrix} \cos \chi & 0 & \sin \chi \\ 0 & 1 & 0 \\ -\sin \chi & 0 & \cos \chi \end{pmatrix} \qquad (7\text{-}21)$$

and

$$\Omega = \begin{pmatrix} \cos\omega & \sin\omega & 0 \\ -\sin\omega & \cos\omega & 0 \\ 0 & 0 & 1 \end{pmatrix} \tag{7-22}$$

Writing the reciprocal-lattice vector **G** of the reciprocal-lattice point of interest as a column vector **h**,

$$\mathbf{h} = \begin{pmatrix} h \\ k \\ l \end{pmatrix}$$

Busing and Levy show that the normal to the selected set of Bragg planes will have the desired orientation for diffraction to occur (and to be observed with the detector) if

$$\mathbf{h}_\theta = \Omega X \Phi \mathbf{UBh} \tag{7-23}$$

has the form

$$\mathbf{h}_\theta = \begin{pmatrix} q \\ 0 \\ 0 \end{pmatrix} \tag{7-24}$$

where

$$q = h_{c1}{}^2 + h_{c2}{}^2 + h_{c3}{}^2 \tag{7-25}$$

and

$$\mathbf{h}_c = \mathbf{Bh} \tag{7-26}$$

The specific attention given to vectors, and their transformation in the Busing and Levy approach, facilitates determination of orientation matrix **U** (note that all the other transformation matrices are known a priori if the lattice parameters are known) from observations of the angle settings at which three indexable reflections are observed. Further, it facilitates least-squares refinements of **U** (and also of **B**, in cases in which the lattice parameters are not well known) on the basis of observations of several reflections. Finally, the approach lends itself to calculation of the angle settings, specifically including $\omega$, required to produce a selected rotation about the diffraction vector. (Busing and Levy refer to this as rotation in azimuth, $\Psi$.) The cited Busing and Levy paper gives full details for the practical use of these procedures. The whole procedure readily can be, and in several cases has been, programmed for implementation on an on-line computer controlling the diffractometer.

**4.** *Axis-positioning and alignment considerations* Range, accuracy, precision, and reliability of axis positioning are factors for which the requirements may vary considerably from one type of crystal study to another. Here, the differing requirements for five types of experiments are considered briefly, including lattice parameter determinations, measurement of reflection intensity by measurement of the peak height, integrated-reflection-intensity measurements, peak shape measurements, and anomalous dispersion measurements.

Very specific and detailed procedures for producing and checking alignment of three- and four-circle diffractometers have been devised by Samson and Schuelke (1967). Hoppe (1969a) discusses sources of mechanical error and methods for detecting them. The discussion here will not deal specifically with these matters; it will be concerned, instead, with estimates of the sizes of errors that may be tolerated for various purposes, rather than how the required precision may be achieved and recognized.

*a. Lattice parameter determinations; Bond's method*  A familiar method developed by Bond (1960) yields both high precision and high accuracy in lattice parameter determinations. In this method, the precise measurement is that of the angle through which the crystal must be rotated to change it from the optimum position for reflection $hkl$, occurring at position $2\theta$, to that for reflection $\bar{h}\bar{k}\bar{l}$, occurring at negative $2\theta$. It is required that the detector window and all other diffracted-beam apertures be large enough not to obstruct any measurable part of the diffracted beams. The angle through which the crystal is rotated then becomes a measure of $2\theta$ free of specimen eccentricity and absorption errors. For a lattice parameter precision of 1 part in $10^7$, at $\theta = 80°$, for example, this angle must be measured with a precision of approximately 0.24 arc sec. For such precision measurements, the alignment of the instrument must also be carefully considered. In particular, the axis of crystal rotation ($\theta$) must be perpendicular to the plane of incidence within a few minutes of arc. This alignment criterion may be discussed in terms of the nonorthogonality of the incident beam and the crystal rotation axis $\theta$ (or $\omega$), called *beam tilt*, and any lack of parallelism of the diffracting planes with the crystal rotation axis, called *specimen tilt*.

A beam tilt of 10 minutes produces an error of the magnitude just considered; i.e., approximately 0.25 second in $2\theta$, corresponding to 1 part in $10^7$ in the lattice parameter for $\theta = 80°$. For a beam tilt of 30 minutes the misalignment corresponds to 1 part in $10^6$ in the lattice parameter measurement. Burke and Tomkeieff (1968) have warned that the two errors are not additive, and have provided the following typical results from their error analysis for specimen tilt and beam tilt:

| $\alpha$ (rad) | $\beta$ (rad) | $\theta = 45°$ | $\theta = 60°$ | $\theta = 75°$ | $\theta = 85°$ |
|---|---|---|---|---|---|
| 0.001 | 0 | $5.00 \times 10^{-7}$ | $5.00 \times 10^{-7}$ | $5.00 \times 10^{-7}$ | $5.00 \times 10^{-7}$ |
| 0.001 | 0.001 | $4.14 \times 10^{-7}$ | $1.54 \times 10^{-7}$ | $3.52 \times 10^{-8}$ | $3.82 \times 10^{-9}$ |
| 0.001 | $-0.001$ | $2.41 \times 10^{-6}$ | $2.15 \times 10^{-7}$ | $2.04 \times 10^{-6}$ | $2.00 \times 10^{-6}$ |

Baker, George, Bellamy, and Causer (1968, 1969) have developed an automated instrument for precision measurement of lattice parameters with the Bond method. Through analysis of the systematic errors, they have developed a strategy which permits the computer-controlled instrument to locate peak positions by fitting a quartic to step-count data and to provide compensation for, or direct assessment of, the more serious errors. For example, their closed-error-loop strategy compensates for eccentricity (typically a source of up to 15 seconds error) of the main angle-measuring gear wheel. This is done by successive and automatic measurement of the crystal rotation angle (between $hkl$ and $\bar{h}\bar{k}\bar{l}$ reflection positions) using different portions of the gear wheel. They obtain, routinely, a precision of 1 part in $10^7$ and, thanks to the closed-error-loop strategy in particular, an accuracy nearly as good. These values compare with about 1 part in $10^5$ routinely obtained with Kossel lines (Yakowitz, 1969) and 1 part in $10^4$ to $10^5$ generally accepted as routinely attainable precision with powder methods. This precision is still low, however, compared to 1 part in $10^9$, or better, attainable in especially favorable cases with x-ray interferometry (Hart, 1969).

*b.   Difference measurements*   For many kinds of studies, interest centers on changes produced in lattice parameters rather than on their absolute values. Thermal expansion studies provide a convenient example for discussion, although irradiation effects might be a more immediately relevant example because of the radiation damage produced in many classes of specimens by the incident x-ray beam.

For measurement of differences, emphasis is placed on precision of angle measurement and specimen alignment, rather than on the accuracy of either (except that the specimen must be reasonably well aligned). Thus, eccentricity of the main angle-measuring gear wheel is not a major consideration, but the precision of sensing the axis position is. Using standard commercially available instruments without modification, Nicklow and Young (1963) measured the thermal expansion coefficients; i.e., the slopes of $a \log d$ vs. $T$ curves, of Al and AgCl over a temperature range of several hundred degrees. Typically, the precision was about 2% even though the temperature was controlled to only $\pm 0.05$ K°. The precision of $d$ value measurement was, therefore, much higher and of the order of 1 part in $10^4$. This corresponded to a precision of average angle measurement (at $\theta \approx 80$) of 0.06° in $2\theta$. In this work, no extrapolation procedures common to precision lattice parameter measurement methods were required, since the absolute value was not required. At each temperature, the apparent $d$ values used were average values obtained on both heating and cooling cycles and for pairs of reflections 180° apart in crystal orientation. The final results, including statements of precision, were based on data from different reflections occurring at different angles, for two different wavelengths, and for several different specimens of the same materials. The precision of an individual angle measurement was approximately 0.01° (2θ). [Angle determinations were made by fitting a parabola to the intensity data obtained at 0.01° (2θ) intervals.]

The precision attainable with care, as opposed to the fairly casual approach used by Nicklow and Young, is, of course, much higher. Simmons (1968) has reviewed most of the studies which have been done with x-ray measurements of thermal expansion. He concluded that routine work with commercial x-ray equipment produces relative expansion measurements with a precision of 1 or 2 parts in $10^4$. Several investigators have reported measurements with a precision of 2 to 4 parts in $10^5$ in the range 4 to 1400°K. With three-crystal instruments and with automatic scanning the expansion of essentially perfect crystals near room temperature has been measured with a precision in the range of 2 to 10 parts in $10^8$. It is evident that the required precision for angle measurement is of the order of (and slightly less than) a tenth of a degree for routine cases, and a hundredth of a degree and one ten-thousandth of a degree for the last two cases, respectively.

*c. Peak-height intensity measurements* The peak-height or stationary-counter-stationary-crystal method of data collection is preferred by some workers for reasons of speed and resolution of neighboring reflections from crystals having large unit cells, e.g., proteins. Clearly a greater emphasis must be placed on angle-setting accuracy with this method than with the integrated intensity method in which the subsequent scanning motion compensates for any small errors in the initial settings. In reciprocal space the problem is to orient the crystal so that the central portion of the chosen reciprocal-lattice point lies within the maximally (and, presumably, approximately uniformly) dense region of the reciprocal-space sampling region.

For the immediate purpose of obtaining some numerical estimates of angle-setting errors that can be tolerated, it is convenient to approach the problem in direct space. The problem, then, is to set the diffractometer for the maximum or other selected portion of a diffraction profile which is itself the convolution of at least three functions, a specimen shape function $\tau(\phi, \omega, \chi, \theta, \Delta\theta)$, a mosaic-spread function $\eta(\Delta\theta)$ (isotropic case), and an x-ray source profile function $g(\lambda, K, \phi, \chi_s)$. For present purposes one can neglect intrinsic diffraction broadening. Any significant amount present would introduce still another profile function $f(\theta, \Delta\theta)$ to be convoluted with the others. The specimen shape function $\tau(\phi, \omega, \chi, \theta, \Delta\theta)$ is meant to include both specimen-size and specimen-absorption effects. To provide an idea of how well the diffractometer must be set, these function are examined one by one.

Take, for example, a spherical specimen of radius $r$ at the distance $l$ from the source of x rays ($l \gg r$ and $l \gg$ source dimension normal to the plane of incidence). Then

$$\tau(\phi, \omega, \chi, \theta, \Delta\theta) = \tau(\theta, \Delta\theta) = \tau\Delta\theta = \frac{3}{4}\left[1 - \frac{l^2}{r^2}(\Delta\theta)^2\right]\frac{l}{r} \qquad (7\text{-}27)$$

for $-r/l \leq \Delta\theta \leq r/l$. Typical values are 0.05 to 0.5 mm for $r$, and 20 to 30 cm for $l$. For the values $r = 0.2$ mm and $l = 24$ cm the full width of $\tau(\Delta\theta)$ at half maximum is $\sim 0.06°$ $(\theta)$. Obviously, this is a rather sharp profile. The effect of absorption can be to sharpen further the profile in the neighborhood of its maximum value.

The mosaic spread function $\eta(\Delta\theta)$ is usually taken to have a Gaussian shape. Although crystals vary greatly in respect to the sharpness of this function, full-width-at-half-maximum (half-width) values of $\frac{1}{4}$ to $\frac{1}{2}$ degree are common. (If this mosaic spread is much less than about 0.1°, it is likely that extinction may be severe.) To provide a numerical example, take 0.24° $(\sim\frac{1}{4})°$ as representative. In this case,

$$\eta(\Delta\theta) = \frac{e^{-\Delta\theta^2/k^2}}{k\sqrt{\pi}} = \frac{e^{(\Delta\theta/0.145°)^2}}{\sqrt{\pi}\,0.145°} \tag{7-28}$$

The convolution $\zeta(\Delta\theta)$ of these first two specimen-related functions,

$$\zeta(\Delta\theta) = \int \eta(\Delta\theta')\tau(\Delta\theta - \Delta\theta')\,d(\Delta\theta') \tag{7-29}$$

is that of a Gaussian with a truncated parabola. Since the Gaussian in this case has the broader profile, the convolution will give a profile of pseudo-Gaussian shape with a half-width somewhat less than the sum of the two half-width values, e.g., 0.30° $(\theta)$.

It should now be evident that no matter how irregular the source intensity function $g(\lambda, K, \phi, \chi_s)$ is, the observed profile, being the convolution of it with $\zeta(\Delta\theta)$, cannot have peaks sharper than that of $\zeta(\Delta\theta)$. [Of course, an irregular source function would be undesirable for other reasons, just as, in most cases, would be one narrow compared to $\zeta(\Delta\theta)$. However, that is another matter and is not dealt with here.]

The question of angle-setting precision can therefore be considered, in this case, as one of setting for the maximum of an approximately Gaussian profile of half-width $\approx 0.30°$ $\theta$. For either a Gaussian or a parabolic profile shape, for setting within 5% of the intensity maximum, the crystal position must be set within approximately 0.05° $(\theta)$ of the correct $(\Delta\theta = 0)$ position; for 1% intensity error the corresponding angle tolerance is approximately 0.02°.

It should be borne in mind that most of the breadth, and hence the tolerance of $\pm 0.02°$ missetting with only 1% intensity error, arises from the mosaic spread for which a "typical" value was assumed. The mosaic spread could be less with the result that the angle-setting tolerance would be less. To a first approximation, the angle tolerance for a given intensity error depends approximately linearly on the mosaic spread [in this example in which $\eta(\Delta\theta)$ is much broader than $\tau(\Delta\theta)$].

The calculation here has been carried out for what is, in effect, a point source; i.e., a source profile $g(\lambda, K, \phi, \chi_s)$ representable by a delta function. The actual source profile will not be a delta function but may well be narrow compared to $\zeta(\Delta\theta)$ or contain peaks that are narrow compared to $\zeta(\Delta\theta)$. For example, a 100-$\mu$m microfocus source subtends an angle of 0.024° at a specimen 24 cm distant from it. The accompanying source profile $g(\lambda, K, \phi, \chi_s)$ would not contribute significant breadth to its convolution, $h(\Delta\theta)$, with $\zeta(\Delta\theta)$. On the other hand, a uniformly intense source $g(\lambda, K, \phi)$ of 1-mm dimension would, its source profile being essentially a rectangle of width $\Delta\theta = 0.24°$.

One concludes from these numerical discussions that crystal-setting accuracy (not merely precision) of $\pm 0.02°$-$\theta$ is an appropriate goal. Contributing to the actual crystal-setting errors are errors in $\omega$, components of those in $\phi$ and $\chi$, those arising from misplacement (off center) of the specimen, and those from mechanical misalignment of the diffractometer axes. It is not a particularly difficult task for a three- or four-circle diffractometer to be constructed such that the maximum run-out on all axes is 0.025 mm ($\sim$0.001 in.); i.e., the sphere of confusion created at the approximate intersection of all axes has a radius of 0.025 mm or less. Again for $l = 24$ cm, this error corresponds to an angular error of approximately 0.006°. Allowing up to 0.04 mm for the mispositioning of the specimen and taking the worst-case position for $\phi$ and $\chi$ axes, one calculates that a precision of $\pm 0.01°$ in setting each of the angles, $\phi$, $\chi$, and $2\theta$, leads to an expected cumulative error $< 0.02°$ in the actual crystal $\theta$ setting. Thus it appears that 0.01 accuracy in all angle settings and 0.025 mm (0.001 in.) error in the intersection of axes are allowable tolerances for attainment of 1% precision (i.e., accuracy of measuring true intensity maximum) in intensity measurements by the stationary-crystal–stationary-counter method.

*d.  Integrated intensity measurements*  The scanning motion ordinarily employed in the measurement of integrated Bragg intensities relaxes the requirements on angle-setting precision. For continuous-scan methods a further accuracy requirement is imposed on constancy, or reproducibility according to a known program, of the scanning speed. The fact that accuracy is less important in the scanning (moving-crystal) case than in the stationary-crystal case can be easily appreciated by a simple direct-space construction relating to the diffraction conditions. For this purpose, imagine first that the crystal specimen diffracts only when the Bragg condition is exactly satisfied. Consider a spherical specimen as shown in Fig. 7-9a whose reflecting planes are perpendicular to the plane of the drawing. Of all possible rays incident from the left at the Bragg angle $\theta$, $a$ and $b$ are the two most widely separated such rays that can be diffracted. Rays $c$ and $d$ play the corresponding role for incidence from the right. One sees that rotation of the specimen about the $y$ axis would not affect these diffraction conditions; thus the rays $a$ and $c$ and the rays $b$ and $d$ can be taken to be the traces of two cones, with axes on $y$ and semiapex angles $90 - \theta$, between or on which all acceptable incident rays must lie. The region bounded by these cones is called the *acceptance region*. Obviously, the source (real or virtual) of these acceptable incident rays traveling through this region must also lie in the region; any part of an

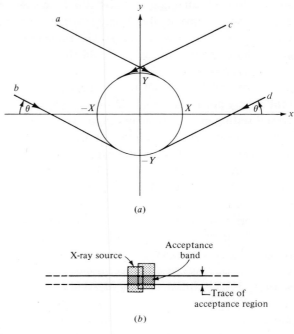

(a)

(b)

FIGURE 7-9
Acceptance region.

x-ray source lying outside of the acceptance region will, in the simplified picture, be ineffective in the diffraction process. Taking $R$ as the radius of the spherical specimen, Fig. 7-9a, note that the displacement of the two defining cones along their common axis is $2R/\cos\theta$, whereas the perpendicular distance between the two defining cone surfaces is simply $2R$. (In fact, of course, the mosaic spread and nonzero width of the intrinsic diffraction profile permit the specimen to accept, for diffraction, incident radiation over a range of angles centered on the nominal $\theta$. However, for present purposes little is gained from such added complexities, and it is better to keep the definition of the acceptance region in terms of the exact nominal diffraction angle, $\theta$.)

Now imagine this acceptance region, which is specific to the reflection of interest, to be rigidly attached to the specimen as it is rotated during the orientation and scanning processes. The question of whether diffraction will occur then can be stated as that of whether the acceptance region passes over the x-ray source. For a correct integrated intensity to be measured directly, every point on the (visible) x-ray source must be moved entirely through the acceptance region from surface to surface. (At the same time, of course, all diffracted rays must be appropriately detected.)

It is a common case that the source size projected perpendicular to the acceptance-region surface is $>2R$. Let the intercept of the acceptance region with the source be called the *acceptance band*, indicated in Fig. 7-9b as seen from the specimen position. The process of scanning through a reflection now becomes the process of moving the acceptance band across the x-ray source. If the crystal scanning motion is in $\theta$ or $\omega$, such motion of the acceptance band is, nominally, perpendicular to its "long" dimension, i.e., vertically in Fig. 7-9b.

Let any errors in specimen orientation be expressed in terms of the angles $\delta$, $\sigma$, and $\Delta\omega$ measured about axes parallel to the incident beam, parallel to the diffraction vector $(S - S_0)$, and parallel to $\theta$, respectively. The effect of a nonzero $\Delta\omega$ will be to shift the whole reflection peak to higher or lower values of the scanning variable without affecting the total intensity diffracted during the scan. The detected intensity can be affected adversely, however, if the detector aperture is set so narrowly as to obscure a portion of the shifted intensity distribution.

The effect of a misorientation about the diffraction vector $(\Delta\sigma \neq 0)$ has no effect for a symmetric specimen in the absence of multiple diffraction effects. If the shape or composition distribution of the specimen (or irradiated portion thereof) does not have rotational symmetry about the diffraction vector, the effect of nonzero $\Delta\sigma$ must, of course, be treated specifically for each individual specimen condition. (Even then, however, source-illumination gradients perpendicular to the scanning motion, along the horizontal direction in Fig. 7-9b, are unlikely to produce any significant errors in the integrated intensity measurements.)

The effect of a nonzero $\delta$ will be to increase the length of the path traveled by each source point, through the acceptance band, by the factor $1/\cos \delta$. Consequently, the integrated intensity measured for that reflection will be increased by the same factor. It is obvious, however, that for any reasonable missettings in the $\delta$, the integrated intensity will not be significantly affected for most purposes. For example, a missetting in $\delta$ of more than $8°$ is required to produce a $1\%$ error in the integrated intensity.

With the help of the acceptance-region concept, for measurement of integrated intensities by scanning in $\theta$ or $\omega$, the requirements on precision in the setting of the specimen orientation are now seen to be very lax indeed, and errors of several degrees can be tolerated. The actual limitations on setting errors are then seen to be imposed by the detector-aperture system; its size, positioning, and motion must be such that effectively all (or a known portion) of the diffracted intensity is actually detected. Obviously, if the aperture system used is slightly larger than the minimum size,[1] no stringent diffractometer-setting requirements need be imposed. To obtain an idea of the tolerance, take as an example a 0.5-mm-diameter specimen 25 cm from an x-ray

---

[1] For discussions of the minimum aperture dimensions required, see Section IV B and also Appendix D of Alexander and Smith (1962), Alexander and Smith (1964), and Ladell and Spielberg (1966).

source of effective (projected) dimension $1 \times 1$ mm$^2$ and a detector aperture placed (1) 5 cm from the specimen and (2) 20 cm from the specimen. In either case, even for the $2\theta$ scan case with no mosaic spread, the aperture dimension measured perpendicular to the plane of incidence must be at least as large as the specimen dimension plus the distance subtending at the specimen the angle that is subtended there by the source; i.e., $0.5 + 0.8 = 1.3$ mm in case (2). If the aperture is made larger than the minimum size by as little as 0.2 mm (as would seem to be prudent in any event in view of the minor increase in background it would produce) the setting tolerance on $\delta$ becomes

$$\pm\Delta\delta = \frac{0.1}{20 \times 2 \sin\theta \cos\theta} = \frac{1}{200 \sin 2\theta} > \tfrac{1}{4}° \qquad (7\text{-}30)$$

in case (2) and greater in case (1). A similar calculation can be carried out for tolerance in $\omega$. As already pointed out, the tolerance on $\sigma$ is ordinarily very large and may be effectively infinite. Thus it is concluded that an angle-setting accuracy of 0.1°, and sometimes poorer, will be adequate for direct (scanning) integrated intensity measurements to better than 1% except, possibly, in the rare case in which strong specimen absorption, highly unsymmetric specimen shape (e.g., protruding whiskers), and strong source-illumination gradients occur together. In reaching this conclusion an assumption has been made which the reader may wish to examine more fully in light of the discussion of multiple scattering in Section IV F. It is that settings about the diffraction vector ($\sigma$ settings) calculated to *avoid* multiple scattering are not critical and, hence, need not be attained with an accuracy better than about 0.1°. While good for nearly all specimens used for crystal-structure analysis, that assumption would not be justified for highly perfect specimens in which a great deal of strong simultaneous diffraction occurred.

The tolerable errors in specimen position (e.g., centering) are, for detector apertures oversize by 0.2 mm, approximately $\pm 0.05$ mm for an extreme of case (2); i.e., displacement either perpendicular to the plane of incidence or in that plane perpendicular to the diffracted-beam path. This estimate suggests that 0.2 mm is probably too small an amount for the detector aperture to be oversized. A better "balance of inconvenience" would obviously be obtained by making the apertures still larger, say 0.4 to 0.6 mm oversize in case (2) and 0.3 to 0.5 mm in case (1), and accepting the relatively small attendant increase in background in return for less difficulty in positioning specimens, particularly those of unsymmetric shape.

While, as just shown, the requirements on specimen orientation and positioning are not demanding, those on the scanning motion are. In order to evaluate diffraction profiles of the various shapes, first consider the errors in measuring a given increment on the profile, such as the one with width $x_0$ in Fig. 7-10. It will turn out that, as also may appear to be obvious, considerable variation in the scanning speed, including stops and starts, may be accommodated if the (arithmetic) average speed over the

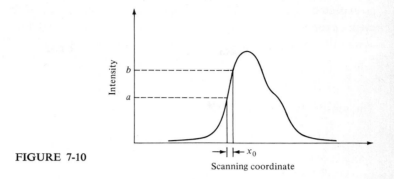

FIGURE 7-10

increment is within acceptable limits and if the increment is taken small enough so that the change in intensity over the width $x_0$ of the increment can be approximated by a straight line. It is perhaps also obvious that the fractional error in measurement of the increment intensity is approximately that in the average scan velocity—or, more directly, the time devoted to measurement of the increment (the latter method of statement also applies directly to the step-scan method). However, the degree of the approximation does depend on both speed and intensity variations, as can be readily shown for the simple case of linear variations in each as the increment is continuously scanned. (Statistical fluctuations, which may be treated separately, are neglected here.) Let the linear variation in intensity $I$ and scanning speed $V$ be represented by

$$I(x) = a + (b - a)\frac{x}{x_0} \qquad (7\text{-}31)$$

and

$$V(x) = V_0\left(1 + k\,\frac{x}{x_0}\right) \qquad (7\text{-}32)$$

where $a$ and $b$ are the intensities (usually in counts per unit time) at the edges of the increment shown in Fig. 7-10, $V_0$ is the scanning velocity at the first edge of the increment and $k$ is the appropriate constant.

The required integral is actually one over time, $I(x)\,dt$. Making the necessary substitutions from $V = dx/dt$ to permit the integration to be carried out in $x$, one finds

$$I_{\text{meas}} = \frac{ax_0}{V_0}\frac{\ln\,(1 + k)}{k} + \frac{(b - a)x_0}{V_0}\frac{k - \ln\,(1 + k)}{k^2} \qquad (7\text{-}33)$$

For $k = 0$ (i.e., velocity constant) (7-33) becomes

$$I_{\text{meas}} = \frac{a + b}{2}\frac{x_0}{V_0} \qquad (k = 0) \qquad (7\text{-}34)$$

as expected. One is interested primarily in the dependence of the intensity-measure-ment error on the speed variation, represented in $k$, and the change of intensity across the increment, $b - a$. Let $b = a + qa$ for convenience. Then (7-33) becomes

$$I_{meas} = \frac{ax_0}{V_0} \frac{\ln(1 + k)}{k} + q \frac{k - \ln(1 + k)}{k^2} \qquad (7\text{-}35)$$

The sensitivity to slope of the profile $(b - a)/x_0$

$$\frac{dI}{dq} = \frac{ax_0}{V_0} \frac{a}{k^2} \frac{k(1 + q)}{1 + k} - \ln(1 + k)\left(1 - \frac{2q}{k}\right) - 2q \qquad (7\text{-}36)$$

For small $k$ this simplifies to

$$\frac{dI}{dk} \xrightarrow[k \to 0]{} \frac{-V_0 a}{x_0}\left(\frac{1}{2} + \frac{q}{3}\right) \qquad (7\text{-}37)$$

(Note that $aV_0/x_0$ would be the correct integrated intensity for an increment for which $b = a$, i.e., a flat-topped increment.) For such an increment, equation (7-37) tells that a 1% speed variation ($k = 0.01$) will produce an error of approximately $\frac{1}{2}\%$ in the integrated intensity measurement, as one might expect almost intuitively. It also follows that, if $b > a$; i.e., $I(x)$ increases in the scanning direction, the error from a given constant acceleration $k$ is increased. On the other hand, partial compensation takes place when $k$ and $q$ are of opposite signs.

What has been calculated here for an assumed trapezoidally shaped increment of the profile is the measurement due to a linear change in scanning speed during the course of that particular measurement. For the total error due to scan-speed error one must also include the full error in the scan speed at the start of the measurement of the increment. With this consideration and (7-37) and with $q < \frac{3}{2}$, it may be concluded that the total error (due to linear scan-speed variation) in measuring the integrated intensity of any profile will be less than the total scan-speed error accumu-lated over the entire measurement interval. In fact, it could be considerably less because the quantity $q$ will be positive on one side of the profile and negative on the other, thus producing some partial compensation of its contribution to the total error, as shown by (7-37).

Since maintenance of scan-speed constancy to much less than 1% variation, e.g., 0.1%, is not unduly difficult, and since any more detailed calculation would have to take into account the detailed shape of the profile and would therefore not be of general applicability, this particular line of inquiry is now concluded. One arrives at the apparently intuitively obvious result that the total measurement error due to a linear scan-speed variation is less than the total scan-speed error accumulated during the measurement. This statement is, of course, sufficiently conservative to cover the case of any monotonic change in scan-speed variations, whether or not it is linear.

If the scanning speed varies in a nonmonotonic but systematic fashion, the error calculation must, almost certainly, be carried out for the specific case in order to be useful. A general statement can be made that the total error will be less than that extrapolated from the greatest speed variation occurring at any point in the scan. Such a statement is not very useful because the actual errors are likely to be very much less than this upper unit, possibly an order of magnitude or more.

If, however, the nonmonotonic scanning-speed errors are random as judged from increment to increment, another calculational method can be used. This model of random speed variation from increment to increment fits best the case of step scanning, the process in which the detector position is changed incrementally. The effective scanning speed for each increment is the width of the increment in the scanning variable divided by the time the detector is permitted to remain at that particular position. The effects of such random errors on the integrated intensity have been discussed in Chap. 6 (Section VII B) and the dependence of the resulting error on the reflection-profile slope is brought out there (6-131). The resulting expressions tend to be sufficiently complicated so that no simply perceived general conclusion can be drawn; instead one must apply them in detail to a specific case of interest.

As for the case of continuous scanning, of course, it can be said that the total percentage error in the integrated intensity measurement will be less than the maximum percentage error in the effective scanning speed; i.e., counting time per step, for any step. Further, for the simple case of an integrated intensity $I$, measured by a step-scanning process involving $N$ steps to cover the entire profile, $I_i$ being the intensity measured for the $i$th step and $\sigma_i$ the standard deviation in the measurement

$$I = \sum_{i=1}^{N} I_i \pm \left( \sum_{i=1}^{N} \sigma_i^2 \right)^{1/2} \qquad (7\text{-}38)$$

Unlike the case for random errors in counting statistics, for random errors in timing at each step the standard deviations expressed in percent would, ordinarily, be similar for all steps. Thus it can be simply concluded that the error in the integrated intensity due to random timing variations in the step-scanning process will be similar to the random error in timing of an individual step.

## III. WAVELENGTH DISCRIMINATION

### A. Need for discrimination

A typical x-ray source ordinarily emits several characteristic radiations superimposed on a continuous spectrum, ranging from the short-wavelength limit $[\lambda_{swl} = 12.35$ A kV$/V_{max}$, where $V_{max}$ is the maximum potential difference applied to the x-ray tube] to the longest wavelength (ordinarily 2 to 3 A) for which absorption by the x-ray tube window is not complete. A further intensity variation in the incident beam spectrum occurs with increased path length because of differing air absorption for differing wavelengths. Even for x rays of a single wavelength incident on the specimen, x rays having different wavelengths may be scattered to the detector; i.e., those due to Compton scattering, to specimen fluorescence, to thermal diffuse scattering (although the wavelength change in this case is not detectable with ordinary equipment) and to elastic scattering, including Bragg scattering.

In any scattering experiment some discrimination between these wavelengths is required because both scattering intensity and scattering angle are wavelength dependent. Historically, for many studies using single-crystal specimens, it was found that the characteristic radiation sufficiently dominated the scattering so that the only wavelength-selection process required was the distinction between the two dominant characteristic wavelengths, $K\beta$ and the combined $K\alpha$ consisting of $K\alpha_1$ and $K\alpha_2$. A simple $\beta$ filter thus served well and also provided the added benefit of eliminating a large part of the white radiation background. This method has several disadvantages. Both in diffractometer scanning and in photographic recording, the Laue streak must be measured in order that its extrapolated value "under" the Bragg peak can be subtracted out to leave only the contribution of the characteristic radiation (see Section IV C). Furthermore the method does little to improve the signal-to-noise ratio, and it is applicable only to crystalline specimens, primarily for measuring the intensity distributions in relatively sharp maxima.

For studies of noncrystalline materials and for studies of either diffuse or inelastic scattering from crystalline materials, it is essential for quantitative work that some wavelength-selection device be used. If not, the x-ray intensity measured at any scattering angle is determined by the convolution of the single-wavelength specimen-scattering function with the incident wavelength distribution. (Obviously if this convolution can be eliminated experimentally, considerable ambiguity in the interpretation of the data will be removed.) If the wavelength discrimination is sufficiently effective and controllable, diffractometry can even be done by scanning in wavelength rather than angle, as has been demonstrated by Giessen and Gordon (1968). In fact, a powder diffractometer with no moving parts, constructed along the lines described by Giessen, is now being offered commercially.

## B.  Discrimination methods

There exists an extensive literature on methods and devices for wavelength discrimination in x-ray detection. For immediate purposes, however, a summary of only certain aspects will suffice. The nondispersive methods of interest are electronic pulse-height analysis, the use of filters, and the use of specific fluorescence from an absorber placed in the scattered beam. The dispersive methods are those involving total reflection or crystal diffraction. After each of these is considered in some detail, their relative effectiveness will be compared at the conclusion of Section III.

**1.** *Pulse-height analysis*  As already indicated in Section V D 5 of the preceding chapter, pulse-height discrimination depends on the existence of a unique relation between the mean amplitude of the electrical pulse put out by the detector plus its immediately associated circuitry and the wavelength of the x-ray beam being detected; i.e., the mean pulse amplitude is a definite function of incident photon energy. Typically, a detector produces one pulse for each quantum (here, x-ray photon) counting event. The pulse passes first through a preamplifier, located adjacent or very close to the detector. Second, it passes through a more remotely located linear amplifier and thence to a pulse-height analyzer (PHA). The PHA contains triggering circuits so arranged (e.g., two Schmidt triggers set at different levels and feeding into an anti-coincidence circuit) that an output pulse is produced only if the input pulse has an amplitude lying between $E$ and $E + \Delta E$. Here $E$ is the base line or threshold level and $\Delta E$ is the window size; $E$ and $\Delta E$ are normally measured in volts.

The practical limitations in pulse-height analysis reside in the counter and the preamplifiers, rather than in the circuitry of the electronic PHA; PHA's with resolving times less than 100 nanoseconds are routinely available—though not always incorporated in commercial equipment. Thus, with the reminder that the resolving time limitation of a PHA applies to the pulse train presented to it rather than that issuing from it, various counter-preamplifier combinations are considered below. Geiger counters, of course, provide pulses with amplitudes essentially independent of the x-ray photon energy and are, therefore, useless for wavelength discrimination. In these discussions of counter-preamplifier combinations, primary interest centers on the pulse-amplitude distribution; i.e., the distribution of amplitudes of output pulses for a fixed x-ray photon energy. The resolving or dead time, i.e., the minimum time between two photon-capture events, if they are to be registered separately, and the quantum counting efficiency are of secondary interest.

A pulse-amplitude distribution is shown schematically in Fig. 7-11 wherein $W$ is the half-width. If $W$ were zero or effectively so, electronic PHA would provide all the wavelength discrimination needed. It is not zero because there is always a certain randomness in the location of the photon-capture event and, particularly, in the events immediately following. In the proportional counter the proximity of the photon capture event to the central wire affects pulse size. In the scintillation counter the output pulse size varies with location of the capture event in relation to the optical path to the first dynode of the photomultiplier tube, with the statistics of the photomultiplier action, and with the statistics of the several light-producing collisions (e.g., as many as 25) that the photoelectron produced in the scintillation crystal undergoes in giving up the energy it has obtained from the x-ray photon. At the end of the

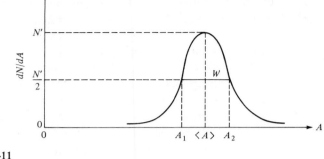

FIGURE 7-11

photoelectron's path, for example, there exists the possibility of a few collisions whose radiation contributes to the scintillation light or for many "small" collisions to produce radiant wavelengths too long to pass efficiently through the required optical path.

In addition, in a scintillation counter there is always the possibility of occurrence of the two-step process leading to an "escape peak." In the first step, the initial photoelectron ejects a bound electron, e.g., from the $K$ or $L$ shell of an atom in the scintillation crystal. In the second step, the shell is refilled and the characteristic photon subsequently emitted escapes from the scintillation crystal without interacting with it. The pulse amplitude for the initial photon is thus reduced by this characteristic amount. The result is to produce in the pulse-amplitude distribution a small second peak shifted from the main one by the pulse amplitude characteristic of the escaped-photon energy. A good PHA instrument, however, discriminates very successfully against this escape peak.

A useful figure of merit for comparing resolutions is the ratio of the width $W$ to the mean amplitude $\langle A \rangle$. Typical values for a scintillation counter and a gas-proportional counter (Miller, 1957) are compared in Table 7-1. The dead times for these two counters are approximately $0.5 \times 10^{-6}$ sec and $1-10 \times 10^{-6}$ sec, respectively (Mirken, 1964). As the table shows, the wavelength discrimination ability of the gas-proportional counter is about twice that of the scintillation counter. In x-ray spectrometry, for example, a gas-proportional counter followed by a PHA unit set for 50% transmission of a desired (characteristic) $K\alpha$ wavelength can discriminate effectively against the $K\alpha$ wavelength of another element for which the atomic number differs by 2 or 3 in the $Z \approx 20$ to 30 range. As is shown in Fig. 7-12, the two types of counters differ markedly in their spectral response; i.e., quantum-counting efficiency, as a function of wavelength. It is clear that the scintillation counter is a much more efficient detector, particularly at the shorter wavelengths, e.g., Mo $K\alpha$ (0.71 A) or Ag $K\alpha$ (0.56 A). The spectral response of the proportional counter, however, of itself and without the use of PHA, gives some very mild discrimination against extra-

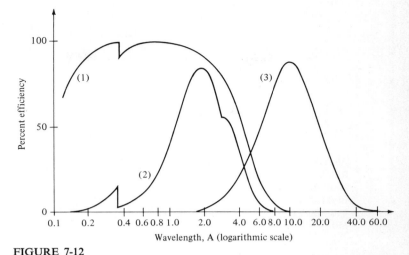

**FIGURE 7-12**
Relative responses of (1) a scintillation counter equipped with a thallium-activated NaI crystal; (2) a xenon-filled proportional counter; (3) proportional counter containing $N_2HeCH_4$. (After Azároff, 1968.)

neous wavelengths. This can be exaggerated by filling the detector with a gas that produces the peak of the response curve (Fig. 7-12) at or near the wavelength of interest.

Much better nondispersive wavelength discrimination is provided by the recently introduced semiconductor detectors. Although their development began around 1959–1960 (Coche, 1968), their application in the x-ray range did not become practical until the end of the following decade. To date, their utilization in x-ray diffraction has been quite limited although they are utilized increasingly in special applications of x-ray spectroscopy to chemical analysis. The principal advantages of semiconductor detectors stem from their high quantum-counting efficiencies coupled with an ability to handle relatively high counting rates ($\simeq 5 \times 10^5$ cps) or high-energy resolution at lower counting rates (e.g., 0.6 keV at 20,000 cps for 14.4-keV radiation).

**Table 7-1  COUNTER CHARACTERISTICS**

| Counter | $W/\langle A \rangle$ for | | | |
|---|---|---|---|---|
| | Al $K\alpha$ | Mn $K\alpha$ | Cu $K\alpha$ | Ag $K\alpha$ |
| Scintillation | — | 0.50 | 0.42 | 0.27 |
| Gas-proportional | 0.47 | 0.25 | 0.21 | 0.13 |

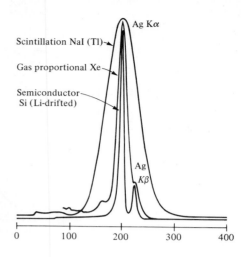

Scintillation NaI (Tl)

Gas proportional Xe

Semiconductor
Si (Li-drifted)

Ag Kα

Ag
Kβ

0    100    200    300    400

**FIGURE 7-13**
Resolution capabilities of three types of x-ray detectors. (Courtesy of Philip G. Burkhalter and William J. Campbell, U.S. Bureau of Mines, College Park, Md.)

The type of semiconductor detector which seems most promising for the x-ray range is the *n-i-p* type now represented by the Li-drifted Ge or Si detectors. The operation of this device as a radiation detector is somewhat like that of a solid-state analog of an ionization chamber. An x-ray photon incident in the Li-drifted zone ionizes Li atoms and thus creates pairs of charged particles which are attracted to opposite electrodes. If all charges created are collected (excepting, of course, the effectively immobile $Li^+$ ions), the pulse size is proportional to the number of charge pairs created and that, in turn, is dependent on the energy of the incident photon. (It should be noted that both the mobility and, particularly, the solubility of Li in Ge are strongly temperature dependent. In consequence, such devices must be kept at cryogenic temperatures both during their manufacture and at all times thereafter, lest the Li atoms migrate and precipitate out as clusters.)

Except for the possible inconvenience of maintenance of cryogenic temperatures, this type of semiconductor detector appears to offer outstanding advantages for x-ray diffractometry. As Fig. 7-13 shows, the energy resolution of these detectors is very much better than that even of a gas-proportional counter. In addition, they are now very stable and have quantum-counting efficiencies as high as or higher than that of the scintillation counter. Finally, when the best energy resolution is not required, the dead time can be less than 100 nanoseconds.

This type of semiconductor detector can provide considerable discrimination against $K\beta$ in the presence of $K\alpha$ x radiation, as is indicated by the resolution of Ag $K\alpha$ and $K\beta$ shown in Fig. 7-13 for a Li-drifted Si detector. At higher energies, of interest in gamma-ray spectroscopy but not in x-ray diffraction, these counters provide better resolution than does a crystal monochromator. At the lower energies of interest

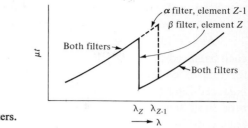

**FIGURE 7-14**
Ideally balanced filters.

in x-ray diffraction, however, the relative resolution of these counters is not presently as good as that. For example, for a particularly carefully fabricated Si(Li) detector, the $W/\langle A \rangle$ value is 0.030 for Fe $K\alpha$ radiation ($\lambda = 1.94$ A) and 0.019 for a 14.4-keV gamma ray ($\lambda = 0.86$ A, somewhat longer than that of Mo $K\alpha$, 0.71 A). Such resolution is remarkable for a nondispersive unit, particularly when the low energy is kept in mind. In the energy range of interest to x-ray diffraction, however, the relative resolution of these counters is less than that of crystal monochromators. Thus, at their present state of development, semiconductor detectors can be employed in various applications, and future developments should enable the utilization of their high efficiency, good stability, good resolution, and low dead time in an even wider range of applications.

**2.  *Balanced filters***   The effective passband (or window) of a pair of balanced filters is sharply defined by the absorption edges of the two filter materials employed, as demonstrated in Fig. 7-14. As their name implies, the two filters are balanced in thickness so that their respective absorption of x rays lying outside the passband, that is, for $< \lambda_Z$ and $> \lambda_{Z-1}$ in Fig. 7-14, ideally is exactly the same. The intensity transmitted through each must be measured independently, but the difference between the two transmitted values is the intensity of x radiation lying between the two absorption edges. Either two separate runs can be employed or automatic alternation of the filters, with approximately separated recording, during a single run (McKinstry and Short, 1960; Jones, 1953). Computer control or other automated operation of a diffractometer lends itself quite easily to such alternation of the filters, so that the intensity data for each particular reflection can be collected through each filter sequentially before the next reflection is measured.

For ideally balanced filters, the wavelength discrimination is as sharply defined as is the absorption edge of either filter. This wavelength cutoff is much sharper than that obtainable with a PHA, and the sharpness is limited only by the fine structure of the absorption edge, usually negligible for x-ray diffraction. Typical choices of anodes, the associated active elements in the balanced filters, and the absorption edge are given in Table 7-2, drawn in part from Furnas (1957) and from vol. III of the *International tables for x-ray crystallography* (1962).

As is evident from Table 7-2, a considerable range of wavelengths is passed by each balanced-filter pair. The difference in the shape of the passband makes it difficult for one to compare the relative passband widths for the various techniques; the balanced-filter analog of the passband profile shown in Fig. 7-11 is a rectangle. Let $p$ be the fractional height at which $W$ is measured on the passband profile. For a balanced-filter pair, $W_p/\langle A \rangle$ ($\approx \Delta\lambda/\lambda$ in Table 7-2) has the same value for all $p$. For $p = \frac{1}{2}$, as it was assumed in Table 7-2, $W_p/\langle A \rangle$ ranges from $\frac{1}{2}$ to $\frac{1}{10}$ that for the scintillation counter, from $\frac{1}{5}$ to 1 for the proportional counter, and from 2 to 20 times that for the Li-drifted Si counter, depending on the wavelengths involved. Obviously, as $p$ increases, the balanced filters appear in a more favorable light. However, a Xe-filled proportional counter used for Cu $K\alpha$ radiation with the PHA set for 50% transmission would provide a wavelength discrimination roughly comparable to (though perhaps not quite as good as) that provided by Co-Ni balanced-filter pair with approximately the same overall intensity loss. The balanced-filter pair does have the advantage that the discrimination is not dependent on the electronic stability of a PHA unit operated under demanding conditions. It has the disadvantage that the filters must be properly prepared and balanced and must be replaced in exactly the same location and orientation for each measurement. The choice of optimum thick-

Table 7-2 SOME BALANCED-FILTER PARAMETERS

| Anode | | $\alpha$ filter | | $\beta$ filter | | |
|---|---|---|---|---|---|---|
| Element | $K\alpha_1$(A) $K\alpha_2$ | Element | abs. edge (A) | Element | abs. edge (A) | Passband relative width $\Delta\lambda/\lambda(K_\alpha)$ |
| Cr | 2.28962 2.29351 | Ti | 2.497 | V | 2.269 | 0.100 |
| Fe | 1.93597 1.93991 | Cr | 2.070 | Mn | 1.896 | 0.090 |
| Cu | 1.54051 1.54433 | Co | 1.608 | Ni | 1.488 | 0.078 |
| Mo | 0.70926 0.713543 | Y Sr | 0.72762 0.76969 | Zr Zr | 0.68877 0.68877 | 0.055 0.114 |
| Ag | 0.559363 0.563775 | Mo Ru Ru Ru | 0.61977 0.560 0.560* 0.560* | Pd Pd Rh Mo | 0.509 0.509 0.533 0.61977 | 0.196 0.092 0.049 0.106 |

* An especially interesting situation occurs here in that the absorption edge of Ru lies between the $K\alpha_1$ and $K\alpha_2$ wavelengths of Ag; hence the choice of Ru and Rh or Ru and Pd as balanced filters eliminates $K\alpha_2$ leaving $K\alpha_1$, while the Ru, Mo choice eliminates $K\alpha_1$, leaving $K\alpha_2$ in the difference.

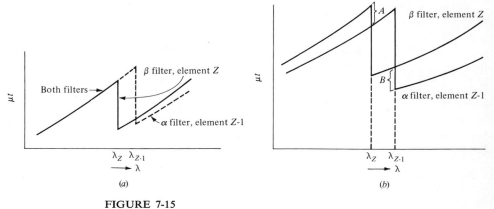

FIGURE 7-15

nesses for balanced filters has been discussed by Kirkpatrick (1944); Soules, Gordon, and Shaw (1956); and Bol (1967), among others.

The matter of balancing the two filters is not quite as straightforward as one might wish. One reason is that the relative sizes of the absorption discontinuities differ for two materials matched on one side of the passbands, as shown in Fig. 7-15$a$. To make the discontinuities equal, so that a match can be obtained on both sides of the passband, a third material having an absorption edge far removed from the passband is required. The absorption coefficient $\mu$ normally varies rather closely as $\lambda^3$ away from the absorption edge. Advantage is taken of this fact to produce the prebalance condition with absorption discontinuities equal, shown in Fig. 7-15$b$, and then to compensate for the differences outside the passband by adding the appropriate thickness of the third material to the $\alpha$-filter side of the filter-pair assembly. In strict fact, with three materials, perfect balance can be achieved at only two points, one on either side of the passband. Such use of a third material, methods for mounting the filters and varying their effective thicknesses, and experimental procedures and methods for producing and verifying the balanced condition have been discussed by Young (1963). In the production of the true balanced condition, there are a number of easily entered pitfalls having to do with inadequate knowledge of the wavelengths actually entering the detector that should be avoided. The wavelengths are most commonly selected by diffraction from a single crystal placed at the specimen position in a diffractometer. The wavelength selected by the Bragg law, however, is not necessarily the only (nor, perhaps, even the most intense!) x radiation being detected. The problems appear to be greatest when the mechanically convenient placement of the filter between the crystal and the detector is used. Because of the low intensity of the selected radiation

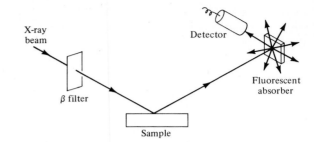

FIGURE 7-16
Arrangement for wavelength discrimination by fluorescence.

on the short-wavelength side of the passband, ordinarily negligible sources of scattering become important. These include fluorescence radiation from the filter materials themselves and, perhaps most often overlooked, air scattering of the $K\alpha$ radiation from that portion of the main beam path which is visible to the detector through the $\beta$ filter. These potential difficulties make desirable the analysis of the wavelength distribution entering the detector and, finally, a test of balance based on pulse-height analysis as well as one based on the dispersive method ($2\theta$ scan) of selecting the nominal wavelength. Results of both kinds of tests have been presented by Young (1963).

**3.  *Fluorescent discriminators*** An interesting arrangement for discriminating against wavelengths that are longer than the one desired was developed by Warren and Mavel (1965) in order to eliminate the Compton scattering from a diffraction intensity measurement. As discussed in Chap. 2, this is an often recurring problem in measurements involving noncrystalline materials.

The arrangement is shown in Fig. 7-16. The x-ray beam scattered from the specimen falls on a metal plate whose absorption edge lies at a wavelength slightly longer than the $K\alpha$ of the incident x-ray beam but shorter than that of most of the Compton-scattered radiation. The metal plate then fluoresces with an intensity proportional to the intensity of $K\alpha$ radiation striking it and this fluorescent radiation constitutes the measured intensity. Proper choice of x-ray tube anode and absorber material is required. Warren and Mavel found the possible choices quite limited; in fact, commercially supplied x-ray diffraction anodes were unsuitable, and a special x-ray tube had to be constructed. It turned out that a rhodium anode ($K\alpha = 0.615$ A) and a molybdenum absorber ($K$ absorption edge at $\lambda = 0.620$ A, $K\alpha = 0.71$ A) worked well. The discrimination against wavelengths longer than the absorption edge is, of course, essentially complete. Discrimination against $K\beta$ and shorter wavelengths was provided by a conventional $\beta$ filter. With this arrangement, it was possible to eliminate nearly all the difficulties previously imposed by the inadequately known Compton-scattered contribution to the total measured scattering intensity.

In essence, this fluorescence method used with a $\beta$ filter is a method for selecting the wavelengths between two absorption edges, much as a balanced-filter pair does, but without need for a second data collection run. The fluorescence yield and efficiency of detecting the fluorescent radiation are, however, quite low (0.5%, Mozzi and Warren, 1969), so that this technique is not recommended for routine use in preference to the balanced filters. Its principal advantage is that of introducing a wavelength discrimination point immediately adjacent to the $K\alpha$ radiation and on the long-wavelength side.

**4. Total reflection** It is possible to devise dispersive devices based on the fact that the index of refraction $n = 1 - \delta$ for x rays is less than unity in solid materials by an amount $\delta$ which is of the order of $10^{-5}$ and is given by

$$\delta = \frac{Ne^2\lambda^2}{2\pi mc^2}f(0) \qquad (7\text{-}39)$$

where $N$ is the total number of electrons per cm$^3$, $e$ is the electronic charge and $m$ the mass, $c$ is the vacuum velocity of light, $\lambda$ is the x-ray wavelength, and $f(0)$ is the effectiveness of the electron for scattering in the forward direction, and differs from unity only to account for anomalous dispersion. (It is, effectively, the resonance term in the Drude-Lorentz theory of anomalous dispersion.) Thus, x rays incident at a sufficiently small glancing angle on a smooth solid surface will be totally reflected at a critical angle (see also Guentert, 1965)

$$\theta_c = \sqrt{2\delta} \qquad (7\text{-}40)$$

where $\theta_c$ is the critical angle for reflection measured from the surface plane.

A slit or limiting edge arrangement which restricts $\theta$ to $\theta \geq \theta_c$ for a particular wavelength then effectively eliminates all shorter wavelengths from the totally reflected spectrum. Even without such a limiting slit (which also serves to block out the main, reflected beam) there is some peaking in the reflected wavelength distribution due to absorption. The absorption also decreases the maximum reflectivity to less than total. Moreover, there is, in fact, some reflectivity at angles greater than the critical angle as shown by the reflectivity curves for Cu $K\alpha$ reflected from nickel and gold in Fig. 7-17. In its effect on spectrum, the total-reflection device differs little from the operation of a filter. In fact, the high degree of collimation required to define the angles and hence the wavelength cutoff may result in more intensity loss (depending on the application) than would a filter.

What makes total-reflection devices valuable is that they can be made up as focusing devices. A number of such devices have been described over the years. Among the most interesting is an ellipsoidal focusing device described by Elliott (1965)

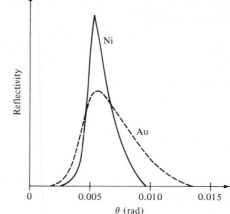

**FIGURE 7-17**

Reflection of Cu $K\alpha$ emanating from a small x-ray source, $1.0 \times 0.1$ mm², by thin films of Au and Ni. (After Witz, 1969.)

which has been found particularly useful in the polymer field (Steward, 1968). Because the Elliott device is patterned on a cavity in the shape of a very prolate ellipsoid of revolution rather than a cylinder, it has a much larger angular aperture ($1.4 \times 10^{-4}$ vs. $7.5 \times 10^{-6}$ steradian) than does the better-known Franks (1958) camera. The surface finish of the walls is chosen for good total-reflection conditions. As is shown schematically in Fig. 7-18, the ends of the ellipsoid are omitted so that the x-ray source (microfocus) and specimen can be placed at, respectively, the left and right loci of the ellipse indicated. A beam stop eliminates the direct beam, ensuring that the radiation reaching the specimen has undergone total reflection (or secondary scattering). The intensity is quite high, being generally an order of magnitude higher than that of the device base on a cylindrical mirror. The result can be quite striking; with a Cu anode microfocus x-ray tube operated at 50 kV and 5 ma, the focused-beam spot is readily visible on a fluorescent screen held at the specimen position in all but brightly lit rooms. For the particular device described by Elliott, the degree of monochromaticity introduced (by the dependence of critical angle on wavelength) is such that, when the device is set for Cu $K\alpha$ radiation (1.54 A), the Cu $K\beta$ radiation (1.39 A) is "only feebly reflected" and the bulk of the white radiation, for the x-ray tube being operated in the range 30 to 50 kV, is "not appreciably reflected."

**5. Crystal monochromators** Crystal monochromators provide the preeminent dispersive method for wavelength discrimination. Flat, singly bent, ground and bent, and doubly bent crystal monochromators—all have been used with success in a great variety of arrangements, some of which already have been discussed in Section II F of the preceding chapter. An excellent eight-page summary-review of the crystal monochromator arrangements reported in the literature through about 1959 is given

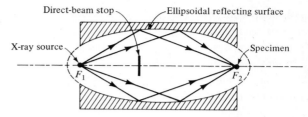

FIGURE 7-18
Focusing, total-reflection device, shown considerably foreshortened for clarity.

in *International tables for x-ray crystallography*, vol. III (1962), pp. 37–46, including two pages of references. A bibliography on crystal monochromatization techniques has been published by the IUCr under the editorship of Herbstein (1967). Witz (1969) has recently reviewed focusing monochromators of both crystal diffraction and total-reflection types. In this section, therefore, following some general comments, attention will be restricted to three more recent developments in crystal x-ray mono-chromators, (1) the use of pyrolytic graphite as a monochromator crystal, (2) the Bonse-Hart arrangement involving multiple, near-total reflection from a highly perfect crystal, and (3) special methods for selecting a wavelength band narrower than the emission profile of the characteristic radiation.

To set crystal monochromators in perspective with the other methods of wavelength discrimination discussed in this chapter, note that $W/\langle A \rangle$ or, rather, $\Delta\lambda/\lambda$ for crystal monochromators can be as small as 0.001 or less, e.g., of the order of $10^{-5}$ for monochromators based on highly perfect crystals. However, it is more frequently of the order of 0.01; for example, monochromator mosaic spread $\geq \frac{1}{4}°$, 1-mm projected target size, target-to-monochromator plus monochromator-to-spec-imen distance $\sim 40$ cm, and monochromator Bragg angle $\Delta_m = 15°$, give $\Delta\lambda/\lambda \sim 0.01$ in the absence of any instrinsic diffraction broadening.

*a. Pyrolytic graphite* This material has recently gained much favorable attention as a crystal monochromator both for x rays (Sparks, 1967) and for neutrons (Riste and Otnes, 1969). In general, x-ray absorption depends approximately on $Z^4$ while maximum possible diffracting power (geometric structure factor = 1) is proportional only to $Z^2$ (mosaic crystal case), where $Z$ is the atomic number. Thus it is that crystals of the lighter elements are expected to form better monochromators than those of heavier elements. In any case it is, of course, desired that there be a strong reflection, e.g., one for which all or nearly all the atoms in the crystal scatter in phase. Together, these two criteria direct attention to light-element single crystals that have a suitable mosaic spread; in the range of $\frac{1}{4}$ to $5°$, depending on the application. Pyrolytic graphite, while not a single crystal, can be made so highly oriented that these conditions are met for the (00.*l*) planes and recent advances in the technology of pyrolytic graphite preparation make possible the control of the "mosaic" spread so that a highly efficient monochromator results (Sparks, 1967). Furthermore, the material can be fashioned quite easily in the form of both singly bent and doubly bent focusing monochromators.

Curiously, the rapidly developing and rather widespread acceptance of pyrolytic graphite as a superior monochromator seems to be based primarily on oral reports and private communications rather than on documentation in the written literature. Oral comments by T. Furnas and by A. E. Smith have been published as part of the discussion following the paper by Witz (1969). For a flat monochromator crystal of pyrolytic graphite, Smith there reports intensity equal to $\sim 80\%$ of that obtained with a $\beta$ filter and significantly improved data quality. With a focusing monochromator of this material, it obviously would be possible to obtain intensities more than an order of magnitude greater than those obtained with a $\beta$ filter or with the unmodified x-ray beam. [With a doubly bent LiF monochromator, Chipman (1956) obtained intensities 40 times those with a singly bent monochromator which, by inference, should give a somewhat—though probably smaller—increase over the intensity from a flat crystal.] It should be noted, however, that the intensity increase from a focusing monochromator is produced by using rays traveling in a wider range of directions. Thus, whether or not the gain will be useful, or will be partially or wholly offset by a decreased signal-to-noise ratio resulting from the broadening of the observed reflection profiles, will depend on the application. For single-crystal work, that portion of the increased beam intensity that arises from cross fire in the plane of incidence and exceeds the mosaic spread of the specimen in that plane is wasted.

*b. Bonse-Hart monochromator*   When weak signals must be detected in the presence of nearby strong signals, such as small-angle diffuse scattering near the main beam or thermal diffuse scattering near a Bragg maximum, wavelength discrimination combined with high angular resolution is needed. A crystal monochromator having a particularly sharp angular and wavelength cutoff at both sides of the passband was devised by Bonse and Hart (1965). Successive symmetric reflections reduce the tails of the resultant reflection profile relative to its maximum, the use of essentially perfect single crystals gives nearly total reflection with consequent small loss at the maximum, and the embodiment of the whole device in one crystal gives it mechanical stability, simplicity, and inherent alignment. The basic element of the device is a rectilinear channel cut parallel to the reflecting planes in a large single crystal of Si or Ge. The successive reflections then take place on the inner surfaces of the channel, as is indicated in Fig. 7-19. (The theory of the operation of this monochromator is presented in Chap. 4.)

FIGURE 7-19
Bonse-Hart monochromator.

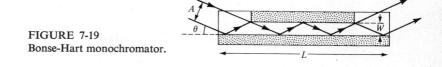

It is clear that if a single reflection reduces the intensity at angle $(\theta_0 + \Delta\theta)$ by the factor $R$ relative to the intensity at $\theta_0$, then $m$ such reflections will reduce the intensity at $(\theta_0 + \Delta\theta)$ by $R^m$. Further, if the reflection is 95% total at $\theta_0$, the final reflection intensity at $\theta_0$ will be $(0.95)^m$. For $m = 5$, this amounts to a loss of only 23% at $\theta_0$, approximately the situation obtaining for the calculations for reflection of Cu $K\alpha$ radiation plotted in Fig. 7-20. Note in this figure that, at only 4 seconds of angle from $\theta_0$, the tails of the reflection curve are depressed by a factor of $10^8$. The half-width is almost unaffected by the successive reflections; it remains approximately

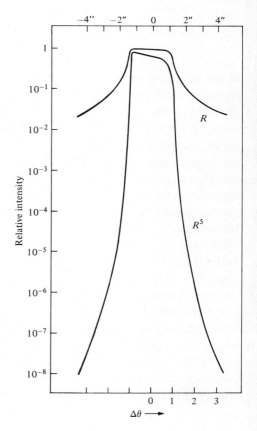

FIGURE 7-20
Reflection curves for a Bonse-Hart monochromator. (After Bonse and Hart, 1965.)

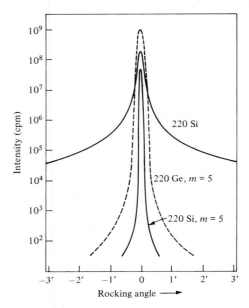

**FIGURE 7-21**
Rocking curves of pairs of grooved crystals of Ge and Si compared with a two-crystal rocking curve of silicon. (After Bonse and Hart, 1965.)

at the relatively small value appropriate to the particular reflection from a perfect crystal. It will be recalled from Chap. 4 that the width of such a reflection is determined by the range of angles at which near-total reflection occurs and that is proportional to the structure factor. Since the atomic scattering factor for Ge exceeds that for Si, a larger integrated intensity (produced by a wider range of near-total reflection) will be obtained with the 220 reflection of Ge than with the 220 reflection of Si. This comparison is shown in Fig. 7-21 which also shows the experimental results corresponding to the calculations plotted in Fig. 7-20. Bonse and Hart found experimentally that with the Si 220 reflection and $m = 5$ the intensity at $\Delta\theta = 9$ seconds was $10^{-5}$ times that at $\theta_0$ while in the two-crystal case the "tail" intensity at this position was 7000 times larger. For the 220 reflections and $m = 5$ (Fig. 7-21) the full width at half-maximum was 5.5 seconds for Si and 15 seconds for Ge. These widths give $\Delta\lambda/\lambda \approx 1 \times 10^{-5}$ and $18 \times 10^{-5}$, respectively, which may be compared with the values in Table 7-2.

Another interesting feature of the Bonse-Hart monochromator is that the cross-sectional dimensions of the beam may be large, e.g., several millimeters, even though it is highly parallel. Thus both extended x-ray sources and extended specimens can be fully utilized. These advantages have led to the construction of a multiple-reflection diffractometer for small-angle x-ray scattering and x-ray spectroscopic studies (Bonse and Hart, 1965b). The diffractometer utilizes two of the channeled

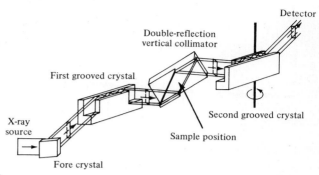

**FIGURE 7-22**
Three-channeled-crystal arrangement limiting both vertical and horizontal divergence. (After Bonse and Hart, 1965.)

perfect crystals, one in the incident beam and one in the detected beam. Such an instrument has a sufficient resolving power to permit measurement of the diffraction maximum of 20,000 A spherical (Dow latex) particles with Cu $K\alpha$ radiation; sufficient intensity was available to permit observation of 20 orders (Bonse and Hart, 1965$b$).

While the Bonse-Hart arrangements do a remarkable job of limiting angular variation in one plane, they do not limit the variation (cross fire) perpendicular to that plane. For some purposes it may be desirable to impose limitations in the second plane; i.e., to limit effective vertical slit height effectively to zero. An arrangement proposed by Bonse and Hart (1965) for this purpose, shown in Fig. 7-22, has a third channeled single crystal inserted in the beam path, this one with its plane of incidence perpendicular to that of the other two. Used simply as a monochromator, the first two channeled crystals alone will give a beam parallel to within less than 1 minute of arc in every direction, yet reasonably high intensities may be expected. Such a device could be very attractive in studies where it is necessary to probe the intensity associated with very small regions of reciprocal space. Taking $\Delta\lambda/\lambda = 18 \times 10^{-5}$, for example, and using the corresponding angular aperture of 15 seconds for the 220 reflection and a Ge monochromator with Cu $K\alpha$ radiation, the sampling region in reciprocal space (see Section II A) becomes a region approximately

$$\frac{73 \times 10^{-6}}{\lambda} \frac{73 \times 10^{-6}}{\lambda} \frac{18 \times 10^{-5}}{\lambda}$$

or approximately $2.6 \times 10^{-13}$ A$^{-3}$.

An interesting application of the perfect-crystal arrangement was made by Renninger (1967). He had sufficient intensity to record the intensities shown in

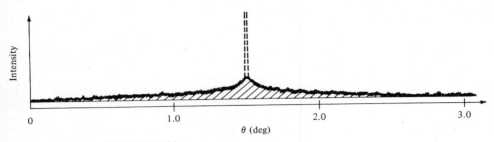

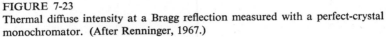

FIGURE 7-23

Thermal diffuse intensity at a Bragg reflection measured with a perfect-crystal monochromator. (After Renninger, 1967.)

Fig. 7-23 with an angular resolution of better than 1 minute of arc. This recording shows how the one-phonon thermal diffuse scattering depends on angle over a region, close to the Bragg maximum, which is usually obscured by the Bragg reflection in instruments having poorer resolution.

*c.  Other devices*    The emission profile of the characteristic wavelength has, typically, $\Delta\lambda_{1/2}/\lambda \approx 10^{-3}$ and Cauchy-like tails that decrease as $(\theta - \theta_0)^{-2}$ where $\theta_0$ is the position of the maximum. A number of methods are available for limiting the passband of a monochromator system to much narrower wavelength ranges. For example, the Bonse-Hart monochromator discussed above provides a reflection profile substantially sharper than this width; as has been pointed out, it effectively eliminates the tails beyond $|\theta - \theta_0| > 1$ minute. If the high degree of parallelism provided by the Bonse-Hart monochromator is not required, a net gain in overall intensity may be realized by use of other, less stringent, methods of passband limitation. Thus Barth (1958) made early use of perfect-crystal diffraction characteristics to produce sharply parallelized and monochromated beams. He described a method involving the use of a single high-index $23\bar{5}4$ reflection from a nearly perfect quartz crystal to obtain a reflection profile width as little as $\frac{1}{30}$ that of the natural x-ray line width.

The smallest value for $\Delta\lambda_{1/2}/\lambda$ has been achieved, probably, by Kohra and co-workers (Kohra, 1962; Kohra and Kikuta, 1968; Kikuta and Kohra, 1970; and Matsushita, Kikuta, and Kohra, 1971) through the use of successive asymmetric Bragg reflections from monolithic perfect crystal. This design makes use of the fact that the intrinsic width of the rocking curve (total reflection region) is decreased by increasing asymmetry, reflections such as 422 with structure factors smaller than that of the 220 reflection are usable, and the refraction effect can be made to cause successive asymmetric reflections to overlap only partially, thus reducing the combined reflection profile to that of the overlap region. Figure 7-24 shows their monochro-

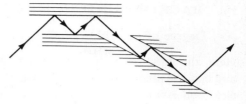

FIGURE 7-24
Multiple-reflection monochromator with
asymmetrically cut crystals. (Kikuta and
Kohra, 1970.)

mator design for which the spread of the beam is calculated to be approximately 0.10 second of arc with a Si crystal, Cu $K\alpha$ radiation, and the 422 reflection. This corresponds to $\Delta\lambda_{1/2}/\lambda \approx 5 \times 10^{-7}$.

A novel approach for limiting $\Delta\lambda/\lambda$ has been put forward by Kottwitz (1968). In common with Bonse and Hart he uses multiple reflections, but in his case the multiple reflections have differing Miller indices and are those involved in the *Umweganregung*, or Renninger, effect. The crystals are not required to be grooved, transmission arrangements can be used, and mosaic crystals as well as perfect crystals can be used. The reflection finally used to produce the exit beam from the monochromator is a forbidden reflection excited by *Unweganregung*. For perfect crystals, $\Delta\lambda_{1/2}/\lambda \approx 10^{-5}$ is readily attained. In a later paper devoted to monochromatization of neutrons, Kottwitz (1969) points out that since $d\lambda/d\theta = 0$ at $\theta = 90°$, a resolution characteristic of perfect crystals can also be obtained with mosaic crystals. The relative effectiveness of the various wavelength discriminators discussed in this section is summarized in Fig. 7-25.

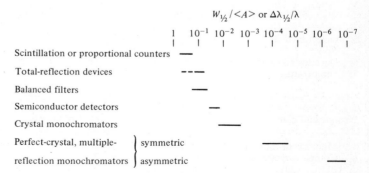

FIGURE 7-25
Wavelength discrimination capabilities of various devices in the x-ray diffraction range (0.5 to 2.0 A).

## IV.  BACKGROUND CONTROL

### A.  Degree of control needed

Here, the question of how much control of background is needed is considered, and how the needed control can be achieved. Answering the latter question requires a clear understanding of the various component contributions to background. Table 7-3 lists the principal components of the background and, also, the principal experimentally controllable instrument features which affect the background intensity.

Although serious systematic errors in measurement of background can arise, it is usually possible to control them to an acceptable degree by using an appropriate instrumental design and procedure. In such cases the significant errors become those due to counting statistics and it is of interest to inquire how much effort—and additional experimental time—should be expended to reduce such errors. An instructive discussion is readily carried through in the context of scanning measurement of Bragg intensities from a small single crystal bathed in a homogeneous x-ray beam. For simplicity, the $2\theta$ scan and equal background levels on the two sides of the Bragg profile are assumed. For such a case, the optimum ratio $t$ of time spent on measuring the Bragg peak to that spent on measuring the background is (Young, 1965):

$$t_{opt} = [\mathscr{S} + 1] \qquad (7\text{-}41)$$

**Table 7-3  BACKGROUND FEATURES***

**Experimentally controllable factors**

1.  Cross fire in incident beam
2.  Effective wavelength spectrum (determined by x-ray tube voltage, filters, target material, monochromator, energy resolution of detector, etc.)
3.  Incident beam size
4.  Scatterers in incident beam (e.g., collimator and beam trap edges, atmosphere, crystal mount, etc.)
5.  Detector (receiving) aperture (dependent on cross fire, mosaic spread, wavelength range effective, etc.)
6.  Detector's "view" (determined by choice of aperture position)
7.  Temperature

| Background components | Most relevant control factors | Angular dependence |
|---|---|---|
| a.  Fluorescent radiation | 2, 5 | Smooth, weak |
| b.  Compton-scattered radiation | 2, 5 | Smooth, strong |
| c.  Elastic scattering from air and obstacles | 3, 4, 5, 6 | Sharp, strong |
| d.  Bragg scattering of harmonic wavelengths | 1, 2, 5 | Sharp, strong |
| e.  Disorder diffuse scattering | 5 | Smooth and sharp |
| f.  Thermal diffuse scattering | 5, 7 | Smooth and sharp |

* Adapted from Young (1965).

where $\mathscr{S}$, the signal-to-noise ratio, is

$$\mathscr{S} = \frac{I_N}{I_B} = \frac{I_N}{tI_b} = \frac{I_p - I_B}{tI_b} \qquad (7\text{-}42)$$

where $I_N$ is the net integrated peak count

$I_p$ is the gross peak intensity measured in time $t_2$

$I_b$ is the background intensity actually measured in time $t_1$

$I_B$ is the background intensity which, on the basis of the $I_b$ measure, would be measured in time $t_2 = t_1 t$.

The total random standard error $\sigma_N$ in the net peak intensity $I_N$ that corresponds to this optimum allocation of time is given by (Young, 1965)

$$\left(\frac{\sigma_N}{I_N}\right)_{\text{opt}} = \frac{\{1 + [1 + (\mathscr{S} + 1)^{1/2}]/\mathscr{S}\}^{1/2}}{(I_N)^{1/2}} \qquad (7\text{-}43)$$

It is clear that for a given value of $I_N$ (presumably the true Bragg intensity) the fractional error $\sigma_N/I_N$ decreases with increasing $\mathscr{S}$. In the limit as $\mathscr{S} \to \infty$, $\sigma_N/I_N \to (I_N)^{-1/2}$, as expected for zero background contribution. It also can be seen from (7-43) that, if $\mathscr{S}$ and $I_N$ are both reasonably large, the fractional error $\sigma_N/I_N$ will tend to be less than some minimum-needed value, e.g., 1%, 2%, or whatever the experimenter has preselected. In that case, less than the optimum time (7-41) can be allotted to the background measurement. Clearly, situations will occur in which, for strong reflections and low backgrounds ($\mathscr{S}$ large), the background need not be measured at all, with obvious savings in the diffractometer setting and measuring time.

Another permutation to be considered is the introduction of an instrumental change, such as addition of a crystal monochromator, which will increase the signal-to-noise ratio $\mathscr{S}$ by the factor $b$ while reducing the net intensity $I_N$ by the factor $a$. The ratio $R$ of initial to final fractional standard deviation is then given by

$$R = \frac{(\sigma_N/I_N)_i^{1/2}}{(\sigma_N/I_N)_f^{1/2}} \left[\frac{b}{a} \cdot \frac{\mathscr{S} + 1 + (\mathscr{S} + 1)^{1/2}}{b\mathscr{S} + 1 + (b\mathscr{S} + 1)^{1/2}}\right]^{1/2} \qquad (7\text{-}44)$$

To take one example, if $\mathscr{S}$ is increased by a factor of 20 while $I_N$ is reduced only by a factor of 4, $R = 0.82$ if the initial $\mathscr{S}$ is 1. For the fractional standard deviation to be reduced ($R > 1$), the signal-to-noise ratio must be less than approximately $\frac{1}{2}$ ($\mathscr{S} = 0.5$ gives $R = 0.97$). The main point of this analysis is now startlingly clear: unless the signal-to-noise ratio is quite poor initially, introducing instrumental changes which cause even a modest intensity reduction while improving the signal-to-noise ratio will represent a net loss rather than a gain in operating efficiency at a chosen level of statistical precision. Using the simpler model of equal times spent on background

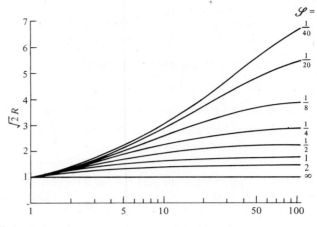

**FIGURE 7-26**
Improvement in $R$ for equal time spent on peak and background.

and on peak, Young (1965) has calculated the $R$ values to be expected for various values of initial $\mathscr{S}$ and ranges of $b$ and $a$ values (Fig. 7-26). The equation analogous to (7-44) but applicable to the equal times model is

$$R' = \left( \frac{b}{a} \cdot \frac{\mathscr{S} + 2}{b\mathscr{S} + 2} \right)^{1/2} \qquad (7\text{-}45)$$

For the previous example of $b = 20$ and $a = 4$, $R' = 0.83$ and 1.02, respectively, for $\mathscr{S} = 1$ and $\frac{1}{2}$. Comparing these with the corresponding values from the optimum time model ($R = 0.82$ and 0.97) indicates that the two models display the same general features concerning gains—or losses—to be anticipated from experimental modifications such as inserting a crystal monochromator. Thus, actual intensity calculations reproduced in Fig. 7-26 serve to emphasize, as a generally applicable point, the fact that intensity is more important than an improvement in signal-to-noise ratio unless the latter is particularly poor.

## B.  Unnecessary background intensity

Unnecessary background intensity is defined as that which can be eliminated without reduction of Bragg peak (or other desired signal) intensity. Fluorescent radiation may, obviously, be controlled by placing the wavelength discrimination device (if one is used) in the scattered beam. Compton radiation is ordinarily not a problem in connection with Bragg intensity measurements. When it is, a narrow passband crystal monochromator placed before the detector can be effective or Warren and Mavel's (1965) fluorescence method could be used, though with an intensity loss. In the presence of large background intensities arising from either extraneous wavelengths or extraneous propagation directions permitted by poor detected-beam collimation (i.e., not emanating from or near the specimen), crystal monochromators can be particularly useful.

Since most background components vary relatively slowly (with angle) compared to the Bragg peak, the measured background intensity is nearly directly proportional to the area of the receiving aperture whereas the Bragg intensity is not. It is therefore worthwhile to calculate or otherwise determine the minimum-size receiving aperture required to permit measurement of the Bragg intensity. Several authors have discussed this calculation (Alexander and Smith, 1964; Ladell and Spielberg, 1963). Furnas (1957), in particular, has provided relevant tabular data. The required minimum size is a function of incident beam cross fire $\tau_x$ (arising from the finite sizes of both specimen and x-ray source), the wavelength range to be used, the diffraction angle (because of dispersion $\tau_d$) scan type (e.g., $\omega$ vs. $2\theta$), and mosaic spread $\tau_m$ both in and perpendicular to the plane of incidence. The aperture size perpendicular to the plane of incidence is dependent on mosaic spread in both the $2\theta$ and $\omega$-scan methods. In the plane of incidence, the minimum aperture dimension is, to a good first approximation, dependent on mosaic spread but not dispersion (e.g., of the $K\alpha_1$ and $K\alpha_2$ wavelengths) in the $2\theta$-scan case and vice versa in the $\omega$-scan case. Roughly, the aperture dimension $D_R$ can be approximated by

$$D_R = L(\tau_x + a\tau_m + b\tau_d) + D_s \qquad (7\text{-}46)$$

where $L$ is the specimen-to-aperture distance, $D_s$ is the dimension of the specimen in the direction being considered (in or out of the plane of incidence), and $a$ and $b$ take on the value 1 or 0 according to the circumstances, as discussed above.

A shortcoming of the simplified expression (7-46) is that the Bragg profile is not sharply defined to have zero value outside of some narrow angular range characterized by a term such as $\tau_m$. In consequence, truncation errors (Kheiker, 1969; Alexander and Smith, 1962) arise from exclusion, by the receiving aperture or the scan range or both, of parts of the tails of the Bragg peaks. Clearly, the extent of such difficulties and the actual size of minimum receiving aperture required differ with each choice of specimen. A practical solution of operational value has been described by Young (1965). A limited selection of receiving apertures is provided. The net integrated Bragg intensity is then measured first with a too-small and then with successively larger apertures (larger in both dimensions). When the measured net Bragg intensity ceases to increase with increasing aperture size, it is recognized that the aperture is then sufficiently large not to introduce experimentally significant

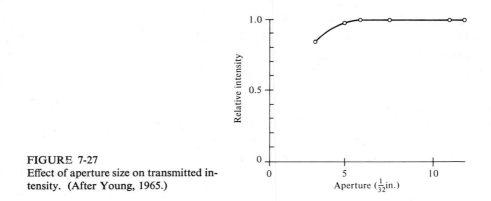

**FIGURE 7-27**
Effect of aperture size on transmitted intensity. (After Young, 1965.)

truncation errors into the measurement of that reflection. A plot of such test results is shown in Fig. 7-27, where it is seen that an aperture size of $\frac{6}{32}$ in. was adequate for that particular case. In the process of setting up to collect data for a particular specimen, one makes such a test for several reflections occurring at various $2\theta$ and orientation values.

One of the most dramatic means for reducing background at no cost in desired intensity is that of reducing the extent of the incident beam path in air (outside of the specimen) which is viewed by the detector. This can be accomplished with a detected-beam tunnel, usually consisting of a metal tube extending from the detector toward the specimen. Figure 7-28 shows the effect of such a tunnel.

Another method for reducing unnecessary background intensity (relative to useful intensity) is the long-recognized one of applying an appropriate potential $V$ to the x-ray tube to produce an optimum ratio of characteristic to white x radiation. The intensity in the white radiation depends approximately on $V^2$ while that of the characteristic radiation depends on $(V - V_K)^m$, where $V_K$ is the excitation potential for characteristic radiation and $m$ is a number larger than unity, $\sim\frac{3}{2}$ for Cu $K\alpha$ radiation. For Cu $K\alpha$ ($V_K = 8.9$ kV), the optimum $V$, giving the maximum ratio of characteristic to total white radiation, is about 36 kV for constant-potential and 51 kVp for full-wave rectification. Much of the short-wavelength white radiation will in fact not pass the detectors normally used, however, and hence the effective optimum $V$ for diffractometry is generally much higher. For heavier target elements, e.g., Mo with $V_K = 20$ kV, the optimum $V$ is very much higher—higher in fact than the 50- to 60-kV maximum available on most commercial diffraction x-ray generators. Figure 7-29 supports this point; recognizing that little of the white radiation at wavelengths less than $\frac{1}{2}$ that of $K\alpha$ will pass the detector, one can easily extrapolate from this figure to see that applied voltages substantially higher than 50 kV, as high as 75 to 100 kV, would be useful.

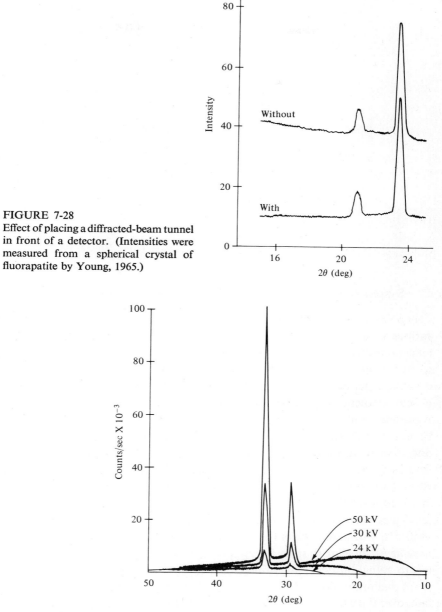

**FIGURE 7-28**
Effect of placing a diffracted-beam tunnel in front of a detector. (Intensities were measured from a spherical crystal of fluorapatite by Young, 1965.)

**FIGURE 7-29**
Effect of increasing the applied tube voltage while keeping the filament current constant. The ratio of the characteristic Mo $K\alpha$ radiation to that of neighboring white radiation goes from 4.7 at 24 kV to 23.0 at 50 kV. (After Young, 1965.)

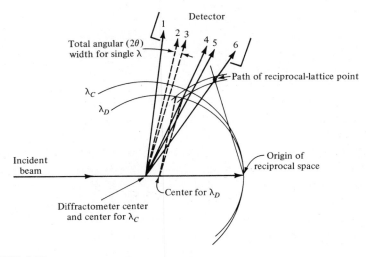

**FIGURE 7-30**
Scanning geometry in reciprocal space. (After Young, 1965.)

## C.   Systematic errors from extraneous-wavelength radiation

The point considered here is the contribution to measured background from the various x radiations present in the white radiation ("Laue") streak underlying Bragg reflections.  The extent of contribution depends on the experimental method used, so that comparisons first will be made of several aspects of the two commonly used diffractometer scan modes, the $2\theta$ scan (moving crystal, moving counter) and the $\omega$ scan (moving crystal, stationary counter).  For this comparison, the analyses by Alexander and Smith (1962) and, particularly, by Burbank (1964) will be used. Figure 7-30 illustrates some of the geometry involved.  For simplicity, we consider only a single incident beam direction; i.e., no cross fire, but permit a range of wavelengths to be present.  For reference, the rays entering the detector are numbered. Rays 4 and 5 show the minimum angular breadth required for the detector aperture in the $2\theta$ scan with a single wavelength $\lambda_C$.  Rays 2 and 3 show the same thing for the particular wavelength $\lambda_D$, after being translated from the actual center of the $\lambda_D$ sphere to a common origin for all scattered rays.  While the reciprocal-lattice point is moving along its path from the $\lambda_C$ sphere to the $\lambda_D$ sphere, the variations in the wavelength being diffracted are strictly a function of the position of the reciprocal-lattice point and are in no way dependent on whether a $2\theta$ or an $\omega$ scan is being made.  In fact, over the whole scan of whatever type, as long as all the diffracted rays lie between rays 1 and 6 (i.e., they do enter the counter aperture), the $2\theta$ scan and the $\omega$ are essentially equivalent insofar as the diffracted rays associated with a given point

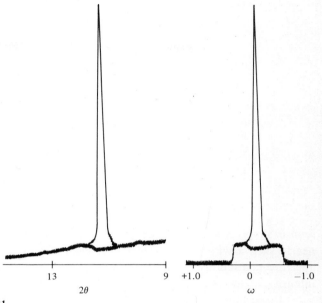

FIGURE 7-31
Comparison of $2\theta$ scan and $\omega$ scans. (After Young, 1965.)

in the reciprocal space of the specimen, e.g., a given reciprocal-lattice point smaller than the uniform portion of the reciprocal-space sampling region, are concerned. A further point of similarity is that in both techniques a certain range of wavelengths $\Delta\lambda$ is to be considered. In the $\omega$-scan case, $\Delta\lambda$ is determined by the limits of the aperture, that is, by the limits indicated by rays 1 and 6. In the $2\theta$-scan case this selection is made by the choice of points at which one measures the background on the two sides of the main peak.

This similarity between the $\omega$ scan and the $2\theta$ scan of a single reciprocal-space point, put forward very effectively by Burbank (1964), is shown in another way in Fig. 7-31. Here, balanced filter scans have been made in both the $\omega$ and $2\theta$ modes. Unnecessarily wide apertures were used so that the various features developed in the discussion of Fig. 7-30 may be clearly seen. All scans in Fig. 7-31 were made under conditions which were identical with the exceptions of the stated selections of filter and of scan mode. Several separate points are noted. (1) The shoulders on the $\omega$ scan occur at the positions equivalent to those of rays 1 and 6 in Fig. 7-30. (2) Let $\lambda_C$ and $\lambda_D$ represent the absorption edges of the two filters. It is clear that the rays corresponding to these points, identifiable in the figure as the positions at which the traces with the two filters diverge, are well inside the shoulder. (3) It is notably clear that the $\omega$

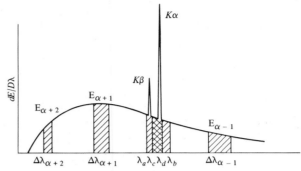

**FIGURE 7-32**
Simultaneously active portions of the incident spectrum.

scan is simply a section of the $2\theta$ scan. (4) Ordinarily, of course, one does not use nearly such a wide aperture in the $\omega$ scan, nor in the $2\theta$ scan, because the wider aperture continuously admits additional background without enhancing the main intensity, as has been discussed in the preceding section.

Regarding Fig. 7-31, one now can ask what wavelengths contribute at each point along the scans and note specifically that it is not the whole white spectrum emitted by the x-ray source. Figure 7-30 would suggest that only a single wavelength, or at most a narrow band of wavelengths, would be active at any one time. In addition to the expected wavelengths indicated in Fig. 7-30, however, there will be certain other wavelengths present which, with other orders of the nominal reflection, also satisfy Bragg's law for this angle of incidence. Figure 7-32 indicates something of the situation. Each wavelength range that satisfies Bragg's law has its own energy $E_n$, its own wavelength spread $\Delta\lambda_n$, its own associated structure factor $F_{n(hkl)}$ (because, of course, it does arise from some other reciprocal-lattice point), etc. Each of these active wavelength ranges contributes a different harmonic wavelength at the same crystal position and same time. Because these harmonics are of $m\lambda/n$ ($m$ and $n$ integers) and not just $\lambda/n$, a scintillation counter followed by a pulse-height analyzer cannot discriminate against all such harmonics. For example, such a detector assembly cannot resolve $4\lambda/5$ from $\lambda$. One must then use sharper wavelength discriminators such as a pair of balanced filters or a crystal monochromator (Fig. 7-25).

One may ask whether the harmonic wavelength contributions under the Bragg peak can be significant. A balanced-filter $\omega$ scan, which may be thought of as a scan across rather than along the Laue streak, shows their effect explicitly, Fig. 7-33. The radiations of harmonic wavelengths pass the $\beta$ filter equally as well as they do the $\alpha$ filter. Therefore they produce equal peaks with both filters as the scan proceeds across the Laue streak. Thus the peaking in the $\alpha$-filter scan is due entirely to the

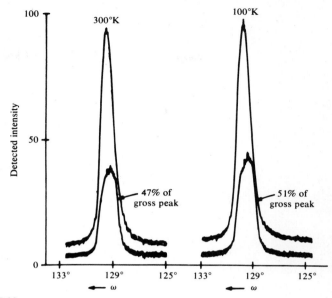

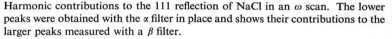

FIGURE 7-33

Harmonic contributions to the 111 reflection of NaCl in an $\omega$ scan. The lower peaks were obtained with the $\alpha$ filter in place and shows their contributions to the larger peaks measured with a $\beta$ filter.

harmonics which, in this case, can account for approximately 50% of the gross peak intensity seen. In another case, the 00.12 reflection of quartz, the entire observable "Bragg" peak turned out to be due to harmonics.

In passing, several small points are noted. The aperture in the scan in Fig. 7-33 is near the minimum size, and hence, the shoulders seen in Fig. 7-31 do not appear here. However, it will be noted that the smaller peak is in fact the wider peak, as it must be if the aperture size is not too small. Also, it is noted that the balanced filters might appear not to be balanced, in that their respective traces (i.e., intensity records) do not come to the same level at the edges of the peak. They should not; this difference in level is due to non-Bragg scattered radiation of passband wavelength. (This is a point to watch for when one is balancing filters.) Finally, note that the data of Fig. 7-33 were obtained with copper radiation. The harmonic contributions are likely to be smaller, though still very significant, if molybdenum radiation is used with conventional equipment.

The fact that the relative contribution of these harmonics is temperature dependent is further evidence that they are indeed harmonic contributions coming from reciprocal-lattice points at different distances from the origin of reciprocal space.

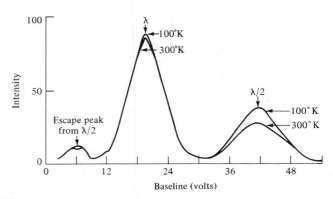

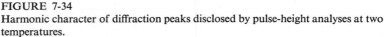

**FIGURE 7-34**
Harmonic character of diffraction peaks disclosed by pulse-height analyses at two temperatures.

This temperature dependence can be used for a nice demonstration of the harmonic character with the aid of a pulse-height analyzer scan made while the diffractometer is fixed at the main peak position, Fig. 7-34. In this case $m$, in the expression $m/n$, equals 1, and therefore the harmonic wavelength nearest to $\lambda$ is $\lambda/2$. In Fig. 7-34 this $\lambda/2$ appears as a second, smaller, peak appearing at twice the energy of the first main peak. The temperature dependence of the second peak is very noticeably greater than that of the first. The reason is that $\lambda/2$ is being diffracted as a 222 reflection, for which the temperature factor is 4 times larger than it is for the 111 reflection responsible for the main peak.

Note that, because the contributions from the individual harmonics depend on structure factors of other reflections, the resulting errors are expected to be systematic. The relative contributions of the harmonics may be estimated in various ways. An approximate relation suggested in a preliminary investigation by Young (1961) is

$$\frac{I_h}{I_N} = 2\Delta\omega \cos \theta \sum_n \frac{\lambda_n^3 d_n E_n'}{\lambda_\alpha^3 (E_\alpha + E_W)} \frac{A_n}{A_\alpha} \frac{\varepsilon_n}{\varepsilon_\alpha} \frac{|F_{n(hkl)}|^2}{|F_{\alpha(hkl)}|^2} \tag{7-47}$$

where   $I_h$ is the total harmonic contribution to the apparent Bragg peak

$I_N$ is the net true Bragg peak intensity arising from passband wavelengths

$\Delta\omega$ is the angle subtended at the crystal by the detector aperture (assumed to be much larger than the angle subtended at the detector by the crystal)

$n$ is the order of the harmonic as referred to unity for the lowest order possible

$\alpha$ is the order of the reflection using the characteristic wavelength

$E_\alpha$ is the energy of the characteristic $\lambda$ in the incident beam

$E_W$ is the energy of the portion of the Bremsstrahlung which falls within the passband containing the characteristic $\lambda$

$E_n' = \partial E_n/\partial\lambda$ is the energy per unit range of $\lambda$ associated with $\lambda_n$

$\varepsilon$ is an extinction factor which is included because, if it is not unity, it is expected to be wavelength dependent

$A$ is the absorption factor.

The other symbols have their usual meanings. For the moment, the main value of this equation is to point out that the harmonic contributions are not random errors. The use of a suitable equation would, in principle, allow calculation by an iterative procedure of all the required harmonic corrections in a set of data. Fitzwater (1965) and Hoppe (1969b), among others, have made this sort of calculation and consider it to be successful.

Experimentally, the elimination of the harmonic contributions originating from the incident white radiation requires that one of several things be done: (1) The harmonic wavelengths may be eliminated from the incident spectrum in the first place, perhaps by use of a crystal monochromator with suitable precautions being taken about the harmonics present in the diffracted beam from it. (2) One may offset the detector along the Laue streak, rather than beside or across it, to determine background, whether doing a continuous scan, using the peak-height method, or reading a film. (3) One may use some sharp wavelength discrimination technique such as the balanced-filter method. It should be noted, however, that with balanced filters four measures are required for each peak, three of the background and one of the gross peak.

It was common practice in many laboratories, some years ago, for workers to read the background by offsetting to the side of the Laue streak, or equivalently, to read a background on a film beside the Bragg reflection spot rather than on the Laue streak adjacent to it. The harmonic contribution was then not eliminated. One example of the effect which this harmonic contribution has may be found in quartz (Young, 1962). Three zones of data were collected, and least-squares refinement gave an $R$ factor based on $|F|^2$ of 7.4%. One of the zones of data had been taken with proper background measurements (all had been taken with $\omega$-scan techniques). Sorting out the reflections of this zone from the same refinement yielded an $R$ factor from that zone alone of 3.7%.

## D.   Background structure

For a variety of reasons (e.g., air scattering at relatively small angles, scattering from obstructions at larger angles) the background from non-Bragg scattered radiation may be markedly nonlinear with angle, increasing sharply with increasing $2\theta$ in the

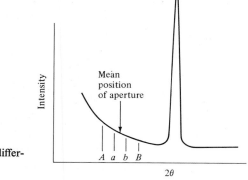

**FIGURE 7-35**
Curved background sampled with different apertures.

vicinity of the peak, as indicated in Fig. 7-35. It is quite clear that, for two apertures centered at the same mean position, the aperture which covers the region only between $a$ and $b$ will measure a lower average intensity per unit area of aperture than will the one extending between $A$ and $B$. Thus, the effect of a concave upward curvature in the background is to cause an overestimation of the background at whatever the nominal position of the aperture is. In this way, the entire background line is artificially increased. In this case, extrapolation even on a measured curved background will not necessarily lead to a correct estimate of the background under a Bragg peak. Of course, if background is measured at only two points on either side of the peak and a straight-line interpolation is used, an obvious error is introduced (Kheiker, 1969).

One of the most common examples of structured background occurs at relatively small $2\theta$ values when a $2\theta$ scan is used with a $\beta$ filter alone. Figure 7-36 shows an example. Here the $2\theta$ angle is small enough, and consequently the dispersion is small enough, so that the absorption edge of the $\beta$ filter is indistinguishable from the edge of the main Bragg peak. It is clear that the true background under the peak is not the average of the values at the two sides but is, rather, an extrapolation of the background trend at the higher side, somewhat as indicated by the dashed lines. As smaller and smaller $2\theta$ values are reached, more and more extreme cases are met until it would be possible, in principle, for the entire white spectrum to be compressed under the main peak, the width of which is then determined by instrumental factors rather than by spectral dispersion.

For the case where there is a structured background the $\omega$ scan with balanced filters offers certain advantages over the $2\theta$ scan. The balanced-filter technique, especially when backed up by pulse-height discrimination, effectively removes all radiation of nonpassband wavelengths (including that which tends to lie under the

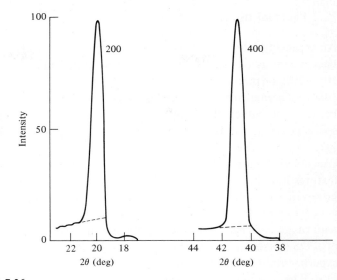

**FIGURE 7-36**
Structured background in $2\theta$ scan at small angles for two orders of the $h00$ reflections.

main peak and to interfere with background measurement in the $2\theta$ scan made with no filter or a $\beta$ filter alone). Thus either the $2\theta$ scan or the $\omega$ scan would obviate the problem shown in Fig. 7-36 if used with balanced filters. Additionally, however, the detector is stationary in the $\omega$ scan and hence continues to receive essentially the same non-Bragg scattered radiation, even at passband wavelengths, throughout the scan. This contribution, regardless of curvature, should then subtract out exactly when the two sets of scan data, one with each member of the balanced-filter pair, are subtracted from each other. (It is understood, of course, that an $\omega$-scan technique will be employed only with balanced filters, or other effective wavelength discrimination, in order that the harmonic contributions shall not be debilitating.)

These considerations of structure in the background have led some workers to choose the $\omega$ scan with balanced filters in the low range of $2\theta$, the $2\theta$ scan with no filter in midrange, and the $2\theta$ scan with a $\beta$ filter in the high range of $2\theta$. The midrange here runs from the $2\theta$ region where a $2\theta$ scan first shows a reasonably straight background between $K\alpha$ and $K\beta$ peaks to the $2\theta$ region where $\beta$ radiation being reflected from higher orders begins to interfere with the background measurement. The balanced-filter $\omega$ scan is then used only when necessary at the small angles because it is inherently slower, since extra measurements have to be made with it, and the aperture width required becomes excessively large at large angles.

## E.   Thermal diffuse scattering

An especially interesting source of systematic error in the experimentally determined background is that due to thermal diffuse scattering, TDS. It can easily amount to 10 to 30% or more of the apparent Bragg peak (Nilsson, 1957; Nicklow and Young, 1964). Thermal diffuse scattering contributes to apparent Bragg intensities largely because the rather slowly varying distribution of one-phonon TDS in reciprocal space peaks at the Bragg position and, hence, is not all included in the measured background. The resulting Bragg-intensity errors can be quite large. At present it appears that corrections for TDS must be based largely on theory, whose salient features have been considered in Section II F, Chap. 2. The design of the experiment, however, can help in several ways, e.g., (1) by operating at reduced temperatures where the TDS is reduced, (2) by using narrow-profile geometry to minimize the scan range and, hence, the TDS correction needed, (3) by permitting easy assessment of TDS anisotropy near the reciprocal-lattice point, and (4) by providing direct experimental tests of the calculated corrections.

The first point rests primarily on the fact that the one-phonon contribution depends approximately linearly on temperature. Whenever possible, therefore, diffraction intensities should be measured at reduced temperature; however, the desired crystalline phase may not exist at the low temperature, or the specimen temperature may be otherwise specified as a requirement of the study. The second point follows from the fact that the one-phonon TDS is intrinsically much less sharply peaked than is the Bragg peak. Thus, as the breadth of the instrumental profile or the width of the scan range is increased, relatively more TDS is included. Chipman and Batterman (1963) have made the point that, for highly perfect crystals and instruments with such narrow instrumental profiles that they sample only the immediate neighborhood of the (very small) reciprocal-lattice point, the TDS contribution is negligibly small. With a two-crystal spectrometer having such an extraordinarily narrow instrumental profile, Renninger (1967) has been able experimentally to display almost separately the TDS (mostly one-phonon) and the Bragg-reflection (Fig. 7-23). An instructive feature of this situation, Fig. 7-23, is that one may readily note the amount of TDS that would be included if the observed Bragg-peak profile (including instrumental broadening effects) had the usual $\frac{1}{4}$ to $\frac{1}{2}°$ half-width instead of being only a few seconds wide. The difference between $\frac{1}{4}$ and $\frac{1}{2}°$ may often be subject to practical control in a single-crystal diffractometer system designed for intensity-data collection. The observed profile breadths can be reduced by the use of small (but brilliant) x-ray focal spots, specimens no larger than necessary, and source-to-specimen distances as large as are consistent with adequate intensity.

The third point above concerns assessment of anisotropy in the TDS. That anisotropy should be expected, even near the reciprocal-lattice point, is clear. It is primarily the one-phonon contribution from the acoustic modes of lattice vibration that will be important, as the optic mode contributions may generally be expected to be much less peaked because of their high dispersion. The dominant TDS contribution not removed in background determinations should, therefore, be that arising from the portions of the acoustic branches close to the origin. There the Debye spectrum approximation is good (for a given direction), and the dispersion is low enough so that the velocity of sound determined from measurements of elastic constants should provide the information needed for calculating the TDS. The diffuse scattering therefore will show anisotropy similar to that shown by the elastic tensor.

Finally, one would like to have an experimental method for spot-checking the accuracy of calculated TDS corrections. The need for such checking for each particular experiment arises because the calculated TDS correction depends strongly on the details of the particular experimental arrangement, e.g., counter aperture, specimen size and aspect presented to the incident beam, incident beam cross fire, and distribution of intensity as a function of ray direction and cross-sectional position in the incident beam. It may often be possible to use the temperature dependence of the background slopes near the peak (but far enough away so that the tails of the Bragg peaks make a negligible contribution) for such an experimental verification. For example, in one study (Nicklow and Young, 1964, 1966) where this method was used, it was shown that the calculations were good to 15% and, hence, the TDS correction was good to 1% of the Bragg intensity.

Very roughly, both the angular and temperature dependence of the one-phonon and two-phonon thermal diffuse scattering contributions may be grasped from the approximate expression

$$I = I_\alpha(1 + K_1 T\varphi + K_2 T^2\varphi^2) \qquad (7\text{-}48)$$

where $I$ is the total intensity including the TDS contribution, $I_\alpha$ is the true Bragg intensity, $T$ is the absolute temperature, $\varphi = \sin^2\theta/\lambda^2$, and $K_1$ and $K_2$ are quantities which are, at most, slowly varying with temperature and which include all the information about the volume of reciprocal space sampled, the elastic constants, etc.

Incorporation of specimen-temperature control in the experimental design is therefore indicated by both points 1 and 4 of the above-enumerated TDS considerations. Obviously, it would be particularly convenient if specimen-temperature control were provided as an integral part of the diffraction experiment under the jurisdiction of, say, the on-line computer used to control the diffractometer.

Calculation of the TDS contribution has engaged the attention of many authors, but no simple exact solutions have been found. The theory of x-ray scattering by lattice vibrations (TDS) has been presented in reviews by several authors (e.g., K. Lonsdale, 1942, 1943; James, 1948; Cochran, 1966; Smith, 1966). The problem is conveniently regarded in two parts, the calculation of the TDS intensity distribution in reciprocal space and the sampling of that distribution, produced by a particular experiment. Each part has its difficulties. In the first part, complete knowledge of the elastic spectrum as a function of direction in the crystal is required. In the second part, detailed knowledge of the composite reciprocal-space sampling-region (as introduced at the end of Section II A) and the weight to be assigned to each point in it is needed.

One may try to calculate this TDS contribution at the Bragg peak, first in reciprocal space and then in the observed profile (Nilsson, 1957; Chipman and Batterman, 1963; Paskin, 1958; Chipman and Paskin, 1959) under certain simplifying assumptions. The first assumption usually made is that TDS is spherically symmetric in reciprocal space over the region of calculation required. Such an assumption was good to about 10% in the immediate neighborhood of the AgCl reciprocal-lattice point studied in one case reported (Nicklow and Young, 1964). The second assumption is that the elastic spectrum is parabolic; i.e., it is a Debye spectrum. This assures that there is no dispersion and, in the region of interest near the origin, the parabolic spectrum can be represented by a straight line of slope obtainable from the elastic constants. Under these assumptions the expected thermal diffuse scattering near the reciprocal-lattice points may be calculated from the elastic constants. As is discussed later, these assumptions with minor modification are realistic for many crystals. The calculation of the distribution of TDS in reciprocal space (near the reciprocal-lattice points) is then straightforward.

As mentioned, the difficult aspect of the calculation of the contribution to the peak intensity is calculation of the volume and proper weighting of the reciprocal-space region actually sampled in the particular case at hand. For each infinitesimal volume element a proper weighting is required which takes into account the cross fire, the distribution of intensity across the x-ray target, and the details of the counter aperture and its motion. Such a calculation is, in fact, the main content of Nilsson's (1957) paper on the subject of TDS contributions to the Bragg peak. Nicklow and Young (1964) made the indicated TDS calculation for the 10,0,0 reflection of an AgCl specimen. By the temperature-dependence method they showed experimentally that their calculation was valid within the experimental error of $\leq 15\%$ in the test. Their results, shown in Fig. 7-37, apply to a geometry which approximates that often used in single-crystal diffractometry. Here the total contribution, to the Bragg peak, of both one- and two-phonon scattering combined is about 30%. Nilsson has quoted 30% for the one-phonon scattering contribution in the case of KCl and NaCl. Chipman and Batterman (1963) mention contributions as high as 70%. Note in Fig. 7-37 that the two-phonon contribution is not so strongly peaked as is the one-phonon contribution and, therefore, may largely be eliminated in the normal procedures of background determination. This fact may be the one somewhat optimistic point about TDS contributions to apparent Bragg peaks; the amount of

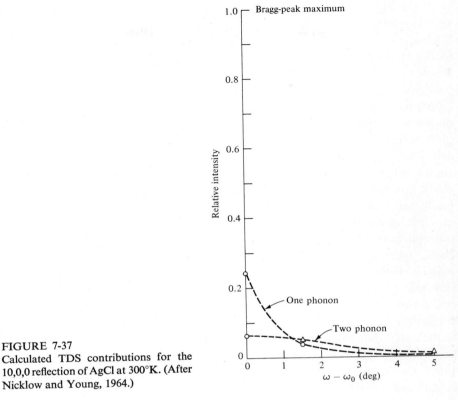

FIGURE 7-37
Calculated TDS contributions for the
10,0,0 reflection of AgCl at 300°K. (After
Nicklow and Young, 1964.)

effort required to calculate the one-phonon contribution in even a simple case is certainly not encouraging. As discussed later, possibly some approximations can be made which will greatly simplify the calculations. In the meantime it is of interest to note that, though tedious, detailed calculations can in fact be carried through at least for the simple cases.

Before proceeding to more complex aspects of TDS, it is of some relevance to consider further the assumptions that went into the calculations in the simple case. Since only the TDS in the immediate neighborhood of the reciprocal-lattice point was of interest, only the long-wavelength region of the elastic spectrum was considered. Generally, this is the region in which the acoustic branch does, in fact, show little or no dispersion and usually shows a fair approximation to a Debye spectrum. Further, one can neglect the contribution of the optic branch, both because of its consistently high frequency and because the dispersion in it is such as to minimize peaking at the Bragg peak.

For more complex crystals the interesting questions include the possibility of approximating acoustic branches of the elastic spectrum with isotropic parabolas (Debye spectrum), in the range of interest, along with the possibility of ignoring optic modes on the ground that they are so dispersive as to show little or no peaking at the Bragg position—hence their contributions would tend to be taken out in normal background subtraction procedures, as are multiple phonon contributions. Cochran (1969) has, in fact, pointed out that contributions from optic modes, in general, do not vary rapidly with phonon wave vector and therefore may be subtracted off by a normal background measurement. Anderson (1963), Schwartz (1964), Rouse and Cooper (1969), Walker and Chipman (1970), and Skelton and Katz (1969), among others, have discussed various rationales for and results of averaging the elastic constants (or wave propagation velocities) to a single value. Generally, the results seem to agree rather well for cubic crystals. Skelton and Katz (1969), for example, calculated less than a 5% difference between the result of the approximate isotropic calculation and one summing explicitly over the actual modes in the known elastic spectrum of copper (511 reflection; summation carried out over 68,920 points). Rouse and Cooper (1969), however, warn strongly against the use of the isotropic approximation for crystals which are not cubic or pseudocubic and they devised an expression for approximating the TDS contribution in terms of the elastic constants of a crystal of any symmetry. A number of other authors have also attested to the applicability of various approximate approaches to calculating TDS contributions to the Bragg peaks from single crystals (e.g., Göttlicher, 1968; Lucas, 1968, 1969). One may, finally, draw a very encouraging conclusion: It now appears that the calculation of the TDS contribution in reciprocal space, to a degree quite satisfactory for correcting single-crystal Bragg intensities, can be successfully carried out for a crystal of any symmetry or complexity on the basis of its elastic constants without reference to details of its elastic spectrum. One could hardly ask for a happier conclusion to a problem that has long stood as a challenge which was not at all certain to be overcome.

The situation with respect to calculating the other part of the problem is not so happy, however; i.e., that part which treats precisely how the TDS intensity in reciprocal space shows up in the actual Bragg-peak scan. The difficulty, as has been pointed out, lies in knowing the weight to assign to each point of the reciprocal-space sampling region (see also Figs. 7-2 to 7-5). The required weighting function (which goes to zero at the outer limits of the sampling region) is called the *resolution function* by several authors (Cooper and Nathans, 1967, 1968a, 1968b; Cooper, 1968; Nielson and Møller, 1969; Cooper and Rouse, 1968; Cochran, 1969; Rouse and Cooper, 1969; Komura and Cooper, 1970). As can be perceived from the discussion in section II, this resolution function will depend in detail on such things as detector (viewing) aperture dimensions, mosaic spread of the crystal, cross fire in the incident beam, specimen (or beam) size, incident beam inhomogeneity, both in intensity and in wavelength, and specimen-to-receiving-aperture distance.

For their neutron diffractometer, Cooper and Nathans (1967) managed to specify these parameters well enough to calculate a resolution function that accounted for their experimental measurements. The neutron diffractometer resolution function has the simplifying feature that the incident beam is reliably homogeneous in a spatial sense and the wavelength distribution can be well approximated by a Gaussian function. Even so, the calculation was tedious. Nilsson (1957) showed that the resolution function (in effect) could be evaluated analytically if slit height were taken to be finite. However, this approximation has been shown to introduce substantial (up to 30%) error (Cooper and Rouse, 1968) in the TDS correction. In the same paper, Cooper and Rouse showed that neglect of the mosaic spread introduces $2\theta$-dependent errors of up to 10% in the TDS correction for a barium fluoride crystal in which the half-width values ranged from 0.2 to 0.8°—indicating a not particularly large mosaic spread.

In an x-ray study, Nicklow and Young (1964) experimentally evaluated the incident beam inhomogeneity and cross fire for their instrument. Directly from these data plus the relevant data on sizes and distances, they calculated the weight to be assigned to the sampling of each point in reciprocal space for each of the several reflections studied. Using the temperature-dependence method mentioned earlier, they showed that all their calculations were correct within the experimental test error of 15% or less.

Because the calculation of the resolution function is so tedious, and must be carried out anew for each reflection and set of specific instrument conditions, there is a substantial incentive for developing experimental approaches which circumvent the need for such calculation. These approaches have taken at least three forms: determination of effective resolution function parameters through their inclusion as adjustable parameters in least-squares refinements of a large body of data, selection of instrumental parameters to produce a convenient resolution function, and direct experimental measurement of the resolution function over the experimental range required.

Lucas (1968, 1969) has suggested a method for correcting TDS entirely after the fact through least-squares crystal-structure refinements based on the apparent Bragg-intensity data. His method also has the advantage that the elastic constants of the crystal need not be well known in advance but may be refined, from some initial estimate, as a part of the iterative process. In the work so far reported (Lucas, 1969), he has tried the method only with cubic crystals. He approximated the true sampling region with a sphere whose radius is a variable parameter in the least-squares refinements. He reports results for Al and KCl in good agreement (<15% difference) with TDS determinations made directly by Mössbauer x rays (Butt and O'Connor, 1967).

The theory has not yet been applied to an anisotropic case. Obviously, if Lucas' approach could be used in the general case, it would be the most desirable way of handling TDS corrections to single-crystal Bragg intensities collected for structure refinement purposes; the addition of a few more adjustable parameters in a least-squares refinement already involving several tens or hundreds of adjustable parameters and several hundreds or thousands of reflections would be a trivial matter in itself. The question of possible correlations with the temperature parameters is, however, a matter requiring further consideration. Certainly the first-order effect of uncorrected TDS is just that of reducing the apparent temperature factor [as has been reported by Nilsson (1957) and numerous later authors]; the separation of TDS parameters and the usual thermal parameters (the $\beta_{ij}$'s) in a least-squares structure refinement will depend on second- and higher-order effects.

Göttlicher (1968) has taken the approach of selecting his instrumental parameters in such a way as to produce a convenient resolution function. By using a small, crystal-monochromated incident beam with small cross fire ("divergence" <15 minutes for 95% of the radiation and $\leq 30$ minutes for all), not-small (172 mm) specimen-to-detector-aperture distance, and large detector apertures ($6 \times 6$ mm$^2$ and $12 \times 12$ mm$^2$), he was able to make the detector aperture the predominant factor in the resolution function. By making his measurements with two different apertures he was able to test his calculations, specifically the simply approximated resolution function in them, experimentally. As might be expected, the effect of a larger angular aperture, in this case increased by moving the detector closer to the specimen, is to allow much more of the background intensity to be recorded. Apparently Göttlicher found his TDS estimates to be good within <10%, for he states that all Bragg intensities were corrected to within 2% and some of the TDS contribution exceeded 30%.

In the same work, Göttlicher (1968) also did a number of other things of interest in the present context. He showed the anisotropy of the TDS contribution directly by comparing $\omega$-scan and $2\theta$-scan results, thus taking advantage of the fact that the two scans sweep out differing paths in reciprocal space. He examined the range over which the scanning needed to be done in order to pick up the TDS contribution that peaked at the Bragg position. This is found to be about 5° (i.e., $\pm 2.5°$) in $\omega$. He used plates (of NaCl) as specimens and, by successively thinning them, he was able to make extinction corrections by the method of Bragg, James, and Bosanquet (1921). Having measured and corrected for both TDS and extinction some 40 reflections, he was able to prepare an electron density map for NaCl which was of a quality superior to any previously reported. But in this high-precision electron density map, he found no discernible effect of the TDS correction on either ionic radii or valence electron density, even though the correction did affect the apparent thermal motion strongly.

Experimental determination of the resolution function can be undertaken in ways. Cochran (1969) recommends that for a spherical specimen it be measured, for a few reflections covering the required range in $2\theta$, by taking a series of photographs, say 10, with the films at the detector aperture and the specimen stationary at 10 successive positions covering the scan range. A microphotometer tracing of the Bragg intensity (characteristic radiation only) is a "vertical" section of the resolution function "in a nearly planar segment of the reflecting sphere." Rouse and Cooper (1969) suggest that a suitable approximation to the resolution function can be obtained by first calculating what it would be if due only to detector aperture size. Then they allow for the other factors in the resolution function "by repeating the calculation with the center of the scan at different points in a partial resolution function centered on the original center of the scan."

Cooper and Nathans (1967) give a thorough analysis of the resolution function and the various factors that enter into a proper calculation of it. They were able to make direct experimental determination of the resolution function (necessarily, however, without inclusion of the specimen's own contributions) by using a highly perfect crystal as specimen and exploring the scan profile as a function of crystal and aperture setting. Pictorially, this corresponds to exploring the composite reciprocal-space sampling region by moving through it along various paths a "reciprocal-lattice point" that is actually nearly pointlike.

# V.  OTHER EFFECTS

## A.  Simultaneous diffraction

Though recognized for many years (Renninger, 1937), simultaneous diffraction (multiple reflection) has only recently begun to be seriously considered as a source of significant errors in "routine precision" measurements of Bragg intensities. When two (or more) reciprocal-lattice points, at $G_1$ and $G_2$, are in contact with the Ewald sphere at the same time, the diffracted beam for $G_1$ is properly oriented to be an incident beam for diffraction with another reciprocal-lattice point at $G_3$ which is related (James, 1948) to the two points at $G_1$ and $G_1$ (simultaneously on the Ewald sphere) by $G_3 = G_2 - G_1$. From $G \equiv h\mathbf{a}^* + k\mathbf{b}^* + l\mathbf{c}^*$, one recognizes that $h_3 = h_2 - h_1$, $k_3 = k_2 - k_1$, and $l_3 = l_2 - l_1$. The net result is that intensity is reflected out of the stronger beams into the weaker ones. The degree to which this occurs depends on the number of reflections operating at once, the strengths of the coupling reflections $h_3 k_3 l_3$, and on much the same parameters, especially, mosaic spread, as are important to secondary extinction (Zachariasen, 1965a, 1967). In a hypothetical extreme case, as Zachariasen (1965a, 1965b) has pointed out, the result could be to make all reflections appear to have the same intensity. In practice, as Zachariasen has shown theoretically and Post (1969) experimentally, the intensity errors resulting from multiple reflection range all the way from trivial ($\ll 1\%$) to very significant, from several percent to several hundred percent for weak reflections. The likelihood that multiple reflection will occur unless deliberately avoided is, in fact, quite high. Obviously, the probability that some second reciprocal-lattice point $h_2 k_2 l_2$ ($= H_2$) will touch the Ewald sphere while measurements are being made on the first, $h_1 k_1 l_1$ ($= H_1$), will increase with increasing thickness of the effective Ewald sphere; i.e., cross fire in the incident beam (see Figs. 7-2 and 7-4). It will also increase with increasing unit-cell size; for very large cells it will be effectively unity (Arndt, 1968). Further, multiple reflection is intrinsic to many commonly used geometries. In the equi-inclination Weissenberg case all intensity measurements are made under conditions of multiple reflection (Yakel and Fankuchen, 1962). With three- and four-circle diffractometers the common practice of mounting the specimen with a symmetry axis coincident with the diffractometer $\theta$ axis produces intrinsic multiple reflection in many cases (Burbank, 1965, plus comment by Willis, 1966; Zocchi and Santoro, 1967).

Irrespective of the instrument or crystal mounting use, the principal thing that the experimenter can do to avoid or to assess multiple reflection is to rotate the specimen about the diffraction vector $\mathbf{S} - \mathbf{S}_0$. This is the rotation that has been designated as either $\sigma$ or $\Psi$, depending on the physical arrangement, in Section II C 3a. Obviously, when the Bragg-reflection condition is satisfied, this axis is parallel to the vector $\mathbf{G}_1$ extending from the origin of the reciprocal space to the reciprocal-lattice point corresponding to the reflection being measured. In the absence of any misalignment and in the absence of specimen shape, absorption, or extinction anisotropies, the only variation of intensity with $\sigma$ then observable will be that caused by the effects of other reciprocal-lattice points entering and leaving the Ewald sphere. Since all the other sources of intensity variation are slowly varying in $\sigma$, certainly all observed sharp variations may be ascribed to simultaneous diffraction (of course, it can happen that an occasional reciprocal-lattice point $H_2$ will move on a path which intersects the sphere at grazing incidence and thus will produce an effect which is not sharp in $\sigma$). Figure 2-36, already encountered in the discussion of this effect in Chap. 2, illustrates the result of such rotation for a particularly graphic case. Figure 7-38 shows a similar test for a case more typical of those encountered in crystal-structure investigations. (The specimen was a fractional-millimeter sphere of fluorapatite.) Here the strong reflection, 11.2, shows only a decrease in intensity as a result of simultaneous diffraction, the weak reflections 45.1 and 52.2 show only intensity increases, and the reflection of intermediate strength, 51.1, appears to exhibit both gains and losses, as would be expected.

Intrinsic multiple reflection can be avoided through straightforward modifica-
ns to the geometric arrangement such as dropping equi-inclination in favor of
ne geometry in the Weissenberg case (Santoro and Zocchi, 1966), mounting

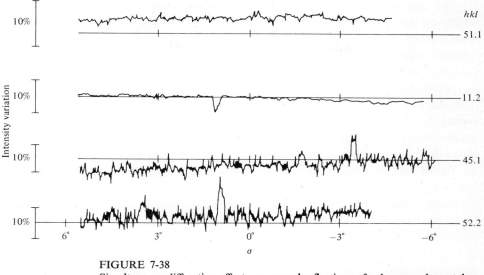

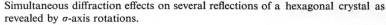

**FIGURE 7-38**
Simultaneous diffraction effects on several reflections of a hexagonal crystal as revealed by $\sigma$-axis rotations.

the crystal in a random orientation or one chosen not to permit any simple crystallographic direction to coincide with the $\Psi$ axis of the diffractometer (Busing and Levy, 1967), or rotating the crystal orientation about $G_1$; i.e., a $\sigma$-axis rotation. Avoidance of accidental multiple reflection to the maximum degree possible, however, requires individual consideration of each reflection; an appropriate rotation angle about $G_1$ must be chosen (if one can be found) corresponding to a flat portion of a plot, for the reflection, like that of Fig. 7-38. Three degrees of freedom for specimen rotation are thus required for the general case and may be instrumentally provided in several ways (Section II C 3).

In the usual four-circle diffractometer they are provided, at least in principle, by the $\chi$, $\phi$, and $\omega$ axes. Busing and Levy (1967) have given the general solution to the problem of choosing $\chi$, $\phi$, and $\omega$ to give a specified rotation about a given reciprocal-lattice vector. (See Section II C 3b.) Santoro and Zocchi (1964) and Powell (1966) have prepared computer programs for the precalculation, for each reflection, of the combinations of settings of $\chi$, $\phi$, and $\omega$ which will place the desired reciprocal-lattice point on the Ewald sphere while keeping all others off (unit-cell size permitting). Willis (1962, 1963), among others, has made use of this feature of the four-circle diffractometer for several years and has used it to produce plots of neutron diffraction intensity vs. rotation about $G_1$, analogous to Fig. 7-38. While very useful for producing rotations about $G_1$, the four-circle diffractometer does have some shortcomings:

(1) The large $\omega$ motions sometimes required produce excessive $\chi$-circle shadowing in reciprocal space and danger of collision with collimation and beam tunnels. (2) Unless a well-programmed computer is on line with the four-circle diffractometer, dynamic use of rotation about $G_1$ is not convenient. (3) For those reflections (reciprocal-lattice points) lying in a central plane perpendicular to $\phi$, rotation about $S - S_0$ is not possible.

As was mentioned in Section II C 3a, an additional mechanical axis can be arranged to be always parallel to the diffraction vector $S - S_0$. This provides the needed rotation directly and avoids most of the objections. This axis has been called the $\sigma$ axis because of its parallelism with $S - S_0$. In the instruments described by Young and coworkers (Young, Goodman, and Kay, 1964; Young, 1969) the $\phi$ and $\chi$ axes are carried by the $\sigma$ axis and, hence, their settings for a given reflection are independent of $\sigma$. With $\omega$ fixed at zero, the standard $\theta - 2\theta$ motion maintains this special axis parallel to $S - S_0$. A particular advantage is that neither change or recalculation of crystal-setting angles initially calculated for some simple (symmetrically aligned) initial orientation nor remounting of the crystal are required to permit scanning of any (and every) reflection for the $\sigma$ angle(s) at which multiple reflection effects are minimized. The results shown in Fig. 7-38 were obtained with such an instrument with a single mounting of the specimen.

The problem of calculating the actual intensity effects of simultaneous diffraction has been treated for neutrons by, for example, Moon and Shull (1964) and, for x rays, by Zachariasen (1965a). The effect which simultaneous diffraction effects may have on an electron density map has been experimentally determined by Panke and Wölfel (1968). They found that the multiple diffraction errors in just five reflections from $Mg_2Si$ were responsible for errors in the electron density map of 0.05 $e/A^3$ on average, 0.1 $e/A^3$ in the most interesting area between ions, and up to 0.5 $e/A^3$ at points of high symmetry. They conclude that for electron density determination from single-crystal measurements, multiple reflection cannot be neglected.

A combined experimental and calculational approach to overcome the multiple reflection problem on the four-circle diffractometer has been devised by Coppens (1968). He first notes that the effect will be significant only if $H_3$, the coupling reflection, is strong and at the same time either $H_1$ or $H_2$ is strong and the other weak. Except in the centrosymmetric case, the coupling from $H_1$ to $H_2$ need not be equal to that from $H_2$ to $H_1$ [i.e., $F(H_3) \neq F(\overline{H}_3)$], and separate terms for the two directions of coupling must be included in an expression for the intensity effect $\Delta I_1$ on $H_1$. In different notation Coppens (1968) cites the expression

$$\Delta I_1 = -kQ(H_1)Q(H_2) - k'Q(H_1)Q(\overline{H}_3) + k''Q(H_2)Q(H_3) \qquad (7\text{-}49)$$

where $Q(H_1)$, for example, is the usual integrated reflectivity per unit volume for reflection $H_1$ and the proportionality factors $k$, $k'$, and $k''$ depend on the appropriate

Lorentz and polarization factors, the mosaic spread, and path lengths in the crystal. Coppens' experimental approach is, then, first to make scans, such as are shown in Fig. 7-38 (also Fig. 2-36), of several strong reflections in order to estimate whether any of the $k$ factors are large enough to be cause for concern. If they are, he then constructs a list of strong reflections and calculates with a computer program whether any one of these will fall on or near the Ewald sphere while any other reflection is being measured. If so, he either calculates new settings of $\omega$, $\phi$, and $\chi$ to provide $\Psi$ rotation sufficient to avoid the situation, or eliminates that reflection from further consideration. He reports that the method has been successful in tests with several crystals, correctly predicting observation of significant multiple reflection effects.

A somewhat different experimental approach has been reported by Young and Mackie (1971). They measured each reflection twice, once at each of two different azimuth ($\Psi$ or $\sigma$) settings, devoting to each measurement only one-half of the time they otherwise would—in order not to extend the total time for data collection. If the two measures differ by more than 3 times the expected standard deviation, that reflection is either omitted or remeasured at a third $\Psi$ value. Since it is highly unlikely that the intensity effect $\Delta I$ from multiple reflection will be the same at two randomly different positions of $\Psi$ (or $\sigma$), this method is believed effectively to rule out all reflections significantly affected by multiple reflection. In an extreme case in which there are always several reciprocal-lattice points on the Ewald sphere at any time (such as would occur for very large unit-cell sizes and large incident-beam cross fire) this method would fail, of course, as would the others mentioned. In such a case, the only approach available would seem to be one of trying to apply calculated corrections, based on theory, to observations after they were made. The only specific contribution of measurement might then be measurements of mosaic spread, incident beam cross fire, and lattice parameters.

## B.  Extinction corrections

For ideally imperfect crystals the diffracted intensity is strictly proportional to $|F|^2$, for ideally perfect crystals it is proportional to $|F|$, and for real crystals the intensity is (ordinarily) intermediate between these two extremes. The departure of the diffracted intensity from the ideally imperfect crystal case is said to be due to the phenomenon of extinction. Physically, extinction may be thought of as arising in two ways. In the first, the coherently diffracting domains in real crystals are large enough so that the incident beam intensity can be significantly reduced by scattering processes on passing through them (primary extinction). In the second, the domains are sufficiently similarly oriented so that either (or both) the incident or diffracted beam for one domain is diminished by diffraction from another nearly parallel oriented domain (secondary extinction). In dealing with these effects, Darwin pictured a crystal as consisting of small coherent (perfect) regions slightly tilted relative to each other, even though he clearly realized that real crystals did not have such "mosaic" structures. As discussed in Chaps. 2 and 3, one can use the kinematical theory to deal with imperfect crystals. It turns out that primary extinction effects become negligible when the coherent regions in a crystal are exceedingly small ($<10^{-5}$ cm), but it is very difficult to prepare samples that meet this condition, although cold-worked metal powders come closest. On the other hand, if a large crystal ($>1$ mm) is perfect, it is necessary to employ the dynamical theory discussed in Chap. 4. As pointed out there, and in Chap. 5, the dynamical theory can be utilized most effectively for crystals that are large compared to the x-ray beam cross section so that the results rather than the details of the theory of extinction are considered here. Treated separately for many years, the two aspects of extinction have now been treated successfully (Zachariasen, 1967) in an integrated fashion, and the distinction between them has diminished.

In crystal-structure refinement studies, it has long been generally recognized that reflections affected in large degree by extinction must either be discarded or corrected for the effect. It is much less widely recognized that correction for even small degrees of extinction becomes important when physically significant precision is demanded. An example occurs in a recent set of refinements based on different hydroxyapatite specimens of the same origin (Sudarsanan and Young, 1969). Before an extinction correction was applied, the largest $R$ (based on $|F|^2$) value was 5.4%, which would have seemed quite satisfactory by the standards of a few years ago. The disagreement among the different results (from different specimens) for the same parameter was $4\sigma$ in some cases; these were anisotropic thermal parameters and $\sigma$ was only $\sim 5\%$ of the parameter value. However, application of Zachariasen's (1963) extinction correction reduced $R$ from 5.4% to 3.5% for the specimen most affected and reduced the disagreements from $4\sigma$ in some cases to $\sim 1\sigma$ in all cases. Thus, in this case at least, the physical significance of the apparent precision was much enhanced by the application of extinction corrections in a case which already appeared to have attained "good" precision.

The practical calculation of small (i.e., $<25\%$) extinction corrections seems to be well in hand, thanks to Zachariasen's (1963) work, and is now widely used. There is also evidence that much larger extinction corrections can be made successfully (Zachariasen, 1967, 1968, 1969). It is of interest to note that Zachariasen's correction works much better for x rays than for neutrons (Cooper and Rouse, 1968; communicated privately).

Many workers have now included Zachariasen's extinction correction procedure directly in their computer program for least-squares crystal-structure refinement (Larson, 1967; plus groups at Oak Ridge National Laboratory, among others). Hamilton and coworkers (W. C. Hamilton, comment following paper by Zachariasen, 1969) have included in their programs allowance for anisotropic extinction corrections.

It is perhaps not yet widely appreciated that extinction depends on the orientation of the plane of incidence, not on $S - S_0$ alone. This anisotropy was experimentally shown for quartz by Pringle (1955). More recently, x-ray topography and ultrasonic standing waves in a quartz plate were used to demonstrate (1) the reality of extinction anisotropy and (2) that its dependence on the orientation of the plane of incidence was in accord with a simplified interpretation of the full theory (Young and Wagner, 1966). It should be expected, therefore, that in a case in which extinction is anisotropic (due, perhaps, to anisotropy in the mosaic spread or in coherent-domain shape) the integrated intensity obtained for a reflection will depend on just how the plane of incidence happens to be oriented about $S - S_0$, as axis, during the scan. (Since different instrument geometries will produce different orientations of the plane of incidence during a scan, it is evident that two experiments could consistently obtain differing results for the intensities of the same reflections from the same specimen if anisotropic extinction is present.) Since this axis is the previously mentioned $\sigma$ axis, it is thus clear how the incorporation of a $\sigma$ axis in the instrument design can contribute to experimental assessment of extinction anisotropy. The experimental demonstration that, for several noncoplanar reflections, the integrated intensity did not depend (in a slowly varying way) on $\sigma$ would constitute proof that extinction anisotropy was not a significant source of error. Practically, it probably would be necessary only to show that the peak intensities, rather than the integrated intensities, did not depend on $\sigma$.

A quantitative experimental method of directly assessing extinction is still much to be desired, both for itself and to complement and guide the theoretical approach. The most notable example of an experimental direct method is Chandrasekhar's method based on the polarization-factor difference between kinematic and dynamical theory. With his coworkers (Chandrasekhar, Ramaseshan, and Singh, 1969), he has mounted a monochromating crystal, selected to give a strong reflection at $2\theta = 90°$, in the diffracted beam in such a way that monochromator and detector can be rotated about that beam as an axis. This monochromator arrangement thus serves as a polarization analyzer. Chandrasekhar et al. show that according to Zachariasen's theory the integrated intensity of reflection from a symmetrically shaped crystal of volume $v$, assumed to consist of nearly spherical domains of radius $r$, is given by

$$R = R_k y \qquad (7\text{-}50)$$

for unpolarized x rays, where the kinematic value

$$R_k = I_0 v A Q_0 \frac{(1 + \cos^2 2\theta)}{2} \qquad (7\text{-}51)$$

and the extinction factor

$$y = \frac{(1 + 2x_0)^{-1/2} + \cos^2 2\theta (1 + 2x_0 \cos^2 2\theta)^{-1/2}}{1 + \cos^2 2\theta} \qquad (7\text{-}52)$$

and

$$x_0 = Q_0 \lambda^{-1} T r^*$$

$$Q_0 \lambda^{-1} = \left| \frac{e^2 \lambda F}{mc^2 V} \right|^2 (\sin 2\theta)^{-1}$$

$$r^* = r \left[ 1 + \left( \frac{r}{\lambda g} \right)^2 \right]^{-1/2} \tag{7-53}$$

$$T = -\frac{1}{A} \frac{dA}{d\mu} = -A^* \frac{dA}{d\mu}$$

$I$ is the incident intensity, $A^* = A^{-1}$ is the absorption factor, $g$ is the factor determining the disorientation of the perfect domains in the crystal, and the other symbols have their usual meanings.

For perpendicular polarization, equations (7-50) to (7-53) give

$$R_\perp = R_{k_\perp} y_\perp \tag{7-54}$$

where

$$R_{k_\perp} = I_0 v A Q_0$$

and

$$y_\perp = (1 + 2x_0)^{-1/2}$$

For parallel polarization

$$R_{\parallel} = R_{k_\parallel} y_{\parallel} \tag{7-55}$$

where

$$R_{k_\parallel} = R_{k_\perp} \cos^2 2\theta$$

$$y_{\parallel} = (1 + 2x_0 \cos^2 2\theta)^{-1/2}$$

From (7-54) and (7-55)

$$R_{k_\perp}^2 = \frac{R_\perp^2 R_\parallel^2 (1 - \cos^2 2\theta)}{R_\perp^2 \cos^4 2\theta - R_\parallel^2 \cos^2 2\theta} \tag{7-56}$$

and

$$x_0 = \frac{1}{2} \left[ \left( \frac{R_{k_\perp}}{R_\perp} \right)^2 - 1 \right] \tag{7-57}$$

A measurement of $R_\perp$ and $R_\parallel$ directly yields $R_{k_\perp}$ (or $|F|^2$) and $x_0$, provided $\theta$ is not too close to 0, 45, or 90°. Observe from (7-53) that a plot of $x_0/Q_0$ against $T$ (for different reflections) gives a straight line of slope $r^* \lambda^{-1}$ passing through the origin. Such a plot may be used to obtain the extinction-free structure factors for those reflections for which (7-56) and (7-57) become ill-conditioned. The value of $x_0/Q_0$ corresponding to any $T$ may be read off from the straight line and, assuming that the data are on an absolute scale, this gives $x_0/R_k = \alpha$, say. Substitution in (7-54) and simplification leads to the relation

$$R_{k_\perp} = R_\perp [\alpha R + (\alpha^2 R + 1)^{-1/2}] \tag{7-58}$$

from which $|F|^2$ may be evaluated. The equations in their present form apply for $r$ and $g$ isotropic, an assumption which may not be valid in all crystals.

With their diffracted-beam polarization analyzer (using the 333 reflection of Ge and Cu $K\alpha$ radiation, for which $2\theta = 90°4'$) Chandrasekhar et al. made experimental determinations of the extinction corrections for several reflections from quartz and compared them with values calculated by Zachariasen's correction method. The results were in excellent agreement, within 1% in most cases, even though extinction reduced the reflection intensity by $\sim 70\%$ in the worst case. The results also agreed very well with structure factor values determined in three cases from *Pendellösung* measurements by others.

Another experimental approach to direct determination of the extinction effect is through the wavelength dependence of the effect, expressed here in equation (7-53). In earlier formulations, the primary and secondary extinction coefficient was readily evaluated experimentally from the approximate relation, for the Bragg arrangement,

$$R_{hkl} = \frac{Qf(A)}{2[\mu + gQf(A)]} \qquad (7\text{-}59)$$

and, for the Laue arrangement,

$$R_{hkl} = Qf(A)\frac{t}{\gamma}\, e^{-[\mu + gQf(A)]t/\gamma}$$

where $Q = Q_0(1 + \cos^2 2\theta)/2$, $t$ is the crystal thickness, and $\gamma$ is the cosine of the angle of incidence. Here $f(A)$ is a function representing the primary extinction effect as defined by Zachariasen (1967).

For no primary extinction, $f(A) = 1$, while for very large primary extinction, $f(A)$ has a small fractional but constant value. Thus a plot of $\ln (R_{hkl}/Q)$ against $F$ or $\lambda$ may show a linear behavior, as was first demonstrated by Bragg, James, and Bosanquet (1921) for NaCl. More recently, Bragg and Azároff (1962) prepared such a plot as a function of incident x-ray wavelength for a fairly perfect silicon crystal cut perpendicular to (111). They found that, apparently, $f(A)$ was not constant (primary extinction was not negligible) but, by assuming various values for the coherent block thickness, they were able to obtain a straight-line plot and a precision of about 15% in the estimate of effective block thickness. With the new, unified view of extinction provided by Zachariasen's general theory of x-ray diffraction, one sees that the quantity to be evaluated from wavelength dependence of the integrated intensity is $r^*$ in equation (7-53).

A discussion of the equations that are applicable to a dynamically diffracting crystal has been given by Weiss (1966, Chap. 2). Until considerably more experimental work has been carried out, particularly a study of primary extinction, it appears unlikely that structure factors can be measured to less than a 1% error. Approximate methods for making corrections in small crystals, primarily for secondary extinction, have been explored by Weiss (1952), Lang (1953), Williamson and Smallman (1955), Vand (1955), Gatineau and Mering (1956), Chandrasekhar (1956), and Zachariasen (1963, 1965b, 1967, 1968, 1969).

## C.   Effective integration of intensities

Most aspects of the problem of measuring an integrated Bragg intensity have been covered in connection with other subjects, especially the discussion of background. There are two additional aspects, however, truncation and incident beam inhomogeneities, that should at least be mentioned. The geometric aspect and the weighting aspect of collecting an integrated intensity are not entirely separable. In what is called the geometric aspect, the problem is that of detecting equally the intensity in all parts of the diffracted beam, of separating contributions from the various reflections (overlap), and of measuring all the intensity including that in the tails of the reflection profile (truncation).

If the reflection profiles were accurately known (see Diamond, 1969, for a discussion of their predictability), truncation corrections could be made as outlined by Alexander and Smith (1962) and scans could be kept short, thereby improving the signal-to-noise ratio, among other things. Figure 7-39 indicates something of the nature of the problem—actual profiles are not always simple. It does seem, however, that a real possibility may exist for use of controlled truncation, followed by calculated truncation corrections, to reduce scanning times and, possibly, to improve precision by restricting measurement to parts of the scan where $\mathscr{S}$ is large.

Overlap of reflections sometimes can be an equally serious problem (Ladell and Spielberg, 1966) and exacerbates truncation problems. Overlap can be minimized by reducing the instrumental profile breadth $b$. The dispersion component of $b$ can be reduced considerably by reducing the effective wavelength spread through crystal monochromatization while the component due to specimen mosaic spread can be reduced by use of an $\omega$ scan.

The weighting aspect of effective integration is the requirement that each part of the specimen shall, except for losses in the specimen itself, receive equal illumination during the course of a scan. This is, in effect, the requirement that the reciprocal-space sampling region have a uniform region large enough to accommodate all significantly intense portions of the reciprocal-lattice point as extended to include specimen features such as mosaicity, finite size, and absorption. In many instruments not equipped with crystal monochromators, this aspect is attended by a design which lets each part of the crystal "see" without obstruction each part of the x-ray source for equal times during the scan. When a crystal monochromator is used, two problems may arise: (1) the different parts of the specimen will receive radiation only from different parts of the source, if the mosaic spread in the monochromator crystal

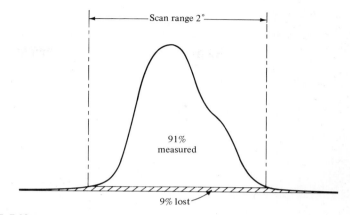

**FIGURE 7-39**
Truncation effect. (After Alexander and Smith, 1962.)

is inadequate to permit each part of the specimen to see equally well each part of the source reflected in the monochromator, and (2) each point on the specimen receives radiation from a different set of points on the monochromator both at one time and in total. Figure 7-40 is designed to help illustrate this point. It is assumed that the intrinsic diffraction profile of a "mosaic block" in the monochromator is narrow compared to the mosaic spread. In Fig. 7-40, $\beta - \beta'$ is the orientation difference of two such blocks which, say, represent the useful extremes. For example, these might correspond to the two 90%-height points (though > 99% would be more appropriate for precision work) on a plot of quantity of blocks as a function of orientation. It is seen that the active mosaic blocks are at different positions in the monochromator and reflect radiation from different parts of the target. Further, the two blocks may have differing reflection efficiencies for several reasons, one of which is difference in extinction. If now another point on the specimen is used, the two corresponding "useful extreme" mosaic-block positions on the monochromator will be shifted along the surface by the difference in $X$ coordinates of the specimen points. In these new positions the monochromator reflectivity may again be different. Without here going into the detail necessary to develop these arguments, one can perhaps accept as plausible the consequence that large (compared to the angle subtended by the source at the monochromator) mosaic spread and uniform reflectivity are important requirements of the monochromator if the weights given different parts of the specimen are not to differ, thereby preventing effective integration for determination of correct integrated intensities.

Many commonly used monochromator crystals do not meet these requirements.

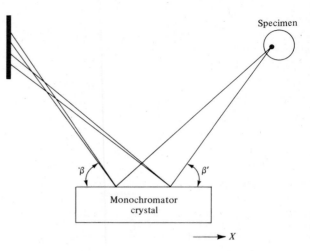

**FIGURE 7-40**
Aspects of monochromator-crystal use.

Of course, better crystals can be selected; the point is that individual selection is necessary (see comments by Azároff and by Chandrasekhar following the paper by Witz, 1969). Clearly, an operational test of the suitability of the crystal is needed. Here equipment design is a factor. If the monochromator mount permits controlled two-dimensional adjustment of the monochromator crystal parallel to its reflecting face, a "good" portion of the crystal may be selected on the basis that small shifts in the monochromator crystal position then do not affect the integrated intensity measured for a particular specimen reflection.

# COMPTON TOTAL CROSS SECTIONS

Calculated from equation (1-104) using Hartree-Fock values in the impulse approximation for $J_i(z)$ and from the Waller-Hartree expression (1-102).

Mo $\lambda = 0.709$

$$(d\sigma/d\Omega)/(e^2/mc^2)/K^2$$

| $\dfrac{\sin\theta}{\lambda}$ | Li† Total Imp. $1s^2$ | Li† Total W.H. $1s^2$ | Li Core Imp. $1s^2$ | Li Total Imp. $1s^2 2s$ | Li Total W.H. $1s^2 2s$ | Be Core Imp. $1s^2$ | Be Total Imp. $1s^2 2s^2$ | Be Total W.H. $1s^2 2s^2$ | B Core Imp. $1s^2$ | B Total Imp. $1s^2 2s^2 2p$ | B Total W.H. $1s^2 2s^2 2p$ |
|---|---|---|---|---|---|---|---|---|---|---|---|
| 0.05 | 0.005 | — | 0.005 | 0.788 | — | 0.000 | 1.417 | — | 0.000 | 1.607 | 0.439 |
| 0.1 | 0.072 | 0.126 | 0.072 | 1.001 | 1.046 | 0.012 | 1.724 | 1.464 | 0.002 | 2.126 | 1.380 |
| 0.2 | 0.492 | 0.447 | 0.491 | 1.473 | 1.437 | 0.189 | 2.113 | 2.222 | 0.076 | 2.731 | 2.772 |
| 0.3 | 0.978 | 0.836 | 0.978 | 1.959 | 1.818 | 0.526 | 2.474 | 2.479 | 0.271 | 3.110 | 3.244 |
| 0.4 | 1.334 | 1.188 | 1.335 | 2.315 | 2.163 | 0.876 | 2.823 | 2.740 | 0.538 | 3.427 | 3.472 |
| 0.5 | 1.558 | 1.452 | 1.559 | 2.536 | 2.424 | 1.163 | 3.105 | 2.984 | 0.808 | 3.705 | 3.668 |
| 0.6 | 1.690 | 1.628 | 1.691 | 2.663 | 2.596 | 1.375 | 3.308 | 3.193 | 1.044 | 3.933 | 3.849 |
| 0.7 | 1.765 | 1.735 | 1.765 | 2.729 | 2.697 | 1.521 | 3.442 | 3.346 | 1.233 | 4.106 | 4.008 |
| 0.8 | 1.802 | — | 1.802 | 2.757 | — | 1.617 | 3.523 | — | 1.376 | 4.227 | — |
| 0.9 | 1.817 | 1.818 | 1.817 | 2.762 | 2.763 | 1.677 | 3.564 | 3.527 | 1.479 | 4.304 | 4.228 |
| 1.0 | 1.817 | — | 1.817 | 2.750 | — | 1.711 | 3.576 | — | 1.551 | 4.346 | — |
| 1.1 | 1.807 | 1.771 | 1.807 | 2.728 | 2.735 | 1.726 | 3.567 | 3.561 | 1.598 | 4.356 | 4.317 |
| 1.2 | 1.790 | — | 1.790 | 2.698 | — | 1.728 | 3.543 | — | 1.625 | 4.344 | — |

Mo $\lambda = 0.709$

| $\dfrac{\sin\theta}{\lambda}$ | F Core Imp. $1s^2$ | F Total Imp. $1s^2 2s^2 2p^5$ | F Total W.H. $1s^2 2s^2 2p^5$ | F− Core Imp. $1s^2$ | F− Total Imp. $1s^2 2s^2 2p^6$ | F− Total W.H. $1s^2 2s^2 2p^6$ | Ne Core Imp. $1s^2 2s^2 2p^6$ | Ne Total Imp. $1s^2 2s^2 2p^6$ | Ne Total W.H. $1s^2 2s^2 2p^6$ |
|---|---|---|---|---|---|---|---|---|---|
| 0.05 | 0.000 | 1.379 | — | 0.000 | 1.466 | — | 0.666 | 0.666 | — |
| 0.1 | 0.000 | 2.403 | 0.860 | 0.000 | 2.802 | 1.345 | 1.963 | 1.963 | 0.815 |
| 0.2 | 0.004 | 3.938 | 2.699 | 0.004 | 4.688 | 3.530 | 3.752 | 3.752 | 2.532 |
| 0.3 | 0.021 | 5.108 | 4.437 | 0.021 | 6.044 | 5.263 | 5.173 | 5.173 | 4.229 |
| 0.4 | 0.070 | 5.844 | 5.656 | 0.070 | 6.831 | 6.463 | 6.158 | 6.158 | 5.567 |
| 0.5 | 0.153 | 6.303 | 6.372 | 0.153 | 7.293 | 7.206 | 6.787 | 6.787 | 6.514 |
| 0.6 | 0.262 | 6.606 | 6.765 | 0.262 | 7.589 | 7.637 | 7.190 | 7.190 | 7.137 |
| 0.7 | 0.388 | 6.821 | 6.984 | 0.388 | 7.794 | 7.880 | 7.459 | 7.459 | 7.516 |
| 0.8 | 0.519 | 6.979 | — | 0.519 | 7.941 | — | 7.645 | 7.645 | — |
| 0.9 | 0.649 | 7.095 | 7.178 | 0.649 | 8.046 | 8.093 | 7.776 | 7.776 | 7.869 |
| 1.0 | 0.770 | 7.178 | — | 0.770 | 8.116 | — | 7.865 | 7.865 | — |
| 1.1 | 0.880 | 7.229 | 7.239 | 0.880 | 8.154 | 8.145 | 7.919 | 7.919 | 7.964 |
| 1.2 | 0.975 | 7.251 | — | 0.975 | 8.162 | — | 7.944 | 7.944 | — |

Mo $\lambda=0.709$

### Na, Mg, Al

| $\frac{\sin\theta}{\lambda}$ | Na Core Imp. $1s^22s^22p^6$ | Na Total Imp. $1s^22s^22p^63s$ | Na Total W.H. $1s^22s^22p^63s$ | Mg Core Imp. $1s^22s^22p^6$ | Mg Total Imp. $1s^22s^22p^63s^2$ | Al Core Imp. $1s^22s^22p^6$ | Al Total Imp. $1s^22s^22p^6$ | Al Total W.H. $3s^23p$ |
|---|---|---|---|---|---|---|---|---|
| 0.05 | 0.280 | 0.322 | — | 0.087 | 0.254 | 0.036 | 0.277 | — |
| 0.1 | 1.312 | 1.921 | 1.503 | 0.711 | 1.884 | 0.414 | 1.465 | 2.278 |
| 0.2 | 3.022 | 3.992 | 2.875 | 2.282 | 4.185 | 1.759 | 4.389 | 4.019 |
| 0.3 | 4.388 | 5.364 | 4.390 | 3.591 | 5.532 | 2.989 | 5.868 | 5.168 |
| 0.4 | 5.459 | 6.436 | 5.739 | 4.699 | 6.646 | 4.063 | 6.964 | 6.311 |
| 0.5 | 6.216 | 7.190 | 6.778 | 5.566 | 7.512 | 4.962 | 7.863 | 7.353 |
| 0.6 | 6.732 | 7.701 | 7.531 | 6.201 | 8.141 | 5.664 | 8.559 | 8.215 |
| 0.7 | 7.083 | 8.044 | 8.035 | 6.648 | 8.576 | 6.189 | 9.069 | 8.888 |
| 0.8 | 7.323 | 8.275 | — | 6.961 | 8.872 | 6.570 | 9.428 | — |
| 0.9 | 7.490 | 8.432 | 8.542 | 7.179 | 9.070 | 6.842 | 9.671 | 9.730 |
| 1.0 | 7.604 | 8.534 | — | 7.330 | 9.197 | 7.033 | 9.830 | — |
| 1.1 | 7.678 | 8.596 | 8.690 | 7.430 | 9.272 | 7.164 | 9.924 | 10.064 |
| 1.2 | 7.717 | 8.622 | — | 7.491 | 9.305 | 7.249 | 9.968 | — |

Mo $\lambda=0.709$

### Si, P, S

| $\frac{\sin\theta}{\lambda}$ | Si Core Imp. $1s^22s^22p^6$ | Si Total Imp. $1s^22s^22p^6$ | Si Total W.H. $3s^23p^2$ | P Core Imp. $1s^22s^22p^6$ | P Total Imp. $1s^22s^22p^63s^23p^3$ | S Core Imp. $1s^22s^22p^6$ | S Total Imp. $1s^22s^22p^63s^23p^4$ |
|---|---|---|---|---|---|---|---|
| 0.05 | 0.015 | 0.564 | — | 0.003 | 0.918 | 0.002 | 1.260 |
| 0.1 | 0.235 | 1.889 | 2.458 | 0.125 | 2.349 | 0.078 | 2.564 |
| 0.2 | 1.322 | 4.669 | 4.558 | 0.939 | 4.934 | 0.701 | 5.184 |
| 0.3 | 2.473 | 6.252 | 5.724 | 1.977 | 6.588 | 1.627 | 7.011 |
| 0.4 | 3.499 | 7.349 | 6.746 | 2.947 | 7.714 | 2.534 | 8.197 |
| 0.5 | 4.392 | 8.248 | 7.717 | 3.820 | 8.614 | 3.367 | 9.099 |
| 0.6 | 5.129 | 8.977 | 8.576 | 4.576 | 9.362 | 4.112 | 9.845 |
| 0.7 | 5.708 | 9.537 | 9.284 | 5.199 | 9.964 | 4.748 | 10.459 |
| 0.8 | 6.148 | 9.950 | — | 5.693 | 10.425 | 5.271 | 10.946 |
| 0.9 | 6.472 | 10.240 | 10.241 | 6.071 | 10.763 | 5.686 | 11.314 |
| 1.0 | 6.707 | 10.433 | — | 6.355 | 10.997 | 6.006 | 11.579 |
| 1.1 | 6.873 | 10.552 | 10.681 | 6.559 | 11.146 | 6.246 | 11.755 |
| 1.2 | 6.984 | 10.612 | — | 6.704 | 11.228 | 6.421 | 11.858 |

Mo $\lambda=0.709$ $\qquad\qquad (d\sigma/d\Omega)/[(e^2/mc^2)K^2]$

| $\dfrac{\sin\theta}{\lambda}$ | Cl Core Imp. $1s^22s^22p^6$ | Cl Total Imp. $1s^22s^22p^63s^23p^5$ | Ar Core Imp. $1s^22s^22p^6$ | Ar Total Imp. $1s^22s^22p^63s^23p^6$ | K Core Imp. $1s^22s^22p^63s^23p^6$ | K Total Imp. Core$+4s$ |
|---|---|---|---|---|---|---|
| 0.05 | 0.001 | 1.217 | 0.000 | 0.574 | 0.332 | 0.368 |
| 0.1 | 0.055 | 2.741 | 0.028 | 2.255 | 1.707 | 2.334 |
| 0.2 | 0.528 | 5.401 | 0.368 | 5.152 | 4.334 | 5.297 |
| 0.3 | 1.346 | 7.374 | 1.050 | 7.472 | 6.642 | 7.617 |
| 0.4 | 2.191 | 8.665 | 1.824 | 8.996 | 8.320 | 9.297 |
| 0.5 | 2.985 | 9.599 | 2.576 | 10.023 | 9.469 | 10.442 |
| 0.6 | 3.707 | 10.346 | 3.274 | 10.806 | 10.303 | 11.269 |
| 0.7 | 4.342 | 10.961 | 3.902 | 11.433 | 10.952 | 11.910 |
| 0.8 | 4.879 | 11.458 | 4.449 | 11.942 | 11.473 | 12.422 |
| 0.9 | 5.318 | 11.843 | 4.910 | 12.344 | 11.887 | 12.825 |
| 1.0 | 5.666 | 12.127 | 5.288 | 12.649 | 12.203 | 13.131 |
| 1.1 | 5.936 | 12.323 | 5.588 | 12.865 | 12.432 | 13.347 |
| 1.2 | 6.143 | 12.445 | 5.820 | 13.004 | 12.580 | 13.481 |

Mo $\lambda=0.709$

| $\dfrac{\sin\theta}{\lambda}$ | Ca Core Imp. $1s^22s^22p^63s^23p^6$ | Ca Total Imp. Core$+4s^2$ | Ca Total W.H. Core$+4s^2$ | Sc Core Imp. $1s^22s^22p^63s^23p^6$ | Sc Total Imp. Core$+3d\,4s^2$ | Ti Core Imp. $1s^22s^22p^63s^23p^6$ | Ti Total Imp. Core$+3d^24s^2$ |
|---|---|---|---|---|---|---|---|
| 0.05 | 0.199 | 0.324 | — | 0.166 | 0.629 | 0.167 | 1.181 |
| 0.1 | 1.163 | 2.370 | 3.088 | 0.842 | 2.527 | 0.812 | 3.069 |
| 0.2 | 3.599 | 5.501 | 5.635 | 3.080 | 5.625 | 2.904 | 6.098 |
| 0.3 | 5.829 | 7.762 | 7.901 | 5.203 | 7.887 | 4.880 | 8.282 |
| 0.4 | 7.610 | 9.558 | 9.664 | 7.016 | 9.791 | 6.625 | 10.178 |
| 0.5 | 8.883 | 10.829 | 10.951 | 8.369 | 11.191 | 7.976 | 11.627 |
| 0.6 | 9.797 | 11.732 | 11.855 | 9.346 | 12.186 | 8.973 | 12.678 |
| 0.7 | 10.489 | 12.408 | 12.524 | 10.074 | 12.915 | 9.721 | 13.449 |
| 0.8 | 11.037 | 12.938 | — | 10.642 | 13.470 | 10.300 | 14.029 |
| 0.9 | 11.474 | 13.355 | 13.488 | 11.095 | 13.902 | 10.762 | 14.475 |
| 1.0 | 11.815 | 13.674 | — | 11.454 | 14.234 | 11.130 | 14.816 |
| 1.1 | 12.072 | 13.906 | 14.147 | 11.730 | 14.478 | 11.419 | 15.068 |
| 1.2 | 12.250 | 14.058 | — | 11.931 | 14.642 | 11.636 | 15.241 |

Mo λ=0.709

| $\dfrac{\sin\theta}{\lambda}$ | V | | Cr | | Mn | | |
|---|---|---|---|---|---|---|---|
| | Core Imp. $1s^22s^22p^63s^23p^6$ | Total Imp. Core$+3d^34s^2$ | Core Imp. $1s^22s^22p^63s^23p^6$ | Total Imp. Core$+3d^54s^2$ | Core Imp. $1s^22s^22p^63s^22p^6$ | Total Imp. Core$+3d^54s^2$ | Total W.H. Core$+3d^54s^2$ |
| 0.05 | 0.170 | 1.766 | 0.164 | 2.643 | 0.149 | 2.608 | — |
| 0.1 | 0.757 | 3.593 | 0.655 | 3.970 | 0.536 | 4.360 | 2.918 |
| 0.2 | 2.730 | 6.559 | 2.509 | 6.616 | 2.276 | 7.250 | 5.764 |
| 0.3 | 4.584 | 8.679 | 4.252 | 8.740 | 3.939 | 9.300 | 8.388 |
| 0.4 | 6.261 | 10.560 | 5.866 | 10.666 | 5.512 | 11.182 | 10.633 |
| 0.5 | 7.603 | 12.048 | 7.201 | 12.242 | 6.853 | 12.767 | 12.523 |
| 0.6 | 8.616 | 13.153 | 8.231 | 13.444 | 7.909 | 13.999 | 14.040 |
| 0.7 | 9.351 | 13.939 | 9.017 | 14.342 | 8.718 | 14.925 | 15.222 |
| 0.8 | 9.944 | 14.550 | 9.625 | 15.014 | 9.344 | 15.617 | — |
| 0.9 | 10.414 | 15.015 | 10.105 | 15.521 | 9.835 | 16.137 | 16.800 |
| 1.0 | 10.790 | 15.368 | 10.488 | 15.902 | 10.228 | 16.525 | — |
| 1.1 | 11.089 | 15.628 | 10.795 | 16.181 | 10.542 | 16.808 | 17.649 |
| 1.2 | 11.318 | 15.808 | 11.032 | 16.375 | 10.788 | 17.004 | — |

Mo λ=0.709

| $\dfrac{\sin\theta}{\lambda}$ | Fe | | | Co | |
|---|---|---|---|---|---|
| | Core Imp. $1s^22s^22p^63s^22p^6$ | Total Imp. Core$+3d^64s^2$ | Total W.H. Core$+3d^64s^2$ | Core Imp. $1s^22s^22p^63s^22p^6$ | Total Imp. Core$+3d^74s^2$ |
| 0.05 | 0.142 | 3.040 | — | 0.136 | 3.612 |
| 0.1 | 0.468 | 4.787 | 2.488 | 0.417 | 5.295 |
| 0.2 | 2.096 | 7.636 | 5.455 | 1.947 | 8.072 |
| 0.3 | 3.671 | 9.654 | 7.960 | 3.453 | 10.059 |
| 0.4 | 5.186 | 11.526 | 10.227 | 4.911 | 11.909 |
| 0.5 | 6.508 | 13.135 | 12.130 | 6.205 | 13.526 |
| 0.6 | 7.573 | 14.414 | 13.679 | 7.268 | 14.838 |
| 0.7 | 8.400 | 15.389 | 14.932 | 8.106 | 15.854 |
| 0.8 | 9.040 | 16.121 | — | 8.759 | 16.623 |
| 0.9 | 9.544 | 16.670 | 16.592 | 9.273 | 17.201 |
| 1.0 | 9.944 | 17.078 | — | 9.681 | 17.630 |
| 1.1 | 10.264 | 17.374 | 17.510 | 10.008 | 17.942 |
| 1.2 | 10.518 | 17.579 | — | 10.268 | 18.158 |

Mo $\lambda = 0.709$

$$(d\sigma/d\Omega) \,/\, (e^2/mc^2)\, K^2$$

| $\sin\theta/\lambda$ | C Core Imp. $1s^2$ | C Total Imp. $1s^22s^22p^2$ | C Total W.H. $1s^22s2p^3$ | N Core Imp. $1s^2$ | N Total Imp. $1s^22s^22p^3$ | N Total W.H. $1s^22s2p^3$ | O Core Imp. $1s^2$ | O Total Imp. $1s^22s^22p^4$ | O Total W.H. $1s^22s^22p^4$ |
|---|---|---|---|---|---|---|---|---|---|
| 0.05 | 0.000 | 1.379 | — | 0.000 | 1.794 | — | 0.000 | 1.204 | — |
| 0.1 | 0.001 | 2.209 | 1.029 | 0.000 | 2.705 | 1.049 | 0.000 | 2.109 | 0.969 |
| 0.2 | 0.032 | 3.092 | 2.504 | 0.011 | 3.688 | 2.816 | 0.007 | 3.644 | 2.789 |
| 0.3 | 0.141 | 3.648 | 3.510 | 0.072 | 4.302 | 4.050 | 0.039 | 4.693 | 4.279 |
| 0.4 | 0.324 | 4.037 | 3.983 | 0.190 | 4.710 | 4.728 | 0.117 | 5.303 | 5.213 |
| 0.5 | 0.542 | 4.338 | 4.357 | 0.351 | 5.009 | 5.089 | 0.235 | 5.679 | 5.742 |
| 0.6 | 0.760 | 4.581 | 4.559 | 0.532 | 5.244 | 5.306 | 0.378 | 5.942 | 6.035 |
| 0.7 | 0.956 | 4.772 | 4.719 | 0.712 | 5.433 | 5.259 | 0.532 | 6.137 | 5.982 |
| 0.8 | 1.121 | 4.915 | — | 0.878 | 5.582 | — | 0.684 | 6.287 | — |
| 0.9 | 1.256 | 5.018 | 4.944 | 1.022 | 5.693 | 5.651 | 0.825 | 6.401 | 6.396 |
| 1.0 | 1.355 | 5.077 | — | 1.143 | 5.770 | — | 0.950 | 6.481 | — |
| 1.1 | 1.431 | 5.106 | 5.054 | 1.239 | 5.814 | 5.758 | 1.057 | 6.530 | 6.485 |
| 1.2 | 1.484 | 5.108 | — | 1.314 | 5.828 | — | 1.146 | 6.550 | — |

Mo $\lambda = 0.709$

| $\sin\theta/\lambda$ | Ni Core Imp. $1s^22s^22p^63s^23p^6$ | Ni Total Imp. Core$+3d^84s^2$ | Cu($3d^9$) Core Imp. $1s^22s^22p^63s^23p^6$ | Cu($3d^9$) Total Imp. Core$+3d^94s$ | Cu($3d^{10}$) Core Imp. $1s^22s^22p^63s^23p^6$ | Cu($3d^{10}$) Total Imp. Core$+3d^{10}4s$ | Cu($3d^{10}$) Total W.H. Core$+3d^{10}4s$ |
|---|---|---|---|---|---|---|---|
| 0.05 | 0.123 | 3.978 | 0.119 | 4.705 | 0.117 | 5.174 | — |
| 0.1 | 0.341 | 5.677 | 0.320 | 5.764 | 0.314 | 6.352 | 1.989 |
| 0.2 | 1.716 | 8.371 | 1.611 | 7.969 | 1.590 | 8.657 | 5.246 |
| 0.3 | 3.169 | 10.358 | 3.010 | 9.884 | 2.997 | 10.665 | 8.100 |
| 0.4 | 4.575 | 12.196 | 4.355 | 11.675 | 4.353 | 12.535 | 10.583 |
| 0.5 | 5.854 | 13.836 | 5.601 | 13.304 | 5.611 | 14.218 | 12.680 |
| 0.6 | 6.928 | 15.196 | 6.661 | 14.676 | 6.676 | 15.618 | 14.460 |
| 0.7 | 7.788 | 16.266 | 7.520 | 15.775 | 7.535 | 16.726 | 15.919 |
| 0.8 | 8.463 | 17.084 | 8.200 | 16.626 | 8.214 | 17.580 | — |
| 0.9 | 8.992 | 17.699 | 8.738 | 17.280 | 8.748 | 18.225 | 18.002 |
| 1.0 | 9.412 | 18.157 | 9.165 | 17.770 | 9.172 | 18.705 | — |
| 1.1 | 9.747 | 18.488 | 9.505 | 18.130 | 9.510 | 19.052 | 19.169 |
| 1.2 | 10.012 | 18.716 | 9.776 | 18.382 | 9.779 | 19.292 | — |

Mo λ=0.709

| $\dfrac{\sin\theta}{\lambda}$ | Zn | | Ga | | Ge | | |
|---|---|---|---|---|---|---|---|
| | Core Imp. $1s^22s^22p^63s^23p^6$ | Total Imp. Core$+3d^{10}4s^2$ | Core Imp. $1s^22s^22p^63s^23d^{10}$ | Total Imp. Core$+4s^24p$ | Core Imp. $1s^22s^22p^63s^23p^63d^{10}$ | Total Imp. Core$+4s^24p^2$ | Total W.H. |
| 0.05 | 0.096 | 4.126 | 2.593 | 3.063 | 1.532 | 2.328 | — |
| 0.1 | 0.254 | 6.211 | 4.276 | 5.964 | 3.510 | 5.718 | 1.589 |
| 0.2 | 1.326 | 8.881 | 6.243 | 8.974 | 5.567 | 9.055 | 4.737 |
| 0.3 | 2.671 | 10.901 | 8.182 | 11.078 | 7.468 | 11.271 | 7.344 |
| 0.4 | 3.972 | 12.736 | 9.978 | 12.892 | 9.233 | 13.090 | 9.911 |
| 0.5 | 5.210 | 14.425 | 11.657 | 14.572 | 10.882 | 14.756 | 12.327 |
| 0.6 | 6.287 | 15.868 | 13.146 | 16.055 | 12.383 | 16.249 | 14.372 |
| 0.7 | 7.174 | 17.037 | 14.394 | 17.287 | 13.673 | 17.522 | 16.044 |
| 0.8 | 7.883 | 17.950 | 15.397 | 18.269 | 14.736 | 18.557 | — |
| 0.9 | 8.442 | 18.645 | 16.182 | 19.026 | 15.583 | 19.369 | 18.542 |
| 1.0 | 8.885 | 19.164 | 16.784 | 19.594 | 16.241 | 19.985 | — |
| 1.1 | 9.238 | 19.542 | 17.235 | 20.009 | 16.742 | 20.438 | 20.183 |
| 1.2 | 9.516 | 19.802 | 17.563 | 20.297 | 17.113 | 20.756 | — |

587

$$(d\sigma/d\Omega)/(e^2/mc^2)K^2$$

Cu $\lambda=1.54$

| $\dfrac{\sin\theta}{\lambda}$ | Li+ | | | Li | | | Be | | | B | | |
|---|---|---|---|---|---|---|---|---|---|---|---|---|
| | Core Imp. $1s^2$ | Total Imp. $1s^2$ | Total W.H. $1s^2$ | Core Imp. $1s^2$ | Total Imp. $1s^2 2s$ | Total W.H. $1s^2 2s$ | Core Imp. $1s^2$ | Total Imp. $1s^2 2s^2$ | Total W.H. $1s^2 2s^2$ | Core Imp. $1s^2$ | Total Imp. $1s^2 2s^2 2p$ | Total W.H. $1s^2 2s^2 2p$ |
| 0.05 | 0.005 | 0.005 | — | 0.005 | 0.786 | — | 0.001 | 1.418 | — | 0.000 | 1.608 | 0.439 |
| 0.1 | 0.073 | 0.073 | 0.126 | 0.073 | 0.998 | 1.045 | 0.016 | 1.724 | 1.463 | 0.005 | 2.127 | 1.379 |
| 0.2 | 0.487 | 0.487 | 0.445 | 0.486 | 1.460 | 1.432 | 0.190 | 2.105 | 2.215 | 0.075 | 2.721 | 2.762 |
| 0.3 | 0.965 | 0.965 | 0.830 | 0.965 | 1.936 | 1.805 | 0.516 | 2.449 | 2.461 | 0.262 | 3.080 | 3.221 |
| 0.4 | 1.310 | 1.310 | 1.173 | 1.310 | 2.275 | 2.136 | 0.855 | 2.775 | 2.705 | 0.516 | 3.369 | 3.428 |
| 0.5 | 1.521 | 1.521 | 1.423 | 1.521 | 2.477 | 2.376 | 1.128 | 3.030 | 2.926 | 0.772 | 3.612 | 3.595 |
| 0.6 | 1.638 | 1.638 | 1.583 | 1.638 | 2.580 | 2.523 | 1.324 | 3.201 | 3.104 | 0.991 | 3.800 | 3.741 |

Cu $\lambda=1.54$

| $\dfrac{\sin\theta}{\lambda}$ | C | | | N | | | O | | |
|---|---|---|---|---|---|---|---|---|---|
| | Core Imp. $1s^2$ | Total Imp. $1s^2 2s^2 2p^3$ | Total W.H. $1s^2 2s^2 2p^3$ | Core Imp. $1s^2$ | Total Imp. $1s^2 2s^2 2p^3$ | Total W.H. $1s^2 2s^2 2p^3$ | Core Imp. $1s^2$ | Total Imp. $1s^2 2s^2 2p^4$ | Total W.H. $1s^2 2s^2 2p^4$ |
| 0.05 | 0.000 | 1.381 | — | 0.000 | 1.796 | — | 0.000 | 1.208 | — |
| 0.1 | 0.002 | 2.209 | 1.028 | 0.000 | 2.704 | 1.048 | 0.000 | 2.109 | 0.969 |
| 0.2 | 0.034 | 3.084 | 2.496 | 0.016 | 3.680 | 2.807 | 0.008 | 3.630 | 2.780 |
| 0.3 | 0.136 | 3.617 | 3.485 | 0.071 | 4.268 | 4.021 | 0.042 | 4.657 | 4.248 |
| 0.4 | 0.306 | 3.972 | 3.932 | 0.178 | 4.639 | 4.668 | 0.110 | 5.224 | 5.147 |
| 0.5 | 0.508 | 4.231 | 4.271 | 0.323 | 4.888 | 4.989 | 0.212 | 5.545 | 5.629 |
| 0.6 | 0.709 | 4.424 | 4.430 | 0.485 | 5.064 | 5.157 | 0.336 | 5.742 | 5.865 |

Cu λ=1.54

| | F | | | F— | | | Ne | | |
|---|---|---|---|---|---|---|---|---|---|
| $\frac{\sin\theta}{\lambda}$ | Core Imp. $1s^2$ | Total Imp. $1s^22s^22p^5$ | Total W.H. $1s^22s^22p^5$ | Core Imp. $1s^2$ | Total Imp. $1s^22s^22p^6$ | Total W.H. $1s^22s^22p^6$ | Core Imp. $1s^22s^22p^6$ | Total Imp. $1s^22s^22p^6$ | Total W.H. $1s^22s^22p^6$ |
| 0.05 | 0.000 | 1.383 | — | 0.000 | 1.471 | — | 0.672 | 0.672 | — |
| 0.1 | 0.000 | 2.403 | 0.860 | 0.000 | 2.802 | 1.344 | 1.963 | 1.963 | 0.815 |
| 0.2 | 0.003 | 3.919 | 2.690 | 0.003 | 4.667 | 3.519 | 3.730 | 3.730 | 2.524 |
| 0.3 | 0.023 | 5.063 | 4.405 | 0.023 | 5.993 | 5.225 | 5.118 | 5.118 | 4.198 |
| 0.4 | 0.067 | 5.757 | 5.584 | 0.067 | 6.733 | 6.381 | 6.059 | 6.059 | 5.496 |
| 0.5 | 0.136 | 6.155 | 6.246 | 0.136 | 7.128 | 7.064 | 6.625 | 6.625 | 6.386 |
| 0.6 | 0.227 | 6.385 | 6.576 | 0.227 | 7.342 | 7.422 | 6.949 | 6.949 | 6.937 |

Cu λ=1.54

| | Na | | | Mg | | Al | | |
|---|---|---|---|---|---|---|---|---|
| $\frac{\sin\theta}{\lambda}$ | Core Imp. $1s^22s^22p^6$ | Total Imp. $1s^22s^22p^63s$ | Total W.H. $1s^22s^22p^63s$ | Core Imp. $1s^22s^22p^6$ | Total Imp. $1s^22s^22p^63s^2$ | Core Imp. $1s^22s^22p^6$ | Total Imp. $1s^22s^22p^6$ | Total W.H. $3s^23p$ |
| 0.05 | 0.285 | 0.326 | — | 0.085 | 0.253 | 0.036 | 0.275 | — |
| 0.1 | 1.309 | 1.918 | 1.502 | 0.702 | 1.873 | 0.406 | 1.455 | 2.277 |
| 0.2 | 2.999 | 3.965 | 2.866 | 2.249 | 4.146 | 1.727 | 4.346 | 4.006 |
| 0.3 | 4.329 | 5.298 | 4.358 | 3.522 | 5.449 | 2.920 | 5.777 | 5.131 |
| 0.4 | 5.354 | 6.319 | 5.666 | 4.582 | 6.503 | 3.945 | 6.808 | 6.230 |
| 0.5 | 6.054 | 7.009 | 6.646 | 5.387 | 7.295 | 4.783 | 7.626 | 7.208 |
| 0.6 | 6.494 | 7.436 | 7.319 | 5.948 | 7.833 | 5.413 | 8.225 | 7.985 |

$$(d\sigma/d\Omega) / (e^2/mc^2) \, K^2$$

Cu $\lambda=1.54$

| $\dfrac{\sin\theta}{\lambda}$ | Si Core Imp. $1s^22s^22p^6$ | Si Total Imp. $1s^22s^22p^6$ | Si Total W.H. $3s^23p^2$ | P Core Imp. $1s^22s^22p^6$ | P Total Imp. $1s^22s^22p^63s^23p^3$ | S Core Imp. $1s^22s^22p^6$ | S Total Imp. $1s^22s^22p^63s^23p^4$ |
|---|---|---|---|---|---|---|---|
| 0.05 | 0.015 | 0.562 | — | 0.002 | 0.916 | 0.001 | 1.258 |
| 0.1 | 0.229 | 1.880 | 2.456 | 0.120 | 2.340 | 0.074 | 2.555 |
| 0.2 | 1.291 | 4.623 | 4.543 | 0.909 | 4.888 | 0.673 | 5.135 |
| 0.3 | 2.404 | 6.153 | 5.683 | 1.909 | 6.484 | 1.561 | 6.900 |
| 0.4 | 3.381 | 7.180 | 6.660 | 2.830 | 7.532 | 2.417 | 8.003 |
| 0.5 | 4.214 | 7.992 | 7.565 | 3.643 | 8.339 | 3.191 | 8.804 |
| 0.6 | 4.880 | 8.617 | 8.336 | 4.328 | 8.976 | 3.865 | 9.431 |

Cu $\lambda=1.54$

| $\dfrac{\sin\theta}{\lambda}$ | Cl Core Imp. $1s^22s^22p^6$ | Cl Total Imp. $1s^22s^22p^63s^23p^5$ | Ar Core Imp. $1s^22s^22p^6$ | Ar Total Imp. $1s^22s^22p^63s^23p^6$ | K Core Imp. $1s^22s^22p^63s^23p^6$ | K Total Imp. Core$+4s$ |
|---|---|---|---|---|---|---|
| 0.05 | 0.001 | 1.215 | 0.000 | 0.572 | 0.329 | 0.366 |
| 0.1 | 0.052 | 2.731 | 0.026 | 2.245 | 1.698 | 2.324 |
| 0.2 | 0.501 | 5.349 | 0.346 | 5.101 | 4.285 | 5.245 |
| 0.3 | 1.280 | 7.256 | 0.988 | 7.349 | 6.522 | 7.489 |
| 0.4 | 2.076 | 8.459 | 1.712 | 8.780 | 8.106 | 9.070 |
| 0.5 | 2.809 | 9.285 | 2.403 | 9.697 | 9.141 | 10.095 |
| 0.6 | 3.461 | 9.906 | 3.031 | 10.342 | 9.844 | 10.783 |

Cu $\lambda = 1.54$

| $\frac{\sin\theta}{\lambda}$ | Ca | | | Sc | | Ti | |
|---|---|---|---|---|---|---|---|
| | Core Imp. $1s^22s^22p^63s^23p^6$ | Total Imp. Core$+4s^2$ | Total W.H. Core$+4s^2$ | Core Imp. $1s^22s^22p^63s^23p^6$ | Total Imp. Core$+3d4s^2$ | Core Imp. $1s^22s^22p^63s^23p^6$ | Total Imp. Core$+3d^24s^2$ |
| 0.05 | 0.198 | 0.322 | | 0.165 | 0.627 | 0.166 | 1.178 |
| 0.1 | 1.154 | 2.359 | 3.085 | 0.833 | 2.515 | 0.804 | 3.056 |
| 0.2 | 3.552 | 5.447 | 5.616 | 3.034 | 5.569 | 2.860 | 6.039 |
| 0.3 | 5.712 | 7.631 | 7.844 | 5.088 | 7.750 | 4.768 | 8.140 |
| 0.4 | 7.397 | 9.320 | 9.541 | 6.807 | 9.543 | 6.418 | 9.920 |
| 0.5 | 8.557 | 10.464 | 10.735 | 8.044 | 10.808 | 7.654 | 11.227 |
| 0.6 | 9.339 | 11.220 | 11.522 | 8.889 | 11.647 | 8.519 | 12.115 |

Cu $\lambda = 1.54$

| $\frac{\sin\theta}{\lambda}$ | V | | Cr | |
|---|---|---|---|---|
| | Core Imp. $1s^22s^22p^63s^23p^6$ | Total Imp. Core$+3d^34s^2$ | Core Imp. $1s^22s^22p^63s^23p^6$ | Total Imp. Core$+3d^54s$ |
| 0.05 | 0.168 | 1.762 | 0.162 | 2.637 |
| 0.1 | 0.748 | 3.578 | 0.646 | 3.952 |
| 0.2 | 2.688 | 6.497 | 2.468 | 6.549 |
| 0.3 | 4.475 | 8.531 | 4.145 | 8.584 |
| 0.4 | 6.058 | 10.291 | 5.666 | 10.385 |
| 0.5 | 7.284 | 11.630 | 6.886 | 11.807 |
| 0.6 | 8.164 | 12.564 | 7.783 | 12.829 |

Cu $\lambda=1.54$

$$(d\sigma/d\Omega)/(e^2/mc^2)\ K^2$$

### Mn

| $\frac{\sin\theta}{\lambda}$ | Core Imp. $1s^22s^22p^63s^22p^6$ | Total Imp. Core$+3d^54s^2$ | Total W.H. Core$+3d^54s^2$ |
|---|---|---|---|
| 0.05 | 0.148 | 2.602 |  |
| 0.1  | 0.528 | 4.341 | 2.916 |
| 0.2  | 2.236 | 7.180 | 5.746 |
| 0.3  | 3.834 | 9.138 | 8.328 |
| 0.4  | 5.315 | 10.891 | 10.497 |
| 0.5  | 6.541 | 12.315 | 12.277 |
| 0.6  | 7.463 | 13.360 | 13.646 |

### Fe

| Core Imp. $1s^22s^22p^63s^22p^6$ | Total Imp. Core$+3d^64s^2$ | Total W.H. Core$+3d^64s^2$ |
|---|---|---|
| 0.141 | 3.033 |  |
| 0.460 | 4.766 | 2.486 |
| 2.057 | 7.561 | 5.437 |
| 3.569 | 9.485 | 7.903 |
| 4.994 | 11.224 | 10.097 |
| 6.201 | 12.666 | 11.892 |
| 7.133 | 13.751 | 13.295 |

### Cu(3d⁹)

| Core Imp. $1s^22s^22p^63s^23p^6$ | Total Imp. Core$+3d^94s$ |
|---|---|
| 0.117 | 4.694 |
| 0.313 | 5.736 |
| 1.574 | 7.881 |
| 2.918 | 9.700 |
| 4.174 | 11.348 |
| 5.308 | 12.839 |
| 6.235 | 13.965 |

Cu $\lambda=1.54$

### Co

| $\frac{\sin\theta}{\lambda}$ | Core Imp. $1s^22s^22p^63s^23p^6$ | Total Imp. Core$+3d^74s^2$ |
|---|---|---|
| 0.05 | 0.134 | 3.604 |
| 0.1  | 0.409 | 5.272 |
| 0.2  | 1.909 | 7.993 |
| 0.3  | 3.355 | 9.884 |
| 0.4  | 4.722 | 11.596 |
| 0.5  | 5.902 | 13.040 |
| 0.6  | 6.832 | 14.150 |

### Ni

| Core Imp. $1s^22s^22p^63s^23p^6$ | Total Imp. Core$+3d^84s^2$ |
|---|---|
| 0.121 | 3.968 |
| 0.334 | 5.651 |
| 1.679 | 8.286 |
| 3.075 | 10.175 |
| 4.390 | 11.870 |
| 5.556 | 13.333 |
| 6.497 | 14.484 |

| $\sin\theta/\lambda$ | Cu(3d¹⁰) | | | Zn | | |
|---|---|---|---|---|---|---|
| | Core Imp. $1s^22s^22p^63s^23p^6$ | Total Imp. Core+$3d^{10}4s$ | Total W.H. Core+$3d^{10}4s$ | Core Imp. $1s^22s^22p^63s^23p^6$ | Total Imp. Core+$3d^{10}4s^2$ | Total W.H. Core+$3d^{10}4s^2$ |
| 0.05 | 0.115 | 5.162 | — | 0.094 | 4.114 | |
| 0.1 | 0.307 | 6.323 | 1.987 | 0.247 | 6.180 | |
| 0.2 | 1.554 | 8.566 | 5.229 | 1.291 | 8.784 | |
| 0.3 | 2.905 | 10.473 | 8.041 | 2.582 | 10.700 | |
| 0.4 | 4.173 | 12.196 | 10.449 | 3.797 | 12.386 | |
| 0.5 | 5.317 | 13.695 | 12.431 | 4.921 | 13.884 | |
| 0.6 | 6.251 | 14.880 | 14.054 | 5.868 | 15.106 | |

| $\sin\theta/\lambda$ | Ga | | Ge | | |
|---|---|---|---|---|---|
| | Core Imp. $1s^22s^22p^63s^23p^63d^{10}$ | Total Imp. Core+$4s^24p$ | Core Imp. $1s^22s^22p^63s^23p^63d^{10}$ | Total Imp. Core+$4s^24p^6$ | Total W.H. Core+$4s^24p^2$ |
| 0.05 | 2.581 | 3.050 | 1.521 | 2.317 | 1.588 |
| 0.1 | 4.247 | 5.931 | 3.480 | 5.684 | 4.722 |
| 0.2 | 6.150 | 8.871 | 5.477 | 8.950 | 7.292 |
| 0.3 | 7.996 | 10.871 | 7.284 | 11.058 | 9.785 |
| 0.4 | 9.657 | 12.533 | 8.914 | 12.720 | 12.084 |
| 0.5 | 11.159 | 14.016 | 10.390 | 14.186 | 13.968 |
| 0.6 | 12.444 | 15.270 | 11.687 | 15.442 | |

# APPENDIX B

## COMPTON PROFILES FOR HARTREE-FOCK WAVEFUNCTIONS

Compton line shape functions $J(q)$ ($q$ in atomic units) calculated from the Hartree-Fock wavefunctions of Clementi. Included also are the expectation values of the kinetic energy $\langle K.E. \rangle$. (Tables prepared by A. Harvey.)

| $\langle K.E.\rangle$ eV / q | 1s Li+ 98.39 | 1s Li 98.22 | 1s Be 184.5 | 1s B 297.3 | 1s C 436.5 | 1s N 602.1 | 1s O 794.4 | 1s F 1013.0 | 1s F- 1013.0 | 1s Ne 1258.0 | 1s Na 1530.0 |
|---|---|---|---|---|---|---|---|---|---|---|---|
| 0.0 | 0.3278 | 0.3286 | 0.2370 | 0.1860 | 0.1532 | 0.1303 | 0.1133 | 0.1003 | 0.1003 | 0.0899 | 0.0815 |
| 0.1 | 0.3263 | 0.3270 | 0.2365 | 0.1857 | 0.1531 | 0.1302 | 0.1133 | 0.1002 | 0.1003 | 0.0899 | 0.0815 |
| 0.2 | 0.3217 | 0.3224 | 0.2347 | 0.1849 | 0.1526 | 0.1299 | 0.1131 | 0.1001 | 0.1001 | 0.0898 | 0.0814 |
| 0.3 | 0.3144 | 0.3149 | 0.2319 | 0.1835 | 0.1518 | 0.1294 | 0.1128 | 0.0999 | 0.0999 | 0.0897 | 0.0813 |
| 0.4 | 0.3044 | 0.3048 | 0.2281 | 0.1816 | 0.1507 | 0.1288 | 0.1123 | 0.0996 | 0.0996 | 0.0894 | 0.0811 |
| 0.5 | 0.2923 | 0.2926 | 0.2232 | 0.1792 | 0.1494 | 0.1279 | 0.1118 | 0.0992 | 0.0992 | 0.0892 | 0.0809 |
| 0.6 | 0.2784 | 0.2785 | 0.2175 | 0.1763 | 0.1477 | 0.1269 | 0.1111 | 0.0987 | 0.0987 | 0.0888 | 0.0806 |
| 0.7 | 0.2631 | 0.2632 | 0.2110 | 0.1730 | 0.1458 | 0.1257 | 0.1103 | 0.0982 | 0.0982 | 0.0884 | 0.0803 |
| 0.8 | 0.2468 | 0.2468 | 0.2038 | 0.1693 | 0.1437 | 0.1243 | 0.1094 | 0.0975 | 0.0975 | 0.0879 | 0.0800 |
| 0.9 | 0.2300 | 0.2300 | 0.1961 | 0.1652 | 0.1413 | 0.1228 | 0.1083 | 0.0968 | 0.0968 | 0.0874 | 0.0796 |
| 1.0 | 0.2131 | 0.2130 | 0.1879 | 0.1608 | 0.1387 | 0.1211 | 0.1072 | 0.0960 | 0.0960 | 0.0868 | 0.0792 |
| 1.2 | 0.1800 | 0.1798 | 0.1707 | 0.1513 | 0.1329 | 0.1174 | 0.1047 | 0.0942 | 0.0942 | 0.0855 | 0.0782 |
| 1.4 | 0.1494 | 0.1492 | 0.1531 | 0.1409 | 0.1265 | 0.1132 | 0.1018 | 0.0922 | 0.0922 | 0.0840 | 0.0770 |
| 1.6 | 0.1223 | 0.1221 | 0.1357 | 0.1302 | 0.1196 | 0.1086 | 0.0986 | 0.0899 | 0.0899 | 0.0823 | 0.0757 |
| 1.8 | 0.0992 | 0.0990 | 0.1191 | 0.1194 | 0.1124 | 0.1038 | 0.0952 | 0.0874 | 0.0874 | 0.0804 | 0.0743 |
| 2.0 | 0.0799 | 0.0798 | 0.1037 | 0.1086 | 0.1051 | 0.0978 | 0.0915 | 0.0847 | 0.0847 | 0.0784 | 0.0728 |
| 2.5 | 0.0460 | 0.0459 | 0.0714 | 0.0837 | 0.0870 | 0.0855 | 0.0818 | 0.0774 | 0.0774 | 0.0729 | 0.0685 |
| 3.0 | 0.0266 | 0.0266 | 0.0481 | 0.0628 | 0.0701 | 0.0724 | 0.0718 | 0.0697 | 0.0697 | 0.0668 | 0.0637 |
| 3.5 | 0.0157 | 0.0156 | 0.0321 | 0.0463 | 0.0555 | 0.0603 | 0.0620 | 0.0618 | 0.0618 | 0.0605 | 0.0586 |
| 4.0 | 0.0095 | 0.0094 | 0.0215 | 0.0338 | 0.0433 | 0.0495 | 0.0529 | 0.0542 | 0.0542 | 0.0542 | 0.0534 |
| 5.0 | 0.0037 | 0.0037 | 0.0099 | 0.0179 | 0.0258 | 0.0324 | 0.0373 | 0.0405 | 0.0405 | 0.0424 | 0.0433 |
| 6.0 | 0.0016 | 0.0016 | 0.0048 | 0.0096 | 0.0153 | 0.0208 | 0.0257 | 0.0295 | 0.0295 | 0.0323 | 0.0342 |
| 7.0 | 0.0008 | 0.0008 | 0.0024 | 0.0053 | 0.0091 | 0.0133 | 0.0174 | 0.0211 | 0.0211 | 0.0241 | 0.0265 |
| 8.0 | 0.0004 | 0.0004 | 0.0013 | 0.0030 | 0.0055 | 0.0086 | 0.0118 | 0.0150 | 0.0150 | 0.0178 | 0.0203 |
| 9.0 | 0.0002 | 0.0002 | 0.0007 | 0.0018 | 0.0034 | 0.0056 | 0.0080 | 0.0106 | 0.0106 | 0.0131 | 0.0154 |
| 10.0 | 0.0001 | 0.0001 | 0.0001 | 0.0011 | 0.0022 | 0.0037 | 0.0055 | 0.0075 | 0.0075 | 0.0096 | 0.0116 |
| 15.0 | 0.0000 | 0.0000 | 0.0001 | 0.0001 | 0.0003 | 0.0006 | 0.0010 | 0.0015 | 0.0015 | 0.0022 | 0.0029 |
| 20.0 |  |  | 0.0000 | 0.0000 | 0.0001 | 0.0001 | 0.0002 | 0.0004 | 0.0004 | 0.0006 | 0.0008 |
| 25.0 |  |  |  |  | 0.0000 | 0.0000 | 0.0001 | 0.0001 | 0.0001 | 0.0002 | 0.0003 |
| 30.0 |  |  |  |  |  |  | 0.0000 | 0.0000 | 0.0000 | 0.0001 | 0.0001 |
| 35.0 |  |  |  |  |  |  |  |  |  | 0.0000 | 0.0000 |

| $q$ | 1s Mg 1820.0 | 1s Al 2155.4 | 1s Si 2508.2 | 1s P 2888.2 | 1s S 3295.1 | 1s Cl 3729.0 | 1s A 4189.5 | 1s K 4677.6 | 1s Ca 5192.5 | 1s Sc 5734.8 | 1s Ti 6304.4 |
|---|---|---|---|---|---|---|---|---|---|---|---|
| ⟨K.E.⟩eV | | | | | | | | | | | |
| 0.0 | 0.0745 | 0.0687 | 0.0636 | 0.0592 | 0.0554 | 0.0520 | 0.0490 | 0.0464 | 0.0439 | 0.0417 | 0.0398 |
| 0.1 | 0.0744 | 0.0687 | 0.0636 | 0.0592 | 0.0554 | 0.0520 | 0.0490 | 0.0464 | 0.0439 | 0.0417 | 0.0398 |
| 0.2 | 0.0744 | 0.0687 | 0.0636 | 0.0592 | 0.0554 | 0.0520 | 0.0490 | 0.0463 | 0.0439 | 0.0417 | 0.0398 |
| 0.3 | 0.0743 | 0.0686 | 0.0635 | 0.0592 | 0.0553 | 0.0520 | 0.0490 | 0.0463 | 0.0439 | 0.0417 | 0.0398 |
| 0.4 | 0.0742 | 0.0685 | 0.0635 | 0.0591 | 0.0553 | 0.0519 | 0.0490 | 0.0463 | 0.0439 | 0.0417 | 0.0398 |
| 0.5 | 0.0740 | 0.0684 | 0.0634 | 0.0590 | 0.0552 | 0.0519 | 0.0489 | 0.0462 | 0.0439 | 0.0417 | 0.0397 |
| 0.6 | 0.0738 | 0.0682 | 0.0632 | 0.0589 | 0.0551 | 0.0518 | 0.0489 | 0.0462 | 0.0438 | 0.0416 | 0.0397 |
| 0.7 | 0.0736 | 0.0681 | 0.0631 | 0.0588 | 0.0551 | 0.0517 | 0.0488 | 0.0461 | 0.0438 | 0.0416 | 0.0397 |
| 0.8 | 0.0733 | 0.0678 | 0.0629 | 0.0587 | 0.0549 | 0.0516 | 0.0487 | 0.0461 | 0.0437 | 0.0415 | 0.0396 |
| 0.9 | 0.0730 | 0.0676 | 0.0628 | 0.0585 | 0.0548 | 0.0515 | 0.0486 | 0.0460 | 0.0437 | 0.0415 | 0.0396 |
| 1.0 | 0.0727 | 0.0674 | 0.0625 | 0.0584 | 0.0547 | 0.0514 | 0.0485 | 0.0459 | 0.0436 | 0.0414 | 0.0395 |
| 1.2 | 0.0719 | 0.0668 | 0.0621 | 0.0580 | 0.0544 | 0.0512 | 0.0483 | 0.0457 | 0.0434 | 0.0413 | 0.0394 |
| 1.4 | 0.0710 | 0.0661 | 0.0615 | 0.0575 | 0.0540 | 0.0509 | 0.0481 | 0.0455 | 0.0432 | 0.0411 | 0.0392 |
| 1.6 | 0.0700 | 0.0653 | 0.0609 | 0.0570 | 0.0536 | 0.0505 | 0.0478 | 0.0453 | 0.0430 | 0.0409 | 0.0391 |
| 1.8 | 0.0689 | 0.0644 | 0.0602 | 0.0564 | 0.0531 | 0.0501 | 0.0474 | 0.0450 | 0.0428 | 0.0407 | 0.0389 |
| 2.0 | 0.0677 | 0.0634 | 0.0594 | 0.0558 | 0.0526 | 0.0497 | 0.0471 | 0.0447 | 0.0425 | 0.0405 | 0.0387 |
| 2.5 | 0.0643 | 0.0607 | 0.0572 | 0.0540 | 0.0511 | 0.0484 | 0.0460 | 0.0438 | 0.0417 | 0.0398 | 0.0381 |
| 3.0 | 0.0605 | 0.0575 | 0.0546 | 0.0519 | 0.0493 | 0.0469 | 0.0447 | 0.0427 | 0.0408 | 0.0390 | 0.0374 |
| 3.5 | 0.0564 | 0.0541 | 0.0518 | 0.0495 | 0.0473 | 0.0452 | 0.0433 | 0.0415 | 0.0397 | 0.0381 | 0.0366 |
| 4.0 | 0.0521 | 0.0505 | 0.0487 | 0.0469 | 0.0452 | 0.0434 | 0.0417 | 0.0401 | 0.0386 | 0.0371 | 0.0357 |
| 5.0 | 0.0435 | 0.0432 | 0.0425 | 0.0416 | 0.0406 | 0.0394 | 0.0383 | 0.0371 | 0.0359 | 0.0348 | 0.0337 |
| 6.0 | 0.0354 | 0.0361 | 0.0363 | 0.0361 | 0.0358 | 0.0352 | 0.0346 | 0.0339 | 0.0331 | 0.0322 | 0.0314 |
| 7.0 | 0.0283 | 0.0296 | 0.0304 | 0.0309 | 0.0311 | 0.0310 | 0.0308 | 0.0305 | 0.0301 | 0.0296 | 0.0290 |
| 8.0 | 0.0223 | 0.0239 | 0.0252 | 0.0260 | 0.0266 | 0.0270 | 0.0272 | 0.0272 | 0.0270 | 0.0268 | 0.0265 |
| 9.0 | 0.0174 | 0.0192 | 0.0206 | 0.0217 | 0.0226 | 0.0232 | 0.0237 | 0.0240 | 0.0241 | 0.0241 | 0.0240 |
| 10.0 | 0.0135 | 0.0152 | 0.0167 | 0.0180 | 0.0190 | 0.0198 | 0.0205 | 0.0210 | 0.0213 | 0.0215 | 0.0216 |
| 15.0 | 0.0038 | 0.0047 | 0.0056 | 0.0065 | 0.0074 | 0.0083 | 0.0091 | 0.0099 | 0.0106 | 0.0112 | 0.0117 |
| 20.0 | 0.0011 | 0.0015 | 0.0019 | 0.0024 | 0.0029 | 0.0034 | 0.0039 | 0.0044 | 0.0049 | 0.0054 | 0.0059 |
| 25.0 | 0.0004 | 0.0006 | 0.0007 | 0.0009 | 0.0012 | 0.0014 | 0.0017 | 0.0020 | 0.0023 | 0.0026 | 0.0029 |
| 30.0 | 0.0001 | 0.0002 | 0.0003 | 0.0004 | 0.0005 | 0.0006 | 0.0007 | 0.0009 | 0.0010 | 0.0012 | 0.0014 |
| 35.0 | 0.0000 | 0.0001 | 0.0001 | 0.0002 | 0.0002 | 0.0003 | 0.0003 | 0.0004 | 0.0005 | 0.0006 | 0.0007 |
| 40.0 | | 0.0000 | 0.0001 | 0.0001 | 0.0001 | 0.0001 | 0.0001 | 0.0002 | 0.0002 | 0.0003 | 0.0003 |
| 45.0 | | | 0.0000 | 0.0000 | 0.0000 | 0.0000 | 0.0000 | 0.0001 | 0.0001 | 0.0001 | 0.0001 |
| 50.0 | | | | | | | | 0.0000 | 0.0000 | 0.0000 | 0.0000 |

| | 1*s* | 1*s* | 1*s* | 1*s* | 1*s* | 1*s* | 1*s* | 1*s* | 1*s* | 1*s* | 1*s* |
|---|---|---|---|---|---|---|---|---|---|---|---|
| | V | Cr | Mn | Fe | Co | Ni | Cu($3d^{10}$) | Cu($3d^9$) | Zn | Ga | Ge |
| ⟨K.E.⟩eV | 6901.0 | 7524.7 | 8175.4 | 8853.6 | 9558.4 | 10290.4 | 11050.0 | 11049.6 | 11835.7 | 12648.4 | 13487.8 |
| *q* | | | | | | | | | | | |
| 0.0 | 0.0379 | 0.0363 | 0.0348 | 0.0333 | 0.0320 | 0.0308 | 0.0296 | 0.0296 | 0.0285 | 0.0275 | 0.0266 |
| 0.1 | 0.0379 | 0.0363 | 0.0348 | 0.0333 | 0.0320 | 0.0308 | 0.0296 | 0.0296 | 0.0285 | 0.0275 | 0.0266 |
| 0.2 | 0.0379 | 0.0363 | 0.0347 | 0.0333 | 0.0320 | 0.0308 | 0.0296 | 0.0296 | 0.0285 | 0.0275 | 0.0266 |
| 0.3 | 0.0379 | 0.0363 | 0.0347 | 0.0333 | 0.0320 | 0.0308 | 0.0296 | 0.0296 | 0.0285 | 0.0275 | 0.0266 |
| 0.4 | 0.0379 | 0.0363 | 0.0347 | 0.0333 | 0.0320 | 0.0308 | 0.0296 | 0.0296 | 0.0285 | 0.0275 | 0.0266 |
| 0.5 | 0.0379 | 0.0362 | 0.0347 | 0.0333 | 0.0320 | 0.0307 | 0.0296 | 0.0296 | 0.0285 | 0.0275 | 0.0265 |
| 0.6 | 0.0378 | 0.0362 | 0.0347 | 0.0333 | 0.0320 | 0.0307 | 0.0296 | 0.0296 | 0.0285 | 0.0275 | 0.0265 |
| 0.7 | 0.0378 | 0.0362 | 0.0347 | 0.0333 | 0.0319 | 0.0307 | 0.0296 | 0.0296 | 0.0285 | 0.0275 | 0.0265 |
| 0.8 | 0.0378 | 0.0362 | 0.0346 | 0.0332 | 0.0319 | 0.0307 | 0.0296 | 0.0296 | 0.0285 | 0.0275 | 0.0265 |
| 0.9 | 0.0378 | 0.0361 | 0.0346 | 0.0332 | 0.0319 | 0.0307 | 0.0295 | 0.0295 | 0.0285 | 0.0275 | 0.0265 |
| 1.0 | 0.0377 | 0.0361 | 0.0346 | 0.0332 | 0.0319 | 0.0306 | 0.0295 | 0.0295 | 0.0284 | 0.0274 | 0.0265 |
| 1.2 | 0.0376 | 0.0360 | 0.0345 | 0.0331 | 0.0318 | 0.0306 | 0.0295 | 0.0295 | 0.0284 | 0.0274 | 0.0264 |
| 1.4 | 0.0375 | 0.0359 | 0.0344 | 0.0330 | 0.0317 | 0.0305 | 0.0294 | 0.0294 | 0.0283 | 0.0273 | 0.0264 |
| 1.6 | 0.0374 | 0.0358 | 0.0343 | 0.0329 | 0.0316 | 0.0305 | 0.0293 | 0.0293 | 0.0283 | 0.0273 | 0.0264 |
| 1.8 | 0.0372 | 0.0356 | 0.0342 | 0.0328 | 0.0315 | 0.0304 | 0.0293 | 0.0293 | 0.0282 | 0.0272 | 0.0263 |
| 2.0 | 0.0370 | 0.0355 | 0.0340 | 0.0327 | 0.0314 | 0.0303 | 0.0292 | 0.0292 | 0.0281 | 0.0272 | 0.0262 |
| 2.5 | 0.0365 | 0.0350 | 0.0336 | 0.0323 | 0.0311 | 0.0300 | 0.0289 | 0.0289 | 0.0279 | 0.0269 | 0.0260 |
| 3.0 | 0.0359 | 0.0345 | 0.0332 | 0.0319 | 0.0307 | 0.0296 | 0.0286 | 0.0286 | 0.0276 | 0.0267 | 0.0258 |
| 3.5 | 0.0352 | 0.0339 | 0.0326 | 0.0314 | 0.0303 | 0.0292 | 0.0282 | 0.0282 | 0.0273 | 0.0264 | 0.0255 |
| 4.0 | 0.0344 | 0.0331 | 0.0320 | 0.0309 | 0.0298 | 0.0288 | 0.0278 | 0.0278 | 0.0269 | 0.0261 | 0.0252 |
| 5.0 | 0.0326 | 0.0316 | 0.0305 | 0.0296 | 0.0287 | 0.0278 | 0.0269 | 0.0269 | 0.0261 | 0.0253 | 0.0245 |
| 6.0 | 0.0306 | 0.0297 | 0.0289 | 0.0281 | 0.0273 | 0.0266 | 0.0258 | 0.0258 | 0.0251 | 0.0244 | 0.0237 |
| 7.0 | 0.0284 | 0.0278 | 0.0272 | 0.0265 | 0.0259 | 0.0252 | 0.0246 | 0.0246 | 0.0240 | 0.0234 | 0.0228 |
| 8.0 | 0.0261 | 0.0257 | 0.0253 | 0.0248 | 0.0243 | 0.0238 | 0.0233 | 0.0233 | 0.0228 | 0.0223 | 0.0218 |
| 9.0 | 0.0239 | 0.0237 | 0.0234 | 0.0231 | 0.0227 | 0.0223 | 0.0219 | 0.0219 | 0.0215 | 0.0211 | 0.0207 |
| 10.0 | 0.0216 | 0.0216 | 0.0215 | 0.0213 | 0.0211 | 0.0208 | 0.0205 | 0.0205 | 0.0202 | 0.1999 | 0.0196 |
| 15.0 | 0.0122 | 0.0126 | 0.0129 | 0.0132 | 0.0134 | 0.0136 | 0.0137 | 0.0137 | 0.0138 | 0.0139 | 0.0139 |
| 20.0 | 0.0063 | 0.0067 | 0.0071 | 0.0075 | 0.0078 | 0.0082 | 0.0084 | 0.0084 | 0.0086 | 0.0089 | 0.0090 |
| 25.0 | 0.0032 | 0.0035 | 0.0038 | 0.0041 | 0.0044 | 0.0046 | 0.0049 | 0.0049 | 0.0051 | 0.0053 | 0.0055 |
| 30.0 | 0.0016 | 0.0018 | 0.0020 | 0.0021 | 0.0023 | 0.0025 | 0.0027 | 0.0027 | 0.0029 | 0.0031 | 0.0032 |
| 35.0 | 0.0008 | 0.0009 | 0.0010 | 0.0011 | 0.0012 | 0.0013 | 0.0014 | 0.0014 | 0.0016 | 0.0017 | 0.0018 |
| 40.0 | 0.0003 | 0.0004 | 0.0005 | 0.0005 | 0.0006 | 0.0006 | 0.0007 | 0.0007 | 0.0008 | 0.0008 | 0.0009 |
| 45.0 | 0.0001 | 0.0001 | 0.0002 | 0.0002 | 0.0002 | 0.0002 | 0.0003 | 0.0003 | 0.0003 | 0.0003 | 0.0003 |
| 50.0 | 0.0000 | 0.0000 | 0.0000 | 0.0000 | 0.0000 | 0.0000 | 0.0000 | 0.0000 | 0.0000 | 0.0000 | 0.0000 |

| $\langle$K.E.$\rangle$eV | 2s Li 5.678 | 2s Be 13.63 | 2s B 26.07 | 2s C 41.83 | 2s N 60.96 | 2s O 84.53 | 2s F 111.5 | 2s F— 105.6 | 2s Ne 141.8 | 2s Na 186.0 | 2s Mg 237.7 |
|---|---|---|---|---|---|---|---|---|---|---|---|
| 0.0 | 1.9356 | 1.3427 | 1.0012 | 0.8057 | 0.6755 | 0.5790 | 0.5079 | 0.5340 | 0.4530 | 0.3902 | 0.3411 |
| 0.1 | 1.6894 | 1.2534 | 0.9627 | 0.7851 | 0.6633 | 0.5712 | 0.5026 | 0.5276 | 0.4492 | 0.3877 | 0.3395 |
| 0.2 | 1.1384 | 1.0245 | 0.8575 | 0.7272 | 0.6282 | 0.5486 | 0.4870 | 0.5091 | 0.4380 | 0.3806 | 0.3348 |
| 0.3 | 0.6124 | 0.7428 | 0.7109 | 0.6417 | 0.5745 | 0.5132 | 0.4624 | 0.4801 | 0.4201 | 0.3691 | 0.3270 |
| 0.4 | 0.2743 | 0.4862 | 0.5523 | 0.5414 | 0.5083 | 0.4682 | 0.4304 | 0.4429 | 0.3965 | 0.3538 | 0.3166 |
| 0.5 | 0.1077 | 0.2928 | 0.4055 | 0.4387 | 0.4361 | 0.4172 | 0.3932 | 0.4003 | 0.3686 | 0.3352 | 0.3038 |
| 0.6 | 0.0411 | 0.1653 | 0.2836 | 0.3428 | 0.3639 | 0.3637 | 0.3529 | 0.3550 | 0.3377 | 0.3144 | 0.2890 |
| 0.7 | 0.0193 | 0.0892 | 0.1907 | 0.2597 | 0.2960 | 0.3107 | 0.3115 | 0.3094 | 0.3051 | 0.2911 | 0.2726 |
| 0.8 | 0.0141 | 0.0475 | 0.1242 | 0.1914 | 0.2355 | 0.2606 | 0.2709 | 0.2656 | 0.2722 | 0.2670 | 0.2549 |
| 0.9 | 0.0134 | 0.0263 | 0.0791 | 0.1379 | 0.1837 | 0.2150 | 0.2323 | 0.2248 | 0.2400 | 0.2425 | 0.2365 |
| 1.0 | 0.0134 | 0.0164 | 0.0499 | 0.0976 | 0.1409 | 0.1748 | 0.1967 | 0.1880 | 0.2093 | 0.2181 | 0.2176 |
| 1.2 | 0.0120 | 0.0108 | 0.0210 | 0.0472 | 0.0796 | 0.1113 | 0.1364 | 0.1277 | 0.1546 | 0.1721 | 0.1804 |
| 1.4 | 0.0096 | 0.0104 | 0.0116 | 0.0231 | 0.0436 | 0.0681 | 0.0912 | 0.0840 | 0.1104 | 0.1316 | 0.1455 |
| 1.6 | 0.0072 | 0.0102 | 0.0093 | 0.0128 | 0.0239 | 0.0480 | 0.0593 | 0.0541 | 0.0768 | 0.0980 | 0.1145 |
| 1.8 | 0.0054 | 0.0094 | 0.0090 | 0.0090 | 0.0140 | 0.0244 | 0.0378 | 0.0344 | 0.0523 | 0.0714 | 0.0882 |
| 2.0 | 0.0040 | 0.0082 | 0.0089 | 0.0079 | 0.0094 | 0.0151 | 0.0241 | 0.0220 | 0.0351 | 0.0510 | 0.0667 |
| 2.5 | 0.0019 | 0.0053 | 0.0077 | 0.0077 | 0.0068 | 0.0070 | 0.0091 | 0.0084 | 0.0133 | 0.0211 | 0.0312 |
| 3.0 | 0.0010 | 0.0032 | 0.0057 | 0.0069 | 0.0067 | 0.0060 | 0.0058 | 0.0055 | 0.0066 | 0.0094 | 0.0143 |
| 3.5 | 0.0005 | 0.0019 | 0.0040 | 0.0056 | 0.0062 | 0.0060 | 0.0054 | 0.0052 | 0.0051 | 0.0057 | 0.0074 |
| 4.0 | 0.0003 | 0.0011 | 0.0027 | 0.0042 | 0.0052 | 0.0056 | 0.0054 | 0.0051 | 0.0049 | 0.0048 | 0.0052 |
| 5.0 | 0.0001 | 0.0005 | 0.0012 | 0.0022 | 0.0033 | 0.0041 | 0.0046 | 0.0043 | 0.0047 | 0.0047 | 0.0046 |
| 6.0 | 0.0000 | 0.0002 | 0.0006 | 0.0012 | 0.0019 | 0.0027 | 0.0033 | 0.0032 | 0.0038 | 0.0042 | 0.0044 |
| 7.0 |  | 0.0001 | 0.0003 | 0.0006 | 0.0011 | 0.0017 | 0.0023 | 0.0022 | 0.0028 | 0.0034 | 0.0038 |
| 8.0 |  | 0.0001 | 0.0002 | 0.0004 | 0.0007 | 0.0011 | 0.0015 | 0.0014 | 0.0020 | 0.0025 | 0.0030 |
| 9.0 |  | 0.0000 | 0.0001 | 0.0002 | 0.0004 | 0.0007 | 0.0010 | 0.0010 | 0.0014 | 0.0018 | 0.0023 |
| 10.0 |  |  | 0.0001 | 0.0001 | 0.0003 | 0.0004 | 0.0007 | 0.0006 | 0.0009 | 0.0013 | 0.0017 |
| 15.0 |  |  | 0.0000 | 0.0000 | 0.0000 | 0.0001 | 0.0001 | 0.0001 | 0.0002 | 0.0003 | 0.0004 |
| 20.0 |  |  |  |  |  | 0·0000 | 0.0000 | 0.0000 | 0.0000 | 0.0001 | 0.0001 |
| 25.0 |  |  |  |  |  |  |  |  |  | 0.0000 | 0.0000 |
| 30.0 |  |  |  |  |  |  |  |  |  |  |  |
| 35.0 |  |  |  |  |  |  |  |  |  |  |  |

| ⟨K.E.⟩eV \ q | $2s$ Al 295.9 | $2s$ Si 260.5 | $2s$ P 431.5 | $2s$ S 509.0 | $2s$ Cl 592.7 | $2s$ A 682.6 | $2s$ K 779.5 | $2s$ Ca 883.0 | $2s$ Sc 993.0 | $2s$ Ti 1110. | $2s$ V 1233. |
|---|---|---|---|---|---|---|---|---|---|---|---|
| 0.0 | 0.3049 | 0.2750 | 0.2507 | 0.2304 | 0.2132 | 0.1984 | 0.1855 | 0.1740 | 0.1639 | 0.1550 | 0.1470 |
| 0.1 | 0.3037 | 0.2741 | 0.2500 | 0.2299 | 0.2128 | 0.1981 | 0.1852 | 0.1738 | 0.1638 | 0.1548 | 0.1469 |
| 0.2 | 0.3004 | 0.2717 | 0.2481 | 0.2281 | 0.2116 | 0.1972 | 0.1844 | 0.1732 | 0.1632 | 0.1544 | 0.1465 |
| 0.3 | 0.2948 | 0.2676 | 0.2450 | 0.2260 | 0.2097 | 0.1956 | 0.1831 | 0.1721 | 0.1623 | 0.1536 | 0.1458 |
| 0.4 | 0.2873 | 0.2620 | 0.2408 | 0.2226 | 0.2070 | 0.1934 | 0.1814 | 0.1706 | 0.1611 | 0.1526 | 0.1449 |
| 0.5 | 0.2780 | 0.2550 | 0.2354 | 0.2184 | 0.2036 | 0.1907 | 0.1791 | 0.1688 | 0.1596 | 0.1513 | 0.1438 |
| 0.6 | 0.2671 | 0.2468 | 0.2291 | 0.2134 | 0.1996 | 0.1874 | 0.1764 | 0.1665 | 0.1577 | 0.1497 | 0.1424 |
| 0.7 | 0.2548 | 0.2375 | 0.2218 | 0.2073 | 0.1950 | 0.1836 | 0.1733 | 0.1639 | 0.1555 | 0.1478 | 0.1408 |
| 0.8 | 0.2415 | 0.2273 | 0.2138 | 0.2013 | 0.1898 | 0.1794 | 0.1698 | 0.1610 | 0.1530 | 0.1457 | 0.1390 |
| 0.9 | 0.2274 | 0.2163 | 0.2051 | 0.1943 | 0.1841 | 0.1747 | 0.1659 | 0.1577 | 0.1502 | 0.1433 | 0.1370 |
| 1.0 | 0.2127 | 0.2047 | 0.1958 | 0.1868 | 0.1780 | 0.1696 | 0.1616 | 0.1541 | 0.1472 | 0.1407 | 0.1347 |
| 1.2 | 0.1826 | 0.1805 | 0.1762 | 0.1708 | 0.1647 | 0.1585 | 0.1523 | 0.1463 | 0.1404 | 0.1349 | 0.1297 |
| 1.4 | 0.1532 | 0.1561 | 0.1559 | 0.1538 | 0.1505 | 0.1464 | 0.1421 | 0.1375 | 0.1329 | 0.1284 | 0.1241 |
| 1.6 | 0.1256 | 0.1324 | 0.1357 | 0.1365 | 0.1357 | 0.1338 | 0.1312 | 0.1282 | 0.1254 | 0.1214 | 0.1179 |
| 1.8 | 0.1013 | 0.1104 | 0.1163 | 0.1196 | 0.1210 | 0.1210 | 0.1201 | 0.1185 | 0.1164 | 0.1139 | 0.1113 |
| 2.0 | 0.0801 | 0.0906 | 0.0982 | 0.1034 | 0.1066 | 0.1083 | 0.1089 | 0.1086 | 0.1076 | 0.1062 | 0.1045 |
| 2.5 | 0.0418 | 0.0519 | 0.0609 | 0.0683 | 0.0742 | 0.0787 | 0.0820 | 0.0843 | 0.0858 | 0.0866 | 0.0868 |
| 3.0 | 0.0207 | 0.0279 | 0.0353 | 0.0423 | 0.0487 | 0.0542 | 0.0588 | 0.0626 | 0.0656 | 0.0680 | 0.0697 |
| 3.5 | 0.0105 | 0.0147 | 0.0197 | 0.0251 | 0.0304 | 0.0356 | 0.0404 | 0.0446 | 0.0483 | 0.0515 | 0.0541 |
| 4.0 | 0.0063 | 0.0082 | 0.0110 | 0.0146 | 0.0185 | 0.0226 | 0.0267 | 0.0307 | 0.0344 | 0.0378 | 0.0408 |
| 5.0 | 0.0045 | 0.0045 | 0.0049 | 0.0058 | 0.0072 | 0.0090 | 0.0112 | 0.0137 | 0.0163 | 0.0189 | 0.0215 |
| 6.0 | 0.0044 | 0.0042 | 0.0041 | 0.0040 | 0.0042 | 0.0046 | 0.0053 | 0.0063 | 0.0076 | 0.0091 | 0.0108 |
| 7.0 | 0.0040 | 0.0041 | 0.0041 | 0.0039 | 0.0038 | 0.0037 | 0.0037 | 0.0039 | 0.0043 | 0.0049 | 0.0057 |
| 8.0 | 0.0034 | 0.0037 | 0.0038 | 0.0038 | 0.0037 | 0.0036 | 0.0035 | 0.0034 | 0.0034 | 0.0035 | 0.0037 |
| 9.0 | 0.0027 | 0.0031 | 0.0034 | 0.0035 | 0.0036 | 0.0035 | 0.0035 | 0.0033 | 0.0032 | 0.0031 | 0.0031 |
| 10.0 | 0.0021 | 0.0025 | 0.0028 | 0.0031 | 0.0032 | 0.0033 | 0.0033 | 0.0033 | 0.0032 | 0.0031 | 0.0030 |
| 15.0 | 0.0005 | 0.0007 | 0.0009 | 0.0011 | 0.0013 | 0.0015 | 0.0017 | 0.0019 | 0.0021 | 0.0022 | 0.0023 |
| 20.0 | 0.0001 | 0.0002 | 0.0003 | 0.0004 | 0.0004 | 0.0006 | 0.0007 | 0.0008 | 0.0009 | 0.0010 | 0.0011 |
| 25.0 | 0.0000 | 0.0001 | 0.0001 | 0.0001 | 0.0002 | 0.0002 | 0.0003 | 0.0003 | 0.0004 | 0.0004 | 0.0005 |
| 30.0 |  | 0.0000 | 0.0000 | 0.0001 | 0.0001 | 0.0001 | 0.0001 | 0.0001 | 0.0002 | 0.0002 | 0.0002 |
| 35.0 |  |  |  | 0.0000 | 0.0000 | 0.0000 | 0.0000 | 0.0000 | 0.0001 | 0.0001 | 0.0001 |

| $\langle K.E.\rangle$ eV $q$ | 2s Cr 1362. | 2s Mn 1497. | 2s Fe 1639. | 2s Co 1786.7 | 2s Ni 1941.0 | 2s Cu(3d$^9$) 2101.6 | 2s Cu(3d$^{10}$) 2101.7 | 2s Zn 2268.3 | 2s Ga 2441.4 | 2s Ge 2620.8 |
|---|---|---|---|---|---|---|---|---|---|---|
| 0.0 | 0.1398 | 0.1333 | 0.1272 | 0.1225 | 0.1174 | 0.1128 | 0.1129 | 0.1086 | 0.1046 | 0.1009 |
| 0.1 | 0.1397 | 0.1332 | 0.1272 | 0.1224 | 0.1174 | 0.1128 | 0.1128 | 0.1085 | 0.1045 | 0.1008 |
| 0.2 | 0.1393 | 0.1329 | 0.1269 | 0.1222 | 0.1172 | 0.1126 | 0.1126 | 0.1084 | 0.1044 | 0.1007 |
| 0.3 | 0.1388 | 0.1324 | 0.1265 | 0.1218 | 0.1169 | 0.1123 | 0.1123 | 0.1081 | 0.1042 | 0.1005 |
| 0.4 | 0.1380 | 0.1317 | 0.1259 | 0.1213 | 0.1164 | 0.1119 | 0.1119 | 0.1077 | 0.1039 | 0.1002 |
| 0.5 | 0.1370 | 0.1309 | 0.1252 | 0.1207 | 0.1159 | 0.1114 | 0.1114 | 0.1073 | 0.1034 | 0.0999 |
| 0.6 | 0.1358 | 0.1298 | 0.1242 | 0.1199 | 0.1152 | 0.1108 | 0.1108 | 0.1067 | 0.1029 | 0.0994 |
| 0.7 | 0.1345 | 0.1286 | 0.1232 | 0.1189 | 0.1143 | 0.1100 | 0.1101 | 0.1061 | 0.1024 | 0.0989 |
| 0.8 | 0.1329 | 0.1272 | 0.1220 | 0.1179 | 0.1134 | 0.1092 | 0.1092 | 0.1053 | 0.1017 | 0.0983 |
| 0.9 | 0.1311 | 0.1257 | 0.1206 | 0.1167 | 0.1123 | 0.1083 | 0.1083 | 0.1045 | 0.1009 | 0.0976 |
| 1.0 | 0.1292 | 0.1240 | 0.1191 | 0.1154 | 0.1112 | 0.1072 | 0.1072 | 0.1035 | 0.1001 | 0.0968 |
| 1.2 | 0.1248 | 0.1202 | 0.1158 | 0.1124 | 0.1085 | 0.1048 | 0.1049 | 0.1014 | 0.0982 | 0.0951 |
| 1.4 | 0.1199 | 0.1158 | 0.1119 | 0.1090 | 0.1055 | 0.1021 | 0.1021 | 0.0990 | 0.0960 | 0.0931 |
| 1.6 | 0.1144 | 0.1110 | 0.1077 | 0.1052 | 0.1021 | 0.0991 | 0.0991 | 0.0962 | 0.0935 | 0.0909 |
| 1.8 | 0.1086 | 0.1059 | 0.1031 | 0.1011 | 0.0984 | 0.0958 | 0.0958 | 0.0932 | 0.0908 | 0.0884 |
| 2.0 | 0.1025 | 0.1005 | 0.0982 | 0.0967 | 0.0945 | 0.0922 | 0.0922 | 0.0900 | 0.0879 | 0.0858 |
| 2.5 | 0.0866 | 0.0861 | 0.0853 | 0.0848 | 0.0837 | 0.0824 | 0.0824 | 0.0811 | 0.0797 | 0.0783 |
| 3.0 | 0.0709 | 0.0716 | 0.0720 | 0.0725 | 0.0723 | 0.0720 | 0.0720 | 0.0715 | 0.0709 | 0.0702 |
| 3.5 | 0.0562 | 0.0579 | 0.0592 | 0.0604 | 0.0611 | 0.0615 | 0.0615 | 0.0618 | 0.0618 | 0.0617 |
| 4.0 | 0.0434 | 0.0456 | 0.0475 | 0.0491 | 0.0504 | 0.0515 | 0.0515 | 0.0523 | 0.0529 | 0.0534 |
| 5.0 | 0.0241 | 0.0264 | 0.0286 | 0.0307 | 0.0326 | 0.0343 | 0.0343 | 0.0357 | 0.0370 | 0.0382 |
| 6.0 | 0.0125 | 0.0143 | 0.0161 | 0.0179 | 0.0197 | 0.0214 | 0.0214 | 0.0229 | 0.0244 | 0.0258 |
| 7.0 | 0.0066 | 0.0077 | 0.0088 | 0.0101 | 0.0114 | 0.0127 | 0.0127 | 0.0140 | 0.0153 | 0.0166 |
| 8.0 | 0.0040 | 0.0045 | 0.0050 | 0.0058 | 0.0066 | 0.0075 | 0.0075 | 0.0084 | 0.0094 | 0.0104 |
| 9.0 | 0.0031 | 0.0032 | 0.0034 | 0.0037 | 0.0041 | 0.0046 | 0.0046 | 0.0052 | 0.0058 | 0.0064 |
| 10.0 | 0.0029 | 0.0028 | 0.0028 | 0.0029 | 0.0030 | 0.0032 | 0.0032 | 0.0035 | 0.0038 | 0.0042 |
| 15.0 | 0.0024 | 0.0024 | 0.0024 | 0.0025 | 0.0024 | 0.0024 | 0.0024 | 0.0023 | 0.0022 | 0.0022 |
| 20.0 | 0.0013 | 0.0014 | 0.0014 | 0.0016 | 0.0017 | 0.0017 | 0.0017 | 0.0018 | 0.0018 | 0.0019 |
| 25.0 | 0.0006 | 0.0007 | 0.0007 | 0.0008 | 0.0009 | 0.0010 | 0.0010 | 0.0010 | 0.0011 | 0.0012 |
| 30.0 | 0.0003 | 0.0003 | 0.0003 | 0.0004 | 0.0005 | 0.0005 | 0.0005 | 0.0006 | 0.0006 | 0.0007 |
| 35.0 | 0.0001 | 0.0001 | 0.0001 | 0.0002 | 0.0002 | 0.0002 | 0.0002 | 0.0003 | 0.0003 | 0.0003 |

| $\langle K.E.\rangle$eV / q | 3s Na 7.268 | 3s Mg 14.56 | 3s Al 25.43 | 3s Si 37.67 | 3s P 51.40 | 3s S 67.20 | 3s Cl 84.51 | 3s A 103.4 | 3s K 130.9 | 3s Ca 162.0 | 3s Sc 189.6 | 3s Ti 217.6 |
|---|---|---|---|---|---|---|---|---|---|---|---|---|
| 0.0 | 2.0651 | 1.5840 | 1.2415 | 1.0368 | 0.8985 | 0.7929 | 0.7126 | 0.6482 | 0.5701 | 0.5082 | 0.4712 | 0.4412 |
| 0.1 | 1.7695 | 1.4399 | 1.1684 | 0.9929 | 0.8693 | 0.7724 | 0.6975 | 0.6367 | 0.5620 | 0.5026 | 0.4666 | 0.4374 |
| 0.2 | 1.1290 | 1.0886 | 0.9766 | 0.8732 | 0.7878 | 0.7145 | 0.6543 | 0.6035 | 0.5389 | 0.4860 | 0.4532 | 0.4263 |
| 0.3 | 0.5564 | 0.6967 | 0.7302 | 0.7079 | 0.6703 | 0.6285 | 0.5888 | 0.5525 | 0.5030 | 0.4598 | 0.4319 | 0.4085 |
| 0.4 | 0.2231 | 0.3873 | 0.4939 | 0.5322 | 0.5374 | 0.5268 | 0.5091 | 0.4889 | 0.4574 | 0.4256 | 0.4040 | 0.3851 |
| 0.5 | 0.0795 | 0.1931 | 0.3067 | 0.3740 | 0.4078 | 0.4221 | 0.4239 | 0.4190 | 0.4045 | 0.3858 | 0.3709 | 0.3571 |
| 0.6 | 0.0319 | 0.0910 | 0.1782 | 0.2480 | 0.2946 | 0.3245 | 0.3407 | 0.3483 | 0.3492 | 0.3424 | 0.3344 | 0.3258 |
| 0.7 | 0.0205 | 0.0499 | 0.0999 | 0.1573 | 0.2041 | 0.2404 | 0.2653 | 0.2815 | 0.2945 | 0.2979 | 0.2961 | 0.2925 |
| 0.8 | 0.0191 | 0.0273 | 0.0570 | 0.0974 | 0.1371 | 0.1728 | 0.2008 | 0.2218 | 0.2430 | 0.2544 | 0.2579 | 0.2587 |
| 0.9 | 0.0190 | 0.0221 | 0.0360 | 0.0609 | 0.0906 | 0.1213 | 0.1484 | 0.1709 | 0.1966 | 0.2133 | 0.2211 | 0.2256 |
| 1.0 | 0.0180 | 0.0212 | 0.0270 | 0.0405 | 0.0604 | 0.0843 | 0.1080 | 0.1294 | 0.1562 | 0.1760 | 0.1867 | 0.1941 |
| 1.2 | 0.0142 | 0.0207 | 0.0233 | 0.0257 | 0.0317 | 0.0427 | 0.0568 | 0.0721 | 0.0947 | 0.1149 | 0.1281 | 0.1386 |
| 1.4 | 0.0100 | 0.0181 | 0.0231 | 0.0238 | 0.0241 | 0.0272 | 0.0331 | 0.0414 | 0.0564 | 0.0725 | 0.0845 | 0.0950 |
| 1.6 | 0.0067 | 0.0143 | 0.0213 | 0.0236 | 0.0232 | 0.0231 | 0.0243 | 0.0275 | 0.0355 | 0.0460 | 0.0550 | 0.0637 |
| 1.8 | 0.0044 | 0.0107 | 0.0180 | 0.0221 | 0.0230 | 0.0227 | 0.0221 | 0.0224 | 0.0257 | 0.0314 | 0.0370 | 0.0430 |
| 2.0 | 0.0029 | 0.0076 | 0.0143 | 0.0194 | 0.0219 | 0.0225 | 0.0219 | 0.0212 | 0.0221 | 0.0245 | 0.0271 | 0.0309 |
| 2.5 | 0.0010 | 0.0030 | 0.0069 | 0.0115 | 0.0154 | 0.0185 | 0.0201 | 0.0205 | 0.0211 | 0.0211 | 0.0205 | 0.0201 |
| 3.0 | 0.0003 | 0.0011 | 0.0029 | 0.0057 | 0.0089 | 0.0122 | 0.0150 | 0.0170 | 0.0191 | 0.0204 | 0.0202 | 0.0195 |
| 3.5 | 0.0002 | 0.0004 | 0.0012 | 0.0027 | 0.0045 | 0.0077 | 0.0096 | 0.0120 | 0.0148 | 0.0172 | 0.0182 | 0.0185 |
| 4.0 | 0.0001 | 0.0002 | 0.0005 | 0.0013 | 0.0022 | 0.0037 | 0.0055 | 0.0076 | 0.0102 | 0.0128 | 0.0145 | 0.0157 |
| 5.0 | 0.0001 | 0.0002 | 0.0002 | 0.0005 | 0.0006 | 0.0010 | 0.0016 | 0.0025 | 0.0039 | 0.0056 | 0.0072 | 0.0087 |
| 6.0 | 0.0001 | 0.0002 | 0.0002 | 0.0004 | 0.0003 | 0.0004 | 0.0006 | 0.0008 | 0.0013 | 0.0021 | 0.0029 | 0.0039 |
| 7.0 | 0.0001 | 0.0001 | 0.0002 | 0.0004 | 0.0003 | 0.0003 | 0.0004 | 0.0004 | 0.0006 | 0.0008 | 0.0011 | 0.0016 |
| 8.0 | 0.0001 | 0.0001 | 0.0001 | 0.0004 | 0.0003 | 0.0003 | 0.0004 | 0.0004 | 0.0004 | 0.0005 | 0.0006 | 0.0007 |
| 9.0 | 0.0000 | 0.0001 | 0.0001 | 0.0003 | 0.0002 | 0.0002 | 0.0003 | 0.0004 | 0.0004 | 0.0004 | 0.0004 | 0.0005 |
| 10.0 | | 0.0001 | 0.0000 | 0.0000 | 0.0001 | 0.0001 | 0.0001 | 0.0002 | 0.0002 | 0.0002 | 0.0004 | 0.0004 |
| 15.0 | | 0.0000 | | | 0.0000 | 0.0000 | 0.0000 | 0.0001 | 0.0001 | 0.0001 | 0.0003 | 0.0003 |
| 20.0 | | | | | | | | 0.0000 | 0.0000 | 0.0000 | 0.0001 | 0.0001 |
| 25.0 | | | | | | | | | | | 0.0000 | 0.0001 |
| 30.0 | | | | | | | | | | | | 0.0000 |
| 32.0 | | | | | | | | | | | | |

| $\langle K.E.\rangle$eV / q | 3s V 246.8 | 3s Cr 277.3 | 3s Mn 308.7 | 3s Fe 342.7 | 3s Co 377.7 | 3s Ni 414.1 | 3s Cu(3d⁹) 451.9 | 3s Cu(3d¹⁰) 445.8 | 3s Zn 490.9 | 3s Ga 538.6 | 3s Ge 590.3 |
|---|---|---|---|---|---|---|---|---|---|---|---|
| 0.0 | 0.4156 | 0.3932 | 0.3737 | 0.3555 | 0.3393 | 0.3246 | 0.3114 | 0.3149 | 0.2993 | 0.2847 | 0.2710 |
| 0.1 | 0.4124 | 0.3905 | 0.3714 | 0.3534 | 0.3375 | 0.3231 | 0.3100 | 0.3135 | 0.2981 | 0.2830 | 0.2701 |
| 0.2 | 0.4031 | 0.3826 | 0.3645 | 0.3475 | 0.3323 | 0.3185 | 0.3059 | 0.3093 | 0.2944 | 0.2805 | 0.2674 |
| 0.3 | 0.3881 | 0.3697 | 0.3534 | 0.3378 | 0.3238 | 0.3110 | 0.2992 | 0.3025 | 0.2886 | 0.2754 | 0.2630 |
| 0.4 | 0.3681 | 0.3526 | 0.3385 | 0.3248 | 0.3124 | 0.3010 | 0.2903 | 0.2931 | 0.2805 | 0.2685 | 0.2570 |
| 0.5 | 0.3441 | 0.3318 | 0.3203 | 0.3090 | 0.2985 | 0.2886 | 0.2793 | 0.2817 | 0.2706 | 0.2599 | 0.2494 |
| 0.6 | 0.3170 | 0.3081 | 0.2995 | 0.2907 | 0.2823 | 0.2742 | 0.2664 | 0.2683 | 0.2590 | 0.2497 | 0.2406 |
| 0.7 | 0.2879 | 0.2826 | 0.2768 | 0.2706 | 0.2644 | 0.2582 | 0.2520 | 0.2534 | 0.2460 | 0.2383 | 0.2305 |
| 0.8 | 0.2579 | 0.2560 | 0.2530 | 0.2493 | 0.2452 | 0.2409 | 0.2364 | 0.2374 | 0.2318 | 0.2258 | 0.2195 |
| 0.9 | 0.2281 | 0.2291 | 0.2287 | 0.2273 | 0.2253 | 0.2228 | 0.2199 | 0.2206 | 0.2169 | 0.2121 | 0.2077 |
| 1.0 | 0.1993 | 0.2027 | 0.2046 | 0.2053 | 0.2052 | 0.2044 | 0.2030 | 0.2034 | 0.2014 | 0.1987 | 0.1953 |
| 1.2 | 0.1470 | 0.1538 | 0.1589 | 0.1630 | 0.1659 | 0.1680 | 0.1693 | 0.1691 | 0.1702 | 0.1704 | 0.1698 |
| 1.4 | 0.1042 | 0.1123 | 0.1191 | 0.1251 | 0.1301 | 0.1341 | 0.1374 | 0.1366 | 0.1400 | 0.1425 | 0.1442 |
| 1.6 | 0.0720 | 0.0797 | 0.0867 | 0.0934 | 0.0992 | 0.1043 | 0.1087 | 0.1077 | 0.1125 | 0.1166 | 0.1198 |
| 1.8 | 0.0494 | 0.0558 | 0.0620 | 0.0683 | 0.0741 | 0.0794 | 0.0843 | 0.0831 | 0.0885 | 0.0934 | 0.0977 |
| 2.0 | 0.0347 | 0.0395 | 0.0443 | 0.0495 | 0.0546 | 0.0595 | 0.0642 | 0.0631 | 0.0685 | 0.0736 | 0.0783 |
| 2.5 | 0.0203 | 0.0211 | 0.0224 | 0.0243 | 0.0266 | 0.0293 | 0.0322 | 0.0315 | 0.0351 | 0.0390 | 0.0429 |
| 3.0 | 0.0186 | 0.0179 | 0.0174 | 0.0173 | 0.0173 | 0.0180 | 0.0188 | 0.0185 | 0.0200 | 0.0219 | 0.0241 |
| 3.5 | 0.0183 | 0.0178 | 0.0171 | 0.0164 | 0.0159 | 0.0154 | 0.0151 | 0.0149 | 0.0150 | 0.0154 | 0.0161 |
| 4.0 | 0.0164 | 0.0166 | 0.0165 | 0.0162 | 0.0158 | 0.0152 | 0.0147 | 0.0145 | 0.0142 | 0.0139 | 0.0138 |
| 5.0 | 0.0101 | 0.0113 | 0.0122 | 0.0129 | 0.0134 | 0.0136 | 0.0137 | 0.0135 | 0.0136 | 0.0136 | 0.0134 |
| 6.0 | 0.0049 | 0.0060 | 0.0070 | 0.0080 | 0.0089 | 0.0097 | 0.0123 | 0.0102 | 0.0108 | 0.0114 | 0.0118 |
| 7.0 | 0.0021 | 0.0028 | 0.0035 | 0.0043 | 0.0051 | 0.0058 | 0.0066 | 0.0065 | 0.0073 | 0.0080 | 0.0087 |
| 8.0 | 0.0009 | 0.0012 | 0.0016 | 0.0021 | 0.0026 | 0.0031 | 0.0037 | 0.0037 | 0.0043 | 0.0050 | 0.0057 |
| 9.0 | 0.0005 | 0.0006 | 0.0008 | 0.0010 | 0.0013 | 0.0016 | 0.0020 | 0.0020 | 0.0024 | 0.0029 | 0.0034 |
| 10.0 | 0.0004 | 0.0004 | 0.0005 | 0.0006 | 0.0007 | 0.0008 | 0.0010 | 0.0010 | 0.0013 | 0.0016 | 0.0019 |
| 15.0 | 0.0003 | 0.0003 | 0.0004 | 0.0004 | 0.0004 | 0.0004 | 0.0003 | 0.0003 | 0.0003 | 0.0003 | 0.0003 |
| 20.0 | 0.0002 | 0.0002 | 0.0002 | 0.0002 | 0.0002 | 0.0003 | 0.0002 | 0.0002 | 0.0002 | 0.0003 | 0.0003 |
| 25.0 | 0.0001 | 0.0001 | 0.0001 | 0.0001 | 0.0001 | 0.0001 | 0.0001 | 0.0001 | 0.0001 | 0.0002 | 0.0002 |
| 30.0 | 0.0000 | 0.0000 | 0.0000 | 0.0000 | 0.0001 | 0.0001 | 0.0000 | 0.0000 | 0.0000 | 0.0001 | 0.0002 |
| 35.0 | 0.0000 | 0.0000 | 0.0000 | 0.0000 | 0.0000 | 0.0000 | 0.0000 | 0.0000 | 0.0000 | 0.0000 | 0.0000 |

| $q$ | $4s$<br>K<br>7.193 | $4s$<br>Ca<br>13.24 | $4s$<br>Sc<br>15.22 | $4s$<br>Ti<br>16.69 | $4s$<br>V<br>18.14 | $4s$<br>Cr<br>19.40 | $4s$<br>Mn<br>20.59 | $4s$<br>Fe<br>22.22 | $4s$<br>Co<br>23.65 | $4s$<br>Ni<br>25.05 | $4s$<br>$Cu(3d^9)$<br>26.48 | $4s$<br>$Cu(3d^{10})$<br>19.04 | $4s$<br>Zn<br>27.79 | $4s$<br>Ga<br>40.78 | $4s$<br>Ge<br>54.04 |
| --- | --- | --- | --- | --- | --- | --- | --- | --- | --- | --- | --- | --- | --- | --- | --- |
| 0.0 | 2.4470 | 1.9496 | 1.8343 | 1.7647 | 1.7110 | 1.6531 | 1.6061 | 1.5542 | 1.5147 | 1.4790 | 1.4389 | 1.7399 | 1.4044 | 1.1776 | 1.0281 |
| 0.1 | 1.9634 | 1.6767 | 1.6074 | 1.5617 | 1.5231 | 1.4845 | 1.4526 | 1.4145 | 1.3842 | 1.3563 | 1.3270 | 1.5425 | 1.3011 | 1.1134 | 0.9848 |
| 0.2 | 1.0308 | 1.0830 | 1.0919 | 1.0917 | 1.0876 | 1.0843 | 1.0818 | 1.0732 | 1.0635 | 1.0537 | 1.0466 | 1.0906 | 1.0393 | 0.9437 | 0.8667 |
| 0.3 | 0.3726 | 0.5461 | 0.5902 | 0.6177 | 0.6390 | 0.6588 | 0.6757 | 0.6903 | 0.6993 | 0.7064 | 0.7163 | 0.6404 | 0.7245 | 0.7228 | 0.7032 |
| 0.4 | 0.1099 | 0.2291 | 0.2662 | 0.2945 | 0.3192 | 0.3434 | 0.3634 | 0.3856 | 0.4028 | 0.4183 | 0.4340 | 0.3308 | 0.4484 | 0.5056 | 0.5291 |
| 0.5 | 0.0483 | 0.0939 | 0.1118 | 0.1282 | 0.1440 | 0.1619 | 0.1752 | 0.1933 | 0.2093 | 0.2245 | 0.2383 | 0.1601 | 0.2516 | 0.3771 | 0.3718 |
| 0.6 | 0.0423 | 0.0533 | 0.0567 | 0.0615 | 0.0676 | 0.0751 | 0.0821 | 0.0924 | 0.1028 | 0.1133 | 0.1226 | 0.0774 | 0.1318 | 0.1986 | 0.2463 |
| 0.7 | 0.0414 | 0.0466 | 0.0434 | 0.0417 | 0.0417 | 0.0425 | 0.0441 | 0.0479 | 0.0526 | 0.0579 | 0.0629 | 0.0405 | 0.0678 | 0.1157 | 0.1559 |
| 0.8 | 0.0369 | 0.0463 | 0.0421 | 0.0384 | 0.0358 | 0.0333 | 0.0319 | 0.0318 | 0.0327 | 0.0341 | 0.0359 | 0.0253 | 0.0377 | 0.0673 | 0.0965 |
| 0.9 | 0.0302 | 0.0440 | 0.0416 | 0.0382 | 0.0353 | 0.0319 | 0.0294 | 0.0276 | 0.0265 | 0.0258 | 0.0256 | 0.0199 | 0.0255 | 0.0419 | 0.0607 |
| 1.0 | 0.0233 | 0.0391 | 0.0389 | 0.0369 | 0.0348 | 0.0318 | 0.0292 | 0.0271 | 0.0253 | 0.0238 | 0.0225 | 0.0186 | 0.0215 | 0.0301 | 0.0410 |
| 1.2 | 0.0125 | 0.0263 | 0.0290 | 0.0296 | 0.0297 | 0.0285 | 0.0273 | 0.0261 | 0.0248 | 0.0234 | 0.0220 | 0.0183 | 0.0205 | 0.0243 | 0.0275 |
| 1.4 | 0.0061 | 0.0154 | 0.0185 | 0.0202 | 0.0216 | 0.0219 | 0.0221 | 0.0221 | 0.0219 | 0.0213 | 0.0207 | 0.0165 | 0.0197 | 0.0241 | 0.0261 |
| 1.6 | 0.0030 | 0.0083 | 0.0108 | 0.0126 | 0.0142 | 0.0152 | 0.0160 | 0.0169 | 0.0173 | 0.0174 | 0.0175 | 0.0131 | 0.0172 | 0.0227 | 0.0257 |
| 1.8 | 0.0016 | 0.0044 | 0.0060 | 0.0074 | 0.0088 | 0.0098 | 0.0108 | 0.0119 | 0.0126 | 0.0132 | 0.0137 | 0.0096 | 0.0139 | 0.0197 | 0.0237 |
| 2.0 | 0.0010 | 0.0025 | 0.0034 | 0.0042 | 0.0052 | 0.0061 | 0.0070 | 0.0080 | 0.0088 | 0.0095 | 0.0101 | 0.0067 | 0.0106 | 0.0159 | 0.0203 |
| 2.5 | 0.0008 | 0.0013 | 0.0013 | 0.0014 | 0.0016 | 0.0019 | 0.0022 | 0.0027 | 0.0031 | 0.0036 | 0.0041 | 0.0025 | 0.0045 | 0.0077 | 0.0111 |
| 3.0 | 0.0008 | 0.0013 | 0.0012 | 0.0011 | 0.0010 | 0.0010 | 0.0010 | 0.0011 | 0.0012 | 0.0014 | 0.0016 | 0.0010 | 0.0018 | 0.0032 | 0.0050 |
| 3.5 | 0.0006 | 0.0011 | 0.0011 | 0.0011 | 0.0010 | 0.0009 | 0.0008 | 0.0008 | 0.0008 | 0.0008 | 0.0008 | 0.0006 | 0.0009 | 0.0014 | 0.0022 |
| 4.0 | 0.0004 | 0.0008 | 0.0009 | 0.0010 | 0.0010 | 0.0009 | 0.0008 | 0.0008 | 0.0007 | 0.0007 | 0.0006 | 0.0005 | 0.0006 | 0.0009 | 0.0012 |
| 5.0 | 0.0002 | 0.0004 | 0.0005 | 0.0006 | 0.0006 | 0.0006 | 0.0007 | 0.0007 | 0.0007 | 0.0006 | 0.0006 | 0.0004 | 0.0006 | 0.0008 | 0.0009 |
| 6.0 | 0.0001 | 0.0001 | 0.0002 | 0.0002 | 0.0003 | 0.0003 | 0.0004 | 0.0004 | 0.0005 | 0.0005 | 0.0005 | 0.0004 | 0.0005 | 0.0007 | 0.0009 |
| 7.0 | 0.0000 | 0.0001 | 0.0001 | 0.0001 | 0.0001 | 0.0002 | 0.0002 | 0.0002 | 0.0003 | 0.0003 | 0.0003 | 0.0002 | 0.0003 | 0.0005 | 0.0007 |
| 8.0 | | 0.0000 | 0.0000 | 0.0000 | 0.0000 | 0.0001 | 0.0001 | 0.0001 | 0.0001 | 0.0002 | 0.0002 | 0.0001 | 0.0002 | 0.0003 | 0.0004 |
| 9.0 | | | | | | 0.0000 | 0.0000 | 0.0000 | 0.0001 | 0.0001 | 0.0001 | 0.0000 | 0.0001 | 0.0002 | 0.0004 |
| 10.0 | | | | | | | | | 0.0000 | 0.0000 | 0.0000 | | 0.0000 | 0.0001 | 0.0003 |
| 15.0 | | | | | | | | | | | | | | 0.0000 | 0.0002 |
| 20.0 | | | | | | | | | | | | | | | 0.0000 |
| 25.0 | | | | | | | | | | | | | | | |
| 30.0 | | | | | | | | | | | | | | | |
| 35.0 | | | | | | | | | | | | | | | |

603

| | 2p | 2p | 2p | 2p | 2p | 2p | 2p | 2p | 2p | 2p | 2p | 2p | 2p | 2p | 2p |
|---|---|---|---|---|---|---|---|---|---|---|---|---|---|---|---|
| | B | C | N | O | F | $F^-$ | Ne | Na | Mg | Al | Si | P | S | Cl | A |
| ⟨K.E.⟩eV | 20.28 | 34.09 | 51.03 | 69.12 | 90.84 | 77.88 | 115.94 | 160.25 | 210.8 | 268.0 | 331.7 | 401.9 | 478.3 | 561.1 | 651.0 |
| q | | | | | | | | | | | | | | | |
| 0.0 | 0.6144 | 0.4799 | 0.3951 | 0.3475 | 0.3070 | 0.3618 | 0.2736 | 0.2246 | 0.1918 | 0.1675 | 0.1490 | 0.1342 | 0.1223 | 0.1124 | 0.1038 |
| 0.1 | 0.6138 | 0.4797 | 0.3951 | 0.3475 | 0.3069 | 0.3617 | 0.2736 | 0.2246 | 0.1918 | 0.1675 | 0.1490 | 0.1342 | 0.1223 | 0.1124 | 0.1038 |
| 0.2 | 0.6062 | 0.4771 | 0.3940 | 0.3468 | 0.3065 | 0.3603 | 0.2734 | 0.2245 | 0.1917 | 0.1675 | 0.1489 | 0.1342 | 0.1223 | 0.1124 | 0.1038 |
| 0.3 | 0.5808 | 0.4677 | 0.3900 | 0.3443 | 0.3050 | 0.3552 | 0.2725 | 0.2242 | 0.1916 | 0.1674 | 0.1489 | 0.1342 | 0.1223 | 0.1124 | 0.1037 |
| 0.4 | 0.5327 | 0.4479 | 0.3810 | 0.3385 | 0.3015 | 0.3448 | 0.2703 | 0.2235 | 0.1913 | 0.1673 | 0.1488 | 0.1341 | 0.1223 | 0.1123 | 0.1037 |
| 0.5 | 0.4667 | 0.4172 | 0.3657 | 0.3286 | 0.2951 | 0.3286 | 0.2663 | 0.2220 | 0.1906 | 0.1670 | 0.1486 | 0.1340 | 0.1222 | 0.1123 | 0.1037 |
| 0.6 | 0.3929 | 0.3781 | 0.3444 | 0.3143 | 0.2857 | 0.3080 | 0.2601 | 0.2196 | 0.1895 | 0.1664 | 0.1483 | 0.1338 | 0.1221 | 0.1122 | 0.1037 |
| 0.7 | 0.3207 | 0.3344 | 0.3183 | 0.2962 | 0.2734 | 0.2846 | 0.2517 | 0.2161 | 0.1879 | 0.1655 | 0.1479 | 0.1336 | 0.1219 | 0.1121 | 0.1036 |
| 0.8 | 0.2559 | 0.2896 | 0.2891 | 0.2752 | 0.2587 | 0.2602 | 0.2414 | 0.2115 | 0.1856 | 0.1643 | 0.1472 | 0.1331 | 0.1217 | 0.1119 | 0.1035 |
| 0.9 | 0.2013 | 0.2468 | 0.2586 | 0.2525 | 0.2421 | 0.2360 | 0.2293 | 0.2057 | 0.1826 | 0.1627 | 0.1462 | 0.1326 | 0.1213 | 0.1116 | 0.1033 |
| 1.0 | 0.1569 | 0.2078 | 0.2286 | 0.2292 | 0.2242 | 0.2127 | 0.2161 | 0.1988 | 0.1789 | 0.1606 | 0.1450 | 0.1318 | 0.1208 | 0.1113 | 0.1031 |
| 1.2 | 0.0943 | 0.1437 | 0.1737 | 0.1841 | 0.1884 | 0.1708 | 0.1878 | 0.1824 | 0.1694 | 0.1551 | 0.1416 | 0.1297 | 0.1193 | 0.1104 | 0.1024 |
| 1.4 | 0.0568 | 0.0977 | 0.1289 | 0.1444 | 0.1545 | 0.1358 | 0.1595 | 0.1637 | 0.1577 | 0.1477 | 0.1369 | 0.1266 | 0.1173 | 0.1089 | 0.1015 |
| 1.6 | 0.0346 | 0.0662 | 0.0945 | 0.1117 | 0.1248 | 0.1074 | 0.1333 | 0.1443 | 0.1443 | 0.1388 | 0.1310 | 0.1226 | 0.1145 | 0.1069 | 0.1001 |
| 1.8 | 0.0215 | 0.0450 | 0.0690 | 0.0858 | 0.0998 | 0.0847 | 0.1103 | 0.1253 | 0.1302 | 0.1288 | 0.1240 | 0.1178 | 0.1110 | 0.1045 | 0.0983 |
| 2.0 | 0.0135 | 0.0309 | 0.0504 | 0.0658 | 0.0795 | 0.0668 | 0.0906 | 0.1075 | 0.1160 | 0.1182 | 0.1162 | 0.1121 | 0.1069 | 0.1015 | 0.0961 |
| 2.5 | 0.0047 | 0.0125 | 0.0234 | 0.0341 | 0.0448 | 0.0372 | 0.0548 | 0.0711 | 0.0834 | 0.0912 | 0.0949 | 0.0957 | 0.0944 | 0.0919 | 0.0888 |
| 3.0 | 0.0018 | 0.0054 | 0.0113 | 0.0180 | 0.0255 | 0.0211 | 0.0332 | 0.0460 | 0.0578 | 0.0673 | 0.0740 | 0.0781 | 0.0800 | 0.0823 | 0.0794 |
| 3.5 | 0.0008 | 0.0025 | 0.0056 | 0.0098 | 0.0148 | 0.0122 | 0.0203 | 0.0297 | 0.0394 | 0.0484 | 0.0559 | 0.0617 | 0.0656 | 0.0679 | 0.0691 |
| 4.0 | 0.0003 | 0.0012 | 0.0029 | 0.0055 | 0.0087 | 0.0072 | 0.0126 | 0.0192 | 0.0267 | 0.0343 | 0.0414 | 0.0475 | 0.0524 | 0.0560 | 0.0586 |
| 5.0 | 0.0001 | 0.0003 | 0.0009 | 0.0018 | 0.0032 | 0.0027 | 0.0051 | 0.0083 | 0.0124 | 0.0171 | 0.0221 | 0.0272 | 0.0319 | 0.0361 | 0.0398 |
| 6.0 | 0.0000 | 0.0001 | 0.0003 | 0.0007 | 0.0013 | 0.0011 | 0.0022 | 0.0038 | 0.0059 | 0.0086 | 0.0118 | 0.0152 | 0.0188 | 0.0224 | 0.0258 |
| 7.0 | | 0.0000 | 0.0001 | 0.0003 | 0.0006 | 0.0005 | 0.0010 | 0.0018 | 0.0030 | 0.0045 | 0.0064 | 0.0086 | 0.0110 | 0.0136 | 0.0163 |
| 8.0 | | | 0.0001 | 0.0001 | 0.0003 | 0.0002 | 0.0005 | 0.0009 | 0.0015 | 0.0024 | 0.0035 | 0.0049 | 0.0065 | 0.0083 | 0.0103 |
| 9.0 | | | 0.0000 | 0.0001 | 0.0001 | 0.0001 | 0.0002 | 0.0005 | 0.0008 | 0.0013 | 0.0020 | 0.0029 | 0.0039 | 0.0051 | 0.0065 |
| 10.0 | | | | 0.0000 | 0.0001 | 0.0000 | 0.0001 | 0.0003 | 0.0005 | 0.0008 | 0.0012 | 0.0017 | 0.0024 | 0.0032 | 0.0041 |
| 15.0 | | | | | 0.0000 | | 0.0000 | 0.0000 | 0.0000 | 0.0001 | 0.0001 | 0.0002 | 0.0003 | 0.0004 | 0.0005 |

The table below lists scattering factors. Column headers give the orbital (2p), element, and ⟨K.E.⟩ in eV; the first column gives $q$.

| ⟨K.E.⟩eV → / $q$ | 2p K 747.0 | 2p Ca 849.7 | 2p Sc 959.2 | 2p Ti 1075. | 2p V 1197. | 2p Cr 1326. | 2p Mn 1462. | 2p Fe 1603. | 2p Co 1751. | 2p Ni 1906. | 2p Cu(3d$^{10}$) 2066.9 | 2p Cu(3d$^9$) 2066.6 | 2p Zn 2233.8 | 2p Ga 2407.3 | 2p Ge 2587.3 |
|---|---|---|---|---|---|---|---|---|---|---|---|---|---|---|---|
| 0.0 | 0.0965 | 0.0902 | 0.0847 | 0.0798 | 0.0755 | 0.0716 | 0.0681 | 0.0650 | 0.0621 | 0.0595 | 0.0572 | 0.0572 | 0.0549 | 0.0529 | 0.0510 |
| 0.1 | 0.0965 | 0.0902 | 0.0847 | 0.0798 | 0.0755 | 0.0716 | 0.0681 | 0.0650 | 0.0621 | 0.0595 | 0.0572 | 0.0572 | 0.0549 | 0.0529 | 0.0510 |
| 0.2 | 0.0965 | 0.0902 | 0.0847 | 0.0798 | 0.0755 | 0.0716 | 0.0681 | 0.0650 | 0.0621 | 0.0595 | 0.0572 | 0.0572 | 0.0549 | 0.0529 | 0.0510 |
| 0.3 | 0.0965 | 0.0902 | 0.0847 | 0.0798 | 0.0755 | 0.0716 | 0.0681 | 0.0650 | 0.0621 | 0.0595 | 0.0572 | 0.0572 | 0.0549 | 0.0529 | 0.0510 |
| 0.4 | 0.0965 | 0.0902 | 0.0847 | 0.0798 | 0.0755 | 0.0716 | 0.0681 | 0.0650 | 0.0621 | 0.0595 | 0.0572 | 0.0572 | 0.0549 | 0.0529 | 0.0510 |
| 0.5 | 0.0965 | 0.0902 | 0.0847 | 0.0798 | 0.0755 | 0.0716 | 0.0681 | 0.0650 | 0.0621 | 0.0595 | 0.0572 | 0.0572 | 0.0549 | 0.0529 | 0.0510 |
| 0.6 | 0.0964 | 0.0902 | 0.0846 | 0.0798 | 0.0755 | 0.0716 | 0.0681 | 0.0650 | 0.0621 | 0.0595 | 0.0572 | 0.0572 | 0.0549 | 0.0529 | 0.0510 |
| 0.7 | 0.0964 | 0.0901 | 0.0846 | 0.0798 | 0.0755 | 0.0716 | 0.0681 | 0.0650 | 0.0621 | 0.0595 | 0.0572 | 0.0572 | 0.0549 | 0.0529 | 0.0510 |
| 0.8 | 0.0963 | 0.0900 | 0.0846 | 0.0797 | 0.0754 | 0.0716 | 0.0681 | 0.0650 | 0.0621 | 0.0595 | 0.0572 | 0.0572 | 0.0549 | 0.0529 | 0.0509 |
| 0.9 | 0.0962 | 0.0900 | 0.0845 | 0.0797 | 0.0754 | 0.0716 | 0.0681 | 0.0650 | 0.0621 | 0.0595 | 0.0572 | 0.0571 | 0.0549 | 0.0529 | 0.0509 |
| 1.0 | 0.0960 | 0.0899 | 0.0844 | 0.0796 | 0.0754 | 0.0715 | 0.0681 | 0.0649 | 0.0621 | 0.0595 | 0.0571 | 0.0571 | 0.0549 | 0.0529 | 0.0509 |
| 1.2 | 0.0956 | 0.0895 | 0.0842 | 0.0794 | 0.0752 | 0.0714 | 0.0680 | 0.0649 | 0.0620 | 0.0594 | 0.0571 | 0.0571 | 0.0549 | 0.0528 | 0.0509 |
| 1.4 | 0.0949 | 0.0890 | 0.0838 | 0.0792 | 0.0750 | 0.0712 | 0.0678 | 0.0648 | 0.0619 | 0.0593 | 0.0570 | 0.0570 | 0.0548 | 0.0528 | 0.0508 |
| 1.6 | 0.0939 | 0.0883 | 0.0833 | 0.0787 | 0.0747 | 0.0710 | 0.0676 | 0.0646 | 0.0618 | 0.0592 | 0.0570 | 0.0570 | 0.0548 | 0.0527 | 0.0508 |
| 1.8 | 0.0926 | 0.0873 | 0.0825 | 0.0782 | 0.0742 | 0.0706 | 0.0674 | 0.0644 | 0.0616 | 0.0591 | 0.0568 | 0.0568 | 0.0547 | 0.0526 | 0.0508 |
| 2.0 | 0.0909 | 0.0861 | 0.0816 | 0.0774 | 0.0737 | 0.0702 | 0.0670 | 0.0641 | 0.0614 | 0.0589 | 0.0567 | 0.0567 | 0.0545 | 0.0525 | 0.0507 |
| 2.5 | 0.0853 | 0.0817 | 0.0782 | 0.0748 | 0.0715 | 0.0685 | 0.0656 | 0.0630 | 0.0605 | 0.0581 | 0.0560 | 0.0560 | 0.0540 | 0.0521 | 0.0503 |
| 3.0 | 0.0778 | 0.0758 | 0.0734 | 0.0709 | 0.0684 | 0.0659 | 0.0635 | 0.0612 | 0.0590 | 0.0569 | 0.0550 | 0.0550 | 0.0531 | 0.0513 | 0.0497 |
| 3.5 | 0.0692 | 0.0686 | 0.0675 | 0.0660 | 0.0643 | 0.0625 | 0.0607 | 0.0588 | 0.0570 | 0.0552 | 0.0535 | 0.0535 | 0.0519 | 0.0503 | 0.0488 |
| 4.0 | 0.0600 | 0.0607 | 0.0607 | 0.0603 | 0.0595 | 0.0584 | 0.0572 | 0.0559 | 0.0545 | 0.0531 | 0.0516 | 0.0516 | 0.0502 | 0.0489 | 0.0475 |
| 5.0 | 0.0428 | 0.0451 | 0.0467 | 0.0479 | 0.0485 | 0.0488 | 0.0488 | 0.0485 | 0.0481 | 0.0474 | 0.0467 | 0.0467 | 0.0459 | 0.0451 | 0.0442 |
| 6.0 | 0.0289 | 0.0317 | 0.0341 | 0.0360 | 0.0376 | 0.0388 | 0.0397 | 0.0403 | 0.0406 | 0.0408 | 0.0407 | 0.0407 | 0.0406 | 0.0403 | 0.0399 |
| 7.0 | 0.0190 | 0.0216 | 0.0239 | 0.0261 | 0.0280 | 0.0297 | 0.0311 | 0.0322 | 0.0331 | 0.0338 | 0.0343 | 0.0343 | 0.0347 | 0.0349 | 0.0350 |
| 8.0 | 0.0123 | 0.0144 | 0.0165 | 0.0185 | 0.0204 | 0.0221 | 0.0236 | 0.0250 | 0.0263 | 0.0273 | 0.0282 | 0.0282 | 0.0289 | 0.0295 | 0.0299 |
| 9.0 | 0.0080 | 0.0096 | 0.0112 | 0.0129 | 0.0145 | 0.0161 | 0.0176 | 0.0191 | 0.0204 | 0.0215 | 0.0226 | 0.0226 | 0.0235 | 0.0243 | 0.0250 |
| 10.0 | 0.0052 | 0.0064 | 0.0076 | 0.0090 | 0.0103 | 0.0117 | 0.0130 | 0.0143 | 0.0156 | 0.0167 | 0.0178 | 0.0178 | 0.0188 | 0.0197 | 0.0205 |
| 15.0 | 0.0007 | 0.0009 | 0.0012 | 0.0015 | 0.0018 | 0.0023 | 0.0027 | 0.0032 | 0.0037 | 0.0043 | 0.0048 | 0.0048 | 0.0054 | 0.0060 | 0.0066 |
| 20.0 | 0.0001 | 0.0002 | 0.0002 | 0.0003 | 0.0004 | 0.0005 | 0.0006 | 0.0008 | 0.0009 | 0.0011 | 0.0013 | 0.0013 | 0.0015 | 0.0018 | 0.0020 |
| 25.0 | 0.0000 | 0.0001 | 0.0001 | 0.0001 | 0.0001 | 0.0001 | 0.0002 | 0.0002 | 0.0003 | 0.0003 | 0.0004 | 0.0004 | 0.0005 | 0.0006 | 0.0006 |
| 30.0 | 0.0000 | 0.0000 | 0.0000 | 0.0000 | 0.0000 | 0.0000 | 0.0001 | 0.0001 | 0.0001 | 0.0001 | 0.0001 | 0.0001 | 0.0002 | 0.0002 | 0.0002 |
| 35.0 | 0.0000 | 0.0000 | 0.0000 | 0.0000 | 0.0000 | 0.0000 | 0.0000 | 0.0000 | 0.0000 | 0.0000 | 0.0000 | 0.0000 | 0.0001 | 0.0001 | 0.0001 |
| 40.0 | 0.0000 | 0.0000 | 0.0000 | 0.0000 | 0.0000 | 0.0000 | 0.0000 | 0.0000 | 0.0000 | 0.0000 | 0.0000 | 0.0000 | 0.0000 | 0.0000 | 0.0000 |

| ⟨K.E.⟩eV | 3p Al 15.54 | 3p Si 25.58 | 3p P 37.09 | 3p S 49.13 | 3p Cl 63.09 | 3p A 78.01 | 3p K 104.5 | 3p Ca 133.7 | 3p Sc 158.9 | 3p Ti 184.5 | 3p V 211.3 |
|---|---|---|---|---|---|---|---|---|---|---|---|
| 0.0 | 0.9159 | 0.7323 | 0.6158 | 0.5535 | 0.4897 | 0.4417 | 0.3758 | 0.3302 | 0.3036 | 0.2828 | 0.2650 |
| 0.1 | 0.9130 | 0.7312 | 0.6153 | 0.5532 | 0.4895 | 0.4416 | 0.3757 | 0.3302 | 0.3035 | 0.2828 | 0.2649 |
| 0.2 | 0.8783 | 0.7178 | 0.6089 | 0.5488 | 0.4872 | 0.4401 | 0.3752 | 0.3299 | 0.3033 | 0.2826 | 0.2648 |
| 0.3 | 0.7775 | 0.6743 | 0.5872 | 0.5338 | 0.4786 | 0.4346 | 0.3729 | 0.3287 | 0.3025 | 0.2820 | 0.2643 |
| 0.4 | 0.6215 | 0.5961 | 0.5447 | 0.5040 | 0.4602 | 0.4225 | 0.3675 | 0.3257 | 0.3005 | 0.2803 | 0.2631 |
| 0.5 | 0.4530 | 0.4941 | 0.4832 | 0.4595 | 0.4308 | 0.4023 | 0.3577 | 0.3202 | 0.2967 | 0.2773 | 0.2608 |
| 0.6 | 0.3074 | 0.3863 | 0.4099 | 0.4039 | 0.3918 | 0.3743 | 0.3427 | 0.3114 | 0.2905 | 0.2724 | 0.2571 |
| 0.7 | 0.1982 | 0.2875 | 0.3335 | 0.3429 | 0.3463 | 0.3401 | 0.3227 | 0.2991 | 0.2817 | 0.2656 | 0.2517 |
| 0.8 | 0.1235 | 0.2057 | 0.2617 | 0.2822 | 0.2981 | 0.3021 | 0.2984 | 0.2835 | 0.2701 | 0.2566 | 0.2446 |
| 0.9 | 0.0752 | 0.1429 | 0.1992 | 0.2262 | 0.2506 | 0.2627 | 0.2711 | 0.2650 | 0.2560 | 0.2456 | 0.2359 |
| 1.0 | 0.0452 | 0.0971 | 0.1480 | 0.1775 | 0.2063 | 0.2243 | 0.2422 | 0.2443 | 0.2398 | 0.2329 | 0.2256 |
| 1.2 | 0.0163 | 0.0431 | 0.0775 | 0.1044 | 0.1333 | 0.1559 | 0.1847 | 0.1994 | 0.2032 | 0.2032 | 0.2010 |
| 1.4 | 0.0067 | 0.0189 | 0.0389 | 0.0591 | 0.0822 | 0.1034 | 0.1338 | 0.1551 | 0.1648 | 0.1706 | 0.1732 |
| 1.6 | 0.0040 | 0.0090 | 0.0194 | 0.0329 | 0.0491 | 0.0663 | 0.0931 | 0.1157 | 0.1286 | 0.1381 | 0.1443 |
| 1.8 | 0.0035 | 0.0054 | 0.0102 | 0.0185 | 0.0289 | 0.0416 | 0.0628 | 0.0834 | 0.0970 | 0.1082 | 0.1167 |
| 2.0 | 0.0035 | 0.0043 | 0.0063 | 0.0110 | 0.0171 | 0.0258 | 0.0414 | 0.0584 | 0.0712 | 0.0824 | 0.0918 |
| 2.5 | 0.0032 | 0.0041 | 0.0044 | 0.0056 | 0.0060 | 0.0085 | 0.0142 | 0.0223 | 0.0300 | 0.0379 | 0.0457 |
| 3.0 | 0.0025 | 0.0036 | 0.0043 | 0.0053 | 0.0044 | 0.0046 | 0.0063 | 0.0091 | 0.0124 | 0.0164 | 0.0210 |
| 3.5 | 0.0018 | 0.0029 | 0.0038 | 0.0051 | 0.0043 | 0.0042 | 0.0047 | 0.0054 | 0.0064 | 0.0079 | 0.0099 |
| 4.0 | 0.0013 | 0.0022 | 0.0031 | 0.0045 | 0.0041 | 0.0042 | 0.0046 | 0.0048 | 0.0049 | 0.0052 | 0.0058 |
| 5.0 | 0.0006 | 0.0012 | 0.0018 | 0.0030 | 0.0030 | 0.0034 | 0.0042 | 0.0046 | 0.0047 | 0.0046 | 0.0044 |
| 6.0 | 0.0003 | 0.0006 | 0.0010 | 0.0018 | 0.0019 | 0.0024 | 0.0031 | 0.0038 | 0.0041 | 0.0043 | 0.0043 |
| 7.0 | 0.0002 | 0.0003 | 0.0005 | 0.0010 | 0.0012 | 0.0015 | 0.0021 | 0.0027 | 0.0032 | 0.0035 | 0.0037 |
| 8.0 | 0.0001 | 0.0002 | 0.0003 | 0.0006 | 0.0007 | 0.0010 | 0.0014 | 0.0018 | 0.0022 | 0.0026 | 0.0029 |
| 9.0 | 0.0000 | 0.0001 | 0.0002 | 0.0003 | 0.0004 | 0.0006 | 0.0009 | 0.0012 | 0.0015 | 0.0019 | 0.0022 |
| 10.0 |  | 0.0001 | 0.0001 | 0.0002 | 0.0003 | 0.0004 | 0.0006 | 0.0008 | 0.0010 | 0.0013 | 0.0015 |
| 15.0 |  | 0.0000 | 0.0000 | 0.0000 | 0.0000 | 0.0000 | 0.0001 | 0.0001 | 0.0002 | 0.0002 | 0.0003 |
| 20.0 |  |  |  |  |  |  | 0.0000 | 0.0000 | 0.0000 | 0.0000 | 0.0001 |
| 25.0 |  |  |  |  |  |  |  |  |  |  | 0.0000 |
| 30.0 |  |  |  |  |  |  |  |  |  |  |  |

| ⟨K.E.⟩eV $q$ | 3p Cr 242.1 | 3p Mn 268.4 | 3p Fe 300.0 | 3p Co 332.9 | 3p Ni 367.2 | 3p Cu($3d^9$) 402.6 | 3p Cu($3d^{10}$) 396.9 | 3p Zn 439.4 | 3p Ga 486.0 | 3p Ge 536.7 | 3p Ga 21.04 | 3p Ge 32.39 |
|---|---|---|---|---|---|---|---|---|---|---|---|---|
| 0.0 | 0.2485 | 0.2357 | 0.2232 | 0.2122 | 0.2019 | 0.1936 | 0.1959 | 0.1856 | 0.1757 | 0.1666 | 0.9112 | 0.7559 |
| 0.1 | 0.2485 | 0.2357 | 0.2232 | 0.2122 | 0.2019 | 0.1936 | 0.1959 | 0.1856 | 0.1757 | 0.1666 | 0.9084 | 0.7547 |
| 0.2 | 0.2484 | 0.2356 | 0.2232 | 0.2122 | 0.2019 | 0.1936 | 0.1959 | 0.1856 | 0.1757 | 0.1665 | 0.8743 | 0.7399 |
| 0.3 | 0.2480 | 0.2354 | 0.2230 | 0.2120 | 0.2018 | 0.1935 | 0.1958 | 0.1855 | 0.1756 | 0.1665 | 0.7752 | 0.6917 |
| 0.4 | 0.2471 | 0.2348 | 0.2225 | 0.2117 | 0.2015 | 0.1933 | 0.1955 | 0.1853 | 0.1755 | 0.1664 | 0.6214 | 0.6051 |
| 0.5 | 0.2452 | 0.2335 | 0.2216 | 0.2109 | 0.2009 | 0.1927 | 0.1950 | 0.1848 | 0.1752 | 0.1661 | 0.4544 | 0.4937 |
| 0.6 | 0.2422 | 0.2315 | 0.2199 | 0.2096 | 0.1999 | 0.1919 | 0.1940 | 0.1841 | 0.1746 | 0.1657 | 0.3094 | 0.3784 |
| 0.7 | 0.2379 | 0.2283 | 0.2174 | 0.2076 | 0.1984 | 0.1905 | 0.1925 | 0.1829 | 0.1737 | 0.1651 | 0.2002 | 0.2754 |
| 0.8 | 0.2322 | 0.2240 | 0.2140 | 0.2048 | 0.1962 | 0.1886 | 0.1905 | 0.1813 | 0.1725 | 0.1641 | 0.1252 | 0.1924 |
| 0.9 | 0.2251 | 0.2184 | 0.2094 | 0.2012 | 0.1932 | 0.1860 | 0.1877 | 0.1791 | 0.1708 | 0.1628 | 0.0675 | 0.1304 |
| 1.0 | 0.2167 | 0.2114 | 0.2038 | 0.1965 | 0.1895 | 0.1827 | 0.1842 | 0.1763 | 0.1687 | 0.1611 | 0.0462 | 0.0863 |
| 1.2 | 0.1963 | 0.1941 | 0.1894 | 0.1845 | 0.1795 | 0.1740 | 0.1750 | 0.1690 | 0.1628 | 0.1565 | 0.0169 | 0.0363 |
| 1.4 | 0.1726 | 0.1731 | 0.1715 | 0.1691 | 0.1666 | 0.1627 | 0.1632 | 0.1592 | 0.1548 | 0.1500 | 0.0072 | 0.0155 |
| 1.6 | 0.1474 | 0.1502 | 0.1514 | 0.1515 | 0.1512 | 0.1492 | 0.1491 | 0.1475 | 0.1449 | 0.1418 | 0.0045 | 0.0080 |
| 1.8 | 0.1225 | 0.1270 | 0.1305 | 0.1328 | 0.1344 | 0.1343 | 0.1338 | 0.1342 | 0.1335 | 0.1320 | 0.0040 | 0.0057 |
| 2.0 | 0.0991 | 0.1050 | 0.1101 | 0.1139 | 0.1172 | 0.1187 | 0.1179 | 0.1201 | 0.1210 | 0.1212 | 0.0040 | 0.0053 |
| 2.5 | 0.0530 | 0.0599 | 0.0663 | 0.0719 | 0.0770 | 0.0813 | 0.0802 | 0.0849 | 0.0889 | 0.0921 | 0.0035 | 0.0051 |
| 3.0 | 0.0257 | 0.0313 | 0.0365 | 0.0416 | 0.0464 | 0.0512 | 0.0502 | 0.0552 | 0.0603 | 0.0648 | 0.0025 | 0.0040 |
| 3.5 | 0.0124 | 0.0157 | 0.0191 | 0.0228 | 0.0263 | 0.0302 | 0.0296 | 0.0338 | 0.0384 | 0.0428 | 0.0016 | 0.0027 |
| 4.0 | 0.0068 | 0.0083 | 0.0102 | 0.0123 | 0.0146 | 0.0172 | 0.0168 | 0.0199 | 0.0233 | 0.0270 | 0.0010 | 0.0017 |
| 5.0 | 0.0045 | 0.0043 | 0.0045 | 0.0049 | 0.0054 | 0.0062 | 0.0061 | 0.0071 | 0.0085 | 0.0101 | 0.0003 | 0.0006 |
| 6.0 | 0.0044 | 0.0041 | 0.0040 | 0.0039 | 0.0038 | 0.0038 | 0.0038 | 0.0039 | 0.0042 | 0.0047 | 0.0001 | 0.0002 |
| 7.0 | 0.0040 | 0.0039 | 0.0039 | 0.0039 | 0.0038 | 0.0036 | 0.0036 | 0.0035 | 0.0035 | 0.0035 | 0.0001 | 0.0001 |
| 8.0 | 0.0033 | 0.0034 | 0.0035 | 0.0036 | 0.0036 | 0.0033 | 0.0035 | 0.0035 | 0.0033 | 0.0034 | 0.0001 | 0.0001 |
| 9.0 | 0.0025 | 0.0027 | 0.0029 | 0.0030 | 0.0032 | 0.0028 | 0.0032 | 0.0033 | 0.0030 | 0.0034 | 0.0001 | 0.0001 |
| 10.0 | 0.0018 | 0.0020 | 0.0022 | 0.0025 | 0.0026 | 0.0008 | 0.0027 | 0.0029 | 0.0030 | 0.0031 | 0.0000 | 0.0001 |
| 15.0 | 0.0003 | 0.0004 | 0.0005 | 0.0006 | 0.0007 | 0.0002 | 0.0008 | 0.0009 | 0.0011 | 0.0012 | 0.0000 | 0.0000 |
| 20.0 | 0.0001 | 0.0001 | 0.0001 | 0.0001 | 0.0002 | 0.0001 | 0.0002 | 0.0003 | 0.0003 | 0.0004 | 0.0000 | 0.0000 |
| 25.0 | 0.0000 | 0.0000 | 0.0000 | 0.0000 | 0.0000 | 0.0000 | 0.0001 | 0.0001 | 0.0001 | 0.0001 | 0.0000 | 0.0000 |
| 30.0 | 0.0000 | 0.0000 | 0.0000 | 0.0000 | 0.0000 | 0.0000 | 0.0000 | 0.0000 | 0.0000 | 0.0000 | 0.0000 | 0.0000 |

| $\langle$K.E.$\rangle$ eV | 3d Sc | 3d Ti | 3d V | 3d Cr | 3d Mn | 3d Fe | 3d Co | 3d Ni | 3d Cu($3d^9$) | 3d Cu($3d^{10}$) | 3d Zn | 3d Ga | 3d Ge |
|---|---|---|---|---|---|---|---|---|---|---|---|---|---|
| q | 83.61 | 106.9 | 129.1 | 151.8 | 176.0 | 198.3 | 222.8 | 248.6 | 275.8 | 256.0 | 304.6 | 356.3 | 410.0 |
| 0.0 | 0.3108 | 0.2701 | 0.2449 | 0.2260 | 0.2093 | 0.1989 | 0.1884 | 0.1792 | 0.1705 | 0.1846 | 0.1624 | 0.1457 | 0.1329 |
| 0.1 | 0.3108 | 0.2701 | 0.2449 | 0.2260 | 0.2093 | 0.1989 | 0.1884 | 0.1792 | 0.1705 | 0.1846 | 0.1624 | 0.1457 | 0.1329 |
| 0.2 | 0.3107 | 0.2701 | 0.2449 | 0.2260 | 0.2092 | 0.1989 | 0.1884 | 0.1792 | 0.1705 | 0.1846 | 0.1624 | 0.1457 | 0.1329 |
| 0.3 | 0.3101 | 0.2698 | 0.2448 | 0.2259 | 0.2092 | 0.1989 | 0.1884 | 0.1792 | 0.1705 | 0.1846 | 0.1624 | 0.1456 | 0.1329 |
| 0.4 | 0.3078 | 0.2689 | 0.2442 | 0.2255 | 0.2090 | 0.1987 | 0.1883 | 0.1791 | 0.1704 | 0.1844 | 0.1624 | 0.1456 | 0.1329 |
| 0.5 | 0.3025 | 0.2665 | 0.2428 | 0.2245 | 0.2084 | 0.1982 | 0.1879 | 0.1788 | 0.1702 | 0.1840 | 0.1622 | 0.1454 | 0.1328 |
| 0.6 | 0.2935 | 0.2621 | 0.2400 | 0.2226 | 0.2071 | 0.1972 | 0.1872 | 0.1782 | 0.1697 | 0.1831 | 0.1618 | 0.1452 | 0.1328 |
| 0.7 | 0.2809 | 0.2553 | 0.2356 | 0.2194 | 0.2050 | 0.1954 | 0.1858 | 0.1771 | 0.1689 | 0.1816 | 0.1612 | 0.1447 | 0.1326 |
| 0.8 | 0.2654 | 0.2462 | 0.2294 | 0.2148 | 0.2018 | 0.1928 | 0.1837 | 0.1753 | 0.1676 | 0.1792 | 0.1602 | 0.1440 | 0.1324 |
| 0.9 | 0.2477 | 0.2351 | 0.2215 | 0.2089 | 0.1975 | 0.1892 | 0.1808 | 0.1729 | 0.1656 | 0.1758 | 0.1586 | 0.1430 | 0.1321 |
| 1.0 | 0.2289 | 0.2223 | 0.2122 | 0.2018 | 0.1921 | 0.1846 | 0.1770 | 0.1697 | 0.1631 | 0.1716 | 0.1565 | 0.1398 | 0.1315 |
| 1.2 | 0.1905 | 0.1939 | 0.1904 | 0.1847 | 0.1787 | 0.1730 | 0.1673 | 0.1614 | 0.1561 | 0.1610 | 0.1507 | 0.1352 | 0.1297 |
| 1.4 | 0.1546 | 0.1646 | 0.1666 | 0.1652 | 0.1627 | 0.1590 | 0.1552 | 0.1509 | 0.1471 | 0.1485 | 0.1430 | 0.1291 | 0.1269 |
| 1.6 | 0.1234 | 0.1368 | 0.1428 | 0.1451 | 0.1455 | 0.1437 | 0.1417 | 0.1391 | 0.1366 | 0.1353 | 0.1339 | 0.1219 | 0.1229 |
| 1.8 | 0.0975 | 0.1123 | 0.1206 | 0.1254 | 0.1282 | 0.1281 | 0.1277 | 0.1267 | 0.1254 | 0.1221 | 0.1239 | 0.1140 | 0.1179 |
| 2.0 | 0.0767 | 0.0913 | 0.1008 | 0.1072 | 0.1116 | 0.1129 | 0.1139 | 0.1142 | 0.1141 | 0.1095 | 0.1135 | 0.1140 | 0.1120 |
| 2.5 | 0.0421 | 0.0538 | 0.0628 | 0.0701 | 0.0762 | 0.0796 | 0.0827 | 0.0852 | 0.0871 | 0.0818 | 0.0866 | 0.0929 | 0.0949 |
| 3.0 | 0.0235 | 0.0317 | 0.0388 | 0.0450 | 0.0508 | 0.0547 | 0.0585 | 0.0619 | 0.0646 | 0.0600 | 0.0671 | 0.0730 | 0.0771 |
| 3.5 | 0.0133 | 0.0189 | 0.0240 | 0.0288 | 0.0336 | 0.0373 | 0.0409 | 0.0443 | 0.0473 | 0.0435 | 0.0501 | 0.0560 | 0.0609 |
| 4.0 | 0.0077 | 0.0113 | 0.0149 | 0.0185 | 0.0222 | 0.0254 | 0.0285 | 0.0315 | 0.0343 | 0.0315 | 0.0370 | 0.0424 | 0.0472 |
| 5.0 | 0.0027 | 0.0042 | 0.0059 | 0.0078 | 0.0099 | 0.0118 | 0.0139 | 0.0159 | 0.0180 | 0.0164 | 0.0201 | 0.0238 | 0.0276 |
| 6.0 | 0.0010 | 0.0017 | 0.0025 | 0.0034 | 0.0045 | 0.0056 | 0.0068 | 0.0081 | 0.0095 | 0.0086 | 0.0109 | 0.0133 | 0.0159 |
| 7.0 | 0.0004 | 0.0007 | 0.0011 | 0.0016 | 0.0021 | 0.0027 | 0.0034 | 0.0042 | 0.0050 | 0.0046 | 0.0059 | 0.0075 | 0.0091 |
| 8.0 | 0.0002 | 0.0003 | 0.0005 | 0.0007 | 0.0010 | 0.0014 | 0.0018 | 0.0022 | 0.0027 | 0.0025 | 0.0033 | 0.0042 | 0.0053 |
| 9.0 | 0.0001 | 0.0002 | 0.0002 | 0.0004 | 0.0005 | 0.0007 | 0.0009 | 0.0012 | 0.0015 | 0.0014 | 0.0019 | 0.0024 | 0.0031 |
| 10.0 | 0.0000 | 0.0001 | 0.0001 | 0.0002 | 0.0003 | 0.0004 | 0.0005 | 0.0007 | 0.0008 | 0.0008 | 0.0011 | 0.0014 | 0.0018 |
| 15.0 | | 0.0000 | 0.0000 | 0.0000 | 0.0000 | 0.0000 | 0.0000 | 0.0000 | 0.0001 | 0.0001 | 0.0001 | 0.0001 | 0.0002 |
| 20.0 | | | | | | | | | 0.0000 | 0.0000 | 0.0000 | 0.0000 | 0.0000 |
| 25.0 | | | | | | | | | | | | | |
| 30.0 | | | | | | | | | | | | | |

# APPENDIX C

## SCATTERING FACTORS FOR HARTREE-FOCK WAVEFUNCTIONS

Tables prepared by A. Harvey.

| $\frac{\sin\theta}{\lambda}$ | 1s Li+ | 1s Li | 1s Be | 1s B | 1s C | 1s N | 1s O | 1s F | 1s F− | 1s Ne | 1s Na |
|---|---|---|---|---|---|---|---|---|---|---|---|
| 0.00 | 1.0000 | 1.0000 | 1.0000 | 1.0000 | 1.0000 | 1.0000 | 1.0000 | 1.0000 | 1.0000 | 1.0000 | 1.0000 |
| 0.05 | 0.9918 | 0.9957 | 0.9957 | 0.9974 | 0.9982 | 0.9987 | 0.9990 | 0.9992 | 0.9992 | 0.9994 | 0.9995 |
| 0.10 | 0.9680 | 0.9679 | 0.9831 | 0.9895 | 0.9929 | 0.9948 | 0.9961 | 0.9969 | 0.9969 | 0.9975 | 0.9980 |
| 0.15 | 0.9303 | 0.9301 | 0.9625 | 0.9767 | 0.9841 | 0.9884 | 0.9913 | 0.9931 | 0.9931 | 0.9945 | 0.9954 |
| 0.20 | 0.8812 | 0.8809 | 0.9349 | 0.9591 | 0.9720 | 0.9796 | 0.9845 | 0.9878 | 0.9878 | 0.9902 | 0.9919 |
| 0.25 | 0.8240 | 0.8236 | 0.9011 | 0.9372 | 0.9567 | 0.9684 | 0.9760 | 0.9811 | 0.9811 | 0.9848 | 0.9875 |
| 0.30 | 0.7617 | 0.7612 | 0.8624 | 0.9115 | 0.9386 | 0.9550 | 0.9657 | 0.9730 | 0.9730 | 0.9782 | 0.9820 |
| 0.35 | 0.6971 | 0.6966 | 0.8198 | 0.8826 | 0.9179 | 0.9396 | 0.9538 | 0.9635 | 0.9635 | 0.9705 | 0.9757 |
| 0.40 | 0.6329 | 0.6322 | 0.7747 | 0.8509 | 0.8949 | 0.9222 | 0.9403 | 0.9528 | 0.9527 | 0.9617 | 0.9684 |
| 0.45 | 0.5707 | 0.5700 | 0.7281 | 0.8171 | 0.8698 | 0.9031 | 0.9253 | 0.9408 | 0.9408 | 0.9519 | 0.9603 |
| 0.50 | 0.5119 | 0.5112 | 0.6810 | 0.7817 | 0.8431 | 0.8825 | 0.9091 | 0.9277 | 0.9277 | 0.9412 | 0.9513 |
| 0.60 | 0.4073 | 0.4068 | 0.5886 | 0.7084 | 0.7859 | 0.8375 | 0.8731 | 0.8984 | 0.8984 | 0.9170 | 0.9310 |
| 0.70 | 0.3215 | 0.3210 | 0.5024 | 0.6346 | 0.7258 | 0.7888 | 0.8334 | 0.8656 | 0.8656 | 0.8896 | 0.9079 |
| 0.80 | 0.2532 | 0.2528 | 0.4252 | 0.5634 | 0.6648 | 0.7378 | 0.7909 | 0.8301 | 0.8301 | 0.8596 | 0.8824 |
| 0.90 | 0.1998 | 0.1994 | 0.3579 | 0.4966 | 0.6047 | 0.6859 | 0.7467 | 0.7924 | 0.7924 | 0.8274 | 0.8547 |
| 1.00 | 0.1582 | 0.1580 | 0.3004 | 0.4355 | 0.5470 | 0.6343 | 0.7016 | 0.7534 | 0.7534 | 0.7936 | 0.8254 |
| 1.10 | 0.1261 | 0.1258 | 0.2520 | 0.3804 | 0.4926 | 0.5839 | 0.6565 | 0.7136 | 0.7136 | 0.7587 | 0.7947 |
| 1.20 | 0.1011 | 0.1009 | 0.2115 | 0.3316 | 0.4419 | 0.5355 | 0.6121 | 0.6737 | 0.6736 | 0.7231 | 0.7631 |
| 1.30 | 0.0816 | 0.0815 | 0.1778 | 0.2886 | 0.3955 | 0.4896 | 0.5688 | 0.6340 | 0.6340 | 0.6873 | 0.7309 |
| 1.40 | 0.0664 | 0.0662 | 0.1499 | 0.2511 | 0.3532 | 0.4464 | 0.5271 | 0.5950 | 0.5950 | 0.6515 | 0.6985 |
| 1.50 | 0.0543 | 0.0543 | 0.1267 | 0.2186 | 0.3151 | 0.4062 | 0.4874 | 0.5572 | 0.5571 | 0.6162 | 0.6660 |

| $\frac{\sin\theta}{\lambda}$ | 1s Mg | 1s Al | 1s Si | 1s P | 1s S | 1s Cl | 1s A | 1s K | 1s Ca | 1s Sc | 1s Ti |
|---|---|---|---|---|---|---|---|---|---|---|---|
| 0.00 | 1.0000 | 1.0000 | 1.0000 | 1.0000 | 1.0000 | 1.0000 | 1.0000 | 1.0000 | 1.0000 | 1.0000 | 1.0000 |
| 0.05 | 0.9996 | 0.9996 | 0.9997 | 0.9997 | 0.9998 | 0.9998 | 0.9998 | 0.9998 | 0.9999 | 0.9999 | 0.9999 |
| 0.10 | 0.9983 | 0.9986 | 0.9988 | 0.9989 | 0.9991 | 0.9992 | 0.9993 | 0.9993 | 0.9994 | 0.9994 | 0.9995 |
| 0.15 | 0.9962 | 0.9968 | 0.9972 | 0.9976 | 0.9979 | 0.9981 | 0.9984 | 0.9985 | 0.9987 | 0.9983 | 0.9989 |
| 0.20 | 0.9933 | 0.9943 | 0.9951 | 0.9957 | 0.9963 | 0.9967 | 0.9971 | 0.9974 | 0.9976 | 0.9978 | 0.9981 |
| 0.25 | 0.9895 | 0.9911 | 0.9923 | 0.9934 | 0.9942 | 0.9949 | 0.9954 | 0.9959 | 0.9963 | 0.9966 | 0.9970 |
| 0.30 | 0.9849 | 0.9872 | 0.9890 | 0.9905 | 0.9916 | 0.9926 | 0.9934 | 0.9941 | 0.9947 | 0.9952 | 0.9956 |
| 0.35 | 0.9796 | 0.9827 | 0.9851 | 0.9870 | 0.9886 | 0.9900 | 0.9911 | 0.9920 | 0.9928 | 0.9935 | 0.9941 |
| 0.40 | 0.9735 | 0.9775 | 0.9806 | 0.9831 | 0.9852 | 0.9869 | 0.9884 | 0.9896 | 0.9906 | 0.9915 | 0.9923 |
| 0.45 | 0.9666 | 0.9716 | 0.9755 | 0.9787 | 0.9813 | 0.9835 | 0.9853 | 0.9868 | 0.9881 | 0.9892 | 0.9902 |
| 0.50 | 0.9591 | 0.9651 | 0.9699 | 0.9738 | 0.9770 | 0.9797 | 0.9819 | 0.9838 | 0.9854 | 0.9867 | 0.9880 |
| 0.60 | 0.9419 | 0.9504 | 0.9572 | 0.9627 | 0.9672 | 0.9710 | 0.9741 | 0.9768 | 0.9791 | 0.9810 | 0.9827 |
| 0.70 | 0.9222 | 0.9334 | 0.9424 | 0.9497 | 0.9558 | 0.9608 | 0.9650 | 0.9686 | 0.9717 | 0.9743 | 0.9766 |
| 0.80 | 0.9002 | 0.9144 | 0.9258 | 0.9351 | 0.9428 | 0.9493 | 0.9547 | 0.9593 | 0.9632 | 0.9666 | 0.9696 |
| 0.90 | 0.8763 | 0.8935 | 0.9075 | 0.9190 | 0.9285 | 0.9364 | 0.9432 | 0.9489 | 0.9538 | 0.9581 | 0.9618 |
| 1.00 | 0.8507 | 0.8711 | 0.8877 | 0.9014 | 0.9129 | 0.9224 | 0.9306 | 0.9375 | 0.9435 | 0.9486 | 0.9531 |
| 1.10 | 0.8237 | 0.8473 | 0.8666 | 0.8827 | 0.8961 | 0.9073 | 0.9169 | 0.9251 | 0.9322 | 0.9383 | 0.9437 |
| 1.20 | 0.7957 | 0.8224 | 0.8444 | 0.8628 | 0.8782 | 0.8912 | 0.9023 | 0.9119 | 0.9201 | 0.9273 | 0.9336 |
| 1.30 | 0.7669 | 0.7966 | 0.8212 | 0.8419 | 0.8594 | 0.8742 | 0.8869 | 0.8978 | 0.9073 | 0.9155 | 0.9228 |
| 1.40 | 0.7375 | 0.7700 | 0.7973 | 0.8203 | 0.8398 | 0.8564 | 0.8707 | 0.8830 | 0.8937 | 0.9030 | 0.9113 |
| 1.50 | 0.7078 | 0.7430 | 0.7728 | 0.7980 | 0.8195 | 0.8379 | 0.8537 | 0.8675 | 0.8794 | 0.8899 | 0.8992 |

| $\frac{\sin\theta}{\lambda}$ | 1s V | 1s Cr | 1s Mn | 1s Fe | 1s Co | 1s Ni | 1s Cu | 1s Zn | 1s Ga | 1s Ge |
|---|---|---|---|---|---|---|---|---|---|---|
| 0.00 | 1.0000 | 1.0000 | 1.0000 | 1.0000 | 1.0000 | 1.0000 | 1.0000 | 1.0000 | 1.0000 | 1.0000 |
| 0.05 | 0.9999 | 0.9999 | 0.9999 | 0.9999 | 0.9999 | 0.9999 | 0.9999 | 0.9999 | 0.9999 | 0.9999 |
| 0.10 | 0.9996 | 0.9996 | 0.9996 | 0.9997 | 0.9997 | 0.9997 | 0.9997 | 0.9997 | 0.9998 | 0.9998 |
| 0.15 | 0.9990 | 0.9991 | 0.9991 | 0.9992 | 0.9993 | 0.9993 | 0.9994 | 0.9994 | 0.9995 | 0.9995 |
| 0.20 | 0.9982 | 0.9984 | 0.9985 | 0.9986 | 0.9987 | 0.9988 | 0.9989 | 0.9990 | 0.9990 | 0.9991 |
| 0.25 | 0.9972 | 0.9975 | 0.9977 | 0.9978 | 0.9980 | 0.9981 | 0.9983 | 0.9984 | 0.9985 | 0.9986 |
| 0.30 | 0.9960 | 0.9964 | 0.9966 | 0.9969 | 0.9971 | 0.9973 | 0.9975 | 0.9977 | 0.9978 | 0.9980 |
| 0.35 | 0.9946 | 0.9950 | 0.9954 | 0.9958 | 0.9961 | 0.9964 | 0.9966 | 0.9968 | 0.9971 | 0.9972 |
| 0.40 | 0.9929 | 0.9935 | 0.9940 | 0.9945 | 0.9949 | 0.9953 | 0.9956 | 0.9959 | 0.9962 | 0.9964 |
| 0.45 | 0.9911 | 0.9918 | 0.9925 | 0.9930 | 0.9936 | 0.9940 | 0.9944 | 0.9948 | 0.9951 | 0.9954 |
| 0.50 | 0.9890 | 0.9899 | 0.9907 | 0.9914 | 0.9921 | 0.9926 | 0.9931 | 0.9936 | 0.9940 | 0.9944 |
| 0.60 | 0.9842 | 0.9855 | 0.9867 | 0.9877 | 0.9886 | 0.9894 | 0.9901 | 0.9908 | 0.9914 | 0.9919 |
| 0.70 | 0.9786 | 0.9804 | 0.9819 | 0.9833 | 0.9845 | 0.9856 | 0.9866 | 0.9875 | 0.9883 | 0.9890 |
| 0.80 | 0.9722 | 0.9745 | 0.9765 | 0.9783 | 0.9799 | 0.9813 | 0.9826 | 0.9837 | 0.9848 | 0.9857 |
| 0.90 | 0.9650 | 0.9679 | 0.9704 | 0.9726 | 0.9746 | 0.9764 | 0.9780 | 0.9795 | 0.9808 | 0.9819 |
| 1.00 | 0.9571 | 0.9606 | 0.9636 | 0.9664 | 0.9688 | 0.9710 | 0.9730 | 0.9747 | 0.9763 | 0.9778 |
| 1.10 | 0.9484 | 0.9526 | 0.9562 | 0.9595 | 0.9624 | 0.9651 | 0.9674 | 0.9695 | 0.9715 | 0.9732 |
| 1.20 | 0.9391 | 0.9440 | 0.9483 | 0.9521 | 0.9555 | 0.9586 | 0.9614 | 0.9639 | 0.9662 | 0.9683 |
| 1.30 | 0.9291 | 0.9347 | 0.9397 | 0.9442 | 0.9481 | 0.9517 | 0.9550 | 0.9579 | 0.9605 | 0.9629 |
| 1.40 | 0.9185 | 0.9249 | 0.9306 | 0.9357 | 0.9402 | 0.9443 | 0.9480 | 0.9514 | 0.9544 | 0.9572 |
| 1.50 | 0.9073 | 0.9145 | 0.9209 | 0.9267 | 0.9319 | 0.9365 | 0.9407 | 0.9445 | 0.9479 | 0.9511 |

| $\frac{\sin\theta}{\lambda}$ | 2s Li | 2s Be | 2s B | 2s C | 2s N | 2s O | 2s F | 2s F | 2s Ne | 2s Na |
|---|---|---|---|---|---|---|---|---|---|---|
| 0.00 | 1.0000 | 1.0000 | 1.0000 | 1.0000 | 1.0000 | 1.0000 | 1.0000 | 1.0000 | 1.0000 | 1.0000 |
| 0.05 | 0.7238 | 0.8576 | 0.9174 | 0.9456 | 0.9613 | 0.9714 | 0.9779 | 0.9761 | 0.9824 | 0.9866 |
| 0.10 | 0.2791 | 0.5495 | 0.7127 | 0.8018 | 0.8554 | 0.8911 | 0.9149 | 0.9085 | 0.9316 | 0.9477 |
| 0.15 | 0.0434 | 0.2687 | 0.4750 | 0.6139 | 0.7073 | 0.7769 | 0.8203 | 0.8083 | 0.8538 | 0.8867 |
| 0.20 | −0.0203 | 0.0950 | 0.2734 | 0.4275 | 0.5462 | 0.6382 | 0.7062 | 0.6896 | 0.7573 | 0.8087 |
| 0.25 | −0.0212 | −0.0163 | 0.1328 | 0.2711 | 0.3953 | 0.5014 | 0.5852 | 0.5660 | 0.6512 | 0.7195 |
| 0.30 | −0.0097 | −0.0201 | 0.0481 | 0.1546 | 0.2680 | 0.3757 | 0.4674 | 0.4481 | 0.5436 | 0.6249 |
| 0.35 | −0.0001 | −0.0146 | 0.0040 | 0.0756 | 0.1689 | 0.2684 | 0.3602 | 0.3424 | 0.4410 | 0.5303 |
| 0.40 | 0.0058 | −0.0068 | −0.0148 | 0.0266 | 0.0966 | 0.1818 | 0.2675 | 0.2524 | 0.3477 | 0.4398 |
| 0.45 | 0.0088 | 0.0001 | −0.0192 | −0.0008 | 0.0470 | 0.1152 | 0.1908 | 0.1787 | 0.2662 | 0.3564 |
| 0.50 | 0.0100 | 0.0089 | −0.0166 | −0.0139 | 0.0151 | 0.0663 | 0.1296 | 0.1204 | 0.1974 | 0.2820 |
| 0.60 | 0.0097 | 0.0124 | −0.0049 | −0.0169 | −0.0134 | 0.0090 | 0.0473 | 0.0429 | 0.0960 | 0.1628 |
| 0.70 | 0.0082 | 0.0128 | 0.0054 | −0.0082 | −0.0167 | −0.0129 | 0.0049 | 0.0034 | 0.0347 | 0.0810 |
| 0.80 | 0.0066 | 0.0119 | 0.0117 | 0.0014 | −0.0103 | −0.0164 | −0.0125 | −0.0125 | 0.0019 | 0.0300 |
| 0.90 | 0.0053 | 0.0105 | 0.0146 | 0.0088 | −0.0019 | −0.0118 | −0.0161 | −0.0154 | −0.0124 | 0.0014 |
| 1.00 | 0.0042 | 0.0091 | 0.0153 | 0.0134 | 0.0056 | −0.0046 | −0.0127 | −0.0119 | −0.0157 | −0.0121 |
| 1.10 | 0.0033 | 0.0077 | 0.0148 | 0.0158 | 0.0111 | 0.0025 | −0.0067 | −0.0061 | −0.0133 | −0.0160 |
| 1.20 | 0.0027 | 0.0065 | 0.0138 | 0.0167 | 0.0147 | 0.0084 | −0.0002 | 0.0001 | −0.0082 | −0.0145 |
| 1.30 | 0.0021 | 0.0055 | 0.0124 | 0.0165 | 0.0167 | 0.0127 | 0.0057 | 0.0056 | −0.0023 | −0.0102 |
| 1.40 | 0.0017 | 0.0046 | 0.0111 | 0.0158 | 0.0175 | 0.0157 | 0.0104 | 0.0101 | 0.0032 | −0.0047 |
| 1.50 | 0.0014 | | 0.0098 | 0.0147 | 0.0176 | 0.0175 | 0.0140 | 0.0134 | 0.0081 | 0.0008 |

| sin θ/λ | 2s Mg | 2s Al | 2s Si | 2s P | 2s S | 2s Cl | 2s A | 2s K | 2s Ca | 2s Sc |
|---|---|---|---|---|---|---|---|---|---|---|
| 0.00 | 1.0000 | 1.0000 | 1.0000 | 1.0000 | 1.0000 | 1.0000 | 1.0000 | 1.0000 | 1.0000 | 1.0000 |
| 0.05 | 0.9895 | 0.9916 | 0.9931 | 0.9942 | 0.9951 | 0.9957 | 0.9963 | 0.9967 | 0.9971 | 0.9975 |
| 0.10 | 0.9589 | 0.9668 | 0.9726 | 0.9770 | 0.9804 | 0.9831 | 0.9853 | 0.9871 | 0.9886 | 0.9898 |
| 0.15 | 0.9101 | 0.9270 | 0.9395 | 0.9491 | 0.9563 | 0.9624 | 0.9672 | 0.9712 | 0.9745 | 0.9772 |
| 0.20 | 0.8465 | 0.8743 | 0.8953 | 0.9114 | 0.9241 | 0.9343 | 0.9425 | 0.9493 | 0.9551 | 0.9599 |
| 0.25 | 0.7717 | 0.8112 | 0.8416 | 0.8653 | 0.8841 | 0.8993 | 0.9117 | 0.9221 | 0.9307 | 0.9380 |
| 0.30 | 0.6900 | 0.7408 | 0.7807 | 0.8123 | 0.8378 | 0.8585 | 0.8755 | 0.8898 | 0.9019 | 0.9121 |
| 0.35 | 0.6053 | 0.6659 | 0.7146 | 0.7541 | 0.7863 | 0.8127 | 0.8347 | 0.8532 | 0.8690 | 0.8823 |
| 0.40 | 0.5211 | 0.5893 | 0.6457 | 0.6923 | 0.7309 | 0.7631 | 0.7900 | 0.8130 | 0.8325 | 0.8493 |
| 0.45 | 0.4403 | 0.5135 | 0.5759 | 0.6286 | 0.6731 | 0.7106 | 0.7424 | 0.7697 | 0.7932 | 0.8134 |
| 0.50 | 0.3650 | 0.4405 | 0.5070 | 0.5645 | 0.6140 | 0.6564 | 0.6927 | 0.7242 | 0.7514 | 0.7751 |
| 0.60 | 0.2361 | 0.3090 | 0.3777 | 0.4404 | 0.4966 | 0.5464 | 0.5902 | 0.6290 | 0.6632 | 0.6933 |
| 0.70 | 0.1387 | 0.2019 | 0.2661 | 0.3281 | 0.3864 | 0.4399 | 0.4885 | 0.5326 | 0.5723 | 0.6079 |
| 0.80 | 0.0707 | 0.1206 | 0.1757 | 0.2324 | 0.2885 | 0.3422 | 0.3925 | 0.4395 | 0.4827 | 0.5223 |
| 0.90 | 0.0269 | 0.0629 | 0.1066 | 0.1550 | 0.2056 | 0.2563 | 0.3056 | 0.3530 | 0.3978 | 0.4396 |
| 1.00 | 0.0013 | 0.0246 | 0.0567 | 0.0954 | 0.1385 | 0.1839 | 0.2298 | 0.2754 | 0.3197 | 0.3620 |
| 1.10 | -0.0116 | 0.0013 | 0.0228 | 0.0517 | 0.0864 | 0.1250 | 0.1659 | 0.2080 | 0.2501 | 0.2914 |
| 1.20 | -0.0161 | -0.0111 | 0.0015 | 0.0214 | 0.0476 | 0.0789 | 0.1138 | 0.1511 | 0.1897 | 0.2287 |
| 1.30 | -0.0154 | -0.0160 | -0.0105 | 0.0017 | 0.0202 | 0.0442 | 0.0726 | 0.1045 | 0.1387 | 0.1743 |
| 1.40 | -0.0119 | -0.0161 | -0.0157 | -0.0098 | 0.0019 | 0.0192 | 0.0413 | 0.0674 | 0.0967 | 0.1282 |
| 1.50 | -0.0069 | -0.0133 | -0.0165 | -0.0154 | -0.0092 | 0.0022 | 0.0183 | 0.0389 | 0.0631 | 0.0901 |

| sin θ/λ | 2s Ti | 2s V | 2s Cr | 2s Mn | 2s Fe | 2s Co | 2s Ni | 2s Cu | 2s Zn | 2s Ga | 2s Ge |
|---|---|---|---|---|---|---|---|---|---|---|---|
| 0.00 | 1.0000 | 1.0000 | 1.0000 | 1.0000 | 1.0000 | 1.0000 | 1.0000 | 1.0000 | 1.0000 | 1.0000 | 1.0000 |
| 0.05 | 0.9977 | 0.9979 | 0.9981 | 0.9983 | 0.9984 | 0.9986 | 0.9987 | 0.9988 | 0.9989 | 0.9990 | 0.9990 |
| 0.10 | 0.9908 | 0.9917 | 0.9925 | 0.9932 | 0.9938 | 0.9943 | 0.9947 | 0.9951 | 0.9955 | 0.9958 | 0.9961 |
| 0.15 | 0.9795 | 0.9815 | 0.9832 | 0.9847 | 0.9860 | 0.9872 | 0.9882 | 0.9890 | 0.9898 | 0.9906 | 0.9912 |
| 0.20 | 0.9639 | 0.9674 | 0.9704 | 0.9730 | 0.9753 | 0.9773 | 0.9790 | 0.9806 | 0.9820 | 0.9833 | 0.9844 |
| 0.25 | 0.9442 | 0.9496 | 0.9542 | 0.9581 | 0.9616 | 0.9647 | 0.9674 | 0.9698 | 0.9720 | 0.9740 | 0.9757 |
| 0.30 | 0.9207 | 0.9282 | 0.9347 | 0.9403 | 0.9453 | 0.9496 | 0.9535 | 0.9569 | 0.9600 | 0.9627 | 0.9652 |
| 0.35 | 0.8938 | 0.9036 | 0.9122 | 0.9197 | 0.9263 | 0.9321 | 0.9372 | 0.9418 | 0.9459 | 0.9496 | 0.9529 |
| 0.40 | 0.8637 | 0.8761 | 0.8870 | 0.8965 | 0.9048 | 0.9122 | 0.9188 | 0.9247 | 0.9300 | 0.9347 | 0.9390 |
| 0.45 | 0.8308 | 0.8460 | 0.8592 | 0.8709 | 0.8812 | 0.8903 | 0.8984 | 0.9057 | 0.9122 | 0.9181 | 0.9234 |
| 0.50 | 0.7956 | 0.8135 | 0.8293 | 0.8432 | 0.8555 | 0.8664 | 0.8762 | 0.8849 | 0.8928 | 0.8999 | 0.9063 |
| 0.60 | 0.7198 | 0.7432 | 0.7640 | 0.7824 | 0.7989 | 0.8136 | 0.8268 | 0.8387 | 0.8495 | 0.8592 | 0.8680 |
| 0.70 | 0.6396 | 0.6681 | 0.6935 | 0.7163 | 0.7369 | 0.7554 | 0.7721 | 0.7873 | 0.8010 | 0.8135 | 0.8249 |
| 0.80 | 0.5582 | 0.5908 | 0.6203 | 0.6471 | 0.6714 | 0.6935 | 0.7136 | 0.7319 | 0.7486 | 0.7639 | 0.7778 |
| 0.90 | 0.4782 | 0.5139 | 0.5466 | 0.5766 | 0.6042 | 0.6294 | 0.6526 | 0.6738 | 0.6933 | 0.7113 | 0.7277 |
| 1.00 | 0.4020 | 0.4395 | 0.4744 | 0.5068 | 0.5369 | 0.5648 | 0.5905 | 0.6143 | 0.6364 | 0.6568 | 0.6756 |
| 1.10 | 0.3312 | 0.3693 | 0.4053 | 0.4392 | 0.4711 | 0.5009 | 0.5287 | 0.5547 | 0.5788 | 0.6014 | 0.6223 |
| 1.20 | 0.2671 | 0.3045 | 0.3406 | 0.3751 | 0.4079 | 0.4390 | 0.4682 | 0.4958 | 0.5217 | 0.5459 | 0.5687 |
| 1.30 | 0.2103 | 0.2461 | 0.2812 | 0.3154 | 0.3484 | 0.3799 | 0.4100 | 0.4387 | 0.4657 | 0.4914 | 0.5155 |
| 1.40 | 0.1610 | 0.1943 | 0.2278 | 0.2608 | 0.2932 | 0.3246 | 0.3549 | 0.3840 | 0.4118 | 0.4383 | 0.4635 |
| 1.50 | 0.1191 | 0.1494 | 0.1805 | 0.2118 | 0.2429 | 0.2735 | 0.3034 | 0.3325 | 0.3605 | 0.3874 | 0.4133 |

| $\frac{\sin\theta}{\lambda}$ | 3s Na | 3s Mg | 3s Al | 3s Si | 3s P | 3s S | 3s Cl | 3s A | 3s K | 3s Ca | 3s Sc |
|---|---|---|---|---|---|---|---|---|---|---|---|
| 0.00 | 1.0000 | 1.0000 | 1.0000 | 1.0000 | 1.0000 | 1.0000 | 1.0000 | 1.0000 | 1.0000 | 1.0000 | 1.0000 |
| 0.05 | 0.6847 | 0.7968 | 0.8647 | 0.9009 | 0.9231 | 0.9386 | 0.9495 | 0.9576 | 0.9659 | 0.9719 | 0.9756 |
| 0.10 | 0.2156 | 0.4058 | 0.5607 | 0.6594 | 0.7266 | 0.7763 | 0.8130 | 0.8411 | 0.8703 | 0.8921 | 0.9058 |
| 0.15 | 0.0051 | 0.1212 | 0.2682 | 0.3898 | 0.4864 | 0.5652 | 0.6275 | 0.6774 | 0.7309 | 0.7728 | 0.7999 |
| 0.20 | −0.0321 | −0.0029 | 0.0798 | 0.1771 | 0.2716 | 0.3589 | 0.4344 | 0.4989 | 0.5710 | 0.6306 | 0.6711 |
| 0.25 | −0.0204 | −0.0321 | −0.0082 | 0.0466 | 0.1164 | 0.1921 | 0.2652 | 0.3330 | 0.4124 | 0.4829 | 0.5334 |
| 0.30 | −0.0056 | −0.0250 | −0.0333 | −0.0153 | 0.0235 | 0.0766 | 0.1359 | 0.1965 | 0.2718 | 0.3443 | 0.3996 |
| 0.35 | 0.0033 | −0.0105 | −0.0289 | −0.0335 | −0.0207 | 0.0083 | 0.0485 | 0.0955 | 0.1583 | 0.2247 | 0.2793 |
| 0.40 | 0.0074 | 0.0013 | −0.0150 | −0.0294 | −0.0334 | −0.0242 | −0.0026 | 0.0284 | 0.0747 | 0.1293 | 0.1785 |
| 0.45 | 0.0085 | 0.0086 | −0.0014 | −0.0168 | −0.0291 | −0.0351 | −0.0267 | −0.0104 | 0.0189 | 0.0590 | 0.0994 |
| 0.50 | 0.0081 | 0.0120 | 0.0085 | −0.0035 | −0.0177 | −0.0287 | −0.0326 | −0.0283 | −0.0139 | 0.0117 | 0.0417 |
| 0.60 | 0.0057 | 0.0121 | 0.0170 | 0.0146 | 0.0060 | 0.0062 | −0.0181 | −0.0271 | −0.0322 | −0.0297 | −0.0196 |
| 0.70 | 0.0033 | 0.0089 | 0.0161 | 0.0202 | 0.0193 | 0.0132 | 0.0036 | −0.0074 | −0.0191 | −0.0283 | −0.0318 |
| 0.80 | 0.0016 | 0.0054 | 0.0120 | 0.0186 | 0.0226 | 0.0228 | 0.0187 | 0.0113 | 0.0012 | −0.0103 | −0.0200 |
| 0.90 | 0.0005 | 0.0028 | 0.0077 | 0.0141 | 0.0202 | 0.0244 | 0.0253 | 0.0228 | 0.0175 | 0.0088 | −0.0013 |
| 1.00 | −0.0001 | 0.0011 | 0.0043 | 0.0094 | 0.0155 | 0.0214 | 0.0255 | 0.0271 | 0.0267 | 0.0227 | 0.0154 |
| 1.10 | −0.0003 | 0.0001 | 0.0019 | 0.0055 | 0.0106 | 0.0166 | 0.0222 | 0.0263 | 0.0297 | 0.0301 | 0.0265 |
| 1.20 | −0.0004 | −0.0004 | 0.0004 | 0.0026 | 0.0065 | 0.0117 | 0.0174 | 0.0227 | 0.0283 | 0.0319 | 0.0319 |
| 1.30 | −0.0003 | −0.0006 | −0.0005 | 0.0008 | 0.0034 | 0.0075 | 0.0125 | 0.0180 | 0.0244 | 0.0300 | 0.0327 |
| 1.40 | −0.0002 | −0.0006 | −0.0008 | −0.0004 | 0.0012 | 0.0041 | 0.0083 | 0.0132 | 0.0195 | 0.0259 | 0.0304 |
| 1.50 | −0.0001 | −0.0004 | −0.0009 | −0.0009 | −0.0001 | 0.0017 | 0.0048 | 0.0089 | 0.0146 | 0.0209 | 0.0263 |

| $\frac{\sin\theta}{\lambda}$ | 3s Ti | 3s V | 3s Cr | 3s Mn | 3s Fe | 3s Co | 3s Ni | 3s Cu($3d^9$) | 3s Cu($3d^{10}$) | 3s Zn | 3s Ga | 3s Ge |
|---|---|---|---|---|---|---|---|---|---|---|---|---|
| 0.00 | 1.0000 | 1.0000 | 1.0000 | 1.0000 | 1.0000 | 1.0000 | 1.0000 | 1.0000 | 1.0000 | 1.0000 | 1.0000 | 1.0000 |
| 0.05 | 0.9785 | 0.9808 | 0.9828 | 0.9844 | 0.9859 | 0.9871 | 0.9882 | 0.9891 | 0.9889 | 0.9899 | 0.9908 | 0.9916 |
| 0.10 | 0.9165 | 0.9264 | 0.9328 | 0.9390 | 0.9446 | 0.9493 | 0.9534 | 0.9571 | 0.9565 | 0.9602 | 0.9636 | 0.9667 |
| 0.15 | 0.8216 | 0.8396 | 0.8548 | 0.8678 | 0.8794 | 0.8894 | 0.8982 | 0.9059 | 0.9047 | 0.9127 | 0.9199 | 0.9265 |
| 0.20 | 0.7041 | 0.7320 | 0.7560 | 0.7767 | 0.7954 | 0.8116 | 0.8259 | 0.8386 | 0.8367 | 0.8499 | 0.8619 | 0.8728 |
| 0.25 | 0.5758 | 0.6125 | 0.6446 | 0.6727 | 0.6984 | 0.7210 | 0.7411 | 0.7591 | 0.7564 | 0.7752 | 0.7924 | 0.8082 |
| 0.30 | 0.4477 | 0.4906 | 0.5289 | 0.5631 | 0.5948 | 0.6230 | 0.6485 | 0.6716 | 0.6682 | 0.6923 | 0.7146 | 0.7353 |
| 0.35 | 0.3289 | 0.3745 | 0.4163 | 0.4545 | 0.4904 | 0.5230 | 0.5528 | 0.5802 | 0.5764 | 0.6050 | 0.6319 | 0.6570 |
| 0.40 | 0.2253 | 0.2702 | 0.3126 | 0.3523 | 0.3904 | 0.4256 | 0.4584 | 0.4888 | 0.4848 | 0.5169 | 0.5473 | 0.5761 |
| 0.45 | 0.1404 | 0.1814 | 0.2216 | 0.2604 | 0.2985 | 0.3346 | 0.3687 | 0.4009 | 0.3968 | 0.4310 | 0.4638 | 0.4954 |
| 0.50 | 0.0748 | 0.1098 | 0.1456 | 0.1814 | 0.2176 | 0.2526 | 0.2865 | 0.3190 | 0.3151 | 0.3499 | 0.3839 | 0.4171 |
| 0.60 | −0.0040 | 0.0160 | 0.0393 | 0.0650 | 0.0929 | 0.1217 | 0.1510 | 0.1804 | 0.1774 | 0.2095 | 0.2420 | 0.2749 |
| 0.70 | −0.0303 | −0.0243 | −0.0141 | −0.0003 | 0.0167 | 0.0361 | 0.0574 | 0.0801 | 0.0781 | 0.1035 | 0.1310 | 0.1597 |
| 0.80 | −0.0268 | −0.0304 | −0.0303 | −0.0270 | −0.0202 | −0.0104 | 0.0020 | 0.0166 | 0.0155 | 0.0329 | 0.0527 | 0.0748 |
| 0.90 | −0.0109 | −0.0192 | −0.0252 | −0.0289 | −0.0299 | −0.0282 | −0.0238 | −0.0169 | −0.0173 | −0.0079 | 0.0041 | 0.0186 |
| 1.00 | 0.0066 | −0.0024 | −0.0108 | −0.0182 | −0.0239 | −0.0276 | −0.0292 | −0.0286 | −0.0285 | −0.0260 | −0.0209 | −0.0134 |
| 1.10 | 0.0206 | 0.0132 | 0.0051 | −0.0030 | −0.0107 | −0.0174 | −0.0226 | −0.0263 | −0.0261 | −0.0284 | −0.0287 | −0.0270 |
| 1.20 | 0.0293 | 0.0247 | 0.0185 | 0.0114 | 0.0039 | −0.0036 | −0.0105 | −0.0165 | −0.0163 | −0.0216 | −0.0255 | −0.0280 |
| 1.30 | 0.0331 | 0.0315 | 0.0278 | 0.0228 | 0.0167 | 0.0099 | 0.0029 | −0.0039 | −0.0037 | −0.0102 | −0.0161 | −0.0212 |
| 1.40 | 0.0331 | 0.0340 | 0.0329 | 0.0302 | 0.0261 | 0.0210 | 0.0150 | 0.0086 | 0.0087 | 0.0022 | −0.0043 | −0.0105 |
| 1.50 | 0.0304 | 0.0332 | 0.0344 | 0.0339 | 0.0320 | 0.0288 | 0.0245 | 0.0193 | 0.0192 | 0.0136 | 0.0075 | 0.0012 |

| $\frac{\sin\theta}{\lambda}$ | 4s K | 4s Ca | 4s Sc | 4s Ti | 4s V | 4s Cr | 4s Mn |
|---|---|---|---|---|---|---|---|
| 0.00 | 1.0000 | 1.0000 | 1.0000 | 1.0000 | 1.0000 | 1.0000 | 1.0000 |
| 0.05 | 0.5552 | 0.6849 | 0.7164 | 0.7378 | 0.7555 | 0.7708 | 0.7833 |
| 0.10 | 0.0613 | 0.2018 | 0.2506 | 0.2876 | 0.3201 | 0.3497 | 0.3750 |
| 0.15 | -0.0415 | -0.0153 | 0.0075 | 0.0286 | 0.0501 | 0.0712 | 0.0914 |
| 0.20 | -0.0182 | -0.0395 | -0.0414 | -0.0395 | -0.0347 | -0.0283 | -0.0203 |
| 0.25 | 0.0046 | -0.0129 | -0.0223 | -0.0288 | -0.0334 | -0.0351 | -0.0373 |
| 0.30 | 0.0122 | 0.0086 | 0.0014 | -0.0053 | -0.0119 | -0.0175 | -0.0223 |
| 0.35 | 0.0114 | 0.0167 | 0.0140 | 0.0100 | 0.0053 | 0.0005 | -0.0044 |
| 0.40 | 0.0079 | 0.0162 | 0.0169 | 0.0158 | 0.0138 | 0.0109 | 0.0077 |
| 0.45 | 0.0044 | 0.0120 | 0.0145 | 0.0155 | 0.0156 | 0.0147 | 0.0133 |
| 0.50 | 0.0017 | 0.0072 | 0.0103 | 0.0123 | 0.0137 | 0.0143 | 0.0143 |
| 0.60 | -0.0008 | 0.0006 | 0.0028 | 0.0049 | 0.0069 | 0.0087 | 0.0101 |
| 0.70 | -0.0011 | -0.0017 | -0.0010 | 0.0002 | 0.0016 | 0.0031 | 0.0045 |
| 0.80 | -0.0004 | -0.0014 | -0.0017 | -0.0015 | -0.0010 | -0.0002 | 0.0008 |
| 0.90 | -0.0003 | -0.0004 | -0.0010 | -0.0014 | -0.0015 | -0.0014 | -0.0010 |
| 1.00 | 0.0008 | 0.0007 | 0.0000 | -0.0005 | -0.0010 | -0.0013 | -0.0014 |
| 1.10 | 0.0010 | 0.0013 | 0.0009 | 0.0004 | -0.0001 | -0.0006 | -0.0009 |
| 1.20 | 0.0010 | 0.0017 | 0.0015 | 0.0011 | 0.0007 | 0.0002 | -0.0002 |
| 1.30 | 0.0009 | 0.0017 | 0.0017 | 0.0015 | 0.0012 | 0.0008 | 0.0004 |
| 1.40 | 0.0008 | 0.0015 | 0.0017 | 0.0017 | 0.0015 | 0.0013 | 0.0010 |
| 1.50 | 0.0006 | 0.0013 | 0.0015 | 0.0016 | 0.0016 | 0.0015 | 0.0013 |

| $\frac{\sin\theta}{\lambda}$ | 4s Fe | 4s Co | 4s Ni | 4s Cu ($3d^9$) | 4s Cu ($3d^{10}$) | 4s Zn | 4s Ga | 4s Ge |
|---|---|---|---|---|---|---|---|---|
| 0.00 | 1.0000 | 1.0000 | 1.0000 | 1.0000 | 1.0000 | 1.0000 | 1.0000 | 1.0000 |
| 0.05 | 0.7970 | 0.8081 | 0.8180 | 0.8271 | 0.7750 | 0.8349 | 0.8759 | 0.9000 |
| 0.10 | 0.4034 | 0.4281 | 0.4506 | 0.4713 | 0.3819 | 0.4893 | 0.5896 | 0.6557 |
| 0.15 | 0.1149 | 0.1368 | 0.1577 | 0.1781 | 0.1184 | 0.1964 | 0.3002 | 0.3824 |
| 0.20 | -0.0099 | 0.0006 | 0.0119 | 0.0242 | -0.0001 | 0.0362 | 0.1010 | 0.1671 |
| 0.25 | -0.0370 | -0.0357 | -0.0331 | -0.0293 | -0.0328 | -0.0244 | -0.0007 | 0.0362 |
| 0.30 | -0.0267 | -0.0300 | -0.0323 | -0.0338 | -0.0281 | -0.0341 | -0.0357 | -0.0243 |
| 0.35 | -0.0095 | -0.0138 | -0.0178 | -0.0215 | -0.0141 | -0.0244 | -0.0361 | -0.0401 |
| 0.40 | 0.0039 | 0.0003 | -0.0036 | -0.0072 | -0.0020 | -0.0108 | -0.0232 | -0.0336 |
| 0.45 | 0.0113 | 0.0090 | 0.0063 | 0.0035 | 0.0054 | 0.0004 | -0.0083 | -0.0190 |
| 0.50 | 0.0139 | 0.0129 | 0.0114 | 0.0097 | 0.0088 | 0.0077 | 0.0036 | -0.0042 |
| 0.60 | 0.0114 | 0.0122 | 0.0126 | 0.0127 | 0.0092 | 0.0124 | 0.0153 | 0.0147 |
| 0.70 | 0.0061 | 0.0074 | 0.0086 | 0.0096 | 0.0064 | 0.0103 | 0.0154 | 0.0191 |
| 0.80 | 0.0019 | 0.0031 | 0.0043 | 0.0054 | 0.0035 | 0.0064 | 0.0109 | 0.0156 |
| 0.90 | -0.0004 | -0.0003 | 0.0011 | 0.0020 | 0.0013 | 0.0029 | 0.0059 | 0.0098 |
| 1.00 | -0.0013 | -0.0010 | -0.0006 | -0.0000 | -0.0001 | 0.0006 | 0.0021 | 0.0047 |
| 1.10 | -0.0012 | -0.0013 | -0.0012 | -0.0010 | -0.0007 | -0.0007 | -0.0002 | 0.0010 |
| 1.20 | -0.0006 | -0.0009 | -0.0011 | -0.0012 | -0.0008 | -0.0011 | -0.0013 | -0.0010 |
| 1.30 | 0.0000 | -0.0003 | -0.0006 | -0.0009 | -0.0006 | -0.0010 | -0.0015 | -0.0018 |
| 1.40 | 0.0006 | 0.0003 | -0.0001 | -0.0004 | -0.0002 | -0.0006 | -0.0012 | -0.0017 |
| 1.50 | 0.0011 | 0.0008 | 0.0005 | 0.0002 | 0.0002 | -0.0001 | -0.0006 | -0.0012 |

| $\frac{\sin\theta}{\lambda}$ | $2p$ B | $2p$ C | $2p$ N | $2p$ O | $2p$ F | $2p$ Fv | $2p$ Ne | $2p$ Na | $2p$ Mg | $2p$ Al |
|---|---|---|---|---|---|---|---|---|---|---|
| 0.00 | 1.0000 | 1.0000 | 1.0000 | 1.0000 | 1.0000 | 1.0000 | 1.0000 | 1.0000 | 1.0000 | 1.0000 |
| 0.05 | 0.8948 | 0.9341 | 0.9546 | 0.9646 | 0.9722 | 0.9609 | 0.9777 | 0.9850 | 0.9891 | 0.9917 |
| 0.10 | 0.6551 | 0.7686 | 0.8345 | 0.8687 | 0.8954 | 0.8594 | 0.9154 | 0.9419 | 0.9572 | 0.9672 |
| 0.15 | 0.4121 | 0.5703 | 0.6763 | 0.7372 | 0.7862 | 0.7283 | 0.8242 | 0.8759 | 0.9073 | 0.9282 |
| 0.20 | 0.2334 | 0.3917 | 0.5162 | 0.5963 | 0.6636 | 0.5948 | 0.7180 | 0.7940 | 0.8431 | 0.8770 |
| 0.25 | 0.1223 | 0.2542 | 0.3765 | 0.4649 | 0.5429 | 0.4738 | 0.6089 | 0.7037 | 0.7694 | 0.8165 |
| 0.30 | 0.0592 | 0.1580 | 0.2653 | 0.3526 | 0.4337 | 0.3708 | 0.5055 | 0.6116 | 0.6906 | 0.7498 |
| 0.35 | 0.0253 | 0.0943 | 0.1817 | 0.2618 | 0.3401 | 0.2864 | 0.4126 | 0.5227 | 0.6107 | 0.6799 |
| 0.40 | 0.0080 | 0.0536 | 0.1214 | 0.1910 | 0.2628 | 0.2189 | 0.3324 | 0.4405 | 0.5330 | 0.6093 |
| 0.45 | -0.0004 | 0.0283 | 0.0788 | 0.1371 | 0.2006 | 0.1657 | 0.2649 | 0.3668 | 0.4598 | 0.5403 |
| 0.50 | -0.0041 | 0.0130 | 0.0494 | 0.0968 | 0.1513 | 0.1242 | 0.2091 | 0.3022 | 0.3926 | 0.4745 |
| 9.60 | -0.0055 | -0.0012 | 0.0161 | 0.0452 | 0.0830 | 0.0675 | 0.1269 | 0.1999 | 0.2787 | 0.3568 |
| 0.70 | -0.0046 | -0.0052 | 0.0016 | 0.0180 | 0.0426 | 0.0344 | 0.0740 | 0.1280 | 0.1918 | 0.2605 |
| 0.80 | -0.0034 | -0.0055 | -0.0039 | 0.0044 | 0.0195 | 0.0156 | 0.0409 | 0.0791 | 0.1283 | 0.1852 |
| 0.90 | -0.0024 | -0.0048 | -0.0055 | -0.0020 | 0.0066 | 0.0051 | 0.0206 | 0.0468 | 0.0832 | 0.1284 |
| 1.00 | -0.0017 | -0.0038 | -0.0055 | -0.0046 | -0.0002 | -0.0004 | 0.0084 | 0.0257 | 0.0519 | 0.0867 |
| 1.10 | -0.0012 | -0.0030 | -0.0049 | -0.0054 | -0.0036 | -0.0031 | 0.0014 | 0.0123 | 0.0305 | 0.0565 |
| 1.20 | -0.0009 | -0.0023 | -0.0041 | -0.0052 | -0.0050 | -0.0042 | -0.0024 | 0.0040 | 0.0163 | 0.0352 |
| 1.30 | -0.0006 | -0.0017 | -0.0033 | -0.0047 | -0.0053 | -0.0045 | -0.0044 | -0.0010 | 0.0069 | 0.0203 |
| 1.40 | -0.0005 | -0.0013 | -0.0027 | -0.0041 | -0.0051 | -0.0043 | -0.0052 | -0.0038 | 0.0009 | 0.0101 |
| 1.50 | -0.0004 | -0.0010 | -0.0022 | -0.0035 | -0.0047 | -0.0040 | -0.0053 | -0.0053 | -0.0028 | 0.0032 |

| $\frac{\sin\theta}{\lambda}$ | $2p$ Si | $2p$ P | $2p$ S | $2p$ Cl | $2p$ A | $2p$ K | $2p$ Ca | $2p$ Sc | $2p$ Ti | $2p$ V |
|---|---|---|---|---|---|---|---|---|---|---|
| 0.00 | 1.0000 | 1.0000 | 1.0000 | 1.0000 | 1.0000 | 1.0000 | 1.0000 | 1.0000 | 1.0000 | 1.0000 |
| 0.05 | 0.9934 | 0.9946 | 0.9955 | 0.9962 | 0.9968 | 0.9972 | 0.9976 | 0.9979 | 0.9981 | 0.9983 |
| 0.10 | 0.9740 | 0.9788 | 0.9824 | 0.9851 | 0.9873 | 0.9890 | 0.9904 | 0.9915 | 0.9924 | 0.9932 |
| 0.15 | 0.9426 | 0.9531 | 0.9609 | 0.9668 | 0.9716 | 0.9754 | 0.9785 | 0.9810 | 0.9831 | 0.9848 |
| 0.20 | 0.9009 | 0.9185 | 0.9318 | 0.9420 | 0.9502 | 0.9567 | 0.9621 | 0.9665 | 0.9702 | 0.9733 |
| 0.25 | 0.8508 | 0.8764 | 0.8960 | 0.9112 | 0.9236 | 0.9334 | 0.9415 | 0.9482 | 0.9538 | 0.9586 |
| 0.30 | 0.7943 | 0.8282 | 0.8545 | 0.8753 | 0.8922 | 0.9059 | 0.9171 | 0.9265 | 0.9344 | 0.9410 |
| 0.35 | 0.7336 | 0.7756 | 0.8087 | 0.8351 | 0.8569 | 0.8746 | 0.8893 | 0.9016 | 0.9120 | 0.9208 |
| 0.40 | 0.6707 | 0.7199 | 0.7595 | 0.7916 | 0.8183 | 0.8402 | 0.8585 | 0.8739 | 0.8869 | 0.8981 |
| 0.45 | 0.6075 | 0.6628 | 0.7082 | 0.7456 | 0.7770 | 0.8031 | 0.8251 | 0.8437 | 0.8595 | 0.8731 |
| 0.50 | 0.5454 | 0.6055 | 0.6558 | 0.6980 | 0.7339 | 0.7640 | 0.7896 | 0.8114 | 0.8301 | 0.8462 |
| 0.60 | 0.4294 | 0.4946 | 0.5518 | 0.6014 | 0.6448 | 0.6820 | 0.7143 | 0.7423 | 0.7666 | 0.7877 |
| 0.70 | 0.3289 | 0.3939 | 0.4535 | 0.5073 | 0.5557 | 0.5984 | 0.6362 | 0.6695 | 0.6988 | 0.7247 |
| 0.80 | 0.2458 | 0.3066 | 0.3651 | 0.4198 | 0.4707 | 0.5167 | 0.5584 | 0.5958 | 0.6294 | 0.6593 |
| 0.90 | 0.1796 | 0.2338 | 0.2883 | 0.3413 | 0.3922 | 0.4396 | 0.4834 | 0.5237 | 0.5603 | 0.5935 |
| 1.00 | 0.1284 | 0.1748 | 0.2237 | 0.2731 | 0.3220 | 0.3688 | 0.4132 | 0.4548 | 0.4934 | 0.5289 |
| 1.10 | 0.0895 | 0.1281 | 0.1706 | 0.2151 | 0.2606 | 0.3054 | 0.3490 | 0.3906 | 0.4300 | 0.4668 |
| 1.20 | 0.0606 | 0.0919 | 0.1278 | 0.1669 | 0.2080 | 0.2498 | 0.2913 | 0.3320 | 0.3711 | 0.4082 |
| 1.30 | 0.0395 | 0.0643 | 0.0939 | 0.1274 | 0.1638 | 0.2018 | 0.2405 | 0.2793 | 0.3172 | 0.3539 |
| 1.40 | 0.0242 | 0.0434 | 0.0675 | 0.0956 | 0.1272 | 0.1610 | 0.1964 | 0.2326 | 0.2687 | 0.3043 |
| 1.50 | 0.0133 | 0.0279 | 0.0471 | 0.0703 | 0.0972 | 0.1269 | 0.1586 | 0.1918 | 0.2256 | 0.2594 |

| $\sin\theta/\lambda$ | 3p Mn | 3p Fe | 3p Co | 3p Ni | 3p Cu($3d^9$) | 3p Cu($3d^{10}$) | 3p Zn | 3p Ga | 3p Gf | 4p Ga | 4p Ge | 3d Sc |
|---|---|---|---|---|---|---|---|---|---|---|---|---|
| 0.00 | 1.0000 | 1.0000 | 1.0000 | 1.0000 | 1.0000 | 1.0000 | 1.0000 | 1.0000 | 1.0000 | 1.0000 | 1.0000 | 1.0000 |
| 0.05 | 0.9821 | 0.9839 | 0.9854 | 0.9867 | 0.9878 | 0.9876 | 0.9888 | 0.9899 | 0.9909 | 0.7771 | 0.8380 | 0.9359 |
| 0.10 | 0.9303 | 0.9371 | 0.9429 | 0.9480 | 0.9522 | 0.9513 | 0.9560 | 0.9603 | 0.9640 | 0.3719 | 0.4967 | 0.7767 |
| 0.15 | 0.8502 | 0.8643 | 0.8763 | 0.8869 | 0.8959 | 0.8940 | 0.9039 | 0.9129 | 0.9209 | 0.1020 | 0.2033 | 0.5868 |
| 0.20 | 0.7498 | 0.7720 | 0.7912 | 0.8082 | 0.8228 | 0.8198 | 0.8358 | 0.8506 | 0.8639 | -0.0060 | 0.0399 | 0.4136 |
| 0.25 | 0.6381 | 0.6679 | 0.6939 | 0.7173 | 0.7377 | 0.7337 | 0.7561 | 0.7769 | 0.7958 | -0.0301 | -0.0232 | 0.2768 |
| 0.30 | 0.5239 | 0.5595 | 0.5913 | 0.6201 | 0.6458 | 0.6410 | 0.6690 | 0.6955 | 0.7198 | -0.0241 | -0.0339 | 0.1773 |
| 0.35 | 0.4142 | 0.4535 | 0.4891 | 0.5218 | 0.5518 | 0.5465 | 0.5790 | 0.6102 | 0.6393 | -0.0120 | -0.0245 | 0.1085 |
| 0.40 | 0.3145 | 0.3548 | 0.3923 | 0.4272 | 0.4598 | 0.4544 | 0.4899 | 0.5245 | 0.5573 | -0.0018 | -0.0109 | 0.0626 |
| 0.45 | 0.2279 | 0.2670 | 0.3042 | 0.3395 | 0.3733 | 0.3680 | 0.4048 | 0.4414 | 0.4766 | 0.0047 | 0.0003 | 0.0326 |
| 0.50 | 0.1559 | 0.1918 | 0.2270 | 0.2612 | 0.2945 | 0.2896 | 0.3262 | 0.3632 | 0.3995 | 0.0080 | 0.0076 | 0.0136 |
| 0.60 | 0.0548 | 0.0809 | 0.1084 | 0.1367 | 0.1654 | 0.1618 | 0.1941 | 0.2279 | 0.2626 | 0.0089 | 0.0127 | -0.0051 |
| 0.70 | 0.0015 | 0.0166 | 0.0344 | 0.0544 | 0.0756 | 0.0733 | 0.0982 | 0.1253 | 0.1545 | 0.0067 | 0.0111 | -0.0104 |
| 0.80 | -0.0185 | -0.0129 | -0.0043 | 0.0072 | 0.0204 | 0.0193 | 0.0358 | 0.0549 | 0.0767 | 0.0040 | 0.0076 | -0.0103 |
| 0.90 | -0.0190 | -0.0200 | -0.0186 | -0.0142 | -0.0081 | -0.0083 | 0.0004 | 0.0117 | 0.0259 | 0.0019 | 0.0042 | -0.0085 |
| 1.00 | -0.0100 | -0.0148 | -0.0180 | -0.0189 | -0.0180 | -0.0180 | -0.0154 | -0.0105 | -0.0031 | 0.0006 | 0.0018 | -0.0065 |
| 1.10 | 0.0019 | -0.0044 | -0.0099 | -0.0141 | -0.0168 | -0.0166 | -0.0182 | -0.0181 | -0.0160 | -0.0002 | 0.0003 | -0.0048 |
| 1.20 | 0.0132 | 0.0070 | 0.0009 | -0.0049 | -0.0095 | -0.0093 | -0.0135 | -0.0164 | -0.0181 | -0.0005 | -0.0005 | -0.0034 |
| 1.30 | 0.0221 | 0.0170 | 0.0114 | 0.0055 | 0.0001 | 0.0002 | -0.0051 | -0.0097 | -0.0136 | -0.0005 | -0.0007 | -0.0024 |
| 1.40 | 0.0281 | 0.0247 | 0.0202 | 0.0150 | 0.0098 | 0.0098 | 0.0044 | -0.0007 | -0.0057 | -0.0003 | -0.0006 | -0.0017 |
| 1.50 | 0.0316 | 0.0297 | 0.0268 | 0.0227 | 0.0183 | 0.0182 | 0.0135 | 0.0085 | 0.0034 | -0.0001 | -0.0003 | -0.0012 |

| $\sin\theta/\lambda$ | 3d Ti | 3d V | 3d Cr | 3d Mn | 3d Fe | 3d Co | 3d Ni | 3d Cu($3d^9$) | 3d Cu($3d^{10}$) | 3d Zn | 3d Ga | 3d Ge |
|---|---|---|---|---|---|---|---|---|---|---|---|---|
| 0.00 | 1.0000 | 1.0000 | 1.0000 | 1.0000 | 1.0000 | 1.0000 | 1.0000 | 1.0000 | 1.0000 | 1.0000 | 1.0000 | 1.0000 |
| 0.05 | 0.9513 | 0.9597 | 0.9655 | 0.9703 | 0.9730 | 0.9757 | 0.9779 | 0.9800 | 0.9761 | 0.9818 | 0.9855 | 0.9880 |
| 0.10 | 0.8242 | 0.8522 | 0.8721 | 0.8888 | 0.8985 | 0.9080 | 0.9162 | 0.9235 | 0.9100 | 0.9301 | 0.9438 | 0.9532 |
| 0.15 | 0.6604 | 0.7079 | 0.7435 | 0.7736 | 0.7921 | 0.8101 | 0.8259 | 0.8402 | 0.8157 | 0.8531 | 0.8798 | 0.8988 |
| 0.20 | 0.4975 | 0.5570 | 0.6041 | 0.6449 | 0.6714 | 0.6973 | 0.7206 | 0.7415 | 0.7087 | 0.7608 | 0.8002 | 0.8296 |
| 0.25 | 0.3571 | 0.4196 | 0.4719 | 0.5187 | 0.5509 | 0.5827 | 0.6117 | 0.6380 | 0.6011 | 0.6626 | 0.7121 | 0.7508 |
| 0.30 | 0.2463 | 0.3048 | 0.3566 | 0.4047 | 0.4401 | 0.4750 | 0.5076 | 0.5374 | 0.5004 | 0.5657 | 0.6215 | 0.6674 |
| 0.35 | 0.1638 | 0.2144 | 0.2618 | 0.3077 | 0.3435 | 0.3792 | 0.4130 | 0.4446 | 0.4102 | 0.4749 | 0.5334 | 0.5838 |
| 0.40 | 0.1045 | 0.1460 | 0.1869 | 0.2282 | 0.2625 | 0.2970 | 0.3304 | 0.3621 | 0.3316 | 0.3928 | 0.4509 | 0.5032 |
| 0.45 | 0.0630 | 0.0956 | 0.1295 | 0.1652 | 0.1965 | 0.2286 | 0.2600 | 0.2906 | 0.2646 | 0.3207 | 0.3760 | 0.4279 |
| 0.50 | 0.0345 | 0.0593 | 0.0865 | 0.1163 | 0.1439 | 0.1727 | 0.2014 | 0.2300 | 0.2084 | 0.2584 | 0.3096 | 0.3592 |
| 0.60 | 0.0032 | 0.0160 | 0.0321 | 0.0511 | 0.0709 | 0.0923 | 0.1146 | 0.1377 | 0.1237 | 0.1613 | 0.2020 | 0.2442 |
| 0.70 | -0.0089 | -0.0036 | 0.0047 | 0.0155 | 0.0283 | 0.0429 | 0.0590 | 0.0762 | 0.0680 | 0.0945 | 0.1247 | 0.1579 |
| 0.80 | -0.0119 | -0.0110 | -0.0077 | -0.0025 | 0.0050 | 0.0141 | 0.0249 | 0.0369 | 0.0326 | 0.0502 | 0.0714 | 0.0959 |
| 0.90 | -0.0112 | -0.0125 | -0.0121 | -0.0105 | -0.0068 | -0.0017 | 0.0049 | 0.0128 | 0.0110 | 0.0219 | 0.0359 | 0.0530 |
| 1.00 | -0.0092 | -0.0114 | -0.0126 | -0.0130 | -0.0118 | -0.0095 | -0.0059 | -0.0013 | -0.0014 | 0.0046 | 0.0132 | 0.0245 |
| 1.10 | -0.0071 | -0.0095 | -0.0113 | -0.0128 | -0.0130 | -0.0126 | -0.0111 | -0.0088 | -0.0081 | -0.0054 | -0.0007 | 0.0061 |
| 1.20 | -0.0053 | -0.0075 | -0.0095 | -0.0113 | -0.0124 | -0.0130 | -0.0130 | -0.0122 | -0.0111 | -0.0106 | -0.0086 | -0.0050 |
| 1.30 | -0.0039 | -0.0057 | -0.0076 | -0.0095 | -0.0110 | -0.0122 | -0.0129 | -0.0132 | -0.0119 | -0.0128 | -0.0127 | -0.0113 |
| 1.40 | -0.0028 | -0.0043 | -0.0060 | -0.0077 | -0.0093 | -0.0107 | -0.0119 | -0.0128 | -0.0116 | -0.0132 | -0.0142 | -0.0144 |
| 1.50 | -0.0021 | -0.0032 | -0.0046 | -0.0061 | -0.0076 | -0.0091 | -0.0105 | -0.0117 | -0.0106 | -0.0126 | -0.0143 | -0.0154 |

| $\dfrac{\sin\theta}{\lambda}$ | 2p Cr | 2p Mn | 2p Fe | 2p Co | 2p Ni | 2p Cu | 2p Zn | 2p Ga | 2p Ge |
|---|---|---|---|---|---|---|---|---|---|
| 0.00 | 1.0000 | 1.0000 | 1.0000 | 1.0000 | 1.0000 | 1.0000 | 1.0000 | 1.0000 | 1.0000 |
| 0.05 | 0.9985 | 0.9986 | 0.9987 | 0.9989 | 0.9989 | 0.9990 | 0.9991 | 0.9992 | 0.9992 |
| 0.10 | 0.9939 | 0.9945 | 0.9950 | 0.9954 | 0.9958 | 0.9961 | 0.9964 | 0.9967 | 0.9969 |
| 0.15 | 0.9863 | 0.9876 | 0.9887 | 0.9897 | 0.9905 | 0.9913 | 0.9920 | 0.9925 | 0.9931 |
| 0.20 | 0.9759 | 0.9781 | 0.9801 | 0.9818 | 0.9833 | 0.9846 | 0.9858 | 0.9868 | 0.9877 |
| 0.25 | 0.9626 | 0.9661 | 0.9691 | 0.9717 | 0.9740 | 0.9761 | 0.9779 | 0.9795 | 0.9809 |
| 0.30 | 0.9467 | 0.9516 | 0.9559 | 0.9596 | 0.9628 | 0.9657 | 0.9683 | 0.9706 | 0.9726 |
| 0.35 | 0.9283 | 0.9348 | 0.9405 | 0.9455 | 0.9498 | 0.9537 | 0.9572 | 0.9602 | 0.9630 |
| 0.40 | 0.9076 | 0.9159 | 0.9232 | 0.9295 | 0.9351 | 0.9401 | 0.9445 | 0.9484 | 0.9520 |
| 0.45 | 0.8848 | 0.8951 | 0.9040 | 0.9118 | 0.9187 | 0.9249 | 0.9304 | 0.9353 | 0.9397 |
| 0.50 | 0.8602 | 0.8724 | 0.8831 | 0.8925 | 0.9008 | 0.9083 | 0.9149 | 0.9208 | 0.9262 |
| 0.60 | 0.8062 | 0.8225 | 0.8369 | 0.8497 | 0.8610 | 0.8711 | 0.8802 | 0.8884 | 0.8958 |
| 0.70 | 0.7476 | 0.7679 | 0.7860 | 0.8021 | 0.8166 | 0.8296 | 0.8412 | 0.8518 | 0.8614 |
| 0.80 | 0.6861 | 0.7102 | 0.7318 | 0.7512 | 0.7687 | 0.7845 | 0.7988 | 0.8118 | 0.8236 |
| 0.90 | 0.6236 | 0.6509 | 0.6756 | 0.6980 | 0.7184 | 0.7369 | 0.7537 | 0.7691 | 0.7832 |
| 1.00 | 0.5615 | 0.5914 | 0.6187 | 0.6438 | 0.6667 | 0.6877 | 0.7069 | 0.7245 | 0.7407 |
| 1.10 | 0.5011 | 0.5328 | 0.5623 | 0.5895 | 0.6145 | 0.6377 | 0.6590 | 0.6787 | 0.6969 |
| 1.20 | 0.4433 | 0.4763 | 0.5072 | 0.5360 | 0.5628 | 0.5878 | 0.6109 | 0.6324 | 0.6524 |
| 1.30 | 0.3891 | 0.4226 | 0.4543 | 0.4842 | 0.5122 | 0.5386 | 0.5632 | 0.5862 | 0.6077 |
| 1.40 | 0.3388 | 0.3722 | 0.4041 | 0.4346 | 0.4634 | 0.4907 | 0.5164 | 0.5406 | 0.5634 |
| 1.50 | 0.2928 | 0.3254 | 0.3571 | 0.3876 | 0.4168 | 0.4446 | 0.4711 | 0.4961 | 0.5199 |

| $\dfrac{\sin\theta}{\lambda}$ | 3p Al | 3p Si | 3p P | 3p S | 3p Cl | 3p A | 3p K | 3p Ca | 3p Sc | 3p Ti | 3p V | 3p Cp |
|---|---|---|---|---|---|---|---|---|---|---|---|---|
| 0.00 | 1.0000 | 1.0000 | 1.0000 | 1.0000 | 1.0000 | 1.0000 | 1.0000 | 1.0000 | 1.0000 | 1.0000 | 1.0000 | 1.0000 |
| 0.05 | 0.7755 | 0.8486 | 0.8895 | 0.9114 | 0.9283 | 0.9411 | 0.9561 | 0.9655 | 0.9706 | 0.9745 | 0.9775 | 0.9800 |
| 0.10 | 0.3693 | 0.5239 | 0.6297 | 0.6936 | 0.7454 | 0.7865 | 0.8364 | 0.8692 | 0.8878 | 0.9019 | 0.9132 | 0.9225 |
| 0.15 | 0.1013 | 0.2352 | 0.3565 | 0.4447 | 0.5219 | 0.5872 | 0.6706 | 0.7301 | 0.7655 | 0.7931 | 0.8156 | 0.8345 |
| 0.20 | -0.0040 | 0.0654 | 0.1556 | 0.2388 | 0.3193 | 0.3932 | 0.4932 | 0.5721 | 0.6223 | 0.6628 | 0.6966 | 0.7255 |
| 0.25 | -0.0262 | -0.0072 | 0.0405 | 0.1007 | 0.1671 | 0.2343 | 0.3314 | 0.4173 | 0.4764 | 0.5260 | 0.5686 | 0.6060 |
| 0.30 | -0.0196 | -0.0262 | -0.0113 | 0.0227 | 0.0679 | 0.1198 | 0.2004 | 0.2810 | 0.3419 | 0.3953 | 0.4430 | 0.4860 |
| 0.35 | -0.0077 | -0.0220 | -0.0260 | -0.0133 | 0.0116 | 0.0458 | 0.1042 | 0.1715 | 0.2277 | 0.2797 | 0.3281 | 0.3731 |
| 0.40 | 0.0018 | -0.0111 | -0.0227 | -0.0239 | -0.0149 | 0.0035 | 0.0402 | 0.0903 | 0.1375 | 0.1839 | 0.2292 | 0.2729 |
| 0.45 |  | -0.0005 | -0.0130 | -0.0212 | -0.0227 | -0.0166 | 0.0020 | 0.0350 | 0.0711 | 0.1093 | 0.1488 | 0.1885 |
| 0.50 | 0.0105 | 0.0072 | -0.0026 | -0.0130 | -0.0203 | -0.0224 | -0.0171 | 0.0010 | 0.0258 | 0.0548 | 0.0869 | 0.1206 |
| 0.60 | 0.0113 | 0.0145 | 0.0120 | -0.0046 | -0.0045 | -0.0135 | -0.0216 | -0.0232 | -0.0171 | -0.0053 | 0.0112 | 0.0313 |
| 0.70 | 0.0095 | 0.0150 | 0.0175 | 0.0154 | 0.0101 | 0.0020 | -0.0077 | -0.0168 | -0.0215 | -0.0216 | -0.0176 | -0.0097 |
| 0.80 | 0.0072 | 0.0128 | 0.0176 | 0.0194 | 0.0184 | 0.0138 | 0.0076 | -0.0014 | -0.0099 | -0.0162 | -0.0197 | -0.0202 |
| 0.90 | 0.0052 | 0.0101 | 0.0153 | 0.0192 | 0.0212 | 0.0202 | 0.0186 | 0.0128 | 0.0049 | 0.0030 | -0.0099 | -0.0149 |
| 1.00 | 0.0036 | 0.0076 | 0.0124 | 0.0170 | 0.0207 | 0.0223 | 0.0243 | 0.0226 | 0.0172 | 0.0104 | 0.0032 | -0.0035 |
| 1.10 | 0.0024 | 0.0054 | 0.0095 | 0.0141 | 0.0184 | 0.0216 | 0.0260 | 0.0276 | 0.0253 | 0.0208 | 0.0149 | 0.0085 |
| 1.20 | 0.0015 | 0.0038 | 0.0071 | 0.0111 | 0.0155 | 0.0194 | 0.0250 | 0.0289 | 0.0293 | 0.0273 | 0.0236 | 0.0186 |
| 1.30 | 0.0009 | 0.0026 | 0.0051 | 0.0085 | 0.0126 | 0.0166 | 0.0226 | 0.0278 | 0.0302 | 0.0305 | 0.0289 | 0.0258 |
| 1.40 | 0.0005 | 0.0017 | 0.0036 | 0.0063 | 0.0098 | 0.0136 | 0.0195 | 0.0253 | 0.0289 | 0.0310 | 0.0313 | 0.0301 |
| 1.50 | 0.0002 | 0.0010 | 0.0024 | 0.0046 | 0.0075 | 0.0109 | 0.0163 | 0.0221 | 0.0264 | 0.0296 | 0.0315 | 0.0320 |

# REFERENCES

## Chapter 1

*Acta Cryst.* (1969): **A25**: 1–276.

ARLINGHAUS, F. J. (1967): Theoretical x-ray scattering factors based on energy-band structure, *Phys. Rev.*, **153**: 743–750.

BATTERMAN, B. W., and CHIPMAN, D. R. (1962): Vibrational amplitudes in germanium and silicon, *Phys. Rev.*, **127**: 690–693.

———, ———, and DEMARCO, J. J. (1961): Absolute measurement of the atomic scattering factors of iron, copper, and aluminum, *Phys. Rev.*, **122**: 68–74.

BENESCH, R. (1972): Algebraic matrices and radial momentum distributions from Hylleras-type wavefunctions, *Phys. Rev.*, **A6**: 573–580.

——— and SMITH, V. H., JR. (1970): Radial momentum distribution for the $^2S$ ground state of the lithium atom, *Chem. Phys. Letters*, **5**: 601–604.

———, WITTE, W., and WÖLFEL, E. (1955): Die Molekuldimensionen von Polymethyl-methacrylat in verschiedenen Lösungsmitteln nach der Streulichtemethode, *Z. Phys. Chem.*, **4**: 65–72.

BERGGREN, K. F., and MARTINO, F. (1971): On the calculation of the Compton profile in crystalline Li H, *Phys. Rev.*, **B3**: 1509–1511.

BISCOE, J., and WARREN, B. (1942): An x-ray study of carbon black, *J. Appl. Phys.*, **13**: 364, 371.

BONHAM, R. A. (1965): Corrections to the incoherent scattering factors for electrons and x rays, *J. Chem. Phys.*, **43**: 1460–1464.

BORIE, B. (1957): X-ray diffraction effects of atomic size in alloys, *Acta Cryst.*, **10**: 89–96.

BRANDT, W. (1970): Compton profile of Li H, *Phys. Rev.*, **B2**: 561, 562.

BROWN, R. E., and SMITH, JR., V. H. (1971): On the discrepancy between theory and experiment for the Compton profile of molecular hydrogen, *Phys. Rev.*, **A5**: 140–143.

CALDER, R. S., COCHRAN, W., GRIFFITHS, D., and LOWDE, R. D. (1962): An x-ray and neutron diffraction analysis of lithium hydride, *J. Phys. Chem. Solids*, **23**: 621–632.

CHIPMAN, D. R. (1960): Temperature dependence of the Debye temperatures of aluminum, lead, and beta brass by an x-ray method, *J. Appl. Phys.*, **31**: 2012–2015.

——— and JENNINGS, L. D. (1963): Measurement of the atomic scattering factor of Ne, Ar, Kr, Xe, *Phys. Rev.*, **132**: 728–734.

——— and PASKIN, A. (1959): Temperature diffuse scattering of x-rays in cubic powders. II. Corrections to integrated intensity measurements, *J. Appl. Phys.*, **30**: 1998–2001.

CLEMENTI, E. (1965): Tables of Hartree-Fock wavefunctions, *IBM J. Res. Dev.*, **9**: 2.

COOPER, M. J. (1965): The effects of porosity and impurity on x-ray integrated intensities for powders, *Phil. Mag.*, **11**: 969–975.

COULSON, C. A. (1941): Momentum distribution in molecular systems, *Proc. Cambridge Phil. Soc.*, **37**: 55–66.

——— (1941): Momentum distribution in molecular systems, *Proc. Cambridge Phil. Soc.*, **37**: 74–81.

COULSON, C. A., and DUNCANSON, W. E. (1941): Momentum distribution in molecular systems, *Proc. Cambridge Phil. Soc.*, **37**: 67–73.

CROMER, D. (1965): Anomalous dispersion corrections computed from self-consistent field relativistic Dirac-Slater wave functions, *Acta Cryst.*, **18**: 17–23.

CURRAT, R., DECICCO, P. D., and WEISS, R. J. (1971): Impulse approximation in Compton scattering, *Phys. Rev.*, **B4**: 4256–4261.

DANIEL, E., and VOSKO, S. H. (1960): Momentum distribution of an interacting electron gas, *Phys. Rev.*, **120**: 2041–2044.

DEMARCO, J. J. (1967): Single crystal measurement of the atomic scattering factor of aluminum, *Phil. Mag.*, **15**: 483–495.

——— and WEISS, R. J. (1965): X-ray determination of the $3d$ orbital population in vanadium metal, *Phys. Rev.*, **140**: A1223–A1225.

——— and ——— (1965): An x-ray determination of the orbital population in iron metal, *Phys. Letters*, **18**: 92–93.

DUNCANSON, W. E., and COULSON, C. A. (1941): Momentum distribution in molecular systems, *Proc. Cambridge Phil. Soc.*, **37**: 406–421.

EISENBERGER, P. (1970): Electron momentum density of the He and $H_2$; Compton x-ray scattering, *Phys. Rev.*, **2**: 1678–1686.

——— and PLATZMAN, P. M. (1970): Compton scattering of x rays from bound electrons, *Phys. Rev.*, **2**: 201–205.

EPSTEIN, I. (1971): (To be published.)

EUWEMA, R. N., WILHITE, D. L., and SURRATT, G. T. (1973): General crystalline Hartree-Fock formalism: Diamond results, *Phys. Rev.*, **B7**: 818–831.

FUKAMACHI, T., and HOSOYA, S. (1970): Electron state in NaF studied by Compton scattering measurements, *Phys. Soc. Japan*, **29**: 736–745.

GOLDBERGER, M. L., and LOW, F. E. (1968): Photon scattering from bound atomic systems at very high energy, *Phys. Rev.*, **176**: 1778–1781.

GOODISMAN, J., and KLEMPERE, W. (1963): On errors in Hartree-Fock calculations, *J. Chem. Phys.*, **38**: 721–725.

GOROFF, I., and KLEINMAN, L. (1968): Charge density of diamond, *Phys. Rev.*, **164**: 1100–1105.

GÖTTLICHER, S., and WÖLFEL, E. (1959): Röntgenographische Bestimmungen der Elektronen-verteilung in Kristallen. VII. Die Elektronendichten in Diamantgitter und in Gitter des Sizicums, *Z. Elekt.*, **63**: 891–901.

GUMMEL, H., and LAX, M. (1957): Thermal capture of electrons in silicon, *Ann. Phys.*, **2**: 28–56.

HALL, G. G. (1966): *Advances in quantum chemistry*. *I*. Academic Press, Inc., New York, and North-Holland Publishing Co., Amsterdam, chap. 4.

HÖNL, H. (1933): Zur Dispersions Theorie der Röntgenstrahlem, *Z. Physik*, **84**: 1–16.

HURST, R. P. (1959): Coherent atomic scattering factors for the lithium hydride crystal field, *Phys. Rev.*, **114**: 746–751.

INKINEN, O., JÄRVINEN, M., and MERISALO, M. (1969): Absolute measurement of the atomic scattering factors of aluminum powder in transmission, *Phys. Rev.*, **178**: 1108–1110.

*International tables for x-ray crystallography* (1962): Kynoch Press, Birmingham, England.

JAUCH, J. J., and ROHRLICH, F. (1955): *Theory of protons and electrons*, Addison-Wesley Publishing Company, Inc., Reading, Mass., p. 229.

JENNINGS, L. D. (1968): Polarization of crystal monochromated x-rays, *Acta Cryst.*, **A24:** 472–474.

KLEIN, O., and NISHINA, Y. (1929): Über die Streung von Strahlung durch freie Elektronen nach der neuen Relativistischen Quantendenmechanik von Dirac, *Z. Phys.*, **52:** 853–868.

MELNGAILIS, J., and DEBENEDETTI, S. (1966): Position annihilation and orthogonalized plane waves in lithium, *Phys. Rev.*, **145:** 400–405.

MERISALO, M., and INKINEN, O. (1966): *Am. Acad. Sci. Fermi*, **6:** 207.

MIJNARENDS, P. E. (1967): Determination of anisotropic momentum distribution in positron annihilation, *Phys. Rev.*, **160:** 512–519.

——— (1971): (To be published.)

NILSSON, N. (1957): *Arkiv Fysik*, **12:** 247–257.

PHILLIPS, W. C. (1973): Compton profile of single-crystal vanadium, *Phys. Rev.*, **B7:** 1047–1051.

——— and WEISS, R. J. (1968): X-ray determination of the electron momentum density in diamond, graphite, and carbon black, *Phys. Rev.*, **176:** 900–904.

——— and ——— (1969): Compton profile of Li H, *Phys. Rev.*, **182:** (3), 923–925.

——— and ——— (1972): (To be published in *Phys. Rev.*)

PLATZMAN, P. M., and EISENBERGER, P. (1971): (To be published in *Phys. Rev.*)

——— and TZOAR, N. (1965): X-ray scattering from an electron gas, *Phys. Rev.*, **139:** A410–A413.

RACCAH, P. M., and HENRICH, V. E. (1969): Absolute experimental x-ray form factor of aluminum, *Phys. Rev.*, **184:** 607–613.

SCHWARTZ, L. H. (1964): Correction of the measured integrated intensities from cubic metallic single crystals for thermal diffuse scattering, *Acta Cryst.*, **17:** 1614–1615.

SUORTTI, P., and JENNINGS, L. D. (1971): Effects of geometrical aberrations on intensities in powder diffractometry, *J. Appl. Cryst.*, **4:** 37–43.

ULSCH, R. C., BONHAM, R. A., and BARTELL, L. S. (1972): Vibrational correction to the calculation of the Compton profile of $H_2$, *Chem. Phys. Letters*, **13:** 6–8.

WAGENFELD, H. (1966): Normal and anomalous photoelectric absorption of x rays in crystals, *Phys. Rev.*, **144:** 216–224.

WALLER, I., and HARTREE, D. R. (1929): On the intensity of total scattering of x rays, *Proc. Roy. Soc.*, **A124:** 119–142.

WEISS, R. J. (1966): *X-ray determination of electron distributions*, North-Holland Publishing Company, Amsterdam.

——— (1967): The absolute x-ray scattering factor of magnesium, *Phil. Mag.*, **16:** 141–146.

——— (1970): Compton profiles of B, $B_4C$, BN, BeO, LiF, and MgO, *Phil. Mag.*, **21:** 1169–1173.

——— and DEMARCO, J. J. (1965): X-ray determination of the 3d-orbital population in vanadium metal, *Phys. Rev.*, **140A:** 1223–1225.

WEPFER, G. G., EUWEMA, R. N., SURRATT, G. T., and WILHITE, D. L. (1973): Electron momentum distribution in diamond, *Intl. J. Quant. Chem.* (Sanibel Proceedings).

## Chapter 2

*General references*

GUINIER, A., and FOURNÉT, G. (1955): *Small-angle scattering of x rays*, John Wiley & Sons, Inc., New York.

JAMES, R. W. (1954): *The optical principles of the diffraction of x rays*, G. Bell & Sons, Ltd., London.

KRIVOGLAZ, M. A. (1969): *Theory of x-ray and thermal neutron scattering by real crystals*, Plenum Press, New York.

WARREN, B. E. (1969): *X-ray diffraction*, Addison-Wesley Publishing Company, Inc., Reading, Mass.


*Special references*

AZÁROFF, L. V., and BUERGER, M. J. (1958): *The powder method in x-ray crystallography*, McGraw-Hill Book Company, New York.

BORIE, B., and SPARKS, C. J., JR. (1971): The interpretation of intensity distributions from disordered binary alloys, *Acta Cryst.*, **A27**: 198–201.

BUERGER, M. J. (1942): *X-ray crystallography*, John Wiley & Sons, Inc., New York.

DELAUNAY, B. (1933): Neu Darstellung der geometrischen Kristallographie, *Z. Krist.*, **84**: 109–149.

FESSLER, R. R., KAPLOW, R., and AVERBACH, B. L. (1966): Pair correlations in liquid and solid aluminum, *Phys. Rev.*, **150**: 34–35.

KAPLOW, R. (1972): Interatomic potentials; aspects which are visible in experimental radial pair distributions, *Proc. Batelle colloq. simulation lattice defects interatomic potentials*, McGraw-Hill Book Company, New York.

——, AVERBACH, B. L., and STRONG, S. L. (1964): Pair correlations in solid lead near the melting temperature, *J. Phys. Chem. Solids*, **25**: 1195–1204.

——, STRONG, S. L., and AVERBACH, B. L. (1965): Radial density functions for liquid mercury and lead, *Phys. Rev.*, **138**: A1336–A1345.

KEATING, D. T. (1963): Interpretation of the neutron or x-ray scattering from a liquid-like binary, *J. Appl. Phys.*, **34**: 923–925.

MILBERG, M. E. (1958): Transparency factor for weakly absorbing samples in x-ray diffractometry, *J. Appl. Phys.*, **29**: 64–65.

SPARKS, C. J., JR., and BORIE, B. (1966): Methods of analysis for diffuse x-ray scattering modulated by local order and atomic displacements, in J. B. COHEN and J. E. HILLIARD (eds.), *Local atomic arrangements studied by x-ray diffraction*, Gordon and Breach, Science Publishers, Inc., New York.

STRONG, S. L., and KAPLOW, R. (1968): The structure of crystalline $B_2O_3$, *Acta Cryst.*, **B24**: 1032–1036.

——, WELLS, A. F., and KAPLOW, R. (1971): On the crystal structure of $B_2O_3$, *Acta Cryst.*, **B27**: 1662–1663.

# Chapter 3

*General references for x-ray diffraction*

AMELINCKX, S., GEVERS, R., RENAUT, G., and VAN LANDUYT, J. (1970): *Modern diffraction and imaging techniques in material science*, North-Holland Publishing Company, Amsterdam.

CLARK, G. L. (1963): *Encyclopedia of x rays and gamma rays*, Reinhold Publishing Co., New York.

EWALD, P. P. (1926): Die Erforschung des Aufbaues der Materie mit Röntgenstrahlen, *Handbuch der Physik*, chap. 4, pp. 207–476, Springer Verlagsgesellschaft, Berlin.

JAMES, R. W. (1962): *The optical principles of the diffraction of x-rays*, 5th ed., G. Bell & Sons, Ltd., London.

LANDAU, L. D., and LIFSHITZ, E. M. (1959): *Electrodynamics of continuous media*, vol. 8, chaps. 14 and 15 (English translation), Academic Press, Inc., New York.

LAUE, M. VON (1960): *Röntgenstrahl-Interferenzen*, Akademische Verlag, Frankfurt.

*General references for electron diffraction*

AMELINCKX, S. (1964): The direct observation of dislocations, in F. SEITZ and D. TURNBULL (eds.), *Solid State Physics*, suppl. 6, Academic Press, Inc., New York.

HIRSCH, P. B., HOWIE, A., NICHOLSON, R. B., PASHLEY, D. W., and WHELAN, M. J. (1965): *Electron microscopy of thin crystals*, Butterworths, London.

LAUE, M. VON (1948): *Materiewellen und ihre Interferenzen*, Akademische Verlagsgesellschaft, Leipzig.

SAADA, G. (1966): *Microscope électronique des lames minces cristallines*, Masson et Cie, Paris.

*General references for optical principles*

BORN, M., and WOLF, E. (1959): *Principles of optics*, 4th ed., Pergamon Press, Ltd., London.

SOMMERFELD, A. (1954): *Optics*, in *Lectures on physics*, vol. 4, Academic Press, Inc., New York.

*Special references*

AFANAS'EV, A. M., and KAGAN, YU. (1968): The role of lattice vibrations in dynamical theory of x rays, *Acta Cryst.*, **A24**: 163–174.

CLEMMOW, P. C. (1959): A symptotic approximation to integrals, appendix 3 in BORN and WOLF (1959).

DEDERICHS, P. H. (1966): *Phys. Condens. Mater.*, **5**: 347–352.

JEFFREYS, H., and JEFFREYS, B. (1956): *Methods of mathematical physics*, p. 503, Cambridge University Press, Cambridge.

KATO, N. (1952): Dynamical theory of electron diffraction for a finite polyhedral crystal. II. Fraunhofer formula, *J. Phys. Soc. Japan*, **7**: 406–414.

——— and UYEDA, R. (1951): Dynamical theory of electron diffraction for a finite polyhedral crystal. II. Comparison with the results of kinematical theory, *Acta Cryst.*, **4**: 229–231.

——— (1953): Dynamical theory of electron diffraction for a finite polyhedral crystal. III. Fresnel diffraction formula, *J. Phys. Soc. Japan*, **8**: 350–359.

LAUE, M. VON (1936): Die äussere Form der Kristalle in ihrem Einfluss auf die Interferenzerscheinungen am Raumgitter, *Ann. Physik*, **26**: 55–68.

MOLIÈRE, G. (1939*a*): Quantenmechanische Theorie der Röntgenstrahlinterferenzen in Kristallen. I. Ableitung und allgemeine Diskussion der dynamischen Grundgleichungen, *Ann. Physik*, **35**: 272–296.

—— (1939*b*): Quantenmechanische Theorie der Röntgenstrahlinterferenzen in Kristallen. II. Dynamische Theorie der Brechung, Reflexion und Absorption von Röntgenstrahlen, *Ann. Physik*, **35**: 297–313.

—— (1939*c*): Aufbau der quantenmechanischen Dispersiontheorie im Sinne eines von M. Laue stammenden Verfahrens, *Ann. Physik*, **36**: 265–274.

OHTSUKI, Y. H. (1964): Temperature dependence of x-ray absorption by crystals. I. Photoelectric absorption, *J. Phys. Soc. Japan*, **19**: 2285–2292.

—— (1965): Temperature dependence of x-ray absorption by crystals. II. Direct phonon absorption, *J. Phys. Soc. Japan*, **20**: 374–380.

SANO, H., OHTAKA, K., and OHTSUKI, Y. (1969): Normal and abnormal absorption coefficients of x-rays, *J. Phys. Soc. Japan*, **27**: 1254–1261.

—— and YANAGAWA, S. (1966): Dynamical theory of diffraction. II. X-ray diffraction, *J. Phys. Soc. Japan*, **21**: 502–506.

SNEDDON, I. N. (1951): *Fourier transforms*, appendix B, McGraw-Hill Book Company, New York.

WAGENFELD, H. (1966): Normal and anomalous photoelectric absorption of x-rays in crystals, *Phys. Rev.*, **144**: 216–224.

## Chapter 4

*General references*
See listing for Chap. 3.

*Special references*

### Review articles

AUTHIER, A. (1970): Ewald waves in theory and experiment, in R. BRILL and R. MASON (eds.), *Advances in structure research by diffraction methods*, **3**: 1–51, Pergamon Press, Oxford.

BATTERMAN, B. W., and COLE, H. (1964): Dynamical diffraction of x rays by perfect crystals, *Rev. Mod. Phys.*, **36**: 681–717.

BORRMANN, G. (1959): Röntgenwellenfelder. Beiträge zur Physik and Chemie des 20 Jahrhunderts, Friedrich Vieweg und Sohn, Brunswick, Germany.

—— (1964): Beugung am Idealkristall, *Z. Krist.*, **120**: 143–181.

EWALD, P. P. (1962): The origin of the dynamical theory of x ray diffraction, *J. Phys. Soc. Japan*, suppl. B-II, **17**: 48–52.

—— (1965): Crystal optics for visible light and x rays, *Rev. Mod. Phys.*, **37**: 46–56.

JAMES, R. W. (1963): The dynamical theory of x-ray diffraction, *Solid State Phys.*, **15**: 55–220.

KATO, N. (1963): Wave-optical theory of diffraction in single crystals, in G. N. RAMACHANDRAN (ed.), *Crystallography and crystal perfection*, pp. 153–173, Academic Press, Inc., New York.

—— (1968): *Pendellösung* fringes in x-ray diffraction, *Acta Geologica et Geographica*, Universitatis Comenianae, Bratislava, **14**: 43–74.

SLATER, J. C. (1958): Interaction of waves in crystals, *Rev. Mod. Phys.*, **30**: 197–222.

WEISS, R. J. (1966): *X-ray determination of electron distributions*, North-Holland Publishing Company, Amsterdam.

## Plane-wave theory

AFANAS'EV, A. M., and PERSTNEV, I. P. (1969): On the Bragg reflection from ideal absorbing crystals, *Acta Cryst.*, **A25**: 520–523.

ARMSTRONG, E. (1946): X-ray studies of surface layers of crystals, *Bell Syst. Tech. J.*, **25**: 136–155.

AUTHIER, A. (1960a): Mise en évidence experimentale de la double réfraction des rayons X, *Compt. Rend. Acad. Sci. Paris*, **251**: 2003–2005.

—— (1960b): Obtension de profiles intrinséques de raies de diffraction des rayons par réflection et par transmission, *Compt. Rend. Acad. Sci. Paris*, **251**: 2502–2504.

—— (1961): Étude de la transmission anomale des rayons x dans des cristaux de silicium, *Bull. Soc. Franc. Minéral. Crist.*, **84**: 51–89.

—— (1962): Trajet des rayons x dans un cristal parfait au voisinage de la reflexion totale, *J. Phys. Rad.*, **23**: 961–969.

BATTERMAN, B. W. (1962): Effect of thermal vibrations on diffraction from perfect crystals. I. The case of anomalous transmission, *Phys. Rev.*, **126**: 1461–1469.

—— and HILDEBRANDT, G. (1967): Observation of x-ray *Pendellösung* fringes in Darwin reflection, *Phys. Stat. Solidi*, **23K**: 147–149.

—— and —— (1968): X-ray *Pendellösung* fringes in Darwin reflection, *Acta Cryst.*, **A24**: 150–157.

BETHE, H. (1928): Theorie der Beugung von Elektronen an Kristallen, *Ann. Physik*, **87**: 55–129.

BONSE, U., and HART, M. (1965): Tailless x ray single-crystal reflection curves obtained by multiple reflection, *Appl. Phys. Letters*, **7**: 238–240.

—— and —— (1966): Small angle x-ray scattering by spherical particles of polystyrene and polyvinyltoluene, *Z. Physik*, **189**: 151–162.

BORRMANN, G. (1941): Über Extinktion der Röntgenstrahlen von Quarz, *Physik Z.*, **42**: 157–162.

—— (1950): Die Absorption der Röntgenstrahlen im Fall der Interferenz, *Z. Physik*, **127**: 297–323.

—— (1951): Die übernormal durchdringende Röntgenstrahlung im Bragg-Fall der Interferenz, *Naturwiss.*, **38**: 330.

—— and HARTWIG, W. (1965): Die Absorption der Röntgenstrahlen im Dreistrahlfall der Interferenz, *Z. Krist.*, **121**: 401–409.

——, HILDEBRANDT, G., and WAGNER, H. (1955): Röntgenstrahl-Fächer im Kalkspat, *Z. Physik*, **142**: 406–414.

BORRMANN, G., and WAGNER, E. H. (1955): Die Absorption interferierender Röntgenstrahlen längs ihrer Wege im dealen Kristallgitter, *Naturwiss.*, **42**: 68.

BROGREN, G., and ADELL, Ö. (1954): The anomalous x-ray transmission in calcite, *Arkiv Fysik*, **8**: 97–112.

BUBÁKOVÁ, R. (1962): The diffraction pattern of Ge (111)—Asymmetrical Bragg case, *Czech. J. Phys.*, **B12**: 776–783.

———, DRAHOKOUPIL, J., and FINGERLAND, A. (1961a): Single crystal diffraction patterns of germanium. Part I., *Czech. J. Phys.*, **B11**: 199–204.

———, ———, and ——— (1961b): A contribution to the theory of the triple crystal diffractometer, *Czech. J. Phys.*, **B11**: 205–222.

———, ———, and ——— (1962a): Single crystal diffraction patterns of germanium. Part II., *Czech. J. Phys.*, **B12**: 538–541.

———, ———, and ——— (1962b): Single crystal diffraction patterns of silicon. Part III., *Czech. J. Phys.*, **B12**: 764–775.

BUCKSCH, R., OTTO, J., and RENNINGER, M. (1967): Die 'Diffraction Pattern' des Idealkristalls für Röntgenstrahlinterferenzen im Bragg Fall, *Acta Cryst.*, **23**: 507–511.

CAMPBELL, H. N. (1951a): X-ray absorption in a crystal set at the Bragg angle, *Acta Cryst.*, **4**: 180–181.

——— (1951b): X-ray absorption in a crystal set at the Bragg angle, *J. Appl. Phys.*, **22**: 1139–1142.

COLE, H., and STEMPLE, N. R. (1962): Effect of crystal perfection and polarity on absorption edges seen in Bragg diffraction, *J. Appl. Phys.*, **33**: 2227–2233.

CORK, J. M. (1932): Laue patterns from thick crystals at rest and oscillating piezoelectrically, *Phys. Rev.*, **42**: 749–752.

COWLEY, J. M., and REES, A. L. G. (1946): Refraction effects in electron diffraction, *Nature*, **158**: 550–552.

——— and ——— (1947): Refraction effects in electron diffraction, *Proc. Phys. Soc. London*, **59**: 287–302.

DARWIN, C. G. (1914a): The theory of x-ray reflection, *Phil. Mag.*, **27**: 315–333 (Kinematical theory).

——— (1914b): The theory of x-ray reflection, Part II, *Phil. Mag.*, **27**: 675–690 (Dynamical theory).

——— (1922): The reflection of x rays from imperfect crystals, *Phil. Mag.*, **43**: 800–829.

DUMOND, J. W. M., and BOLLMANN, V. L. (1936): New and unexplained effects in Laue x-ray reflection in calcite, *Phys. Rev.*, **50**: 97.

EWALD, P. P. (1916a): Zur Begründung der Kristalloptik. I. Theorie der Dispersion, *Ann. Physik*, **49**: 1–38.

——— (1916b): Zur Begründung der Kristalloptik. II. Theorie der Reflexion und Brechung, *Ann. Physik*, **49**: 117–143.

——— (1917): Zur Begründung der Kristalloptik. III. Röntgenstrahlen, *Ann. Physik*, **54**: 519–597.

——— (1937): Zur Begründung der Kristalloptik. IV. Aufstellung einer allgemeinen Dispersions bedingung, inbesondere für Röntgenfelder, *Z. Krist.*, **97**: 1–27.

—— (1958): Group velocity and phase velocity in x-ray crystal optics, *Acta Cryst.*, **11**: 888–891.

—— and HÉNO, Y. (1968): X-ray diffraction in the case of three strong rays. I. Crystal composed of non-absorbing point atoms, *Acta Cryst.*, **A24**: 5–15.

FINGERLAND, A. (1962): On practical calculation with reflection curves for perfect crystals, *Czech. J. Phys.*, **12**: 264–277.

—— (1971): Some properties of the single-crystal rocking curve in the Bragg case, *Acta Cryst.*, **27A**: 280–284.

—— and DRAHOKOUPIL, J. (1970): A comment on the paper: Coherent crystal radiation affects the measurement of the x-ray linewidth (by Das Gupta and Welch, 1968), *Acta Cryst.*, **A26**: 569–571.

HALL, C. R., and HIRSCH, P. B. (1965): Effect of thermal diffuse scattering on propagation of high energy electrons through crystals, *Proc. Roy. Soc.*, **A286**: 158–177.

HAMILTON, W. C. (1957): The effect of crystal shape and setting on secondary extinction, *Acta Cryst.*, **10**: 629–634.

HART, M., and MILNE, A. D. (1968): Direct observation of plane wave and spherical wave *Pendellösung* fringes, *Phys. Stat. Solidi*, **26**: 185–189.

HASHIZUME, H., NAKAYAMA, K., MATSUSHITA, T., and KOHRA, K. (1970): Variation of Bragg case diffraction curves of x rays from a thin silicon crystal with crystal thickness, *J. Phys. Soc. Japan*, **29**: 806.

HEIDENREICH, R. D. (1942): Electron reflections in MgO crystals with the electron microscope, *Phys. Rev.*, **62**: 291–292.

—— and STURKEY, L. (1945): Crystal interference phenomena in electron microscope images, *J. Appl. Phys.*, **16**: 97–105.

HÉNO, Y., and EWALD, P. P. (1968): Diffraction des rayons x dans ler ca de trois rayons forts. II. Influence de l'absorption et du facteur de diffusion atomique, *Acta Cryst.*, **A24**: 16–42.

HILDEBRANDT, G. (1959*a*): Gekrümmte Röntgenstrahlen im schwach verformten Kristallgitter. A. Laue Fall der Interferenz, *Z. Krist.*, **112**: 312–339.

—— (1959*b*): Gekrümmte Röntgenstrahlen im schwach verformten Kristallgitter. B. Bragg-Fall der interferenz, *Z. Krist.*, **112**: 340–361.

—— (1966): Die Absorption von Röntgenstrahlen im Germanium in einem Dreistrahlfall der Interferenz, *Phys. Stat. Solidi*, **15**: K131–134.

HIRSCH, P. B., and RAMACHANDRAN, G. N. (1950): Intensity of x-ray reflection from perfect and mosaic absorbing crystals, *Acta Cryst.*, **3**: 187–194.

—— and HOWIE, A. (1947): Refraction effects in electron diffraction, *Proc. Phys. Soc.* (London), **71**: 287–302.

——, HOWIE, A., NICHOLSON, R. B., PASHLEY, D. W., and WHELAN, M. J. (1965): *Electron microscopy of thin crystals*, Butterworths, London.

HONJO, G., and MIHAMA, K. (1954): Fine structure due to refraction effect in electron diffraction pattern of powder sample. Part II. Multiple structures due to double refraction given by randomly oriented smoke particle of magnesium and cadmium oxide, *J. Phys. Soc. Japan*, **9**: 184–198.

HUNTER, L. P. (1959): X-ray measurement of microstrains in germanium single crystals, *J. Appl. Phys.*, **30**: 874–884.

KATO, N. (1949): Refractive index of electron in crystal medium, *Proc. Japan Acad.*, **25**: 41–44.

——— (1952*a*): Dynamical theory of electron diffraction for a finite polyhedral crystal. I. Extension of Bethe's theory, *J. Phys. Soc. Japan*, **7**: 397–406.

——— (1952*b*): Dynamical theory of electron diffraction for a finite polyhedral crystal. II. Fraunhofer formula, *J. Phys. Soc. Japan*, **7**: 406–414.

——— (1953): Dynamical theory of electron diffraction for a finite polyhedral crystal. III. Fresnel diffraction formula, *J. Phys. Soc. Japan*, **8**: 350–359.

——— (1955): Integrated intensities of the diffracted and transmitted x rays due to ideally perfect crystal (Laue case), *J. Phys. Soc. Japan*, **10**: 46–55.

——— (1958): The flow of x rays and material waves in ideally perfect single crystals, *Acta Cryst.*, **11**: 885–887.

——— (1960): The energy flow of x rays in an ideally perfect crystal: Comparison between theory and experiment, *Acta Cryst.*, **13**: 349–356.

——— (1964): *Pendellösung* fringes in distorted crystals. II. Application to two-beam cases, *J. Phys. Soc. Japan*, **19**: 67–77.

———, KATAGAWA, T., and SAKA, T. (1971): A few remarks on the Bragg cases in the dynamical theory of crystal diffraction, *Krystallographia*, **16**: 1110–1116 (in Russian).

KIKUTA, S. (1971): X-ray crystal collimators using successive asymmetric diffractions and their applications to measurements of diffraction curves. II. Type I collimator, *J. Phys. Soc. Japan*, **30**: 222–227.

——— and KOHRA, K. (1968): Variation with thickness in the profile of Laue-case diffraction curve of x rays from a thin Si crystal, *J. Phys. Soc. Japan*, **25**: 924.

——— and ——— (1970): X-ray crystal collimators using successive asymmetric diffractions and their applications to measurements of diffraction curves. I. General considerations on collimators, *J. Phys. Soc. Japan*, **29**: 1322–1328.

———, KAWASHIMA, K., and KOHRA, K. (1970): Weak extra peak in rocking curves of x-ray reflection for incident beams obtained by successive asymmetric reflections, *Acta Cryst.*, **A26**: 694–696.

KINDER, E. (1943): Magnesiumoxydkristalle im electronenmikroskop, *Naturwiss.*, **31**: 149.

KNOWLES, J. W. (1956): Anomalous absorption of slow neutrons and x rays in nearly perfect single crystals, *Acta Cryst.*, **9**: 61–69.

KOHRA, K. (1962): An application of asymmetric reflection for obtaining x-ray beams of extremely narrow angular spread, *J. Phys. Soc. Japan*, **17**: 589–590.

——— and KIKUTA, S. (1968): A method of obtaining an extremely parallel x-ray beam by successive asymmetric diffractions and its applications, *Acta Cryst.*, **A24**: 200–205.

KOSSEL, W. (1943): Linien gleicher Dicke im Elektronenmikroskop, *Naturwiss.*, **31**: 323–324.

KURIYAMA, M., and MIYAKAWA, T. (1970): Primary and secondary extinctions in the dynamical theory for an imperfect crystal, *Acta Cryst.*, **A26**: 667–673.

LANG, A. R., and HART, M. (1961): Direct determination of x-ray reflection phase relationships through simultaneous reflection, *Phys. Rev. Letters*, **4**: 120–121.

LAUE, M. VON (1931): Die dynamische Theorie der Röntgenstrahlinterferenzen in neuer Form, *Ergeb. Exakt. Naturwiss.*, **10**: 133–158.

────── (1940): Interferenz-doppelbrechung von Röntgenstrahlen in Kristallprismen, *Naturwiss.*, **28**: 645–646.

────── (1949): Die Absorption der Röntgenstrahlen in Kristallen im Interferenzfall, *Acta Cryst.*, **2**: 106–113.

────── (1952): Die Energieströmung bei Röntgenstrahlinterferenzen in Kristallen, *Acta Cryst.*, **5**: 619–625.

────── (1953): Der Teilchenstrom bei Raumgitterinterferenzen von Materiewellen, *Acta Cryst.*, **6**: 217.

LEHFELD-SOSNOWSKA, M., and MALGRANGE, C. (1968): Observation of oscillations in rocking curves of the Laue reflected and refracted beams from thin Si single crystals, *Phys. Stat. Solidi*, **30**: K23–25.

LING, D., and WAGENFELD, H. (1965): Anomalous transmission of x rays in perfect single germanium crystals at liquid nitrogen temperature, *Phys. Letters*, **15**: 8–10.

MALGRANGE, C., and AUTHIER, A. (1965): Interferences entre les champs d'ondes créés par double réfraction des rayons x, *Compt. Rend. Acad. Sci. Paris*, **261**: 3774–3777.

MATSUSHITA, T., KIKUTA, S., and KOHRA, K. (1971): X-ray crystal collimators using successive asymmetric diffractions and their applications. III. Type II collimator, *J. Phys. Soc. Japan*, **30**: 1136–1144.

MENTER, J. W. (1956): The direct study by electron microscopy of crystal lattices and their imperfections, *Proc. Roy. Soc.*, **A236**: 119–135.

MOLIÈRE, K., and NIEHRS, H. (1954): Interferenzbrechung von Elektronenstrahlen. I. Zur Theorie der Elektroneninterferenzen an parallelepipedischen Kristallen, *Zeit. f. Physik*, **137**: 445–462.

NAKAYAMA, K., HASHIZUME, H., and KOHRA, K. (1970): Equal-thickness interference fringes in the Bragg-case diffraction of x rays, *J. Phys. Soc. Japan*, **30**: 897.

NIEHRS, H. (1954): Das Strahlungsfeld auf der Kristallrückseite bei Elektroneninterferenzen, *Zeit. f. Physik*, **138**: 570–597.

OKKERSE, B. (1962): Anomalous transmission of x rays in germanium. Part I. *Philips Res. Rept.*, **17**: 464–478.

PENNING, P. (1967): Dynamical theory for simultaneous x-ray diffraction, *Advan. X-Ray Anal.*, **10**: 67–79.

────── and POLDER, D. (1968a): Dynamical theory for simultaneous x-ray diffraction. Part I. Theorems concerning the *n*-beam case, *Philips Res. Rept.*, **23**: 1–11.

────── and ────── (1968b): Dynamical theory for simultaneous x-ray diffraction. Part II. Application to the three-beam case, *Philips Res. Rept.*, **23**: 12–24.

PRINS, J. A. (1930): Die Reflexion von Röntgenstrahlen an absorbierenden idealen Kristallen, *Z. Physik*, **63**: 477–493.

RAMACHANDRAN, G. N. (1954): X-ray anti-reflections in crystals. Part II. Calculation of the integrated reflection and integrated anti-reflection for an internal reflection, *Proc. Indian Acad. Sci.*, **A39**: 65–80.

RENNINGER, M. (1937): "Umweganregung," eine bisher unbeachtete Wechselwirkungserscheinung bei Raumgitterinterferenzen, *Z. Physik*, **106**: 141–176.

────── (1955): Messungen zur Röntgenstrahl-Optik des Idealkristalls. I. Bestätigung der Darwin-Ewald-Prins-Kohler-Kurve, *Acta Cryst.*, **8**: 597–606.

RENNINGER, M. (1961): Asymmetrische Bragg-Reflexion am Idealkristall zur Erhöhung des Doppelspektrometer-Auflösungvermögens, *Z. Naturforsch.*, **169**: 1110–1111.

——— (1967): The asymmetric Bragg reflection and its application in double diffractometry, *Advan. X-Ray Anal.*, **10**: 32–41.

——— (1968): Messungen zur Röntgenstrahl-Optik des Idealkristalls. II. Diffraction Pattern mit Pendellösung gleicher Neigung, *Acta Cryst.*, **A24**: 143–149.

SAKA, T., KATAGAWA, T., and KATO, N. (1972a): The theory of x-ray crystal diffraction for finite polyhedral crystals. I. The Laue-Bragg cases, *Acta Cryst.*, **A28**: 102–112.

———, ———, and ——— (1972b): The theory of x-ray crystal diffraction for finite polyhedral crystals. II. The Laue-(Bragg) cases, *Acta Cryst.*, **A28**: 113–119.

SCHWARZ, G., and ROGOSA, G. L. (1954): Transmission of x rays through calcite in Laue diffraction, *Phys. Rev.*, **95**: 950–953.

STURKEY, L. (1948): Index of refraction for electrons in crystalline media, *Phys. Rev.*, **73**: 183.

VINEYARD, G. H. (1954): Multiple scattering of neutrons, *Phys. Rev.*, **96**: 93–98.

WAGENFELD, H. (1962): Comments on anomalous absorption of x rays, *J. Appl. Phys.*, **33**: 2907–2908.

WAGNER, E. H. (1959a): Group velocity and energy (or particle) flow density of waves in a periodic medium, *Acta Cryst.*, **12**: 345–346.

——— (1959b): Über Gruppengeschwindigkeit, Energiestromdichte und Energiedichte in der Röntgen bzw. Lichtoptik der Kristalle, *Z. Physik*, **154**: 352–360.

WAGNER, H. (1956): Röntgenstrahlung anomaler Schwächung im Falle Braggscher Reflexion, *Z. Physik*, **146**: 127–168.

WERNER, S. A., and ARROTT, A. (1966): Propagation of Bragg-reflected neutrons in large mosaic crystals and the efficiency of monochrometers, *Phys. Rev.*, **140**: 675–686.

———, ———, KING, J. S., and KENDRICK, H. (1966): Propagation of Bragg-reflected neutrons in bounded mosaic crystals, *J. Appl. Phys.*, **37**: 2343–2350.

WHELAN, M. J. (1965): Inelastic scattering of fast electrons by crystals. II. Phonon scattering, *J. Appl. Phys.*, **36**: 2103–2110.

YOSHIOKA, H. (1954): On the anomalous absorption of x rays by crystals at the setting of Bragg reflection, *J. Phys. Soc. Japan*, **9**: 636–640.

ZACHARIASEN, W. H. (1945): *Theory of x-ray diffraction in crystals*, John Wiley & Sons, Inc., New York.

——— (1952): On the anomalous transparency of thick crystals to x rays, *Proc. Natl. Acad. Sci.*, **38**: 378–382.

——— (1967a): A general theory of x-ray diffraction in real crystals, *Phys. Rev. Letters*, **18**: 195–196.

——— (1967b): A general theory of x-ray diffraction in crystals, *Acta Cryst.*, **A23**: 558–564.

——— (1968): Experimental tests of the general formula for the integrated intensity of a real crystal, *Acta Cryst.*, **A24**: 212–216.

*Spherical-wave theory*

AUSTERMAN, S. B., and NEWKIRK, J. B. (1967): Experimental procedures in x ray diffraction topography, *Advan. X-Ray Anal.*, **10**: 134–152.

AZÁROFF, L. V. (1964): X ray diffraction studies of crystal perfection, *Progr. Solid State Chem.*, **1**: 347–379.

BARRETT, C. S. (1945): New microscopy and its potentialities, *Trans. AIME*, **161**: 15–64.

BONSE, U., and HART, M. (1965): An x-ray interferometer, *Appl. Phys. Letters*, **6**: 155.

——— and ——— (1966): Principles and design of Laue-case x-ray interferometers, *Z. Physik*, **188**: 154–164.

———, ———, and NEWKIRK, J. B. (1967): X-ray diffraction topography, *Advan. X-Ray Anal.*, **10**: 1–8.

BORRMANN, G., and LEHMANN, K. (1963): Some problems of x-ray optics: Partial reflection and superposition of wave fields, in G. N. RAMACHANDRAN (ed.), *Crystallography and crystal perfection*, pp. 101–108, Academic Press, London.

DEMARCO, J. J., and WEISS, R. J. (1964): An x-ray search for ionicity in GaAs, *Phys. Letters*, **13**: 209–212.

——— and ——— (1965a): Absolute x-ray scattering factors of silicon and germanium, *Phys. Rev.*, **137A**: 1869–1870.

——— and ——— (1965b): The integrated intensities of perfect crystals, *Acta Cryst.*, **19**: 68–72.

EWALD, P. P. (1925): Die Intensitäten der Röntgenreflexe und der Strukturfaktor, *Phys. Zeit.*, **26**: 29–32.

HART, M., and LANG, A. R. (1965): The influence of x-ray polarisation on the visibility of *Pendellösung* fringes in x-ray diffraction topographs, *Acta Cryst.*, **19**: 73–77.

——— and MILNE, A. D. (1968): Direct observation of plane wave and spherical wave *Pendellösung* fringes, *Phys. Stat. Solidi*, **26**: 185–189.

——— and ——— (1969): An accurate absolute scattering factor for silicon, *Acta Cryst.*, **A25**: 134–138.

——— and ——— (1970): Absolute measurement of structure factors using a new dynamical interference effect, *Acta Cryst.*, **A26**: 223–229.

HATTORI, H., and KATO, N. (1966): An experimental study on the form of x-ray dispersion surface by means of *Pendellösung* fringes, *J. Phys. Soc. Japan*, **21**: 1772–1775.

———, KURIYAMA, H., and KATO, N. (1965): Effects of x-ray polarization on *Pendellösung* fringes, *J. Phys. Soc. Japan*, **20**: 1047–1050.

———, ———, KATAGAWA, T., and KATO, N. (1965): Absolute measurement of structure factors of Si single crystal by means of x-ray *Pendellösung* fringes, *J. Phys. Soc. Japan*, **20**: 988–996.

HOERNI, J. (1950): Diffraction des électrons par le graphite, *Helv. Phys. Acta*, **23**: 587–622.

HOMMA, S., ANDO, Y., and KATO, N. (1966): Absolute positions of *Pendellösung* fringes in x-ray cases, *J. Phys. Soc. Japan*, **21**: 1160–1165.

KAMBE, K. (1957a): Study of simultaneous reflection in electron diffraction by crystals. I. Theoretical treatment, *J. Phys. Soc. Japan*, **12**: 13–25.

——— (1957b): Study of simultaneous reflection in electron diffraction by crystals. II. Experimental confirmation, *J. Phys. Soc. Japan*, **12**: 25–31.

KATO, N. (1960): Dynamical x-ray diffraction theory of spherical waves, *Z. Naturforsch.*, **15a**: 369–370.

——— (1961a): A theoretical study of *Pendellösung* fringes. I. General considerations, *Acta Cryst.*, **14**: 526–532.

KATO, N. (1961b): A theoretical study of *Pendellösung* fringes. II. Detailed discussion based upon a spherical-wave theory, *Acta Cryst.*, **14**: 627–636.

——— (1968a): Reciprocity theorem in optics and its application to x-ray diffraction topographs, *Acta Cryst.*, **A24**: 157–160.

——— (1968b): Spherical-wave theory of dynamical x-ray diffraction for absorbing perfect crystal. I. The crystal wave fields, *J. Appl. Phys.*, **39**: 2225–2230.

——— (1968c): Spherical-wave theory of dynamical x-ray diffraction for absorbing perfect crystal. II. Integrated reflection power, *J. Appl. Phys.*, **39**: 2231–2237.

——— (1969): The determination of structure factors by means of *Pendellösung* fringes, *Acta Cryst.*, **A25**: 119–128.

——— and LANG, A. R. (1959): A study of *Pendellösung* fringes in x-ray diffraction, *Acta Cryst.*, **12**: 787–794.

——— and TANEMURA, S. (1967): Absolute measurement of structure factor with high precision, *Phys. Rev. Letters*, **19**: 22–24.

KIKUTA, S., MATSUSHITA, T., and KOHRA, K. (1970): A new method of determining structure factors of x rays using half-valued width of diffraction curves from perfect crystal, *Phys. Letters*, **33A**: 151–152.

KOSSEL, W. (1943): Linien gleicher Dicke im Elektronenmikroskop, *Naturwiss.*, **31**: 323–324.

KURIKI-SUONIO, K. (1970): Charge density. Experimental, *Report Series in Physics, Univ. Helsinki*, no. 11.

LANG, A. R. (1957): A method for the examination of crystal sections using penetrating characteristic x radiation, *Acta Met.*, **5**: 358–364.

——— (1958): Direct observation of individual dislocations by x-ray diffraction, *J. Appl. Phys.*, **29**: 597.

——— (1959): The projection topograph. A new method in x-ray diffraction micrography, *Acta Cryst.*, **12**: 249–250.

——— (1963): Topography, x-ray diffraction, pp. 1053–1058, and Topography, x-ray diffraction in dislocation studies, pp. 1058–1063, in G. L. CLARK (ed.), *Encyclopedia of x-rays and gamma rays*, Reinhold Publishing Co., New York.

——— (1970): Recent application of x-ray topography, in S. AMELINCKX et al. (eds.), *Modern diffraction and imaging techniques in material science*, pp. 407–479, North-Holland Publishing Company, Amsterdam.

LAUE, M. VON (1935a): Der optische Reziprozitätsatz in Anwendung auf die Röntgenstrahl-interferenzen, *Naturwiss.*, **23**: 373.

——— (1935b): Die Fluoresenzröntgenstrahlung von Einkristallen, *Ann. Physik*, **23**: 705–746.

——— (1960): *Röntgenstrahl Interferenzen*, Akademische Verlag, Frankfurt.

LEHMANN, K., and BORRMANN, G. (1967): Zur Umlenkung und Überlagerung von Röntgenwellenfeldern, *Z. Krist.*, **125**: 234–248.

LEHMPFUHL, G., and REISSLAND, A. (1968): Photographical record of the dispersion surface in rotating crystal electron diffraction pattern, *Z. Naturforsch.*, **23a**: 544–549.

LORENTZ, H. A. (1905): On the radiation of heat in a system of bodies having a uniform temperature, *Proc. Acad. Amsterdam*, 401–425.

SAKA, T., KATAGAWA, T., and KATO, N. (1972a): The theory of x-ray crystal diffraction for finite polyhedral crystal. I. The Laue-Bragg cases, *Acta Cryst.*, **A27**: 102–112.

———, ———, and ——— (1972b): The theory of x-ray crystal diffraction for finite polyhedral crystal. II. The Laue-Bragg$^m$ cases, *Acta Cryst.*, **A27**: 113–119.

———, ———, and ——— (1973): The theory of x-ray crystal for finite polyhedral crystal. III. The Bragg cases, *Acta Cryst.*, **A28**: 192–200.

SCHIFF, L. I. (1948): *Quantum mechanics*, p. 251, McGraw-Hill Book Company, New York.

TAKAGI, S. (1969): A dynamical theory of diffraction for a distorted crystal, *J. Phys. Soc. Japan*, **27**: 1239–1253.

TANEMURA, S., and KATO, N. (1972): Absolute measurement of structure factors of Si by using *Pendellösung* and interferometry fringes, *Acta Cryst.*, **A28**: 69–80.

URAGAMI, T. S. (1969): *Pendellösung* fringes in Bragg case, *J. Phys. Soc. Japan*, **27**: 147–154.

——— (1970): *Pendellösung* fringes in Bragg case in a crystal of finite thickness, *J. Phys. Soc. Japan*, **28**: 1508–1527.

——— (1971): *Pendellösung* fringes in a finite crystal, *J. Phys. Soc. Japan*, **31**: 1141–1161.

WEISS, R. J. (1966): A measurement of the dispersion correction, *Acta Cryst.*, **20**: 457–458.

## Chapter 5

*General references*
See listing for Chap. 3.


*Special references*

ANDO, Y., and KATO, N. (1966): X-ray diffraction topographs of an elastically distorted crystal, *Acta Cryst.*, **21**: 284–285.

——— and ——— (1970): X-ray diffraction phenomena in elastically distorted crystals, *J. Appl. Cryst.*, **3**: 74–89.

AUTHIER, A. (1968): Contrast of a stacking fault on x-ray topographs, *Phys. Stat. Solidi*, **27**: 77–93.

——— and LANG, A. R. (1964): Three-dimensional x-ray topographic studies of internal dislocation sources in silicon, *J. Appl. Phys.*, **35**: 1956–1959.

——— and SAUVAGE, M. (1966): Dislocations de macl dans la calcite: interferences entre les champs d'onde créés a la traversée d'une lamelle de macl, *J. Phys. Radium*, **27**: 137–142.

———, MILNE, D., and SAUVAGE, M. (1968): X-ray dynamical contrast of a planar defect, *Phys. Stat. Solidi*, **26**: 469–484.

BATTERMAN, B. W. (1961): Imaginary part of x-ray scattering factor for germanium. Comparison of theory and experiment, *J. Appl. Phys.*, **32**: 998–1001.

——— (1962a): Effect of thermal vibration on diffraction from perfect crystals. I. The case of anomalous transmission, *Phys. Rev.*, **126**: 1461–1469.

——— (1962b): Effect of thermal vibration on diffraction from perfect crystals. II. The Bragg case of reflection, *Phys. Rev.*, **127**: 686–690.

BATTERMAN, B. W., and CHIPMAN, D. R. (1962): Vibration amplitudes in germanium and silicon, *Phys. Rev.*, **127**: 690–693.

BLECH, I. A., and MEIERAN, E. S. (1967): Enhanced x-ray diffraction from substrate crystals containing discontinuous surface films, *J. Appl. Phys.*, **38**: 2913–2919.

BONSE, U. (1964*a*): Theorie der Ausbreitung von Röntgen-Wellenfeldstrahlen im schwach deformierten Kristallgitter, *Z. Physik*, **177**: 385–423.

—— (1964*b*): Starke Ablenkung von Röntgen-Wellenfeldstrahlen im elastisch gebogenen Kristallen, *Z. Physik*, **177**: 529–542.

—— (1964*c*): Zum Kontrast an Versetzungen im Röntgenfeld, *Z. Physik*, **177**: 543–561.

—— and HART, M. (1965*a*): An x-ray interferometer, *Appl. Phys. Letters*, **6**: 155–156.

—— and —— (1965*b*): An x-ray interferometer with long separated interfering beam paths, *Appl. Phys. Letters*, **7**: 99–100.

—— and —— (1965*c*): Principles and design of Laue-case x-ray interferometers, *Z. Physik*, **188**: 154–164.

—— and —— (1966*a*): Moiré patterns of atomic planes obtained by x-ray interferometry, *Z. Physik*, **190**: 455–467.

—— and —— (1966*b*): An x-ray interferometer with Bragg case beam splitting and beam recombination, *Z. Physik*, **194**: 1–17.

—— and —— (1968): Combined Laue- and Bragg-case x-ray interferometer, *Acta Cryst.*, **A24**: 240–245.

—— and TE KAAT, E. (1971): The defocussed x-ray interferometer, *Z. Physik*, **243**: 14–45.

BORRMANN, G., and HILDEBRANDT, G. (1959): Absorption und Weg interferierender Röntgenstrahlen im schwach deformierten Kristallgitter, *Z. Physik*, **159**: 189–199.

BRADLER, J., and LANG, A. R. (1968): Use of the Ewald sphere in aligning crystal pairs to produce x-ray Moiré fringes, *Acta Cryst.*, **A24**: 246–247.

CHIKAWA, J. (1965): X-ray Moiré measurement of small lattice variations due to doped impurity in CdS, in S. PEISER (ed.), *Crystal growth*, pp. 817–823, Pergamon Press, Oxford.

—— and AUSTERMAN, S. B. (1968): X-ray diffraction contrast of inversion twin boundaries in BeO crystals, *J. Appl. Cryst.*, **1**: 165–171.

COLE, H., and STEMPLE, N. R. (1962): Effect of crystal perfection and polarity on absorption edges seen in Bragg diffraction, *J. Appl. Phys.*, **33**: 2227–2233.

COSTER, D., KNOL, K. S., and PRINS, J. A. (1930): Unterschiede in der Intensität der Röntgenstrahlen-reflexion an den beiden 111-Flächen der Zinkblende, *Z. Physik*, **63**: 345–369.

DESLATTES, R. D. (1969): Optical and x-ray interferometry of silicon lattice spacing, *Appl. Phys. Letters*, **15**: 386–388.

FRANK, F. C., LAWN, B. R., LANG, A. R., and WILKS, E. M. (1967): A study of strains in abraded diamond surfaces, *Proc. Roy. Soc. London*, **A301**: 239–252.

FUKUSHIMA, E. (1954): A boundary effect on the intensity of x-rays reflected from a quartz plate, *Acta Cryst.*, **7**: 459–460.

HART, M. (1966): *Pendellösung* fringes in elastically deformed silicon, *Z. Physik*, **189**: 269–291.

HARUTA, K., and SPENCER, W. J. (1966): Strain in thin metal films on quartz, *J. Appl. Phys.*, **37**: 2232–2233.

HASHIMOTO, H., HOWIE, A., and WHELAN, M. J. (1960): Anomalous electron absorption effects in metal films, *Phil. Mag.*, **5**: 967–974.

HASHIZUME, H., and KOHRA, K. (1970): Propagation of x-rays in a slightly distorted silicon crystal, *J. Phys. Soc. Japan*, **29**: 805.

—— and —— (1971): Some studies on x-ray wave fields in elastically distorted single crystals. I. Experimental observations, *J. Phys. Soc. Japan*, **31**: 204–216.

HILDEBRANDT, G. (1959a): Gekrümmte Röntgenstrahlen im schwach verformten Kristall-gitter. A. Laue-Fall der Interferenz, *Z. Krist.*, **112**: 312–339.

—— (1959b): Gekrümmte Röntgenstrahlen im schwach verformten Kristallgitter. B. Bragg-Fall der Interferenz, *Z. Krist.*, **112**: 340–361.

IKENO, S., MARUYAMA, M., and KATO, N. (1968): X-ray topographic studies of NaCl crystals grown from aqueous solution with Mn ions, *J. Crystal Growth*, **3–4**: 683–693.

KAMBE, K. (1965a): Dynamische Theorie der Röntgen-Strahl-Interferenzen an schwach verzerrten Kristallgittern. I. Herleitung einer strahlenoptischen Näherung aus Maxwellschen Gleichungen, *Z. Naturforsch.*, **20a**: 770–786.

—— (1965b): Der Debye-Waller Faktor in der dynamischen Theorie von Röntgen und Elektroneninterferenzen an Kristallen, *Z. Naturforsch.*, **20a**: 1730–1732.

—— (1968): Dynamische Theorie der Röntgenstrahl Interferenzen an schwach verzerrten Kristallgitter. II. Strahlenoptik von Bloch-Wellen im allgemeinen Fall und im Zweistrahlfall, *Z. Naturforsch.*, **23a**: 25–43.

KATO, N. (1952a): Dynamical theory of electron diffraction for a finite polyhedral crystal. I. Extension of Bethe's theory, *J. Phys. Soc. Japan*, **7**: 397–406.

—— (1952b): Dynamical theory of electron diffraction for a finite polyhedral crystal. II. Fraunhofer formula, *J. Phys. Soc. Japan*, **7**: 406–414.

—— (1953): Dynamical theory of electron diffraction for a finite polyhedral crystal. III. Fresnel diffraction formula, *J. Phys. Soc. Japan*, **8**: 350–359.

—— (1956): Anomalous enhancement of x-ray reflection intensity at the boundary of ground and etched regions on crystal surface, *J. Phys. Soc. Japan*, **11**: 748–754.

—— (1963a): Dynamical diffraction theory of waves in distorted crystals. I. General formulation and treatment for perfect crystals, *Acta Cryst.*, **16**: 276–281.

—— (1963b): Dynamical diffraction theory of waves in distorted crystals. II. Perturbation theory, *Acta Cryst.*, **16**: 282–290.

—— (1963c): *Pendellösung* fringes in distorted crystals. I. Fermat's principle of Bloch waves, *J. Phys. Soc. Japan*, **18**: 1785–1791.

—— (1964a): *Pendellösung* fringes in distorted crystals. II. Application to two beam cases, *J. Phys. Soc. Japan*, **19**: 67–77.

—— (1964b): *Pendellösung* fringes in distorted crystals. III. Application to homo-geneously bent crystals, *J. Phys. Soc. Japan*, **19**: 971–985.

—— (1965): Dynamical diffraction theory for highly distorted crystals, *Proc. Intern. Conf. Melbourne*, **1A**: 3–4.

—— and ANDO, Y. (1966): Contraction of *Pendellösung* fringes in distorted crystals, *J. Phys. Soc. Japan*, **21**: 964–968.

—— and PATEL, J. R. (1968): A computer calculation of x-ray diffraction topographs for distorted crystals, *Appl. Phys. Letters*, **13**: 42–44.

KATO, N., and PATEL, J. R. (1973): X-ray diffraction topographs of silicon crystals with super-posed oxide film. I. Theory and computational procedures, *J. Appl. Phys.*, **44**: 965–970.

—— and TANEMURA, S. (1967): Absolute measurements of structure factor with high precision, *Phys. Rev. Letters*, **19**: 22–24.

——, USAMI, K., and KATAGAWA, T. (1967): The x-ray diffraction image of a stacking fault, *Advan. X-Ray Anal.*, **10**: 46–66.

—— and UYEDA, R. (1951*a*): Dynamical theory of electron diffraction for a finite poly-hedral crystal. I. Extension of Bethe's theory, *Acta Cryst.*, **4**: 227–229.

—— and UYEDA, R. (1951*b*): Dynamical theory of electron diffraction for a finite poly-hedral crystal. II. Comparison with results of kinematical theory, *Acta Cryst.*, **4**: 229–231.

KOHRA, K., KIKUTA, S., and ANNAKA, S. (1965): Temperature effect on profile of x-ray diffrac-tion of the Bragg case from a germanium single crystal, *J. Phys. Soc. Japan*, **20**: 1965–1966.

——, ——, ——, and NAKANO, S. (1966): Study of temperature effect on x-ray diffrac-tion curves from single crystals by a triple-crystal spectrometer, *J. Phys. Soc. Japan*, **21**: 1565–1572.

——, UYEDA, R., and MIYAKE, S. (1950): An exception to Friedel law in electron diffraction. II. Theoretical consideration, *Acta Cryst.*, **3**: 479–481.

—— and YOSHIMATSU, M. (1962): X-ray observations of lattice defects, in particular stacking faults in the neighborhood of a twin, *J. Phys. Soc. Japan*, **17**: 1041–1045.

KURIYAMA, M. (1971): On the principle of x-ray interferometry, *Acta Cryst.*, **A27**: 273–280.

LANG, A. R. (1968): X-ray Moiré topography of lattice defects in quartz, *Nature, London*, **220**: 652–657.

LAUE, M. VON (1940): Interferenz-doppelbrechung von Röntgenstrahlen in Kristalprismen, *Naturwiss.*, **28**: 645–646.

MEIERAN, E. S., and BLECH, I. A. (1965): X-ray extinction contrast topography of silicon strained by thin surface films, *J. Appl. Phys.*, **36**: 3162–3167.

—— and —— (1968): Contrast asymmetries in Lang topographs of crystals strained by thin films, *Phys. Stat. Solidi*, **29**: 653–667.

MITSUISHI, T., NAGASAKI, H., and UYEDA, R. (1951): A new type of interference fringes observed in electron-micrograph of crystalline substances, *Proc. Japan Acad.*, **27**: 86–87.

MIYAKE, S., and UYEDA, R. (1950): An exception to Friedel's law in electron diffraction, *Acta Cryst.*, **3**: 314.

—— and —— (1955): Friedel's law in the dynamical theory of diffraction, *Acta Cryst.*, **8**: 335–342.

NISHIKAWA, S., and MATSUKAWA, K. (1928): Hemihedry of zinc blende and x-ray reflection, *Proc. Imp. Acad. Tokyo*, **4**: 96–97.

OKAYA, Y., SAITO, Y., and PEPINSKY, R. (1955): New method in x-ray crystal structure deter-mination involving the use of anomalous dispersion, *Phys. Rev.*, **98**: 1857–1858.

PARTHATHARASY, R. (1960): The temperature factor in the dynamical theory of x-ray inter-ference for a perfect crystal with heat motion, *Acta Cryst.*, **13**: 802–806.

PATEL, J. R., and KATO, N. (1968): X-ray dynamical diffraction effects of oxide films on silicon substrates, *Appl. Phys. Letters*, **13**: 40–42.

————, and ———— (1973): X-ray diffraction topographs of silicon crystals with superposed oxide films. II. *Pendellösung* fringes: Comparison of experiment with theory, *J. Appl. Phys.*, **44**: 971–977.

———— and WAGNER, R. (1972): (In preparation.)

PENNING, P., and POLDER, D. (1961): Anomalous transmission of x-rays in elastically deformed crystals, *Philips Res. Rept.*, **16**: 419–440.

SCHWUTTKE, G. H., and HOWARD, J. K. (1968): X-ray stress topography of thin films on germanium and silicon, *J. Appl. Phys.*, **39**: 1581–1591.

TAKAGI, S. (1962): Dynamical theory of diffraction applicable to crystals with any kind of small distortion, *Acta Cryst.*, **15**: 1311–1312.

———— (1969): A dynamical theory of diffraction for a distorted crystal, *J. Phys. Soc. Japan*, **27**: 1239–1253.

TAUPIN, D. (1964): Théorie dynamique de la diffraction des rayons x par les cristaux déformés, *Bull. Soc. Franc. Minéral. Crist.*, **87**: 469–511.

WHELAN, M. J., and HIRSCH, P. B. (1957a): Electron diffraction from crystals containing stacking faults. I, *Phil. Mag.*, **2**: 1121–1142.

———— and ———— (1957b): Electron diffraction from crystals containing stacking faults. II, *Phil. Mag.*, **2**: 1303–1325.

————, ————, HORNE, R. W., and BOLLMAN, W. (1957): Dislocations and stacking faults in stainless steel, *Proc. Roy. Soc. London*, **240**: 524–538.

WILKENS, M. (1964a): Zur Theorie des Kontrasts von elektronenmikroskopisch abgebildeten Gitterfehlen, *Phys. Stat. Solidi*, **5**: 175–186.

———— (1964b): Streuung von Blochwellen schneller Elektronen in Kristallen mit Gitterbaufehlern, *Phys. Stat. Solidi*, **6**: 939–956.

———— (1966): Modifizierte Bloch-Wellen und ihre Anwendung auf den elektronenmikroskopischen Beugungskontrast von Gitterfehlern, *Phys. Stat. Solidi*, **13**: 529–542.

YOSHIMATSU, M. (1965): A new type of x-ray *Pendellösung* fringes observed in a quartz single crystal, *Japan J. Appl. Phys.*, **4**: 619–620.

ZENER, C., and BILINSKY, S. (1936): Theory of the effect of temperature on the reflection of x-rays by crystals. III. High temperature. Allotropic crystals, *Phys. Rev.*, **50**: 101–104.

ZIMAN, J. M. (1964): *Principles of the theory of solids*, chap. 6, Cambridge University Press, Cambridge.

# Chapter 6

ABRAMOWITZ, M., and STEGUN, I. A. (1968): *Handbook of mathematical functions*, U.S. Government Printing Office, Washington.

BARTH, H. (1960): Möglichkeiten der Präzisionsgitterkonstantenmessung mit hochmonochromatischer Röntgen-Strahlung, *Acta Cryst.*, **13**: 830–832.

BEU, K. E. (1962): Report no. 2 of the apparatus and standards committee of the American Crystallographic Association, p. 11.

BOOM, G. (1966): *Accurate lattice parameters and the LPC method*, van Denderen, Groningen.

BRAGG, W. H. (1921): Application of the ionisation spectrometer to the determination of the structure of minute crystals, *Proc. Phys. Soc.*, **33**: 222–224.

BRENTANO, J. (1917): Monochromateur pour rayons Röntgen, *Arch. Sc. Phys. et Nat.*, **44**: 66–68.

—— (1919): Sur un dispositif pour l'analyse spectrographique de la structure des substances à l'état de particules désordonnées par les rayons Röntgen, *Arch. Sc. Phys. et Nat.*, **1**: 550–552.

BRINDLEY, G. W. (1945): Effect of grain or particle size on x-ray reflections from mixed powders and alloys, considered in relation to the quantitative determination of crystalline substances by x-ray methods, *Phil. Mag.*, **36**: 347–369.

BURGER, H. C., and VAN CITTERT, P. H. (1932): Wahre und scheinbare Intensitätsverteilung in Spektrallinien, *Z. Physik*, **79**: 772–780.

COLE, H. (1970): Bragg's law and energy sensitive detectors, *J. Appl. Cryst.*, **3**: 405–406.

DEMARCO, J. J., and WEISS, R. J. (1962): U.S. Army Materials Research Laboratory, no. 114, Watertown, Mass. Quoted in R. J. Weiss (1966): *X-ray determination of electron distributions*, North-Holland Publishing Company, Amsterdam.

DOI, K. (1961): Interpretation of line profiles by the Fourier method, with special reference to the size distribution of carbon black crystallites, *Acta Cryst.*, **14**: 830–834.

EASTABROOK, J. N. (1952): Effect of vertical divergence on the displacement and breadth of x-ray powder diffraction lines, *Brit. J. Appl. Phys.*, **3**: 349–352.

—— and WILSON, A. J. C. (1952): The diffraction of x-rays by distorted crystal aggregates. III. Remarks on the interpretation of the Fourier coefficients, *Proc. Phys. Soc.*, **B65**: 67–75.

EDWARDS, H. J., and LANGFORD, J. I. (1971): A comparison between the variances of the Cu $K\alpha$ and Fe $K\alpha$ spectral distributions, *J. Appl. Cryst.*, **4**: 43–50.

—— and TOMAN, K. (1969): The intensity distribution and variance of the iron $K\alpha$ multiplet, *J. Appl. Cryst.*, **2**: 240–246.

—— and —— (1970a): Correction for the satellite group in the variance method of profile analysis, *J. Appl. Cryst.*, **3**: 157–164.

—— and —— (1970b): The additivity of variances in powder diffraction profile analysis, *J. Appl. Cryst.*, **3**: 165–171.

ERGUN, S. (1968): Direct method for unfolding convolution products—Its application to x-ray scattering intensities, *J. Appl. Cryst.*, **1**: 19–23.

FINGERLAND, A. (1960): Method of moments in analysis of x-ray diffraction lines, *Czech J. Phys.*, **B10**: 233–239.

FUKAMACHI, T., HOSOYA, S., and TERASAKI, O. (1972): The precision of the interplanar distances measured by an energy-dispersive diffractometer, *Technical Report of ISSP*, Series A, no. 549, pp. ii + 28.

——, ——, and —— (1973): The precision of the interplanar distances measured by an energy-dispersive diffractometer, *J. Appl. Cryst.*, **6**: 117–122.

FURNAS, T. C. (1965): Paper given at ACA Meeting, Gatlinburg, Tenn., June 27.

GALE, B. (1963): The positions of Debye diffraction line peaks, *Brit. J. Appl. Phys.*, **14**: 357–364.

——— (1968): The aberrations of a focusing x-ray diffraction instrument: second-order theory, *Brit. J. Appl. Phys. (J. Phys. D)*, **1**: 393–408.

GIESSEN, B. C., and GORDON, G. E. (1968): X-ray diffraction: New high speed technique based on x-ray spectroscopy, *Science*, **159**: 973–975.

GILLHAM, C. J. (1971): Centroid shifts due to axial divergence and other geometrical factors in Seemann-Bohlin diffractometry, *J. Appl. Cryst.*, **4**: 498–506.

——— and KING, H. W. (1972): Measurements of centroid and peak shifts due to dispersion and the Lorentz factor at very high Bragg angles, *J. Appl. Cryst.*, **5**: 23–27.

HARDY, G. H., and ROGOSINSKI, W. W. (1944): *Fourier series*, Cambridge University Press, Cambridge.

HARRISON, R. J., and PASKIN, A. (1964): The effects of granularity on the diffracted intensity in powders, *Acta Cryst.*, **17**: 325–333.

KILLEAN, R. C. G. (1967a): Least-squares weighting schemes for diffractometer-collected data, *Acta Cryst.*, **23**: 54–56.

——— (1967b): A note on the *a priori* estimation of *R* factors for constant-count-per-reflexion diffractometer experiments, *Acta Cryst.*, **23**: 1109–1110.

——— (1969): Least-squares weighting schemes for diffractometer-collected data. III. Optimization process, *Acta Cryst.*, **B25**: 977–978.

KUKOL', V. V. (1962): O nesostojatel'nosti primenjaemykh metodov i polucennykh resul'tatov pri issledovanii substruktury kristallov putem Fur'eanaliza formy odnoj difrakcionnoj linii, *Fiz. Tverdogo Tela*, **4**: 724–735 (in Russian).

KUNZE, G. (1964a): Korrekturen höherer Ordnung für die mit Bragg-Brentano- und Seemann-Bohlin-Systemen gewonnenen Messgrössen (unter Berücksichtigung der Primärstrahldivergenz), *Z. angew. Phys.*, **17**: 412–421.

——— (1964b): Intensitäts-, Absorptions- und Verschiebungsfaktoren von Interferenzlinien bei Bragg-Brentano- und Seemann-Bohlin-Diffraktometern. I, *Z. angew. Phys.*, **17**: 522–534.

——— (1964c): *Idem*, II, *Z. Physik*, **18**: 28–37.

LADELL, J. (1961): Interpretation of diffractometer line profiles distortion due to the diffraction process, *Acta Cryst.*, **14**: 47–53.

———, MACK, M., PARRISH, W., and TAYLOR, J. (1959): Dispersion, Lorentz and polarization effects in the centroid method of precision lattice parameter determination, *Acta Cryst.*, **12**: 567–570.

LANG, A. R. (1956a): Effect of dispersion and geometric intensity factors on x-ray back-reflection line profiles, *J. Appl. Phys.*, **27**: 485–488.

——— (1956b): Diffracted-beam monochromatization techniques in x-ray diffractometry, *Rev. Sci. Instr.*, **27**: 17–25.

LANGFORD, J. I. (1962): Counter diffractometer: The effect of axial divergence on the breadth of powder lines, *J. Sci. Instr.*, **39**: 515–516.

——— (1968a): The variance and other measures of line broadening in powder diffractometry. I. Practical considerations, *J. Appl. Cryst.*, **1**: 48–59.

——— (1968b): The variance and other measures of line broadening in powder diffractometry. II. Determination of particle size, *J. Appl. Cryst.*, **1**: 131–138.

LANGFORD, J. I., and WILSON, A. J. C. (1962): Counter diffractometer: the effect of specimen transparency on the intensity, position and breadth of x-ray powder diffraction lines, *J. Sci. Instr.*, **39**: 581–585.

——— and ——— (1963): On variance as a measure of line broadening in diffractometry: some preliminary measurements on annealed aluminium and nickel and on cold-worked nickel, in G. N. RAMACHANDRAN (ed.), *Crystallography and crystal perfection*, pp. 207–222, Academic Press, London.

LAURIAT, J. P., and PÉRIO, P. (1972): Adaptation d'un ensemble de détection Si(Li) à un diffractometre X, *J. Appl. Cryst.*, **5**: 177–183.

MACK, M., and PARRISH, W. (1967): Seeman-Bohlin x-ray diffractometry. II. Comparison of aberrations and intensity with conventional diffractometer, *Acta Cryst.*, **23**: 693–700.

——— and SPIELBERG, N. (1958): Statistical factors in x-ray intensity measurements, *Spectrochim. Acta,* **12**: 169–178.

MILBERG, M. E. (1958): Transparency factor for weakly absorbing samples in x-ray diffractometry, *J. Appl. Phys.*, **29**: 64–65.

MITCHELL, C. M., and DE WOLFF, P. M. (1967): Elimination of the dispersion effect in the analysis of diffraction line profiles, *Acta Cryst.*, **22**: 325–328.

MITRA, G. B. (1964): The fourth moment of diffraction profiles, *Brit. J. Appl. Phys.*, **15**: 917–921.

——— and WILSON, A. J. C. (1960): Variation with particle size of the effective x-ray absorption coefficient of heterogeneous slabs, *Brit. J. Appl. Phys.*, **11**: 43–45.

OGILVIE, R. E. (1963): Parafocusing diffractometry, *Rev. Sci. Instr.*, **34**: 1344–1347.

PARRISH, W. (1962): X-ray intensity measurements. Geiger, proportional and scintillation counters, *International Tables for X-Ray Crystallography*, Vol. III, pp. 144–156, Kynoch Press, Birmingham, England.

———, MACK, M., and VAJDA, I. (1967): Seeman-Bohlin linkage for Norelco x-ray diffractometer, *Norelco Reporter*, **14**: 56–59.

——— and WILSON, A. J. C. (1959): Precision measurement of lattice parameters of polycrystalline specimens, *International Tables for X-Ray Crystallography*, Vol. II, pp. 216–234, Kynoch Press, Birmingham, England.

PATERSON, M. S. (1950): Calculation of the correction for instrumental broadening in x-ray diffraction lines, *Proc. Phys. Soc.*, **A63**: 477–482.

PIKE, E. R. (1957): Counter diffractometer—the effect of vertical divergence on the displacement and breadth of powder diffraction lines, *J. Sci. Instr.*, **34**: 355–361.

——— (1959a): Counter diffractometer—the effect of vertical divergence on the displacement and breadth of powder diffraction lines. Errata, *J. Sci. Instr.*, **36**: 52–53.

——— (1959b): Counter diffractometer—the effect of dispersion, Lorentz and polarisation factors on the positions of x-ray powder diffraction lines, *Acta Cryst.*, **12**: 87–92.

——— (1962): Focusing geometry in x-ray powder diffractometers, *J. Sci. Instr.*, **39**: 222–223.

——— and LADELL, J. (1961): The Lorentz factor in powder diffraction, *Acta Cryst.*, **14**: 53–54.

——— and WILSON, A. J. C. (1959): Counter diffractometer—the theory of the use of centroids of diffraction profiles for high accuracy in the measurement of diffraction angles, *Brit. J. Appl. Phys.*, **10**: 56–68.

RIETVELD, H. M. (1969): A profile refinement method for nuclear and magnetic structures, *J. Appl. Cryst.*, **2**: 65–71.

RULAND, W. (1964): The separation of coherent and incoherent Compton x-ray scattering, *Brit. J. Appl. Phys.*, **15**: 1301–1307.

SCHÄFER, K. (1933): Atomfaktorbestimmungen im Gebiet der anomalen Dispersion. II, *Z. Physik*, **86**: 738–759.

SHOEMAKER, D. P. (1968): Optimization of counting times in computer-controlled x-ray and neutron single-crystal diffractometry, *Acta Cryst.*, **A24**: 136–142.

SOLLER, W. (1924): A new precision x-ray spectrometer, *Phys. Rev.*, **24**: 158–167.

SPENCER, R. C. (1931): Additional theory of the double x-ray spectrometer, *Phys. Rev.*, **38**: 618–629.

——— (1949): Discussion of 'Geometrical factors affecting x-ray spectrometer maxima,' *J. Appl. Phys.*, **20**: 413–414.

STOKES, A. R. (1948): A numerical Fourier-analysis method for the correction of widths and shapes of lines on x-ray powder photographs, *Proc. Phys. Soc.*, **61**: 382–391.

STRONG, S. L., and KAPLOW, R. (1966): Elimination of incoherent x-rays with diffracted-beam monochromators, *Rev. Sci. Instr.*, **37**: 1495–1496.

SUORTTI, P., and JENNINGS, L. D. (1971): Effects of geometrical aberrations on intensities in powder diffractometry, *J. Appl. Cryst.*, **4**: 37–43.

SZÁNTÓ, I. S., and VARGA, L. (1969): An x-ray method for the determination of domain size from the tails of diffraction profiles, *J. Appl. Cryst.*, **2**: 72–76.

TAYLOR, A. (1944): The influence of crystal size on the absorption factor as applied to Debye-Scherrer diffraction patterns, *Phil. Mag.*, **35**: 215–229.

THOMSEN, J. S., and YAP, F. Y. (1968a): Effect of statistical counting errors on wavelength criteria for x-ray spectra, *J. Res. Natl. Bur. Std.*, **72A**: 187–205.

——— and ——— (1968b): Simplified method of computing centroids of x-ray profiles, *Acta Cryst.*, **A24**: 702–703.

TOURNARIE, M. (1956a): Utilisation du deuxième moment comme critère d'élargissement des raies Debye-Scherrer. Élimination de l'effet instrumental, *Compt. Rend. Acad. Sci. Paris*, **242**: 2016–2018.

——— (1956b): Utilisation du deuxième moment comme critère d'elargissement des raies Debye-Scherrer. Signification physique, *Compt. Rend. Acad. Sci. Paris*, **242**: 2161–2164.

WARREN, B. E. (1959): X-ray studies of deformed metals, *Progr. Metal Phys.*, **8**: 147–202.

——— (1960): X-ray measurement of grain size, *J. Appl. Phys.*, **31**: 2237–2239.

WASER, J., and SCHOMAKER, V. (1953): The Fourier inversion of diffraction data, *Rev. Mod. Phys.*, **25**: 671–690.

WASSERMANN, G., and WIEWIOROWSKY, J. (1953): Über ein Geiger-Zählrohr-Goniometer nach dem Seemann-Bohlin-Prinzip, *Z. Met.*, **44**: 567–570.

WEISS, R. J. (1966): *X-ray determination of electron distributions*, North-Holland Publishing Company, Amsterdam.

WILCHINSKY, ZIGMOND W. (1951): Effect of crystal, grain, and particle size on x-ray power diffracted from powders, *Acta Cryst.*, **4**: 1–9.

WILSON, A. J. C. (1940): On the correction of lattice spacings for refraction, *Proc. Cambridge Phil. Soc.*, **36**: 485–489.

——— (1950): Geiger-counter x-ray spectrometer—influence of size and absorption coefficient of specimen on position and shape of powder diffraction maxima, *J. Sci. Instr.*, **27**: 321–325.

——— (1958): Effect of absorption on mean wavelength of x-ray emission lines, *Proc. Phys. Soc.*, **72**: 924–925.

——— (1962a): Refraction broadening in powder diffractometry, *Proc. Phys. Soc.*, **80**: 303–305.

——— (1962b): On variance as a measure of line broadening in diffractometry. I. General theory and small particle size, *Proc. Phys. Soc.*, **80**: 286–294.

——— (1963a): *Mathematical theory of x-ray powder diffractometry*, Centrex Publishing Co., Eindhoven.

——— (1963b): On variance as a measure of line broadening in diffractometry. II. Mistakes and strain, *Proc. Phys. Soc.*, **81**: 41–46.

——— (1963c): On variance as a measure of line broadening in diffractometry. III. A note on dislocations, *Proc. Phys. Soc.*, **82**: 986–991.

——— (1964): Aberrations and line broadening in x-ray powder diffractometry, in G. N. RAMACHANDRAN (ed.), *Advanced methods of crystallography*, pp. 221–250, Academic Press, Ltd., London.

——— (1965a): On variance as a measure of line broadening in diffractometry. IV. The effect of physical aberrations, *Proc. Phys. Soc.*, **85**: 171–176.

——— (1965b): The location of peaks, *Brit. J. Appl. Phys.*, **16**: 665–674.

——— (1967): Statistical variance of line-profile parameters: Measures of intensity, location and dispersion, *Acta Cryst.*, **23**: 888–898.

——— (1968): Statistical variance of line-profile parameters: Measures of intensity, location and dispersion: Corrigenda, *Acta Cryst.*, **A24**: 478.

——— (1969a): The moments of a powder diffraction profile in the kinematic approximation, *Acta Cryst.*, **A25**: S 15.

——— (1969b): Statistical variance of line-profile parameters: Addendum, *Acta Cryst.*, **A25**: 584–585.

——— (1970): Limitations on the additivity of moments in line-profile analysis, *J. Appl. Cryst.*, **3**: 71–73.

——— (1971a): Influence of sample position on intensity, width, and position of a diffraction line in x-ray powder diffractometry (in Russian), *Kristallografija*, **16**: 1127–1130.

——— (1971b): The moments of a powder diffraction profile in the kinematic tangent-plane approximation, *Acta Cryst.*, **A27**: 599–604.

——— (1971c): Some further considerations in particle-size broadening, *J. Appl. Cryst.*, **4**: 440–443.

——— (1972): Note on the aberrations of a fixed-angle energy-dispersive powder diffractometer, *Technical Report of ISSP*, Series A, no. 548, pp. ii + 30.

——— (1973): Note on the aberrations of a fixed-angle energy-dispersive powder diffractometer, *J. Appl. Cryst*, **6**: 230–237.

——— and DELF, B. W. (1961): Effect of variations in the quantum counting efficiency of detectors on the mean wavelength of x-ray emission lines, *Proc. Phys. Soc.*, **78**: 1256–1258.

———, THOMSEN, J. S., and YAP, F. Y. (1965): Minimization of the variance of parameters derived from x-ray powder diffractometer line profiles, *Appl. Phys. Letters*, **7**: 163–165.

WOLFF, P. M. DE (1947): A theory of x-ray absorption in mixed powders, *Physica*, **13**: 62–78.

——— (1956): Measurement of particle absorption by x-ray fluorescence, *Acta Cryst.*, **9**: 682–683.

———, TAYLOR, J., and PARRISH, W. (1959): Experimental study of effect of crystallite size statistics on x-ray diffractometer intensities, *J. Appl. Phys.*, **30**: 63–69.

WOOSTER, W. A. (1962): *Diffuse x-ray reflexions from crystals*, Clarendon Press, Oxford.

YOUNG, R. A., GERDES, R. J., and WILSON, A. J. C. (1967): Propagation of some systematic errors in x-ray line-profile analysis, *Acta Cryst.*, **22**: 155–162.

ZEVIN, L. S., UMANSKIJ, M. M., KHEJKER, D. M., and PANČENKO, JU. M. (1961): The question of diffractometer methods of precision measurement of unit cell parameters, *Soviet. Phys. Cryst.*, **6**: 277–283 (*Kristallografija*, **6**: 348–356).

## Chapter 7

ABRAHAMS, S. C. (1964): Evaluation of digital automatic diffractometer systems, *Acta Cryst.*, **17**: 1190–1195.

ALEXANDER, L. E., and SMITH, G. S. (1962): Single crystal intensity measurements with the three-circle diffractometer, *Acta Cryst.*, **15**: 983–1004.

——— and ——— (1964): Receiving aperture widths in single crystal diffractometry, *Acta Cryst.*, **17**: 447–448.

ANDERSON, O. L. (1963): A simplified method for calculating the Debye temperature from elastic constants, *J. Phys. Chem. Solids*, **24**: 909–917.

ARNDT, U. W. (1964): Analogue and digital single-crystal diffractometers, *Acta Cryst.*, **17**: 1183–1190.

——— (1968): The optimum strategy in measuring structure factors, *Acta Cryst.*, **B24**: 1355–1357.

———, CROWTHER, R. A., and MALLETT, J. F. W. (1968): A computer-linked cathode-ray tube microdensitometer for x-ray crystallography, *J. Sci. Instr.*, (2) **1**: 510–516.

——— and PHILLIPS, D. C. (1959): In A. HARGREAVES and E. STANLEY, Summarized proceedings of a conference on x-ray analysis, Manchester, April 1958, *Brit. J. Appl. Phys.*, **10**: 116–124.

——— and ——— (1961): The linear diffractometer, *Acta Cryst.*, **14**: 807–818.

——— and ——— (1963): Conference on x-ray diffractometry, *Brit. J. Appl. Phys.*, **14**: 229–236, London, November 1962.

——— and WILLIS, B. T. M. (1966): *Single crystal diffractometry*, Cambridge University Press, Cambridge.

AZÁROFF, L. V. (1968): *Elements of x-ray crystallography*, p. 98, McGraw-Hill Book Company, New York.

BAKER, T. W., GEORGE, J. D., BELLAMY, B. A. and CAUSER, R. (1968): Fully automated high precision x-ray diffraction, *Advan. X-Ray Anal.*, **12**: 359–374, and, in longer form, *AERE Harwell Rept.*, AERE-R5152 (February 1969).

BARTH, H. (1958): Monochromatisierung einer Rontgen-Spektrallinie, *Z. Naturforsch.*, **13a**: 680–698.

BINNS, J. V. (1964): Setting accuracy in point-drive, single-crystal, x-ray diffractometers, *J. Sci. Instr.*, **41**: 715–721.

BOL, W. (1967): The use of balanced filters in x-ray diffraction, *J. Sci. Instr.*, **44**: 736–739.

BOND, W. L. (1960): Precision lattice constant determination, *Acta Cryst.*, **13**: 814–818.

BONSE, U., and HART, M. (1965*a*): Tailless x-ray single-crystal reflection curves obtained by multiple reflection, *Appl. Phys. Letters*, **7**: 238–240.

——— and ——— (1965*b*): A new tool for small angle x-ray scattering and x-ray spectroscopy: The multiple reflection diffractometer, *Proc. conf. x-ray small-angle scattering*, H. BRUMBERGER (ed.), Gordon and Breach, Science Publishers, Inc., New York.

BRAGG, R. H., and AZÁROFF, L. V. (1962): Direct study of imperfections in nearly perfect crystals, in J. B. NEWKIRK and J. H. WERNICK (eds.), *Direct observation of imperfections in crystals*, pp. 415–429, Interscience Publishers, Inc., New York.

BRAGG, W. L., JAMES, R. W., and BOSANQUET, C. H. (1921): The intensity of reflexion of x-rays by rocksalt. Part II, *Phil. Mag.*, **42**: 1–17.

BUERGER, M. J. (1942): *X-ray crystallography*, John Wiley & Sons, Inc., New York.

BURBANK, R. D. (1964): A comparison of $\omega$ and $2\theta$ scans for integrated intensity measurements, *Acta Cryst.*, **17**: 434–442.

——— (1965): Intrinsic and systematic multiple reflection, *Acta Cryst.*, **19**: 957–962.

BURKE, J., and TOMKEIEFF, M. V. (1968): Specimen and beam tilt errors in Bond's method of lattice parameter determination, *Acta Cryst.*, **A24**: 683–685.

BUSING, W. R., and LEVY, H. A. (1967): Angle calculations for 3- and 4-circle x-ray and neutron diffractometers, *Acta Cryst.*, **22**: 457–464.

BUTT, N. M., and O'CONNOR, D. A. (1967): The determination of x-ray temperature factors for aluminum and potassium chloride single crystals using nuclear resonant radiation, *Proc. Phys. Soc.*, **90**: 247–252.

CHANDRASEKHAR, S. (1956): A first-order correction for extinction in crystals, *Acta Cryst.*, **9**: 954–956.

———, RAMASESHAN, S., and SINGH, A. K. (1969): Experimental determination of the extinction factor by the use of polarized x rays, *Acta Cryst.*, **A25**: 140–142.

CHIPMAN, D. R. (1956): Improved monochromator for diffuse x-ray scattering measurements, *Rev. Sci. Instr.*, **27**: 164–165.

——— and BATTERMAN, B. W. (1963): Contribution of thermal diffuse scattering to integrated Bragg reflections from perfect crystals, *J. Appl. Phys.*, **34**: 912–913.

——— and PASKIN, A. (1959): Temperature diffuse scattering of x-rays in cubic powders. II. Corrections to integrated intensity measurements, *J. Appl. Phys.*, **30**: 1998–2001.

CLIFTON, D. F., FULLER, A., and MCLACHLAN, D. (1951): The adaptation of a Geiger counter to the Weissenberg camera, *Rev. Sci. Instr.*, **22**: 1024–1025.

COCHE, A. (1968): Les détecteurs a semiconduteurs: évolution, performances, perspectives d'avenir, *Nucleus*, **9**: 114–121.

CÒCHRAN, W. (1966): X-ray scattering by phonons [1], in R. W. H. STEVENSON (ed.), *Phonons in perfect lattices and in lattices with point imperfections*, pp. 153–160, Oliver & Boyd, Ltd., Edinburgh.

—— (1969): The correction of measured structure factors for thermal diffuse scattering, *Acta Cryst.*, **A25**: 95–101.

COLE, H. (1952): Approximate elastic spectrum of elastic waves in AgCl from x-ray scattering, *J. Appl. Phys.*, **24**: 482–487.

——, CHAMBERS, F., and WOOD, G. C. (1961): An x-ray polarizer, *J. Appl. Phys.*, **32**: 1942.

COOPER, M. J. (1968): The resolution function in neutron diffractometry. IV. Application of the resolution function to measurement of Bragg peaks, *Acta Cryst.*, **A24**: 624–627.

—— and NATHANS, R. (1967): The resolution function in neutron diffractometry I. The resolution function of a neutron diffractometer and its application to phonon measurements, *Acta Cryst.*, **23**: 357–367.

—— and —— (1968a): The resolution function in neutron diffractometry. II. The resolution function of a conventional two-crystal neutron diffractometer for elastic scattering, *Acta Cryst.*, **A24**: 481–484.

—— and —— (1968b): The resolution function in neutron diffractometry. III. Experimental determination and properties of the 'elastic two-crystal' resolution function, *Acta Cryst.*, **A24**: 619–624.

—— and ROUSE, K. D. (1968): The correction of measured integrated Bragg intensities for first-order thermal diffuse scattering, *Acta Cryst.*, **A24**: 405–410.

COPPENS, P. (1968): The elimination of multiple reflection on the four-circle diffractometer, *Acta Cryst.*, **A24**: 253–257.

DAVIS, M. F., GROTER, C., and KAY, H. R. (1968): On choosing off-line automatic x-ray diffractometers, *J. Appl. Cryst.*, **1**: 209–217.

DIAMOND, R. (1969): Profile analysis in single crystal diffractometry, *Acta Cryst.*, **A25**: 43–55.

ELLIOTT, A. (1965): The use of toroidal reflecting surfaces on x-ray diffraction cameras, *J. Sci. Instr.*, **42**: 312–316.

EVANS, H. T. (1953): Use of a Geiger counter for the measurement of x-ray intensities from small single crystals, *Rev. Sci. Instr.*, **24**: 156–161.

FITZWATER, D. R. (1965): Comment in published discussion of paper by Young (1965).

FRANKS, A. (1958): Some developments and applications of microfocus x-ray diffraction techniques, *Brit. J. Appl. Phys.*, **9**: 349–352.

FURNAS, T. C. (1957): *Single crystal orienter instruction manual*, General Electric Company, Milwaukee.

GATINEAU, L., and MERING, J. (1956): Une methode de correction des effects d'extinction affectant les intensites des rayons x reflectis par un crystal unique, *Compt. Rend.*, **242**: 2018–2020.

GIESSEN, W. C., and GORDON, G. E. (1968): X-ray diffraction: new high speed technique based on x-ray spectroscopy, *Science*, **159**: 973–975.

GÖTTLICHER, S. (1968): Der Beitrag der thermisch diffusen Streustrahlung zur Intensität der Röntgeninterferenzen und die Elektronendichtverteilung im NaCl, *Acta Cryst.*, **B24**: 122–129.

GUENTERT, O. J. (1965): Study of the anomalous surface reflection of x-rays, *J. Appl. Phys.*, **36**: 1361–1366.

HART, M. (1969): High precision lattice parameter measurements by multiple Bragg reflexion diffractometry, *Proc. Roy. Soc.*, **A309**: 281–296.

HERBSTEIN, F. H. (ed.) (1967): *Methods of obtaining monochromatic x-rays and neutrons*, Bibliography 3, A. Oosthoek's Uitgevers Mij N.V. Utrecht, The Netherlands.

HIRSHFELD, F. L., and SCHMIDT, C. M. J. (1953): Geiger counter measurements of single crystal Bragg reflections. The geometrical problem, *Bull. Res. Council Israel*, **3**: 37–39.

HOPPE, W. (1969*a*): Electronic and mechanical sources of error in diffractometry, *Acta Cryst.*, **A25**: 67–76.

——— (1969*b*): Comment in published discussion of paper by Kheiker (1969), *International tables for x-ray crystallography*, Kynoch Press, Birmingham, England.

JAMES, R. W. (1948): The optical principles of the diffraction of x-rays, G. Bell & Sons, Ltd., London.

JEFFREY, J. W. (1969): X-ray photography as a means of accurate intensity measurement, *Acta Cryst.*, **A25**: 153–157.

——— and WHITAKER, A. (1965): Experimental requirements for accurate x-ray intensity measurements by photographic means, *Acta Cryst.*, **19**: 963–967.

JONES, C. K. (1953): A highly stable automatic x-ray diffraction apparatus with only one moving part, *Rev. Sci. Instr.*, **24**: 380–387.

KHEIKER, D. M. (1969): The geometry of integrated intensity measurement and errors due to non-monochromatic radiation, *Acta Cryst.*, **A25**: 82–88.

KIKUTA, S., and KOHRA, K. (1970): X-ray collimators using successive asymmetric diffractions and their applications to measurements of diffraction curves. I. General considerations on collimators, *J. Phys. Soc. Japan*, **29**: 1322–1328.

KIRKPATRICK, PAUL (1944): Theory and use of Ross filters, *Rev. Sci. Instr.*, **15**: 223–229.

KOHRA, K. (1962): An application of asymmetric reflection for obtaining x-ray beams of extremely narrow angular spread, *J. Phys. Soc. Japan*, **17**: 589–590.

——— and KIKUTA, S. (1968): A method of obtaining an extremely parallel x-ray beam by successive asymmetric diffractions and its applications, *Acta Cryst.*, **A24**: 200–205.

KOMURA, S., and COOPER, M. J. (1970): The resolution function of a twin-rotor neutron time-of-flight spectrometer, *Japan. J. Appl. Phys.*, **9**: 866–874.

KOTTWITZ, D. A. (1968): High-resolution monochromatization of neutrons and x-rays by multiple Bragg reflection, *Acta Cryst.*, **A24**: 117–126.

——— (1969): High resolution monochromatization of neutrons by multiple Bragg reflection in hexagonal close-packed crystals, *Acta Cryst.*, **A25**: 459–464.

LADELL, J., and CATH, P. G. (1963): PAILRED II. An automatic diffractometer for crystal structure analysis, *Norelco Reptr.*, *Special Issue*.

——— and LOWITZSCH, K. (1960): Automatic single crystal diffractometry: The kinematic problem, *Acta Cryst.*, **13**: 205–215.

——— and SPIELBERG, N. (1966): Theory of the measurement of integrated intensities obtained with single-crystal counter diffractometers, *Acta Cryst.*, **21**: 103–118.

LANG, A. R. (1953): Extinction in x-ray diffraction patterns of powders, *Proc. Phys. Soc. London*, **B66**: 1003–1008.

LARSON, A. C. (1967): Inclusion of secondary extinction in least-squares calculations, *Acta Cryst.*, **23**: 664–665.

LONSDALE, K. (1942/3): Experimental study of x-ray scattering in relation to crystal dynamics, *Rept. Progr. Phys.*, **9**: 256–293.

LUCAS, B. W. (1968): Correction for second-order diffuse x-ray scattering in the determination of the elastic constants of crystals, *Acta Cryst.*, **A24**: 336–338.

——— (1969): On the contribution of thermal diffuse x-ray scattering to the integrated Bragg intensities of single crystals, **A25**: 627–631.

MATSUSHITA, T., KIKUTA, S., and KOHRA, K. (1971): X-ray crystal collimators using successive asymmetric diffraction and their applications to measurements of diffraction curves. II. Type II collimator, *J. Phys. Soc. Japan*, **30**: 1136–1144.

MCKINSTRY, H. A., and SHORT, M. A. (1960): Automatically operated balanced filters for a counter diffractometer, *J. Sci. Instr.*, **37**: 178.

MILLEDGE, H. J. (1969): Real crystals as a source of error, *Acta Cryst.*, **A25**: 173–180.

MILLER, D. C. (1957): Some considerations in the use of pulse height analysis with x-rays, *Norelco Reptr.*, **4**: 37–40.

MIRKEN, L. S. (1964): *Handbook of x-ray analysis of polycrystalline materials*, p. 59, J. E. S. BRADLEY (trans.), Consultants Bureau, New York.

MOON, R. M., and SHULL, G. G. (1964): The effects of simultaneous reflections on single-crystal neutron diffraction intensities, *Acta Cryst.*, **17**: 805–812.

MOZZI, R. L., and WARREN, B. E. (1969): The structure of vitreous silica, *J. Appl. Cryst.*, **2**: 164–172.

NICKLOW, R. M., and YOUNG, R. A. (1963): Thermal expansion of AgCl, *Phys. Rev.*, **139**: 1936–1943.

——— and ——— (1964): A study of lattice vibrations through temperature dependence of x-ray Bragg intensities, *Tech. Rept. 3*, DDC Document No. AD605501.

——— and ——— (1966): Lattice vibrations in aluminum and the temperature dependence of Bragg intensities, *Phys. Rev.*, **152**: 591–596.

NIELSON, N., and MØLLER, H. B. (1969): Resolution of a triple axis spectrometer, *Acta Cryst.*, **A25**: 547–550.

NILSSON, N. (1957): On the corrections of the measured integrated Bragg reflections due to thermal diffuse scattering, *Arkiv Fysik*, **12**: 247.

PANKE, D., and WÖLFEL, E. (1968): The effect of multiple diffraction on the determination of electron density distributions, *J. Appl. Cryst.*, **1**: 255–257.

PASKIN, A. (1968): Contributions of one and two phonon scattering to temperature diffuse scattering, *Acta Cryst.*, **11**: 165–168.

POST, B. (1969): The intensities of multiple diffraction effects, *Acta Cryst.*, **A25**: 94.

POWELL, M. J. D. (1966): Citation in Arndt and Phillips (1966).

PRINGLE, G. E. (1955): The sandwich theory of crystal texture, Oral presentation, ACA Meeting, Pasadena, Calif.

PRYOR, A. W. (1966): Debye-Waller factors in crystals of the sodium chloride structure, *Acta Cryst.*, **20**: 138–140.

RENNINGER, M. (1937): 'Umweganregung', eine bisher unbeachtete Wechselwirkungserscheinung bei Raumgitterinterferenzen, *Z. Phys.*, **106**: 141–176.

——— (1967): Experimental determination of the integrated contribution of temperature diffuse scattering in x-ray reflections, *Advan. X-Ray Anal.*, **10**: 42–45.

RISTE, T., and OTNES, K. (1969): Oriented graphite as a neutron monochromator, *Nucl. Instr. Methods*, **75**: 197–202.

ROUSE, K. D., and COOPER, M. J. (1969): The correction of measured integrated Bragg intensities for anisotropic thermal diffuse scattering, *Acta Cryst.*, **A25**: 615–621.

SAMSON, S., and SCHUELKE, W. W. (1967): Accurate and straightforward alignment procedures for x-ray diffractometers equipped with an Eulerian cradle, *Rev. Sci. Instr.*, **38**: 1273–1283.

SANTORO, A., and ZOCCHI, M. (1964): Geometrical properties of a four-circle neutron diffractometer for measuring intensities at an 'optimum' azimuth of the reflecting planes, *Acta Cryst.*, **17**: 597–602.

——— and ——— (1966): Multiple diffraction in the Weissenberg method, *Acta Cryst.*, **21**: 293–297.

SCHWARTZ, L. H. (1964): Corrections of the measured integrated intensities from cubic metallic single crystals for thermal diffuse scattering, *Acta Cryst.*, **17**: 1614–1615.

SIMMONS, R. O. (1968): Oral communication, *Symposium on Thermal Expansion in Solids*, NBS, Gaithersburg, Md.

SKELTON, E. F., and KATZ, L. J. (1969): Analytical evaluation of thermal diffuse scattering contributions to integrated x-ray intensities in the vicinity of a Bragg reflection, *Acta Cryst.*, **A25**: 319–329.

SMITH, T. (1966): X-ray scattering by phonons [2], in R. W. H. STEVENSON (ed.), *Phonons in perfect lattices and in lattices with point imperfections*, pp. 161–169, Oliver & Boyd, Ltd., Edinburgh.

SOULES, J. A., GORDON, W. L., and SHAW, C. H. (1956): Design of differential x-ray filters for low-intensity scattering experiments, *Rev. Sci. Instr.*, **27**: 12–14.

SPARKS, C. J. (1967): Oral communication, ACA Meeting, Minneapolis, Minn.

SPIELBERG, N., and LADELL, J. (1963): On minimum receiving apertures in single crystal diffractometry, *Acta Cryst.*, **16**: 1057–1058.

STEWARD, E. H. (1968): Private communication.

SUDARSANAN, K., and YOUNG, R. A. (1969): Significant precision in crystal structural details: Holly Springs hydroxyapatite, *Acta Cryst.*, **B25**: 1534–1543.

VAND, V. (1955): Methods for the correction of x-ray intensities for primary and secondary extinction in crystal structure analyses, *J. Appl. Phys.*, **26**: 1191–1194.

WALKER, C. B., and CHIPMAN, D. R. (1970): Thermal diffuse scattering in integrated intensities of Bragg reflections, *Acta Cryst.*, **A26**: 447.

WARREN, B. E., and MAVEL, G. (1965): Elimination of the Compton component in amorphous scattering, *Rev. Sci. Instr.*, **36**: 196–197.

WEISS, R. J. (1952): Extinction effects in powders, *Proc. Phys. Soc. London*, **B65**: 553–555.

——— (1966): *X-ray determination of electron distributions*, North-Holland Publishing Company, Amsterdam.

WILLIAMSON, G. K., and SMALLMAN, R. E. (1955): X-ray extinction and the effect of cold work on integrated intensities, *Proc. Phys. Soc. London*, **B68**: 577–585.

WILLIS, B. T. M. (1962): Use of a three-circle goniometer for diffraction measurements, *Brit. J. Appl. Phys.*, **13**: 548–550.

——— (1963): Neutron diffraction studies of the actinide oxides. I. Uranium dioxide and thorium dioxide at room temperature, *Proc. Roy. Soc.*, **A274**: 122–133.

—— (1966): A note on Burbank's paper on 'Intrinsic and systematic multiple diffraction,' *Acta Cryst.*, **21**: 175.

WITZ, J. (1969): Focusing monochromators, *Acta Cryst.*, **A25**: 30–41.

WOOSTER, W. A. (1967): The interpretation of x-ray photographs with the use of computers, *Acta Cryst.*, **23**: 714–717.

XUONG, N. H., KRAUT, J., SEELY, O., FREER, S. T., and WRIGHT, C. (1968): Rapid measurements of large numbers of reflections for proteins, *Acta Cryst.*, **B24**: 289–291.

YAKEL, H. L., and FANKUCHEN, I. (1962): Systematic multiple diffraction in equi-inclination Weissenberg geometry, *Acta Cryst.*, **15**: 1188.

YAKOWITZ, H. (1969): The divergent beam x-ray technique, in *Electron probe microanalysis*, pp. 361–431, Academic Press, Inc., New York.

YOUNG, R. A. (1961): Background intensities in single crystal diffractometry, *Tech. Rept. 2*, DDC Document No. AD262912.

—— (1962): Mechanism of the phase transition in quartz, final rept., DDC Document No. AD276235.

—— (1963): Balanced filters for x-ray diffractometry, *Z. Krist.*, **118**: 233–247.

—— (1965): Background factors and technique design, *Trans. Am. Cryst. Assoc.*, **1**: 42–66.

—— (1969): Present problems and future opportunities in precise intensity measurements with single-crystal diffractometers, *Acta Cryst.*, **A25**: 55–66.

——, GOODMAN, R. M., and KAY, M. I. (1964): Oral presentation, ACA Meeting, Bozeman, Mont.

—— and MACKIE, P. E. (1971): Oral presentation, ACA Meeting, Columbia, S.C.

—— and WAGNER, C. E. (1966): Intensity contrast in diffraction from nearly perfect crystals, *Brit. J. Appl. Phys.*, **17**: 723–727.

ZACHARIASEN, W. H. (1945): *Theory of x-ray diffraction in crystals*, John Wiley & Sons, Inc., New York.

—— (1963): The secondary extinction correction, *Acta Cryst.*, **16**: 1139–1144.

—— (1965a): Multiple diffraction in imperfect crystals, *Acta Cryst.*, **18**: 705–710.

—— (1965b): Extinction, *Trans. Am. Cryst. Assoc.*, **1**: 33–41.

—— (1967): A general theory of x-ray diffraction in crystals, *Acta Cryst.*, **23**: 558–564.

—— (1968): Experimental tests of the general formula for the integrated intensity of a real crystal, *Acta Cryst.*, **A24**: 212–216.

—— (1969): Theoretical corrections for extinction, *Acta Cryst.*, **A25**: 102.

ZOCCHI, M., and SANTORO, A. (1967): Simultaneous diffraction with the three-circle diffractometer, *Acta Cryst.*, **22**: 331–334.

# INDEXES

# NAME INDEX

Abrahams, S. C., 509
Abramowitz, M., 476
Adell, Ö., 286, 289
Afanas'ev, A. M., 182, 292
Aldred, E., 338
Alexander, L. E., 525, 551, 554, 578, 579
Allison, S. K., 510
Amenlinckx, S., 384
Anderson, O. L., 566
Ando, Y., 329, 414, 423, 431
Annaka, S., 438
Arlinghaus, F. J., 64
Armstrong, E., 277
Arndt, V. W., 509, 510, 570
Arrott, A., 295
Austerman, S. B., 320, 371

Authier, A., 274, 275, 281, 288, 371, 436
Averbach, B. L., 94, 95, 101, 113, 114
Azároff, L. V., 142, 320, 515, 533, 577, 580

Baker, T. W., 510, 520
Barrett, C. S., 320
Bartell, L. S., 45
Barth, H., 546
Batterman, B. W., 39, 64, 281, 438, 562, 564
Bellamy, B. A., 520
Benesch, R., 12, 21, 64, 320
Berggren, K. F., 56, 57
Bethe, H., 222, 271

Beu, K., 484
Bilinsky, V., 438
Binns, J. V., 509
Biscoe, J., 52
Blech, I. A., 429, 432
Bol, W., 537
Bollmann, W., 277, 358
Bond, W. L., 509, 519
Bonham, R. A., 2, 45
Bonse, U., 288, 320, 337, 433–436, 542, 544, 545, 547
Borie, B., 39, 132
Born, M., 217, 389, 394
Borrmann, G., 278, 279, 281, 284, 335, 408
Bosanquet, C. H., 568, 577
Bradler, J., 437
Bragg, R. H., 577
Bragg, W. H., 444
Bragg, W. L., 510, 568
Brandt, W., 56, 57
Brentano, J., 444
Brindley, G. W., 490
Brogren, G., 286, 288
Brown, R. E., 45
Bubáková, R., 286
Bucksh, R., 249
Buerger, M. J., 142, 143, 510
Burbank, R. D., 554, 555, 570
Burger, H. C., 475
Burke, J., 519
Burkhalter, P. G., 534
Busing, W. R., 516–518, 571
Butt, N. A., 567

Calder, R. S., 57
Campbell, H. N., 278
Campbell, W. J., 534
Cath, P. G., 510
Causer, R., 520
Chambers, F., 503
Chandrasekhar, S., 575, 577, 578, 580
Chikawa, J., 371, 436
Chipman, D. R., 39, 43f., 438, 542, 562, 564, 566

Cittert, P. H. von, 475
Clementi, E., 62
Clifton, D. F., 510
Coche, A., 533
Cochran, W., 563, 566, 568
Cole, H., 249, 424, 470, 503
Compton, A. H., 510
Cooper, M. J., 39, 566, 567, 569, 574
Coppens, P., 572, 573
Cork, J. M., 277
Coster, D., 423
Coulson, C. A., 59, 60
Cowley, J. M., 263
Cromer, D., 17
Crowther, R. A., 509
Currat, R., 2, 27

Dai, K., 476
Daniel, E., 45, 64
Darwin, C. G., 222, 295, 573
Davis, M. F., 509
De Benedetti, S., 45
De Cicco, P. D., 2, 27
Dederichs, P. H., 182
Delaunay, B., 142
Delf, B. W., 470
De Marco, J. J., 32, 63, 64, 336, 486
Deslattes, R. D., 432
de Wolff, P. M., 467, 471, 486
Diamond, R., 578
Drahokaupil, J., 286
DuMond, J. W. M., 277
Duncanson, W. E., 59, 60

Eastabrook, J. N., 460, 461, 476, 478, 480, 493
Edwards, H. J., 478, 479
Eisenberger, P., 2, 23, 44
Elliot, A., 539, 540
Epstein, I., 60
Ergun, S., 475
Euwema, R. N., 50
Evans, H. T., 510
Ewald, P. P., 222, 225, 228, 234, 267, 284, 295

Fankuchen, I., 570
Fessler, R. R., 94, 95, 114
Fingerland, A., 262, 286, 480
Fitzwater, D. R., 559
Fournét, G., 69
Franck, F. C., 432
Franks, A., 540
Freer, S. T., 509
Fukamachi, T., 31, 471
Fukushima, E., 432
Fuller, A., 510
Furnas, T. C., 447, 514, 536, 542, 551

Gale, B., 450
Gatineau, L., 578
George, J. D., 520
Gerdes, R. J., 477, 478, 493
Giessen, W. C., 470, 530, 537
Gillham, C. J., 450, 451, 468
Goldberger, M. L., 44
Goodisman, J., 39
Goodman, R. M., 516, 572
Gordon, G. E., 470, 530
Gordon, W. F., 537
Goroff, I., 49
Göttlicher, S., 49, 566, 568
Groter, C., 509
Guentert, O. J., 539
Guinier, A., 69
Gummel, H., 22

Hall, C. R., 284
Hall, G. G., 39
Hamilton, W. C., 295, 574
Hardy, G. H., 477
Harrison, R. J., 486
Hart, M., 271, 288, 320, 331, 334, 337,
   418, 419, 434–436, 520, 542, 544, 545,
   547
Hartwig, W., 284
Haruta, K., 432
Hashizume, H., 288, 432
Hattori, H., 331, 332, 334, 337
Heidenreich, R. D., 271, 337
Héno, Y., 225, 284

Henrich, V. E., 64
Herbstein, F. H., 541
Hildebrandt, G., 278, 281, 284, 288, 292,
   408
Hirsch, P. B., 234, 282, 284, 292, 293,
   358, 383
Hirshfeld, F. L., 510
Hoerni, J., 333
Homma, S., 310, 329
Honjo, G., 263
Hönl, H., 17
Hoppe, W., 519, 559
Horne, R. W., 358
Hosoya, S., 31, 471
Howard, J. K., 432
Hunter, L. P., 281
Hurst, R. P., 56, 57

Ikeno, S., 389
Inkinen, O., 55, 64

James, R. W., 69, 563, 568, 569, 577
Jauch, J. J., 2, 28
Jeffrey, J. W., 509
Jeffreys, B., 217
Jeffreys, H., 217
Jennings, L. D., 39, 43f., 490
Jones, C. K., 535

Kagen, Yu., 182
Kajimura, S., 322
Kambe, K., 333, 409, 433, 438
Kaplow, R., 94, 95, 101, 113, 114, 139,
   140, 445
Katagawa, T., 234, 257, 262, 311, 371
Kato, N., 209, 234, 257, 262, 263, 271,
   277, 279, 291, 295, 311, 325, 327, 329,
   331, 332, 334, 337, 351, 371, 389, 401,
   407, 408, 414, 423, 427, 431, 433, 434,
   436
Katz, L. J., 566
Kawashima, K., 286
Kay, H. R., 509
Kay, M. I., 516, 572

Keating, D., 128
Kelvin, D., 217
Kheiker, D. M., 493, 551, 560
Kikuta, S., 286, 288, 336, 438, 546, 547
Killean, R. C. G., 483, 494
Kinder, E., 271
King, H. W., 468
Kirkpatrick, P., 537
Kleinman, L., 49
Klemperer, W., 39
Knol, K. S., 423
Knowles, J. W., 277
Kohra, K., 286, 288, 336, 366, 424, 432, 438, 546, 547
Komura, S., 566
Kossel, W., 271
Kottwitz, D. A., 547
Kraut, J., 509
Krivoglaz, M. A., 69
Kukol, V. V., 493
Kunze, G., 450
Kuriyama, M., 295, 334, 434
Kurki-Suonio, K., 336

Ladell, J., 469, 507, 509, 510, 512, 525, 551, 578
Lang, A. R., 295, 320, 334, 337, 432, 437, 448, 463, 467, 469, 471, 578
Langford, J. I., 458, 461, 477–479, 486
Larson, A. C., 574
Laue, M. von, 208, 222, 225, 234, 263, 278, 279, 295, 353
Lauriat, J. P., 471
Lawn, B. R., 432
Lax, M., 22
Lehfeld, Sosnowska M., 288
Lehmann, K., 335
Lehmpfuhl, G., 333
Levy, H. A., 516–518, 571
Ling, D., 284
Lonsdale, K., 563
Low, F. E., 44
Lowitzsch, K., 510, 512

Lucas, B. W., 566–568

Mack, M., 443, 450, 455, 469, 493
Mackie, P. E., 573
McKinstry, H. A., 535
McLachlan, D., 510
Malgrange, C., 288
Mallett, S. F. W., 509
Martino, F., 56f., 57
Maruyama, M., 389
Matsukawa, K., 423
Matsushita, T., 286, 288, 336, 546
Mavel, G., 538, 551
Meieran, E. S., 427, 429, 432
Melngailis, J., 45
Menter, J. W., 282
Mering, J., 578
Merisalo, M., 55
Mihama, K., 263
Mijnarends, P. E., 52
Milberg, M. E., 89, 458, 486
Milledge, H. J., 509
Miller, D. C., 532
Milne, A. D., 271, 331, 337, 436
Mirken, L. S., 532
Mitchell, C. M., 467, 471
Mitra, G. B., 480, 486
Mitsuishi, T., 436
Miyakawa, T., 295
Miyake, S., 424
Moliére, G., 182, 263
Moller, H. B., 566
Moon, R. M., 572
Mozzi, R. L., 539

Nagasaki, H., 436
Nakayama, K., 288
Nathans, R., 566, 567, 569
Newkirk, J. B., 320
Nicklow, R. M., 520, 521, 562–565, 567
Niehrs, H., 263, 271
Nielson, N., 566
Niggli, P., 142
Nilsson, N., 39, 562, 564, 567, 568

Nishikawa, S., 423

O'Connor, D. A., 567
Ogilvie, R. E., 450
Ohtsuki, Y. H., 182
Okaya, Y., 424
Okkerse, B., 281, 282, 284, 292
Otnes, K., 541
Otto, J., 249

Pančenko, Ju. M., 493
Panke, D., 572
Parathatharasy, R., 438
Parrish, W., 442, 443, 450, 455, 470, 473, 485
Paskin, A., 39, 486, 564
Patel, J. R., 427, 428
Paterson, M. S., 475
Penning, P., 284, 433
Pepinsky, R., 424
Pério, P., 471
Perstnev, I. P., 292
Phillips, D. C., 509, 510
Phillips, W. C., 46, 52, 55–57, 63
Pike, E. R., 443, 461, 469, 493
Platzman, P. M., 2, 23
Polder, D., 284, 433
Post, B., 569
Powell, M. J. D., 571
Pringle, G. E., 575
Prins, J. A., 284, 423

Raccah, P. M., 64
Ramachandran, G. N., 291, 292
Ramaseshan, S., 575
Rees, A. L. G., 263
Reissland, A., 333
Renninger, M., 249, 286, 288, 545, 546, 562, 569
Rietveld, H. M., 474
Riste, T., 541
Rogosa, G. L., 286
Rogosinski, W. W., 477

Rohrlich, F., 2, 28
Rouse, K. D., 566, 567, 569, 574
Ruland, W., 445

Saada, G., 292
Saito, Y., 424
Saka, T., 234, 257, 262, 311, 335
Samson, S., 519
Santora, O., 570, 571
Sauvage, T., 371, 436
Schäfer, K., 490
Schiff, L. I., 298
Schmidt, C. M. J., 510
Schomaker, V., 480
Schuelke, W. W., 519
Schwartz, G., 286
Schwartz, L. H., 39, 566
Schwuttke, G. H., 432
Seely, O., 509
Shaw, C. H., 537
Shoemaker, D. P., 480
Short, M. A., 535
Shull, G. G., 572
Simmons, R. O., 521
Singh, A. K., 575
Skelton, E. F., 566
Smallman, R. E., 578
Smith, A. E., 542
Smith, G. S., 525, 551, 578, 579
Smith, T., 563
Smith, V. H., Jr., 12, 14
Sneddon, I. N., 217
Soller, W., 450
Sommerfeld, A., 389
Soules, J. A., 537
Sparks, C. J., Jr., 132, 541, 542
Spencer, W. J., 432, 479
Spielberg, N., 493, 507, 525, 551, 578
Stegun, I. A., 476
Stemple, N. R., 249, 424
Steward, E. H., 540
Stokes, A. R., 474, 475, 479
Strong, S. L., 101, 113, 139, 445
Sturkey, L., 263, 271, 337
Sudarsanan, K., 574

Suortti, P., 39, 490
Surratt, G. T., 50
Szántó, I. S., 480


Takagi, S., 311, 335, 434
Tanemura, S., 337, 436
Taupin, D., 434
Taylor, A., 469, 485, 490
TeKaat, E., 434, 436
Terasaki, O., 471


Weiss, R. J., 2, 27, 32, 36, 37, 44, 46, 49,
    52, 55–57, 63, 336, 486, 489, 577, 578
Wells, A. F., 139
Wepfer, G. G., 50
Werner, S. A., 295
Whelan, M. J., 284, 358
Whitaker, A., 509
Wiewiarowsky, J., 442
Wilchinsky, Z. W., 486
Wilhite, D. L., 50
Wilkens, M., 434
Wilks, E. M., 432
Williamson, G. K., 578
Willis, B. T. M., 509, 570, 571
Wilson, A. J. C., 442, 449, 450, 452, 454,
    455, 457, 458, 461, 465, 467–471, 473,
    476–480, 484, 486, 493
Witte, W., 584

Witz, J., 540–542, 580
Wolf, E., 217, 389, 394
Wölfel, E., 49, 572, 584
Wolff, P. M. De, 485
Wood, E., 277
Wood, G. C., 503
Wooster, W. A., 475, 509
Wright, C., 509


Xuong, N. H., 509


Yakel, H. L., 570
Yakowitz, H., 520
Yanagawa, S., 182
Yap, F. Y., 493
Yoshimatsu, M., 366, 371
Yoshioka, H., 286
Young, R. A., 477, 478, 493, 516, 520,
    521, 537, 538, 548–555, 559, 562–567,
    572–575


Zachariasen, W. H., 278, 289, 291, 295,
    569, 572, 574, 577, 578
Zener, C., 438
Zevin, L. S., 493
Ziman, J. M., 433
Zocchi, M., 570, 571

# SUBJECT INDEX

Absorption, 188, 399
  coefficient, 469
  by crystals, 241
Acceptance band, 525
Acceptance region, 523, 524
Accuracy of wavefunction, 8
Action integral, 396
Air scattering, 487
Alignment considerations, 518
Aluminum, 64
Amplitude region, 203
Analysis of line profiles, 474
Angle-setting errors, 521
Angle-setting procedures, 516
Angular divergence, 297
Anomalous dispersion, 16

*Anpassung,* 248
Atomic charge distributions, 116
Atomic form factor, 199
Atomic pair distribution, 95
Atomic vibrations, 101, 113
Autocorrelation function, 79
Axial divergence, 460
Axis positioning, 518

$B_4C$, 57, 58
Background, 548
Background components, 548
Background structure, 559
Balanced-filter parameters, 536
Balanced filters, 535

Beam inhomogeneities, 578
Beam tilt, 519
Benzene, 59, 61, 62
BeO, 57, 58
Berg-Barrett arrangement, 321, 322
Berg-Barrett topographs, 290
$\beta$ filter, 530
Bloch wave, 224, 225, 230, 234
BN, 57, 58
Bond's method, 519
Bonse-Hart arrangement, 541
Bonse-Hart interferometer, 337, 435
Bonse-Hart monochromator, 542, 543, 546
Boron, 57, 58
Borrmann absorption, 234, 263, 278, 282–284, 292, 335
Borrmann fan, 321, 409
Borrmann-Lehmann fringes, 337
Bragg-Brentano arrangement, 442, 444, 449
Bragg-Brentano diffractometer, 448
Bragg-Brentano focusing, 440, 441, 444
Bravais lattices, 142
Bright-field image, 271
Brillouin zones, 143, 149
Broadening functions, 102, 472
Buerger equi-inclination, 510

Carbon, 49
Carbon black, 49
Cauchois aberration, 464
Caustics, 310
Centroid, 469
Charge density, 178
Clustering, 130
Coherence, 197
Coherent scattering, 71
Compton cross sections, 35
Compton effect, 18, 21
Compton profile, 2, 9, 10, 21, 32–34, 37, 66
  helium, 72
  hydrogenic atoms, 24
  lithium, 40
  measurements, 38

Compton profile:
  polycrystalline titanium, 65
  vanadium metal, 27
Compton scattering, 2, 488
Configuration interactions, 8
Continuous-scan methods, 523
Contraction factor, 398
Convergence, 503
Convoluted functions, 472
Correction for background, 487
Correlation energy, 7
Cosmic rays, 488
Coulson-Duncanson theory, 61
Counter characteristics, 533
Counting statistics, 482
Crystal monochromators, 540, 580
Crystal-shape function, 200
Crystal-structure factor, 199, 203, 204, 224
Crystal wave, 190
Crystal wave fields, 299, 307
Current density, 178, 181
Curved-crystal arrangements, 447, 463, 464
Cyclohexane, 59, 61, 62

Dark-field image, 271
Darwin-Prins rocking curve, 286
Darwin rocking curve, 249, 250
Dead times, 532
Debye-Waller factor, 109
Diamond, 49, 53
Dielectric constant, 177
Diffracted intensity, 487
Diffraction broadening, 480
Diffraction profile, 474
Diffraction vector, 74, 148
Diffractometer:
  motions, 509
  task, 508
Diffuse intensity, 110
Diffuse scattering, 489
Dirac transformation, 19
Dislocations, 213
Dispersion, 466, 467
Dispersion correction, 199

Dispersion point, 228, 252
Dispersion relation, 185, 222
Dispersion surface, 191, 228, 266
Distorted crystals, 214
Divergence, 503
Double refraction, 263, 264, 271, 272
Dynamical theory, 195, 222

Effective spectrum, 464
Eikonal surfaces, 391
Eikonal theory, 233, 389, 400
Elastic spectrum, 566
Electromagnetic wave fields, 180
Electron transmission diffraction, 157
Electronic troubles, 488
Energy density, 179, 187
Energy flow, 187
    spherical wave, 197
Energy flux, 179
Equal-thickness fringes, 271
Equatorial divergence, 459
Escape peak, 532
Eulerian cradle, 514
Eulerian geometry, 515
Euler's equation, 395
Ewald construction, 202, 203
Ewald rocking curves, 250, 260, 289
Extinction corrections, 573
Extinction effect, 70, 469
    primary, 573, 577
    secondary, 573, 577

Fault plane, 354
Fermat's principle, 392, 403
Field vectors, 226
Film methods, 509
Finiteness of crystal, 205
Fixed-count timing, 482
Fixed-time counting, 483
Flourescent discriminators, 538
Fluorescent radiation, 488
Form factor, 197, 199
Fourier method, 475
Fourier transform, 473, 478
Fraunhofer approximations, 196

Fraunhofer diffraction, 208, 309
Fraunhofer equation, 195
Fresnel diffraction, 209
Fresnel integral, 218
Fresnel theory, 190
Friedel's law, 423
Fringe phenomena, 335
Fringe position, 330
Fringe shape, 331
Fringes of equal inclination, 271

Generalized diffraction function, 215
Geometrical aberrations, 449, 463, 490
Goniostat, 514
Graphite, 49, 53

$H_2$ molecule, 44
Hart and Milne fringes, 337
Hartree-Fock method, 7
Hartree-Fock wavefunctions, 8, 11, 35
Heitler-London scheme, 8
Helium, 44
Huang diffuse scattering, 134
Hund-Mullikan scheme, 8
Hydrogen atom, 9
Hydrogenic solution, 4, 22

Imperfect crystal, 205
Impulse approximation, 23, 24, 27
Incident wave, 190
Independent-axes diffractometers, 514
Inert gases, 43
Integrated intensity, 70, 290, 326, 501
Integrated-intensity measurements, 523
Integrated power ratios, 323
Integrated reflecting power, 49, 290, 323
Intensity of x-ray diffraction maxima, 480
Intensity profiles, 277
Intrinsic line profiles, 471
Iron, 60

Johann aberration, 464

Kelvin's stationary-phase method, 300
Kikuchi patterns, 444
Kinematical theory, 69, 195
Klein-Nishina formula, 2, 35
Kossel figures, 295
Kossel lines, 520

Lang method, 320, 321
Lang topographs, 290
Lattice bending, 213
Lattice extinction, 144
Lattice image, 282
Lattice-parameter determinations, 519
Laue-Bragg reflection, 202
Laue pattern, 150
Lehmann-Borrmann fringes, 335
LiF, 55, 56
LiH, 55, 56, 126
LiMg, 46, 49
Line centroid, 491
Line profiles, 473
    synthesis, 474
Linear absorption coefficient, 189, 234
Linear diffractometer, 510, 511
Linear-tracking diffractometers, 510
Liquid aluminum, 93, 94
Lithium, 45
Localized defects, 159
Lorentz factor, 467, 469, 513

Magnesium, 46
Magnetic permeability, 178
Material equations, 177
Maxwell equations, 176, 181, 191, 225
Measurement of Bragg intensity, 486
Mechanical analog instruments, 510
Method of steepest descent, 295
Micro-densitometry, 509
Miller indices, 143
Misfit boundaries, 380, 383, 388
Moiré fringes, 436
Momentum space, 18
Momentum wavefunctions, 19
Monochromator:
    curved crystal, 463

Monochromator:
    plane crystal, 463
    pyrolytic graphite, 541
Multiple reflection, 569
Multiple scattering, 70

Neon, 44
*Netzebene* fringes, 278, 282, 283

Optimization of effort, 493
Oscillator density, 16
Oscillator strength, 17

PAILRED, 510–512
Pair distributions, 115, 139
Particle orientation, 484
Particle-size effects, 85, 87, 155
Peak-height intensity, 521
*Pendellösung* fringes, 265, 267, 268, 288,
    295, 328, 337, 409, 577
Perfect coherence, 198
Phase factor, 198
Phase relationships, 72
Physical aberrations, 464, 469
Planar defects, 210
Plane of incidence, 501
Plane-wave approximation, 71
Plane-wave solution, 185
Plane-wave theory, 296
Poisson distribution function, 482
Polarization factor, 198, 467, 468
Polyethylene, 59, 61, 62
Postreflection, 444, 449
Powder diffractometer, 444
Power ratio, 290
Poynting vector, 230
Prereflection, 444, 449
Primary extinction, 573, 577
Primitive cell, 142
Prin's rocking curve, 249
Propagation of error, 490
Proportional counter, 470, 532, 533
Pulse-height analysis, 470, 531
Pulse-size discrimination, 470, 488

Pyrolytic graphite, 541

Quantum-counting efficiency, 470

Radioactivity, 488
Random setting errors, 483
Ray-optical considerations, 279
Reciprocal lattice, 144
Reciprocal-lattice vectors, 200
Reciprocal-space sampling region, 501, 506, 508
Reciprocity theorem, 399
Reduced cell, 142
Reduced intensity, 84, 114
Reflected wave, 190
Reflection, 189, 192
Refraction, 189, 192, 465
Relativistic corrections, 28
Renninger effect, 547
Resolution function, 566–568
Resonance factor, 225
Resonanzfehler, 240, 253
Riemann sheet, 248
Rocking curve, 255, 286

Scattering factor, 10–12, 33, 37, 116, 124, 199
Scattering of x-rays, 69
Scattering power, 197
Scintillation counter, 470, 532, 533
Secondary extinction, 573, 577
Section topograph, 321, 322, 327–329
Seemann-Bohlin arrangement, 442, 444, 449
Semiconductor detector, 533, 534
Short-range order, 129, 130
Simultaneous diffraction, 150, 569, 571
Single-crystal diffractometer, 508, 510
Size-effect coefficients, 132
Small-angle boundary, 386, 387
Small-angle intensity, 91
Small-angle reduced intensity, 92
Small-angle scattering, 89

Snell's law, 191
Sodium, 45
Solid state detector, 470
Soller slits, 443, 450, 461–463
Space wavefunctions, 19
Specimen displacement, 457
Specimen tilt, 458, 519
Specimen transparency, 457
Spherical-wave theory, 295
Stächlen, 208, 212
Stacking fault, 210, 211, 364, 367, 370
image, 365
Stationary-counter-stationary-crystal method, 521
Stationary-phase method, 215, 217, 218, 295
Statistical absorption effects, 485
Statistical fluctuation, 491
Statistical matters, 481
Step-scanning mode, 486
Stokes method, 478, 479
Structure factor, 147
Systematic absences, 144
Systematic errors, 554

Tangential continuity, 190
Thermal diffuse scattering, 108, 109, 489, 562
Thomson scattering, 197
Three-dimensional pair distribution, 75
Tie point, 228
Titanium, 65
Topograph:
section, 321, 322, 327–329
silicon, 322
traverse, 322, 327, 328
Topographic methods, 321
Topographic observations, 320
Total reflection, 194, 539
Total scattering, 16
Transverse wave, 186, 225
Truncation effect, 578, 579
Twin boundary, 380
Twin plane, 372
Twinned crystals, 371
Two-beam approximation, 227

Two-dimensional pair distribution function, 77

*Umweganregung,* 547
Unit-cell vectors, 141
Unit cells, 141

Vacancies, 213
Vanadium, 60
Variance methods, 478
Variances of counting rates, 483
Variational principle, 394
Vertical truncation, 477
Voronoi zone, 142, 150

Waller-Hartree expression, 35

Waller-Hartree theory, 2
Waller integral, 292
Warren coefficient, 132, 134
Wave fields, 180, 244, 248, 250, 256
Wavefunction, 10
Wavelength discrimination, 529
Weissenberg camera, 510
Width:
    of receiving slit, 456
    of source, 456
Wigner-Seitz cell, 142, 150

X-ray diffraction topographs, 320
X-ray interferometer, 434
X-ray scattering factors, 34

Zachariasen's theory, 575